The Origin and Evolution of Birds

The Origin and Evolution of Birds

Alan Feduccia

Yale University Press New Haven and London

Designed by James J. Johnson and set in Aster Roman type by The Marathon Group, Durham, North Carolina. Printed in the United States of America by Vail-Ballou Press, Binghamton, New York.

The paper in this book meets the guidelines for permanence and durability of the Committee on Production Guidelines for Book Longevity of the Council on Library Resources.

Library of Congress Cataloging-in-Publication Data

Feduccia, Alan.
The origin and evolution of birds / Alan Feduccia.
p. cm.
Includes bibliographical references (p.) and index.
ISBN 0-300-06460-8 (hardcover : alk. paper)

1. Birds—Evolution. 2. Birds—Origin. 3. Birds, Fossil.
I. Title.
QL677.3.F43 1996
598.2'38—dc20 95-46758

A catalogue record for this book is available from the British Library.

10 9 8 7 6 5 4 3 2 1

CONTENTS

Preface

Charles Darwin devoted his life to uncovering the primary mechanism of evolution. In *On the Origin of Species*, published in 1859, he elucidated that mechanism: the principle of natural selection, which provided the motivation that propelled generations of scientists for the quest to uncover evolutionary relations of both plants and animals. Of particular interest here, there began the search for the evolutionary relations of birds, using evidence from paleontology and comparative anatomy. Over the past century evidence from myriad fields—behavior, ecology, physiology, comparative biochemistry, DNA hybridization—have contributed greatly to our understanding of avian relations, but the bedrock of the field has remained paleontology and comparative biology; indeed, paleontology remains the external check for molecular data. This book focuses on the testimony for avian evolution provided from paleontology and comparative biology.

The Origin and Evolution of Birds treats two major themes. The first, presented in chapters 1 through 3, is the discovery of the earliest bird, the *urvogel, Archaeopteryx*, and the origin and early evolution of birds and avian flight. The remaining chapters cover the evolution of archaic Mesozoic birds (chapter 4) and modern Tertiary birds (chapters 5 through 8), organized around such themes as the evolution of filter-feeding, flightlessness, raptorial birds, and land birds. The organization is similar to that of my earlier, popular book *The Age of Birds* (Harvard University Press, 1980), and that book served more or less as an outline for this volume.

The field of avian evolution has undergone revolutionary changes in both discovery and theory during the past two decades, often pitting opposing sides in bitter controversy. Nowhere has the vitriole been greater than between those paleontologists who advocate a dinosaurian origin of birds, coupled with the counter-intuitive ground-up, or cursorial, origin of avian flight, and those ornithologists, such as myself, who favor the time-honored and intuitively facile arboreal, or tree-down, theory of the evolution of flight. Many vertebrate paleontologists have accepted a formal, rigid cladistic methodology as the sine qua non for reconstructing phylogenies and, using this scheme in an almost religious manner, have discarded geological time as a tool in deciphering evolution, with the result that superficial resemblance often dictates relatedness. To such workers it is inconsequential that birdlike dinosaurs occur some 75 million or more years after the origin of birds. Yet to those more classically trained, like myself, this time gap is a sure sign of convergent evolution, defined simply as the acquisition of similar adaptations in unrelated groups of organisms. Even the names first given to these primarily late Cretaceous birdlike dinosaurs—*Ornithomimus*, the bird-mimic, *Struthiomimus*, the ostrich-mimic, and *Gallimimus*, the fowl-mimic—indicate that their original describers thought that these dinosaurs had independently evolved birdlike features.

Convergent evolution is a central theme of vertebrate history, and it is a predominant thread throughout the avian ranks. Everywhere we look birds from different evolutionary backgrounds have come to look alike: Northern Hemisphere auklets and Southern Hemisphere diving-petrels, great auks and penguins, Old World hornbills and New World toucans, swifts and swallows, Malagasy false sunbirds and true sunbirds, Old World oscine flycatchers and New World suboscine flycatchers—to mention but a few. Although it is relatively easy to uncover superficial, whole-animal resemblance caused by convergence, such as the examples of birds just mentioned, or in vertebrates—fish and dolphins, and bats and pterosaurs—convergence is often so complete, so elusive and subtle, that clues can be difficult to ascertain. One case of massive or whole-animal convergence is that of Mesozoic, toothed divers (*Hesperornis*) and modern loons, and indeed the different development of the proximal end of the tibia was the first clue of nonrelationship, combined with the fact that the first primitive loon, *Colymboides*, occurs in the Eocene, some 15 or more million years after *Hesperornis*. In early classifications many convergent pairs were placed together, for example, swifts and swallows, and hawks and owls, illustrating that convergence is an insidious and treacherous trap, baited and waiting for the unsuspecting worker.

Nowhere has the trap been more successful than in luring paleontologists to the theropod dinosaurian origin of birds. Theropods, with their bipedal gait and foreshortened arms, superficially resemble flightless birds, especially the ratites, through convergence, and this first led Darwin's champion, Thomas Huxley, to note that "the Dinosaurians are in many important characters intermediate between certain reptiles and certain birds—the birds referred to being the ostrich-tribe" (Darwin 1959). Much more recently, Allison Andors, in an article concerning the convergence between *Tyrannosaurus* and the great flightless bird *Diatryma* (*Natural History*, June 1995), quoted the novelist Flannery O'Connor: "Everything that rises must converge." This is the most important theme of bird evolution.

Just as important, and a discourse of the first part of this book, is that a dinosaurian origin of birds is inextricably linked with the cursorial, or ground-up origin, of avian flight, which is a biophysical impossibility. How could a large, deep-bodied, obligately bipedal reptile, with a long, heavy, balancing tail and greatly foreshortened forelimbs, fighting gravity all the way, give rise to avian flight? The answer, of course, is that it could not. Yet such prestigious institutions as the American Museum of Natural History in New York have recently revised their entire dinosaur hall to conform to this theory, and their accompanying book *Discovering Dinosaurs* (Norell, Dingus, and Gaffney, 1995), proclaimed that "the smallest dinosaur is the bee hummingbird . . . found only on Cuba" (25). They, like many others who adhere to a rigid cladistic methodology, have been led into that perfidious trap of convergence that lies camouflaged and waiting for the unsuspecting victim. I join many others in generally accepting the tenets of cladistic theory, that only shared, derived features can indicate relatedness, but the implementation of cladistic methodology is where the intractable problems are encountered. Often the procedure is more like a board game than a careful scientific examination of anatomy, and simple tallies of characters by computer are used to determine who is related to whom. As a consequence, convergence is almost impossible to weed out. Within birds, cladistic methods have grouped such unrelated and disparate forms as the ancient, toothed hesperornithiform divers and loons, and even hawks and owls. It is not enough to state simply that the closest relatives of hawks remain unknown, there must be a hypothesis of relationship, and until another better scheme is published, then it remains as the supposed best genealogy. But does science really work that way? Often there is simply insufficient evidence to offer such bold hypotheses, and cladograms confer an aura of truth to what is in reality speculation. For example, we do not know the closest relatives of hawks; but does this mean that they are related to owls by default? Such is the case for the origin of birds. Paleontological cladists claim that opponents of the theropod origin must produce a more suitable ancestor, but alas, we simply don't have sufficient evidence. We can only say, as dictated by science and logic, that the ancestor was surely a small, quadrupedal, arboreal archosaur, a pre-dinosaur in the overall scheme of the genealogy.

The idea of a theropod origin of birds originally became coupled with the idea, now largely disproven, of hot-bloodedness or endothermy, in dinosaurs, the idea being that hot-blooded dinosaurs became coated with feathers for insulation, sprouted wing feathers on the arms, and flew off into the sunset. Proclaimed one author of the first known bird, "*Archaeopteryx* supports two theories: warm-blooded dinosaurs and the dinosaurian ancestry of birds"! However, although the theory of hot-bloodedness has fallen from favor, the advocates of a theropod origin of birds still reconstruct earthbound dinosaurs with a feathered coat, defying logic, as the feathers of birds that have lost the ability to fly invariably degenerate and become hairlike in appearance.

The Mesozoic era, we have just begun to understand and as I discuss in chapter 4, was a period of adaptive radiation of many archaic groups of birds that became extinct at the end of the Cretaceous. Predominant in this radiation were the recently discovered "opposite," or enantiornithine, birds, so named because the three metatarsal bones fused from proximal to distal, the opposite direction of that of modern, ornithurine birds. The enantiornithines were quite diverse, but were all characterized by a well-developed, precocious flight apparatus but a primitive, urvogel-style pelvic region and hindlimbs. These opposite birds are included in the same subclass, the Sauriurae, with the urvogel, *Archaeopteryx*, but the lineage of modern-type birds, the Ornithurae, as current evidence tells us, evolved at about the same time as the Sauriurae, the early Cretaceous, and is represented by such early forms as the Mongolian *Ambiortus* and the Chinese wader *Gansus*. Other Mesozoic ornithurines included the toothed divers, the hesperornithiforms, and the ichthyornithiforms, but all of these groups became extinct at the close of the Cretaceous, along with the myriad groups of reptiles and other organisms.

The fossil record of birds now tells us clearly that:

(1) the Cretaceous-Tertiary boundary event was very significant for birds; this is a recent discovery, but surely, with all the life-forms that became extinct, the analogy of the miner's canary should tell us that birds would be the first to be affected; (2) the diversification of modern orders was post-Mesozoic; almost all modern orders thought to be present in the Cretaceous have now been shown to be enantiornithines, yet all modern orders are known from Eocene deposits; (3) the early Tertiary diversification was explosive: no modern orders are present at the close of the Cretaceous, but all are present by the Eocene; if this evidence is valid, then all modern orders may have arisen explosively during a period of some 5 to 10 million years, and with this many-stranded genealogy, relationships of modern orders may well be lost to the past; and (4) the mammalian fossil record provides a good model for avian diversification; that the mammalian radiation should serve as a model for bird evolution is eminently logical.

As I mentioned above, I treat the Tertiary radiation of birds in an unconventional manner, by exploring such themes as filter-feeding and wading birds, the evolution of flightlessness, the evolution of birds of prey, and the rise of land birds. In each chapter specific topics illustrate themes of evolution—for example, flightless birds illustrate such concepts as the mode and tempo of evolution, as well as many features of flight, which can be studied by examining what happens in the flightless state.

One outcome of the revelation that the modern avian orders are post-Cretaceous is the recognition that there exist numerous living fossils within the modern avifauna ranging across the spectrum of the class Aves. To mention a few, tinamou-like birds are an offshoot of the lineage that led to the ratites, tropicbirds are primitive, basal pelecaniforms, the Australian freckled duck, *Stictonetta*, is primitive, as is the Australian magpie-goose, *Anseranas*, and the South American screamers are an early offshoot of the anseriform birds. Continuing, the South American seriamas are primitive gruiforms and were no doubt basal to the carnivorous phorusrhacids, and many of the monotypic genera within the Gruiformes, such as the finfoots and sunbitterns, are surely ancient relicts. Rollers and Malagasy ground-rollers (Coraciidae and Brachypteracidae) were common birds of the Eocene, as were primitive piciform relatives of the puffbirds, mousebird-like birds, and varied caprimulgiforms. Indeed, the island of Madagascar may well be an avian refugium of the early Tertiary, with its ground-rollers, cuckoo-roller, and mesites, just as it is for such mammalian relicts as the lemurs, and its preservation, seriously threatened today, is absolutely essential. The conservation of many of these avian relicts all over the world should indeed be of great concern, as the modern avifauna provides us with a living laboratory of the early Tertiary avian radiation.

This book owes a large debt to a large number of individuals, all of whom cannot be mentioned here. However, I would like to acknowledge the following for their invaluable help: Herculano Alvarenga, Allison Andors, Philip Ashmole, Ian Atkinson, Herman Berkhoudt, Walter Bock, Walter Boles, José Bonaparte, Michael Brett-Surman, Alan Brush, Eric Buffetaut, Paul Bühler, Kenneth Campbell, Robert Carroll, Robert Chandler, Sankar Chatterjee, Jacques Cheneval, Luis Chiappe, Anusuya Chinsamy, Leslie Christidis, Richard Cifelli, Nicholas and Elsie Collias, Charles Collins, A. H. Cramer, Arthur Cruickshank, Nina Cummings, Michael Daniels, Ron Dorfman, Carla Dove, Steve Edinger, Andrzej Elzanowski, Robert Farrar, C. Hilary Fry, Stephen Gatesy, Ed Gerken, Paul Germain, Frank Gill, John Goss-Custard, Lance Grande, Mark Hallett, Alan Harris, Dave Harris, Bernd Haubitz, Hartmut Haubold, Max Hecht, W.-D. Heinrich, Angelika Hesse, Richard Hink, Lian-hai Hou, Peter Houde, Helen James, Robert Johnson, Philip Kahl, Bill Kelley, Lloyd Kiff, Evgeny Kurochkin, Roxie Laybourne, Karel Liem, David Ligon, Larry Marshall, Beverly McCulloch, Gordon Maclean, Don Merton, David Meyer, Carlo Morandini, Cécile Mourer-Chauviré, Martin Moynihan, Giuseppe Muscio, Åke Norberg, Ulla Norberg, Rory O'Brien, John Ostrom, Kenneth Parkes, Raymond Paynter, Jr., Colin Pennycuick, Stefan Peters, François Poplin, Robert Raikow, Pat Rich, Siegfried Rietschel, John Ruben, Dale Russell, Peter Ryan, José Sanz, Gary Schnell, Paul Sereno, Christopher Shaw, Bernard Sigé, Robin Simpson, David Steadman, Burkhard Stephan, Peter Stettenheim, Robert Storer, Stuart Strahl, Clark Sumida, Sam Tarsitano, Tim Tokaryk, Michael Tove, James Vanden Berge, Günter Viohl, Alick Walker, Glenn Walsberg, Doug Wechsler, Peter Wellnhofer, Rupert Wild, Wolfgang Wiltschko, Lawrence Witmer, Derek Yalden, Zhonghe Zhou, and Gart Zweers. I thank the many museums and institutions that contributed illustrations, including the American Museum of Natural History, Bavarian State Collection of Paleontology and Historical Geology, Bishop Museum (Honolulu), Black Hills Institute of Geological Research, Bürgermeister Müller Museum (Solnhofen), Canterbury Museum (Christchurch), Carnegie Museum of Natural History, Field Museum of Natural History, George C. Page Museum (Los Angeles), Humboldt University Museum für Naturkunde (Berlin), Jura-Museum (Eichstätt), Museum of Comparative Zoology (Harvard), Museum of New Zealand (Wellington), National Geographic Society, Natural History Museum of Los Angeles County, New York Zoological Society,

Senckenberg Museum (Frankfurt), National Museum of Natural History (Smithsonian), and Peabody Museum of Natural History (Yale).

I especially acknowledge my highly esteemed colleagues Larry Martin of the University of Kansas and Storrs Olson of the Smithsonian Institution, who over the years have stoked the fires of enthusiasm and helped clarify many important issues. I also wish to thank Dr. Dorothy S. Fuller, sister of George Miksch Sutton, who kindly arranged for her late brother's pen-and-ink bird drawings to be used in this book. Much credit is also due Susan Whitfield, artist-illustrator in the Department of Biology at the University of North Carolina, Chapel Hill, who rendered many of the illustrations, particularly line drawings and cladograms. My friend and colleague John P. O'Neill of Louisiana State University skillfully rendered the frontispieces for the eight chapters, as well as other pen-and-inks. Yale University Press also deserves credit for putting together a difficult project, in particular Science Editor Jean Thomson Black, who enthusiastically sought the manuscript, and Laura Jones Dooley, whose skilled editing led this book to completion.

The Origin and Evolution of Birds

Archaeopteryx displaying in a late Jurassic ginkgo. Although *Archaeopteryx* has been envisioned as a cursorial predator, most evidence indicates that it was primarily a primitive arboreal bird and a trunk climber and that it does not represent a terrestrial stage in the evolution of avian flight and feathers. (Drawing by John P. O'Neill)

1

Feathered Reptiles

At a time when tropical temperatures warmed much of the Northern Hemisphere and low, palmlike vegetation covered what is now central Europe, a feathered creature the size of a crow met its death in a shallow lagoon. Of the event itself this is all we can know, separated from us as it is by approximately 150 million years. But the death is recorded nonetheless, chronicled by sediments that, throughout the millennia, settled and consolidated into lithographic limestone, a fine-grained limestone that preserved not only the shape of the bones but the delicate impression of feathers. The creature thus memorialized was *Archaeopteryx lithographica*, and, though indisputably birdlike, it could with equal truth be called reptilian. The forearms that once held feathers ended in three fingers with sharp, recurved claws. The *Archaeopteryx* fossil is, in fact, the most superb example of a specimen perfectly intermediate between two higher groups of living organisms—what has come to be called a "missing link," a Rosetta stone of evolution. Its discovery in 1861, just two years after publication of Charles Darwin's *On the Origin of Species* (1859), seemed an unparalleled act of cosmic good will toward science, for by fulfilling the Darwinian expectation that such intermediate forms existed, this one fossil had a profound influence on the ultimate acceptance of the concept of evolution through natural selection. And for students of avian evolution, *Archaeopteryx* became the focal point of efforts to determine the descent of birds.

In attempting to construct a genealogical history, or phylogeny, for birds, we must look both to living forms and to the fossil record of ancient birds. *Archaeopteryx*, the oldest avian form yet discovered, is represented by a feather and seven skeletal specimens, several of which, notably the famous Berlin specimen, are almost perfectly preserved. But the fossil record is rarely so cooperative. Most bird bones are hollow and thin walled—adaptations that lighten the skeleton for flight—and are therefore not easily preserved. An extreme example of this is seen in the frigatebird, whose feathers, it is said, probably weigh more than the dried skeleton. True or not, it does illustrate the point that bird fossils are formed from fragile bones, are often fragmentary, or consist of single bones. Although bird fossils are sometimes abundant, only the rare find of an associated skeleton allows major advances in the effort to establish evolutionary relationships.

One could, technically, establish a phylogeny of birds, or any other group, exclusive of the fossil record, and perhaps have a reasonably good idea of the major lineages using evidence from such diverse areas as anatomy and biochemical and genetic (DNA) comparisons. Yet, even then, problems are legion. Not only is there considerable argument about the methodology that should be employed, but the search for meaningful anatomical features (known as characters) that elucidate relationships is laden with problems because, beneath their feathers, birds tend to look very much alike anatomically. In order for flight to be possible, flight architecture was invented early on and the forelimbs had to assume an elongated form in the first birds. That basic form has been retained with very little modification in all modern birds. Mammals, by contrast, possess many features, such as teeth, that differ substantially among the various groups and indicate evolutionary relationships. Traits of the major avian groups, as we shall see, present few such clues.

Because of the extremely restrictive physiological and anatomical demands of flight, birds are finely tuned metabolic machines, with very high metabolic rates and constrained aerodynamic morphologies. This restrictive aerodynamic design means that birds are the most structurally uniform of all vertebrate orders. Evolutionary biologist Walter Bock (1963a) has argued convincingly that if birds are to fly, they must conform to a narrow set of structural and physiological requirements, and so slight morphological divergence has been the characteristic feature

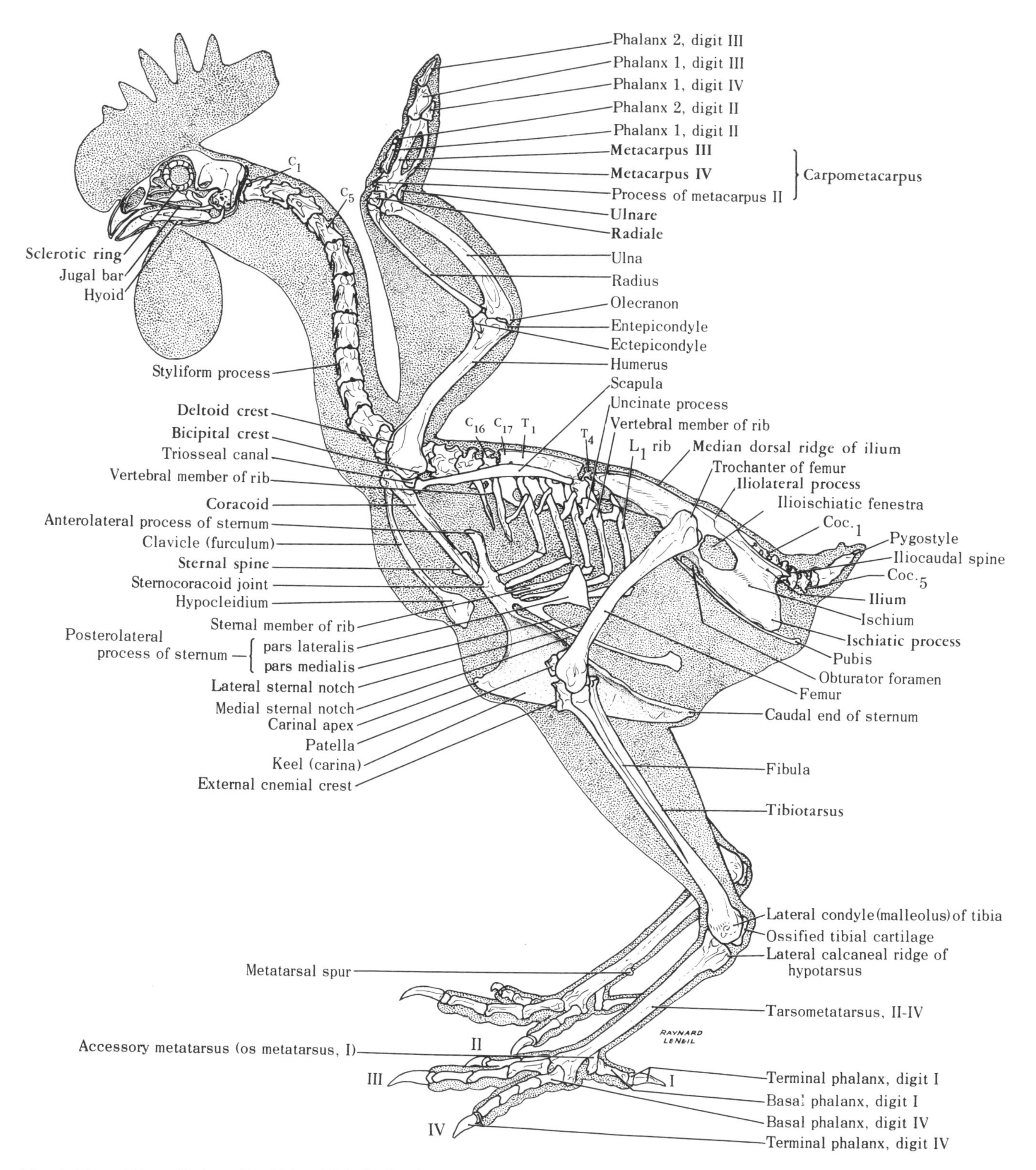

The skeleton of the male domestic chicken (*Gallus*), showing the major features of the avian skeleton. (Adapted from Lucas and Stettenheim 1972; courtesy Peter Stettenheim)

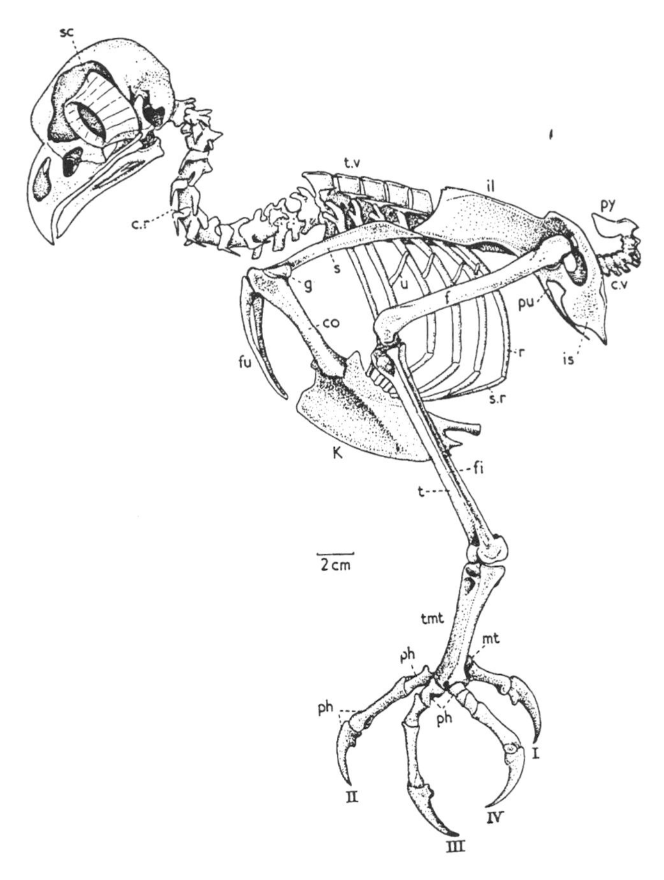

Skeleton of the eagle owl (*Bubo bubo*) with the wings removed. Note that the fibula extends the entire length of the tibia. Structures on the left side only are shown except for the right clavicle. Abbreviations: *co,* coracoid; *c.r.,* cervical rib; *c.v.,* caudal vertebrae; *f,* femur; *fi,* fibula; *fu,* furcula (fused clavicles); *g,* glenoid; *il,* ilium; *is,* ischium; *k,* sternal keel; *mt,* first metatarsal; *ph,* phalanx; *pu,* pubis; *py,* pygostyle; *r,* vertebral rib; *s,* scapula; *sc,* sclerotic rings (scleral ossicles); *s.r.,* sternal rib; *t,* tibiotarsus; *tmt,* tarsometatarsus; *t.v.,* thoracic vertebrae; *u,* uncinate process; *I–IV,* digits. (From Bellairs and Jenkin 1960; courtesy Academic Press)

of their primary and secondary adaptive radiations. Flight is, in a morphological sense, the biomechanically and physiologically most restrictive vertebrate locomotor adaptation, permitting little latitude for new designs. Ground-dwelling tetrapods, such as terrestrial mammals, and swimming forms, such as fishes, have not been so constrained in their structure by extreme physiological demands or aerodynamic constraints, and as a consequence more morphological divergence has characterized their adaptive radiations. As an analogy, an engineer can construct a terrestrial vehicle in diverse configurations, but there is really only one basic design for fixed-wing aircraft. Unlike mammals, which possess ample features such as teeth that vary considerably between the various groups, the higher categories of birds present few such clues to their interrelations. As a result, many of the major lineages of birds are defined on one or a handful of diagnostic anatomical features.

Small Size and Light Weight

As we shall see, small size is essential for the evolution of flight, for only in small animals could the original, rudimentary protofeathers have had an aerodynamic effect, and small size can be viewed as a general characteristic of birds. Although such conspicuous birds as ostriches and penguins capture the attention of the public, in reality, the vast majority of the some 9,700 living bird species (Sibley and Monroe 1990) weigh much less than a kilogram (2.2 lbs.) and are fewer than 50 centimeters (20 in.) in length (Proctor and Lynch 1993). In fact, although there were some exceptionally large flying birds, such as the extinct New World teratorns, which probably depended on winds to get aloft, the upper limit of mass in normal flying birds is represented by such types as the large bustards, storks, and swans, approaching some 16 or so kilos (35 lbs.). As well as being lightweight, the avian body is compact. Body mass is concentrated between the wings and the center of gravity, which hangs below the wings to convey aerodynamic stability. This is accomplished by lodging the major flight muscle mass—comprising up to 40 percent of body weight—on the sternum.

Because it is dogma that the avian body is characterized by light weight, Prange, Anderson, and Rahn surprised the ornithological world when they presented data to support a conclusion that, "contrary to common expectations, the skeletons of birds are not lighter than those of mammals" (1979, 105). But they merely showed that the allometry of skeletal mass to body mass in birds is very similar to that in mammals. Allometry refers to relative growth in which the proportions of a particular structure change as overall size changes. The more interesting question revolves around the relation of weight to volume. As Paul Bühler has noted, "At the same time that selection pressure diminishes the weight of a skeleton, it probably also diminishes, to a certain degree, the weight of the soft parts of the body. So, the constant skeleton mass to body mass relation does not indicate that the avian skeleton in general is as heavy as the mammalian skeleton" (1992, 391–392). Indeed, one need only pick up the skull of a nightjar in one hand and the skull of a mammal of similar body size in the other to confirm this expectation. As we shall see, in addition to overall lightening, weight has been redistributed in the avian body: with birds' loss of teeth, the gizzard gained importance as the masticatory organ, and as flight was perfected, the major flight muscles comprised a disproportionate amount of the body's weight.

Many anatomical traits are unique to modern birds

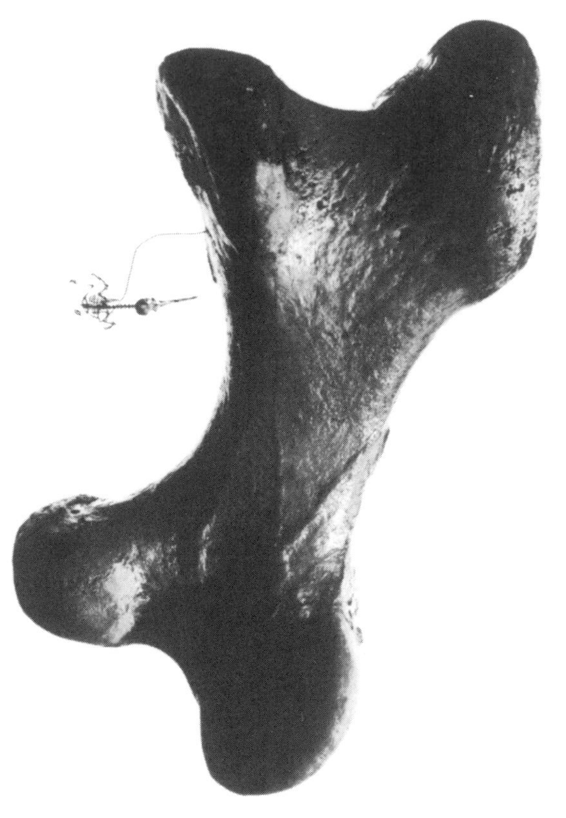

Hummingbird skeleton and the hollow femur (thigh bone) of an extinct Malagasy elephantbird (*Aepyornis*), illustrating each end of the avian size spectrum. The bee hummingbird (*Mellisuga helenae*), at 6.4 cm (2.5 in.) and 2.5 gm (1/10 oz.), is the smallest extant bird, whereas elephantbirds are estimated to have been up to 3 m (9.8 ft.) tall and to weigh up to 450 kg (1,000 lbs.). (Courtesy Department of Library Services, American Museum of Natural History)

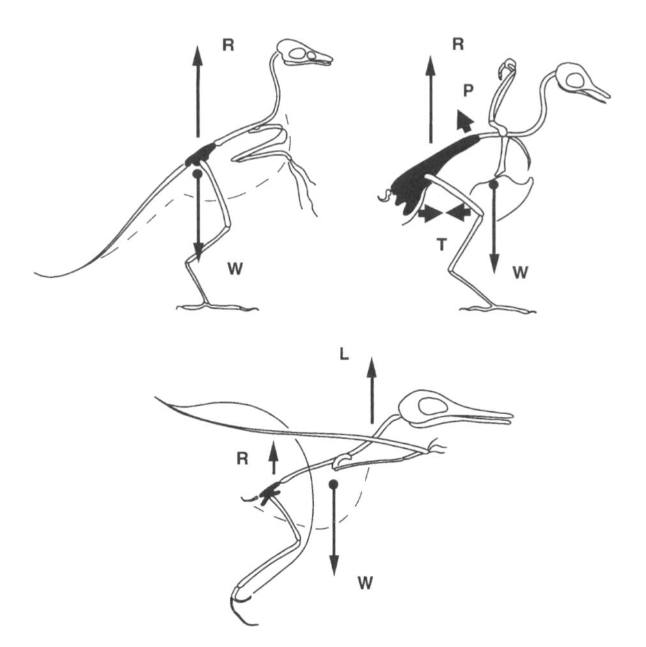

In *Archaeopteryx* (*top left*), the long tail allowed the body weight (*W*) to act along approximately the same line as the upward reaction from the hip joint (*R*). Modern birds (*top right*) have lost the tail, and the body weight therefore acts ahead of the hip joint, tending to topple the body forward. This is resisted by the enormously expanded, shoehorn-shaped synsacrum, which works as a lever. Tonic muscles (*T*) pull the posterior end of the synsacrum toward the femur, and this in turn holds the anterior end of the body up (*P*). The later pterosaurs (*bottom*) also lost the tail but did not develop a similarly expanded synsacrum. They had no adaptation to permit movements to be balanced about the hip joint in bipedal standing. A running take-off might have been possible, as shown, with part of the weight supported aerodynamically (*L*) by the wings. (Modified from Pennycuick 1986)

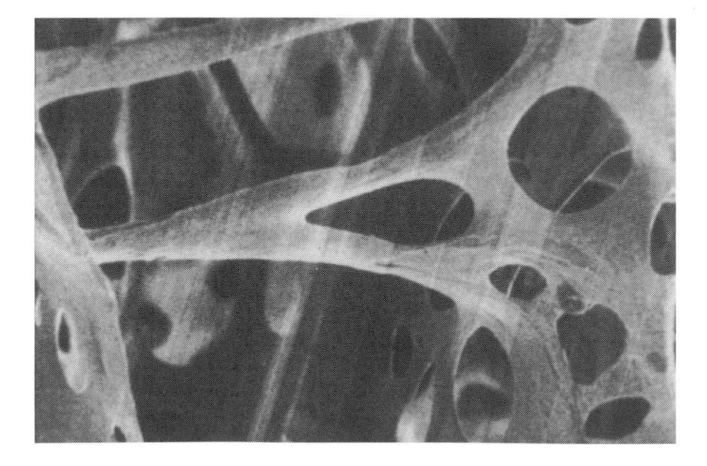

Scanning electron micrograph of a forked bony beam from the interior of the braincase wall of the tawny owl (*Strix aluco*), illustrating the best-known weight-reducing strategy in birds: the substitution of air for marrow or marrowlike tissue in the cavities of bones as well as in the skull. Section in image 1 mm wide. (From Bühler 1992; courtesy *Los Angeles County Museum of Natural History, Science Series* 36)

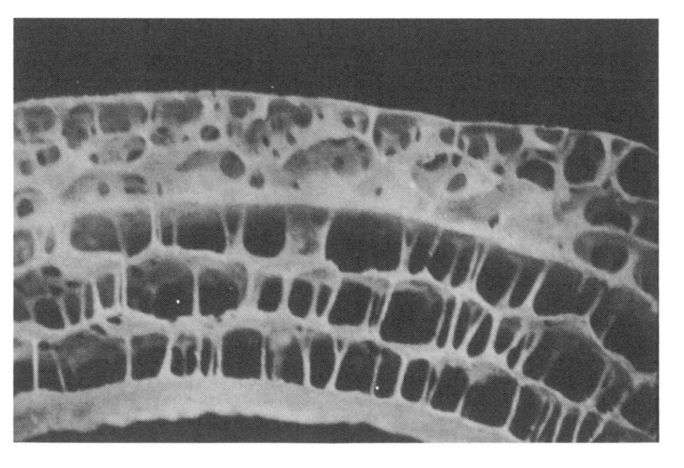

Four-story construction of the braincase of a long-eared owl (*Asio otus*), illustrating the "sandwich" construction in the walls of the avian braincase. This is one of the most visually impressive weight-reducing strategies in birds. (From Bühler 1992; courtesy *Los Angeles County Museum of Natural History, Science Series* 36)

and have evolved as adaptations for flight. The major bones are hollow and pneumatized, with direct connections to an extensive air-sac and respiratory system; such bones as the lightweight, hollow humerus are exemplary of this structural complexity. In addition, such bones as the vertebrae, skull, and jaws have undergone pneumatization. Lightweight, pneumatized skull bones are perhaps best exemplified in owls and nightjars, where broadening of the braincase is accomplished by air-filled "sandwich" bone (Bühler 1992). Many bones of modern birds have also been fused or deleted, more than in any other group of living vertebrates (but closely paralleling the extinct pterosaurs); these adaptations are also associated with lightening of the skeleton to permit flight. In most birds, the fibula, too, is reduced to a splint along the outer tibiotarsus. Bird embryos have a full-length fibula, but it is generally reduced through ontogeny, or embryological development (owls are an exception); birds also parallel pterosaurs in this feature. As one might expect, the first known bird, *Archaeopteryx*, had a fibula that ran the full length of the tibia. However, like modern birds, *Archaeopteryx* had thin-walled hollow bones with a thickness within the range of variation of modern birds, and it exhibited pneumatization in the braincase (Bühler 1992).

Distinctive Attributes of Birds

Small size and light weight are general features of birds, but their salient features include feathers, the avian-type furcula (the wishbone, or fused clavicles), and a reversed first toe known as the hallux. Other diagnostic features are anatomical traits involved in feeding and locomotion—the skull, the beak, and skeletal structures involved in flight, perching, and running, including the feet, the wings, and a miscellany of odds and ends.

The Avian Skull. The first birds had a full complement of uniform reptilian teeth in upper and lower jaws, but over time these were reduced and eventually lost. Cretaceous toothed divers, the hesperornithiform birds (see chapter 4), illustrate an intermediate stage of avian reduction of teeth in which the premaxilla lacked teeth and the teeth of the upper jaw were restricted to the maxilla; ultimately, the entire tooth row and maxilla were lost. The toothed jaws were transformed into the lighter bill or beak. The modern beak is covered with a horny sheath, a keratinized thickening of epidermal corneum called the rhamphotheca, a usually hard and rigid, highly adaptable, often plastic structure transformed into everything from a nutcracker in finches and grosbeaks to an instrument for tearing flesh in hawks and owls. Much of the work accomplished in mammals by the mammalian jaw, with its complex heterodont dentition and grinding molars, is taken over in birds by the muscular gizzard, often augmented by grinding, with the ingestion of grit and gizzard stones. As much as 2.3 kilos (5 lbs.) of grit and stones have been recovered from the individual gizzards of fossilized moas.

The bird skull is reptilian in many respects, with its single occipital condyle and movable quadrate, which articulates with the lower jaw. However, the general proportions of the avian skull are governed by the much larger relative size of the brain and eyes, and the avian brain fills the cavity of the braincase quite tightly, rather than being separated from it by loose tissue, as in reptiles. Also, the avian skull has a unique form of cranial kinesis—the movement of all or part of the upper jaw relative to the braincase, a universal feature in birds (Bock 1964). The dorsoventral swing of the upper jaw relative to the braincase results in, among other things, a larger gape and a faster closing, more efficient jaw mechanism and permits the bill to maintain a more or less stationary axis as the jaws open and close, thus enhancing birds' ability to capture prey. Cranial kinesis as such is not unique to birds and is also found among fishes, fossil amphibians, and reptiles.

Genital Organs. Aside from skeletal modification, myriad adaptations of the bird's anatomy reduce overall weight, including the sex organs: most females have but one ovary, and many males lack a penis, although a penis-like intromittent organ has been reacquired by some large flightless birds, such as the ostrich, as well as by tinamous, most waterfowl, storks, and curassows, and a smaller organ is found in chickens and turkeys. Because most birds lack external genitalia, copulation normally involves only a brief cloacal contact in which sperm is transferred. All birds lay eggs—that is, they are oviparous—and, in fact, birds are the only vertebrate class to have no members that are viviparous, giving birth to live young (Blackburn and Evans 1986).

Fusion, Deletion, and Skeletal Rigidity. The avian body is extremely compact. Almost all the major skull bones are fused into a single rigid structure, the wing and leg bones are reduced in number, and many elements are fused. The wing, for example, has three instead of the normal five fingers of the vertebrate hand. From this three-fingered hand emerge the primary flight feathers, numbering from nine to twelve in flying birds; from the ulnar region come the secondary flight feathers, sometimes, but by no means always, attached to the ulna by individual bony knobs known as ulnar quill knobs (Edington and Miller 1941; Yalden 1985). Collectively, the flight feathers are called remiges (singular, remex).

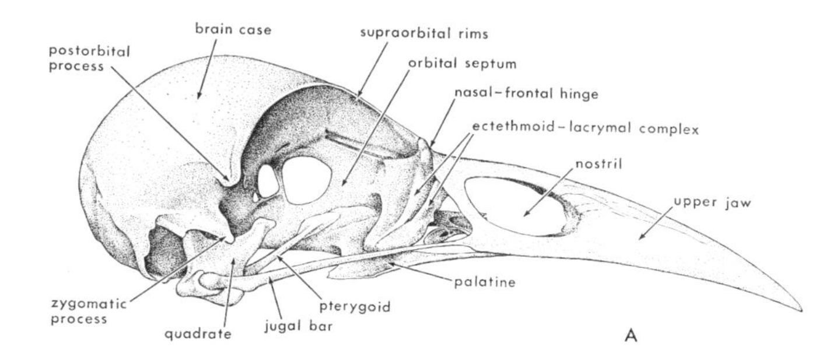

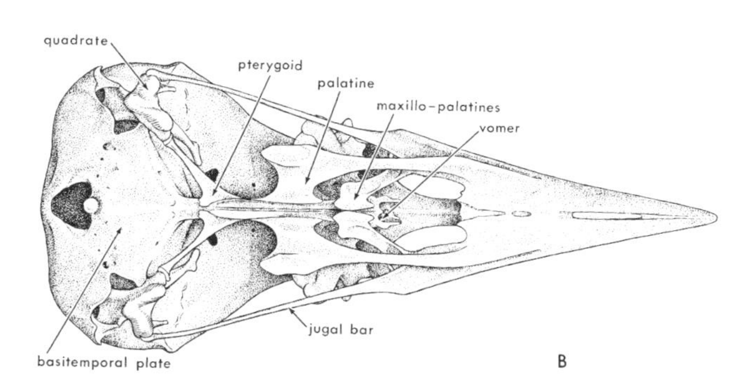

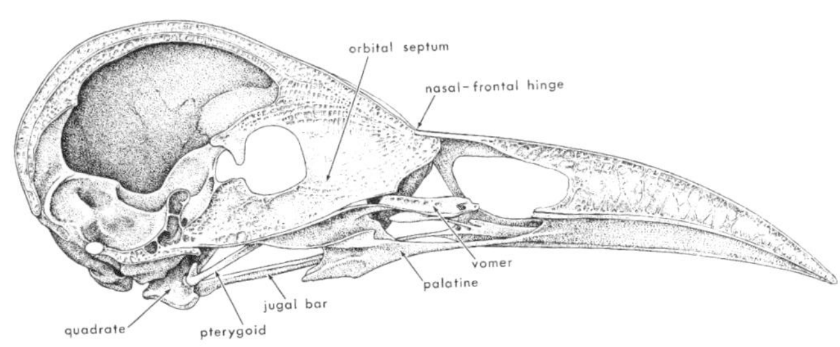

Left, Skull of a crow (*Corvus*) seen from the side (*A*) and from beneath (*B*). *Above,* skull of a crow cut along the midsagittal plane. (From Bock, *Journal of Morphology,* copyright © 1964; reprinted by permission of John Wiley and Sons)

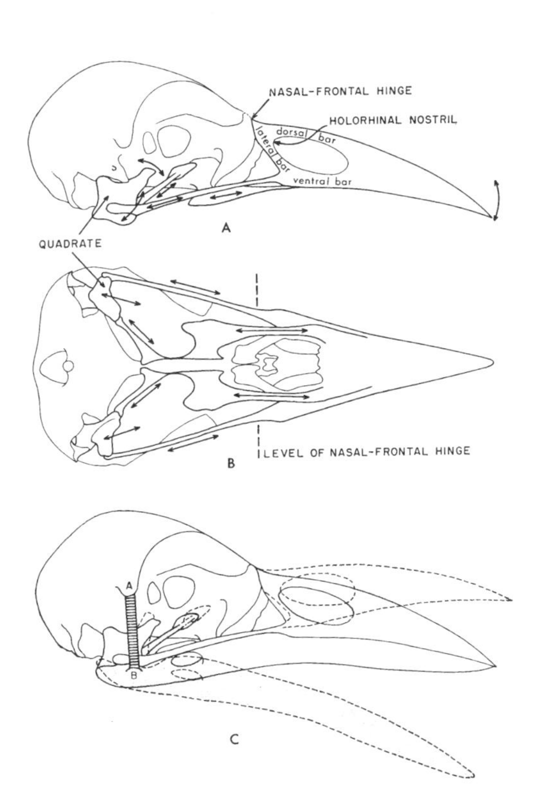

Schematic figures of a crow skull seen from the side (*A*), from beneath (*B*), and from the side with the mandible and the postorbital ligament in place (*C*), illustrating the movement of the upper jaw in cranial kinesis. The anterior and posterior movements of the various elements of the palate associated with the up and down movements of the upper jaw in prokinesis are shown in *A* and *B*. The positions of the open jaws (stippled) relative to the closed jaws are shown in *C*. (From Bock, *Journal of Morphology,* copyright © 1964; reprinted by permission of John Wiley and Sons)

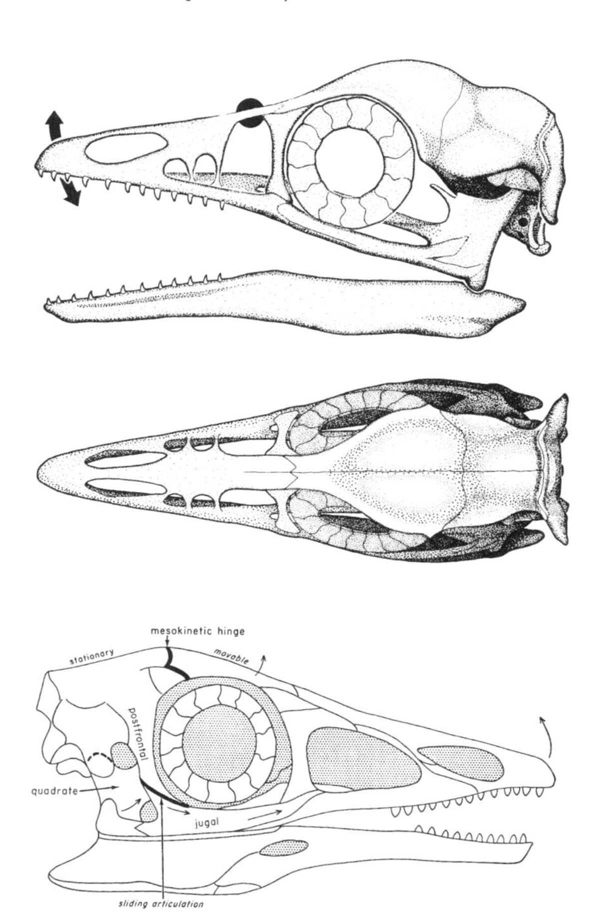

Reconstruction of *Archaeopteryx* skull by Paul Bühler, showing prokinesis, as in modern birds. The round dot marks the hypothetical craniofacial flexion zone, between the nasal and frontal bones, termed a nasofrontal hinge. The braincase is larger than it had been supposed, and Bühler's studies indicate that *Archaeopteryx* was much more birdlike than had been assumed previously, having among other avian features a prokinetic skull with an "inflated" and much more voluminous braincase. (From Bühler 1985; courtesy Academic Press.) Before Bühler's studies, the skull of *Archaeopteryx* was assumed to be of a reptilian, mesokinetic type, as in illustration at bottom, in which the hinge is between the parietal and frontal bones. It has been assumed that the modern prokinetic avian skull is derived from a reptilian mesokinetic type. (From Bock, *Journal of Morphology,* copyright © 1964; reprinted by permission of John Wiley and Sons.) Modern birds have evolved diverse forms of cranial kinesis, including rhynchokinesis (ratites and shorebirds) and amphikinesis (specialized shorebirds).

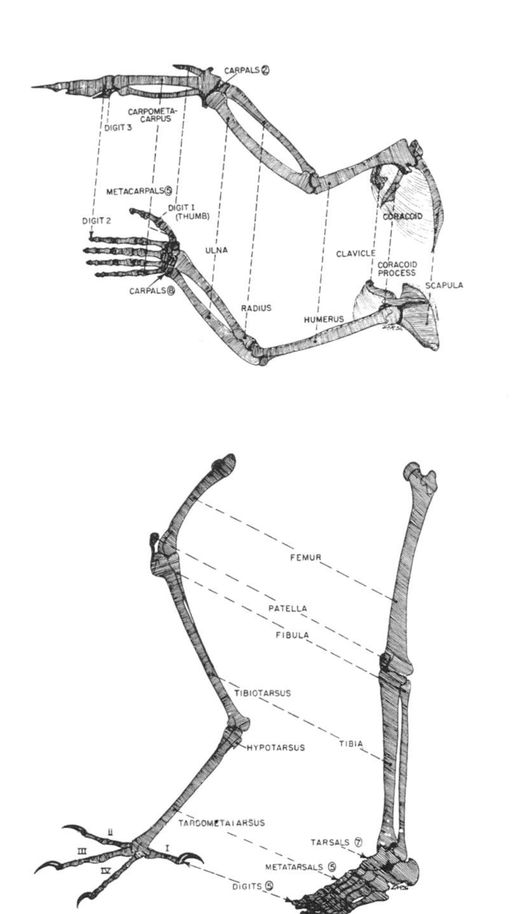

Homologies of the forelimbs and hindlimbs of a bird and a human. (From Van Tyne and Berger, *Fundamentals of ornithology,* copyright © 1976; reprinted by permission of John Wiley and Sons)

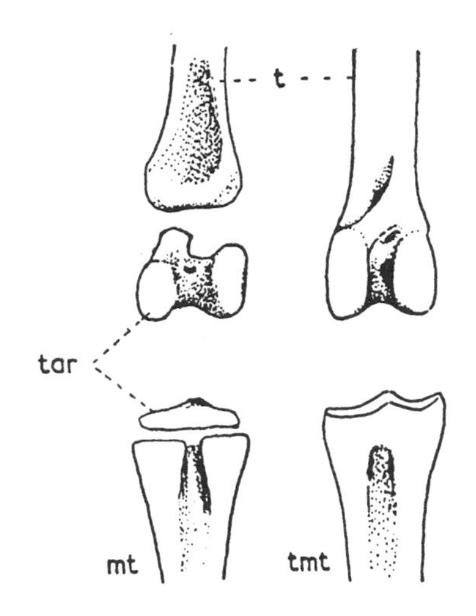

Right ankle region of (*left*) young chicken (*Gallus*) and (*right*) mature chicken, showing contribution of the tarsals to the tibiotarsus and tarsometatarsus. Front view not to scale. Bird embryos have at least three tarsal elements, but their identities remain obscure (Holmgren 1953). (Modified after Bellairs and Jenkin 1960; courtesy Academic Press)

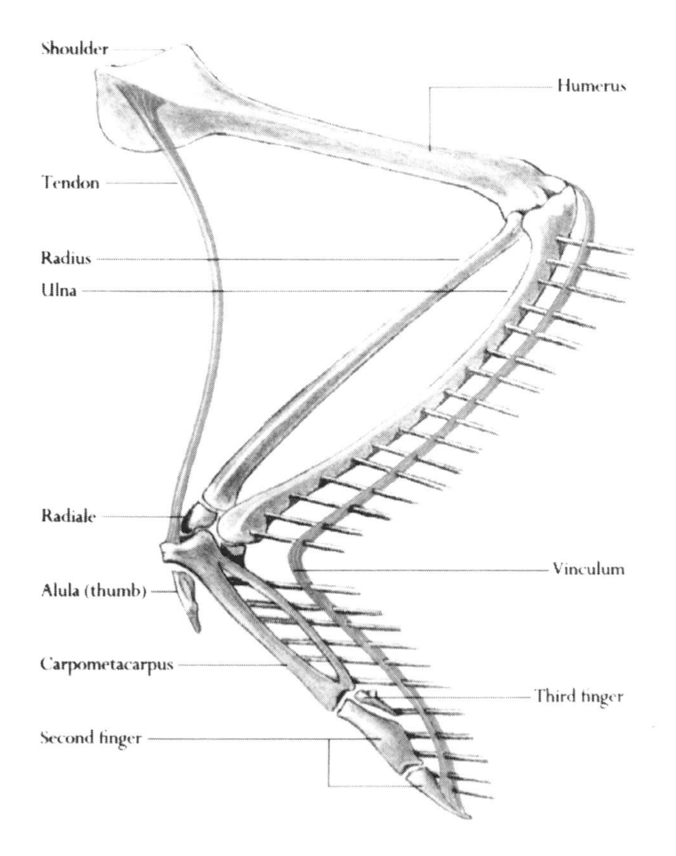

The main bones of the wing. The three avian metacarpals are fused to two carpals to form the avian carpometacarpus, the main bone of the avian hand. The two carpal elements, the ulnare and radiale, at the distal ends of the ulna and radius, respectively, lie between the forearm and manus and play an important role in supporting and controlling wing movements in active flapping flight. These carpal elements appear to restrict the movements of the manus and thus keep the primary feathers in proper alignment during the upstroke and downstroke (Vasquez 1992). The tendon running from the shoulder to the wrist stops the wing from opening too far and stretches the web of skin that forms the leading edge of the wing. A ligament, the vinculum, holds the flight feathers in place. (From Burton 1990; drawing by Sean Milne; courtesy Eddison Sadd Editions)

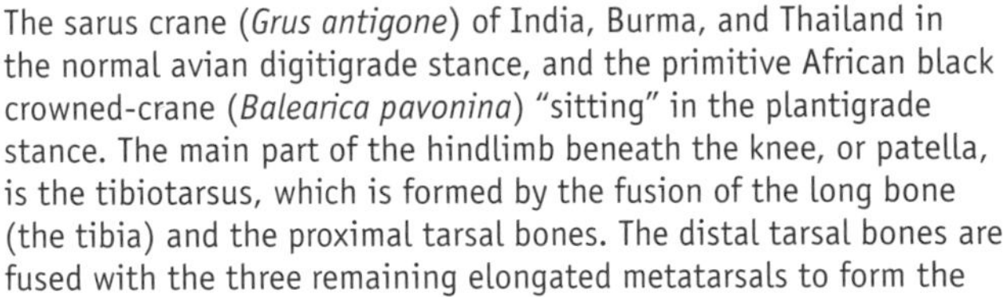

The sarus crane (*Grus antigone*) of India, Burma, and Thailand in the normal avian digitigrade stance, and the primitive African black crowned-crane (*Balearica pavonina*) "sitting" in the plantigrade stance. The main part of the hindlimb beneath the knee, or patella, is the tibiotarsus, which is formed by the fusion of the long bone (the tibia) and the proximal tarsal bones. The distal tarsal bones are fused with the three remaining elongated metatarsals to form the tarsometatarsus; thus, the three fused metatarsals, which end distally in three pulley-shaped processes, the trochleas, correspond to the three anterior toes, 2, 3, and 4. The only free metatarsal is the small rudimentary, distal metatarsal one that is posteriorly directed with its toe composed of a basal and terminal phalanx; this is an adaptation for perching called the hallux. (Photos by author)

Compared to the human, a plantigrade vertebrate with a foot composed of tarsals, or ankle bones, metatarsals, or foot bones, and phalanges, or toes, birds are digitigrade—that is, they walk on their toes. Many young birds as well as the adults of especially long-legged birds, when resting, will extend the foot in a fashion analogous to humans. The joint that is formed between the rounded medial and lateral condyles of the tibiotarsus and the head of the tarsometatarsus is called a mesotarsal joint, because it passes between the proximal (upper) and distal (lower) tarsal elements. This type of joint is found in a number of descendants of archosaurs, including birds, pterosaurs, and dinosaurs, as well as in diverse Triassic thecodonts.

Still other fusion has taken place. In the vertebral column, for example, the sacral vertebrae have fused with other thoracic, lumbar, and urosacral vertebrae to form the synsacrum (fused sacral region vertebrae), and in a number of bird families, the thoracic vertebrae are fused to form a notarium, as in the pterosaurs. Ribs are equipped with horizontal bony connecting flaps called uncinate processes, struts that extend from the vertical upper ribs to overlap and bind the adjacent ribs, in effect binding the entire rib cage together. Paired coracoids, the anterior (forward) elements of the pectoral girdle, act as mechanical struts to prevent collapse of the anterior chest region during flight. The clavicles are fused to form the furcula, or wishbone; this uniquely avian trait stabilizes the shoulder joint and helps form a rigid chest region. The paired clavicles are typically fused in the midline, where their distal ends form a flattened plate known as the hypocleidium; however, many modern birds lack a hypocleidium, as does *Archaeopteryx*. The furcula serves biomechanically as a spring-spacer, bending and recoiling during flight, may play a role in breathing (Jenkins, Dial, and Goslow 1988), and, along with a special membrane, may serve as a partial site for the attachment of the pectoralis flight muscles that power the downstroke of the wing in flight (Berger 1960; Raikow 1985a).

The first bird, *Archaeopteryx*, had a long reptilian tail with numerous vertebrae, but in most modern birds, the distal four to seven embryonic caudal vertebrae are fused into a tapered, single bony pygostyle (or plowshare bone) from which the tail feathers (rectrices) emanate. A true pygostyle is absent in most ratites and the tinamous, and embryonic ratites, especially the kiwi, have a long, unmodified reptilelike tail. The pygostyle also accommodates the bilobed oil gland (also called the uropygial, or preen, gland) on either side, which secretes a rich fatty, waxy oil that birds apply externally with the bill to clean feathers

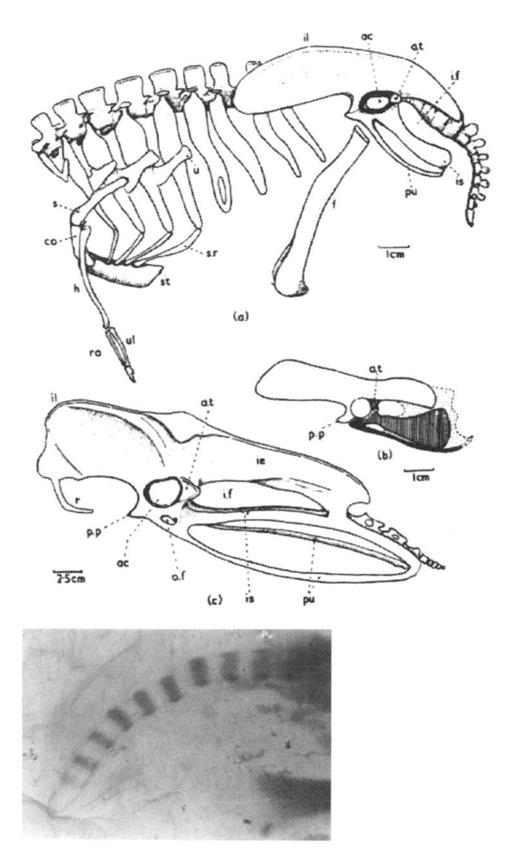

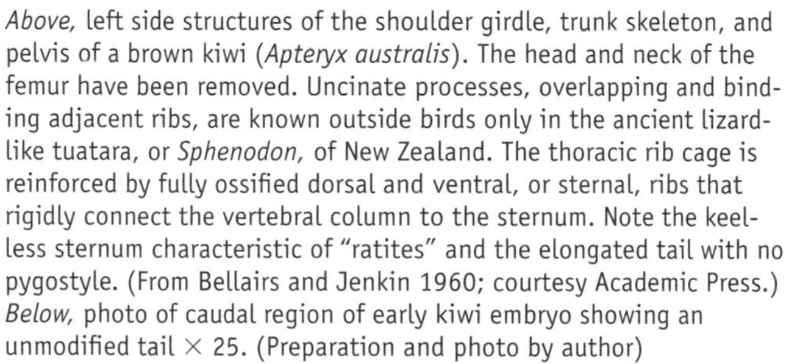

Above, left side structures of the shoulder girdle, trunk skeleton, and pelvis of a brown kiwi (*Apteryx australis*). The head and neck of the femur have been removed. Uncinate processes, overlapping and binding adjacent ribs, are known outside birds only in the ancient lizard-like tuatara, or *Sphenodon,* of New Zealand. The thoracic rib cage is reinforced by fully ossified dorsal and ventral, or sternal, ribs that rigidly connect the vertebral column to the sternum. Note the keelless sternum characteristic of "ratites" and the elongated tail with no pygostyle. (From Bellairs and Jenkin 1960; courtesy Academic Press.) *Below,* photo of caudal region of early kiwi embryo showing an unmodified tail × 25. (Preparation and photo by author)

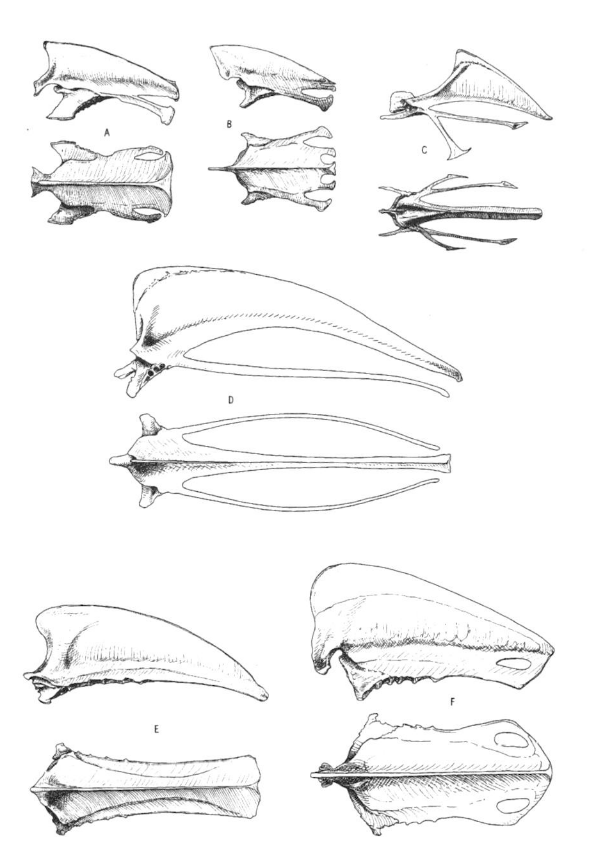

Examples of diversity in the structure of the avian sternum. *A,* russet-backed oropendola (*Psarocolius alfredi,* Icteridae); *B,* belted kingfisher (*Megaceryle alcyon,* Alcedinidae); *C,* scaled quail *(Callipepla squamata,* Phasianidae); *D,* great tinamou (*Tinamus major,* Tinamidae); *E,* limpkin (*Aramus guarauna,* Aramidae); *F,* scarlet macaw (*Ara macao,* Psittacidae). Sternal structure, including the pattern of notching on posterior margin, has long been an aid in classifying birds (Heimerdinger and Ames 1967; Ames et al. 1968), but the functional significance of these notches remains unclear (Feduccia 1972). (From Van Tyne and Berger, *Fundamentals of ornithology,* copyright © 1959; reprinted by permission of John Wiley and Sons)

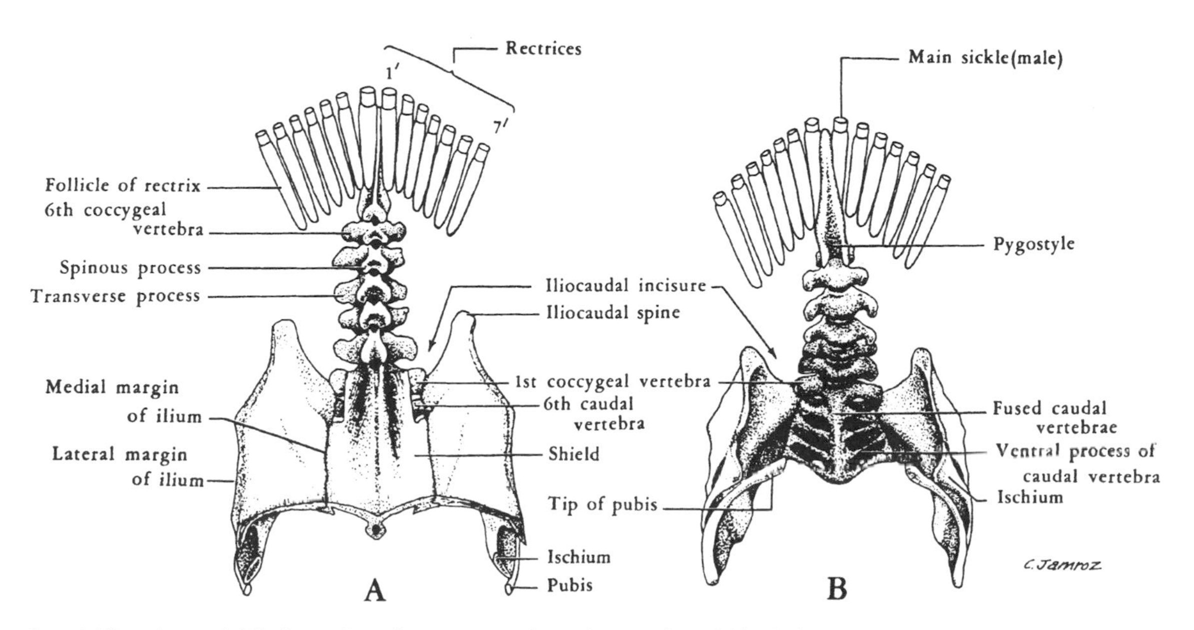

Dorsal (*A*) and ventral (*B*) views of rectrices, coccygeal vertebrae, and caudal half of the pelvic girdle to illustrate the relation of the rectrices to the pygostyle in the chicken (*Gallus*). (From Lucas and Stettenheim 1972; courtesy Peter Stettenheim; modified by J. Cagle)

and preserve their moistness and flexibility and sustain the plumage as an insulatory and waterproofing layer (Jacob and Ziswiler 1982). This gland is absent in certain birds, such as the ostrich and emu, bustards, and some parrots, pigeons, and woodpeckers, to mention a few. This branched, alveolar oil gland is the only conspicuous gland derived from the skin of birds, and demonstrates that, as with other organs, the skin, too, has become greatly reduced in weight and is paper-thin in most species of flying birds.

The Sternum and Flight Apparatus

The keeled, or carinate, bird sternum is highly modified to accommodate the extensive flight musculature. In a sense, the carinate sternum is analogous to the mammalian skull's sagittal crest, which serves for the attachment of massive jaw muscles. Its most striking feature is a large, bony expanse along the midline called the keel, or carina (Latin, keel of a ship), which provides a broad area for the attachment of the two major flight muscles. The main muscle arising from the keel and responsible for raising the wing for the recovery stroke in modern birds is the large supracoracoideus, and it has unusual features that allow it to perform this function. Originating in the lower region of the carina, it extends up the sternum and is connected to the humerus by a tendon that passes through the triosseal canal, so called because it is formed by the conjunction of the coracoid, the scapula, and the furcula. The supracoracoideus inserts by tendinous fibers on the dorsal surface of the humerus, just distal to its articular head. The supracoracoideus muscle is thus firmly anchored to both the carina and the humerus, and by running a tendon over a pulley in the shoulder, it does most of the work needed to raise the wing in flight. If it did not, the dorsal elevators, small muscle on the back of the body associated with the scapula, would have to perform this task by themselves. Indeed, in bats, which lack the keeled sternum and the supracoracoideus, and in early birds, such as *Archaeopteryx*, that had not yet evolved the supracoracoideus, the dorsal elevators of the scapula were probably solely responsible for effecting the recovery stroke. In birds that literally fly through the water, and are therefore among the "best" fliers, such as penguins and auks, these dorsal elevators increase in mass, with a concomitant increase in the size of the blade of the scapula. The keeled or unkeeled condition of the sternum was for many years the primary criterion used to define the major groupings of living birds, respectively, the carinate and ratite birds. The flightless ratites (Latin, *ratis*, raft) include the living kiwis, ostriches, emus, cassowaries, and rheas; these birds have lost the carina and their sternum is a more or less flat breast plate. This keel-less sternum is also characteristic of other birds that have lost the power of flight, including the Cretaceous foot-propelled diver *Hesperornis*, and the keel is all but absent in New Zealand's flightless owl parrot, the kakapo *(Strigops habroptilus)*.

The massive pectoralis muscle (sometimes called the pectoralis major) pulls the wing down for the power stroke in flight. The pectoralis may be as little as 1.7 times as large as the supracoracoideus in hummingbirds and as much as 15 to 22 times as large in gulls and kites (George and Berger 1966, 22). It originates in part from the sternum and in part from the furcula and adjacent membranes, the coracoclavicular membranes, the ribs, and the coracoid (Berger 1960; Raikow 1985a), and it inserts by a large tendinous area on the deltoid crest of the humerus, thus pulling the wing downward. The powerful pectoralis and supracoracoideus muscles, together comprising about 15 percent of body weight, may range from 7.8 percent in the white-throated rail (*Laterallus albigularis*) to 30 percent in certain hummingbirds and up to 36.7 percent in Cassin's dove (*Leptotila rufinucha*) (George and Berger 1966; Hartman 1961). The magnitude of these muscles singularly illustrates that avian architecture is designed around one theme: flight. All other structural considerations are subservient.

Bills and Feet

The features of birds most likely to be modified rapidly through natural selection are the bill and the feet, the primary implements for feeding and locomotion. The bills replace the lost reptilian teeth with an epidermally derived, keratinized beak. This horny sheath, or rhamphotheca, as mentioned, may be equipped with lateral cutting ridges, a hooked tip for tearing flesh, even a "horny palate" for holding food. The lateral cutting edges of the upper beak are known as the upper tomia. The keratinized beak grows throughout a bird's life and maintains its characteristic form through wear caused by eating and other activities. The horny beak may also perform many of the functions of the mammalian forelimb, the avian forelimb being preempted by flight. In addition to eating, bills are used in nest building, fighting, and grooming. Birds kept in zoological parks, which often cannot perform their normal functions with the beak and are given foods they would not eat in the wild, may exhibit long, deformed bills, especially the tips.

Differences in beak shape thus directly reflect different species' way of life, as exemplified by the adaptive radiation of the Hawaiian honeycreepers and the Galápagos, or Darwin's, finches. Bird beaks are infinitely varied; each variant is a slightly different modification for a particular

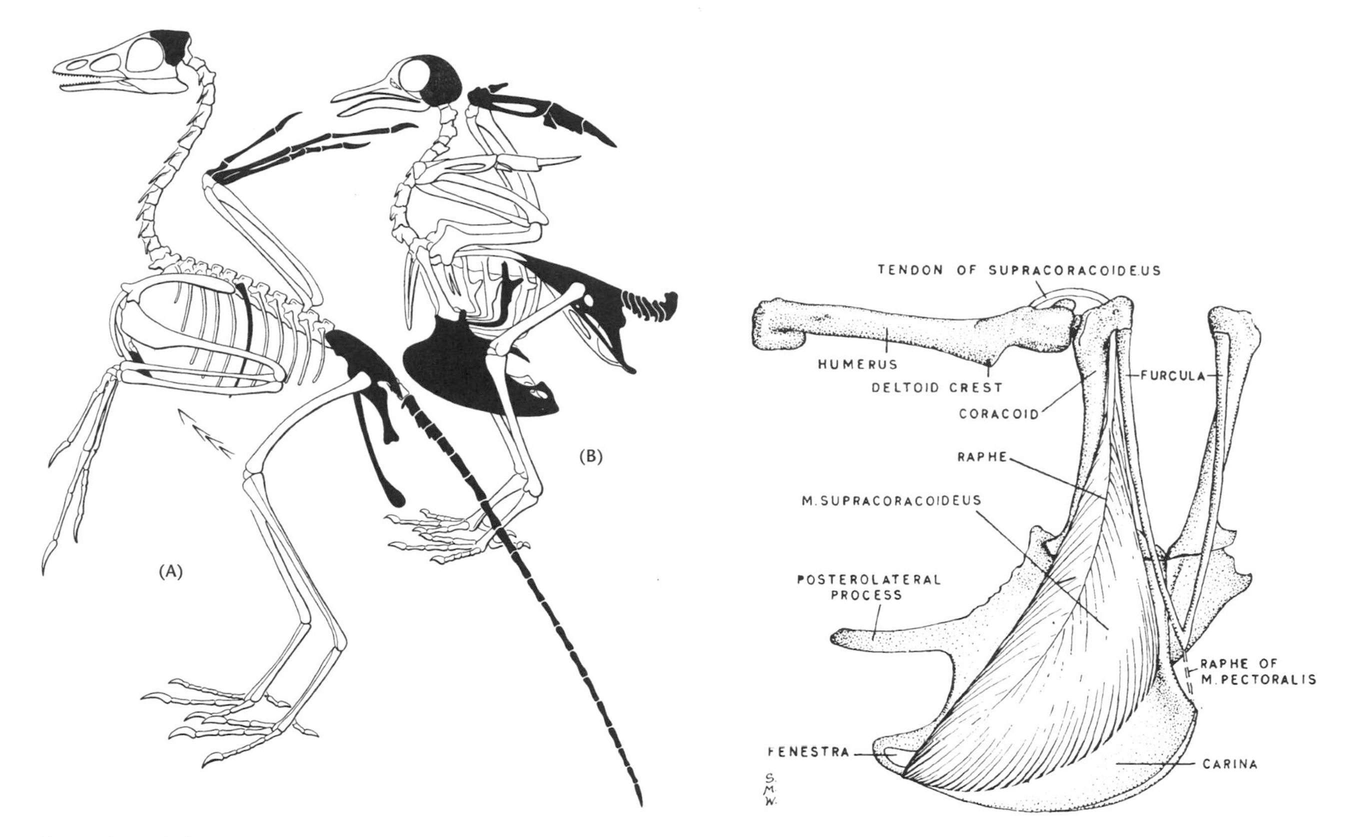

Comparison of the skeletons of *Archaeopteryx* (*A*) and a modern domestic pigeon (*B*). Comparable regions of the skeleton (braincase, hand, sternum, rib, pelvis, tail) are shaded black. In modern birds, the tail is reduced and several terminal vertebrae are fused to form the pygostyle; the braincase is expanded and the head bones are fused; the hand bones are fused to form a carpometacarpus and the fingers are reduced; the pelvic girdle is joined to the fused sacral vertebrae (the synsacrum); the ribs are bound rigidly by uncinate processes; and the sternum is greatly expanded and keeled for the attachment of the major flight muscles. All are adaptations associated with perfecting the mechanism of flight. (From Colbert and Morales, *Evolution of the vertebrates,* copyright © 1991; reprinted by permission of John Wiley and Sons)

A front (anterior and lateral) view of the sternum and pectoral girdle of the domestic pigeon, showing the general relation of the supracoracoideus and its tendon of insertion. The tendon passes through the triosseal (three-bone) canal formed by the articulation of the furcula (fused clavicles), coracoid, and scapula (not shown here). The supracoracoideus is the major muscle used to raise the wing in flight in modern birds. The pectoralis muscle, which powers the wing in flight, has been removed; it arises partly from the area of the sternum not occupied by the supracoracoideus, but mostly from the front and side surfaces of the furcula and adjacent membranes. (From George and Berger 1966; courtesy Academic Press)

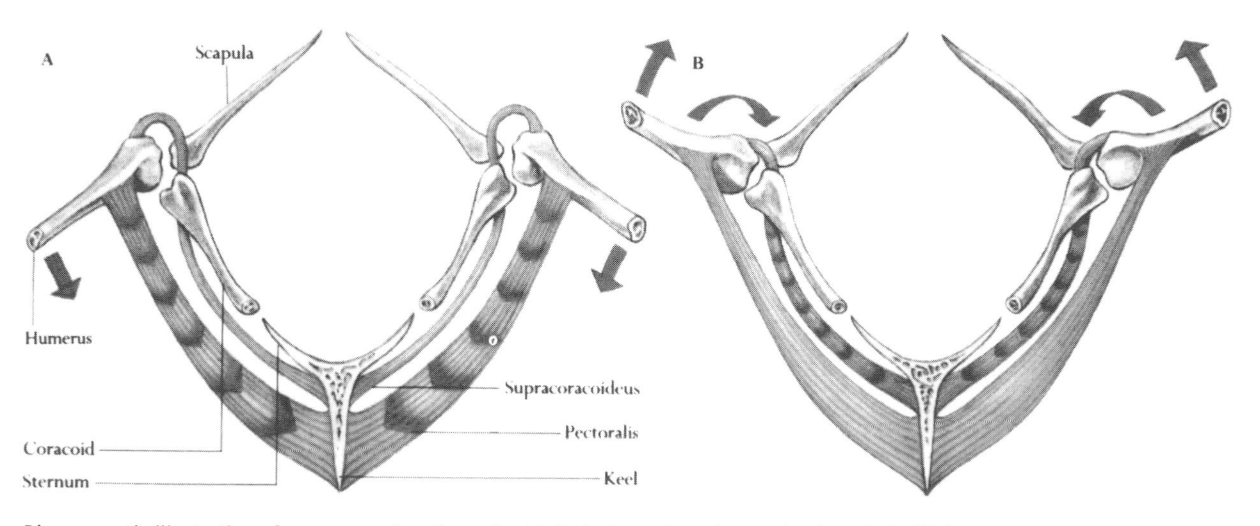

Diagrammatic illustration of a cross-section through a bird's body to show the mechanism of the flight muscles. The muscles are anchored to the breastbone and pull on the humerus of the wings. *A,* contraction of the large pectoralis muscles pulls the wings down; *B,* contraction of the smaller supracoracoideus muscles acts through "pulleys" to pull the wings up, or effect the recovery stroke. (From Burton 1990; drawing by Sean Milne; courtesy Eddison Sadd Editions)

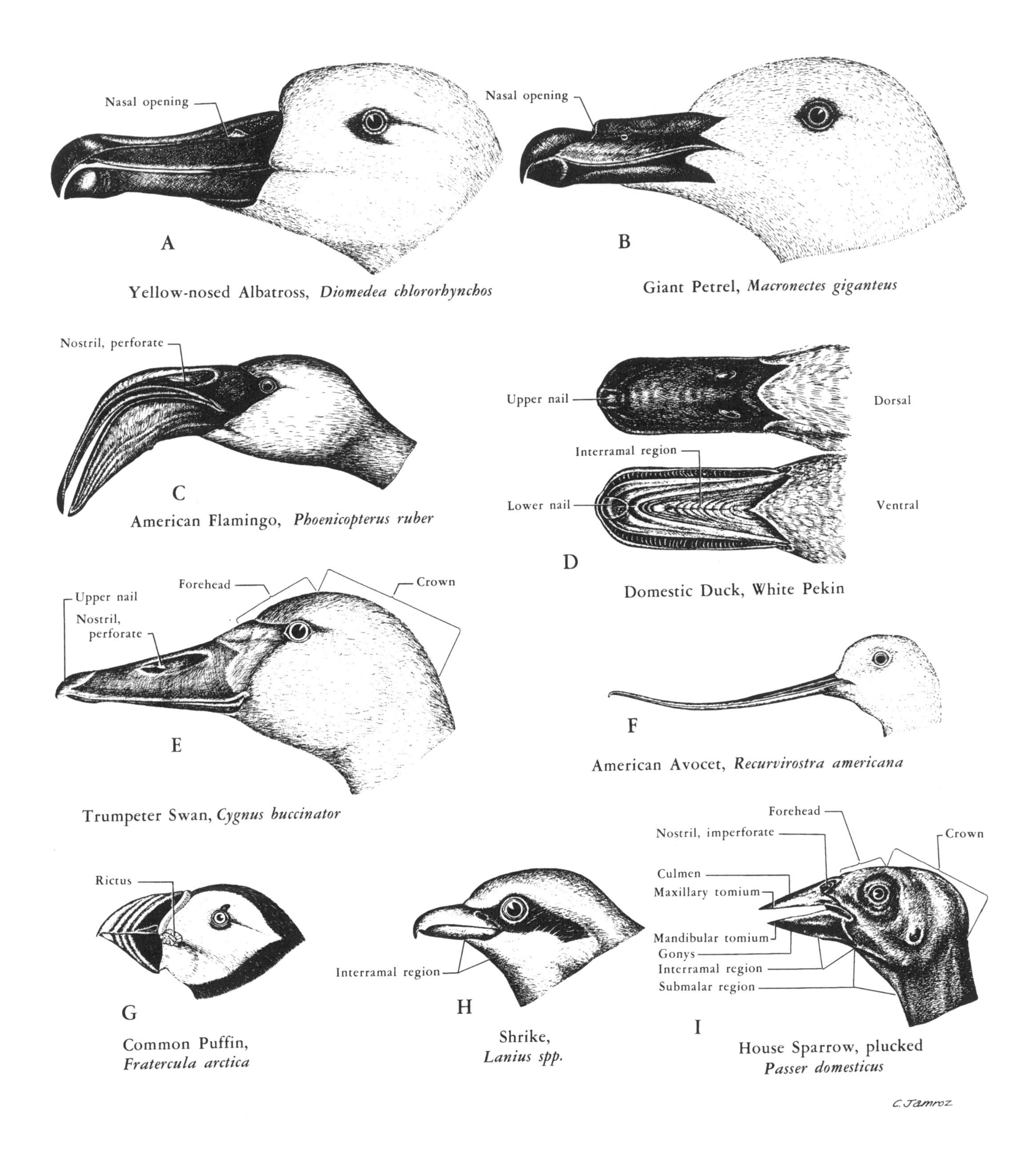

Examples of the diversity in shape and structure of bird bills produced by adaptive radiation. (From Lucas and Stettenheim 1972; courtesy Peter Stettenheim)

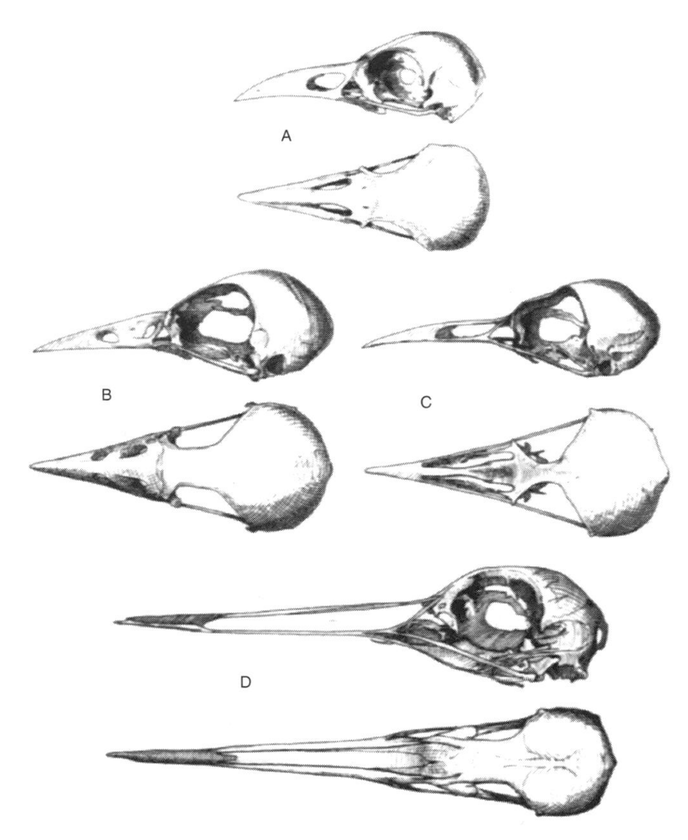

Dorsal and lateral views of skulls to show the four main types of bony nostrils, revealed by removal of the cornified beak, or rhamphotheca. *A,* holorhinal: American crow (*Corvus brachyrhynchos,* Corvidae); *B,* amphirhinal: white-cheeked antbird (*Gymnopithys leucaspis,* Formicariidae); *C,* pseudoschizorhinal: blackish cinclodes (*Cinclodes antarcticus,* Furnariidae); *D,* schizorhinal: sandhill crane (*Grus canadensis,* Gruidae). These nasal types generally relate to the type of cranial kinesis. The holorhinal and its variant, the amphirhinal, are characteristic of prokinetic skulls; whereas the schizorhinal, or shorebird, type is associated with bending at the tip and a form of kinesis known as rhynchokinesis. (From Van Tyne and Berger, *Fundamentals of ornithology,* copyright © 1959; reprinted by permission of John Wiley and Sons)

style of feeding, compromised with adaptations for other activities. There are the soft, lamellated, leathery, and flexible structures of ducks and flamingos designed for straining algae and invertebrates from mud and water. Other dramatic adaptive types include the raptorial tearing bills of owls and hawks, the conical bills of seedeaters, the broad gapes of nightjars, frogmouths, swifts, swallows, and others, the flower-probing bills of hummingbirds, the chisel-shaped bills of woodpeckers, the elongate, slender bills of waders and shorebirds, adapted for probing in mudflats, the daggerlike bills of herons and anhingas, used for impaling fish, and the serrated bills of the fish-eating mergansers. Among the most bizarre beaks are the boat-shaped bills of the African shoebill and tropical American boat-billed heron, adapted to catch slippery lungfish and frogs, respectively; the spatulalike bills of spoonbills; and the trowel-like bills of skimmers, in which the lower mandible overshoots the upper, so that it is plowed beneath the water's surface to catch fish.

The Perching Foot. Found in no other group of vertebrates, the original perching foot of birds is a unique structure that has become modified into a marvelous variety of adaptive types. The reptiles ancestral to birds, whatever they were, had a pamprodactyl foot with all five digits present and directed anteriorly. In the avian lineage the fifth digit was lost and the first digit (the hallux, homologue of our big toe) rotated to the posterior position to oppose the remaining toes, thus producing an anisodactyl foot.

Archaeopteryx and other ancestral birds had anisodactyl feet with three front and one opposing hind toe adapted for perching in trees. But the ancient anisodactyl foot was less efficient anatomically than the foot of modern birds, and the hallux of *Archaeopteryx* was slightly elevated above the plane of the anterior toes and decidedly shorter. In modern perching birds (the passerines), the hallux is not elevated and is usually longer than the second toe, almost as long as the third toe. By the early Cretaceous period, birds such as the Chinese *Sinornis* and *Cathayornis* appear to have had modern anisodactyl feet with highly curved claws, possibly as efficient as those of modern perching birds, although the soft anatomy is unknown. Nevertheless, a number of other perching bird foot types have evolved from the ancestral anisodactyl foot, including the syndactyl foot, in which the anterior toes are partially encased by a common sheath of skin, and the zygodactyl and heterodactyl feet, in which either the fourth or the second toe, respectively, is reversed (see table 1.1). Climbing feet have also evolved in many avian lineages, and the anisodactyl, zygodactyl, and pamprodactyl feet all appear to be efficient climbing structures, although the pamprodactyl is perhaps the most highly specialized climbing foot (Bock and Miller 1959).

Drastic structural changes to the avian foot have also occurred in many water birds, especially in the strong swimmers: the anterior toes have become webbed and the hallux has been reduced. Most swimmers, such as penguins, loons, tube-nosed swimmers, ducks and allies, and the shorebirds and allies, including gulls and terns, have anisodactyl palmate feet (some have semipalmate feet), usually with the hallux small, elevated, or lacking. In ducks, the hallux may be lobed. The avian order Pelecaniformes, composed of aquatic birds, is defined primarily by a medially directed hallux and all four toes webbed, or the totipalmate foot. Wading birds and birds that walk on soft substrates have adaptations that spread their weight to prevent sinking (Raikow 1985a). Birds that wade in

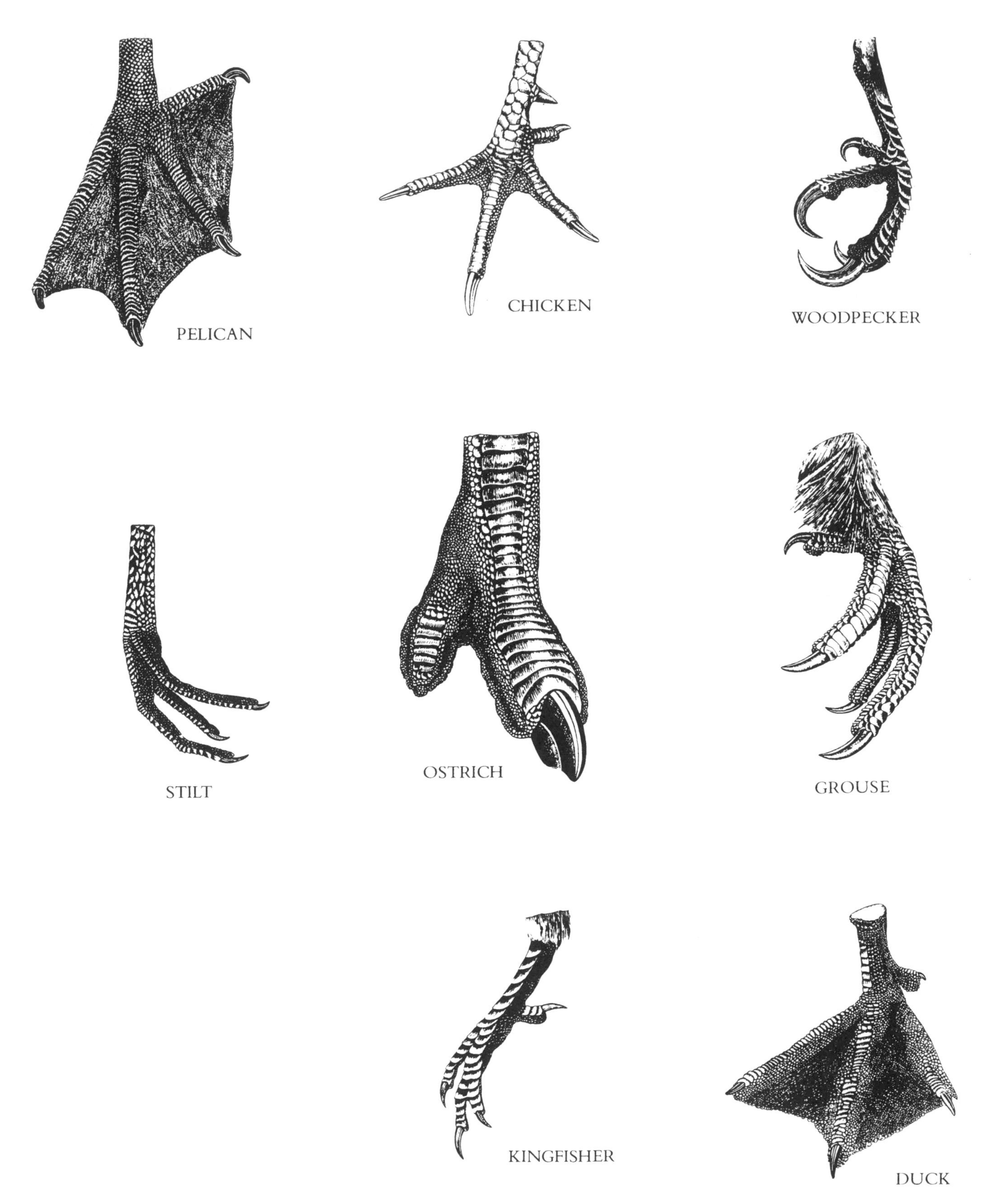

Examples of the diversity in shape and structure of bird feet, all modified from the original anisodactyl foot, with three anterior toes and one opposing posteriorly directed toe, the hallux, digit 1. (From Lucas and Stettenheim 1972; courtesy Peter Stettenheim)

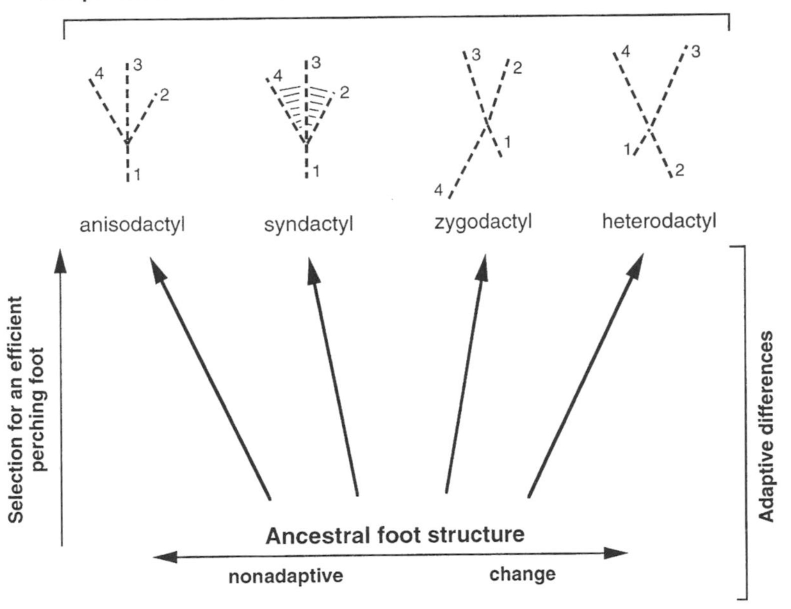

Schematic diagram illustrating the pattern of multiple pathways of evolution of perching feet in, *from left to right,* anisodactyl, syndactyl, zygodactyl, and heterodactyl birds. The evolution of the four arrangements from the ancestral condition was under the control of the same selection force for a more efficient perching foot. Differences observed in the vertical comparisons are adaptive, whereas horizontal differences represent multiple pathways of adaptation. (Modified from Bock and Miller 1959 and Bock 1967)

Table 1.1 The five major foot types as defined by toe arrangement

Anisodactyl. Second, third, and fourth toes anterior; first toe (hallux) posterior (to the tarsometatarsus) and usually long. This is the perching foot in primitive birds (*Archaeopteryx*) and the most common perching foot in modern birds (passerines); it is also, however, a perfectly suitable climbing foot.

Syndactyl. Second, third, and fourth toes anterior and partly encased by a common sheath of skin; first toe (hallux) posterior and usually long. This type of anisodactyl foot is found primarily in many coraciiform birds.

Zygodactyl. Second and third toes anterior; first and fourth toes posterior. This is the characteristic foot of the woodpeckers and allies, parrots, and cuckoos, but many birds, such as some owls, the mousebirds, and the musophagids, have a temporary zygodactyl foot, in which the fourth toe can be held either anterior or posterior when the bird is perching.

Heterodactyl. Third and fourth toes anterior; first (usually short) and second toes posterior. This foot is characteristic only of the Neotropical trogons.

Pamprodactyl. All four toes anterior; first toe on the inside of the foot, next to the second toe, but the position of the first and fourth toes is not fixed. This is one of the more specialized climbing feet and has evolved independently in different groups, such as one genus of parrot (*Micropsitta*), certain swifts, the oilbird, and the mousebirds.

Note: Other more specialized foot types include ectropodactyl, tridactyl (involving the loss of the hallux), and didactyl, the foot of the ostrich.

Source: After Bock and Miller 1959; see also Raikow 1985

muddy or soft-bottomed waters may have anisodactyl feet with elongated toes and a long, incumbent hallux, as do herons (family Ardeidae). Storks and ibises have basally webbed toes. Stilts and avocets have webbed feet for wading and swimming, with a greatly reduced or absent hallux. The marsh-dwelling rails (Rallidae) have long toes and claws, and the bizarrely lengthened toes and claws of the jacanas, or lily trotters (Jacanidae), are adapted for walking on floating lily pads. Among the Galliformes (chickens and allies) and the Columbiformes (pigeons and allies), foot adaptations include densely feathered snowshoe feet in grouse, particularly ptarmigan (*Lagopus*) and sandgrouse (*Syrrhaptes*) for walking on snow and sand, respectively.

Although loons and grebes are both foot-propelled divers, they derive from different ancestral stock and have developed different foot structures. Loons have an anisodactyl palmate foot with three webbed anterior toes. Grebes, in contrast, have an anisodactyl foot with lobate webbing and flattened claws. In both loons and grebes the hallux is small and elevated. Lobate, webbed toes are also present among gruiform birds (rails and relatives) in the finfoots, or sungrebes (Heliornithidae), and the coots (Rallidae), and they occur among the shorebirds within the phalaropes (Phalaropidae); the extinct, toothed, hesperornithiform birds of the Cretaceous also had lobate, webbed toes.

Terrestrial or semiterrestrial birds occur in a variety of avian orders, including ratites and the galliform, gruiform, and charadriiform orders. The most highly modified feet, such as those of ostriches, emus, cassowaries, and rheas, lack a hallux, and in the kiwis and tinamous, the hallux is small and elevated. These facts make it extremely improbable that cursorial dinosaurs would have evolved a foot that featured a reversed hallux, as frequently illustrated, which would be unlike their running counterparts among birds and would only have encumbered their running abilities. In fact, most ratites are tridactyl, and ostriches, having reduced their number of toes to two, are didactyl, approaching the extreme adaptations of running hoofed mammals. As Storer (1971) has argued, problems of balance inherent in the avian bipedal stance have likely prevented birds from developing the most extreme cursorial adaptations observed in such hoofed mammals as the horses.

The Avian Wing. Of equal prominence with bills and feet are wings, which reflect birds' ability to fly. Through natural selection, the size and shape of wings have been optimized to minimize the energy required to fly at the speed and style characteristic for a given bird. One can tell much about a bird's mode of living by its wing, which makes wing shape a central diagnostic feature of birds. Wing shapes vary tremendously: from the broad spans of the albatross, which can soar for hours with minimal energy expenditure, to the tiny fanlike wings of hummingbirds, which hover as they forage, requiring enormous expenditure of energy. Each type requires a dramatically different wing morphology.

The energy expenditure of flight can be expressed as the relation between the total wing area and body mass, or wing loading, which is calculated by the number of grams of bird weight that a given area of bird wing must carry and is expressed by grams per square centimeter of wing surface area. Wing loading is thus the ratio of bird weight to wing area.

Another expression of wing functional morphology is called aspect ratio—simply stated, the ratio of wing length to wing width. Thus, long and narrow wings, designed for high speed, have a high aspect ratio; that is, the length divided by the width produces a high number. Conversely, short, broad wings, designed for low speed and maneuverability, have a low aspect ratio; the length divided by the width gives a low number.

Ornithologists tend to recognize four basic wing types (Savile 1957), but this number is arbitrary, and there are many more styles of wings. Birds such as gallinaceous species, doves, woodcock, woodpeckers, and most passerines tend to have short, broad *elliptical wings* that are designed for maneuvering in and out of thick brush and dense, woodland vegetation and are characterized by low aspect ratios, uniform pressure distribution over the wing surface, and pronounced camber, or convex curvature. Although this wing is less effective than some other designs at high sustained speeds, it is splendidly suited for twisting flight in dense vegetation (Savile 1957). Interestingly, this is the wing type of the earliest known bird, *Archaeopteryx*.

Among the birds characterized by long, narrow *high-speed wings* are swifts, swallows, falcons, hummingbirds, sandpipers, and plovers; their wings have moderately high aspect ratios, lower camber than most birds, and taper to a slender, elliptical tip, giving a swept-back appearance. These slim, unslotted wings permit fast flight in open habitats. Seabirds such as albatrosses, shearwaters, and tropicbirds have *high-aspect-ratio wings*, designed for long-distance gliding; their wings are long, narrow, and flat, with no slotting on the primary feathers. In such birds, wing length greatly exceeds width, and aspect ratios as high as 15 and 18 have been calculated for two species of albatrosses; by comparison, the catbird's aspect ratio is 4.7 (Savile 1957). Last, vultures, eagles, and storks have *high-lift* or *slotted soaring wings*, of moderate aspect ratio with pronounced camber and slotting, which in large birds produces a very efficient soaring wing. These rela-

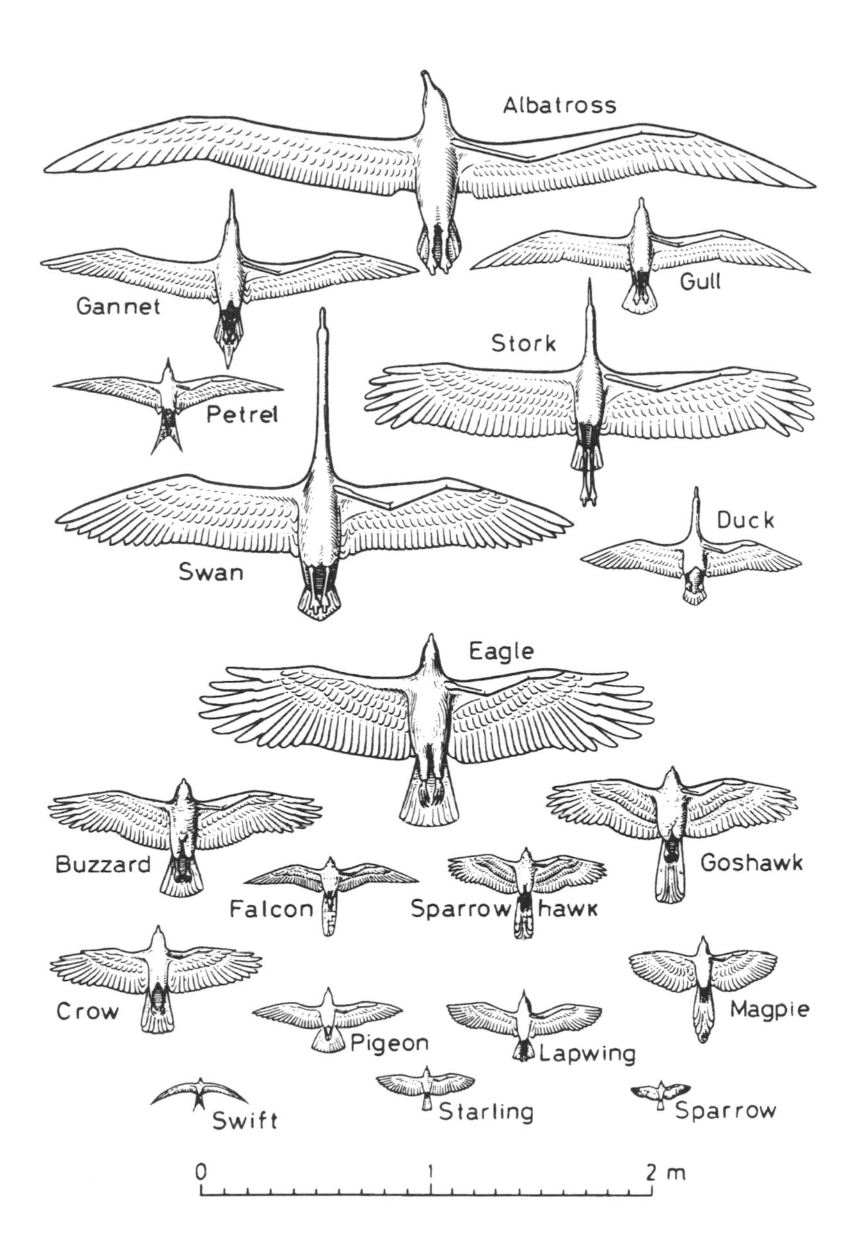

Diversity of wing shapes in birds. The four basic types of wings are illustrated by: upper, high aspect ratio (albatross); lower left, high-speed wing (swift); lower right, elliptical wing (magpie); middle, soaring wing (eagle). All levels of intermediacy are seen in the various birds. (From U. M. Norberg 1990, after Herzog 1968; courtesy Springer-Verlag)

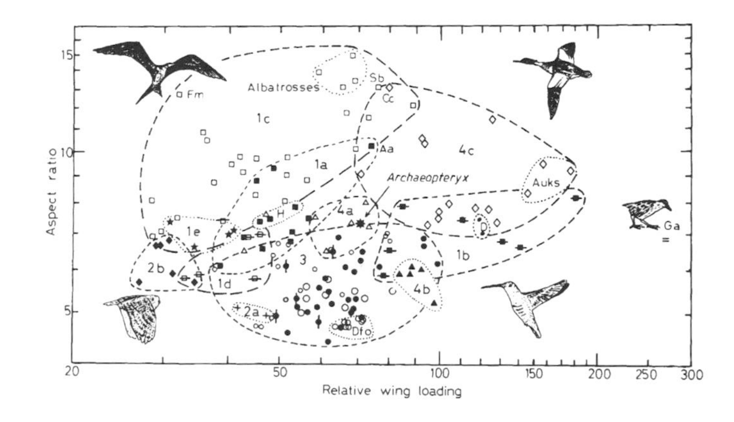

Aspect ratio versus relative wing loading in various birds. The encircled foraging groups are defined solely by the similarity of flight and foraging modes, not by systematic affinity. Aa = *Apus apus* (swift); Cc = *Cygnus cygnus* (swan); D = diving petrels; Df = Darwin's finches; Fm = *Fregata magnificens* (frigatebird); Ga = *Gallirallus australis* (flightless New Zealand wood rail); H = swallows; Sb = *Sula bassana* (booby). Note particularly the position of the urvogel *Archaeopteryx*. Data from sources compiled in U. M. Norberg and R. Å. Norberg (1989) and from Pennycuick (1987). (From U. M. Norberg 1990; after U. M. Norberg and R. Å. Norberg 1989)

tively broad, short wings provide for light loading, and because the primaries are slotted, each can be used as an individual airfoil, a major feature of the high-lift wing.

A triangular fold of skin known as the propatagium (or prepatagium) considerably increases the surface area of the wing. This web of skin stretches between the shoulder region and the wrist. Its leading edge is supported by a tendon, which may also control its stretch, prevent the wing from opening too far, and automatically extend the wrist and spread the flight feathers. A smaller fold of skin, the metapatagium, between the trunk and the elbow, adds more wing surface area. Wing patagia like these may have been critical in the early evolution of flight, for jumping and gliding, before true powered flight originated.

The tail is in fact an aerodynamic adjunct of the wing (Thomas 1993); it is an indispensable adaptation for generating lift and providing stability in pitch. In flying birds, the tail provides additional wing surface area and is, as would be expected for increasing lift, spread and depressed at very low speeds, particularly during take-off and landing (Pennycuick 1972). Among other aerodynamic advantages, tails give birds added control at low-speed flight. In *Archaeopteryx*, a pair of tail feathers emanates laterally from each caudal vertebra, whereas in modern birds the tail feathers, or rectrices, attach to the soft tissue of the pygostyle region and can be spread transversely or moved up and down. Tail morphology varies tremendously, reflecting the myriad of variations on styles of flight.

Urvogel—*Archaeopteryx*

The phylogeny of a group of organisms would be incomplete without the study of fossils as well as living organisms. The fossil record allows us to place organisms and their lineages on the geological time scale and permits us to see changes in diversity and morphology through time, including extinctions and rates of evolution. *Archaeopteryx*, for example, living in the late Jurassic period, was preserved in what geologists call Upper Jurassic limestone—the terms *early*, *medial*, and *late* indicate spans of time, whereas *Lower*, *Middle*, and *Upper* refer to the actual rock strata. The illustration, above, comparing *Archaeopteryx* with the modern pigeon points up dramatic differences in the two species' shape and structure. Paleontological data also provide evidence on rates of evolution, extinction, and times of major adaptive innovations. The fossil heritage of birds is incomplete and difficult to interpret. Yet, we do have an excellent starting point, *Archaeopteryx*, and this good fortune appears all the more exceptional when we consider the circumstances of the specimens' fossilization and discovery.

Each of the seven known specimens of *Archaeopteryx* was recovered from the fine-grained Solnhofen limestone of the Altmühl region of Bavaria. Although fossils are rare in this limestone, the meticulous mining of this fine stone over the ages has produced a vast array of amazingly well-preserved fossils and thus provides a rare great window to the past. During the late Jurassic (Lower Tithonian), the region was within the subtropical zone at 25–30° North latitude. A great variety of plants, invertebrates, fishes, and reptiles shared this resting place, and they allow us to form some idea of the late Jurassic habitat and the conditions that led to this wealth of fossils. As Barthel, Swinburne, and Morris note in their book *Solnhofen* (1990), the region is a plateau just north of Munich known as the Southern Franconian Alb. Cutting through the plateau are the Danube and Altmühl valleys, their rivers slicing through hundreds of feet of Upper Jurassic *Plattenkalk*, or platy limestone. The most famous unit of limestone, Solnhofen, is named for the nearby village of Solnhofen. These limestones are predominantly calcium carbonate ($CaCO_3$) deposited as limy, calcareous mud (*calx*, lime) derived from both inorganic processes and lime-secreting algae and settling to the bottom as very fine material. Much calcium carbonate derives from the shells of invertebrates in the form of the mineral calcite, and the tiny shells, or tests, of these microscopic foraminifera make up much of the limestone, forming as limy muds of considerable thickness, particularly along tropical shoals.

In some regions the Solnhofen limestone attained a thickness of about 90 meters (295 ft.). These deposits are marine backreef sediments (Buisonjé 1985) that are today found in a longitudinal belt some 80 kilometers (50 mi.) long and 30 kilometers (19 mi.) wide and during deposition were partially separated from the Tethys Sea by discontinuous coral reefs along eastern and southern regions of the sediments. Paul de Buisonjé (1985) has envisioned the paleoecology as indicating an arid, warm, tropical paleoclimate along the coast, with less arid or semiarid conditions prevailing inland. A modern example of such a semienclosed marine basin with an estuarinelike circulation is the 70-kilometer-long (43 mi.) Gulf of Cariaco on the northern coast of Venezuela. In this situation, as in the Solnhofen basin, a combination of wind and upwelling water produces a high concentration of inorganic nutrients and a high productivity of plankton and causes seasonal, midsummer seawater blooms of coccolithophorids (calcareous planktonic algae) in the lagoonal waters. When upwelling seawater rich in inorganic nutrients combines with relatively high temperatures, fish and other macroorganisms may be killed. In the Gulf of Cariaco, thousands of small fish die in this way every summer, and it is probable that much of the water in the backreef basin of

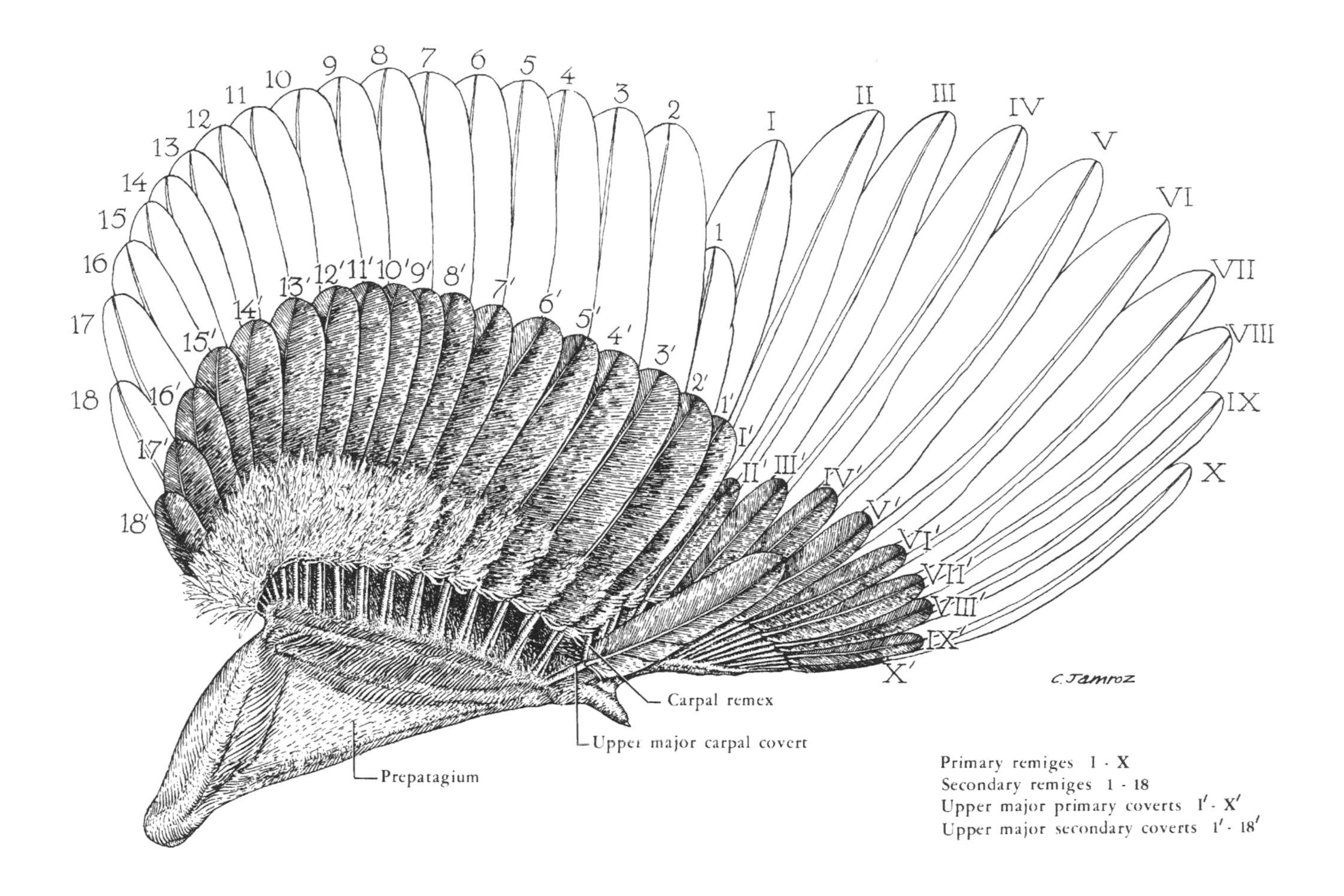

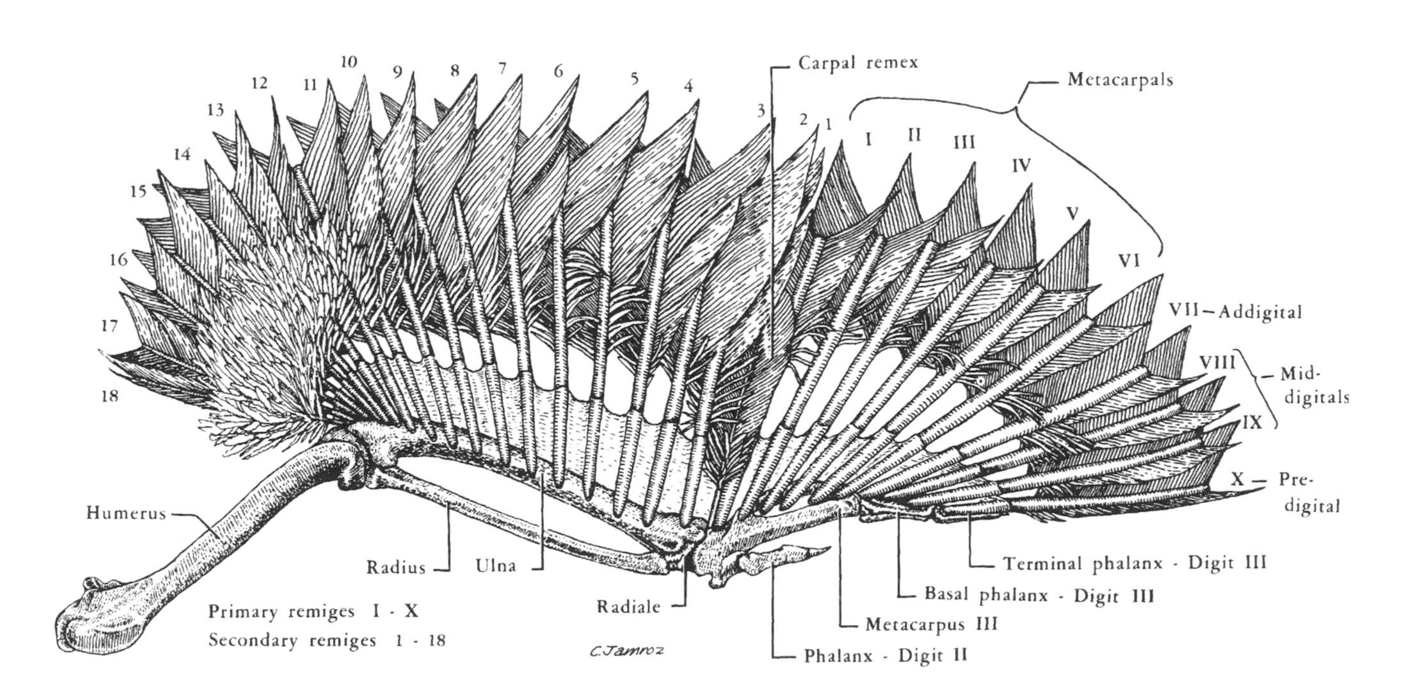

Dorsal view of, *top,* the extended left wing of a white leghorn chicken and, *bottom,* of the skeletal attachments of the primaries and secondaries of the same wing. (From Lucas and Stettenheim 1972; courtesy Peter Stettenheim)

Reconstruction of the late Jurassic shoreline of European lagoons that produced the Solnhofen limestone of Bavaria. Here *Archaeopteryx* is shown with the chicken-sized coelurosaurian dinosaur *Compsognathus* and the tailed, Jurassic pterosaur *Rhamphorhynchus*. The Solnhofen deposits preserved twenty-nine described species of pterosaurs (many may be growth stages, and as few as five species may have existed), ranging from the size of a sparrow to 1.2 m. (4 ft.) in length (*Rhamphorhynchus*), a variety of insects from moths to flies and dragonflies, crustaceans, ammonites, and soft-bodied jellyfish, rarely preserved as fossils. (Mural by Charles R. Knight; courtesy Field Museum of Natural History, neg. no. 75017)

Table 1.2 Geological time scale in millions of years before the present (dates from Harland et al. 1990). Duration of time periods given in parentheses, with rounded values. Some text figures may have slightly different dates depending on author's source.

Era	Period		Epoch	Million years before present
CENOZOIC (65) (Age of Birds and Mammals)	QUATERNARY (1.64)		Recent (Holocene) (0.01)	
			Pleistocene (1.63)	1.64
	TERTIARY	Neogene (22)	Pliocene (3.5)	5.2
			Miocene (18.3)	23.5
		Paleogene (42)	Oligocene (12.0)	35.5
			Eocene (21.0)	56.5
			Paleocene (8.5)	65
MESOZOIC (180) (Age of Reptiles)	CRETACEOUS (81)		Late (32)	97
			Early (49)	146
	JURASSIC (62)		Late (12)	157
			Middle (21)	178
			Early (30)	208
	TRIASSIC (37)			245

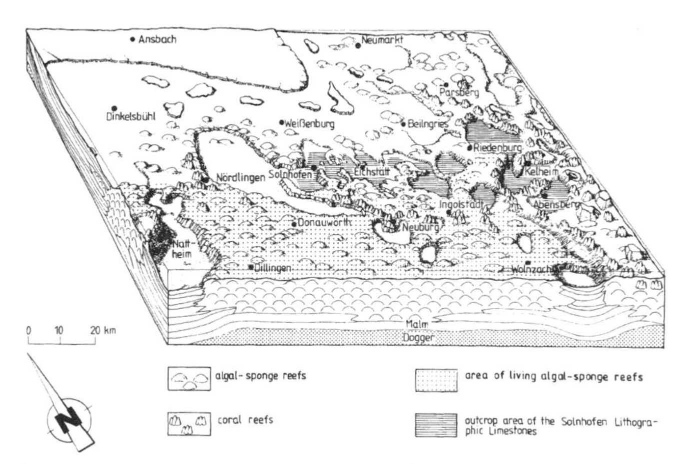

Paleogeographic reconstruction of the sedimentation area of the Solnhofen limestone. (From Viohl 1985; courtesy G. Viohl and Jura-Museum)

Solnhofen was similarly "poisoned": the dead macroorganisms sank quickly to the ocean floor and were soon covered by more slowly sinking masses of dead coccoliths, the microscopic calcified disc remains of the dead algae.

Perhaps the most widely accepted model to explain what made the Solnhofen waters so poisonous, however, is the excessive concentration of salt caused by intense evaporation. This would certainly kill macroorganisms, and it correlates with the shriveled appearance of many Solnhofen jellyfish. In addition, fine preservation of organic material is another consequence of hypersalinity (Barthel, Swinburne, and Morris 1990). Either model nicely explains the otherwise unexplainable: a fish (*Caturus*) preserved with small fish-prey halfway in its mouth; a horseshoe crab (*Limulus*) dead at the end of a spiral death-trail of footmarks in the sediment.

Complementing these details of Solnhofen paleoecology, Günter Viohl (1985a), curator at Eichstätt's Jura-Museum, has provided more information on paleoclimate, proposing that the region had an arid or semiarid monsoonal climate with alternating dry and rainy seasons. Viohl estimates the remarkable Solnhofen fauna to compose some six hundred fossil species from various localities (Barthel, Swinburne, and Morris 1990, 13). The most abundant fossils of this exceedingly rich assemblage include crinoids, ammonites, fish, and crustaceans, but more fragile, soft-bodied animals, normally void in the fossil record, such as jellyfish and squid, are also preserved, as are many insects. In fact, the remarkably diverse insect fauna, including mayflies, thirteen genera of dragonflies (the most known from any fossil deposit), cockroaches, water skaters, locusts and crickets, bugs and water scorpions, cicadas, lacewings, beetles, bees and wasps, and caddis flies and dipteran flies, in combination with abundant morphs of pterosaurs and the primitive bird *Archaeopteryx*, indicate a volant fauna that was preserved near the shore. The myriad of insects could not have been far from land; their diversity and numbers would dramatically decrease farther out to sea. And many insects, including mayflies, caddis flies, and most dragonflies, could not have been permanently away from freshwater, or at least brackish coastal water, where they would have to deposit their eggs and their larvae would undergo metamorphosis.

Some paleontologists have envisioned sea cliffs up which pterosaurs and even *Archaeopteryx* may have climbed before launching themselves into strong winds (Rayner 1991; Peters and Gorgner 1992), but as Yale University's John Ostrom has pointed out, the Solnhofen limestone contains "almost no detrital sediments; this fact suggests that nearby terrestrial relief was low and that any high cliffs would have been many miles away from the Solnhofen lagoons" (1986a, 80). Günter Viohl, in contrast,

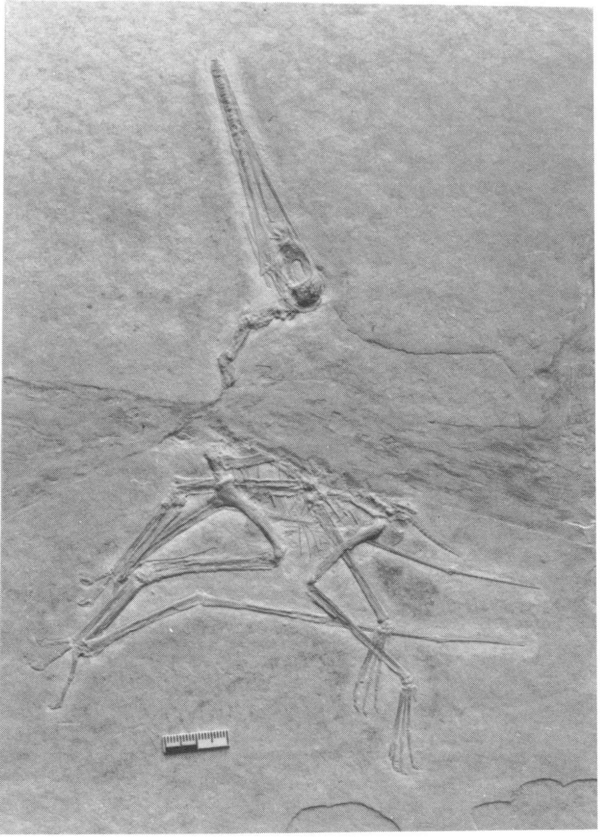

Pterodactylus elegans, left, and *Pterodactylus kochi,* the smallest known pterosaurs, both from the Solnhofen limestone. Scale in cm. (Photos courtesy G. Viohl and Jura-Museum)

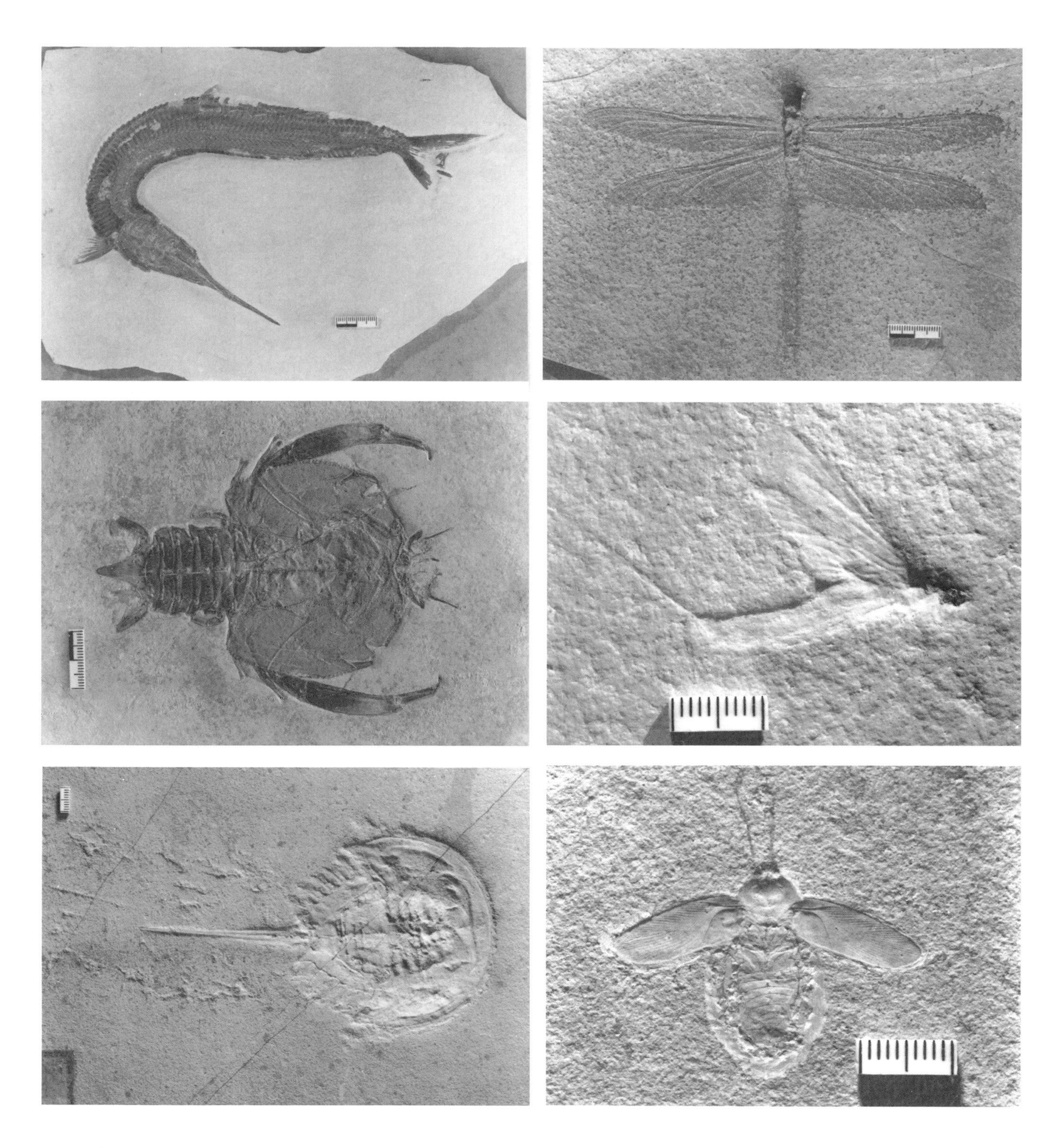

Spectacular Solnhofen fossils include: *left, upper to lower,* aspidorhynchiform holostean fish *Aspidorhynchus acutiformis,* decapod crustacean *Cycleryon propinquus,* and horseshoe crab (xiphosuran chelicerate) *Mesolimulus walchi,* at the end of its track; *right, upper to lower,* dragonfly *Aeschnogomphus intermedius,* mayfly *Hexagenites cellulosus,* and cockroach *Lithoblatta lithophila*. The presence of insects indicates a volant fauna preserved near the shoreline, with freshwater for breeding. (Photos courtesy G. Viohl and Jura-Museum)

has envisioned the habitat of *Archaeopteryx* as "on the island formed by the Central German Swell and the London-Brabant-Massif, on the Bohemian Island, and perhaps some smaller offshore islands. Along the shoreline of the slowly emerging land in the NW, and the offshore islands, marine abrasion probably formed cliffs, though we have no direct evidence" (1985a, 41).

An intriguing aspect of the Solnhofen paleogeography is the different morphotypes of *Archaeopteryx* occurring on offshore islands. As Storrs Olson of the Smithsonian Institution has remarked, "Perhaps not enough has been made of the fact that the known individuals [of *Archaeopteryx*] were inhabitants of islands" (1987a, 74). Could it be that the dramatic size and morphological differences in the seven urvogel specimens (Houck et al. 1990) is best explained by geographic isolation on these secluded Solnhofen landmasses? As mentioned earlier, the Hawaiian honeycreepers and Galápagos finches provide spectacular examples of insular adaptive radiation, and a recent example of microevolutionary change in beak size has been documented in the medium ground finch (*Geospiza fortis*) on Daphne Major Island as an outcome of the exceptionally strong El Niño of 1982–1983 (Grant and Grant 1993; Greenwood 1993). In the case of these finches, small beak sizes were selectively favored when larger seeds became scarce. Microevolutionary events may thus be exacerbated in island situations, and climatic events that may seem relatively minor for broadly adapted continental species may have dramatic effects in an insular world such as the Solnhofen lagoons.

In addition to the insects and other invertebrates, Solnhofen fossils include a vast array of vertebrates, including chondrichthian and osteichthian fishes, turtles, ichthyosaurs, plesiosaurs, lizards, rhynchocephalians, crocodiles, a single, small saurischian dinosaur *Compsognathus*, seven described genera of pterosaurs (now thought to be five species exhibiting indeterminate growth; Bennett 1993a), and the earliest known bird, *Archaeopteryx*. It is a picture of a very diverse fauna of aquatic, terrestrial, and volant organisms in and around this arid, late Jurassic lagoonal region. Given strong offshore winds, creatures flying over the water would occasionally die or fall in and drift to the bottom, where they would be covered by viscous sediments. Storms would have blown flying insects, pterosaurs, and an occasional *Archaeopteryx* into the lagoons (de Beer 1954). The fine sediments sealed their fate and, with each passing age, formed a new layer of lithographic limestone.

Caves dating from the late Stone Age are indelibly inscribed with scratched drawings and colored murals on the Solnhofen Plattenkalk, and later the Romans mined the stone for bath linings, for buildings, and for inscription tablets (Barthel, Swinburne, and Morris 1990). The limestone was mined extensively during the Middle Ages and became a valuable export; the famed church of Hagia Sofia in Istanbul has a medieval mosaic floor crafted from Solnhofen limestone. Even today, all but the most modern buildings in the Altmühl region are finished with floors and roofs of this fine stone. But the modern era of mining the quarries began with the invention by Alois Senefelder in 1793 of the printing process called lithography. The stone began to be quarried for lithographic use in the nineteenth century. New sites were opened, and more important, workers began to check each slab meticulously for its intrinsic quality. Only this careful mining process made the discovery of *Archaeopteryx* possible.

Urvogel Discoveries. In 1861, Hermann von Meyer of Frankfurt wrote to H. G. Brown, German publisher of the *New Yearbook of Mineralogy*, that a fossil bird feather had been discovered the previous year in the quarry of the village of Solnhofen, not far from Munich. This unexpected discovery caused a sensation, for although the first specimen was merely a feather, it clearly meant that there was a bird in rocks dating from the Age of Reptiles. The blackish feather was 60 millimeters (2.5 in.) long and 11 millimeters (0.44 in.) wide, with its vane on one side of the quill that was roughly half as wide as that of the other—the same contours as the flight feathers of modern birds and conforming nicely in size and shape to a secondary feather of *Archaeopteryx*. The implications of this avian hint from the late Jurassic began to resonate. And within a month Meyer was reporting the discovery of a complete fossil skeleton in another quarry from the Langenaltheim region of Bavaria, not far from the original site. It had a long, reptilian tail exhibiting many vertebrae, but attached to each there was what appeared to be a pair of short feathers. The limestone had captured these subtle impressions, as well as the startling image of feathered wings. Here clearly was a mosaic of avian and reptilian characteristics. Had it not been for the serendipitous role of the uniquely fine-grained Plattenkalk, the creature would have been classified as a reptile. Instead, Meyer gave it the avian genus name "ancient wing" (*archaos*, ancient; *pteryx*, wing) and the species designation *lithographica*, in reference to the lithographic limestone in which it was preserved.

The main slab of the single feather impression went to the Museum of the Academy of Sciences in Munich; the counterslab made its way to the Humboldt Museum für Naturkunde in Berlin. But in 1861, with Victorian science on the march and *On the Origin of Species* two years in print, the discovery of the complete skeleton, a specimen purported to be an intermediate form between two higher

Left, one of the Solnhofen limestone quarries near Eichstätt, Bavaria, still active today. *Above,* a worker meticulously separating slabs of limestone. (Photos by author)

Gerhard Heilmann's well-known reconstruction of *Archaeopteryx.* (From Heilmann 1926)

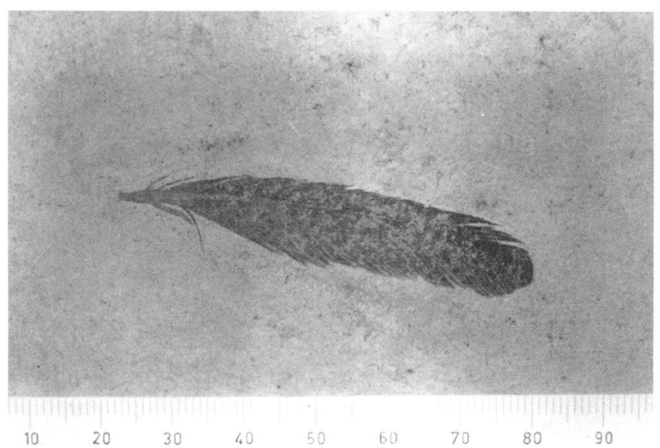

The first piece of evidence that birds existed in late Jurassic times was the single secondary flight feather, 6 cm (2.5 in.) in length, reported by Hermann von Meyer in 1861. Note that the two sides (vanes) of the feather are asymmetric, an arrangement found elsewhere only in the primary and secondary flight feathers and outer tail feathers of modern flying birds. Evidence of *Archaeopteryx*'s ability to fly has thus been available for more than a hundred years. (Photo by author)

groups of animals, would not be so quietly reinterred in dusty packing cases and cataloged into obscurity.

The Plattenkalk fossils have been highly prized by local people as far back as the late Stone Age, when they were kept as ornaments, but by the late 1700s, the fossils had become quite valuable, and by the mid-nineteenth century, many lay collectors were amassing extensive holdings of fossils from the Solnhofen quarries. Quarry hands likewise were making profits selling their finds. Dr. C. F. Häberlein, a Royal Bavarian District medical officer as well as an obstetrician in Pappenheim, was a collector who accepted fossils in payment for medical services (see Viohl 1985b). The complete skeleton made its way into his hands. A speculator at heart, he offered the fossil for sale after only three months. All across Europe a mad scramble began, involving royal councillors and learned professors. The German court tried to obtain the curious specimen for the Bavarian State Collection of Paleontology and Historical Geology in Munich. This effort met with the objections of an influential professor of zoology at Munich University named J. Andreas Wagner. Wagner was a staunch opponent of the new Darwinian theory, and he vehemently opposed the idea that there might be a transitional link between reptiles and birds. He expressed his sentiments at the Munich Academy of Sciences in 1861 in a paper aptly titled "A New Reptile Supposedly Furnished with Bird Feathers" (Über Ein neues, augeblich mit Vogelfedern versehenes Reptil . . . 1861a). Wagner also ignored Meyer's avian name for the fossil and gave it the reptilian designation *Griphosaurus* (*gryps*, mythical beast; *sauros*, lizard).

Meanwhile, Dr. Häberlein was protecting his investment by preserving the aura of mystery that shrouded *Archaeopteryx*. At first he allowed no one to make any drawings of the fossil, but he did let a few people inspect it to dispel rumors that it might be a fake. One of the inspectors sketched the fossil and produced an accurate lithograph. This was shown to Wagner, but he remained adamant that birds did not exist as far back as the late Jurassic.

The significance of *Archaeopteryx* was not wasted on the British. In spite of his confirmed antievolutionist position, Sir Richard Owen, superintendent of the British Museum (Natural History), along with George Robert Waterhouse, keeper of the Geology Department, recommended that the trustees of the British Museum bid for the prize. After haggling for several months, Häberlein agreed to sell *Archaeopteryx* along with 1,703 other Solnhofen specimens from his collection for the sum of 700 pounds (two years' acquisition budget for the museum), which became the dowry for one of Häberlein's six living daughters, and the controversial fossil arrived in London in November 1862.

The image of *Archaeopteryx* as a link between reptiles and birds was not readily embraced either by the public or by the scientific community. The obvious implications for a theory of evolution were far too disturbing, especially to the entrenched ecclesiastical view that reptiles and birds were static groups placed on earth in their present unchanging forms.

The debate in England squared off between two main protagonists: Owen, the acknowledged leader of British science and an intimate of the royal family, and Thomas Henry Huxley, the most eloquent defender of Darwin and evolutionary theory. Owen (1862, 1863) carefully studied the fossil and in 1863 insisted on giving it yet another name, *Archaeopteryx macrura*, meaning long-tailed (Owen 1863). While he was at work, three brief papers supporting his views were being published in England, one by Henry Woodward in the *Intellectual Observer*, one translation of a paper by Hermann von Meyer, and a translation of Wagner's paper of 1861. Never having seen the actual fossil, Wagner remained confident in his classification of it as a pterosaur, a long-tailed flying reptile: "In conclusion, I must add a few words to ward off Darwinian misinterpretation of our new saurian. At the first glance of the Griphosaurus [*Archaeopteryx*] we might certainly form a notion that we have before us an intermediate creature, engaged in the transition from the saurian to the bird. Darwin and his adherents will probably employ the new discovery as an exceedingly welcome occurrence for the justification of their strange views upon the transformation of animals but in this they will be wrong" (Wagner 1862, 266).

Richard Owen, the famed anatomist who at the age of thirty-eight had coined the term *Dinosauria*, was in fact not a creationist, as was popularly thought, but rather had great disdain for Darwin's explanation of evolution by natural selection and was, as mentioned, at odds with "Darwin's Bulldog," Thomas Huxley. Huxley had previously demolished Owen in the great "hippocampus debate" and was to discredit Owen again by showing that he had described the London specimen of *Archaeopteryx* as lying belly-up on the slab, instead of belly-down, as observed by Huxley (1868a).

Owen believed in something called "continuous creation," in which new life-forms were created from time to time, from modifications on a basic, archetypal plan, and his view of *Archaeopteryx* was indeed bizarre. Owen insisted that the famed urvogel was a bird, the earliest bird, transmuted from a long-tailed pterosaur such as the Solnhofen *Ramphorhynchus*. Owen argued that although scientists did not as yet know the causes of evolution they could nevertheless see the general sequence of change, and here we clearly see the evolutionist in Owen. "Neither

The first complete fossil skeleton of *Archaeopteryx,* known as the London specimen. It was discovered in 1861, shortly after the isolated feather was uncovered, and it proved the existence of a reptile-bird. The London specimen shows clearly the furcula, or wishbone (fused clavicles), indicated by the arrow, characteristic of modern birds. (Courtesy the Natural History Museum, London)

Skeleton and restoration of *Compsognathus longipes*. (From Witmer 1991, modified from Ostrom 1978; courtesy Bayerischen Staatssammlung für Paläontologie und historische Geologie, Munich)

has the biologist been able, as yet, to explain how the *Ramphorhynchus* became transmuted into the *Archaeopteryx*. . . . Every bone in the bird was antecedently present in the framework of the Pterodactyle. . . . Some pterodactyles had long tails and all had toothed jaws. A bird of the oolitic period [=*Archaeopteryx*] combined a long tail of many vertebrae with true avian wings" (Owen, 1863).

In contrast, Huxley, who in 1868 published his first paper on bird origins, focused more on the "proof" of evolution, and he clearly regarded *Archaeopteryx* as a real bird. In studying the London specimen he made comparisons with a small theropod dinosaur, the chicken-sized *Compsognathus* (*kompsos*, elegant; *gnathos*, jaw), which also had been described from the Solnhofen limestone by Wagner (1861b). Referring not to *Archaeopteryx* but to the small Solnhofen dinosaur, Huxley noted that "but a single specimen, obtained from those Solnhofen slates, . . . affords a still nearer approximation to the 'missing link' between reptiles and birds" (1868b, 73). Huxley's focus was on the dinosaurs' birdlike features, particularly the chickenlike pelvis and hindlimb, and, in fact, his paper on bird origins was really directed toward proof of evolution; *Archaeopteryx* for Huxley exemplified the existence of "reptilian birds," which he viewed as a side branch off the base of the main avian tree and having little to do with the mainsprings of modern flying birds. He believed birds to have originated from dinosaurs and modern flying birds to have arisen through the flightless ratites, the ostriches and their allies. Huxley's influence on Darwin is apparent when Darwin (1859) refers to *Archaeopteryx* as "that strange Secondary bird, with a long lizard-like tail" (Secondary here referring to the Mesozoic deposits).

Even in the United States, professors of natural history were coming around to the Darwinian view, bolstered by this new evidence. As Professor Othniel Charles Marsh of Yale University wrote in 1877: "The classes of Birds and Reptiles, as now living, are separated by a gulf so profound that a few years since it was cited by opponents of evolution as the most important break in the animal series, and one which that doctrine could not bridge over. Since then, as Huxley has clearly shown, this gap has been virtually filled by the discovery of bird like Reptiles and reptilian Birds. *Compsognathus* and *Archaeopteryx* of the Old World . . . are the stepping stones by which the evolutionist of to-day leads the doubting brother across the shallow remnant of the gulf, once thought impassable" (1877, 352).

But even Thomas Huxley, who later proposed a complete classification of all living birds based on the structure of the bony palate, could not accept the presence in the jaws of *Archaeopteryx* of teeth, a reptilian carry-over. The London specimen was somewhat disarticulated, and as late as 1868, Huxley wrote that the skull was lost. But in fact there remained part of the upper jaw with four teeth, as Sir John Evans had pointed out in 1865 in an article on the cranium and jaw preserved in the fossil slab. Evans fully accepted *Archaeopteryx* as a link between reptiles and birds, but in the same article he quoted a letter Hermann von Meyer had written him expressing the anti-Darwinian view that still prevailed in many circles:

> It would appear that the jaw really belongs to the *Archaeopteryx* and arming the jaw with teeth would contradict the view of the *Archaeopteryx* being a bird or embryonic form of bird. But after all, I do not believe that God formed His creatures after the systems devised by our philosophical wisdom. Of the classes of birds and reptiles as we define them, the

> Creator knows nothing, and just as little of a prototype, or of a constant embryonic condition of the bird, which might be recognized in the *Archaeopteryx*. The *Archaeopteryx* is of its kind just as perfect a creature as other creatures, and if we are not able to include this fossil animal in our system, our shortsightedness is alone to blame. (1865, 421)

The controversy over *Archaeopteryx* continued.

An Avian Rosetta Stone. In 1877, while scientists all over the world were still debating the significance of the London *Archaeopteryx*, news of the discovery in the autumn of 1876 of another skeletal specimen was announced, this one from a quarry on the Blumenberg River, just outside Eichstätt, some 30 kilometers (19 mi.) east of the Langenaltheim sites where the first feather and London specimen were discovered. Again it fell into the hands of a speculator—none other than the son of Dr. C. F. Häberlein. Ernst Otto Häberlein, a tax consultant, bought the fossil, which the quarry owner regarded as a *Pterodactylus*, for 2,000 marks. Along with a Dr. Redenbacher from Pappenheim, he had assembled an extensive collection of Solnhofen fossils. In this case, however, Häberlein offered this fossil for sale, and this time, the Germans, embarrassed at having lost the first skeleton to the British Museum, were determined to come up with the funds. Four years of negotiations followed, involving the Bavarian State Collection in Munich, private endowments, the Prussian Ministry of Culture, and numerous men of science. Ultimately, the industrial magnate Dr. Werner von Siemens himself bought the specimen, reselling it to the Prussian ministry in 1881 for 20,000 marks, the purchase price. The fossil and counterslab were presented to the Humboldt Museum für Naturkunde in Berlin, where they reside today. Later, the Prussian state also bought the remainder of E. O. Häberlein's Solnhofen collection.

The Berlin *Archaeopteryx* may well be the most important natural history specimen in existence, comparable perhaps in scientific and even monetary value to the Rosetta stone. Beyond doubt, it is the most widely known and illustrated fossil animal—a perfectly preserved Darwinian intermediate, a bird that has anatomical features of a reptile, feathers, and a long, lizard-like tail. In contrast to the London specimen, the skeleton of the Berlin fossil is articulated in a natural pose with the wings extended. The Berlin specimen, as it is known, is preserved in much the same anatomical position as the small Solnhofen dinosaur *Compsognathus*, with the head arched back over the neck, a preservational pose also exhibited often by pterosaurs. The skull, with upper and lower teeth, is complete, and attached to the outspread arms and hands are complete impressions of primary and secondary flight feathers, nearly identical in detail to those of modern birds. The long tail shows a pair of tail feathers attached symmetrically to each vertebra.

At first, scientists thought that the Berlin specimen belonged to the same species as the London specimen. But in 1897, Wilhelm Barnim Dames, the German paleontologist who first described the Berlin fossil, dubbed it *Archaeopteryx siemensii*. Later, in the 1920s, Bronislav Petronievics, greatly overemphasizing the differences between the London and Berlin specimens, named the Berlin bird *Archaeornis siemensii*, placing it in a different genus from *Archaeopteryx* (1925, 1927). He believed that *"Archaeornis"* gave rise to all modern birds except the ostriches and allies, which he thought were descended from the London *Archaeopteryx*. Today, however, most students regard all the specimens as belonging to the single genus *Archaeopteryx*, but whether one or more fossil species existed is still questioned.

An interesting twist to the history of the Berlin specimen was revealed by Yale University's dinosaur authority John Ostrom (1985a), who discovered through the work of his secretary, Miriam Schwartz, in 1983, that O. C. Marsh had attempted to obtain the Berlin specimen for Yale's Peabody Museum of Natural History. Schwartz discovered an envelope containing some letters, a newspaper clipping, and two tracings. One of the tracings was of the Berlin specimen, and it was accompanied by a letter from one F. A. Schwartz dated from Nuremberg on 7 March 1879, offering to sell the Berlin specimen (with other Solnhofen fossils) for 10,000 dollars. This offer predated any offer to sell the specimen elsewhere, but there has yet to be found any documentation as to why Marsh did not follow through on this rare opportunity. Nevertheless, this earliest known drawing of the Berlin specimen clearly shows the presence of contour feathers along the back and legs and a tuft of feathers behind the skull, all now prepared away. And so the drawing provides valuable new information, illustrating that the famous urvogel was probably fully feathered. Lamenting the loss of the "Yale specimen," Ostrom calls it "the one that flew the coop."

Later Discoveries. Remarkably, it was almost a hundred years later before another specimen of *Archaeopteryx* came to light. The third specimen, discovered in 1956 in a quarry shed by a student at the University of Erlangen, came from the same quarry at Langenaltheim that produced the London specimen. Klaus Fesefeldt, a geologist from Erlangen, determined that it was from an area only 229 meters (750 ft.) from the site of the recovery of the London bird, but some 6 meters (20 ft.) higher in the

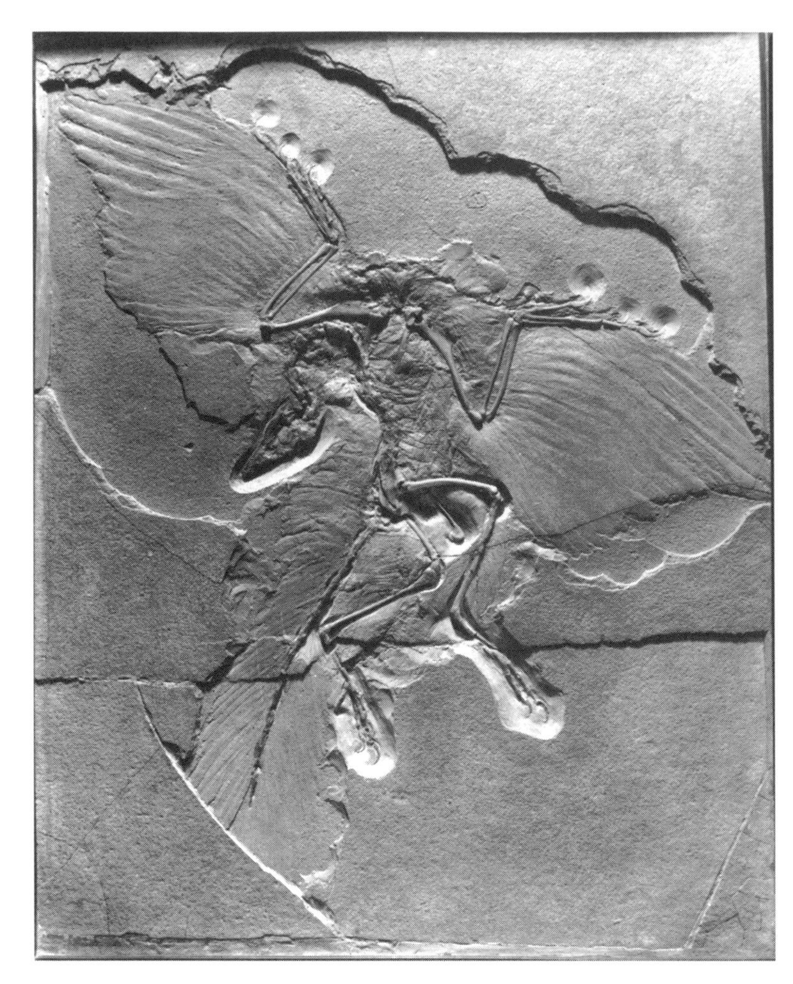

The Berlin specimen of *Archaeopteryx*, found in 1877 in a limestone quarry near the Bavarian town of Eichstätt, main slab and counterslab. (Courtesy Humboldt Museum für Naturkunde, Berlin)

Reconstruction of Berlin *Archaeopteryx*, by Gillard u. Steinbacher, 1959. (Courtesy Department of Library Services, American Museum of Natural History, neg. no. 325288)

Wing as preserved on the counterslab of the Berlin specimen. Note the asymmetry of the primary feathers, indicated by the arrow, proving that the wing was aerodynamic. (Courtesy Humboldt Museum für Naturkunde, Berlin)

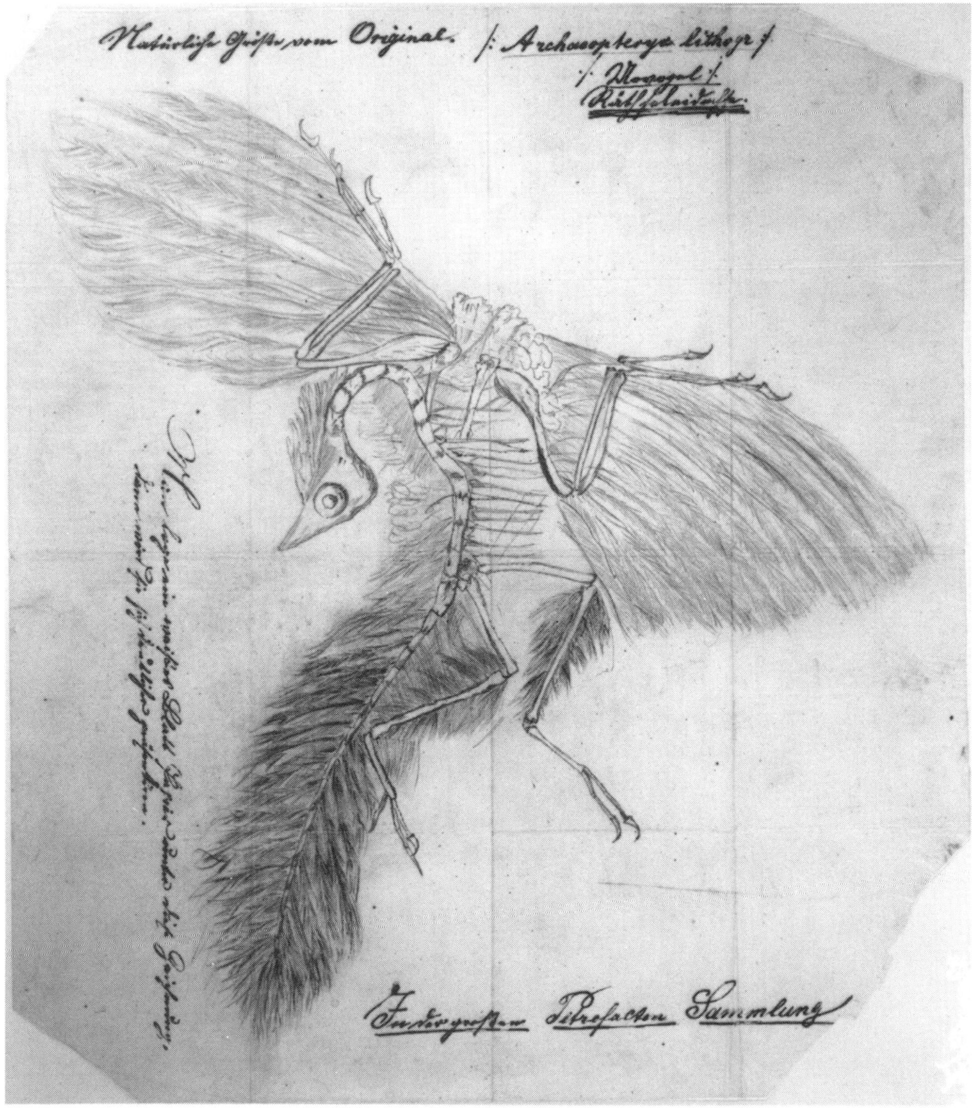

Pencil tracing of the Berlin specimen of *Archaeopteryx* in the Yale University Archival Collection, showing the presence of contour feathers on various parts of the body, including the throat, back, breast, and legs, which were apparently later prepared away. The specimen is labeled at the upper right as Urvogel/"Ratseleidesche" (original bird/"riddle lizard"). This sketch was included with a letter dated 7 March 1879, from F. A. Schwartz of Nuremberg, offering to sell the "Berlin *Archaeopteryx*" and a collection of Solnhofen fossils to Yale's Peabody Museum. This letter and sketch were discovered in August 1983 by Miriam Schwartz, secretary to John Ostrom. This specimen shows that *Archaeopteryx* most likely had a complete covering of contour body feathers. (Courtesy John Ostrom and Peabody Museum of Natural History, Yale University)

quarry. This specimen, which was mainly a torso, showed some feather impressions, but it was so badly disarticulated and, presumably, decomposed that its identification took two years. It was described by Florian Heller in 1959 and was on display for some four years near Solnhofen in the Maxberg Museum of the Solenhofer Aktienverein, owned by one of the Solnhofen quarry companies, being dubbed the "Maxberg specimen." Eduard Opitsch of Pappenheim removed it from display in 1982, and it resided safely in his home, though he allowed no one to see it. Fortunately, before Opitsch reclaimed the fossil, John Ostrom was permitted to study it, and he was able to determine that some of the foot bones, the metatarsals, were partly fused. This anatomical condition presages the complete fusion in modern birds. The story ends sadly: when Opitsch, a bachelor, died in February 1991 at age ninety-one, his heir, a nephew, was unable to find the slab and counterslab, which are now presumed to have been stolen shortly after his death (Wellnhofer 1992).

In 1970, Ostrom was studying pterosaurs in Europe and happened on a fourth specimen of *Archaeopteryx* that had been displayed as a distinctive species of pterosaur (Ostrom 1970). The fossil, consisting of parts of one hand, miscellaneous arm and leg bones, feet, and some feather impressions, had been recovered a century earlier, in 1855, in a large quarry at Jachenhausen, near Kelheim, well east of the other localities. Amazingly, then, this specimen had been recovered six years before the discovery of the feather and the London specimen and had been announced in 1857 as a flying reptile or pterosaur by none other than Hermann von Meyer, who described it as *Pterodactylus crassipes*. The specimen has been on display since 1860 in the Teyler Museum in Haarlem, Netherlands, and has been variously dubbed the "Teyler" or "Haarlem" specimen. The faint impressions of feathers around the body on the slab led Ostrom to his remarkable discovery. But it is the extraordinary preservation of a finger claw with the horny sheath that has brought attention to the Teyler *Archaeopteryx*.

A beautifully preserved, small specimen of *Archaeopteryx* was reported in 1973 by F. X. Mayr of Eichstätt, founder of the Jura-Museum. Discovered in 1951 in a quarry near Workerszell, just north of Eichstätt, the fossil was initially thought to be a juvenile specimen of the small Solnhofen dinosaur *Compsognathus*. Twenty years passed before Mayr studied it and noticed the faint feather impressions, marking it as the fifth specimen of *Archaeopteryx;* it was described by Peter Wellnhofer of the Bavarian State Collection in Munich in 1974. It is some one-third smaller than the well-preserved London specimen and therefore has led some to believe that it represents a distinctive genus; Michael Howgate (1984, 1985) proposed that it be known as *Jurapteryx*. However, species and generic limits of these Jurassic birds are impossible to determine (see also Houck et al. 1990), so the proposal has not been widely accepted, and the fossil remains today simply a small specimen of *Archaeopteryx*. The Eichstätt specimen is now on prominent display in the Jura-Museum and shows exceptionally well-preserved bones with a nearly perfect skull, but there is no furcula (it may have been lost in preservation), and the feather impressions are so faint that they are barely discernible as outlines of wings and tail.

In November 1987, Günter Viohl of the Jura-Museum discovered yet another specimen of *Archaeopteryx* in the private collection of the former mayor of Solnhofen, Friedrich Müller; the exact quarry of its unearthing is unknown. It has been studied in detail and described by Peter Wellnhofer (Wellnhofer 1988a, 1988b, 1992; also Ostrom 1992). The specimen is now the property of the Bürgermeister Müller Museum, Solnhofen, and has been dubbed the "Solnhofen" specimen. This, the sixth skeletal specimen of the famous urvogel, is the largest of all the specimens and is some 10 percent larger than the next largest example, the London specimen, based on total length of wing skeleton. The nearly complete skeleton is nicely preserved in its natural articulation, and one can see the impressions of flight feathers of the left wing (mainly shafts). In osteological features, it agrees favorably with the London specimen, particularly in tooth morphology and skeletal proportions.

Just as the excitement over the Solnhofen specimen was dying down, Wellnhofer announced through the German newspapers in late April 1993 the sensational discovery of still another specimen of urvogel, this one smaller, with longer limb bones and strongly curved, arboreally adapted claws. The most exciting aspect of the discovery, however, is that this specimen has a sternum (without keel) for the attachment of the flight muscles, the only specimen to exhibit ossified sternal development. The specimen was discovered in the Langenaltheim quarry district, where the London and Maxberg specimens were recovered. From the known stratigraphy, it appears to be the youngest, the London specimen being the oldest. Wellnhofer (1993) has described the specimen in a beautiful monograph and considers it sufficiently different from the other examples to be named a new species, *Archaeopteryx bavarica*, characterized by relatively longer tibiae and hind legs. It is dubbed the "Solenhofer Aktien-Verein" specimen in honor of the quarry company that owns it. Feather impressions of primary and secondary flight feathers as well as tail feathers are preserved and are typical of other forms of *Archaeopteryx*. The lower jaws are preserved, showing the lingual surface, which exhibits in-

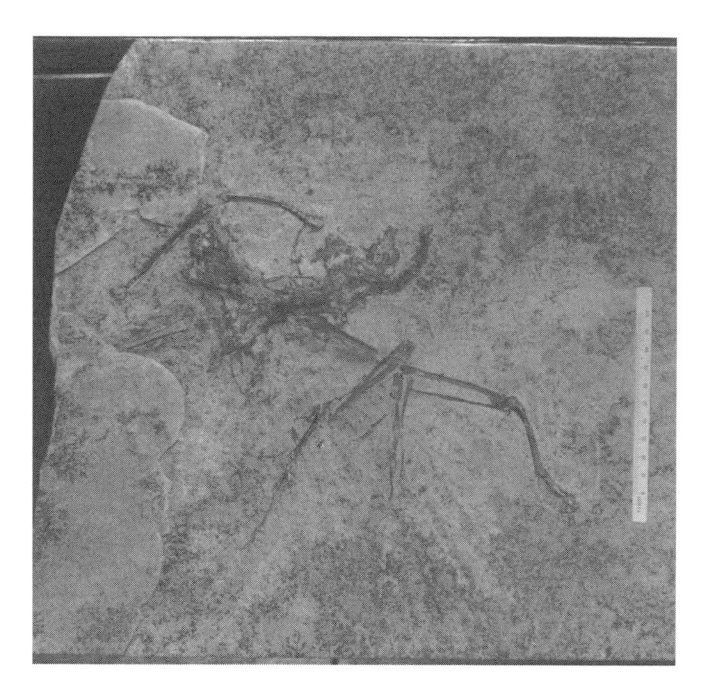

Left, The decomposed and disarticulated remains of the Maxberg specimen of *Archaeopteryx,* the third specimen to be discovered. This specimen, found in 1956, clearly shows the metatarsal bone partly fused, foreshadowing the fused metatarsals of modern birds. (Courtesy John Ostrom and Peabody Museum of Natural History, Yale University)

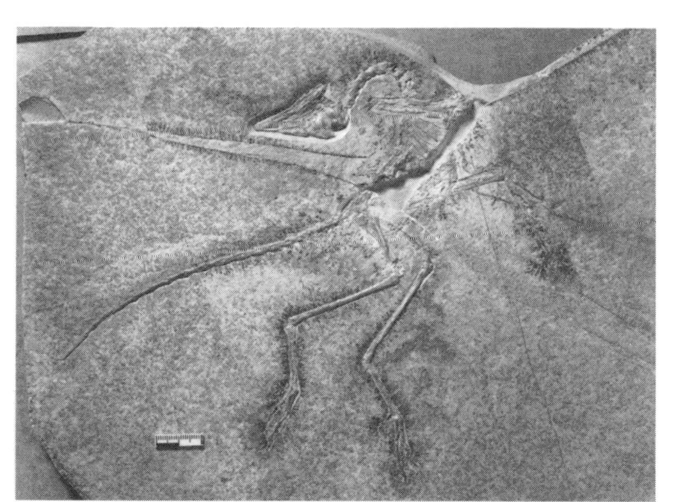

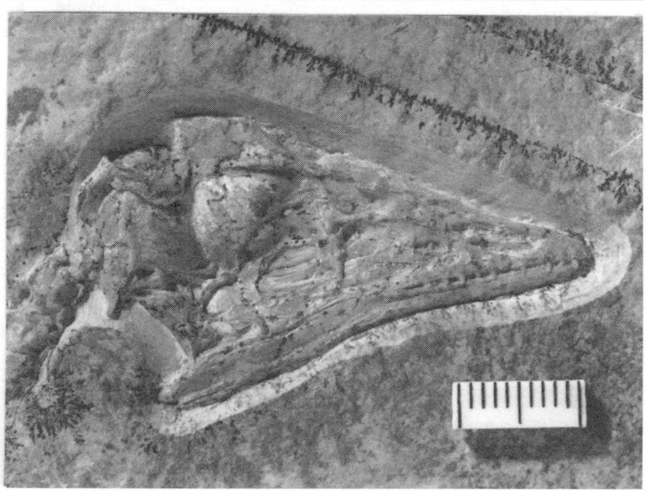

The Eichstätt specimen of *Archaeopteryx*. Unearthed in 1951, it was thought for some twenty years to be the small dinosaur *Compsognathus*. The outlines of the wing and tail are very faint. *Below,* enlargement of the skull (length 39 mm, 1.56 in.); the teeth are clearly visible. Cm scale. (Photo by Franz Höck, Munich; courtesy G. Viohl and Jura-Museum)

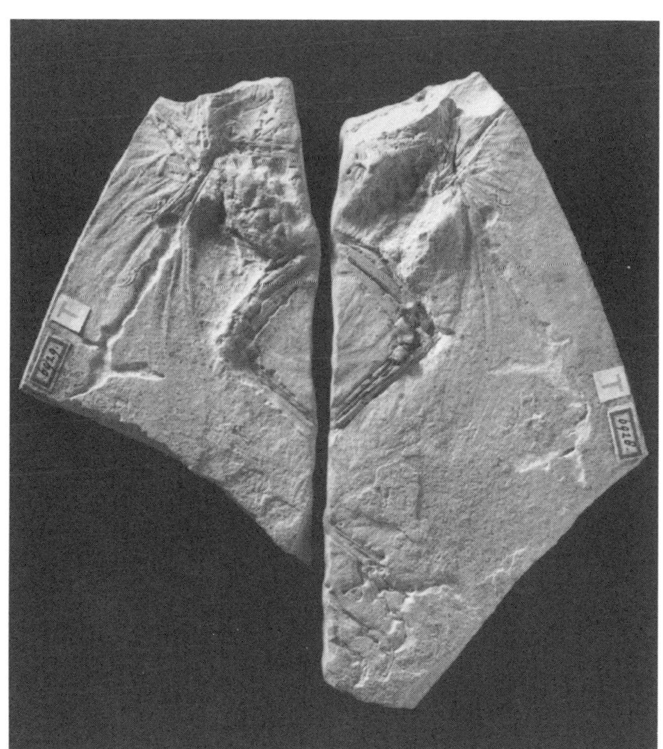

The Teyler specimen of *Archaeopteryx*. Found in 1855, it was long thought to be a pterosaur. In 1970 John Ostrom recognized it as a specimen of *Archaeopteryx*. This fossil is particularly interesting because one finger claw is preserved with its horny sheath. (Courtesy John Ostrom and Peabody Museum of Natural History, Yale University)

The Solnhofen specimen of *Archaeopteryx*. The well-preserved sixth skeletal specimen of the urvogel resides in the collection of the Bürgermeister Müller Museum in Solnhofen, Bavaria. The largest known *Archaeopteryx,* it is some 10 percent larger than the London specimen. (Courtesy Peter Wellnhofer and Bürgermeister Müller Museum)

Archaeopteryx bavarica, a new species of *Archaeopteryx* discovered in 1992, which represents the seventh skeletal specimen of the urvogel. The new bird is called the "Solenhofer Aktien-Verein" specimen in honor of the quarry company that owns it. (Photo by Franz Höck, Munich; courtesy Peter Wellnhofer)

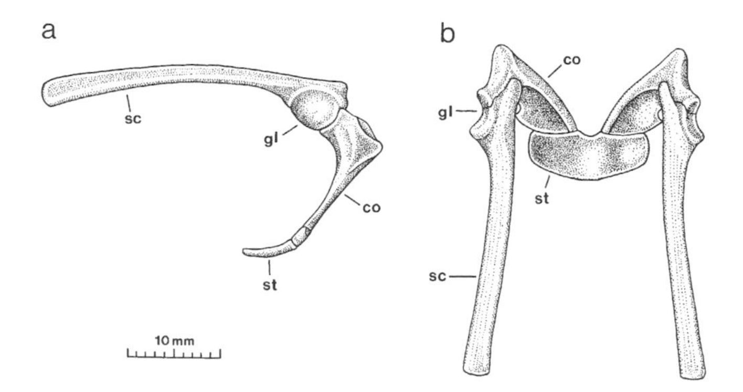

Wellnhofer's tentative restoration of the shoulder girdle and sternum of *Archaeopteryx bavarica: a,* right scapula and coracoid with sternum in lateral view; note the nearly 90° of articulation of the scapula and coracoid, a character unique to flying birds; *b,* scapulae and coracoids in articulation with the sternum, in dorsal view. Abbreviations: *co,* coracoid, *gl,* glenoid fossa, *sc,* scapula, *st,* sternum. (Modified after Wellnhofer 1993)

terdental plates, found also in thecodonts and theropod dinosaurs. The ungual phalanges of both manus (hand) and pes (foot) are extremely well preserved, still covered with their horny claws with sharply pointed tips; some claws are as well preserved as those of the Teyler specimen. Wellnhofer interprets the strongly curved claws of the pes to indicate a perching function and the presence of an ossified sternum to suggest that "*Archaeopteryx* was capable of powered flight and could not have been such a poor flyer as often has been stated" (1993, 2).

These, then, are the seven specimens of *Archaeopteryx*. By comparison with the Solnhofen *Archaeopteryx*, which is the largest example, the others, in size, are as follows: Solnhofen 100 percent, London 90 percent, Maxberg 87 percent, Berlin 77 percent, *Archaeopteryx bavarica* 66 percent, and Eichstätt 50 percent.

The Eichstätt Conference. Three years before the Solnhofen specimen was discovered, an International *Archaeopteryx* Conference, hosted by Günter Viohl of the Jura-Museum, was held in Eichstätt in 1984, to discuss problems involved with the morphology of *Archaeopteryx* and the origin of avian flight and feathers. Thirty-four of the world's specialists spoke on the famous urvogel, and two field trips were made to the Solnhofen quarries where the fossils were unearthed. The proceedings were published in the form of a nicely edited book entitled *The Beginnings of Birds*, edited by Max Hecht, Ostrom, Viohl, and Wellnhofer. Highlighting the conference was the presence of actual specimens of the Solnhofen dinosaur *Compsognathus* and three *Archaeopteryx* specimens—the main slab of the single feather and the Teyler and Eichstätt specimens—as well as a detailed video of the London specimen, considered too fragile to travel. A variety of topics such as the true anatomy of the urvogel, its possible relation to various groups of archosaurian reptiles, and the origin of bird flight—from the ground-up or trees-down—were covered. Also discussed was the sticky issue of the origin of feathers, the paramount feature defining the class of birds (see Dodson 1985 and Howgate 1985). There was general agreement that birds must have evolved via an arboreal route, but the question of avian ancestry appeared much more controversial.

Rivals for *Archaeopteryx*. Without the presence of feathers, the identification of the bones of early birds would in fact be impossible; the skeletons would be classified as reptilian, so that Jurassic and Triassic fossils that have been thought to be avian have been traditionally viewed with extreme skepticism; feathers or compelling osteological evidence must be present. Considerable interest was attached, therefore, to the description by Russian paleontologist A. S. Rautian in 1978 of a feather from an early bird named *Praeornis sharovi*, from Upper Jurassic lake deposits in Kazakhstan. The fossil has since been cited as an early stage in the evolution of typical avian feathers, intermediate between a reptilian scale and a feather. Through the courtesy of Evgeny Kurochkin, I was able to examine the feather in Moscow in 1982. I cannot confirm its avian status, however; in fact, I agree with Walter Bock (1986, 60) that the fossil appears to conform nicely to the pinnate leaf of a cycad (Nessov 1992a) or similar plant material. Further, its large size means that it would have come from an animal at the upper limits of known flying birds. *Praeornis* cannot be regarded as avian.

Several additional rivals for *Archaeopteryx* have emerged over the years. In 1881, O. C. Marsh described what he considered to be a Jurassic bird from Wyoming, *Laopteryx prisca*, which, following study by John Ostrom (1986b), turned out to be a pterosaur. In 1978, James Jensen of Brigham Young University, known as "dinosaur Jim," announced what he called a possible rival for *Archaeopteryx* from the late Jurassic of Colorado (J. L. M. 1978, 284; Jensen 1981). But the main specimen on which he based the claim is a disarticulated femur, and single limb elements from the Jurassic simply cannot be identified with any certainty as avian. The material has since been restudied; none was shown to be demonstrably bird, and most has been referred to pterosaurs (Jensen and Padian 1989). Few skeletal features of the limb bones of *Archaeopteryx* are not similar to other Jurassic reptiles, although, as we shall see in chapter 2, the skull, pelvis, and some other bones are more birdlike than once thought.

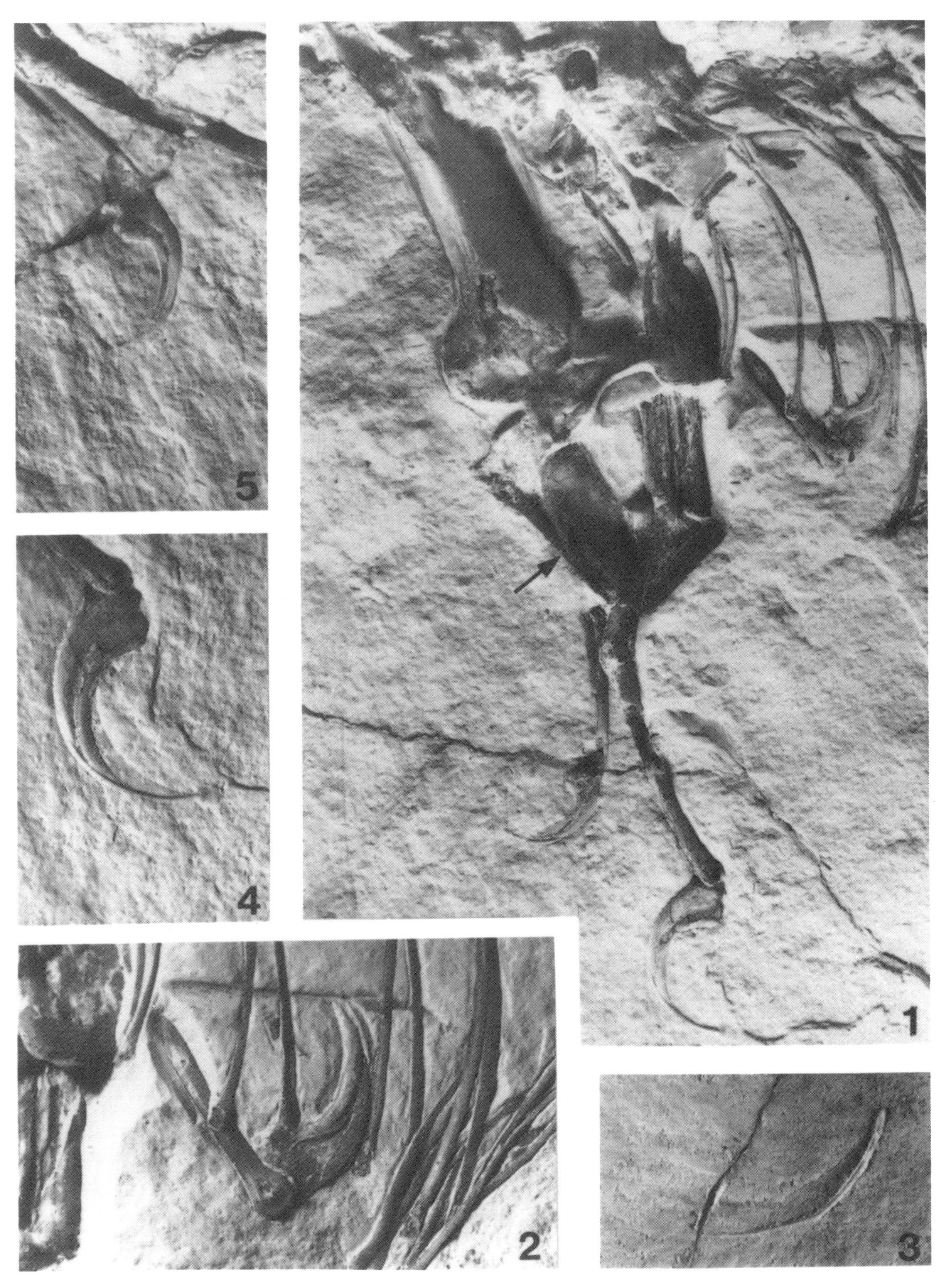

Archaeopteryx bavarica. 1, detail of shoulder girdle area showing the proximal parts of both humeri, scapulae, coracoids, the sternum (arrow), and right digits; × 2.3 natural size. *2,* detail of *1,* showing first right digit with bony claw, right coracoid, ribs, and gastralia; × 3.9. *3,* horny claw of first right digit, lying dislocated at the lower margin of the slab; × 3.5. *4,* claw with horny sheath or second right digit; × 3.7. *5,* claw with horny sheath of third right digit; × 3.2. (Photos by Franz Höck, Munich; courtesy Peter Wellnhofer)

Late Jurassic and Early Cretaceous Chinese Birds under Study. In spite of the controversy surrounding the above supposed Jurassic birds, Hou Lian-Hai and Zhou Zhonghe of Beijing's Academia Sinica are currently studying interesting fossil remains of birds from the late Jurassic and early Cretaceous of China, collected by themselves, farmers, and colleagues, which include not only skeletal elements but feathers. Although these are mostly very small birds, they closely resemble *Archaeopteryx* in many features, especially the skull and pelvic girdle. These discoveries have opened a new era in paleornithology in China, and to date, Mesozoic birds have been recovered from five regions in northern, eastern, and northeastern China: Gansu, Hebei, Liaoning, and Shandong Provinces and Inner Mongolia. The most productive horizon is the Jiufotang Formation (Lower Cretaceous) in Liaoning Province. Generally speaking, the Mesozoic bird fossil-bearing beds of China include four horizons, ranging from possibly Upper Jurassic to Lower Cretaceous.

These Chinese avian fossils represent one of the most dramatic breakthroughs in paleornithology since the discovery of *Archaeopteryx*, and well-preserved specimens of more than twenty individuals have been recovered, representing diverse morphological types, particularly in the specimens from Liaoning. At least three major adaptive types having been recovered, ranging from shore to terrestrial and predatory types, indicating an early ecological adaptive radiation. In addition, all have been recovered in terrestrial or lacustrine sediments, whereas *Archaeopteryx* has been found only in the lagoonal coastal Solnhofen deposits. And the wide geographic distribution of the Chinese fossils, as well as their occurrence in diverse geologic strata, illustrate not only a long history of birds that preceded *Archaeopteryx* but also considerable geographic distribution. With the exception of *Gansus*, which is an ornithurine, or modern type, bird, the others are grouped within the enantiornithine, or "opposite," birds, discussed in chapter 4.

Of special interest is a new urvogel, *Confuciusornis sanctus*, possibly close in age to *Archaeopteryx*, from the Chinese late Jurassic, that represents the oldest evidence for a beaked, edentulous (toothless) bird (Hou, Zhou, Gu, and Zhang 1995; Hou 1995; Hou, Zhou, Martin, and Feduccia 1995). Also under study is another Chinese specimen equivalent in age to *Confuciusornis* and intermediate in morphology between *Archaeopteryx* and the enantiornithines. *Confuciusornis*, which is about the size of the Eichstätt specimen of *Archaeopteryx*, was collected by farmers in 1994 from freshwater deposits in the Yixian Formation in Liaoning Province of northeastern China and is associated with numerous fishes, insects, amphibians, reptiles, and mammals. This bird has been described as late Jurassic, the same age as *Archaeopteryx*, and it shares many primitive characters with the famous urvogel. However, the dating of discontinuous lake deposits, especially the late Jurassic–early Cretaceous sequence in China, is problematic, and a broad range of dates has been obtained. Given the primitive morphology of this urvogel, however, it is at least tentatively considered to be approximately coeval with *Archaeopteryx*, and it would certainly be of equal or more interest if an urvogel with unreduced manal (hand) digits could be shown to have persisted into the realm of the Cretaceous. The wing skeleton exhibits the primitive pattern of *Archaeopteryx*, featuring a manus with unfused carpal elements, long digits, strongly curved, arboreally adapted claws, and a particularly strongly recurved and hypertrophied claw on the first digit. Two leg skeletons from the same site provide evidence for a clothing of body contour feathers.

The hindlimbs and pelvis are very similar to that of the primitive *Archaeopteryx*, and the fifth metatarsal is present, a condition found only in the Solnhofen urvogel. The tarsometatarsus is fused proximally, as in the Cretaceous opposite birds, and the hallux is reflexed and slightly elevated, as in *Archaeopteryx*. The three anterior toes are typically avian, with the middle toe longest, and exhibit highly curved, arboreally adapted claws. The skull combines primitive *Archaeopteryx*-like features with advanced characters, some strongly resembling modern birds, and especially the late Cretaceous opposite bird *Gobipteryx*, particularly in the large, broad, and heavy premaxilla. There is no question that *Confuciusornis* has a well-developed nasofrontal hinge for a modern avian prokinetic skull. *Confuciusornis* points to the strong possibility of a

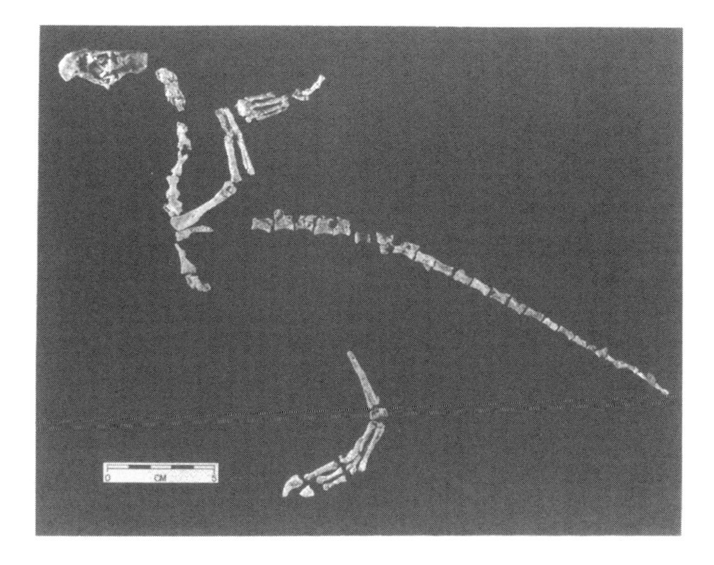

Protoavis texensis, small individual. (From Chatterjee 1991, fig. 7; courtesy Royal Society, London)

previously undiscovered pre-*Archaeopteryx*, or rapid post-*Archaeopteryx*, evolution of birds, as well as the rapid loss of teeth and acquisition of a keratinous beak in some early birds. Its preservation in deposits that represent a freshwater lake surrounded by a lush forest indicate that much of early avian evolution must have occurred in continental interiors.

Remains of still another urvogel of late Jurassic age have been reported recently from North Korea. The specimen, found in Shinniju City, is said to include a skull and articulated forelimb with feather impressions and is housed at the Kinnissei University Natural Sciences Museum. The find was announced in the North Korean newspaper *Nodong Sinmum* on 22 October 1993 and was written up in a Japanese magazine entitled *Korean Pictorial*, which carried a color photo of the specimen (1994, no. 2). We must await the detailed descriptions, determination of precise geologic ages, and naming and analyses of relations among these and other recently discovered primitive birds; however, for now it is clear that Chinese and Korean urvogels existed and that they were widely distributed geographically by the late Jurassic.

The Protoavis Controversy. A new chapter in the rivals for *Archaeopteryx* was written in August 1986, when paleontologist Sankar Chatterjee of Texas Tech University announced the discovery of a fossil he called *"Protoavis"* from the 225-million-year-old west Texas Dockum Formation (late Triassic), predating *Archaeopteryx* by some 75 million years. If confirmed, this find would push the origin of birds back to the dawn of the dinosaurs. The two crow-sized creatures, whose discovery startled the paleontological community, are claimed by Chatterjee to be closer to the ancestry of modern birds than is *Archaeopteryx* and are supposed to possess a furcula, a keeled sternum, strutlike coracoids, a birdlike quadrate bone, and quill nodes on the ulna for attachment of flight feathers (Beardsley 1986, 677). The discovery created a tremendously bitter and acrimonious controversy. Among other problems, the announcement was made by the National Geographic Society, which sponsored the research, but without the usual documentation; and it was not until June 1991 that Chatterjee published some of the documentation in a monograph entitled "Cranial Anatomy and Relationships of a New Triassic Bird from Texas" in the *Philosophical Transactions of the Royal Society* (also Chatterjee 1994, 1995), the remainder, on the postcranial skeleton and flight origins, to be published later. Because the material is quite fragmentary, possibly not associated, and there are no feather remains, a discovery that would extend the stratigraphic range of birds by half was destined to spark the vitriolic debate that ensued, and the controversy appeared in many major scientific journals, often with sharp personal attacks (Ostrom 1987b, 1991b; Anderson 1991; Monastersky 1991; Zimmer 1992). The questions surrounding the *Protoavis* discovery will not soon be settled. Chatterjee firmly maintains that *Protoavis* is the oldest bird, with a lightly built, pneumatized skull exhibiting an inflated braincase and an enormous orbit, and he infers the presence of feathers from quill knobs. The posterior teeth are lacking; the only remaining teeth are at the tip of the jaws. Presumed to be like that of modern birds, the skull is of a modified diapsid type (see chapter 2), with the orbit confluent with the upper and lower temporal openings and with loss of the bony bar that once separated them. The pelvis was quite birdlike, with an avian antitrochanter, and there was fusion of the ilium and ischium and lack of a distal symphysis, or junction, on the ischium and pubis. The cervical vertebrae were of the avian heterocoelous (saddle-shaped) type, and the pectoral apparatus was quite birdlike. According to Chatterjee, the humerus exhibited both bicipital and pectoral crests, and there was a triosseal canal, and the scapula and coracoid met at an acute angle, as in modern birds. In addition, the keeled sternum and a V-shaped furcula, together with other features, indicated to Chatterjee that *Protoavis* was capable of powered flight and could take off from the ground. The features Chatterjee illustrates are without question quite birdlike, and an early bird from the late Triassic is certainly possible. Yet, with the evolution of late Triassic thecodonts one might wonder if in *Protoavis* we are seeing an early experiment in thecodonts becoming somewhat birdlike in morphology. However, the major questions shrouding Chatterjee's discovery will likely remain, at least for the time being, because of the controversial nature of the evidence, not to mention the controversial nature of the problem itself, and the implications of finding a real bird from the Triassic.

The Archaeopteryx Flap. Perhaps the greatest challenge to *Archaeopteryx* came in 1986, when the eminent British cosmologist Fred Hoyle and his colleague Chandra Wickramasinghe charged in their book *Archaeopteryx, the Primordial Bird: A Case of Fossil Forgery* that the London specimen is a forgery. Hoyle and Wickramasinghe claim that the urvogel in the British Museum (and presumably the other specimens) was manufactured by pressing modern bird feathers into a layer of artificial cement (containing pulverized limestone from the Solnhofen quarries) that was painted on the two surfaces of a split slab containing a small dinosaur. Hoyle speculates that someone in Bavaria, possibly Carl Häberlein,

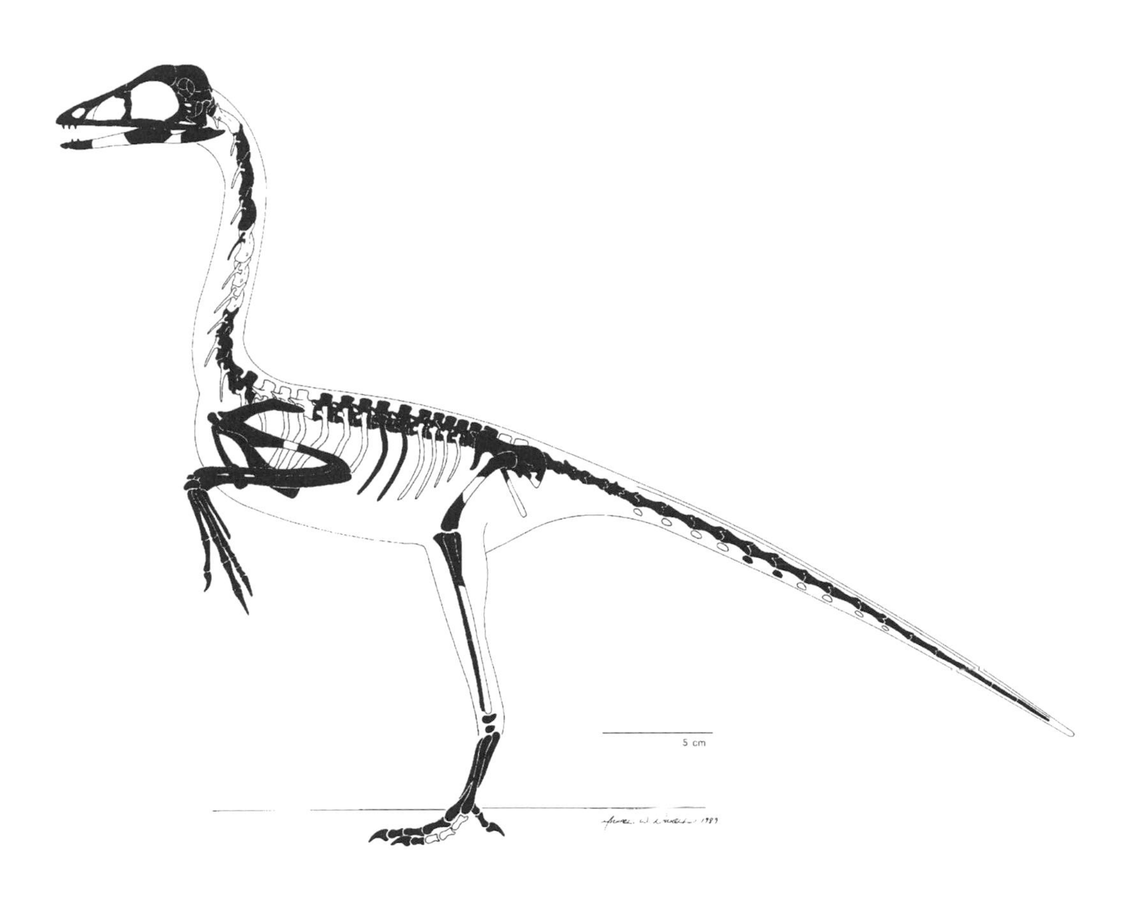

Composite skeletal restoration of Sankar Chatterjee's *Protoavis texensis* from the late Triassic of Texas. If on the avian family tree, it would predate *Archaeopteryx* by some 75 million years and, most likely, preclude a dinosaurian origin of birds. (From Chatterjee 1991, fig. 5; courtesy Royal Society, London)

produced the forgery for simple monetary gain and that Sir Richard Owen, knowing the fossil to be a forgery, purchased it as part of an elaborate scheme to undermine Darwin and his theory of evolution. Harvard University's Stephen Jay Gould (1986a), in a brilliant essay on this fiasco, pointed out that Hoyle's arguments make no sense at all, in that Owen, who fostered his own version of evolution, was not an antievolutionist but a fierce anti-Darwinian, quite possibly for personal reasons. Hoyle's arguments have been painstakingly demolished by Alan Charig, curator of fossil reptiles and birds at the British Museum, and his colleagues (Greenaway et al. 1986; see also Courtice 1987 and Dickson 1987). Perhaps the most devastating of their arguments (if one must choose) against forgery comes from the formation of manganese dendrites on the Solnhofen slabs. Small treelike patterns are formed in the limestone layers by the precipitation of manganese dioxide, and each "dendrite" is unique, just like snowflakes. These unique dendritic patterns cover both the slab and the counterslab of the London specimen, and some are directly on top of the feather impressions. Alan Charig and colleagues photographed the dendrites on the counterslab, then printed the negative backward and compared it with the corresponding dendrites on the main slab; they match perfectly. This can only mean that the dendrites must have formed on the bedding plane before the limestone slabs were split.

The most challenging aspect of the affair lies in explaining the motive for Hoyle and Wickramasinghe to place their reputations on the line for such a ludicrous proposal. According to Tom Kemp (1986), the reason relates to the authors' strange theory of evolution, by which major evolutionary change occurs as the result of "genetic storms." During such periods, the authors argue, showers of virus particles from outer space invade the planet and are incorporated into the genomes of living plants and animals. These periodic showers result in concomitant periods of mass extinction as well as in major evolutionary change. Apparently, a late Jurassic *Archaeopteryx* does not fit this picture, and the specimen must therefore be a forgery!

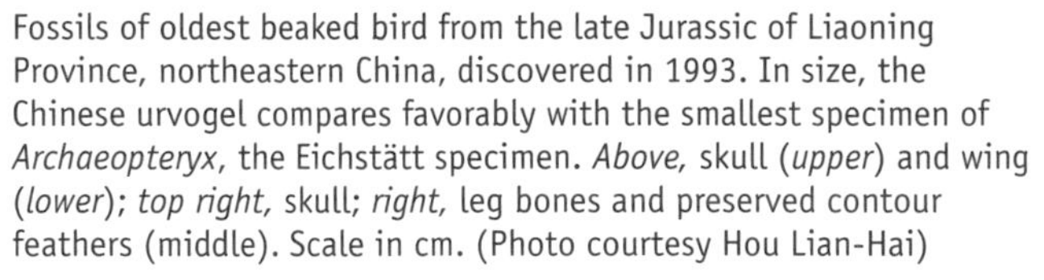

Fossils of oldest beaked bird from the late Jurassic of Liaoning Province, northeastern China, discovered in 1993. In size, the Chinese urvogel compares favorably with the smallest specimen of *Archaeopteryx*, the Eichstätt specimen. *Above,* skull (*upper*) and wing (*lower*); *top right,* skull; *right,* leg bones and preserved contour feathers (middle). Scale in cm. (Photo courtesy Hou Lian-Hai)

Table 1.3 The orders of living birds of the world

Order	Families	Number of taxa Genera	Species	Members
Tinamiformes	1	9	47	Tinamous
Rheiformes	1	1	2	Rheas
Struthioniformes	1	1	1	Ostrich
Casuariiformes	2	2	4	Cassowaries, emu
Apterygiformes	1	1	3	Kiwis
Dinornithiformes†	2	6	11	Moas (Pleistocene–Recent, New Zealand)
Aepyornithiformes†	1	1	7(?)	Elephantbirds (Pleistocene–Recent, Madagascar)
Dromornithiformes†	1	5	8	Mihirungs (Oligocene–Recent, Australia, Tasmania, New Guinea—possibly neognaths)
Sphenisciformes	1	6	17	Penguins
Procellariiformes	4	22	114	Tube-nosed seabirds: petrels, shearwaters, albatrosses, storm-petrels, diving-petrels
Pelecaniformes	6	8	66	Water birds with totipalmate feet: cormorants, pelicans, anhingas, boobies, gannets, frigatebirds, tropicbirds
Ciconiiformes	4	29	86	Heterogeneous assemblage of long-legged wading birds: storks, herons, hamerkop, shoebill
Falconiformes	5	82	309	Raptors: falcons, caracaras, hawks, eagles, Old World vultures, kites, osprey, secretarybird, New World vultures
Galliformes	5	74	282	Gallinaceous birds: grouse, quails, pheasants, chickens, curassows, guans, chachalacas, guineafowl, moundbuilders: excludes hoatzin
Gruiformes	11	55	214	Diverse terrestrial and marsh birds: rails, coots, sungrebes, cranes, sunbittern, kagu, limpkin, seriemas, bustards, buttonquails, trumpeters, mesites
Podicipediformes	1	6	22	Grebes
Charadriiformes	18	84	349	Shorebirds and allies: sandpipers, plovers, phalaropes, stilts, jacanas, painted-snipes, pratincoles, gulls and terns, seed snipes, sheathbills, skimmers, skuas, auks
Pteroclidiformes	1	2	16	Sandgrouse
Threskiornithiformes	1	14	33	Ibises, spoonbills
Anseriformes	2	48	161	Waterfowl: ducks, geese, swans, screamers
Phoenicopteriformes	1	1(3)	5(6)	Flamingos
Gaviiformes	1	1	5	Loons
Columbiformes	1	42	316	Pigeons, doves
Psittaciformes	3	80	360	Parrots, macaws
Coliiformes	1	2	6	Mousebirds
Musophagiformes	1	5	23	Turacos, plaintain-eaters
Cuculiformes	5	29	142	Cuckoos
Opisthocomiformes	1	1	1	Hoatzin
Strigiformes	2	25	173	Owls, barn owls
Caprimulgiformes	5	20	116	Nightjars, potoos, frogmouths, oilbird, owlet-nightjars
Apodiformes	3	127	425	Swifts, crested-swifts, hummingbirds
Trogoniformes	1	6	39	Trogons, quetzals
Coraciiformes	10	46	219	Kingfishers and allies: todies, motmots, bee-eaters, rollers, hoopoe, woodhoopoes, hornbills
Piciformes	8	66	407	Woodpeckers and allies: wrynecks, piculets, barbets, toucans, honeyguides, jacamars, puffbirds
Passeriformes	84±	1,168	5,739	Perching birds, songbirds, passerines

Table 1.3 The orders of living birds of the world (*continued*)

Order	Families	Number of taxa Genera	Species	Members
Suborder Tyranni	(17±	293	1,161)	"Suboscines," Old World lyrebirds, scrub-birds, rifleman, pittas, broadbills, etc., New World flycatchers, manakins, ovenbirds, antbirds, etc., and allies
Suborder Passeri	(67±	875	4,578)	Oscines, crows and allies, thrushes and allies, Old World insect-eaters and weavers and allies (including nine-primaried oscines)
Totals		2,063	9,702	

Sources: Following primarily Gill (1995); tentative numbers of genera and species from Monroe and Sibley (1993).
Notes: Numbers of genera and species given later may not add precisely because various birds have not been assigned to specific families. Most avian orders date from the Eocene, some date from the Paleocene, and passerines may be somewhat younger. The classification of the major orders of birds remains controversial, so here, pending new evidence, a more or less traditional classification and sequence is employed, with the exception of the early Tertiary shorebird mosaics and the ducks, flamingos, and ibises (see chapter 5). As Ernst Mayr has so appropriately put it, "There is no simple solution for the problem of finding the perfect classification of birds. No instructions exist that would tell us how to convert a phylogenetic bush into a linear sequence of the higher taxa of birds. I believe that the only way this problem can ever be solved is by international agreement. And that is the current status of the problem of avian classification" (1980, 122).
†Extinct

Table 1.4 Abbreviated classification of major groups of reptiles and higher categories of extinct birds

Class Reptilia
- Subclass Diapsida
  - Infraclass Lepidosauromorpha (Permian and Triassic eosuchians; rhynchocephalians, including the living genus *Sphenodon*; lizards, monitors [varanids], and snakes, etc.)
  - Infraclass Archosauromorpha
    - Superorder Protorosauria
      - Order Prolacertiformes (*Protorosaurus*), Upper Permian; (*Cosesaurus* and *Macrocnemius*, small, birdlike forms, Middle Triassic)
    - Superorder Archosauria (Upper Permian–Recent)
      - Order Thecodontia (Upper Permian–Upper Triassic, basal archosaurs)
        - Suborder Proterosuchia (*Chasmatosaurus*, Lower Triassic)
        - Suborders Rauisuchia, Aetosauria, Phytosauria, etc.
        - Suborder Ornithosuchia* (*Euparkeria*, Lower Triassic; *Ornithosuchus*, Upper Triassic; *Lagosuchus*, Middle Triassic, etc.); also called "pseudosuchians"
        - Suborder(s) Incertae Sedis* (avimorph thecodonts, including *Scleromochlus*, Upper Triassic, *Longisquama*, Upper Triassic, *Megalancosaurus*, Upper Triassic, etc.); also called "pseudosuchians"
      - Order Crocodylia (Triassic–Recent)
      - Order Pterosauria (Triassic–Cretaceous)
        - Suborder Rhamphorhynchoidea (Upper Triassic–Upper Jurassic)
        - Suborder Pterodactyloidea (Upper Jurassic–Upper Cretaceous)
      - Order Saurischia (Upper Triassic–Upper Cretaceous; "reptile-hip dinosaurs")
        - Suborder Staurikosauria (Middle–Upper Triassic early dinosaurs, including *Eoraptor, Staurikosaurus*)
        - Suborder Theropoda (early forms, including *Herrerasaurus, Coelophysis, Podokesaurus*, Upper Triassic; well-known carnosaurs, such as *Allosaurus*, Upper Jurassic, and *Tyrannosaurus*, Upper Cretaceous; coelurosaurs, including such dromaeosaurs as *Deinonychus*, Lower Cretaceous; birdlike Upper Cretaceous forms, such as *Troödon* and *Stenonychosaurus;* and Lower–Upper Cretaceous Ornithomimidae, including *Struthiomimus* and possibly *Mononykus*)
        - Suborder Sauropodomorpha (Upper Triassic–Upper Cretaceous)
      - Order Ornithischia (Upper Triassic–Upper Cretaceous; "bird-hip dinosaurs")

Table 1.4 Abbreviated classification of major groups of reptiles and higher categories of extinct birds (*continued*)

Class Aves
- Subclass Sauriurae
  - Infraclass Archaeornithes (Upper Jurassic)
    - Order Archaeopterygiformes (*Archaeopteryx*, Upper Jurassic, Germany)
    - Order Confuciusornithiformes (*Confuciusornis*, Upper Jurassic, China)
  - Infraclass Enantiornithes (Lower–Upper Cretaceous; ordinal divisions uncertain)
    - Order Sinornithiformes (*Sinornis*, Lower Cretaceous, China)
    - Order Cathayornithiformes (*Cathayornis*, Lower Cretaceous, China)
    - Order Iberomesornithiformes (*Iberomesornis*, *Noguerornis*, *Concornis*, Lower Cretaceous, Spain)
    - Order Gobipterygiformes (*Gobipteryx*, etc., Upper Cretaceous, Mongolia)
    - Order Alexornithiformes (*Alexornis*, Upper Cretaceous, Baja California)
    - Order Enantiornithiformes (*Enantiornis*, *Avisaurus*, *Nanantius*, etc., Lower–Upper Cretaceous, worldwide)
    - Order Patagopterygiformes (*Patagopteryx*, Upper Cretaceous, Argentina, incertae sedis)
- Subclass Ornithurae
  - Infraclass Odontornithes or Odontoholcae (Lower–Upper Cretaceous)
    - Order Hesperornithiformes (*Enaliornis*, *Baptornis*, *Hesperornis*, etc., Lower–Upper Cretaceous, probably worldwide)
  - Infraclass Neornithes or Carinata (Lower Cretaceous–Recent)
    - Superorder Ambiortimorphae
      - Order Ambiortiformes (*Ambiortus*, Lower Cretaceous, Mongolia)
      - Order Ichthyornithiformes (*Ichthyornis*, etc., Upper Cretaceous, possibly worldwide)
      - Order Apatornithiformes (*Apatornis*, Upper Cretaceous, North America)
    - Superorder Incertae Sedis
      - Order Gansuiformes (*Gansus*, Lower Cretaceous, China)
      - Order Chaoyangiaformes (*Chaoyangia*, Lower Cretaceous, China)
    - Superorder Palaeognathae (Paleocene–Recent)
      - Order Lithornithiformes (*Lithornis*, *Pseudocrypturus*, etc., Paleocene–Eocene, Europe and North America)
      - Order Remiornithiformes (*Remiornis*, Upper Paleocene, France)
    - Superorder Neognathae (Upper Cretaceous–Recent)
      - Order Charadriiformes (Upper Cretaceous–Paleocene "transitional shorebirds," *Graculavus*, *Telmatornis*, *Cimolopteryx*, etc.)
      - Order Gastornithiformes (Diatrymiformes) (incertae sedis, *Diatryma*, *Gastornis*, Paleocene–Eocene, Europe and North America)
      - Order Sandcoliiformes (*Sandcolius*, *Eobucco*, etc., Middle Eocene, Wyoming, probably Eocene of Europe)

Notes: The term *thecodontian* is generally used to refer to late Permian and Triassic archosaurs that do not fit into the category of crocodilians, pterosaurs, or dinosaurs (Weishampel et al. 1990).
* These groups are often included in the paraphyletic taxon Pseudosuchia

Rendering of the oldest beaked bird from the late Jurassic of northeastern China, exhibiting the primitive *Archaeopteryx*-like wing with three clawed fingers. The leg skeleton shows contour body feathers. (Drawing by John P. O'Neill)

2

The Descent of Birds

If *Archaeopteryx* is the oldest known bird, does it necessarily follow that *Archaeopteryx* is ancestral to all subsequent birds? Many authors, going back to Huxley and even Darwin, have argued that *Archaeopteryx* is really a sideline of avian evolution and not on the direct line leading to modern birds. Huxley considered *Archaeopteryx* somewhat irrelevant to the issue of bird origins (Desmond 1982), writing "that, in many respects, *Archaeopteryx* is more remote from the boundary-line between birds and reptiles than some living Ratitae [ostriches and kin] are" (1868a, 248). Probably owing to Huxley's influence, Darwin made little mention of *Archaeopteryx*, terming it "that strange Secondary bird," in reference to its stratigraphic occurrence. In recent years, Larry Martin of the University of Kansas has argued, as have others, that *Archaeopteryx* had many unique features and that the famous urvogel was not on the main line of avian evolution (Martin 1983a, 1985, 1987, 1991). *Archaeopteryx* remains central to all discussions on bird origins because it is unique in being the oldest known complete fossil bird, feathers and all. But its well-developed flight feathers, with differentiated primary and secondary feathers, much as in modern birds, among other avian features, tell of a long avian history that preceded it, and recent discoveries of possibly coeval avian remains from China confirm this assertion. So, *Archaeopteryx* is not the ultimate avian prototype, although it is very close in most of its structure to the ancestry of all subsequent birds.

The question remains, though, what antedated *Archaeopteryx?* What were the reptilian ancestors from which it and other birds descended? Most of the various groups of ruling reptiles of the Mesozoic era have, at one time or another, been considered the ancestors of birds, from lizards to pterosaurs and crocodiles to dinosaurs. After much more than a century of investigation and a fairly satisfactory fossil record of reptiles, the answer remains highly controversial. Nonetheless, just two major theories are widely accepted today. They differ with respect to specific lines of descent, and equally important, they differ widely in terms of the time when the first bird appeared.

By tracing the genealogy of reptiles, we can see where the two theories diverge. Ancestral to all other reptilian groups were the stem reptiles, scientifically termed cotylosaurs. The cotylosaurs appeared long before the Age of Reptiles, in late Paleozoic time, and differed from their descendants in having solid, or anapsid, skulls that were not perforated by openings in the back of the skull for the jaw musculature called temporal fenestrae (windows).

The Archosaurs

The next major group in our scheme is represented by the archosaur radiation, which includes diverse, diminutive forms, up to the most spectacular reptiles to ever exist, the dinosaurs. The somewhat catchall term *archosaur* refers to the thecodonts, or basal archosaurs, and their various derivatives; they are Mesozoic reptiles with diapsid skulls (with two temporal openings) and a large opening in front of the eye, the antorbital opening, or antorbital fenestra. Ignoring the birds that are archosaur derivatives, five orders are included within the Archosauria: Crocodylia (crocodiles and alligators), Saurischia (reptile-hip dinosaurs), Ornithischia (bird-hip dinosaurs), Pterosauria (flying reptiles), and the ancestral Thecodontia, which appeared in late Permian to early Triassic times, approximately 245 million years ago, and had their own evolutionary heyday. In addition, archosaurs probably shared a common ancestry with such diverse Mesozoic reptiles as the protorosaurs, rhynchosaurs, and trilophosaurids. Benton (1985) used the all-encompassing term *Archosauromorpha* to accommodate the entire group. The other diapsids were to become the lizards, snakes (derivatives of fossorial varanid, or monitor, lizards), and their kin and are termed

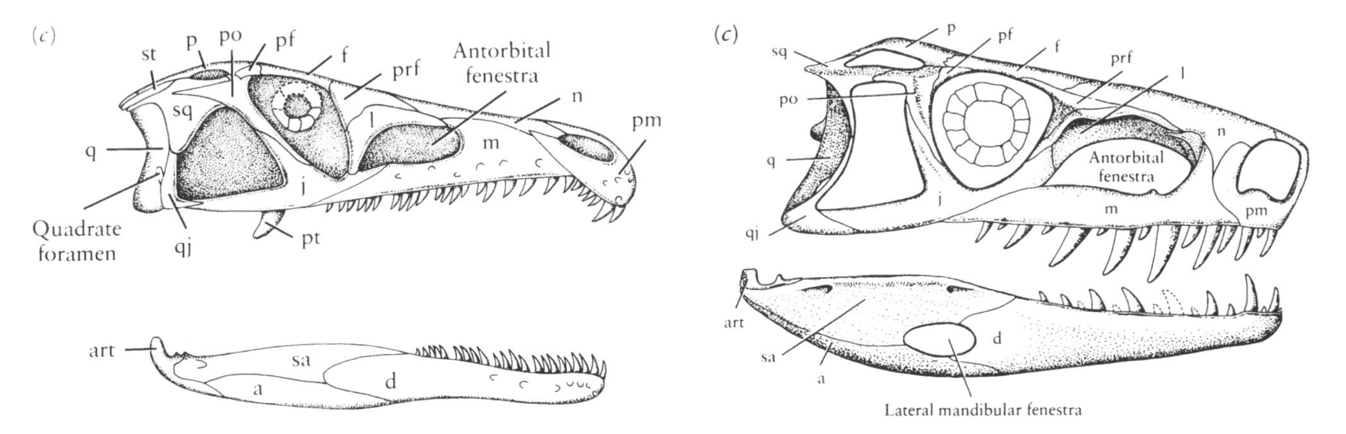

Left, skull of the primitive archosaur *Proterosuchus* [*Chasmatosaurus*] from the Lower Triassic of South Africa. Note the diapsid skull with two temporal openings behind the orbit and the antorbital fenestra anterior to the orbit. *Right,* skull of the gracile Lower Triassic thecodont *Euparkeria*. (*Left,* after Cruickshank 1972, courtesy A. R. I. Cruickshank; *right,* after Ewer 1965; courtesy Royal Society, London)

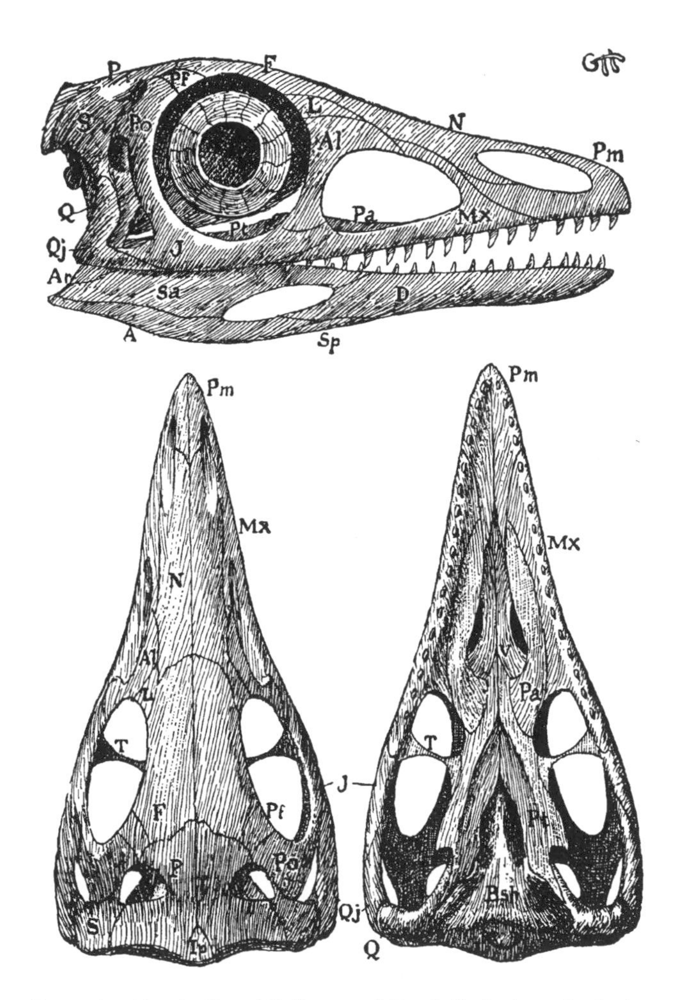

Reconstruction by Gerard Heilmann of the skull of a hypothetical proavis in side, top, and palatal views. Note that Heilmann has shown the skull with the typical diapsid condition, but with reduced temporal fenestrae; *Archaeopteryx* does not exhibit these types of temporal openings but is considered, like modern birds, a modified diapsid. (From Heilmann 1926)

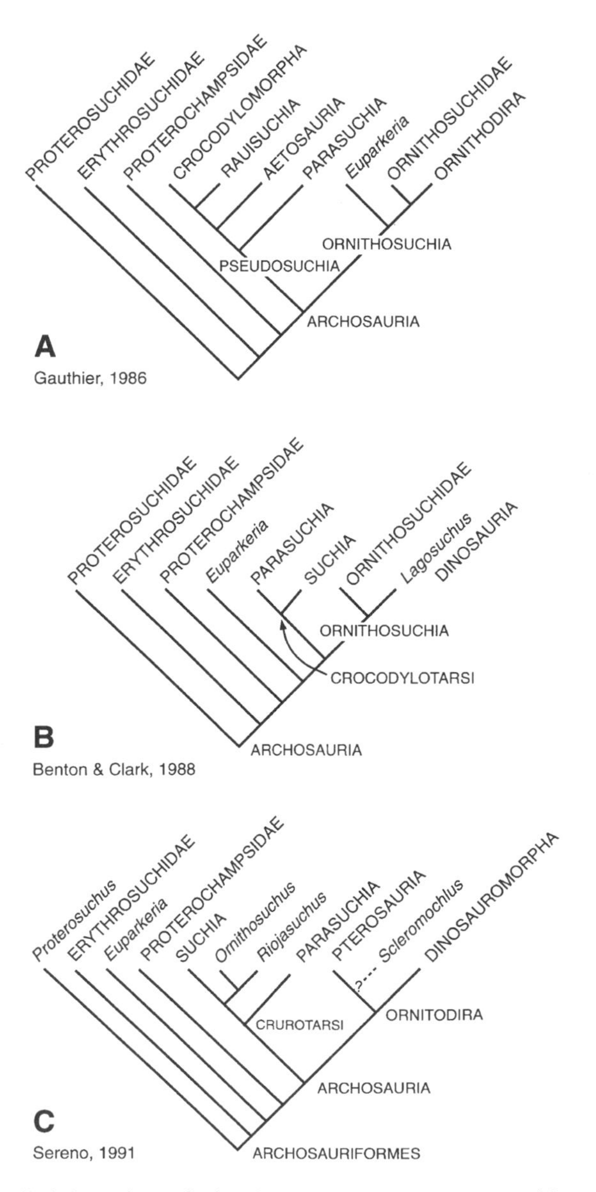

Cladistic hypotheses for basal archosaur phylogeny. *A,* Gauthier 1986; *B,* Benton and Clark 1988; *C,* Sereno 1991. (Modified from Sereno 1991; courtesy Society of Vertebrate Paleontology)

the lepidosauromorphs. According to this scheme, primitive archosauromorphs such as the protorosaurs (the lizardlike Prolacertiformes), rhynchosaurs, and trilophosaurids extended back to the late Permian and Triassic as the earliest members of the archosauromorph assemblage and were preceded by earlier, more primitive forms known as the eosuchians. Here, however, we are interested in the thecodonts and their descendants because it is within the thecodont assemblage that one finds the earliest reptiles that might qualify for bird ancestry.

We should not be deluded by these sophisticated names; archosaurs are diverse reptiles with few clues in their morphology to unite them into nice, neat groups. For example, the thecodonts are themselves an assorted, mixed group with few uniting features, so that although we refer to an order Thecodontia, it includes most Triassic archosaurs, an incongruous assemblage indeed. The name *thecodont* refers to the fact that the teeth are set in sockets, yet this character is common to other archosauromorphs and evolved convergently in mosasaurs (seagoing monitor lizards) and therapsids (mammal-like reptiles) (Carroll 1988). The thecodonts, like many basal phyletic groups, are not easily definable by modern standards but are characterized by the presence of diapsid skulls, thecodont teeth, and an antorbital fenestra (Witmer 1987), the last character uniting them with such later archosaurs as dinosaurs and pterosaurs.

Radiation of Basal Archosaurs

Thecodonts or basal archosaurs were dominant during Triassic times, enjoyed a worldwide distribution, and are grouped into four or so suborders by many authors. The Proterosuchia were short-legged quadrupeds that represented the most primitive forms; they emerged early in the Triassic period and quickly diverged into a number of more advanced groups comprising a heterogeneous assemblage of up to ten families, plus a number of forms of uncertain affinity. José Bonaparte's (1975) revelation that the archosaurian ankle joints are organized according to two basic morphological types, "crocodile-normal" and "crocodile-reversed," has gone a long way toward clarifying our understanding of this enigmatic group of reptiles (Chatterjee 1982). Yet tremendous confusion remains, as exemplified by the cladograms, showing the phylogenetic sequence of branching, of basal archosaurs, although a recent study by Paul Sereno (1991) has helped to interpret the relations of a number of forms.

Most authors still recognize two of the highly specialized thecodont groups, the heavily armored herbivores known as the Aetosauria and the convergently crocodile-like Phytosauria, which, unlike later crocodiles, had their nostrils in front of the eyes instead of at the tip of the snout. Other groups include forms that became dinosaur-like in various characters, including the Ornithosuchia, Lagosuchia, and Rauisuchia. Our focus, however, is not on the systematics of basal archosaurs, so here I use the term *thecodont* to describe this heterogeneous group of late Permian to late Triassic basal archosaurs. Historically, the term *pseudosuchian* (literally, false crocodile) was used, but today that term has no taxonomic reality. Many of the more interesting forms are especially well known in South Africa and South America, and various species showed a trend away from the original quadrupedal stance toward structural features associated with the attainment of a more obligatory bipedal, upright posture that would also characterize the dinosaurs. The thecodonts or ornithosuchians were thus ancestral to all the Mesozoic ruling reptiles, including the dinosaurs; through one forebear or another, thecodonts gave rise to birds.

Two Divergent Theories

It is at this point, the early to medial Triassic, that proponents of one of the two principal theories place the entry of the first bird, holding that birds descended directly from thecodonts about 230 or more million years ago. The competing theory postulates a much later entry of birds into the evolutionary arena after the line of descent had continued from thecodonts to the saurischian dinosaurs and their subsequent split into distinctive lineages. According to the dinosaur theory of bird origins, birds are directly descended from the later carnivorous theropod lineage of dinosaurs, a group that itself derives from a lineage of bipedal thecodonts with foreshortened forelimbs, close in time and structure to the Middle Triassic lagosuchid genera *Lagosuchus* (Bonaparte 1975, 1978) and *Lagerpeton* (Sereno and Arcucci 1993) but in appearance more like the late Triassic *Ornithosuchus* or *Postosuchus*. More specifically, most recent advocates of the dinosaurian theory for bird origins picture birds being derived from the dromaeosaurs, typified by such genera as the early Cretaceous *Deinonychus* and the late Cretaceous *Velociraptor*. But before we can compare the two theories, we must continue our reptilian genealogy.

Pterosaurs: Flying Reptiles Convergent on Birds

Among the main lineages of archosaurs are the pterosaurs, or Mesozoic flying reptiles, a group characterized by large membranous wings that are supported by a single greatly elongated fourth digit. Like other archosaurs, the skulls of these dragons of the sky had two

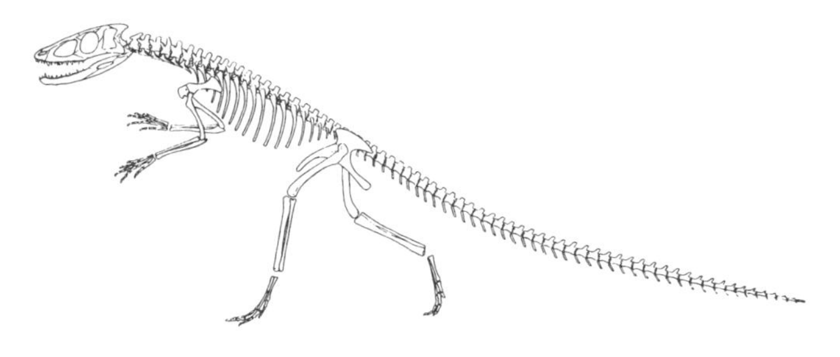

A generalized thecodont (*Hesperosuchus*) from Upper Triassic deposits in Arizona. This reptile was slightly more than 1 m (3 ft.) long. (From Colbert, *Evolution of the Vertebrates,* copyright © 1969; reprinted by permission of John Wiley and Sons)

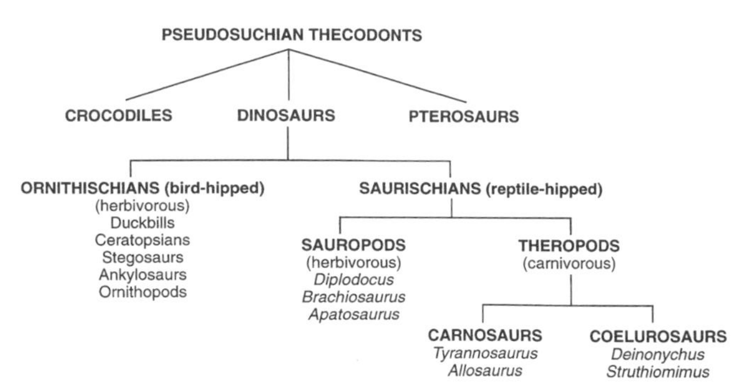

The descent of ruling reptiles from thecodonts. The descent of pterosaurs and the branching of the dinosaurs shown here is generally accepted, regardless of views on the origin of birds.

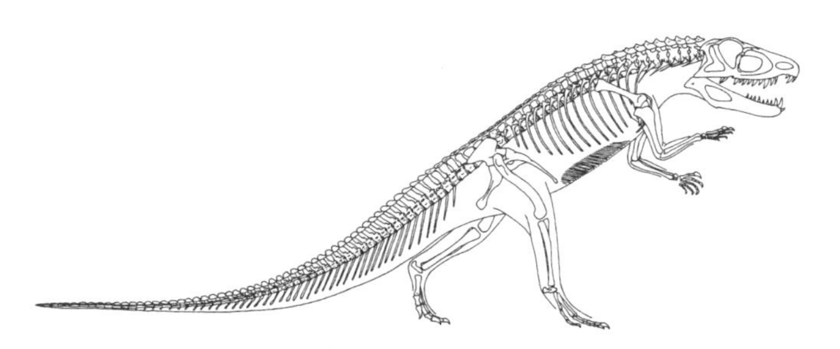

Above, the bipedal, dinosaurlike or ornithosuchian thecodont *Ornithosuchus* from the Upper Triassic of Scotland. *Ornithosuchus* was about 4 m (13 ft.) long, with forelimbs about two-thirds the length of the hindlimbs. *Right,* reconstruction of *Ornithosuchus*. (*Above,* from Walker 1964; courtesy Royal Society, London; *right,* from Heilmann 1926)

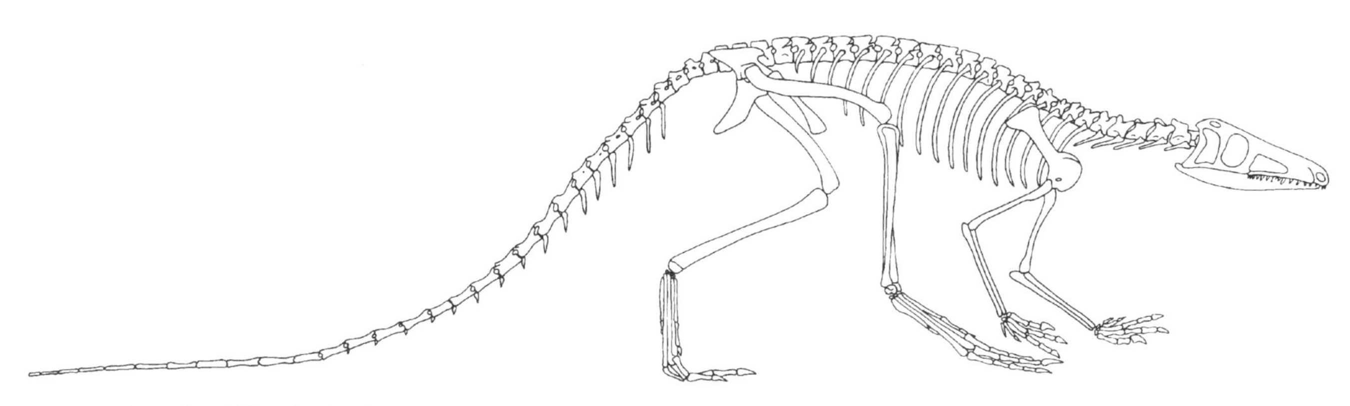

Lagosuchus, from the Middle Triassic of South America, is often considered a close ancestor of dinosaurs. This small, lightly built thecodont was about 30 cm (1 ft.) in length. Some believe that *Lagosuchus* represents the sister-group of both dinosaurs and pterosaurs. (From Bonaparte 1978; courtesy José Bonaparte)

openings in the temporal region and an antorbital fenestra. Because of their ability to fly and other flight characteristics similar to birds, pterosaurs were once considered to be ancestral to birds, but the theory never gained wide acceptance. The pterosaur wing is formed by completely different bones, there are no clavicles (wishbone), and the scapula and coracoid are fused to form a distinctive L-shaped structure. It is now clear that pterosaurs were at best distant cousins to birds, but they do provide interesting evolutionary comparisons. Pterosaurs are divided into two suborders, the more primitive, tailed rhamphorhynchoids, which appeared in the late Triassic and continued throughout the Jurassic, and the pterodactyloids, with shortened tails and reduced or absent teeth, which occurred from the late Jurassic to the late Cretaceous. The pterosaur lived on earth for some 150 million years, a span equal to that enjoyed so far by the birds, a group they paralleled in the evolution of many structural features associated with flight.

The Upper Triassic of the Italian Alps (some 220 million years old) has produced some of the most primitive pterosaurs found, including the nearly complete skeleton of *Eudimorphodon* and the even more primitive *Peteinosaurus;* much of the fossil material has been described by paleontologist Rupert Wild of the Staatliches Museum für Naturkunde in Stuttgart (Wild 1978). Although there have been great disputes over the exact relations of pterosaurs, many believe that they belong within the archosaurs (Wellnhofer 1991). However, Wild (1983) has argued that similarities in the skulls of *Eudimorphodon,* the best-known of the primitive Triassic pterosaurs, and those of the eosuchians, primitive Permian reptiles that preceded the archosauromorphs, indicate that the pterosaurs are direct descendants of the eosuchians. Following this interpretation, Peter Wellnhofer has stated that "the origin of pterosaurs probably lay with Permian eosuchians or certain transitional forms between eosuchians and prolacertilians, but not with thecodonts. Pterosaurs and thecodonts are branches of the diapsids which developed independently of one another from Upper Permian or Lower Triassic eosuchians" (1991, 44). Michael Benton argued in 1985 that the pterosaurs were not archosaurs but stated, in a note to his article, "I now accept that the pterosaurs are archosaurs, and a close sister-group of the Dinosauria, as argued by Padian (1983, 1984) and Gauthier (thesis, 1984)" (Wellnhofer 1991, 44). As with the thecodonts, then, controversy continues as to the correct systematic position of the pterosaurs. Given the current evidence, we must for now regard pterosaurs as an independent branch of the diapsids and assign them to their own subclass, the Pterosauromorpha.

Aside from the usual systematic difficulties, one thing is certain—pterosaurs became very birdlike in structure (Galton 1980; Ostrom 1986b); by late Cretaceous time the pteranodontid pterosaurs exhibited entire suites of convergent, birdlike features, including an expanded braincase, a cerebellum greatly elaborated with large floccular lobes, very large optic lobes, an enlarged orbit and eye, an ossified sternum, a keeled sternum, a hollow, light skeleton with thin-walled bones, pneumatized bones, a reduced, splintlike fibula, a birdlike tibia, a mesotarsal joint, the loss of a tail, a notarium (fused thoracic vertebrae), and the loss of teeth.

Although pterosaurs originated in the Triassic, millions of years before *Archaeopteryx,* they were predominant only in the later Mesozoic era, through the Jurassic and Cretaceous periods, and went through two waves of evolution in their invasion of the air. For flight, instead of feathers, pterosaurs depended on a wing membrane, or patagium, that was somewhat like that of bats, but has been compared favorably in aerodynamic terms with a hang-glider. Being volant, the pterosaurs evolved features that converged on birds in many respects, as outlined above. The skeleton was exceptionally light in weight and the bones were hollow. The sternum had a slight keel for the attachment of the flight muscles that stretched from the keel to an enlarged bony structure on the humerus. The wing bones, though, were highly specialized and different from those of birds. The three inner fingers had short, clawed hooks, possibly used for clinging to sea cliffs; or perhaps some pterosaurs hung upside down, in bat fashion.

Pterosaurs are most frequently preserved in marine sediments, many laid down in the great epicontinental oceans that covered much of the Americas and Europe during Cretaceous times, and it seems likely that the great majority of flying reptiles lived near Mesozoic coasts and fed on fish and other marine organisms. The early pterosaurs had teeth adapted for feeding on fish, squid, and similar prey, but the pterosaur jaw shows incredible diversity, and by the late Cretaceous, the pteranodontids had lost their teeth altogether. Illustrating this diverse adaptive radiation is a most unusual pterosaur from the early Cretaceous of Argentina, discovered in 1970 by José Bonaparte of the Museo Argentino de Ciencias Naturales of Buenos Aires; he named it *Pterodaustro.* Dubbed the "flamingo pterosaur" (see chapter 4), this creature was a truly remarkable filter-feeder that sieved small organisms from the shallows with hundreds of flexible "teeth" that were borne on its upwardly curved lower jaw (Wellnhofer 1991). Another noteworthy pterosaur was described in 1971 by the Russian paleontologist A. G. Sharov, who called it *Sordes pilosus,* which means, roughly translated, "hairy evil spirit." This late Jurassic pterosaur from the

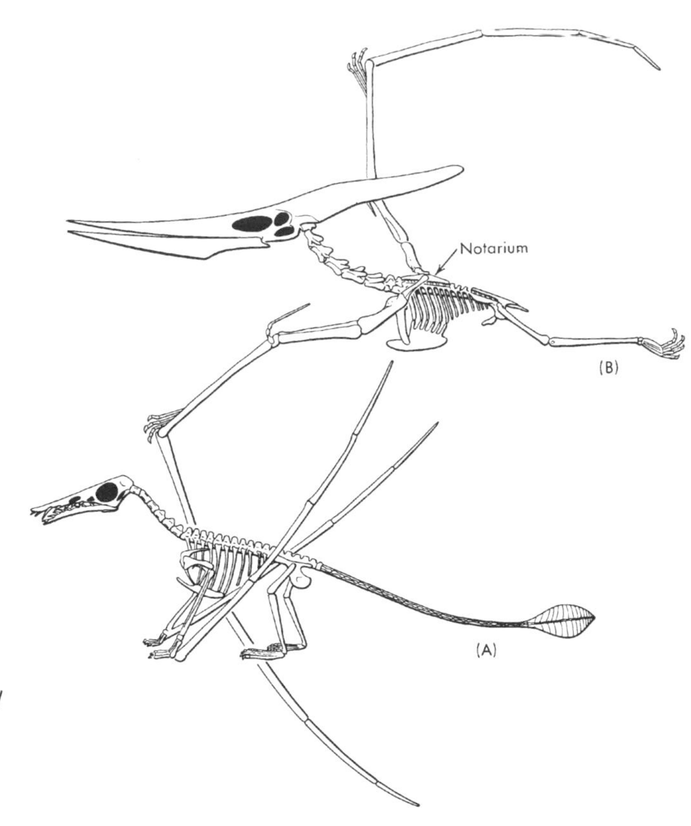

Pterosaurs: *A, Rhamphorhynchus* of late Jurassic age, with a wing spread of about 1 m (3 ft.); *B, Pteranodon* of late Cretaceous age, with a wing spread of slightly less than 7 m (23 ft.). The notarium (fused thoracic vertebrae) served to attach the upper edge of the scapula to the backbone, making a strong base for the great wing. Note the loss of the tail and teeth by Cretaceous time and the greatly increased keel on the sternum, which served as the place of attachment for the flight muscles. (From Colbert and Morales, *Evolution of the vertebrates,* copyright © 1991; reprinted by permission of John Wiley and Sons)

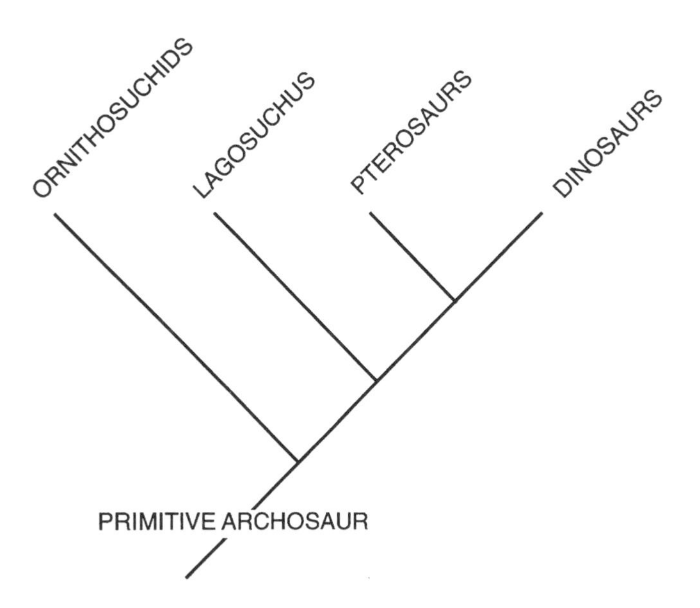

Cladogram showing the hypothetical phylogenetic relations among the pterosaurs, dinosaurs, *Lagosuchus,* and the ornithosuchids as suggested by Kevin Padian (1984). In this view, dinosaurs and pterosaurs are regarded as sister-groups sharing a common ancestor with *Lagosuchus* and the ornithosuchids. Paleontologist Rupert Wild of the Staatliches Museum in Stuttgart believes, however, that pterosaurs and thecodonts shared a common ancestor in the ancient Permian eosuchians, which preceded the archosaurian radiation. (Modified after Wellnhofer 1991)

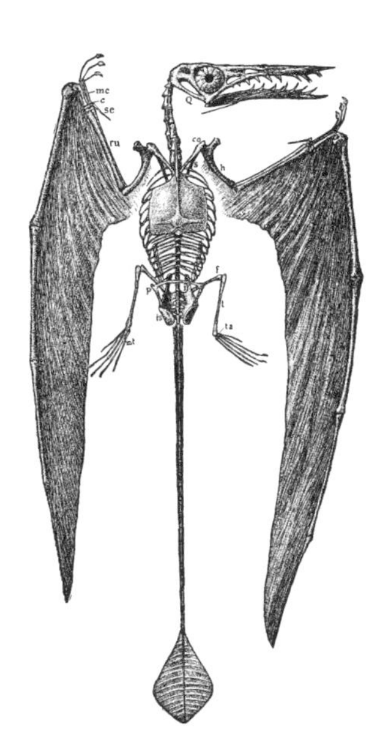

Skeleton of *Rhamphorhynchus gemmingi,* with wing membranes preserved. Although the pterosaur *Sordes* and several rhamphorhynchoid pterosaurs have been discovered with some type of integumentary, hairlike structures, now known to be stiffening rods in the wing membranes, reconstructing pterosaurs with fur, as is currently popular, is totally unfounded. (From Heilmann 1926)

Karatua Mountains of Kazakhstan was remarkable in having hairlike structures: "Long, dense and fairly thick hair covers the whole body, and the curvature of individual hairs suggests ample elasticity. There was also hair on the flight membrane . . . although it was sparser here, and shorter" (Sharov 1971, translation in Wellnhofer 1991, 102). The longest hairs were about 0.61 centimeters (0.24 in.). Traces of "hair" covering had been found previously in pterosaurs from several localities, including Solnhofen, but not the hairs or hairlike structures themselves (Wellnhofer 1991). Tremendous significance has been attached to the occurrence of "hair" in pterosaurs, and many workers have inappropriately reconstructed all pterosaurs with thick hair, proof to them that all pterosaurs regulated their temperature like modern birds and mammals. It should be noted that of the scores of pterosaurs recovered with the wing membrane preserved, only a few have been discovered with hairlike integumentary structures, and these structures have been variously interpreted, by those who have examined *Sordes*, including myself (also Bock 1986), as something other than hair, perhaps supportive collagenous structures of some sort. In addition, Shilov and Stephan (1976) appropriately point out that there is no reason to assume that there is a direct link between the presence of hair, which can control temperature, and basal metabolic rate. To assume that pterosaurs were exactly like bats because of the presence of these integumentary structures is to grossly overinterpret the evidence. Interestingly, more than a hundred years earlier, H. G. Seeley in 1870 suggested that the pterosaurs were covered with hair and were warm-blooded, like birds!

Most recently, David Unwin and Natasha Bakhurina (1994) studied additional, well-preserved specimens of *Sordes* from Kazakhstan and concluded that pterosaurs had no fur; what had appeared to be fur were in fact two kinds of fibers that served as stiffening rods in the wing membranes—long, ridged fibers in the outer sections and thin, curly fibers closer to the body, where more flexibility was needed. They also found evidence that the wing membrane was much more extensive than previously thought, attaching along the body from the front limbs to the hind feet, in batlike fashion, and that there was a smaller wing, or uropatagium, attaching to both legs and controlled by a fifth toe. This new view, which they see as possibly universal among pterosaurs, would picture flying reptiles with severely impeded movement on the ground and poor terrestrial ability, pointing to a "gravity-assisted" (arboreal) origin, as in birds, and a highly manipulable flight surface with low wing-loading, which indicates slow, maneuverable flight consistent with a mode of life as aerial insectivores or surface piscivores.

The flying reptiles came in a great variety of sizes, ranging from the late Jurassic, sparrow-sized rhamphorhynchoid *Pterodactylus elegans* of the Solnhofen limestone to the largest, a pterodactyloid named *Quetzalcoatlus* from the late Cretaceous of Texas, with an estimated wingspan of some 10 to 11 meters (33–36 ft.) (Langston 1981). The amazing radiation of pterosaurs is just now coming to light, and new discoveries from South America are being made practically every year (Zimmer 1994). The Texas pterosaur was the largest of all known flying vertebrates and, as deduced from the geology of the region, is thought to have lived as far as several hundred miles from the nearest Cretaceous sea. It had an unusually long neck and is speculated perhaps to have been a scavenger, feeding on the carcasses of large dinosaurs.

We see, therefore, that the first evolutionary radiation of pterosaurs produced small to moderate-sized forms with long tails and teeth set in sockets. Many varieties of these tailed pterosaurs are preserved in the Solnhofen lagoons where *Archaeopteryx* lived (numerous species in seven genera [Barthel, Swinburne, and Morris 1990], although Christopher Bennett [1993a] thinks there may be as few as five medium to large species represented by various growth stages). The second evolutionary radiation occurred during the Cretaceous, when the pteranodontid pterosaurs became very large and, as in the avian lineage, lost their teeth and bony tail and strengthened the dorsal vertebrate by fusion, producing a notarium. By that time the large, familiar North American *Pteranodon* had attained a wingspan of some 6 meters (20 ft.) and had a large bony cranial crest thought to have been variously a forward rudder or an aerodynamic counterbalance. Bennett (1992) has shown that *Pteranodon* was markedly sexually dimorphic, with the larger males probably using the hypertrophied crests for sexual display. Immature individuals of *Pteranodon* do not differ significantly in size from adults, and this, combined with the extensive fusion of mature skeletons, suggests that these late Cretaceous pterosaurs had determinate growth (Bennett 1993b), unlike their Jurassic predecessors (Bennett 1993a) the rhamphorhyncoids and pterodactyloids, which exhibited indeterminate growth as evidenced by well-preserved growth series. Pterosaurs probably depended on sea breezes and associated thermals for soaring. When the epicontinental seas retreated at the close of the Cretaceous period, the pterosaurs were doomed to extinction.

Dinosaur Divergence. The dinosaurs, too, were doomed, but according to one of the two prevailing theories, one of their lineages managed to give rise to the birds. Very early on, as the chart on page 48 shows, dinosaurs branched into two major groups delineated by the structure of their hips: the reptile-hipped saurischians and the bird-hipped ornithischians. There has long been a strong temptation to try to derive birds from the ornithischians because of this amazing but anatomically superficial resemblance of the hips. Gerard Heilmann in 1926 wrote that "the mere fact that [the pubis] was directed backward, like that of the birds, has evidently so hypnotized several scientists that they have overlooked, or tried to set aside, the many conspicuous differences between the birds and the Predentates [Ornithischia]" (148). (Ornithischian dinosaurs are called Predentates because they are united as a monophyletic group, or clade, primarily by having a "predentary bone" in the lower jaw.) The possible ornithischian origin of birds was dropped for some time, but Peter M. Galton advocated this now unpopular view as recently as 1970. Ornithischians were highly specialized herbivores, too far removed from the main line of dinosaurian evolution to have given rise to any major group. Their line dead-ended with such forms as the duckbills, armored ankylosaurs, plated stegosaurs, and horned ceratopsians like *Triceratops*.

The reptile-hipped saurischians themselves branched into separate herbivorous and carnivorous evolutionary lines. The herbivores, called sauropods, became highly specialized, particularly in adaptations associated with feeding, and many attained gigantic sizes. Giant forms such as *Diplodocus*, *Brachiosaurus*, and *Apatosaurus* (*Brontosaurus*) are among the favorites of museumgoers, but that is where their story ends. It is the carnivorous line of saurischians, called theropods, that, according to many paleontologists, continues the theoretical lineage toward the birds.

Two groups of theropods emerged, distinguishable by size and other skeletal features. The larger group, called carnosaurs, led, once again, to prized museum pieces—*Tyrannosaurus* and *Allosaurus*. The smaller sized group, however, termed coelurosaurs, included the ornithomimid (bird mimic) *Struthiomimus*, a late Cretaceous dinosaur convergent on living ostriches (*Struthio*), and the small, chicken-sized coelurid *Compsognathus*, previously encountered in the Upper Jurassic Solnhofen limestone, as well as such well-known dromaeosaurs as *Deinonychus* and *Velociraptor*. Many authors do not distinguish between carnosaurs and coelurosaurs, attributing the differences to size alone rather than to a phyletic split.

Old Theories of Bird Origins

Thomas Huxley, as we saw in chapter 1, found the resemblance between *Compsognathus* and birds very telling: "Surely there is nothing very wild or illegitimate in the hypothesis that the phylum of the Class of Aves has its foot in the Dinosaurian Reptiles—that these, passing through a series of such modifications as are exhibited in one of their phases by *Compsognathus*, have given rise to [birds]" (1868b, 74). Of *Compsognathus*, Huxley wrote that "a single specimen, obtained from those Solenhofen slates . . . affords a still nearer approximation to the 'missing link' between reptiles and birds" (1868b, 73). Thus was born the theory that birds originated from dinosaurs. Yet Huxley went much further and made extensive comparisons with the many large dinosaurs from both the ornithischians (*Scelidosaurus* and *Iguanodon*) and the carnosaur saurischians (*Megalosaurus*), noting particularly the similarities between them and birds in the large number of sacral vertebrae and the birdlike nature of the pelvis and hindlimbs. Huxley was clearly enamored with *Compsognathus*, finding in its hindlimb a very birdlike (chickenlike) skeleton. In 1870, Huxley seemed to drift toward the ornithischians, pointing to the pelvis of the ornithischian *Hypsilophodon* and claiming that it "affords unequivocal evidences of a further step towards the bird" (1870a, 28). But later that year, in his classification of the dinosaurs, Huxley created a separate group for *Compsognathus* and noted its resemblances to birds in "the peculiarities of the hind limb and pelvis" (1870b, 38). Huxley's theories were augmented by Georg Baur (1883, 1884) who argued for a dinosaurian origin of birds based particularly on the ontogeny of the avian tarsus, which he claimed recapitulated the dinosaur phylogeny. Yet Baur felt that, based on the pubis, it is "in the herbivorous Dinosaurs and especially in the ornithopod-like forms we must seek for the ancestry of birds" (1884, 1275).

Baur's hypothesis brings up an important point: that in many of these early writings, including those of Huxley, it is often difficult to ascertain exactly what the phylogenetic hypothesis is. Huxley's writings appear sometimes to advocate a dinosaurian origin of birds, at other times to propose a common ancestry of dinosaurs and birds—namely, what would now be interpreted as a thecodont origin of birds.

Huxley's contemporary Benjamin Mudge, of the University of Kansas, among others, found the dinosaurs far too diverse and highly specialized to have been the progenitors of birds: "The dinosaurs vary so much from each other that it is difficult to give a single trait that runs throughout the whole. But no single genus or set of genera have many features in common with the birds, or a single persistent, typical element of structure which is found in both" (1879, 226). However, in the same volume, Samuel Wendell Williston, also at the University of Kansas, countered Mudge, noting that "scarcely a single trait of structure runs through the whole of Dinosauria; but that fact does not affect the relation existing between the most

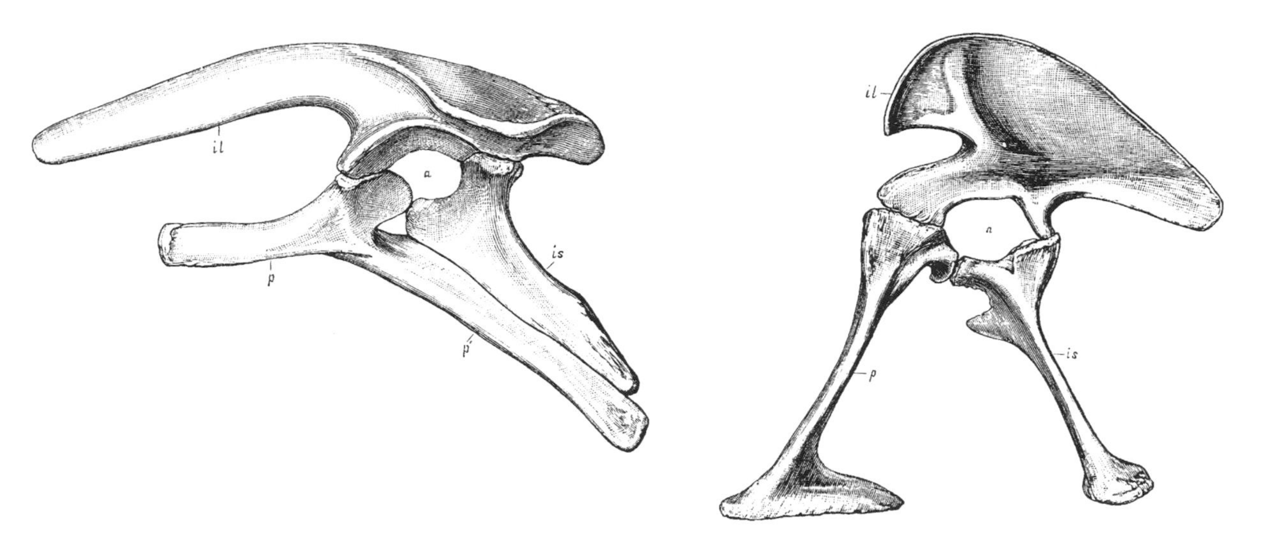

Pelves of, *left,* an ornithischian (bird-hipped) dinosaur, *Stegosaurus,* and, *right,* a saurischian (reptile-hipped) dinosaur, *Allosaurus*. (From Marsh 1896)

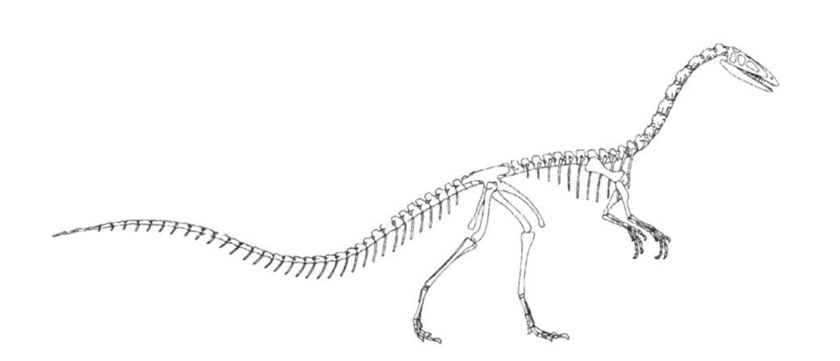

Skeleton of the North American late Triassic theropod *Coelophysis,* illustrating the structure of a fairly early and generally primitive dinosaur. It is about 2.5 m (8 ft.) long. (From Colbert, *Evolution of the vertebrates,* copyright © 1969; reprinted by permission of John Wiley and Sons)

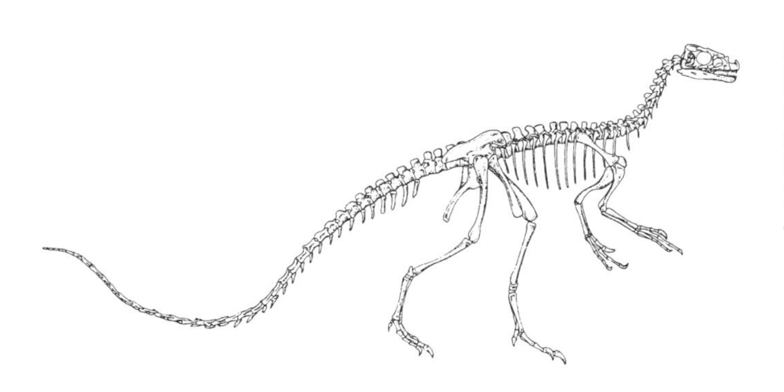

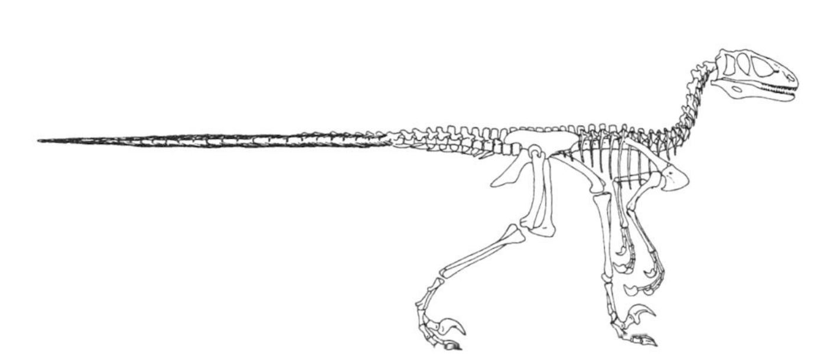

Left, skeleton reconstruction of the late Jurassic *Ornitholestes,* the best known of the family Coeluridae, occupying a central position within the coelurosaurs. It is about 2 m (6.5 ft.) long. *Below left,* skeletal reconstruction of the large, rapacious coelurosaur *Deinonychus,* about 3 m (10 ft.) long. (*Ornitholestes* from Osborn 1916; *Deinonychus* from Ostrom 1976c and Carroll 1988, courtesy John Ostrom and Yale University Peabody Museum of Natural History)

Skeletal reconstructions of two late Cretaceous dinosaurs of the family Ornithomimidae, the bird mimics. *Left,* the emu mimic *Dromiceiomimus* (2.9 m; 10 ft.); *right,* the ostrich mimic *Struthiomimus* (3.5 m; 12 ft.). As Dale Russell has remarked, "A combination of adaptations seen in ratite birds and anteaters, superimposed upon the skeletal framework of a theropod must have made the ostrich dinosaurs one of the most peculiar and interesting groups of terrestrial Mesozoic vertebrates" (1972, 401). (From Russell 1972; courtesy Dale Russell)

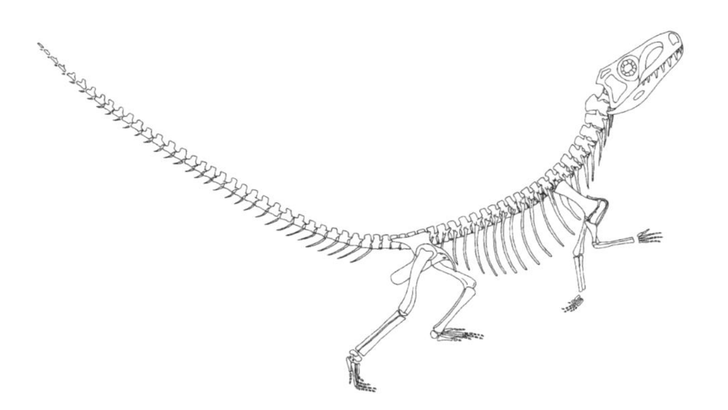

Skeletal reconstruction of the early Triassic thecodont *Euparkeria* from South Africa. *Euparkeria* is specialized in having the crocodiloid type of ankle joint, and dermal armor. Yet it is important in illustrating a stage in which basal archosaurs or thecodonts were still quadrupedal, before the commitment to an obligatory bipedal stance. *Euparkeria* could probably run for short distances on its hindlimbs. (Modified from Ewer 1965; courtesy Royal Society, London)

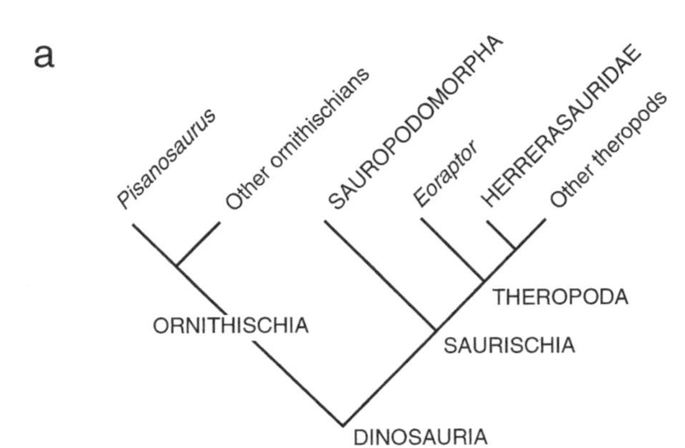

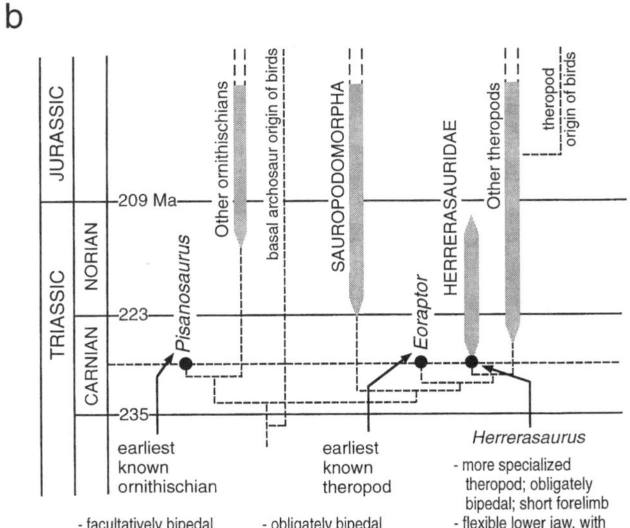

Hypothesis of early dinosaurian phylogeny (*a*) based on numerical cladistic analysis of 132 anatomical characters in 12 terminal taxa, by Paul Sereno (1993). Corresponding phylogram (*b*) showing recorded temporal ranges; dashed horizontal line is 230 million years. *Pisanosaurus* is the earliest known ornithischian, and *Eoraptor* and *Herrerasaurus* are the earliest known theropodlike dinosaurs. All three are known from the Ischigualasto Formation of Argentina, dated at approximately 230 million years old. (Modified after Sereno et al. 1993; reprinted with permission from *Nature* 361:64–66, © 1993 Macmillan Journals Ltd.; hypotheses on bird origins added to *b*, as well as bottom section and silhouettes, which are not to scale)

Lesothosaurus
- 1 to 2m. (up to 6.5 ft.
- ascending process of astragalus
- 5 m. (15 ft.)-too large for origin of flight

avian dinosaurs and the most reptilian birds" (1879, 458–459). Williston further derived birds from a specific group of dinosaurs: "True dinosaurs [apparently referring to theropods] . . . may have given off branches that developed upwards into birds" (458). Later, Williston (1925) retreated from his position, supporting instead a hypothesis of a common ancestry of birds and dinosaurs.

Still others had entered the arena from time to time, often with conflicting ideas. For example, Huxley's adversary Sir Richard Owen (1875) served up an extensive criticism of the Huxley view of dinosaurs and birds. Owen saw similarities between birds and mammals, particularly the egg-laying monotremes, but in the final analysis, he considered the pterosaurs to be the reptiles closest to birds. Then there was Robert Wiedersheim, who published a series of papers on bird origins, always invoking a common reptilian ancestor for *Archaeopteryx* and pterosaurs (see Witmer 1991). To continue, Carol Vogt considered the resemblances between dinosaurs and birds to be "only related to the development of the power of keeping an upright position upon the hind-feet," that is, he felt that birds and dinosaurs were convergent (1880, 448). Vogt is credited with his own version of avian origins; he viewed "the ancestors of the *Archaeopteryx* as terrestrial Reptiles in the form of lizards" (1880, 456).

Broom's Pseudosuchian Thecodont Hypothesis. The debate continued, but the arguments began to shift when Robert Broom, a prominent South African paleontologist, first proposed what has come to be known as "the pseudosuchian thecodont hypothesis" for bird origins. In 1913, Broom described from the rich Lower Triassic deposits of South Africa the pseudosuchian *Euparkeria* (redescribed by Ewer 1965), which he believed was ancestral not only to birds but also to the ruling reptiles:

> There cannot, I think, be the slightest doubt that the Pseudosuchia have close affinities with the Dinosaurs, or at least with the Theropoda. This has been recognized by Marsh, v. Huene and others. In fact there seems to me little doubt that the ancestral Dinosaur was a Pseudosuchian. . . . There is still another group for which some Pseudosuchian has probably been ancestral, namely the Birds. For a time one or other of the Dinosaurs was regarded as near the avian ancestor. . . . Seven years ago . . . I argued that the bird had come from the groups immediately ancestral to the Theropodous Dinosaurs. The Pseudosuchia, now that it is better known, proves to be just such a group as is required. In those points where we find the Dinosaur too specialized, we see the Pseudosuchia still primitive enough. (1913, 631)

In *Euparkeria*, the small, 230-million-year-old thecodont, still quadrupedal but tending toward bipedality, there appeared to be all the necessary anatomical qualifications for the ancestor of birds. No longer was it necessary to deal with the problem of dinosaurian specialization. Broom, then, was arguing for a common-ancestry hypothesis for birds and theropods, based on new finds of less specialized thecodonts (see addendum).

Heilmann's *Origin of Birds*. And so when Gerhard Heilmann, a Dane, wrote the first major book on avian evolution, *The Origin of Birds* (1926), he considered Broom's *Euparkeria* the key to avian ancestry. Heilmann meticulously described how *Archaeopteryx* could have arisen from *Euparkeria* and how flight could have evolved from small bipedal thecodonts. But his argument works equally well to validate a much later descent of birds from a small coelurosaurian dinosaur. Indeed, Heilmann himself seems to have had misgivings; he found many similarities between *Archaeopteryx* and the small coelurosaurian dinosaurs, just as Huxley had found similarities between *Compsognathus* and birds, particularly chickenlike ground-dwellers. His ambivalence permeates the book, but in the end he argued for early thecodontian ancestry:

> It would seem a rather obvious conclusion that it is amongst the coelurosaurs that we are to look for the bird ancestor. And yet, this would be too rash, for the very fact that clavicles are wanting would in itself be sufficient to prove that these saurians could not possibly be the ancestors of birds. . . . We have therefore reasons to hope that in a group of reptiles closely related to the coelurosaurs we shall be able to find an animal wholly without the shortcomings here indicated for bird ancestors . . . . such a group is possibly the pseudosuchians. . . . All our requirements of a bird ancestor are met in the pseudosuchians, and nothing in their structure militates against the view that one of them might have been the ancestor of the birds. (1926, 183–185)

Heilmann's *Origin of Birds* was engaging and well documented; it had all the earmarks of authority. It also had a lasting influence. Although Heilmann argued for a thecodontian ancestry of birds, he did not reject a relation with theropods, and in fact he argued for something close to Broom's common-ancestry hypothesis. For the next fifty years, although some authors proposed alternative theories (Holmgren 1955), Heilmann's early thecodontian ancestor was supported in virtually every subsequent textbook and paper on avian origins, perhaps, as has been noted, "more by default than by direct demonstration" (Bock 1969, 148).

What's the problem? We have some of the best-preserved fossils in the entire vertebrate series in the seven skeletons of *Archaeopteryx*, we have a wonderfully preserved array of fossil reptiles from the Triassic, Jurassic, and Cretaceous periods, and we have scores of well-educated scientists working on the problems of avian relations. So why isn't the problem resolved after much more than a century? Simply put, the matter of resolving phylogenetic relations is just not easy.

Methodology in Systematics

Establishing family trees or phylogenetic relations among vertebrates, both living and fossil, has been one of the most challenging aspects of the study of organisms since the time of the ancient Greeks, who devised a classificatory system based on their view of overall similarity. However, in spite of the advent of evolutionary theory, "natural groups fashioned by nonevolutionary systematists survived evolutionary reinterpretation largely untouched" (Sereno 1990, 9), and since that time the methodology of establishing evolutionary relations, especially in the past fifty years, has been one of the most debated issues in systematic biology.

Until recently few systematists or paleontologists were seriously concerned with the details of the methodology used in establishing phylogenies and hence classifications. Most were content with the methodology termed *evolutionary systematics*, which uses an eclectic spectrum of evidence from all lines of biological information to establish relations. Such lines of evidence might derive from a blend of such disciplines as comparative anatomy and embryology, functional anatomy and biomechanics, biochemistry, physiology, and behavior; Walter Bock has expressed the view that "greater phenotypical similarity implies greater genetical similarity and hence closer relationship" (1973, 377). By this eclectic methodology, workers have analyzed characters in great detail with respect to, for example, biomechanics, in an attempt to evaluate the relative usefulness, or weight, of each character in establishing relations. But the school of evolutionary systematics has been criticized because it lacks a set of procedures or a defined methodology by which quantitative guidelines can be followed and reproduced, as in the more sophisticated, "hard" sciences. Indeed, much of the selection of characters used in systematics is based on an individual's experience with a particular group of organisms, and characters are given various weights based on the studies of functional anatomy and the systematist's intuitive feelings concerning their relative importance. By this methodology the evolutionary systematist may choose a particular key character that he or she feels provides the answer to relations, ignoring a suite of other characters as of lesser importance or possibly due to convergent evolution. The school of evolutionary systematics thus suffers from being part science, part art, and has undergone considerable criticism as a consequence.

The phenetic, or numerical taxonomic, school of systematics emerged during the late 1960s and 1970s in an attempt to define a precise set of procedures that all systematists could easily reproduce, with the idea that "decreasing the amount of art in taxonomy is desirable" (Hull 1970, 49). The pheneticists proclaimed that organisms should be classified simply by their degree of "overall similarity," based on measuring as large a number of phenotypic characters as possible and giving each character equal weight (Sokal and Sneath 1963). Organisms were to be treated as objects, or operational taxonomic units (OTUs), and placed in classifications that would have the advantages of being convenient and stable. Proponents seemed to ignore the fact that the resultant classifications might be wrong, however, and many of the phenetic schemes and classifications appeared to be based on size, ecology, or some parameter other than evolution.

The general lack of confidence in systematic methodological procedures led to the search for a course of action to deal with systematic theory on a more scientific basis, and a new approach has emerged in the past twenty or so years; it is known variously as cladistics (cladism), phylogenetic systematics, or Hennigian systematics, and derives from the writings of the German scientist Willi Hennig (1966). The general methods of phylogenetic systematics have been outlined by numerous authors, including a lucid account by Wiley (1981) and a brief but well-written summary article by Sereno (1990). The general argument for this methodological paradigm is that "no methodology (phenetics included) has succeeded in specifying a procedure that would direct systematists, working independently, to subdivide a given organism into the same morphologic units—characters and character states—all of which carry equivalent information" (Sereno 1990, 10). The Hennigian methodology is concerned only with the establishment of *clades*, or monophyletic groups, which include the common ancestor and all its descendants. Hence, phylogenetic systematics expresses only the branching pattern of the phylogeny in its classifications. The term *grade* is used in contrast to clade, being defined as a paraphyletic group that is delineated on the basis of morphologic distance—that is, a group of organisms that are similar in their level of structural organization or, more simply, a level of anagenetic advance, or progressive "upward" evolution.

Describing the methodology of phylogenetic systematics will be easier if several terms are defined. As in other approaches to the problem of phylogeny, the term *character* refers to any recognizable attribute of an organism, in paleontology usually a feature or conformation of a bone or bones. The term *plesiomorphy* refers to a primitive character within a group, and such a character carries no information about relations. For example, feathers represent a plesiomorphous character in birds because they were present in the ancestral bird. Therefore, the presence of feathers in robins and cardinals does not imply that the two groups shared an immediate common ancestor. By contrast, both robins and cardinals possess a type of palate found only in passerine birds, the aegithognathous palate (in which the vomers are large, completely fused, and anteriorly truncated and separate the maxilopalatines), which represents a shared, derived character, or *synapomorphy*, thereby uniting them in common ordinal status, the order Passeriformes, within the class Aves. The term *apomorphy* is used to refer to such derived or specialized characters, and a synapomorphy is such a specialized, or derived, character shared by two or more groups. The term *symplesiomorphy* is used to designate shared primitive characters. Finally, Hennig used the term *sister-group* to refer to two groups united by the presence of one or more synapomorphies. It is important to note that "nearly all systematists now agree that phylogenetic affinity can only be inferred from synapomorphies and not from symplesiomorphies (shared primitive character states), and that this realization is due primarily to the influence of Hennig" (Sereno 1990, 10).

In the general practice of phylogenetic systematics, the validity of synapomorphies is determined by so-called outgroup comparison, by which one can assess the distribution of character states. For example, if a character exhibits two different states within a group (the ingroup), the state that occurs in near relatives (outgroup) is considered primitive, or *plesiomorphic*. The character state that occurs only within the ingroup is considered derived, or *apomorphic*. Character polarity is thus established by outgroup comparisons. Sereno (1990) used as an example the predentary bone in the ornithischian dinosaurs, a derived character that defines the group. Within the ingroup, the dinosaurs, the predentary bone is found only within the ornithischians, is absent in the sauriscians, but also is absent in the outgroup, the archosaurs, which are closely related to dinosaurs. The predentary is therefore a synapomorphy for the Ornithischia, a monophyletic group, but is plesiomorphic within the ornithischians. In other words, a character state that represents a synapomorphy at one level in the hierarchy is a symplesiomorphy at another level. Interestingly, the recognition that the predentary was evidence of a monophyletic Ornithischia was recognized as early as 1891 by Georg Baur, some seventy-five years before the publication of Hennig's *Phylogenetic Systematics*.

Establishing character polarities is a difficult procedure, as exemplified by a recent study of the ancient amphibians the caecilians. Jenkins and Walsh (1993) reported an early Jurassic caecilian, extending the geologic range of the group back from the late Cretaceous. Previous cladistic analyses had concluded that the two genera of rhinatrematid caecilians were primitive and had the primitive condition of an open (zygokrotaphic) skull roof in the temporal region, a condition shared with modern salamanders and frogs. However, newly discovered early Jurassic caecilians have a solidly roofed (stegokrotaphic) skull, and are primitive in a number of other cranial characters, a strong argument that the zygokrotaphic condition may be a derived condition. Another example, as we shall see in chapter 6, involves a specialized type of palate, the paleognathous type, that characterizes the large, flightless ratites and the tinamous. This was thought by numerous workers—evolutionary systematists and cladists alike—to be a derived condition, and therefore a synapomorphy indicative of evolutionary affinity or monophyly of the various ratites. Yet the discovery by Houde and Olson in 1981 of Paleocene and early Eocene tinamou-like birds with paleognathous palates provided compelling evidence that the ratite palate represented the primitive condition and was therefore not evidence of a monophyly assemblage.

A key problem with all of these methodologies is that the analysis of *homologous* characters, characters that are similar as a result of derivation from common ancestry, is not simple; identification of characters is often in the eyes of the beholder, and there is a great deal of character incongruence within given taxa. Such incongruence is termed *homoplasy*, a word that is often used as a synonym for convergence. Thus, characters that are similar in nature but have arisen from different ancestry are convergent. Convergence is common within vertebrates and results from the evolution of similar attributes in unrelated animals that are adapted for a particular way of life or habitat. Some prime avian examples are the Northern Hemisphere auks and the Southern Hemisphere penguins, which are unrelated but which share many common characters because both evolved adaptations for "flying" underwater. Among marine vertebrates a classic example of massive convergence is apparent between whales and porpoises and the extinct marine reptiles, the ichthyosaurs. The phenomenon known as parallelism is another evolutionary pattern in which nonhomologous characters are developed separately in two or more lineages of common

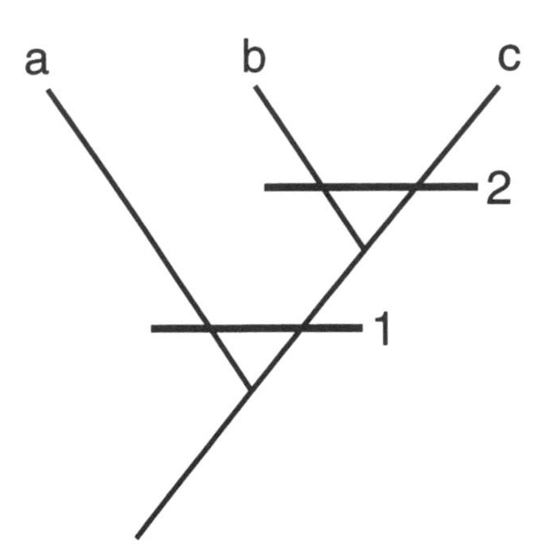

Phylogenetic hypothesis expressed by a cladogram. *a, b,* and *c* belong to a monophyletic group if they all possess some derived characters, or synapomorphies (1), not shared with other taxa. Taxa *b* and *c* are considered to be more closely related to each other than either is to *a* if *b* and *c* share other derived characters (2) not found in *a*. Taxon *a* is thus considered to be the sister-group of *b* and *c*. Taxa *b* and *c* are sister-groups of each other. Cladistic theory is based in part on the assumption that derived characters or evolutionary novelties will exhibit a nested pattern, which is a natural pattern of evolutionary theory. (Diagram modified from Carroll 1988)

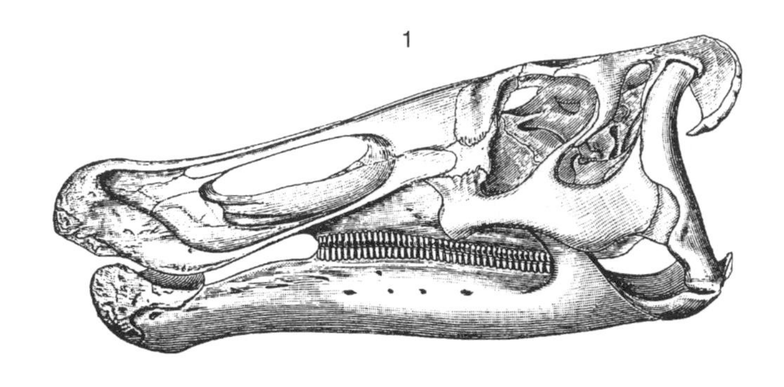

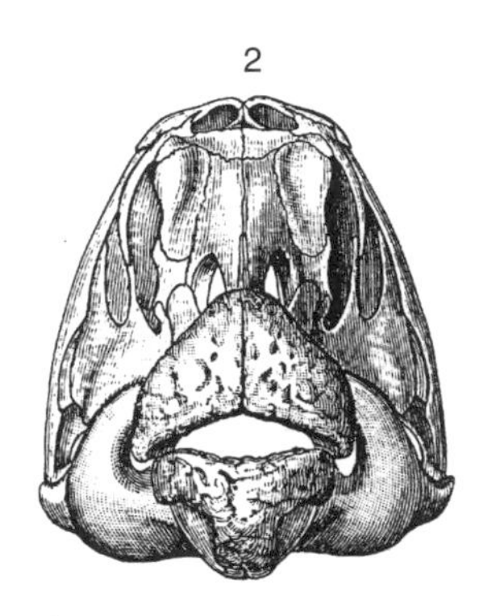

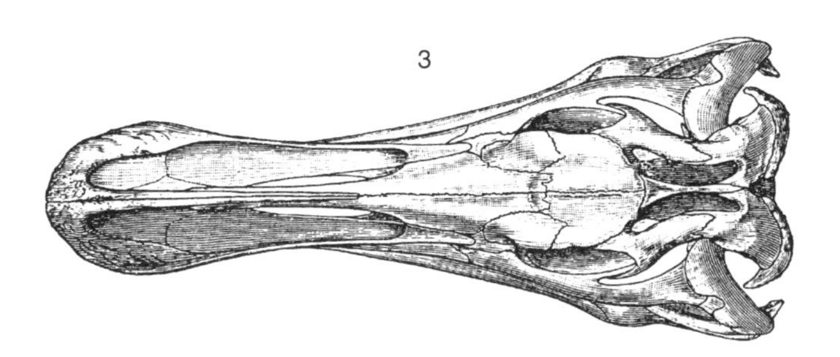

Skull of the Upper Cretaceous hadrosaur *Claosaurus,* in lateral, frontal, and dorsal views (*top to bottom*), illustrating the predentary bone, a synapomorphy that unites the ornithischian dinosaurs and is not found in any theropod dinosaurs. (From Marsh 1896)

ancestry on the basis of, or channeled by, characteristics of that ancestry (Carroll 1988, Simpson 1961).

To overcome the difficulties posed by convergence and parallelism, cladists typically invoke parsimony as the principal criterion to decide among alternative phylogenies—that is, it is more logical to accept a phylogenetic hypothesis that is dependent on the smallest number of changes. In cladistic analysis it is generally assumed that convergence is relatively rare and that most characters evolved only once. However, as Carroll has pointed out, "Biologists working with both modern and extinct groups argue that convergence is very common. . . . Arguments for the close relationship of groups based only on the common presence of derived features are of little value, if convergence is equally or more common than the unique origin of derived characters" (1988, 8).

Cladistics is thus not without criticism, and many of the criticisms are lodged at the notion that cladistics is a panacea for all systematics. L. B. Halstead (1982) has criticized cladistic methodology for the apparent precision and respectability that cladistics confers by its cladograms on what is in reality no more than speculation. He points out that the procedures of selecting synapomorphies, inferring polarity, and resolving conflicts are subjective and unreliable. As an example, one person might examine a set of skulls and discover three synapomorphies, whereas another might discover six from the eye region alone, though all six might be interrelated as a "character complex" and therefore represent but a single true character. Because all characters are given equal weight, drastically divergent phylogenies might result. Cladistic methodology, like its predecessors, is replete with subjectivity. As Michael T. Ghiselin has noted:

> Cladistic techniques often give ambiguous results. Different data imply different genealogies. As the cladists see it, the only permissible solution under such circumstances is to opt for the most "parsimonious" tree, in the sense of the one that invokes the smallest number of changes. One cannot use one's understanding of the organisms to decide which changes have occurred more than once. This is called "character weighting" and is not allowed. We are forbidden, for example, to say that a vestigial part represents the ancestral condition, or to consider what would be physiologically advantageous in a new environment. We are told that invoking multiple changes means an ad hoc hypothesis—even when we know that multiple changes have in fact occurred. Again the Popperian philosophy can be invoked against such views. Popper clearly distinguishes between ad hoc hypotheses, intended to preclude refutation, and auxiliary hypotheses, which enrich the system and narrow down the range of acceptable possibilities. There need be nothing ad hoc in phylogenetics about invoking stratigraphy, biogeography, genetics, embryology, or ecology. Popper calls not for naive parsimony, but for stringency. (1984, 219)

Still, cladistics does represent the most rigorous method for the analysis of morphology, even if its proponents tend to exaggerate its results. Unfortunately, because of the simplicity of the methodology, many workers unfamiliar with the particular group of organisms under study have relied on the literature for character analyses rather than studying the actual specimens. They have inferred character polarities that are often very difficult to assess and have eliminated the time component of the geologic record in their analyses (as well as any other pertinent information). Once the computer establishes the phylogeny, they insist on its correctness. In a book review of *Origins of the Higher Groups of Tetrapods* (Schultze and Trueb 1991), Clark argues that "the book again and again demonstrates that similarity lies in the eye of the beholder, and that the particular hypothesis being advocated strongly colors perceptions of morphological resemblance" (1992, 533).

In addition, many recent cladistic analyses strikingly resemble the phenetic analyses of the late 1960s and early 1970s, in which large numbers of characters, many of trivial nature and dubious character polarity, are tabulated and a phylogenetic analysis is performed using the computer program PAUP (Phylogenetic Analysis Using Parsimony; Swafford 1991). A recent example is a phylogeny of the Neotropical woodcreepers by Robert J. Raikow (1994), in which, using PAUP, Raikow performs a numerical cladistic analysis on thirty-six characters of some forty-two species representing thirteen genera. Characters and their coding for computer analysis include: M. caudofemoralis femoral segment of belly: (0) narrow, (1) wide; M. flexor cruris lateralis origin from caudal aponeurosis: (0) present, (1) absent; nostril shape: (0) elongate, (1) roundish; and extent of tendon ossification: (0) almost none, (1) extensive. It is inconceivable that a compelling case for character polarity, or direction of evolutionary change, for any of these features could be established. Raikow claims that "the overall phylogeny is well supported," even though he states, "The problem is . . . that the characters used in this study will not distinguish species-level differences in these genera" (113).

We need to understand cladistic analysis here because it lies at the core of the debate concerning bird origins. Examining examples of cladistic theory in action,

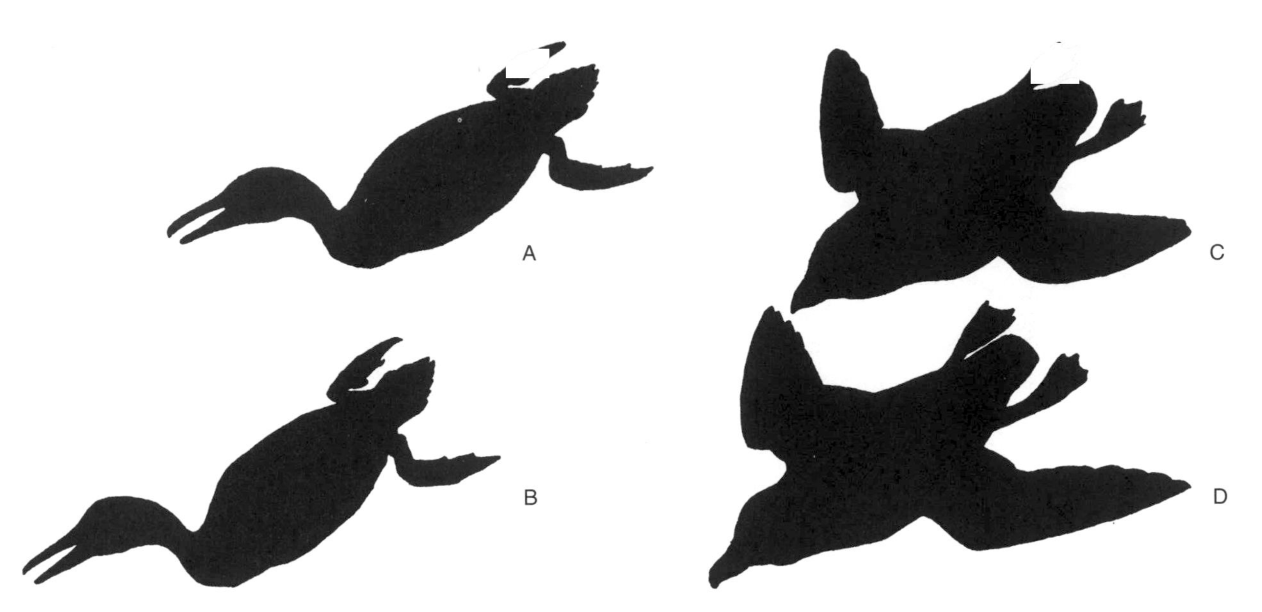

Convergent evolution, the acquisition of similar adaptations in unrelated groups of organisms, is an insidious trap that lies baited and waiting for those attempting phylogenetic reconstruction. Here, whole animal convergence is illustrated by two sets of unrelated look-alikes. The example on the left concerns look-alikes among the foot-propelled divers: *A,* the Cretaceous toothed bird *Hesperornis* and *B,* the modern loon. The example on the right concerns look-alikes among the wing-propelled divers, *C,* a Southern Hemisphere diving-petrel and *D,* a Northern Hemisphere auk. (Adapted from Fisher and Peterson 1964)

Homobatrachotoxin
in *Pitohui* (class Aves)

Homobatrachotoxin
in *Phyllobates* (class Amphibia)

Convergent evolution of an individual, complex character. Homobatrachotoxin, thought to be restricted to the Neotropical poison-dart frogs of the genus *Phyllobates,* is a complex steroidal alkaloid that is used by South American Indians for their poison blowgun darts. Homobatrachotoxin has also evolved in a poison bird, the genus *Pitohui,* endemic to New Guinea, recently discovered by Dumbacher et al. (1992). (Modified from Dumbacher et al. 1992)

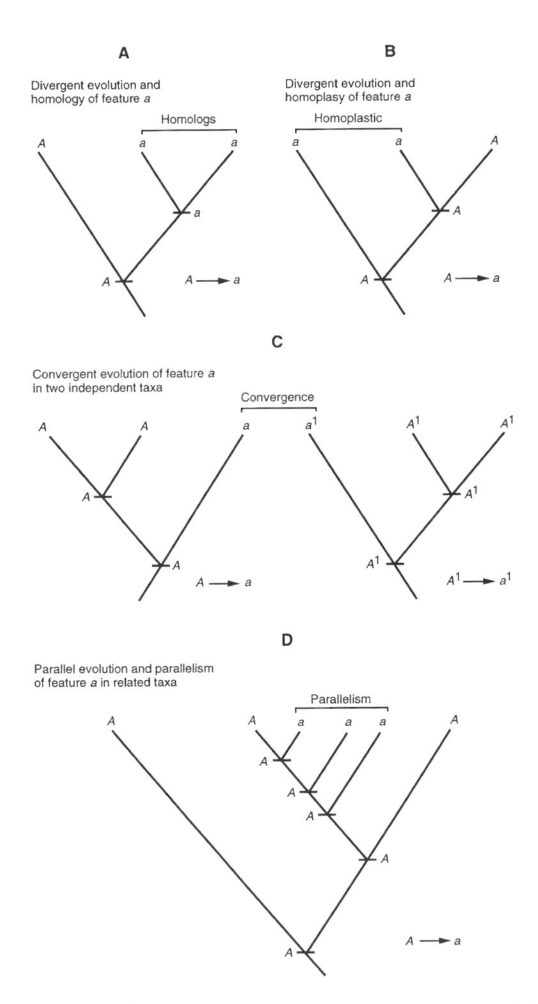

Diagrams of phylogenies showing different types of relations of structures symbolized by *a: A,* Divergent evolution and homology of features *a; B,* divergent evolution and homoplasy of features *a; C,* convergent evolution; *D,* parallel evolution. Features *a* are homoplastic in both convergent and parallel evolution. (From Walker and Liem 1994; courtesy Saunders College Publishing)

from both reptilian and avian groups, will indicate where cladistic methodology has been shown to have difficulties.

Case 1: *Hupehsuchus*, Enigmatic Triassic Aquatic Reptile from China. The case of *Hupehsuchus* (Carroll and Dong 1991) provides an example in which the simple tabulation of total numbers of perceived, derived characters shared with other groups gives misleading results because "most of derived characters appear to be subject to convergence among secondarily aquatic reptiles" (152). *Hupehsuchus* was an aquatic, diapsid reptile (skull with two temporal openings) with a girdle and limbs highly adapted for aquatic locomotion. In attempting to assess its phylogenetic position, Robert Carroll and Dong Zhi-Ming attempted to use standard cladistic methodology—calculating the relative number of derived features shared with other animals. Cladistic methodology places *Hupehsuchus* as the sister-group of highly specialized marine reptiles, the ichthyosaurs and the nothosaurs. As Carroll and Dong pointed out, "These data might be interpreted as indicating that all early Mesozoic aquatic reptiles shared a common ancestry, distinct from that of the lepisosauromorphs and archosauromorphs. This seems unlikely, however, because ichthyosaurs, nothosaurs, placodonts and thalattosaurs exhibit very different proportions of the trunk and limbs, indicating divergent modes of aquatic locomotion, as well as having very different patterns of skull and vertebral morphology" (149). They further noted, "The problem is to establish whether or not the large number of derived characters shared by these groups are actually homologous. . . . The difficulty of identifying unique derived characters uniting the Hupehsuchia with ichthyosaurs, despite the great number of derived similarities, raises the possibility that skeletal features that are common to secondarily aquatic reptiles might have evolved convergently in each of these groups" (149–150).

To test their hypothesis, the authors made further comparisons with two other groups of secondarily aquatic diapsid reptiles, the plesiosaurs and the mosasaurs. These groups are ideal because the specialized body plan of plesiosaurs is not known before the early Jurassic, and it is well documented that mosasaurs are derived from varanid (monitor) lizards, which evolved in the medial to late Cretaceous; indeed, the living Southeast Asian *Varanus salvator*, a large coastal and seafaring monitor, is analogous to a veritable proto-mosasaur. Yet the plesiosaurs shared twenty-six derived features of the skeleton with *Hupehsuchus*, and mosasaurs shared twenty-nine. According to the standard approach of phylogenetic systematics, convergent or homoplastic characters should be identifiable by the principle of parsimony (Patterson 1982). That is, if any two groups are united by many shared derived characters, but fewer characters support an alternative relation, the smaller number of characters are assumed to be the result of convergence. "Comparison of *Hupehsuchus* and mosasaurs, however," wrote the authors, "suggests that *most* of the derived characters exhibited by *Hupehsuchus* were convergently acquired by mosasaurs. How could this be recognized if we did not know from other evidence that mosasaurs evolved separately from terrestrial ancestors [varanid lizards] that retained the primitive character state for these features? The principle of parsimony cannot be directly applied in this situation" (151).

Carroll and Dong noted further that most of the changes associated with an aquatic mode of existence are "of such general nature that it is not possible to refute the possibility of their common origin by direct observation," and Rieppel has noted that many of the similarities among various marine diapsids are attributable to paedomorphosis: "The same tissues, the same developmental processes, and the same functional explanations may be involved in all groups, but the changes are not strictly homologous because they have occurred independently in each group" (1989, 151). Carroll and Dong ask, "Is there any category of characters that is not commonly subject to convergence? Is it ever possible to establish phylogenetic relationships without some specific information regarding the strict homology of the characters in question? The problem of establishing the relationships among the aquatic diapsids suggests that it is not" (151). Last, they argue, *"The principle of parsimony cannot be used directly to identify homologous characters if most of the derived characters are convergent"* (131, emphasis in original).

Case 2: The Supposed Monophyly of Loons, Grebes, and the Extinct Hesperornithiforms. This case provides an example from birds in which highly specialized foot-propelled divers give the illusion of being a monophyletic assemblage by phylogenetic analysis using the simple tabulation of supposed shared, derived characters (Cracraft 1982a). Careful examination of the embryology of certain aspects of their anatomy and DNA-DNA comparisons, however, show that these disparate groups have a very low probability of being a single phylogenetic lineage.

Using cladistic methodology, Joel Cracraft (1982a) proposed that the ancient toothed hesperornithiform birds (restricted to the Cretaceous) and the modern loons (which are thought to extend back to late Cretaceous but are likely of early Tertiary origin) and grebes (first recorded in the Miocene) were monophyletic. His analysis was based primarily on the shapes of various skeletal characters that he considered to be shared, derived conditions, or synapomorphies. This case is similar to that of case 1: the subjects are animals that have undergone extreme modification for a particular mode of life—in this case, foot-propelled diving. Massive convergence would therefore be expected to have occurred. Using cladistic methodology, however, it is all but impossible to ascertain which characters are homoplasious—caused by convergence—and which are homologous—shared and derived.

Most of the characters used by Cracraft are associated in one way or another with the locomotor apparatus, pelvic girdle and hind limbs. As Robert Storer has pointed out, "Adaptive modifications of such foot-propelled diving birds as the Hesperornithes, loons, and grebes include a long and extremely narrow pelvis, a short femur with a double, hingelike articulation at and above the acetabulum, a long tibiotarsus provided with a long cnemial crest, a laterally compressed tarsometatarsus, and the tying in of the leg with the body musculature nearly to the ankle joint. . . . The various groups of foot-propelled diving birds have long been recognized as prime examples of convergence" (1971, 163).

An examination of just a few of Cracraft's skeletal characters point up the problem. One example is character 5, defining the Gaviomorphae: "The cnemial crests of the tibiotarsus are raised proximally into a more or less sharp point compared to other birds"(42). Yet, careful examination of the cnemial crest of hesperornithiform birds shows it to be formed by the kneecap, or patella, only, whereas in loons it is an exclusive extension of the tibiotarsus, the patella being an extremely small structure embedded in the tendons of the knee. In contrast, grebes form the cnemial crest by compounding a large patella and a long extension of the tibiotarsus. Other differences indicative of an independent origin of the three groups include the webbed feet of loons and the lobate webbing on the feet of grebes, as well as major differences in the structure and mechanism of the ankle and toe joints (Stolpe 1935, Storer 1971). Cracraft discussed Stolpe's (1935) evidence and dismissed it, stating that it "*contains no relevant information bearing on the problem*" (51, emphasis in original).

Extensive DNA-DNA comparisons by Charles Sibley and Jon Ahlquist "support a relationship among loons, penguins, and procellariids, but not a close relationship between loons and grebes" (1990, 550). Sibley and Ahlquist further state that "the grebes have no close living relatives" (551) and that "the errors in Cracraft's reconstruction of the phylogeny of the diving birds are due to the difficulties of interpreting morphological characters, not to the principles he used as the basis for his analysis" (150).

The case of the association of loons and grebes is of particular interest because it follows a general historical pattern involving cladistic analysis—that is, a reversion to classifications of the past. Early classification associated these avian groups, but after Stolpe's famous paper of 1935, most authors accepted the view that the similarities between loons and grebes, as well as hesperornithiforms, were caused by convergence. Then cladistic analysis, generally unable to deal with massive convergence, led to the groups being placed back together as a monophyletic assemblage.

Ernst Mayr (1981) used Cracraft's analysis of loon and grebe phylogeny to point out the deficiencies of cladistic analysis in discerning homologous from nonhomologous characters. "Both grebes and loons, two orders of diving birds, have a prominent spur on the knee and were therefore called sister-groups by one cladist. However, other anatomical and biochemical differences between the two taxa indicate that the shared derived feature was acquired by convergence" (512).

As Storrs Olson has stated, "The evidence for this [Cracraft's monophyletic grouping] was derived almost entirely from the same convergent specializations of the hindlimb and pelvis that led earlier workers to the same erroneous conclusions. Cracraft's phylogeny requires, among other things, that the Hesperornithiformes re-evolved teeth from an edentulous ancestor. I can only hope to distance myself from those who would accept the likelihood of such nonsense" (1985a, 91).

In 1986, Cracraft modified his view, stating that "the hypothesis that *Hesperornis* and some neognathus (loons, grebes) might be closely related (Cracraft 1982a) cannot be accepted" (393). However, the disparate loons and grebes still remain as cladistic sister-groups.

Case 3: The Supposed Monophyly of the Hawks and Owls. This example compounds problems of case 2. In this case, similarities in various types of diurnal birds of prey, which have, by virtually all twentieth-century systematists, been viewed as convergent, are, through the methodology of phylogenetic systematics, considered as monophyletic (Cracraft 1981). This is very similar philosophically to the situation of case 2. According to Cracraft, "What is not often realized by contemporary ornithologists is that numerous 19th century workers saw the many similarities of hawks and owls as evidence of close relationship; it was 20th century systematists who overemphasized the differences and concluded convergence. . . . Owls can be included in the Falconi because they possess a derived tarsometatarsal and pelvic morphology shared with pandionids and acciptrids" (694).

Sibley and Ahlquist (1990), in reviewing the history of the classification, stated of Cracraft's approach that "the

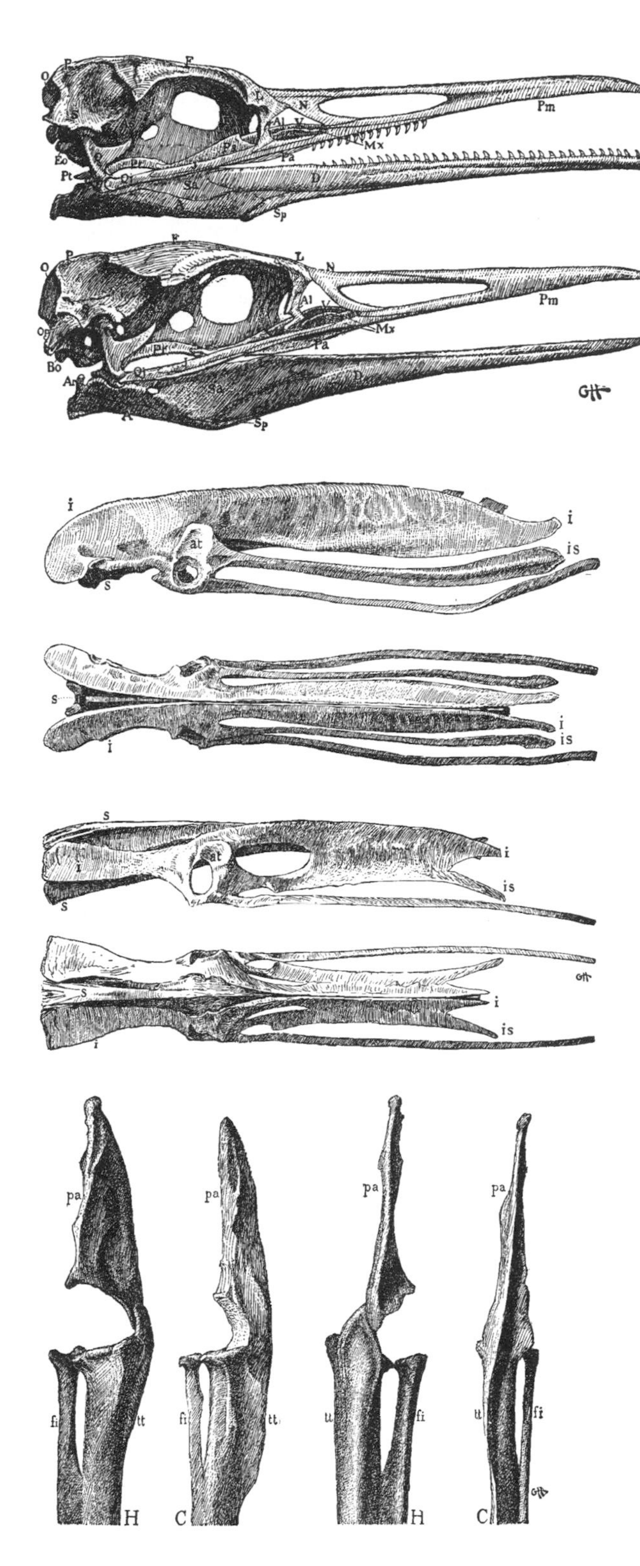

A remarkable example of whole animal convergent evolution that cladistic analysis was unable to detect involves the loons, grebes, and hesperornithiform birds, now shown to be convergent on the basis of embryology and biochemical comparisons. *Top,* the skulls of the toothed, Cretaceous *Hesperornis* (*above*), compared to the common loon (*Gavia immer*) (*below*). *Center,* pelves in lateral and dorsal views of *Hesperornis* (*above*), compared to that of the great crested grebe (*Podiceps cristatus*) (*below*). *Bottom,* cnemial crest of the tibiotarsus, which houses the powerful swimming muscles, in two views of *Hesperornis* and the common loon. In *Hesperornis* the cnemial crest is formed embryologically from the patella, in loons, by a projection of the tibia, and in grebes, by contributions from the patella and the tibia. (From Heilmann 1926)

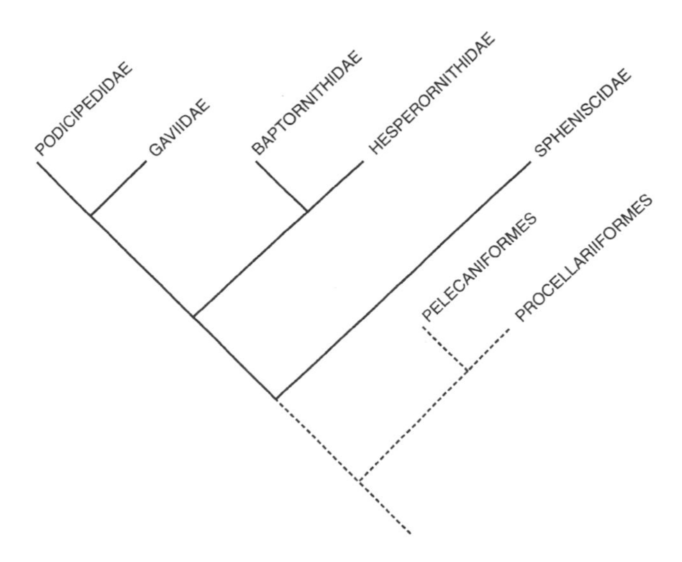

Cladogram resulting from cladistic analysis of loons, grebes, and hesperornithiform birds by Cracraft (1986). Among other problems, this scheme would require that *Hesperornis* evolve teeth on the maxilla from edentulous ancestors. (Modified from Cracraft 1986)

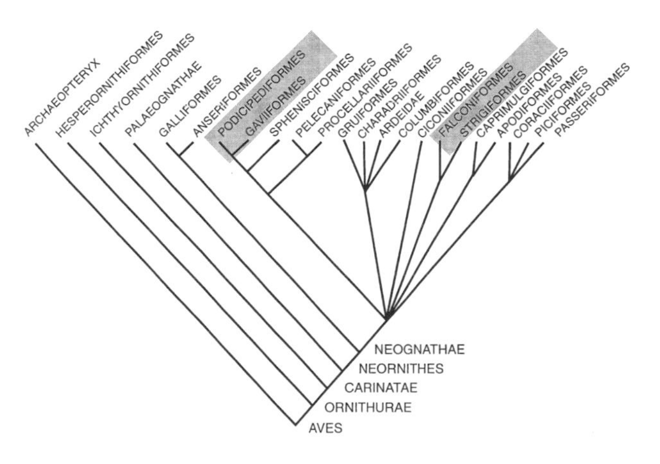

Postulated cladistic relations of some major avian higher taxa by Cracraft. This analysis shows monophyly of loons and grebes, but Cracraft at this point had abandoned the idea that the clade included the hesperornithiforms. In this scheme, hawks and owls also form a monophyletic group. (Modified from Cracraft 1988)

convergent similarities between owls and falconiforms caused them to be placed together in many later classifications, and this treatment was recently revived by Cracraft (1981)" (473). However, they argue, their DNA-DNA hybridization evidence clearly shows that "the Falconides and owls (Strigi) are not closely related to one another. These conclusions are congruent with substantial morphological evidence" (484). Again, as in the above cases, the simplistic tabulation of shared, derived characters proves unequal to the task of sorting out relations where large suites of convergent characteristics have evolved for a particular mode of life. It is crucial to understand such problems and the pitfalls of modern systematic methodology in order in discuss recent theories on bird origins, which are, for the most part, rooted in cladistic methodology.

Recent Theories of Bird Origins

Going back to the question at hand—the origin of birds—it is interesting to trace the development of the various theories throughout the 1970s to the present (for more in-depth treatment, see Witmer 1991, an excellent detailed review). The era officially began when the ornithischian theory for avian origins was resurrected by Peter M. Galton (1970), who, following Georg Baur (1883) and J. E. V. Boas (1930), advocated the relation of the two groups based mainly on their common possession of a retroverted, or opisthopubic, pubis. Instead, however, of deriving birds directly from the known ornithischians, Galton supported a view of a mid-Triassic common ancestry of birds and ornithischians from a cursorial opisthopubic bipedal dinosaur lacking the more refined ornithischian specializations. Galton's view was not readily accepted, however. He eventually abandoned the hypothesis and sided with John Ostrom's revival of the theropod-bird hypothesis (Bakker and Galton 1974).

It has been commonly accepted that among the living amniotes, birds and crocodiles are nearest relatives. But it was a surprise when Alick D. Walker of Newcastle University (1972, 1977) advocated a crocodilian hypothesis for bird origins, by which he suggested that a common ancestor of birds and crocodylomorphs was at "a higher level of organization than that of the Thecodontia." In a sense, this is a twist on Robert Broom and Gerard Heilmann's theory of a pseudosuchian ancestry, since by this view both birds and crocodiles would arise in the Triassic from a common thecodont ancestor somewhat more specialized than Broom's *Euparkeria*. Walker derived his theory from the morphology of *Sphenosuchus*, a primitive, probably arboreal relative of crocodiles from the late Triassic of South Africa, in which Walker found avianlike cranial similarities in the pneumaticity of the bone, kinesis, palatal structure, inner ear morphology, and quadrate articulation. Walker's crocodylomorph hypothesis had its advocates. Kenneth Whetstone and Larry Martin (1979, 1981) confirmed the similarities of quadrate articulation between crocodylomorphs and birds and noted additional similarities in the tympanic pneumaticity. Martin, Stewart, and Whetstone (1980) offered tooth morphology and replacement as still another important feature that might unite birds and crocodylomorphs. Although debate continued on the possible relation of birds and crocodylomorphs, Walker himself announced in 1985 that "the original concept of a particularly close relationship between birds and crocodiles has become so tenuous that it is very difficult to sustain" (133). Study of the aetosaur *Stagonolepis* and *Sphenosuchus* had led Walker to believe that the two groups had acquired their quadrate-braincase articulations in different ways.

Although Walker has recanted his long-held view, Martin (1991) still considers a crocodylomorph hypothesis of bird origin to be viable, and Sam Tarsitano (1991) holds the view that the avian ancestor is to be found among the thecodonts, but not too distantly removed from the crocodylomorphs. Tarsitano (1985, 1991) has held a more eclectic view than most, posing the possibility that birds may be part of a monophyletic assemblage that includes the crocodylomorphs, the theropods, and at least some of the thecodonts. At first glance, this proposal may appear distorted, but over the past decade it has become clear that the Triassic archosaurs lumped together as thecodonts are a heterogeneous assemblage, with some forms related to dinosaurs and others related to crocodiles (Carroll 1988, 269), and numerous relationships between the proterosuchian thecodonts and various advanced groups have been proposed by Romer (1966), Sill (1974), Thulborn (1982), Chatterjee (1982), and Bonaparte (1984).

Many workers consider thecodonts to be undefinable, a heterogeneous assemblage (Gauthier 1986), Triassic archosaurs with thecodont teeth (set in sockets) and an antorbital fenestra. In other words, they are Triassic or basal archosaurs that are not crocodiles, pterosaurs, or dinosaurs. In essence, then, "the crocodylomorph hypothesis can be considered a specific version of the pseudosuchian thecodont hypothesis" (Witmer 1991, 451). Yet, the essence of all the variations of the thecodont hypothesis is that paleontologists are seeking a bird ancestor with few specializations earlier in time than theropods and that they are ascribing all of the synapomophic resemblances of theropods and *Archaeopteryx* (and birds) to homoplasy—convergence or parallelism.

We shall return to the crocodylomorph similarities to birds shortly but must mention the announcement by

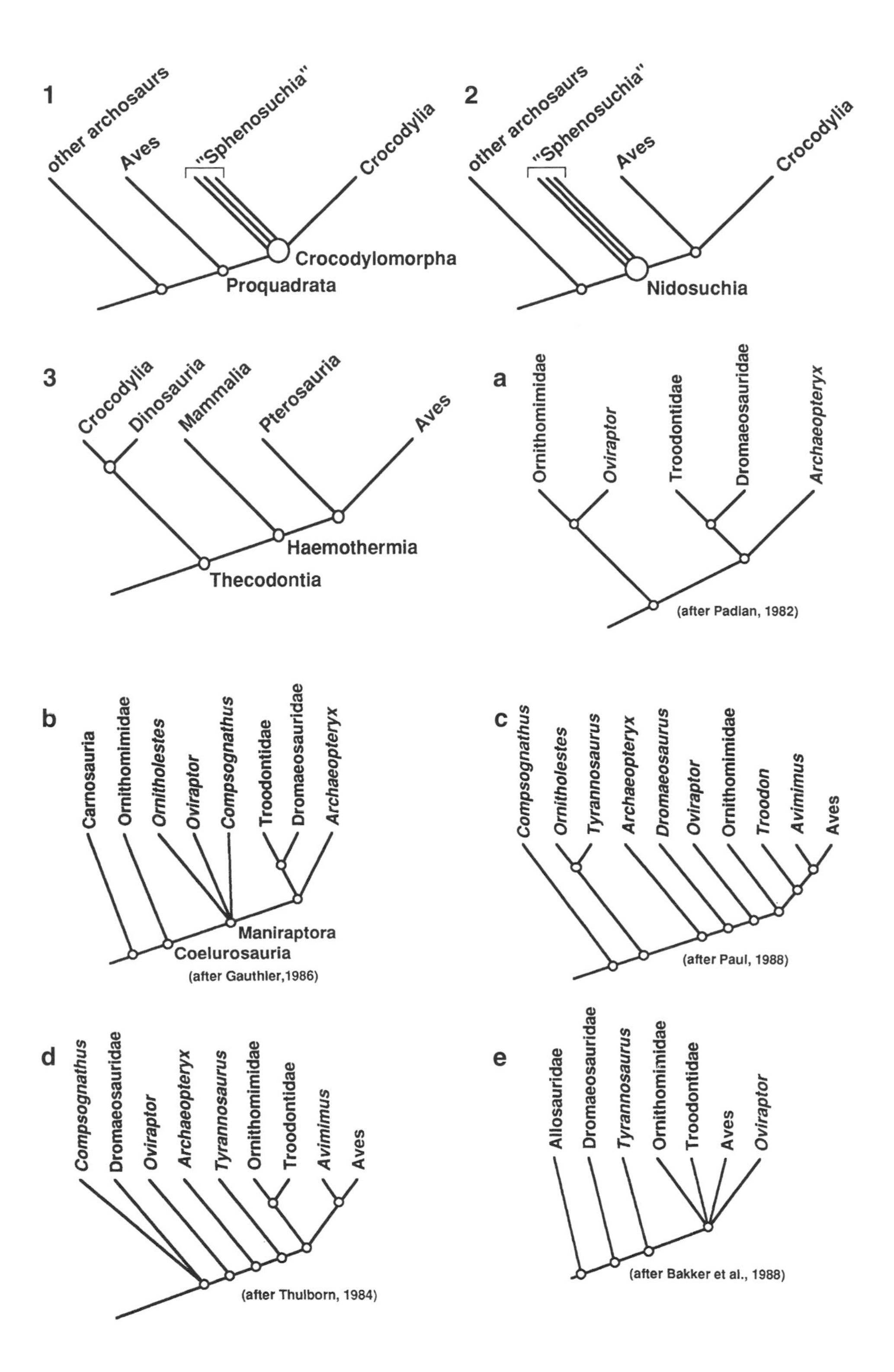

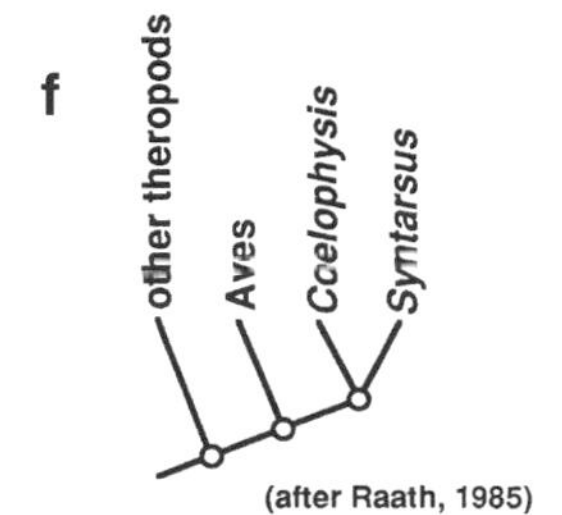

1–3, hypothetical phylogenetic relations of birds and other Amniota (reptiles and mammals) based on the analyses of *1,* Walker 1972, 1977; *2,* Whetstone and Whybrow 1983 and Martin 1983a, 1983b; taxon names are from Whetstone and Whybrow 1983 and *3,* Gardiner 1982. In *1* and *2,* "Sphenosuchia" is shown as paraphyletic on the basis of the analysis of Benton and Clark 1988. *a–f,* hypothetical phylogenetic relations of birds and various theropod dinosaurs on the basis of the analyses of *a,* Padian 1982; *b,* Gauthier 1986; *c,* Paul 1988; *d,* Thulborn 1984; *e,* Bakker et al. 1988; and *f,* Raath 1985. In most cases, these authors included other taxa in their cladograms; only the genera of major higher taxa are included here. (Modified from Witmer 1991; reprinted from Hans-Peter Schultze and Linda Trueb, eds., *Origins of the higher groups of tetrapods: controversy and consensus,* copyright © 1991 by Cornell University, used by permission of the publisher, Cornell University Press)

Brian Gardiner (1982), based on cladistic analysis, that birds shared a common origin with mammals and were therefore part of Richard Owen's group the Haemothermia. According to this scheme, which gained support from S. Løvtrup (1985), birds were not archosaurs, the Archosauria consisting of dinosaurs and crocodiles, and the Archosauria was thus the sister-group of the Haemothermia. Jacques Gauthier, Arnold Kluge, and Timothy Rowe (1988) have refuted Gardiner's proposal, but it illustrates many of the difficulties inherent in phylogenetic analysis.

Most of the recent debate on bird origins, however, has centered on the new version of Thomas Huxley's dinosaurian hypothesis, initiated and since advocated by John H. Ostrom. In 1973, in a page-long paper in *Nature*, Ostrom outlined the basis of his new theory and rapidly followed this up with a series of papers on the origin of birds and avian flight (see especially 1974 and 1976a), many of which are taken up in the next chapter. Ostrom's hypothesis, simply put, is that birds are not only descended directly from dinosaurs as proposed by Thomas Huxley but are descended from a specific subset of the theropods, the small coelurosaurian dinosaurs similar to the splendid *Deinonychus*, the lightly built, early Cretaceous (110-million-year-old) coelurosaur that he discovered back in 1964 (Ostrom 1969a). In *Deinonychus*, Ostrom saw remarkable overall similarity to the first bird *Archaeopteryx*. Ostrom's theory gained considerable momentum during the ensuing years and is now practically dogma among vertebrate paleontologists. The human-sized *Deinonychus* ("terrible-claw") was named for the large, curved, cutting claws on the feet that were apparently used to rip open prey. Although this animal was some 40 million years younger than *Archaeopteryx*, Ostrom became convinced by studying and comparing the arm and hand, the hips, and the ankles and shoulder bones that *Archaeopteryx* and therefore birds must be derived from this group of gracile dinosaurs:

> It has been repeatedly observed that the *Archaeopteryx* specimens are very birdlike, but also possess a number of reptilian features . . . . the actual fact is that these specimens are not particularly like *modern* birds at all. If feather impressions had not been preserved in the London and Berlin specimens, they never would have been identified as birds. Instead, they would unquestionably have been labelled as coelurosaurian dinosaurs . . . . the last three specimens [of *Archaeopteryx*] to be recognized were all misidentified at first, and the Eichstätt specimen for twenty years was thought to be a small specimen of the dinosaur *Compsognathus*. (1975, 61)

Of course, the point is seldom mentioned that the Teyler specimen of *Archaeopteryx* was also misidentified—but as a pterosaur.

Most of Ostrom's comparisons of coelurosaurs were made directly with *Archaeopteryx* and, more specifically, with the bones of the postcranial skeleton. Interestingly, most of the characters that allied birds with a crocodylomorph ancestry were associated with the skull. Also intriguing is that on the discovery of *Deinonychus antirrhopus* in the mid-1960s (1969a) and the subsequent revelation of its avian similarities (1976a), Ostrom based his comparisons on overall similarity, much like paleontological analyses that predated cladistics. Nevertheless, Ostrom saw strong comparisons between coelurosaurs and *Archaeopteryx*, particularly in: the presence of a "semilunate carpal"; the phalangeal formula and proportions and pattern of digital reduction in the hand, or manus; the supposed elongation of the forelimb and particularly the manus; a "dinosaurian" pubis with a distally expanded pubic foot; a mesotarsal joint; an ascending process of the astragalus (ankle bone, Rieppel 1993); a reduced metatarsal 1, with the loss of a connection between metatarsal V and the tarsus; and a reversed hallux.

Here we must step back for a moment, because Ostrom's theory for the dinosaurian origin of birds is inextricably linked to his early proposal that dinosaurs were hot-blooded, or endothermic, and to his later theory that, because the coelurosaurs were earthbound, cursorial animals, avian flight and feathers originated from the ground up. This is the so-called cursorial theory of avian flight. Ostrom's view is perhaps most vividly revealed in a *National Geographic* article on dinosaurs: "*Archaeopteryx* supports two theories: warm-bloodedness in dinosaurs and dinosaurian ancestry of birds" (1978b, 168). In other words, dinosaurs were earthbound and were certainly not going up trees; hence, feathers would have to originate first in a thermoregulatory context, to insulate hot-blooded dinosaurs—that is, in a context other than an aerodynamic one. These ideas are tackled in the next chapter.

With the concomitant rise of new ideas on dinosaur physiology and behavior, along with the rise of phylogenetic systematics, Ostrom's ideas immediately gained near complete acceptance within the world of paleontology. Robert Bakker and Peter Galton (1974) used the new, radical view as the basis for a new class of vertebrates, the Dinosauria, that included the birds and was based almost entirely on speculative physiological parameters—namely, endothermy. A year later, Bakker (1975) published an article in *Scientific American* entitled "Dinosaur Renaissance": dinosaurs were now on the same activity plane as birds and mammals, and birds were a simple offshoot of dinosaurs; put more simply, birds are living dinosaurs.

The implications of a dinosaurian origin of birds were startling and provocative: birds and therefore avian flight originated from the ground up from hot-blooded dinosaurs insulated with feathers. Illustrating the wide acceptance of this radical notion, we are now told that "the smallest dinosaur is the bee hummingbird, *Mellisuga helenae*, found only in Cuba," and that "dinosaurs like condors are more closely related to *Tyrannosaurus rex* than *T. rex* is to familiar dinosaurs such as *Apatosaurus*" (Norell et al. 1995, 25, 73).

Phylogenetic analyses, however, are based solely on the validity of the characters being analyzed, not on speculative scenarios. The hypothesis of a dinosaurian origin of birds must ultimately either rest on the homology of the supposed shared, derived character similarities between birds and dinosaurs or fall on the proposition that the characters in birds and dinosaurs acquired a similar appearance through parallelism or convergence. So, not unexpectedly, as early as 1976, Max Hecht published a paper in which he claimed that Ostrom's characters were of "low weight" or were convergent based on the independent origins of bipedality in birds and dinosaurs. Bipedality in birds would have arisen from quadrupedal archosaurs under a different set of selective pressures resulting from the obligate release of the forelimbs for wings and flight; quadrupedality would be biophysically impossible under such circumstances. Contrarily, dinosaurs would have acquired their bipedality as an adaptation for bipedal locomotion, directly inherited from similar precursor thecodonts. Dinosaurs and birds, though sharing the common theme of bipedality, are therefore characterized by a fundamentally different groundplan, or *bauplan*.

The basic question, then, is whether birds are similar to dinosaurs because they are derived from dinosaurs or because they became similar through convergence. Hecht's proposal that the shared similarities might be caused by convergence takes us right back to Thomas Huxley's presentation before the Geological Society of London in 1869, "Further Evidence of the Affinity between the Dinosaurian Reptiles and Birds" (published in 1870). After Huxley read his presentation, Harold Grovier Seeley stood up and said that he "thought it possible that the peculiar structure of the hinder limbs of the Dinosauria was due to the functions they performed rather than to any actual affinity with birds" (in Discussion to Huxley 1870a, 31). Seeley's view, which was virtually identical to that of Karl Vogt of the University of Geneva (1880), held that the similarities resulted from convergence. His theory (1866, 1881) of bird origins linked birds with the pterosaurs, and he considered *Archaeopteryx* an intermediate form. And so Huxley answered Seeley with a dose of the same medicine, stating that the similarities between birds and pterosaurs were due to "physiological action, and not to affinity" (1870b, 38). But it was Huxley who presented the first hint of convergence: "And if the whole hind limb quarters from the ilium to the toes, of a half-hatched chicken could be suddenly enlarged, ossified, and fossilized as they are, they would furnish us with the last step of the transition between Birds and Reptiles; for there would be nothing in their characters to prevent us from referring them to the Dinosauria" (1870a). In this intrepid statement, Huxley tells us of the remarkable apparent similarity between the hindlimb quarters of bipedal running birds such as the Galliformes and the bipedal running dinosaurs. In fact, much of the confusion in bird embryology results from the fact that almost all studies have been done on the chicken, which is in fact rather atypical of the class Aves as a whole. The hindlimbs of the chickenlike dinosaur *Compsognathus* and a chicken are superficially comparable because both are bipedal ground dwellers. Yet superficial comparison to other birds, including *Archaeopteryx*, is much less favorable. The assertion that the similarities between dinosaurs and birds were the result of convergence was issued time and time again after Seeley's proclamation, by Mudge (1879), Dollo (1882, 1883), Dames (1884), Parker (1887), Fürbringer (1888), Osborn (1900), Broom (1913), Heilmann (1926), Simpson (1946), de Beer (1954), and Romer (1966), as well as in numerous more recent articles. Perhaps the most influential evolutionary biologist of his time, George Gaylord Simpson, emphatically asserted: "Almost all the special resemblances of some saurischians to birds, so long noted and so much stressed in the literature, are demonstrably parallelisms and convergences" (1946, 94). It was not, and is not, an idea devoid of supporters or reason.

Thus, Ostrom's discovery of *Deinonychus* and the revitalization of the dinosaurian hypothesis for bird origins seemed like a case of déjà vu, for just as Seeley had attributed Huxley's list of dinosaurian-bird similarities to convergence, so have Max Hecht, Sam Tarsitano, and Larry Martin become the most vocal opponents of the theropod hypothesis. Tarsitano and Hecht (1980) and Hecht and Tarsitano (1982) published a lengthy argument against a theropod-bird derivation, questioning the homology of many of the characters used (also Tarsitano 1991). Martin has also published extensively to refute many of the characters used to unite birds and theropods (see Martin 1983a, 1983b, 1991). Martin's view, perhaps first stated in Whetstone and Martin (1979), does not differ dramatically from the pseudosuchian thecodont hypothesis, that birds and crocodylomorphs shared thecodontian ancestry before the advent of either ornithischian or saurischian dinosaurs.

Supposed Theropod-Bird Synapomorphies: Notes and Comments. Although the paleontological community has overwhelmingly accepted Ostrom's views and cladistic methodology, Ostrom's most vocal advocates have been Kevin Padian of the University of California at Berkeley and Jacques Gauthier of the California Academy of Sciences in San Francisco. Ostrom did not use cladistics in his paper on bird origins (1976a), but he did indicate that he had attempted to use only shared, derived characters; related birds closely to the dromaeosaurids, including *Deinonychus;* and mentioned that he thought birds were derived from some unknown coelurosaurian dinosaur similar to *Ornitholestes*. It was Kevin Padian (1982), however, who first attempted to overcome the "ambiguities" of Ostrom's approach by providing the first true cladogram of *Archaeopteryx* and related archosaurs, which showed the deinonychosaurs, Troödontidae and Dromaeosauridae, as the sister-group of birds. Later, Gauthier and Padian (1985) provided a more complete cladogram with diagnoses of archosaurian and saurischian taxa, and Gauthier (1986) provided a detailed character analysis (using some eighty-four characters of varying importance) of the entire archosaur assemblage, with emphasis on the saurischians and bird origins. However, although Gauthier performed an analysis on the characters, he never analyzed the characters themselves.

Gauthier's scheme considers birds derived coelurosaurs, a group composed of the Ornithomimidae (bird-mimics) and the Maniraptora (hand-raptors), which includes *Archaeopteryx*. Within the Maniraptora, *Archaeopteryx* and the deinonychosaurs are sister-taxa, although the monophyly of the Deinonychosauria is seriously questioned (Witmer 1991). Many of the characters appear trivial, and the analyses are difficult to understand and ambiguous, raising serious questions about many of the supposed synapomorphies between birds and coelurosaurs (Tarsitano 1991; Martin 1991). In addition, many of the characters are known to be primitive, and some have criticized the nature of Gauthier's analysis on this basis (Martin 1988). Yet for paleontologists who support a dinosaurian origin of birds, Gauthier's paper is the benchmark and is now referred to as the standard work on the topic.

Gauthier defines "theropoda . . . to include birds and all saurischians that are closer to birds than they are to sauropodomorphs" (1986, 18). In his phylogenetic analysis, therefore, *Procompsognathus* comes out at the level of the Ceratosauria. Since then, Paul Sereno and Rupert Wild have shown that *Procompsognathus* was incorrectly identified: "*Procompsognathus triassicus*, long held to be a primitive theropod, is actually a paleontological chimera composed of the postcranial skeleton of a *Segisaurus*-like ceratosaur and the skull of a basal crocodylomorph" (1992, 435)—in other words, it is a composite, with an archosaur body and crocodylomorph head. According to Gauthier's analysis, in "*Theropoda* . . . in contrast to the ancestral condition, the lacrimal forms much of the skull roof anterior and lateral to the prefrontal above the orbit in *Procompsognathus* (Pers. obs.), Ceratosauria, Carnosauria, Ornithomimidae, [and] Deinonychosauria. . . . The presence of the apomorphic condition in theropods as diverse as *Procompsognathus* . . . , *Ceratosaurus* . . . , *Deinonychus* . . . , *Allosaurus* . . . , *Tyrannosaurus* . . . , *Oviraptor* . . . , *Gobipteryx* . . . , and the Ratitae and Tinami, suggests that the vomers are fused anteriorly in Theropoda generally" (18). Such definitive analyses of poorly known taxa do not lend confidence to the overall phylogenetic conclusions. In addition, many of the key characters seen as uniting birds and theropods are disputed, among them the nature of the pelvis (Martin 1991; Tarsitano 1991), the homology of the digits (Hinchliffe and Hecht 1984; Hinchliffe 1985; Martin 1991; Tarsitano 1991), the nature of the teeth (Martin, Stewart, and Whetstone 1980; Martin 1991), the hallux (Tarsitano and Hecht 1980; Martin 1991; Feduccia 1993a), the ascending process of the astragalus (Martin, Stewart, and Whetstone 1980; Martin 1991; also see McGowan 1984, 1985, and reply by Martin and Stewart 1985), the pubis (Martin 1983a, 1983b, 1991; Tarsitano 1991; also see Wellnhofer 1985), and even the supposed unique semilunate carpal thought to be shared by *Deinonychus* and *Archaeopteryx* (and modern birds) (Martin 1991; Tarsitano 1991).

Problems of Digital Homology. Although the consensus among paleontologists today is that *Archaeopteryx* and therefore birds are derived coelurosaurs, the truth is that if one can show that one key synapomorphy, such as that of the homology of the digits of the manus, is falsified, then the remainder of the "synapomorphies" must be considered to be convergent. As James Clark stated in a review of a book dealing in part with bird origins, "The only substantive problem with the theropod-bird hypothesis remains the discrepancy between the homology of the digits of the manus as indicated by the fossils and the development of modern birds" (1992, 534).

In the nineteenth century, Owen (1836) identified the bird digits as 2–3–4, whereas Parker (1888a) numbered them 1–2–3, and in modern times, developmental biologists have numbered them 2–3–4, while paleontologists have tended to use 1–2–3 because of the supposed dinosaur-bird link. Tarsitano and Hecht (1980), who reexamined the problem, believed that the pattern of digital re-

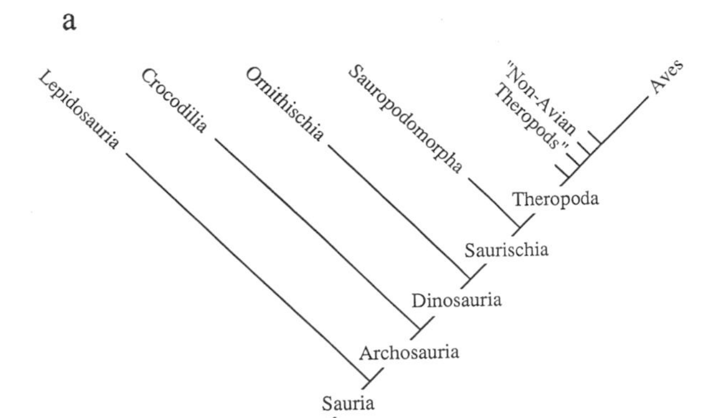

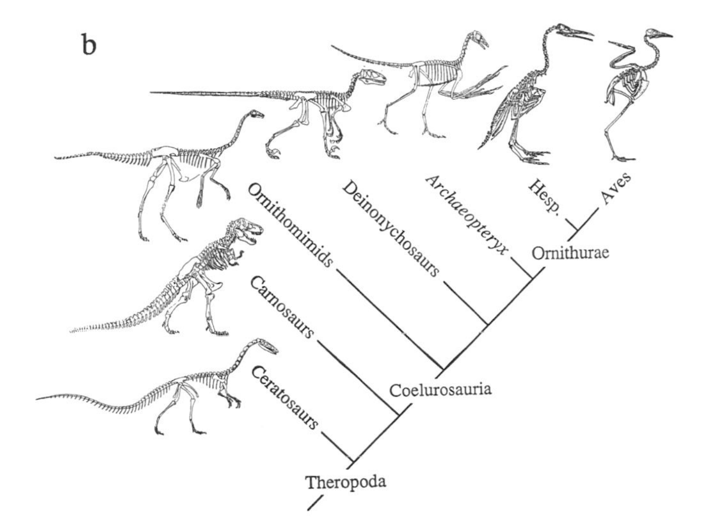

Simplified cladograms illustrating the current dogma from the paleontological world concerning bird origins from theropod dinosaurs, specifically the dromaeosaurs, exemplified by *Deinonychus:* (*a*) selected taxa of Sauria; (*b*) major modes of the clade Theropoda; representative theropods figured, *from lower left to upper right, Coelophysis, Tyrannosaurus, Struthiomimus, Deinonychus, Archaeopteryx, Hesperornis,* and a Paleocene lithornithid (paleognathous) bird (all from Carroll 1988). Cladograms simplified from Gauthier 1984, 1986. (From Gatesy 1990; courtesy S. M. Gatesy and editor, *Paleobiology*)

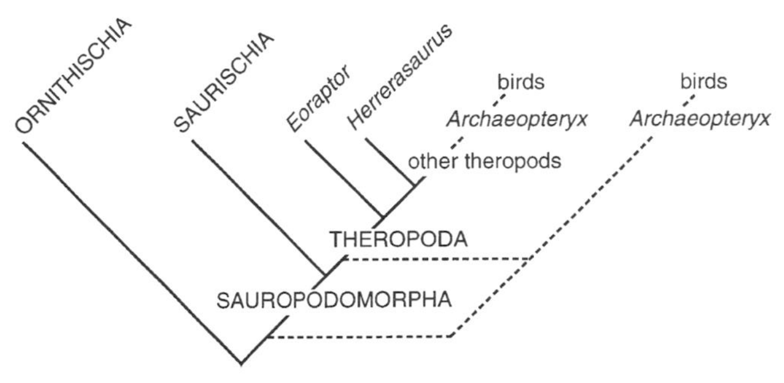

Cladogram of postulated relations of dinosaur groups and birds. Solid lines following Sereno (1993). Dotted lines are postulated relations of birds. The dotted lines on the left represent relations based on the 1–2–3 hypothesis of digital homology, whereas the dotted lines on right represent relations based on the 2–3–4 hypothesis of digital homology. (Modified after Hecht and Hecht 1994)

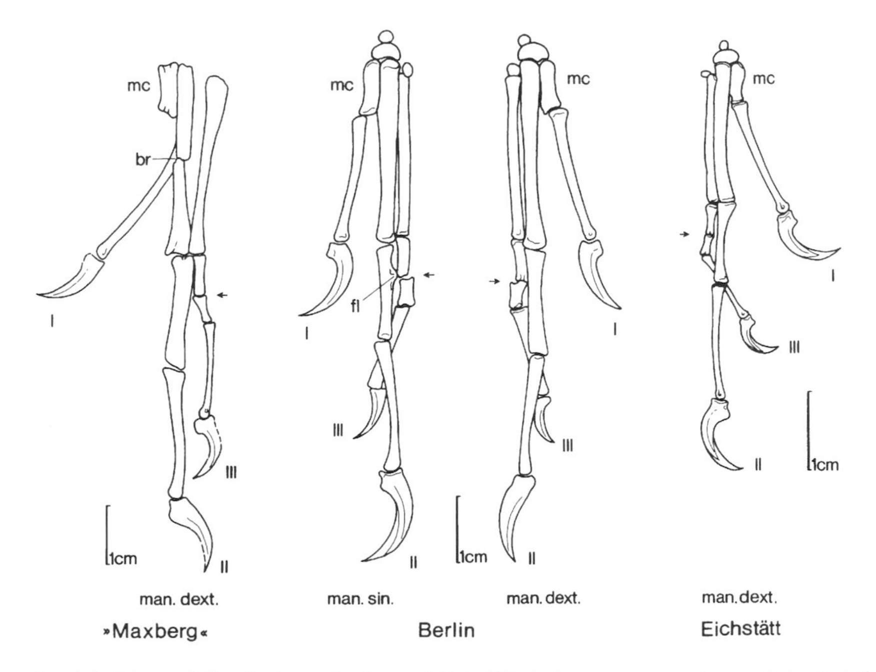

Hand skeletons of the Maxberg, Berlin, and Eichstätt *Archaeopteryx* specimens. Note that the claws of the hand have toppled or twisted to lie on their sides (Yalden 1985) and that the third and fourth fingers (here numbered 2 and 3 by Wellnhofer) are crossed; this condition is unknown in the dinosaur. Arrows indicate the joints between the first and second phalanx of digit 3, as numbered here. Abbreviations: *br,* break of metacarpal 2 in the Maxberg specimen; *fl,* flange on the proximal phalanx of left digit 2 in the Berlin specimen; *mc,* metacarpals. (From Wellnhofer 1985; courtesy G. Viohl and Jura-Museum)

duction was different in birds and dinosaurs and that therefore birds and dinosaurs could be related only at the level of ancestral thecodonts. J. R. Hinchliffe of the University of Wales has presented substantial evidence (1985) from experimental embryology that the remaining three digits in the bird wing (and therefore presumably that of *Archaeopteryx*) are the middle three, 2–3–4, exactly what one would expect from normal symmetrical digital reduction, a necessary outcome of the digital reduction pattern of the primitive amniote pattern (developmental bauplan of Shubin 1991). In contrast, the derived dinosaur pattern, which is seen in such early theropods as *Herrerasaurus*, with digits 1–2–3 retained, is illustrative of an adult morphology unique to amniotes and would require reduction of digits 4 and 5 as well as the fifth metacarpal—digital reduction from the postaxial side, not the primitive amniote reduction pattern. This type of reduction is visible in early dinosaurs; in such forms as the late Triassic *Herrerasaurus*, digits 4 and 5 are reduced, with a phalangeal formula of 2–3–4–1–0 and a hand that is adapted for "grasping and raking" (Sereno and Novas 1992). In other words, paleontological evidence favors the view that the dinosaur hand is a derived amniote condition, characterized by the reduction of two digits, 4 and 5, leaving 1–2–3, and therefore that it is a different type of amniote hand than the bird manus. To derive the bird hand from that of dinosaurs, assuming the correct numbering of the bird digits is 2–3–4, would require a mutation affecting symmetry and central axis to reverse the process (Hecht and Hecht 1994), and it is highly improbable that a developmental program could be altered with all evidences of prior history erased.

To add to the confusion, the phalangeal formula for dinosaurs is 2–3–4 for digits 1–2–3, the primitive formula of the first three digits of the pentameral (five-digit) manus, and *Archaeopteryx* has the same phalangeal formula (Wellnhofer 1985, 1991). To simplify, to have acquired a phalangeal formula of 2–3–4 with the three middle digits (2–3–4) retained, *Archaeopteryx* would have to have lost digits 1 and 5, as well as one, and only one, phalanx from each remaining digit (2–3–4). Just how important these formulas are is questionable, because the phalangeal foot formula is 2–3–4 in the Solnhofen specimen and 2–3–4 in the Berlin and Eichstätt specimens. At present, then, the digital relation of birds and dinosaurs remains elusive, with some evidence to support both views, and eventual resolution of this apparently intractable problem may be difficult. The relations of birds, as Hecht and Hecht point out, are "further complicated by the similarity of the developmental bauplan of the amniote manus. In order to relate birds to the saurischian or theropod clade, it is necessary to deny the reductive amniote sequence which has been inferred from many studies on the development of the manus of modern birds" (1994, 335–336).

Neil Shubin has come down on the side of the 1–2–3 hypothesis, basing his conclusion largely on "phylogenetic analysis that relies on the total evidence"—that is, the theropod derivation of birds. Yet he has stated that "without fossil theropods, and the phylogenetic interpretations that they imply, the II–III–IV interpretation of the homologies of the avian digits would, perhaps, be a more likely interpretation" (1994, 264).

To complicate the matter further, the *Archaeopteryx* manus is not a grasping, raking hand and does not closely resemble that of a theropod dinosaur, and as the digits have been preserved, the claws are toppled or twisted to lie on their sides, indicating that the claws pointed ventrally from the wing in life (Yalden 1985), as would be expected for trunk climbing. In addition, in most specimens, the middle finger is crossed over the outer finger in a strange style of preservation not found in any known dinosaur, theropod or ornithischian.

The Semilunate Carpal. There is little agreement regarding the homologies of the bones of the avian wrist and manus (Hinchliffe and Hecht 1984; Hinchliffe 1985, 1989a, 1989b, 1991). *Archaeopteryx* exhibits a typical avian carpus of four bones, as illustrated in the Eichstätt specimen, and contrary to popular belief (Vasquez 1992), the cuneiform conforms to the typical V-shape of modern birds (Martin 1991, 1995b). In *Archaeopteryx*, as in all birds, the semilunate bone forms a unit with the metacarpals, and the scapholunar and cuneiform, homologized with the radiale and ulnare of reptiles, articulate with the radius and ulna, respectively. However, the true ulnare appears to be lost in ontogeny, while the avian "cuneiform" is, according to British embryologist Richard Hinchliffe, a pisiform and the radiale is probably the avian "scapholunar." The two carpal bones that fuse with the metacarpals are a distal carpal element, designated "Bone X" by Hinchliffe, and the semilunate bone (distal carpal 3), which represents a single center of ossification that ultimately joins with the metacarpals to form the trochlea carpalis of the carpometacarpus. Thus, in modern birds the wrist consists of two carpal elements, a small carpal at the distal end of the ulna, the ulnare (cuneiform), and a carpal at the distal end of the radius, the radiale (scapholunar). Their role in supporting and controlling wing movements in active, flapping flight has been elucidated recently by Rick Vasquez (1992), who showed, not surprisingly, that the *Archaeopteryx* manus lacked the kinematic sophistication of the modern bird hand.

The semilunate carpal element in *Archaeopteryx* has

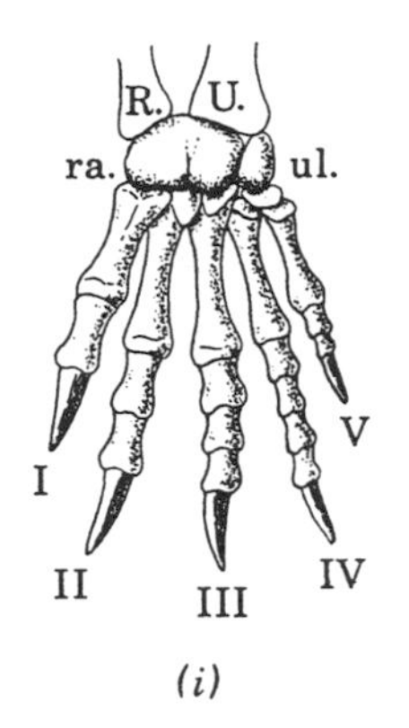

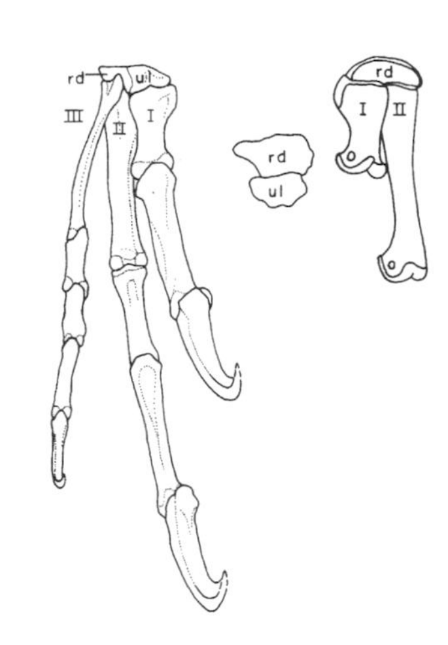

Left, restoration of left manus of *Postosuchus* showing the position of the radiale (*ra.*) and ulnare (*ul.*) relative to the radius and ulna. *Lower right,* restoration of the wrist and manus of the Eichstätt specimen of *Archaeopteryx, left,* showing the semilunate carpal, compared to that of *Deinonychus,* in palmar (open hand) view, and anconal (back of hand) view, with a proximal view of the radiale and ulnare in articulation. (*Postosuchus* from Chatterjee 1985; courtesy Royal Society, London; *Archaeopteryx* and *Deinonychus* from Martin 1991; reprinted from Hans-Peter Schultze and Linda Trueb, eds., *Origins of the higher tetrapods: controversy and consensus,* copyright © 1991 by Cornell University, used by permission of the publisher, Cornell University Press)

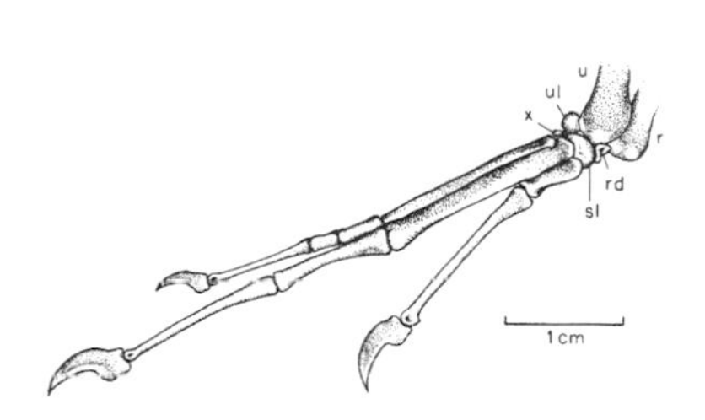

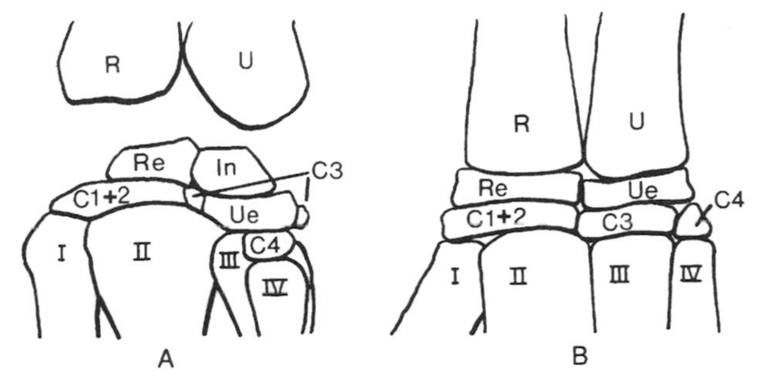

Right carpus, palmar view, of *Syntarsus rhodensiensis* (*A*) and *Coelophysis bauri* (*B*), late Triassic theropods that illustrate the typical theropod condition of having two rows of carpal elements, the proximal elements being the radiale and ulnare, which articulate with the two bones of the forearm. "*Deinonychus,* from the Lower Cretaceous, would seem to have only two carpals, the radiale and ulnare, an unusual condition in a theropod" (Colbert 1989, 94). If birds derive from dinosaurs, it would have to be late in time from some specialized dromaeosaurids. (From Colbert 1989; courtesy E. H. Colbert and Museum of Northern Arizona)

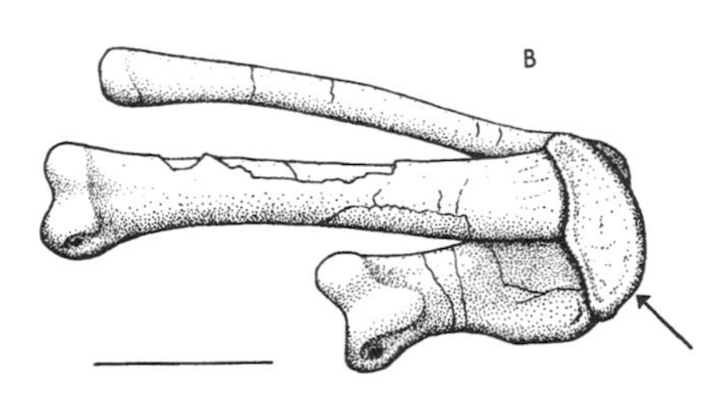

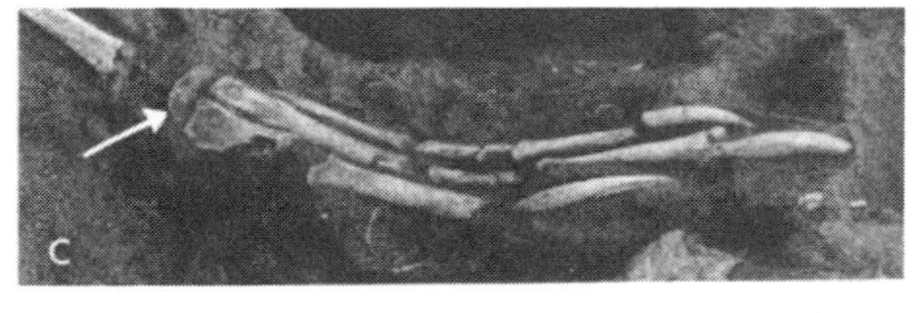

Carpus and metacarpals of the Eichstätt specimen of *Archaeopteryx* (*A*), compared with those of the early Cretaceous *Deinonychus antirrhopus* (*B*) and the late Cretaceous *Velociraptor mongoliensis* (*C*). Arrows point to the half-moon–shaped distal carpal in each, considered to be synapomorphy of theropods and *Archaeopteryx,* but this superficial resemblance is found only in four dinosaurs, and all but one are from the Cretaceous period, two from the late Cretaceous. *A* and *B* represent right wrists; *C* depicts the left wrist and hand. Scale division in *A* = 0.5 mm; scale line in *B* = 3 cm. (From Ostrom 1976; courtesy John Ostrom and Academic Press)

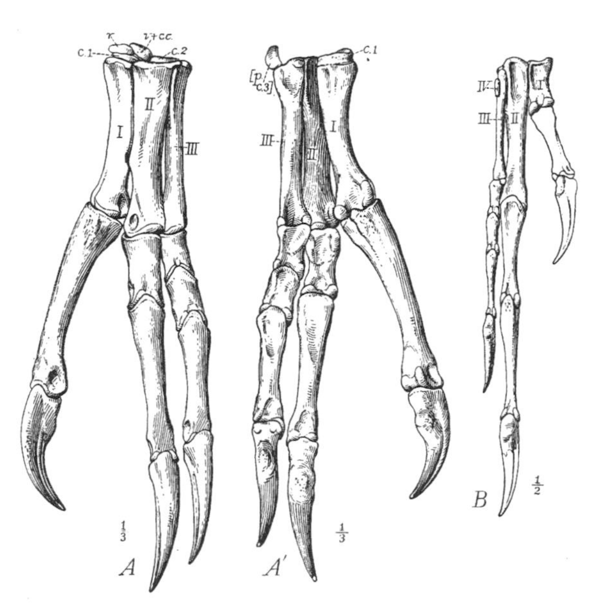

Manus of *Struthiomimus altus,* dorsal aspect (*A*), palmar aspect (*A*'); and manus of *Ornitholestes hermanni* (B), in palmar view. Abbreviation: *r,* radiale. In the ornithomimids the carpal elements are tremendously reduced. (From Osborn 1916)

assumed great significance because a superficially similar element has been found in a handful of theropod dinosaurs and has thus been hailed by some paleontologists as the definitive character linking birds to a theropod ancestry. But this theropod semilunate carpal element is known in only four types of dinosaurs, two of them from the late Cretaceous: two dromaeosaurs, *Deinonychus* (early Cretaceous) and *Velociraptor* (late Cretaceous); the troödont *Sinornithoides* (=*Stenonychosaurus* and *Saurornithoides?* exact taxa unclear; late Cretaceous); and the poorly known coelurosaur *Coelurus* (late Jurassic) (Ostrom 1985b, 1995). It is unknown in any primitive Triassic theropods (Ostrom 1985b, 167). There is also considerable doubt about the possible homology of the element in these theropods and *Archaeopteryx* (Martin 1991; Tarsitano 1991). Adding to the confusion, dinosaurs typically have two rows of carpal elements, as seen in the late Triassic *Coelophysis* and *Syntarsus*, and the often-cited birdlike Jurassic theropod *Ornitholestes* lacks any carpal element that remotely resembles the semilunate structure. In the ornithomimids and other theropod dinosaurs, the carpal elements are greatly reduced in size and number, again showing no such semilunate carpal structure.

Attempting to homologize the semilunate bone of *Archaeopteryx* with that of a superficially similar element in four theropods proves extremely difficult, and Tarsitano (1991) claims that a specimen of a juvenile *Deinonychus* housed at Harvard University's Museum of Comparative Zoology exhibits a semilunate carpal composed of two unfused carpi, whereas the avian semilunate represents a single unit of ossification.

The actual problem is that some dinosaurs have lost an entire row of carpal elements that are still present in birds. And it is not known for certain which row is lost, as in, for example, *Deinonychus*, making it almost impossible to homologize structures of the avian and theropod wrist. If, for example, the distal carpals are missing and the element represents the radiale, a proximal element, as Ostrom assumed (1976a, 1991), then there is no homology between the semilunate carpal of theropods and the semilunate carpal of birds, a distal carpal element (Martin 1991, 511). Writes Ostrom, "Of special importance is a single carpal bone (= radiale?; the identity of the element is of less importance than its function)" (1991b, 479). But the identity of the element is the *only* point of importance to establish homology with any similar avian structure.

This problem of homology, combined with the erratic distribution of the character and the probable nonhomology of the digits of birds and theropods, renders this character of dubious utility. That certain theropods would independently develop an avianlike "semilunate carpal element" that would provide for "automatic carpal supination-pronation synchronized with carpal flexion-extension" (Ostrom 1995, 262), given their diverse adaptive radiation and probable diverse hand movements, is not surprising, and a possible precursal element is present in the hand of the Triassic thecodont *Postosuchus* (Chatterjee 1985, 421).

The Pelvis. Another dramatic character is the footed, or booted, pubes that are united distally. This character is seen in a number of theropods and superficially resembles the condition in *Archaeopteryx*. However, the footed pubis is also known among thecodonts, for example *Postosuchus*, and as Sankar Chatterjee notes, "It appears that an expanded pubic foot has evolved independently on several occasions among archosaurs, as first pointed out by A. D. Walker (1969), in poposaurids, *Herrerasaurus*, and *Staurikosaurus* during the Triassic, and in several theropods and in *Archaeopteryx* during Jurassic and Cretaceous times" (1985, 449). However, although the footed pubis is a primitive archosaur character, *Archaeopteryx* does not have a pubic foot in the sense of the same character in theropods. In *Archaeopteryx* and a number of early Cretaceous enantiornithine birds, the pubes meet distally to form a "hypopubic cup," resembling cupped hands, the forearms analogous to the pubic arm. The typical theropod pubis has a prominent, hooked-back pubic foot or boot, which is a massive bony structure resembling an inverted anvil. Because the pubis extends vertically downward, when the animal was seated or stretched on the ground, the pubic foot would augment the limbs in receiving the weight of the body. Whatever the case, the condition appears to be primitive within the archosaurs.

Much dispute has centered around the proper orientation of the pubis. Advocates of the dinosaurian origin of birds have oriented the pubis more vertically, and Wellnhofer (1985) and Ostrom (1976a, 1991b) propose that the Berlin pubis was rotated backward after death. The dispute arises in part because in *Archaeopteryx* the pubis is loosely connected to the ilium and ischium, so that fossilization somewhat displaces the pubis, and thus the orientation of the pubes is slightly different in all the *Archaeopteryx* specimens. In spite of all this, there can be little doubt that the pubis is at least moderately retroverted (opisthopubic), or oriented backward, and that the appropriate term for the structure observed in *Archaeopteryx* is a "pubic apron," a primitive character, rather than a true pubic foot of the thecodont *Postosuchus* and theropod dinosaurs, which is thought to have partially borne the weight of the seated or stretched-out animal (Chatterjee 1985, 439). In addition, if, as we shall see in chapter 3, the

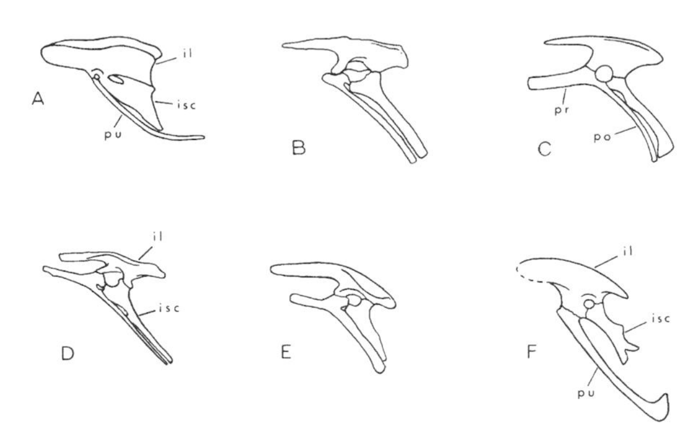

Pelves, in left lateral view, of several ornithischians compared with the pelvic organization of a modern bird and that as preserved in the Berlin *Archaeopteryx*. The postpubic rod (*po*) of ornithischians has been equated with the avian pubis (*pu*). The prepubic process (*pr*) is considered a new structure. *A, Columba; B, Scelidosaurus; C, Camptosaurus; D, Thescelosaurus; E, Stegosaurus; F, Archaeopteryx.* Ostrom restored the Berlin specimen's pubis in a more vertical, dinosaurian orientation, arguing that it was displaced during preservation. Abbreviations: *il,* ilium; *isc,* ischium; *po,* postpubic ramus; *pr,* prepubic ramus; *pu,* pubis. Sketches not to scale. (From Ostrom 1976; courtesy John Ostrom and Academic Press)

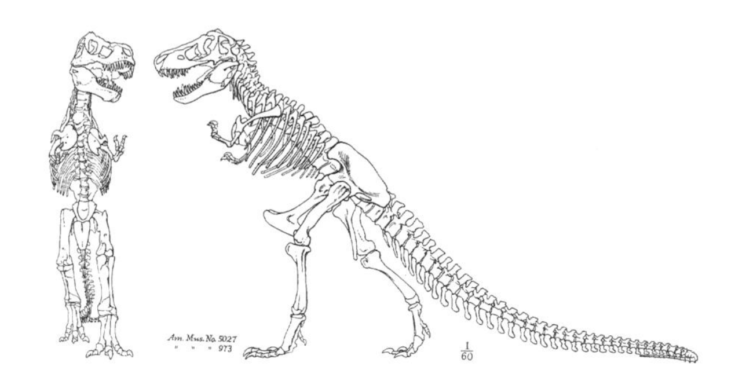

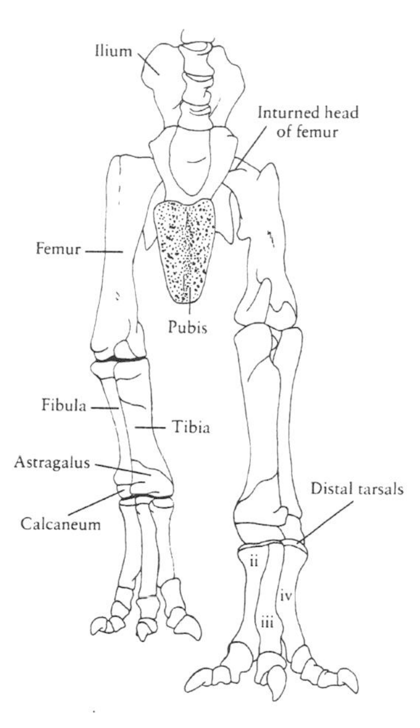

Above, skeletal reconstruction of the late Cretaceous *Tyrannosaurus,* illustrating the structure of a typical theropod pubis, with its vertical orientation and distal foot. The distal end is a massive bony structure somewhat resembling an inverted anvil, with a rough, pitted surface; it is thought to have partly borne the weight of the animal when seated. Note the near coequal length of the pubes and the femora. *Left,* anterior view of the rear limbs of *Tyrannosaurus,* showing the massive, "inverted anvil" appearance of the fused pubes, totally unlike the "hypopubic cup" formed by the fused *Archaeopteryx* pubes.The vertical posture and transverse knee and ankle joints are characteristic of both saurischians and ornithischians. (Modified from Osborn 1916 and Carroll 1988)

Ostrom's skeletal reconstruction of two "Late Jurassic bipedal predators," *Ornitholestes hermanni* (*above*), a coelurosaurian dinosaur from the Morrison Formation (Kimmeridgian age) of North America, and *Archaeopteryx* (*below*) from the Solnhofen Limestone (Kimmeridgian age) of Europe. Note that Ostrom reconstructs the pubis of *Archaeopteryx* in a more or less theropod orientation but that it was most likely retroverted in a more avian position. Note also the very different proportions of both fore- and hindlimb elements. Scales = 5 cm. (From Ostrom 1976; courtesy John Ostrom and Academic Press)

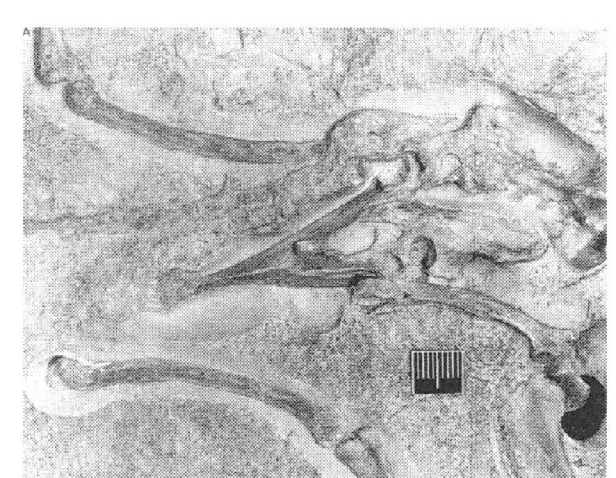

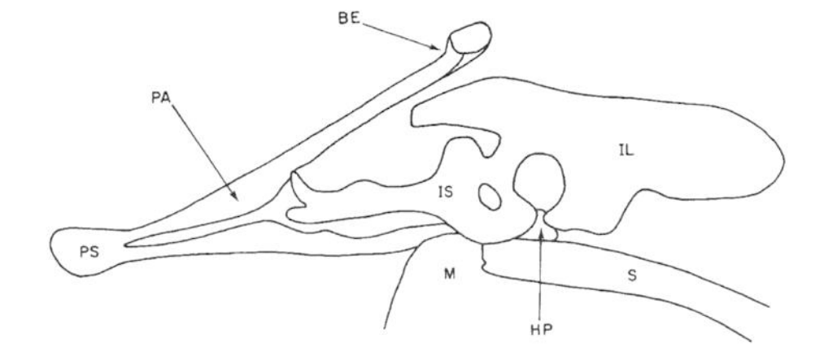

Above, overall view of the pubic structure of the London specimen of *Archaeopteryx,* illustrating the pubic apron. *Below,* interpretive diagram of the pelvic region. Abbreviations: *BE,* bend in left pubis below its head; *HP,* head of the right pubis; *IL,* ilium; *IS,* ischium; *M,* matrix; *PA,* pubic apron; *PS,* pubic symphysis; *S,* right scapula. The pubic symphysis forms a primitive pubic apron with a distal "hypopubic cup," resembling two cupped hands, the forearms analogous to the pubes. This is totally unlike the theropod pubic foot or boot. (From Tarsitano and Hecht 1980; photo courtesy S. Tarsitano and Academic Press)

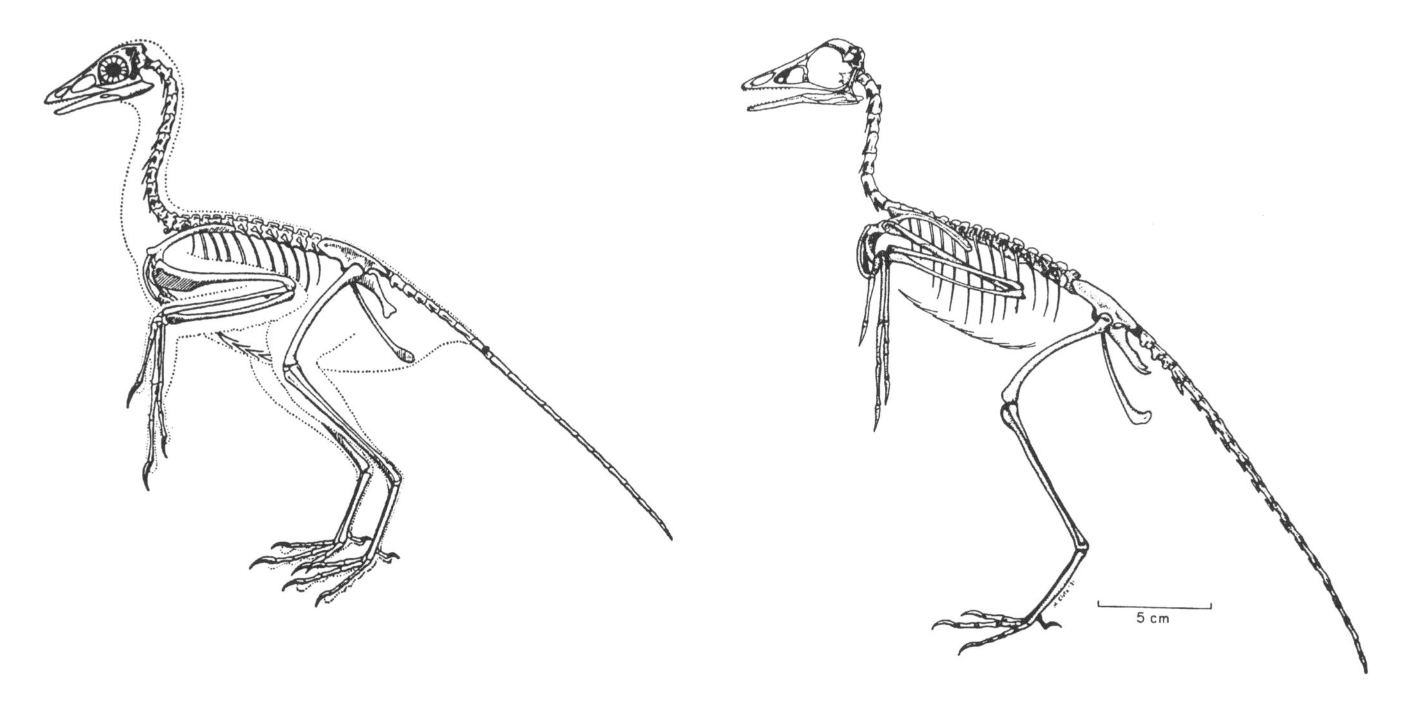

Restoration, by Larry Martin, of the skeleton of *Archaeopteryx,* with the pubis in opisthopubic (retroverted), or avian, orientation (*right*), compared to that of W. E. Swinton (*left*). The pubis has been reconstructed as reflexed anywhere from near the theropod 110° to an avian opisthopubic 140–150°. (From Martin 1991; reprinted from Hans-Peter Schultze and Linda Trueb, eds., *Origins of the higher tetrapods: controversy and consensus,* copyright © 1991 by Cornell University, used by permission of the publisher, Cornell University Press; Swinton 1960, courtesy Academic Press)

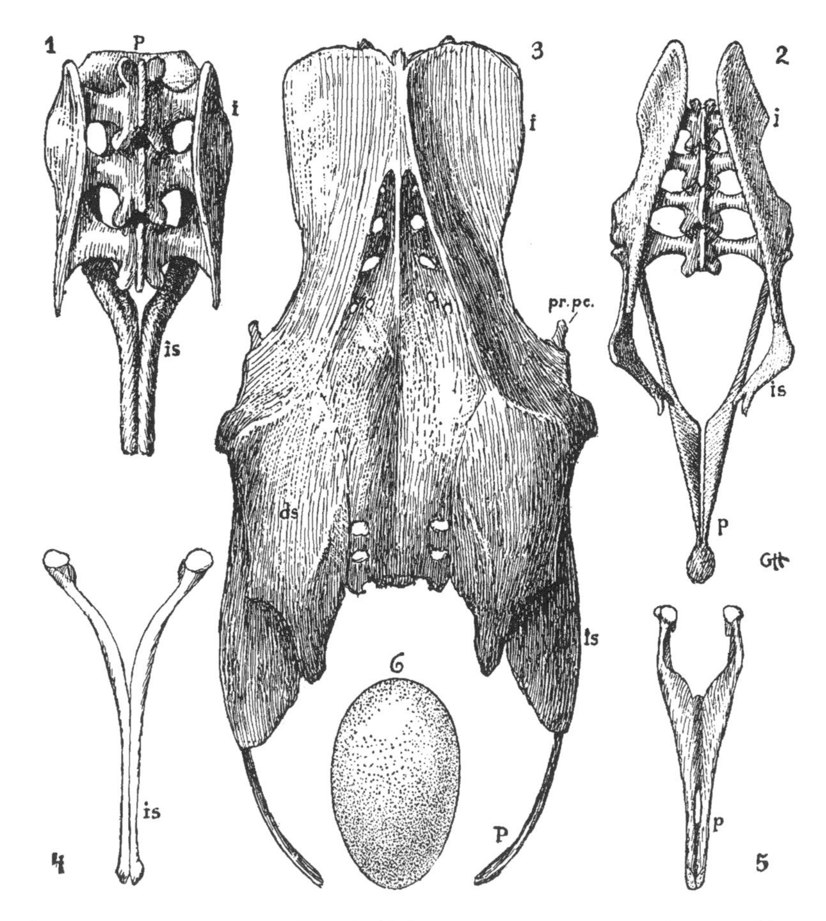

1–5, restored pelves: *1,* late Triassic thecodont *Ornithosuchus; 2, Archaeopteryx; 3, Gallus; 4,* ischial symphysis of *Dryosaurus altus; 5,* pubic symphysis of late Jurassic *Coelurus; 6,* natural size of *Archaeopteryx* egg in accordance with pelvic width. The pubic symphysis was lost with the advent of the large eggs characteristic of modern birds, providing large quantities of yolk for an advanced embryonic development. (From Heilmann 1926)

evidence points to an arboreal, tree-climbing *Archaeopteryx*, a vertically oriented pubis would be maladaptive, a hindrance in trunk climbing.

Thomas Huxley was struck by the similarity of the pelvis of birds and such ornithischian dinosaurs as *Hypsilophodon* and *Iguanodon*, and he noted that the opisthopuby of *Hypsilophodon* "affords unequivocal evidences of a further step towards the bird" (1870a, 28). However, Wellnhofer concluded that "the pelvis of *Archaeopteryx* is not bird-like at all. There is, however, a trend of ornithization, compared to the coelurosaurian pelvis, by the enlargement of the preacetabular portion of the ilium and by the elongation of the pubes. The same trend occurred in other theropods as well, without any closer relationship to the ancestral stock of *Archaeopteryx*, for example in the Upper Cretaceous *Adasaurus* and *Segnosaurus* (Barsbold 1983)" (1985, 121). Contrarily, the *Archaeopteryx* pelvis is very birdlike, and many early Cretaceous birds exhibit very similar anatomy. And again, the ornithization of the theropod pelvis in the late Cretaceous illustrates once more that dinosaurs converged on birds by the Cretaceous period.

Partially reflexed pubes are known in several additional theropods, including *Deinonychus* (early Cretaceous), *Velociraptor* (late Cretaceous), *Segnosaurus* (late Cretaceous), and *Erlikosaurus* (late Cretaceous) (Ostrom 1985). Wellnhofer (1985, 120) notes that the only birdlike pelvic feature is the elongation of the preacetabular (anterior) portion of the ilium. But the pubis is directed more or less ventrally in many early archosaurs, for example, in the early Triassic *Euparkeria*, and reflexed pubes of early ornithischians indicate multiple evolution of the opisthopubic condition, meaning that the reflexed pubis is a very homoplastic feature. And finding the dinosaurian pubis in the late Triassic thecodont *Postosuchus* argues for the primitive nature of this character and means that it is of limited value in determining relations. Nevertheless, as one can see in the illustration on page 151, in at least one specimen of an early Cretaceous enantiornithine bird, the pubic apron is preserved much like the London specimen of *Archaeopteryx*, and in a lateral view of another similar specimen, it is clear that the enantiornithines had attained an avian opisthopubic pelvis.

The Ascending Process of the Astragalus. Among the principal characters marshaled to support the bird-theropod connection are the ascending process of the ankle bone (astragalus) (Rieppel 1993) and the mesotarsal joint. Like reptiles, birds have two proximal tarsal elements, a lateral calcaneum (heel) and a medial astragalus; these two proximal tarsals are fused or articulated with the tibia, and in Mesozoic forms with the fibula. The distal tarsals either fuse or loosely articulate with the metatarsus. Debate on the homologies of this type of ankle, the bones of the mesotarsal joint, goes back to Huxley (1870a), who noted that bird and theropod dinosaur ankles share a triangular prominence in front of the distal tibia known as the ascending process of the astragalus. The homology of this element in birds and theropods is still debated, however (Martin 1991). In 1984, McGowan presented evidence from cleared and stained embryos that indicated to him that the ratites have an ascending process of the astragalus similar to theropod dinosaurs that is continuous with the astragalus and therefore that this condition was primitive for birds; carinates, in contrast, have a separate ossification in the same position, the pretibial bone, which fuses partly with the bony calcaneum. Martin and Stewart (1985) challenged this argument. In brief, Martin and colleagues (1980), following other workers (Morse 1872; Jollie 1977), had earlier argued that what appeared to be the ascending process was in fact a separate pretibial bone, and in this case, therefore, the process would be homologous in ratites and carinates but not homologous to that of theropods. Earlier yet, Welles and Long (1974) recognized five types of theropod tarsi, each with an ascending process of the astragalus but each of independent origin, with the more derived types approaching a more birdlike state. In addition, similar structures have evolved in the ornithischian dinosaurs, such as *Hypsilophodon* (Galton 1974) and the Triassic thecodont *Lagerpeton* (Wells and Long 1974); even *Lagosuchus* has a similar arrangement. An ascending process of any kind braces the small tarsal bones, in a mesotarsal joint, against the long bone, the tibia, to provide support and to solve problems of torque generation about the ankle (Tarsitano 1991, 561) and is a functional requisite for any mesotarsal foot. Even if the theropod ankle were to be proved homologous to that of ratites, the presence of similar structures in the ornithischians and late thecodonts makes them of limited importance and negates their use as synapomorphies linking birds and theropods. Further complicating the issue, tibiotarsi of a variety of pterosaurs are even more birdlike and have frequently been misidentified as bird fossils. In many pterosaurs the astragalus and calcaneum are completely fused both to each other and to the tibia, the fibula is greatly reduced, and pterosaurs exhibit a very birdlike distal end with attachment areas for an anterior transverse ligament, lateral and medial ligamentous prominences, and an anteroposteriorly expanded pulley-like articular surface (Galton 1980). There can be little doubt that if these pterosaur tibiae were found

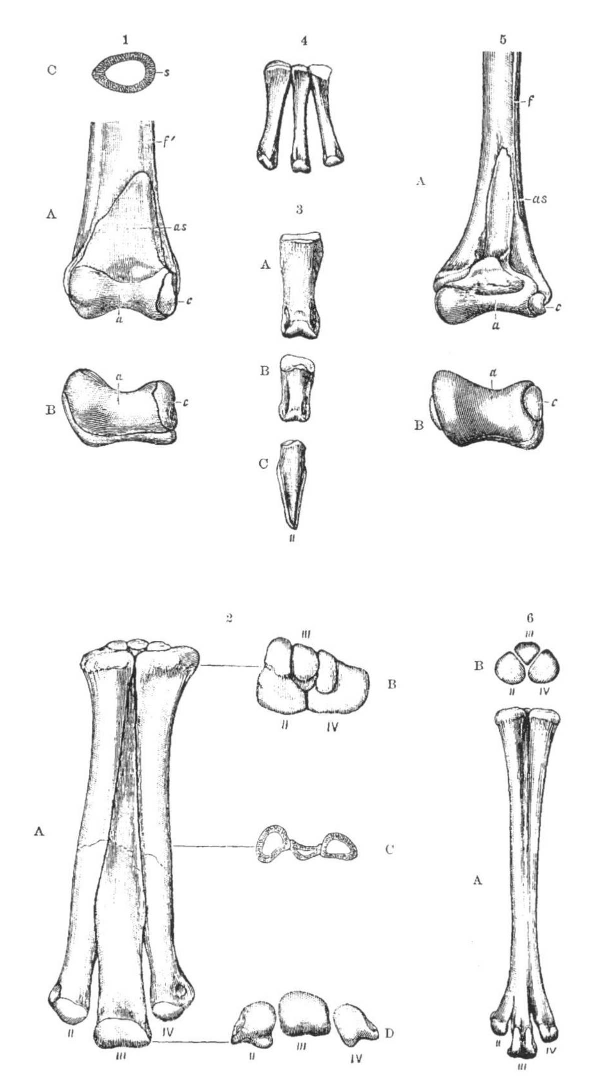

Illustration of the supposed homology of the theropod and avian foot: *1,* left tibia of the late Cretaceous *Ornithomimus velox,* in (*a*) anterior view, (*b*) distal view, and (*c*) transverse section; *2,* left metatarsals of same specimen; *3,* phalanges of second digit of same foot; *4,* left metatarsals of *Ornithomimus,* smaller individual; *5,* left tibia of young ostrich, *Struthio; 6,* left metatarsals of young turkey, *Meleagris*. Note that the comparison is of a bird and the late Cretaceous birdlike dinosaur *Ornithomimus*. Abbreviations: *a,* astragalus; *as,* ascending process of astragalus; *c,* calcaneum; *f,* fibula; *f* ', face for fibula; *II,* second metatarsal; *III,* third metatarsal; *IV,* fourth metatarsal. (From Marsh 1896)

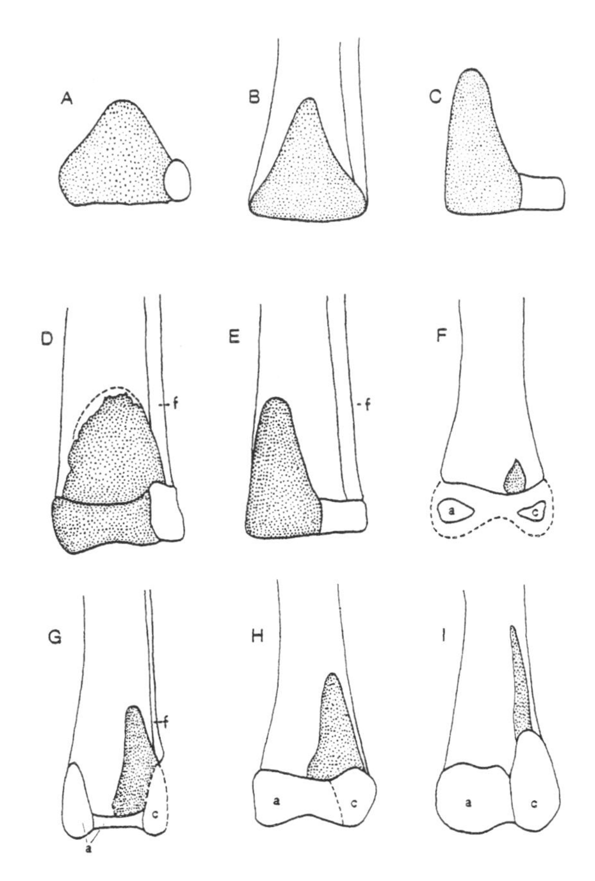

The complexities of possible homologies of the tarsal regions of birds and theropods: tibia and tarsus of dinosaurs and birds. *A–C,* restorations of the tarsal region in *Archaeopteryx; D,* the theropod *Deinonychus; E, Archaeopteryx,* as described and illustrated by Ostrom (1976), a combination of *B* and *C,* above; *F,* the domestic chicken, *Gallus; G, Archaeopteryx* restored after the London, Berlin, and Eichstätt specimens; *H, Baptornis advenus* (extinct hesperornithiform); *I,* the hoatzin (*Opisthocomus hoazin*). The coarsely stippled area represents the so-called ascending process of the astragalus. Abbreviations: *f,* fibula; *a,* astragalus; *c,* calcaneum. (From Martin, Stewart, and Whetstone 1980; courtesy *Auk*)

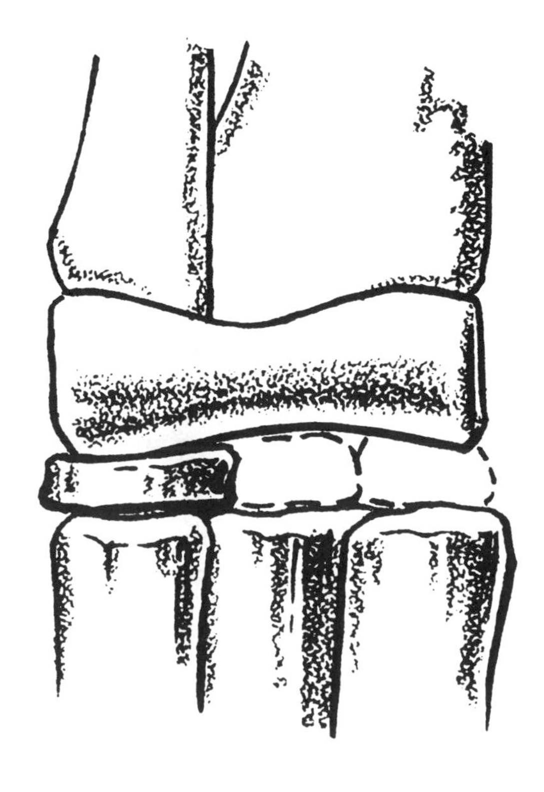

Right tarsus of the late Triassic *Coelophysis* with distal ends of tibia-fibula, fused astragalocalcaneum, and proximal ends of metatarsals. *Coelophysis* typically has a low dorsal ascending process, but the ascending process may be low, high, or absent (Colbert 1989, 108). Even if the ascending process of the astragalus was homologous with the ratites, it also matches with late theropods and the ornithischian *Hypsilophodon*. (From Colbert 1989; courtesy E. H. Colbert and Museum of Northern Arizona)

today as isolated elements in Tertiary deposits, they would be classified as birds and the characters homologized with those of modern carinates.

The Furcula. Great importance has been attributed to the furcula, or fused clavicles, in birds and *Archaeopteryx*, and it has been considered one of the two salient features of the class Aves (the other being feathers). Many advocates of the theropod-bird nexus have made much of Heilmann's statement that "it would seem a rather obvious conclusion that it is amongst the Coelurosaurs that we are to look for the bird-ancestor" (1926, 183). Yet, noted Heilmann, because "the clavicles are wanting" in coelurosaurs, "these saurians could not possibly be the ancestors of the birds" (183). He did, however, suggest that the ancestor of birds was "closely akin to the Coelurosaurs" but "wholly without the shortcomings" (185). As we have seen, Heilmann argued for a close relation of birds and theropods, similar to the common-ancestry hypothesis of such workers as Robert Broom (1913).

Although the presence of clavicles in a number of dinosaurs has been known for many years, in recent years structures thought to be birdlike furculae have been reported from various dinosaurs. However, the structure is disputed in many of these dinosaurs; the supposed furcula of the late Cretaceous *Troödon* was first reported as "gastralia," for example, and many of these structures are difficult to interpret. Nevertheless, the late Cretaceous theropods *Oviraptor* and *Ingenia* (Barsbold 1983) do appear to have furculalike structures, and clavicles are known in such theropods as the ceratosaurs *Segisaurus* and *Carnotaurus*, although similar structures in various other theropod taxa have been discounted (Bryant and Russell 1993). Because the architecture of the dinosaur shoulder is so dramatically different from that of *Archaeopteryx* as well as modern birds, it seems unlikely that any of these structures could have articulated or functioned in a manner similar to the bird furcula or the hypertrophied furcula of the first bird, *Archaeopteryx* (Martin 1991), which is a large, flat, U-shaped structure lacking the hypocleidium found in many, but by no means all, modern carinate birds (see illustration on p. 2). In modern birds the furcula serves as a transverse spacer, an integral part of the bony architecture of the triosseal canal, and the partial point of origin of the pectoralis muscle. In addition, the ventral symphysis, often represented as a flattened hypocleidium, meets the sternal keel to form a more rigid pectoral architecture. The furcula is examined in more detail in chapter 6, but we should note here that when birds become secondarily flightless, the furcula normally degenerates into two clavicular splints; this is true in both the large ratites and other orders where flightlessness evolved (p. 264). Degeneration of the furcula with the evolution of flightlessness proves that it is intimately involved in the flight apparatus and argues strongly that whatever these structures are that are found in late Cretaceous dinosaurs, they are very likely not homologous with the furcula of birds. In studying the distribution of clavicles with "Dinosauria," Bryant and Russell reached a similar conclusion, that "the furcula of birds may be a neomorph, or may represent the reappearance of a 'lost' structure" (1993, 171).

Other Characters. In describing the seventh specimen of *Archaeopteryx*, Wellnhofer (1993) noted that it had interdental plates that were shared only with theropods, and this was duly reported in the *New York Times* (Browne 1993, C9). However, again we find that interdental plates are widespread among Triassic archosaurs, including the well-known early Triassic *Euparkeria* and Chatterjee's late Triassic *Postosuchus*, and *Postosuchus* even has hollow, thin-walled bones, often cited as a dinosaur-bird link. In addition, the presence of cranial air sacs has been used to bolster various arguments, but cranial air sacs appear to be almost universally present in archosaurs (Witmer 1987, 1990). Chatterjee's *Postosuchus* thus illustrates two important points: the difficulty in defining dinosaur, and the fact that most of the characters used to unite dinosaurs and birds are found in this and other late Triassic thecodonts. Finally, features such as the dinosaurian teeth of *Postosuchus* are totally unlike those of *Archaeopteryx*, which lack the serrated edges characteristic of theropods.

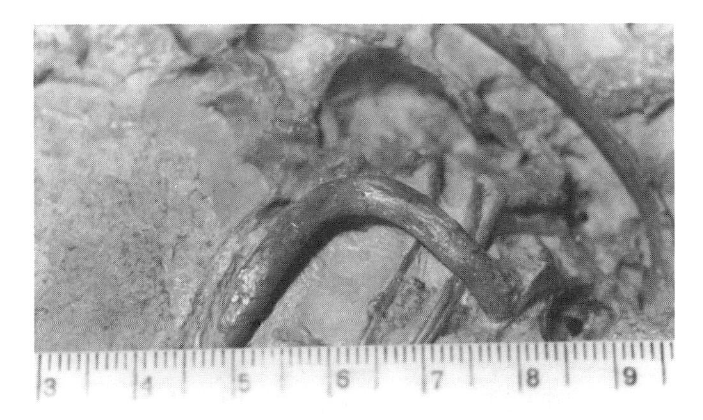

Hypertrophied furcula of the London specimen of *Archaeopteryx;* there is no such homologous structure in dinosaurs. (Photo courtesy S. Tarsitano)

Above, skeleton and life reconstruction of the late Triassic thecodont *Postosuchus,* a large, facultatively bipedal reptile from the late Triassic of Texas that is thought to be close to the ancestry of carnosaurs and illustrates a mosaic of thecodont and theropod characters. Length, about 4 m (13 ft.). *Opposite,* skull, lower jaw, and pubis. *Postosuchus* has an advanced pelvis with a slightly open acetabulum, footed or "booted" pubes that are united distally, and hollow, thin-walled bones. Its forelimb is some 64% the length of the hindlimb, and the skull exhibits interdental plates, which have been identified in the latest discovered specimen of *Archaeopteryx* and are mistakenly thought to link the first bird to theropod dinosaurs. *Postosuchus* thus illustrates many of the characters thought to be important in uniting birds and theropods. (From Chatterjee 1985, figs. 2, 8, 15, 19, 20; courtesy Royal Society, London)

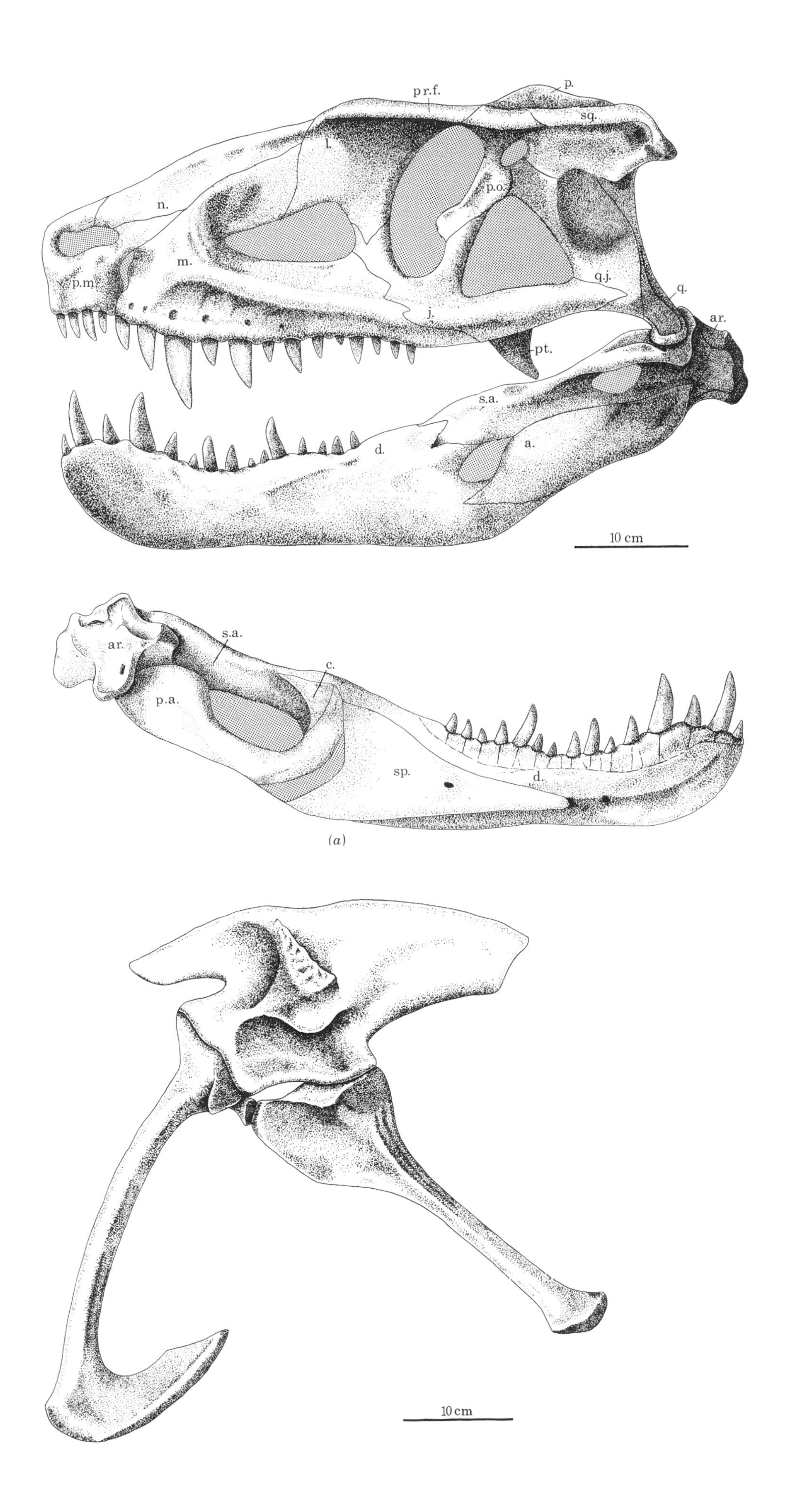

(a)

Dorsal view of the pes of *Postosuchus,* illustrating the thecodont level of organization, without a mesotarsal joint. *Postosuchus* has a sigmoidal femur, as opposed to the dinosaurian femur, which has a straight shaft and strongly inturned head; it was probably facultatively bipedal. (From Chatterjee 1985, fig. 18; courtesy Royal Society, London)

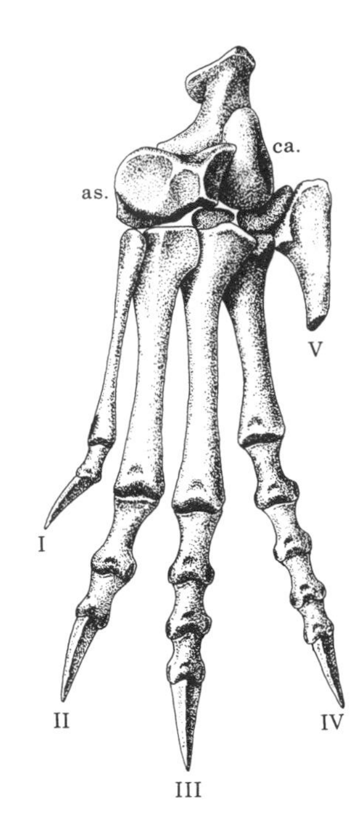

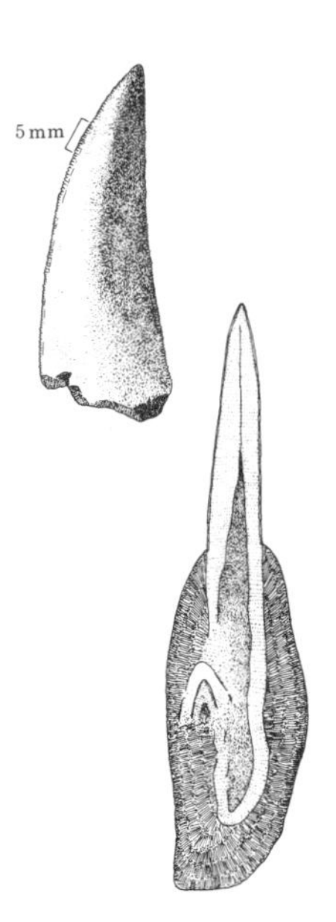

Above, tooth structure of the crown of *Postosuchus,* with serrations, and transverse section of the right dentary, showing replacement tooth lingual to the old tooth. Note the lateral compression of teeth, characteristic of theropods. *Below,* a comparison of dental serration counts of *Postosuchus* with some theropods and the first known bird, *Archaeopteryx*.

Genus	Dental serrations per 5 mm
Coelophysis	34–36
Allosaurus	10–12
Ceratosaurus	10
Deinonychus	16–18
Dromaeosaurus	16
Postosuchus	12
Indosuchus	12
Albertosaurus	9–12
Tyrannosaurus	7–9
Archaeopteryx	**none**

(From Chatterjee 1985, fig. 10; courtesy Royal Society, London)

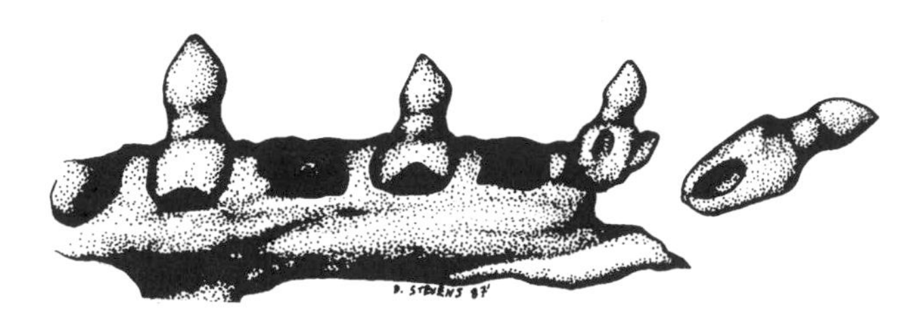

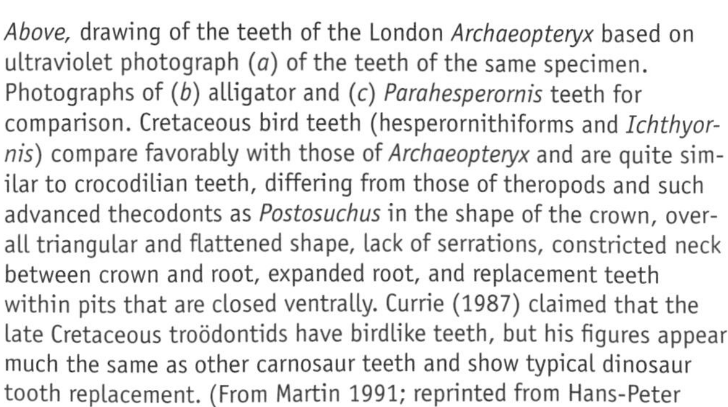

Above, drawing of the teeth of the London *Archaeopteryx* based on ultraviolet photograph (*a*) of the teeth of the same specimen. Photographs of (*b*) alligator and (*c*) *Parahesperornis* teeth for comparison. Cretaceous bird teeth (hesperornithiforms and *Ichthyornis*) compare favorably with those of *Archaeopteryx* and are quite similar to crocodilian teeth, differing from those of theropods and such advanced thecodonts as *Postosuchus* in the shape of the crown, overall triangular and flattened shape, lack of serrations, constricted neck between crown and root, expanded root, and replacement teeth within pits that are closed ventrally. Currie (1987) claimed that the late Cretaceous troödontids have birdlike teeth, but his figures appear much the same as other carnosaur teeth and show typical dinosaur tooth replacement. (From Martin 1991; reprinted from Hans-Peter Schultze and Linda Trueb, eds., *Origins of the higher tetrapods: controversy and consensus,* copyright © 1991 by Cornell University, used by permission of the publisher, Cornell University Press)

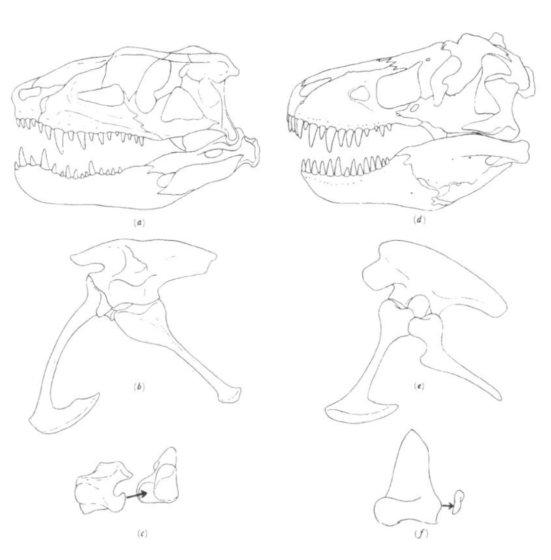

Comparisons between the advanced late Triassic thecodont *Postosuchus* and *Tyrannosaurus,* illustrating the difficulty in defining "dinosaur." *a–c,* skull, pelvis, and ankle joint of *Postosuchus; d–f,* skull, pelvis, and ankle joint of *Tyrannosaurus* (modified from Osborn 1916). In the ankle joint, the arrow indicates the position of the socket opposite the corresponding peg. (From Chatterjee 1985, fig. 30; courtesy Royal Society, London)

The Birdlike Structure of *Archaeopteryx*. Ostrom (1991b) and other bird-theropod advocates have attempted to answer most of the objections raised by various authors to the theropod origin of birds, but most of the serious questions still remain and probably will persist.

An intriguing point that is often omitted by those who advocate the dinosaur-bird nexus is that most recent workers who have studied various anatomical features of *Archaeopteryx* have found the creature to be much more birdlike than previously imagined, engendering Darwin's notion of a strange "secondary" bird instead of a reptile with feathers. Kenneth Whetstone, for example, restudied the braincase of the London specimen after it was newly prepared and concluded that *Archaeopteryx* differs from theropods in the structure of the quadrate, the prootic regions, and the morphology of the occiput. As he put it, "The skull is much broader and more bird-like than earlier interpreted" (1983, 439). Alick Walker reinterpreted the otic region of the London specimen and found that it was "of a primitive, basically avian type" (1985, 123); and Paul Bühler (1985), who studied the skull of the Eichstätt, Berlin, and London specimens, discovered that the quadrate was movable against the skull and that the upper jaw was movable in the avian prokinetic manner. He further discovered that the forebrain and cerebellum were enlarged in relation to the braincase and that the optic lobes were separated by the cerebellum, indicating "that *Archaeopteryx* developed to a much more birdlike level than it has been assumed" (135). Further, Haubitz and colleagues, using computed tomography of *Archaeopteryx*, have concluded that, "contrary to earlier interpretations of these specimens, the results . . . show the presence of an avian-like double-headed quadrate bone in this earliest true bird" (1988, 206). And, in describing the skull of the seventh specimen of *Archaeopteryx*, Elzanowski and Wellnhofer found the palatine to be distinctively avian and "different from the bone of theropods and early dinosaurs. . . . the palatine structure supports strongly that *Archaeopteryx* is a primitive bird rather than a feathered preavian archosaur" (1994, 331).

Yalden (1985) and I (1993a) have shown that the claws of *Archaeopteryx* are quite birdlike—specifically, that the manus claws are like those of trunk-climbing birds and the pes claws are like those of perching birds. The manus claws most closely resemble those of the modern Neotropical passerine trunk-climbing woodhewers of the family Dendrocolaptidae (Feduccia 1993a).

Archaeopteryx also has the unique avian feature of the reversed hallux, a perching adaptation in which the hind toe opposes the front three in grasping branches. Ostrom (1978a, 1991b) claimed that the small late Jurassic dinosaur *Compsognathus* exhibited a distally articulated and reversed hallux as in modern birds, but his argument hinged on the assumption that the ungual claw was not preserved in the correct anatomical position, whereas his own photographs show decisively that this is not the case. Detailed preservation of feet in the theropods *Deinonychus*, *Velociraptor*, *Saurornithoides*, and *Coelophysis* unequivocally show a nonreversed dinosaurian hallux, and the metatarsal scars for the hallical attachment of myriad other dinosaurs show that the first metatarsal and therefore the first toe were universally medial in the dinosaurs. And, as I noted in chapter 1, it is impossible to imagine why a cursorial dinosaur would have a reversed first toe; it could only be a hindrance in running, and no known dinosaur can be shown to have had one.

Finally, as seen in chapter 1, the coracoids meet the avian, straplike scapula at a 90° angle, a condition unique to volant birds. In addition, the glenoid (socket for the humerus) faces laterally or outward in *Archaeopteryx* and modern birds (Jenkins 1993); this condition does not exist in any known dinosaur, "and there is really no reason that it should" (Martin 1991, 532). In theropods the orientation of the glenoid is ventral, as one might expect. There can be little doubt that the resemblance of *Archaeopteryx* to theropod dinosaurs has been grossly overestimated, often incorrectly, and that *Archaeopteryx* was correctly interpreted by Huxley and Darwin as a primitive bird.

Recast Theories of Bird Origins

Everyone agrees that *Archaeopteryx* was an archosaur, and the essential question is whether birds derive from basal archosaurs (thecodonts) or from their descendants, the dinosaurs. Gauthier's criticism of the "thecodont" ancestry hypothesis is based on the notion that because the thecodonts, or basal archosaurs, are simply an undefinable (lacking specific synapomorphies) heterogeneous assemblage of Triassic archosaurs, the thecodont hypothesis can not be considered legitimate: "From a phylogenetic perspective, 'Thecodontia' and Archosauria are diagnosed by the same synapomorphies. Thus, these taxa are redundant, and when one says that birds evolved from 'thecodonts' one is simply reiterating that birds are part of Archosauria" (1986, 2). This critique vividly illustrates the difference between the classical school of evolutionary systematists and modern phylogenetic systematics; it ignores the fact that at the base of any lineage the collective forms may not have large suites of synapomorphies to provide a concise definition. Thecodonts are Triassic archosaurs that have teeth set in sockets and the only early archosauromorphs with an antorbital fenestra. In addition, thecodonts have a lower temporal bar, a relatively short neck, and a single row of conical teeth in the premaxilla as well

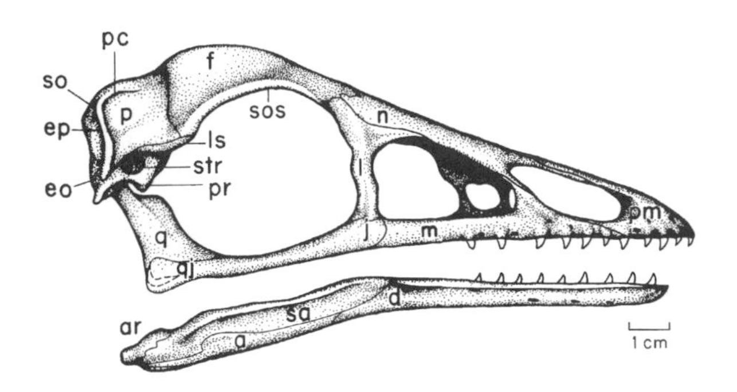

Restoration of the Eichstätt *Archaeopteryx* skull and mandible based on study of the specimens and comparison with the London specimen and Wellnhofer 1974. (From Martin 1991; reprinted from Hans-Peter Schultze and Linda Trueb, eds., *Origins of the higher tetrapods: controversy and consensus,* copyright © 1991 by Cornell University, used by permission of the publisher, Cornell University Press)

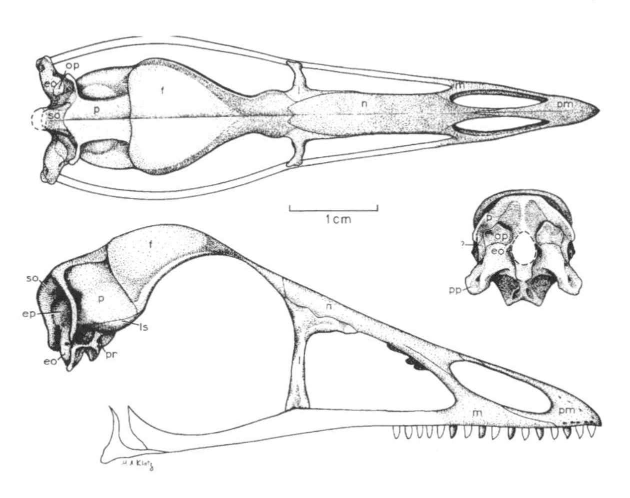

Restorations of the London *Archaeopteryx* skull in dorsal, lateral, and posterior views. Note that it lacks the two diapsid temporal openings characteristic of theropods and *Postosuchus,* and is, overall, birdlike and dramatically different from the dinosaur skull. (From Martin 1991; reprinted from Hans-Peter Schultze and Linda Trueb, eds., *Origins of the higher tetrapods: controversy and consensus,* copyright © 1991 by Cornell University, used by permission of the publisher, Cornell University Press)

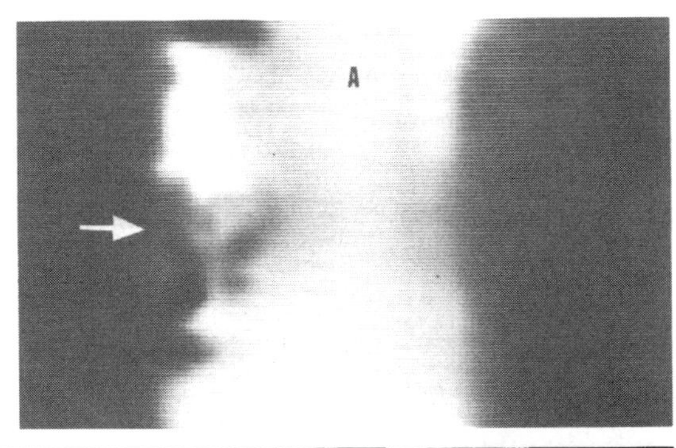

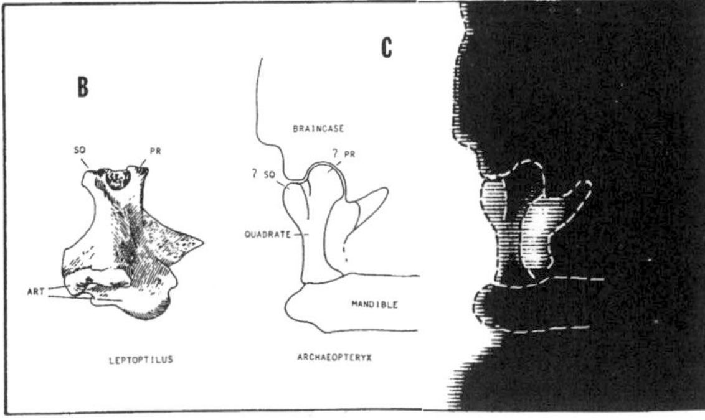

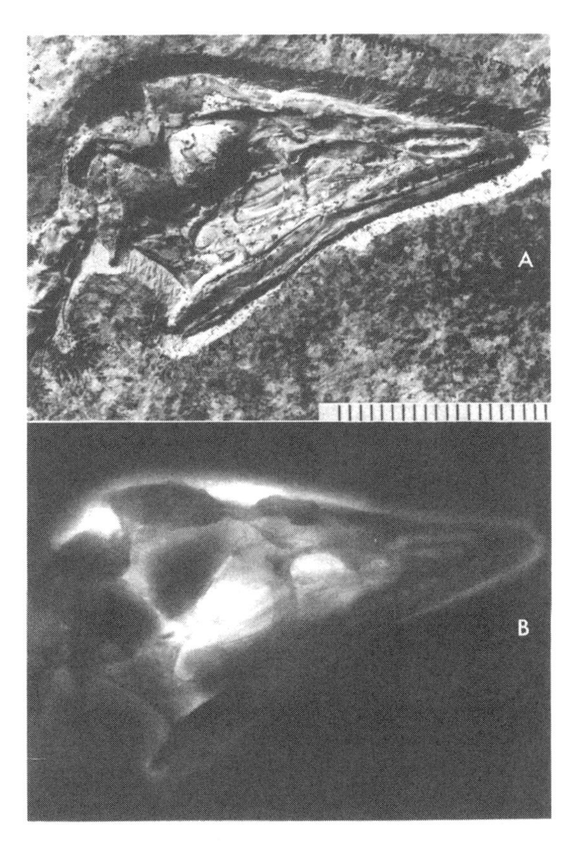

Right, skull of the Eichstätt *Archaeopteryx: A,* photograph and *B,* conventional X-ray image, by W. Stürmer, Erlangen. *Left, A,* computed tomography X-ray image of the right quadrate region of the Eichstätt *Archaeopteryx* skull reconstructed in an oblique plane 32° to the plane of the skull, appearing to show a bifurcated upper extremity of the quadrate (arrow); *B,* left quadrate of a modern stork, *Leptoptilus,* showing the avian double-condyle articulation with the squamosal (*SQ*) and the prootic (*PR*) and the articular surfaces for the mandible (*ART*); *C,* negative of the same image with dashed outline interpretation and line drawing explanation. The "fuzzy" resolution results from "blending" narrow adjacent stripes of computer data acquired from several planes within the specimen to produce the computerized image in a plane oblique to the specimen surface. Contrary to earlier interpretations of these specimens, the results of computed tomography appear to show the presence of an avianlike double-headed quadrate bone in *Archaeopteryx.* (From Haubitz et al. 1988; courtesy B. Haubitz and *Paleobiology*)

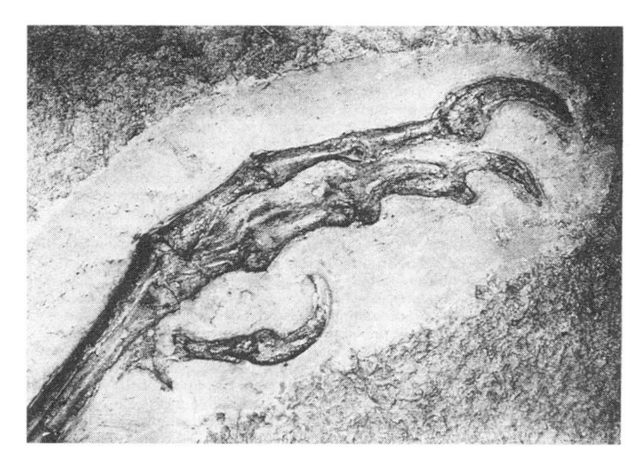

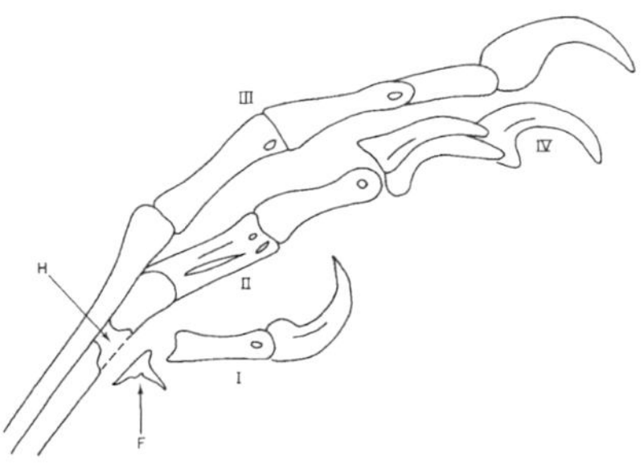

Foot of the London *Archaeopteryx* illustrating the strongly curved claws of a perching bird (see chapter 3) and the fully reversed hallux, digit 1, a character unique to birds and found in no theropod dinosaur. (Photo courtesy S. Tarsitano)

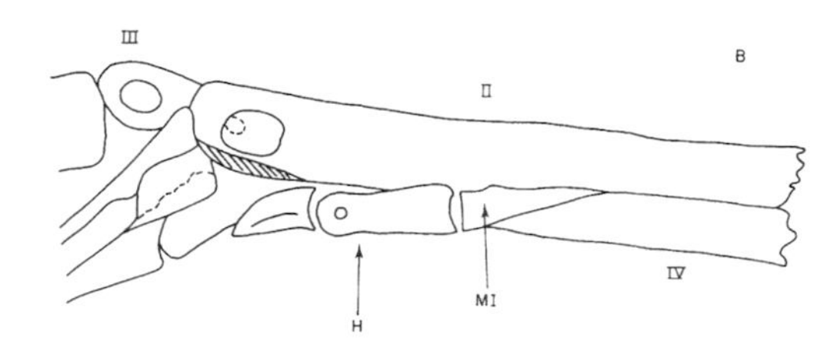

A, medial view of the right pes of *Compsognathus,* said to have a reversed hallux. *B,* interpretative diagram of right pes of *Compsognathus*. Abbreviations: *H,* hallux; *MI,* metatarsal 1. The hallux is preserved in articulation and is clearly not reversed. The theropod *Saurornithoides mongoliensis* (American Museum #6516) was likewise preserved in articulation, with the hallux parallel to the other digits. No articulated theropod has ever been found with a reversed hallux. (Photo courtesy S. Tarsitano and Academic Press)

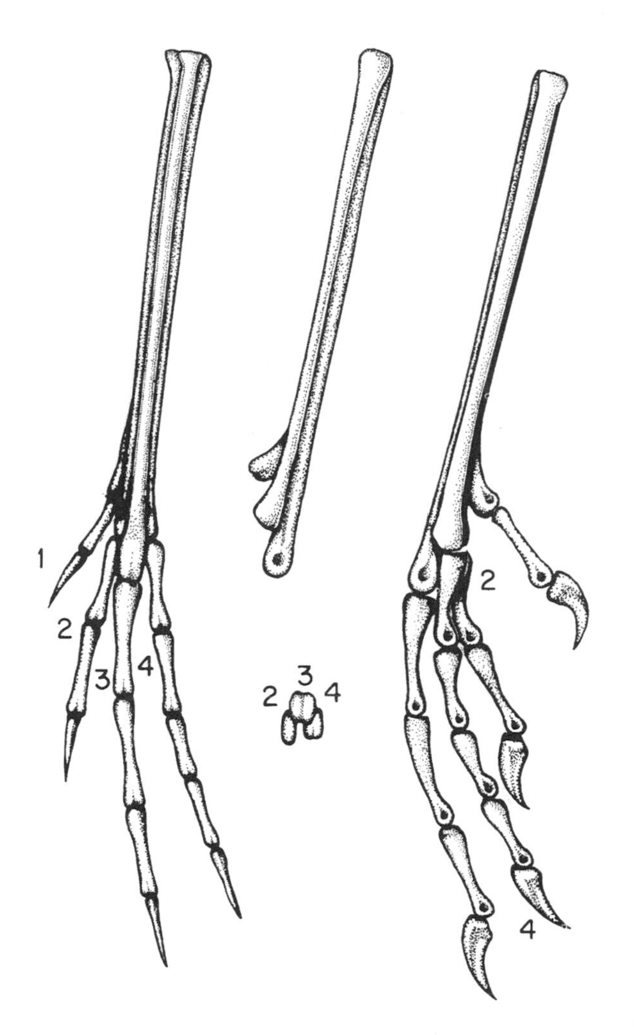

Left to right, restoration of the left foot of the Eichstätt *Archaeopteryx* in anterior view, tarsometatarsus in medial view with metatarsal 1 attached (distal view below), and lateral view. (From Martin 1991; reprinted from Hans-Peter Schultze and Linda Trueb, eds., *Origins of the higher tetrapods: controversy and consensus;* copyright © 1991 by Cornell University; used by permission of the publisher, Cornell University Press)

as in the maxilla and dentary (Carroll 1988, 269). One can extricate oneself from the rigid format of cladistics and argue effectively that the "thecodont" hypothesis is legitimate—that is, that birds arose from the basal archosaur lineage before the advent of forms defined as dinosaurs or before the split of the dinosaurian lineages Ornithischia and Saurischia. We thus have two hypotheses, as stated earlier: Did birds arise from dinosaurs or did they arise from their ancestors, the thecodonts (or basal archosaurs)? Indeed, I would prefer to divide the possible bird derivations into three instead of two entities: (1) birds derive from small Triassic, quadrupedal, arboreal, basal archosaurs (or their allies); (2) birds derive from a common stem with the dinosaurs, from terrestrial, bipedal, transitional thecodonts; (3) birds derive directly from the theropod dinosaurs, as Ostrom and most paleontologists believe, later in time.

Basal archosaurs are characterized by two distinctive ankle joint articulation arrangements (Carroll 1988; Sereno and Arcucci 1990; Sereno 1991). In addition, the lineage or lineages that produced the dinosaurs as well as pterosaurs and birds tended toward what is called a mesotarsal joint—that is, an ankle joint in which the major joint passes between the proximal and distal tarsals. Within this assemblage of thecodonts some forms from the Middle Triassic of South America had clearly approached the dinosaur grade of morphological organization, while the Middle Triassic *Lagosuchus*, redescribed by José Bonaparte (1975), and the Middle Triassic *Lagerpeton* (Romer 1972; Sereno and Arcucci 1993) are considered to be very close to the definition of "dinosaur." *Lagosuchus* is a small, lightly built thecodont about 30 centimeters (12 in.) long, with the posterior limb longer than the forelimb, and with a mesotarsal hinge between the astragalus and calcaneum proximally and between the distal tarsals and metatarsals distally. Kevin Padian (1984) has proposed that *Lagosuchus* is closely related to both dinosaurs and pterosaurs and that the ancestor of both dinosaurs and pterosaurs was a biped with a fully erect stance and gait. By expressing this view, Padian has run into the same problem as advocates of the dinosaur-bird nexus—that is, he is strapped with the ground-up origin of pterosaur flight. Padian thus presupposes that pterosaurs were bipeds and could move along the ground on two legs and that pterosaur flight evolved from the ground up, a view endorsed by Sereno (1991). This model has been criticized and refuted on biophysical grounds by Ulla Norberg (1990), however, and is taken up in chapter 3.

Although the basal archosaurs may lack adequate definition based on synapomorphies, so, too, do the early lineages of dinosaurs, and it is not even clear exactly what constitutes a dinosaur! In almost all cases, including the earliest known dinosaur, *Eoraptor*, from the late Triassic of Argentina (Sereno, Forster, Rogers, and Monetta 1993), one can ask the question, is it a late thecodont or an early dinosaur? It is, in fact, the near vertical posture, and concomitant complete fenestration of the acetabulum assumed by dinosaurs, that perhaps best defines the early forms from their thecodont predecessors. In dinosaurs a mesotarsal joint is present with the line of flexion between the proximal and distal tarsals. The astragalus and calcaneum are reduced and integrated with the ends of the tibia and fibula. The above features are, in fact, where dinosaurs exhibit their closest resemblance to birds, and the question, of course, is, are the resemblances due to homology or to convergence, since a bipedal stance is assumed in both forms, albeit for different reasons.

Walter Coombs reviewed the physical constraints on cursorial animals and produced a list of "adaptations that are ubiquitous among phylogenetically diverse animals which run and may be regarded as inevitable in any cursor" (1978, 343). Coombs's list includes, among other more detailed features, relatively long limbs, small forelimbs, hingelike joints, short and massive proximal limb elements, long and slender distal limb elements, manus and pes with pronounced medial symmetry, digitigrade to unguligrade stance, interlocked or fused metapodials, and reduced or lost inner or outer digits. With this in mind, Huxley's comparisons of running dinosaurs and chickens were sure to be exceptionally favorable for a dinosaur-bird link. Among the thecodonts, ornithosuchids and lagosuchids brought the rear limbs close to a vertical orientation, but *Lagosuchus* was probably facultatively (by choice), not obligately, bipedal. A number of other thecodonts, including *Ornithosuchus* and the remarkable late Triassic *Postosuchus* (Chatterjee 1985), approached dinosaur "grade" and are for all practical purposes "dinosaurs" except for lacking the trivial characters that define a "dinosaur." In the case of birds, their bipedalism was achieved in a totally different manner, by the eventual release of the forelimbs from a climbing to a flying function. However, the structural form of bipedalism in birds, particularly in secondarily ground-dwelling forms, such as chickens, came to resemble closely that of the bipedal cursorial theropods.

The question of relative length of forelimb to hindlimb has never been adequately addressed. Theropods have been variously described as balanced seesaws and the like, with the body pivoting about the hip joint and the long, heavy tail serving as a balancing organ. This anatomical bauplan is generally combined with large size, a trademark of the dinosaurian radiation. In the earliest dinosaurs, and even in their thecodontian precursors, the forelimbs were exceedingly short, the worst possible

Silhouettes of various theropod dinosaurs illustrating the proportions of forelimbs and hindlimbs. *Above, left to right,* the earliest known saurischian, the late Triassic *Eoraptor* (forelimb to hindlimb ratio = 50%); late Triassic basal saurischian dinosaur *Herrerasaurus* (50%); late Jurassic derived "coelurosaur" *Compsognathus* (37%); *below, left to right,* late Triassic–early Jurassic theropod *Syntarsus* (35–45%); late Triassic basal theropod *Coelophysis* (35–45%); late Cretaceous "ornithomimid?" *Mononykus* (20+%). In most basal archosaurs the forelimbs range from approximately 60 to 75% of hindlimb length (Sereno 1991); and in theropods the forelimbs are 50% or less of the hindlimb length. Birds probably arose in a line of archosaurs that had not yet made the evolutionary commitment to foreshorten the forelimbs. (Drawing by Susan Whitfield)

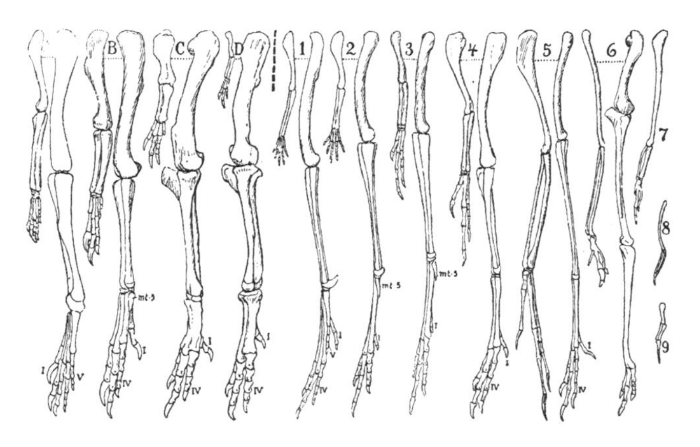

Proportions of forelimbs and hindlimbs in thecodonts, theropods, and birds. *A and 1,* thecodonts: (*A*) *Ornithosuchus woodwardi,* (*1*) *Saltoposuchus longipes*. *B–D,* theropod dinosaurs: (*B*) *Archisaurus colurus,* (*C*) *Ceratosaurus nasicornis,* (*D*) *Tyrannosaurus rex*. *2–4,* "coelurosaurus": (*2*) *Procompsognathus triassicus,* (*3*) *Compsognathus longipes* (known to have two digits), (*4*) *Ornitholestes*. *5,* flying bird: *Archaeopteryx*. *6–9,* secondarily flightless birds: (*6*) rhea, (*7*) ostrich, (*8*) kiwi, (*9*) emu. (From Heilmann 1926, 171)

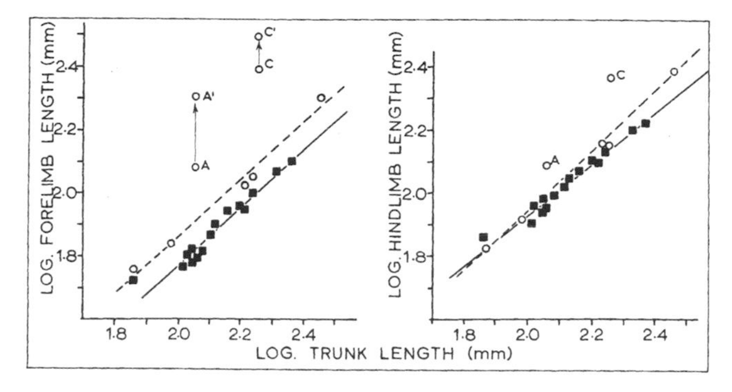

Relative limb lengths for *Archaeopteryx* and the flying lemur *Cynocephalus* plotted on the graphs (from Thorington and Heaney 1981) of limb lengths in modern squirrels. Lower points for *Archaeopteryx* and *Cynocephalus* are those calculated (as by Thorington and Heaney) with forelimb length as (length of humerus plus length of radius); upper points are plotted with the length of the longest digit added. (From Yalden 1985; courtesy G. Viohl and Jura-Museum)

anatomical architecture for flight evolution. For example, while in the basal archosaurs the forelimbs average some 60 to 75 percent of the length of the hindlimbs, in basal theropods, such as the late Triassic *Eoraptor* and *Herrerasaurus*, the forelimbs are some 50 percent or less that of the hindlimbs. In the late Triassic *Syntarsus* and *Coelophysis*, the forelimbs are some 35 to 45 percent of the hindlimbs, and in *Compsognathus*, the figure is 37 percent; it is even less in many late Cretaceous forms, such as *Tyrannosaurus* (26 percent).

Gliding and flying vertebrates are characterized by elongated forelimbs (Yalden 1985), and Thorington and Heaney (1981) have documented this by comparing gliding squirrels with their nongliding relatives. Gliding squirrels do not have longer hindlimbs than nongliders, but the flying lemur (*Cynocephalus*) does have elongated hindlimbs. *Archaeopteryx* has, in its limb proportions (expressed in relation to trunk length), more elongate limb bones than flying squirrels, but its limb bones are less elongate than those of *Cynocephalus*. As Yalden (1985) has shown, if the longest finger is added to limb length, then both *Cynocephalus* and *Archaeopteryx* have forelimbs that are 1.7 times their trunk length.

As we have seen, by the late Triassic, Jurassic, and Cretaceous periods, the theropods are classically divided into two groups, the Carnosauria and the Coelurosauria, the latter group thought by many to have given rise to birds. Yet, as we have noted, these two groups are as poorly defined as basal archosaurs, with the term *carnosaur* generally applying to larger forms and *coelurosaur* to the smaller species, and many authors do not accept this taxonomic division. To illustrate the problem, the classic Solnhofen coelurosaur *Compsognathus* has the hand reduced to two digits and has interdental plates, both characteristics of carnosaurs, although interdental plates are also a feature of thecodonts; and Ostrom's *Deinonychus* combines an almost equal number of carnosaur and coelurosaur characters. Carroll has suggested that "the 'carnosaur' families may each have evolved separately from different groups that have been classified as coelurosaurs. Or perhaps several separate lineages of both large and small theropods evolved independently from among primitive forms such as *Staurikosaurus* and *Herrerasaurus*" (1988, 290).

Tarsitano and Hecht (1980) have projected a logical, new version of Heilmann's pseudosuchian thecodont hypothesis, concluding that birds originated from unarmored mesotarsal thecodonts, a group also ancestral to the dinosaurs. Their cladogram has birds branching off somewhere between *Euparkeria* and *Lagosuchus*. Interestingly, their proposal has a striking similarity to those of many other authors, such as Broom, and in fact, these authors' views do not differ drastically from Huxley's "common ancestry" hypothesis of birds and dinosaurs. Many authors have interpreted Heilmann as rejecting any avian relation to theropods, but this is not true. Heilmann mainly rejected a theropod hypothesis because the "clavicles are wanting" in the coelurosaurs, and in fact he argued for a hypothesis not too distant from the type of common ancestry hypothesis. According to Heilmann, "All the bird-like features we met with in the Coelurosaurs may be considered as a further development of some of the possibilities inherent in the Pseudosuchians" (1926, 185). Heilmann thus made a very important point. Most would agree that birds are derived within the archosaur assemblage, and in spite of all the current bluster, the crux is, in reality, a much more refined question: at exactly what point in the archosaur lineage and at what time did birds branch off? In addition, Heilmann was basing his arguments on the incipiently or facultatively bipedal *Euparkeria*, and not on fully quadrupedal arboreal thecodonts or basal archosaurs.

A Miscellany of Avimorph Thecodonts. Tarsitano and Hecht (1980) and other subsequent authors advocating a basal archosaur or thecodontian origin of birds have sought the avian ancestors among reptiles that would be ideal for evolving flight from the arboreal setting. That is, the ancestral bird would hypothetically be small, quadrupedal, and arboreal. Sam Tarsitano (1985, 1991) in particular has rejected dinosaurs as bird ancestors because they are too large, are bipedal runners, and could not possibly have evolved flight from the trees down. Robert Broom (1906) was also bothered by the problem of size and argued for miniaturization in the origin of birds, feeling that the ancestor of birds must have been very small (Witmer 1991).

These authors point to small Triassic basal archosaurs such as *Longisquama* (Sharov 1970; Bakker 1975; Haubold and Buffetaut 1987), *Scleromochlus* (Woodward 1907), and *Megalancosaurus* (Calzavara, Muscio, and Wild 1980; Tarsitano 1991; Feduccia and Wild 1993) as the most probable types of avian progenitors. Another possible proavian progenitor is the enigmatic Spanish Middle Triassic *Cosesaurus aviceps*, described by Ellenberger and De Villalta (1974). This small reptile, 14 centimeters long (5.5 in.), is represented only by a negative impression of the skeleton with little anatomical detail; yet its skull, braincase, and beaklike jaws give a very birdlike impression, and Ellenberger and De Villalta indicated that traces of feather impressions were present. After restudying the specimen, however, Sanz and Lopez-Martinez (1984) argued that this juvenile archosauromorph was best placed within the Prolacertiformes, primitive relatives of the archosaurs. Nevertheless, as Andrew Milner has pointed out, the specimen does show "how progenetic dwarfing may have reshaped the typical archosaur skull to result in the avian skull" (1985, 544). Even though *Cosesaurus* may not be a valid archosaur, it vividly illustrates how birds may have evolved through neoteny (or arrested development) somewhere in the tangle of the early archosaur radiation.

Longisquama insignis is a remarkable, tiny (some 50 millimeters [2 in.] long) presumed thecodont from the late Triassic of Turkestan and the only known reptile to possess scales that show a possible intermediacy with feathers. The tiny specimen is preserved in its entirety, crushed on a slab. It exhibits an antorbital fenestra, a well-developed furcula, and long keeled, overlapping body scales. Another enigmatic thecodont is *Scleromochlus taylori* from the Upper Triassic Elgin sandstone of Scotland. Described and named by Arthur Smith Woodward of the British Museum in 1907, the tiny creature, which was only some 23 centimeters (9 in.) long, was preserved only as the imprints of bones in the coarse sandstone. Friedrich von Huene (1914) studied the creature and became convinced that it dwelled in trees, could jump from limb to limb, perhaps even glided. He became confident that the creature closely approximated the prototype of pterosaurs and on this basis suggested an arboreal origin of flying reptiles from a passive parachuting and gliding phase. However, the hindlimbs of *Scleromochlus* are considerably longer than the forelimbs, a condition one would not expect in the ancestor of either birds or pterosaurs.

By far the most birdlike of these small arboreal thecodonts is the Upper Triassic *Megalancosaurus preonensis* from the Preone Valley in Italy (Calzavara, Muscio, and Wild 1981; Feduccia and Wild 1993). Estimated to be about 32 centimeters (12.6 in.) long, *Megalancosaurus* is remarkable in having elongated forelimbs and an exceptionally birdlike head and elongated cervical vertebrae. The birdlike skull has a relatively enormous orbit and a beaklike snout with tiny isodont teeth set in sockets. In addition, this arboreal thecodont has a hand with five digits, each with deeply hooked, bladelike claws, and a distinctive avian straplike scapula.

Silvio Renesto performed a complete analysis of the osteology of this interesting reptile and concluded that its phylogenetic position, though somewhat enigmatic, places it near both the Prolacertiformes, a possible sister-group of the Archosauria, and the Archosauria itself: "This reptile may represent an early lineage of the Archosauromorpha, related both to the Prolacertiformes and the Archosauria" (1994, 51). However, *Megalancosaurus* does have the salient antorbital fenestra of the archosaurs, and as Renesto concludes, "*Megalancosaurus* was a small reptile with a high degree of adaptation toward arboreal life, and is probably an archosauromorph" (38).

At the base of any lineage one would expect few shared, derived characters to link ancestral-descendent groups, and therefore the synapomorphies linking basal archosaurs with possible descendants like pterosaurs and birds are limited. One could question whether the shared similarities of these thecodonts and birds indicate any true relation, but as Carroll has correctly emphasized, "When attempting to establish relationships of any group with a fossil record, we must emphasize the earliest known members, because they have had the shortest amount of time to evolve new characters since their initial divergence. Hence, they should provide us with the best opportunity to identify the derived features that they share with their closest sister group" (1988, 8).

The Triassic thecodont *Megalancosaurus*, particularly in combination with *Longisquama* and *Scleromochlus*, thus show considerable similarity with birds, even providing a glimpse into the earliest known furcula and featherlike scales. Their morphological diversity also illustrates that there was an extensive adaptive radiation of small arboreal basal archosaurs that in combination exhibit many synapomorphies with birds. In Tarsitano and Hecht's view, our current knowledge does not permit us to go beyond representing the thecodonts as a grade, and "the relationships of *Archaeopteryx* and other derived taxa are placed within that polyphyletic and/or paraphyletic taxon" (1980, 178). The derivation of birds (and, independently, pterosaurs) from small arboreal Triassic thecodonts is intuitively pleasing and fits very nicely with the arboreal theory for the origin of flight, the only theory of flight origins that can easily be explained in biophysical terms.

New Birdlike Dinosaurs. In recent years the advocates of a dinosaur-bird genealogy have discovered birdlike characters in a number of dinosaurs and have described a number of forms that appear to be birdlike dinosaurs, most of them of late Cretaceous age. Philip Currie of the Tyrrell Museum of Paleontology (1987) reported birdlike characters in the late Cretaceous *Troödon* and in troödontid theropods in general (including *Stenonychosaurus*, *Pectinodon*, and *Saurornithoides*). Then, in 1991, Angela Milner, paleontologist at the British Museum, and Susan Evans described a single maxilla from the Upper Jurassic of Europe as *Lisboasaurus estesi*, a so-called maniraptoran theropod, close to the avian line. And in 1992 and 1993, two new provocative small, late Cretaceous Mongolian archosaurs were described, one as a theropod, *Archaeornithoides deinosauriscus*, the "closest of the known non-avian relatives of *Archaeopteryx* and other

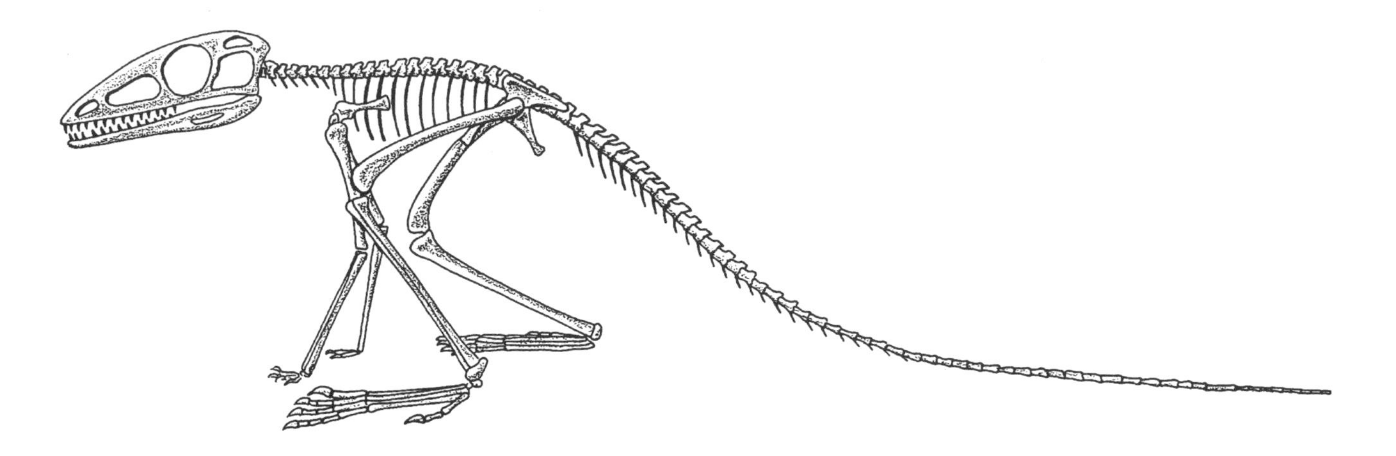

Skeletal reconstruction of the basal archosaur *Scleromochlus taylori* from the Upper Triassic Elgin sandstone of Scotland, approximately 23 cm (9 in.) long. Freidrich von Huene (1914) classified *Scleromochlus* as a pseudosuchian thecodont and thought it to be an arboreal creature that jumped from branch to branch, possibly gliding using skinfolds on the forelimbs, and close to the actual ancestry of pterosaurs. (From von Huene 1914; modified after Wellnhofer 1991)

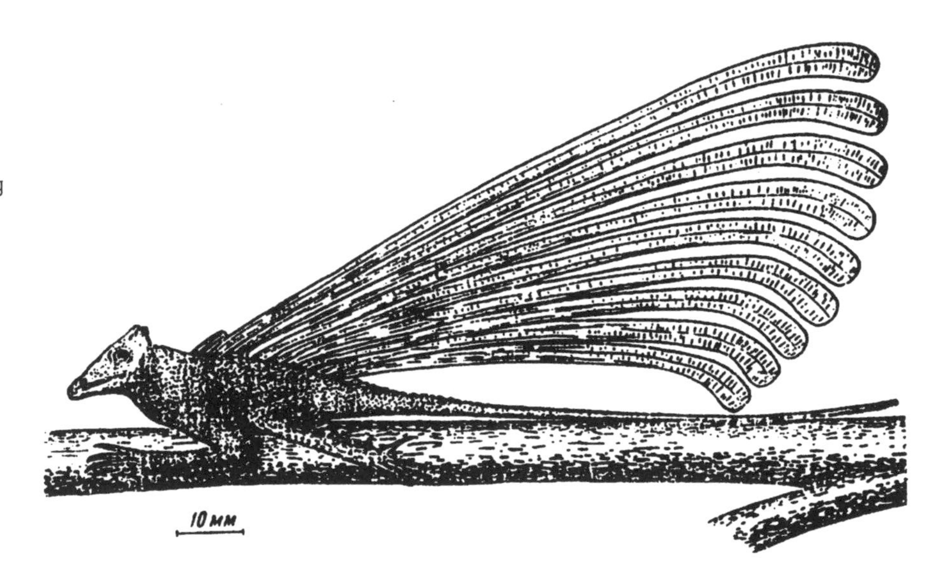

Life reconstruction of the small arboreal thecodont *Longisquama* ("long scale"), described by A. G. Sharov from the late Triassic of Kirghizia. The 10–12.5-cm-long (5 in.) creature is thought by Haubold and Buffetaut (1987) to possess a unique gliding adaptation, a double series of long scalelike structures that were unfolded in butterfly fashion to form a gliding wing. (From Haubold and Buffetaut 1987; courtesy L'Académie des Sciences, Paris)

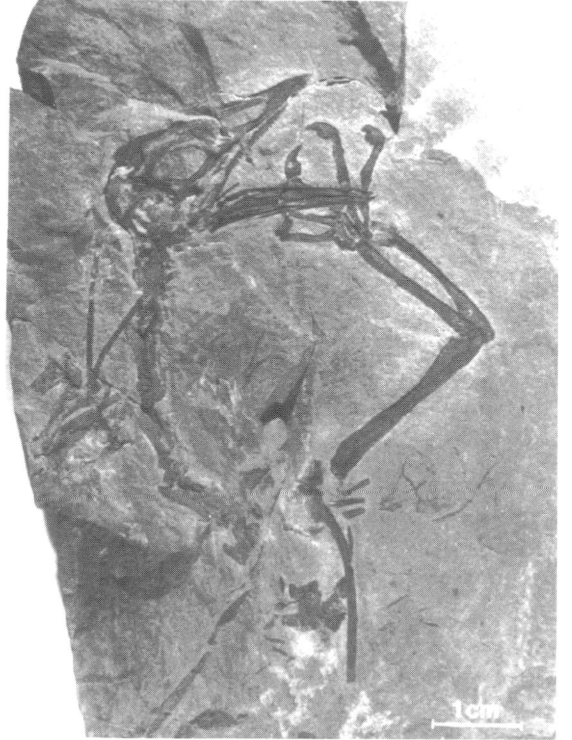

Life reconstruction of the late Triassic avimorph thecodont *Megalancosaurus* (total length approximately 32 cm; 12.6 in.) and actual fossil specimen. (Drawing by John P. O'Neill; photo courtesy Museo Friulano di Storia Naturale, Udine)

birds" (Elzanowski and Wellnhofer 1992, 1993); the other as a bird, *Mononykus*, a bizarre, flightless, turkey-sized creature with very short forearms and a long dinosaurian tail (Perle, Norell, Chiappe, and Clark 1993a; and correction, 1993b; Perle et al. 1994). Milner has stated that *Mononykus* is of "great importance to understanding primitive birds" and that it appears to occupy a "transitional position between *Archaeopteryx* and all other birds" (1993, 589). Yet Storrs Olson, who has probably examined more fossil birds that any other living paleontologist, studied the specimen and has stated that "it's a very interesting dinosaur, but there's no way that thing could be a bird" (Noble 1993, 81), and Yale University's dinosaur expert John Ostrom notes, "*Mononykus* was not a bird, . . . it clearly was a fleet-footed fossorial theropod. . . . Reasoning of such dubious quality demonstrates a fundamental flaw in the cladistic methodology" (1994, 172). The handful (four or five) of so-called synapomorphies used to make the *Mononykus*-bird nexus are trivial features, and employing the same methodology, one could make a much better case that pteranodontid pterosaurs, with some eight or more good avian synapomorphies, including their enlarged orbit and eye, reduced fibula, notarium, loss of teeth, hollow, thin-walled bones, and keeled sternum, were birds! With its small head and large tail with well-developed vertbral haemal (blood) arches, *Mononykus* will no doubt prove to be a theropod, perhaps an aberrant member of the Ornithomimidae. To clothe *Mononykus* with feathers with absolutely no evidence is contrary to all standards of fossil reconstruction.

Yet what we now see clearly is that small coelurosaurian dinosaurs were becoming very birdlike during the late Cretaceous, some 75 or so million years after the appearance of *Archaeopteryx*. Very interestingly, the skeletons of the two earliest known dinosaurs, *Eoraptor* and *Herrerasaurus*, "show that dinosaurs had developed many specialized anatomical features early, long before the reptiles became common. By the time of *Eoraptor*, during the mid-Carnian stage of the Triassic period, dinosaurs had already split into the major groups of carnivorous theropods and herbivorous ornithischians" (Monastersky 1993, 61). In other words, although the major lineages of dinosaurs diverged during the Triassic, we see none of the key dinosaur-bird synapomorphies in early dinosaurs; we do see these characters in dinosaurs of the latest Cretaceous. One could argue that this is rather what one would expect if, in fact, small dinosaurs were converging on birds during the Cretaceous. Early paleontologists clearly recognized this when they named the group the ornithiomimids (bird-mimics) and gave the creatures such generic names as the ancient bird-mimic (*Archaeornithomimus*), bird-mimic (*Ornithomimus*), fowl-mimic (*Gallimimus*), emu-mimic (*Dromiceiomimus*), ostrich-mimic (*Struthiomimus*), and Garuda-mimic (*Garudimimus*), this last creature named by Rinchen Barsbold for the Garuda, an Indian mythological bird. All of these bird-mimics are found in the late Cretaceous.

Another vision advanced by the discovery of the late Cretaceous birdlike theropod *Mononykus* is the view that feathers were present in dinosaurs, a fantasy resurrected from the wild speculations of Robert Bakker (1986) and Gregory Paul (1988). This problem is addressed in the next chapter, but suffice it to say here that Paul has adopted the extreme position that coelurosaurs are derived from archaeopterygids. He has illustrated numerous feathered dinosaurs, including a plumed *Ornitholestes* perched in the branches of a tree!

Complications for a Theropod Origin. Aside from the numerous problems with the synapomorphies uniting birds and dinosaurs, the theory of the dinosaurian origin of birds is replete with other complications. First, the timing is off. *Archaeopteryx* occurs in the late Jurassic, some 150 million years ago, and presumably birds originated much earlier, say, possibly medial to late Triassic. However, the earliest dinosaurs, *Eoraptor* (if it is a dinosaur) and *Herrerasaurus* (Sereno, Forster, Rogers, and Monetta 1993; Sereno and Novas 1992; Novas 1993; Sereno 1993; Sereno and Novas 1993), both from the late Triassic of Argentina, lack the synapomorphies that are used to unite birds and dinosaurs. Birds are supposed to be derived from the coelurosaurian dinosaurs, and the form initially used to relate dinosaurs to birds, *Deinonychus*, is from the early Cretaceous, some 40 million years after *Archaeopteryx*, and the most birdlike dinosaurs are from the late Cretaceous, some 75 or more million years after the appearance of the first bird. In fact, one could interpret the temporal evidence as indicating that birds and dinosaurs are indeed examples of convergent evolution.

Second, the ancestors of birds must be suited to flight, and dinosaurs, even the early ones: (1) are large (turkey-sized; most much larger, beginning in the 3-meter [10 ft.] range—too large for feathers to have any aerodynamic effect), deep-bodied, bipedal, cursorial, earthbound predators; and (2) had greatly shortened forelimbs, approximately one-half the length of the hindlimbs in both *Eoraptor* and *Herrerasaurus*, the earliest two known dinosaurs, as well as in the classical late Triassic "coelurosaur" *Coelophysis*. This represents the worst possible morphological arrangement for the evolution of flight: a large, bipedal runner, with shortened forelimbs, and it has strapped the advocates of the dinosaur origin of birds with the cursorial origin of avian flight and the notion of hot-blooded dinosaurs clothed with feathers for insulation.

Are birds derived from dinosaurs? This would depend entirely on what one defines as a dinosaur, or Dinosauromorpha, as defined by Sereno (1991; Sereno and Arcucci 1990; Sereno and Novas 1990; Sereno and Arcucci 1993), which includes such basal archosaurs as *Lagerpeton*, *Lagosuchus*, and *Pseudolagosuchus*. The real discussion over the past several decades, however, is whether birds derive from advanced theropods, such as the dromaeosaurs. The evidence against such an avian origin is overwhelming. Not only extensive morphological evidence, but the geological timescale and the biophysical attributes of these earthbound running creatures make the theropod theory of bird origins extremely unlikely. We find ourselves back to the common-ancestry hypothesis of Broom and Heilmann, which certainly at this stage remains the most plausible scenario for a derivation of birds from within the Dinosauromorpha.

The coelurosaurs had already attained the erect posture and greatly reduced forelimbs suggestive of terrestrial rather than arboreal locomotion. Many thecodonts, although structurally close to the dinosaur "grade" in morphology, were small, had not yet become fully erect and obligately bipedal, and were arboreal, with little reduction in their forelimbs. These thecodonts could easily have retained or evolved the elongation of the forelimbs seen in *Archaeopteryx* and presumably present in protobirds, but for coelurosaurs to develop lengthened forelimbs from their reduced and somewhat vestigial hands would demand a drastic reversal in evolution. As we shall see in chapter 3, *Archaeopteryx* was well on its way to becoming a true bird and could well be the descendant of some form intermediate between reptiles and birds that occurred much earlier in time than coelurosaurs. Birds originated in the trees.

Hypothetical late Triassic or early Jurassic arboreal "protoavis "at a post-parachuting and early gliding stage in a coniferous forest. Somewhere in pre-Solnhofen sediments a similar animal awaits unearthing. (Drawing by John P. O'Neill)

3

The Genesis of Avian Flight

When they began to fly, birds either lifted themselves up from the ground or glided down from high places; it is difficult to imagine an alternative. In either case, the anatomical changes needed for flying must have evolved in a sequence of very small steps, because nothing we know about evolution allows us to believe that feathered wings could have appeared abruptly as an innovation in avian anatomy. Wings must have evolved over a very long time span. And each new modification of body plan or limbs during that period must have made some contribution to fitness long before the day when a jumping or gliding creature gave the first strong beat of its forelimbs and ceased simply falling back to earth.

So, to reconstruct the evolution of modern birds, we must account for the sequence of changes that replaced reptilian scales with feathers, and, along the way, we must answer certain questions: Were the reptilian ancestors of birds runners, jumpers, or parachuters and then gliders? Was *Archaeopteryx* itself at home on the ground or in trees? Could it only glide or did it already have the ability to sustain flight? What was the original advantage of feathers or their epidermal precursors?

Vertebrates Take to the Air

Before speculating about how flight originated in birds, we should consider other large animals that have taken to the air. All other flying or gliding vertebrates began their evolution in trees, with the possible exception of the pterosaurs, which may have begun flying by gliding out from sea cliffs. As a general rule, then, beginning fliers use the energy that gravity provides: they climb up and coast down. They do not start their flight by expending the burst of effort needed to rise off the ground, because if they did, once off the ground, there would be no energy to sustain the lift.

Several kinds of reptile or amphibian have achieved periods of airborne existence one way or another, but all in the arboreal setting. Although few living vertebrates are pure parachuters—that is, with a descent angle greater than 45°—myriad vertebrates are good gliders (with a descent angle less than 45°), "traveling from tree to tree and voluntarily steepening their descent angle" (U. M. Norberg 1990, 2). Early examples were the late Permian and late Triassic "dawn lizards," represented by the late Permian genera *Coelurosauravus* and *Daedalosaurus* and the late Triassic genera *Kuehneosaurus* and *Icarosaurus*, which were related to the precursors of true lizards and, like birds, are first known as fossils in the late Jurassic period. Some dawn lizards developed a pair of horizontal sails in which a modified set of ribs grew straight out to form struts lying in a horizontal plane instead of curving around to enclose the chest. The skin stretched over these ribs provided a sail surface, and this allowed the animal to volplane, descending without power, from tree to tree.

The Russian zoologist A. G. Sharov (1970, 1971) described several gliding reptiles from the late Triassic of Kirghizia. One species was the lizard-like Triassic *Sharovipteryx (Podopteryx) mirabilis*. (The fossil was renamed in 1981 [Cowen 1981] in honor of Sharov when it was discovered that *Podopteryx* was already claimed by a fish described a hundred years earlier.) *Sharovipteryx* exhibited an elongated head and hindlimbs, but it was unique in having a wing membrane, or patagium, that stretched across the hindlimbs to the base of the tail and perhaps a smaller flap between the forelimbs and body (Gans et al. 1987). Even stranger was the other arboreal reptile from the same deposits described by Sharov as *Longisquama* (mentioned in chapter 2), so named because of a row of some ten elongated and highly modified scales along the back, each apparently corresponding to a dorsal vertebra. This tiny thecodont was only about 10–12.5 centimeters (3.9–4.9 in.) long, and the highly modified, elongated dorsal scales were actually longer than the body. According to

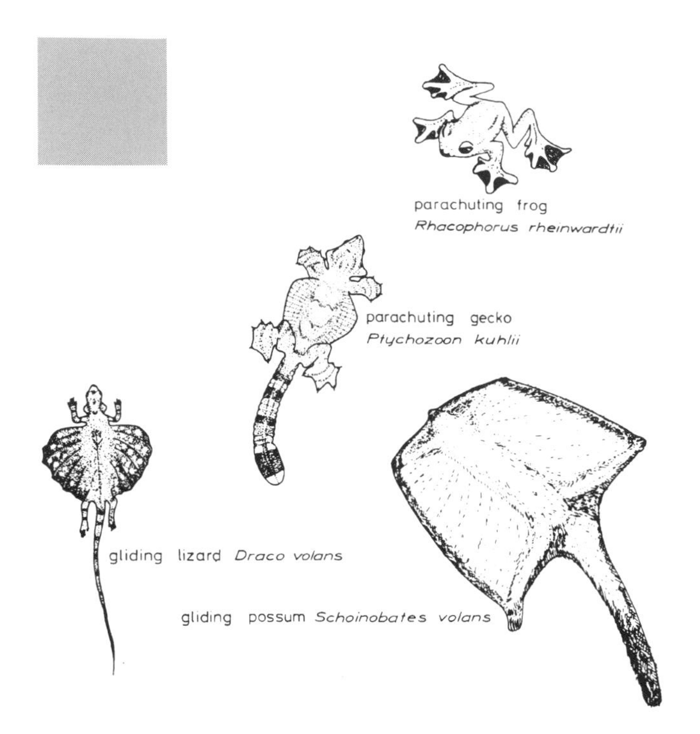

Sampling of gliding and parachuting vertebrates: parachuting frog (*Rhacophorus rheinwardtii*), parachuting gecko (*Ptychozoon kuhlii*), flying dragon (*Draco volans*), gliding possum (*Schoinobates volans*). *Inset:* examples of vertebrates that originated flight from the ground up. (Drawings from U. M. Norberg 1990; courtesy Springer-Verlag)

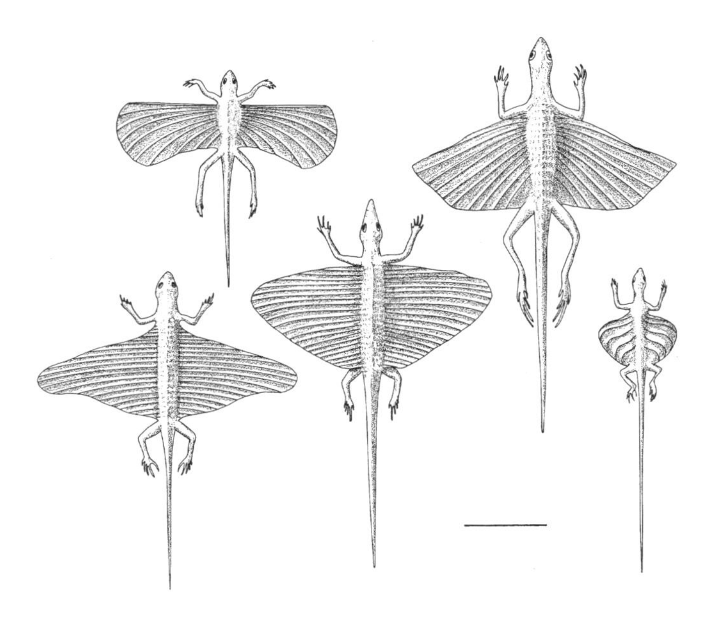

Ancient gliding reptiles. *Top row,* the late Triassic *Icarosaurus* (*left*) and *Kuehneosaurus* (*right*). *Bottom row,* the earlier late Permian *Daedalosaurus* and *Coelurosauravus* compared with the modern flying agamid lizard, the flying dragon *Draco*. (From Wellnhofer 1991; courtesy Peter Wellnhofer)

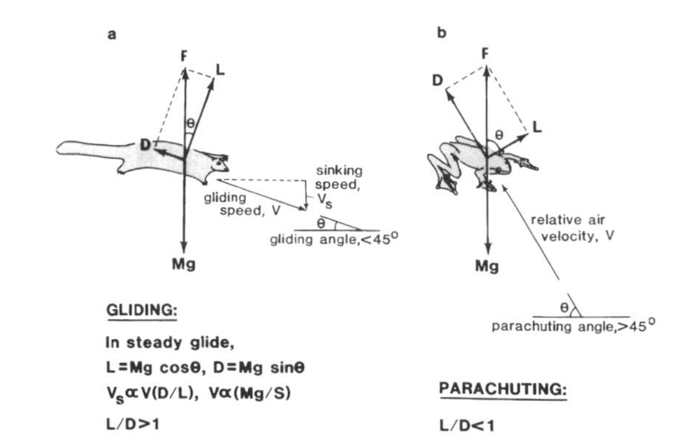

Aerodynamics of gliding flight and parachuting. In parachuting the relations between the lift:drag ratio and the equilibrium angle of descent are as for gliding, but because the area supporting weight is so small, lift becomes small in relation to the combined drag of the body and the gliding surfaces. Therefore, the overall lift to drag ratio (L/D) is lower, resulting in a correspondingly steeper glide path. (Reprinted by permission of the publishers from Dennis M. Bramble, Karel F. Liem and David B. Wake, *Functional vertebrate morphology,* The Belknap Press of Harvard University Press, copyright 1985 by the President and Fellows of Harvard College)

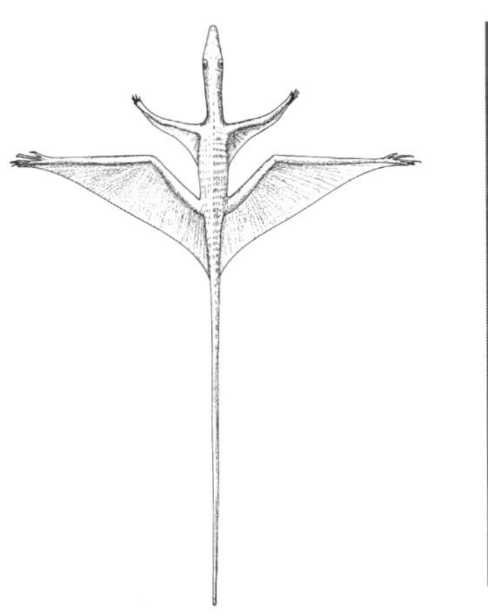

Above left, Sharovipteryx, in life reconstruction, from the late Triassic of Kirghizia, represents another early experiment in gliding, with the main gliding membrane between the hindlimbs, and a tentative reconstruction of a gliding membrane behind the forelimbs. *Above right,* the actual fossil, showing the impression of the gliding membrane, long hind legs, and long tail. Total length, approximately 19.6 mm (7.7 in.). (Reconstruction from Wellnhofer 1991; courtesy Peter Wellnhofer; photo courtesy E. N. Kurochkin)

Haubold and Buffetaut (1987), these dorsal scales could be folded upward in butterfly-wing fashion and then down, where they formed horizontal gliders. *Longisquama* thus illustrates a bizarre and unique solution to the problem of gliding. In conquering a new adaptive zone—gliding—the dawn lizards and other gliders also demonstrate that the forest environment of the late Permian and Triassic periods presented a new ecological niche to be exploited and that reptiles were among the first to do so.

A group of extant lizards of the family Agamidae, at a much later time, also acquired the art of gliding. Living in trees of the Malay Peninsula and islands of the western Pacific, these "flying dragons" number some two dozen species of the genus *Draco*, the best known being *Draco volans*, which can glide up to 60 meters (197 ft.). The flight structure of these lizards is a membrane stretched between the forelimb and hindlimb that is stiffened by six very long ribs, somewhat in the manner of the dawn lizards. Although this equipment seems meager, *Draco* lizards are highly maneuverable and can steer well enough to select a landing site on a particular tree. The family Gekkonidae includes the "flying geckos," *Ptychozoon*, parachuters equipped with broad skin flaps on either side of the body; during flight, these flaps are spread out by the limbs and the body is flattened. Flying geckos also have smaller skin flaps bordering the neck and tail, as well as webbed feet.

Among snakes, which are descended from varanid lizards, there are tree-dwellers that have developed adaptations permitting them some travel time in the air. Members of the Bornean colubrid snake genus *Chrysopelea*, for example, engage in a kind of parachuting. These snakes can hurl themselves from high in a tree, pulling in their abdomens to create a ventral concavity, so that their bodies become airfoils, enabling them to sail rather than drop to a lower point. And among amphibians there exist "flying frogs" of the families Rhacophoridae and Hylidae of Central and South America, Australasia, and Southeast Asia, which rely on webbed feet to achieve a flight surface, in some cases coupled with the ability to flatten the body. In flight, the parachuting Malay frog *Rhacophorus* extends its limbs, which support broad flaps of skin, and its fully webbed toes. Although these parachuting frogs have little steering ability, they are at least able to check the full impact of their fall, and the ability is clearly adaptive: the animal can quickly escape from predators and is protected from injury if it falls accidentally.

The only reptiles to develop true powered flight were the pterosaurs. Kevin Padian (1983) has proposed that pterosaurs were bipedal and evolved flight from the ground up. He argued that these reptiles could not have walked quadrupedally by moving the forelimbs parasagittally, alongside the median plane, over the ground. However, Colin Pennycuick (1986), using a simple system of weight balances, has shown convincingly that pterosaurs must have walked quadrupedally; they would not have been able to support their weight on the hindlegs without support from the wings, and their system of balance would not permit bipedal locomotion (U. M. Norberg 1990).

Gliding adaptations have evolved independently in three diverse orders of mammals (Vaughan 1978). Among the Australian pouched mammals, or marsupials, gliders have evolved in three genera of flying phalangers, including the pygmy glider *Acrobates*, the greater glider *Schoionobates*, and the sugar glider *Petaurus*. The colugos, members of the Dermoptera of Southeast Asia that are thought, like bats, to be derivatives of ancient insectivores, are called flying lemurs because of their lemurlike faces. The two species of the Dermoptera genus *Cynocephalus* have a gliding membrane that stretches between the forelimbs and hindlimbs and even between the digits and including the tail region, allowing them to glide up to 100 meters (328 ft.) between forest trees. And gliding is common among the Rodentia, with the Sciuridae, or flying squirrels, having evolved gliding in some twelve genera and the West and Central African Anomaluridae, or scaly-tailed squirrels, having developed gliding in three genera. In flying squirrels, the gliding membrane is stretched by a short cartilaginous brace that arises from the wrist; in scaly-tailed squirrels, the cartilaginous spine arises from the elbow.

Gliding has tremendous advantages over running because it is relatively fast and economical in terms of energy. A squirrel cannot run faster than 2 meters (6.6 ft.) per second on the ground, but when gliding it can attain a speed of 15 meters (49 ft.) per second (Scholey 1986). The combination of climbing and gliding is an attractive mode of transportation for many reasons, and although the glider expends greater energy when climbing instead of running, the high speeds it attains when gliding outweigh the time it expends in climbing, thus reducing the energy cost of movement (Rayner 1986, 1991; Scholey 1986). In other words, it is less expensive in terms of energy to climb up a tree trunk and then fly down to the next one than to climb in both directions (R. Å. Norberg 1981). As an interesting aside, gliding mammals are similar in size to bats, ranging from 10 grams to 1.5 kilos (0.35 oz.–3.3 lbs.) (U. M. Norberg 1990).

Some of the most interesting mammalian parachutists are the two Malagasy lemurs of the genus *Propithecus*, called sifakas for their distinctive cries. *Propithecus verreauxi* and *P. diadema* not only have a small "gliding membrane," or patagium, analogous to that of birds, between the forearm and body but their entire forearm is

Longisquama, the small arboreal thecodontian reptile from the late Triassic of Kirghizia, is the only known Mesozoic reptile with featherlike scales. It is interpreted as having had a unique gliding adaptation, with a double series of long scalelike appendages that were developed along its back. These could be folded and unfolded in butterfly fashion to form a continuous wing. (From Haubold and Buffetaut 1987; courtesy L'Académie des Sciences, Paris)

Adult New World monkeys known as sakis (*Pithecia monachus*), or voladors, with the top saki seen from below in full flight. (From Moynihan 1976; courtesy Princeton University Press)

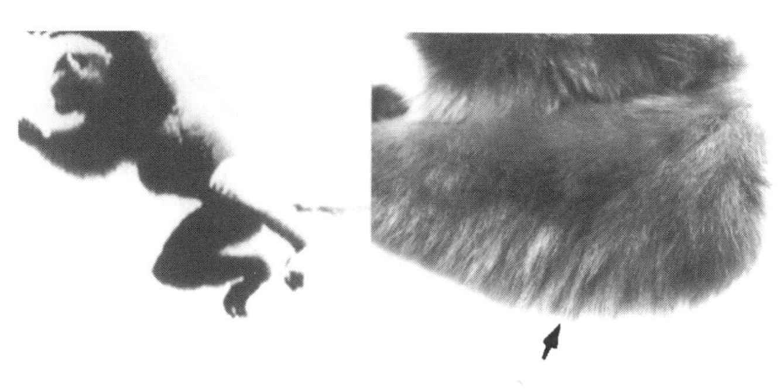

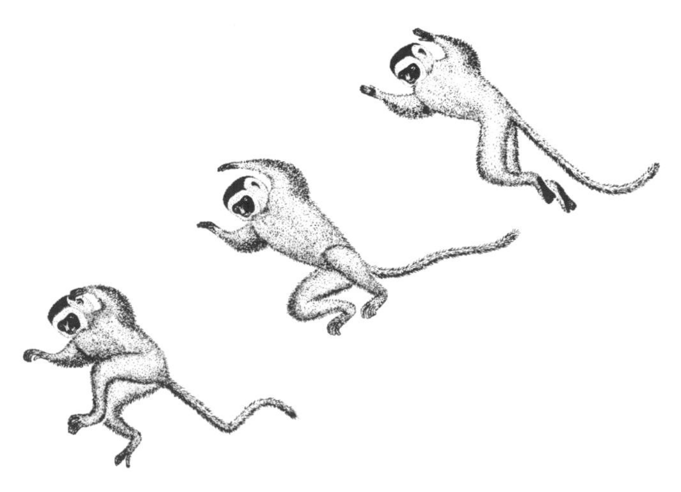

Right, leaping sequence for Verreaux's sifaka (*Propithecus verreauxi*), modified from actual footage; *Above left,* photo taken from action sequence of leaping for *P. verreauxi* (both species average approximately 100 cm (39 in.) in length; *Above right,* photo of left arm of the diadem sifaka (*P. diadema*), showing the extension of hair projecting rearward; *D,* arrow shows point at which measurements in text were taken. (From Feduccia 1993c; courtesy *Journal of Theoretical Biology*)

covered by a thick, rearward-projecting mat of hair that can measure 50 percent or more of the forearm (antebrachium) width. These diurnal lemurs inhabit deciduous and evergreen forests and use their powerful hindlimbs to propel themselves from tree to tree in leaps of up to 10 meters (33 ft.). With forearms outstretched, and using their gliding patagium and rearward projecting forearm hair as an adaptive "proto-wing," these arboreal parachutists are able to increase the gliding area, reduce the wing loading, and accomplish safe landings (Feduccia 1993b). Although these lemurs are too large to represent an actual analogy of flight origins, wind-tunnel experiments on these and other arboreal aerialists could greatly refine the precise aerodynamic adaptations involved (Steve Edinger and Jeff Thomason, pers. comm.).

Among New World primates, which lack prehensile tails, there are numerous semi-aerial adaptations, and the sakis of the genus *Pithecia* are notable among them. *Pithecia monachus*, known locally as "volador" (flier), has a long body, with moderately long arms and legs, and its pelage, or body hair, is a distinctive long, rather coarse, grizzled blackish fur. Sakis prefer mature moist forest with tall trees and mixed forest with moderately tall trees and are usually seen at about 10–35 meters (33–115 ft.) above the ground (Moynihan 1976), where they frequently glide in flying squirrel fashion.

Unlike the lemur parachutists, these vertebrate "gliders" are able to select a tree trunk at a great distance from their takeoff point, steer accurately through the intervening air and maneuver as necessary before striking the target tree, often gliding up at a steep angle just before landing. Parachuting and gliding are advantageous evolutionary adaptations in the context of escaping from predators, but an even more important selective pressure may be to gain optimal foraging ability over a vertical zone in the forest (R. Å. Norberg 1981).

Among mammals, only the bats have developed true flight. Derived from arboreal insectivores, bats have formed their wing from a membrane stretched between their skeletal "fingers" and their hind appendage. The oldest bat, *Icaronycteris index*, is known from the Eocene of North America, some 50 million years ago, and bats are among the most commonly preserved mammals in the Middle Eocene Messel oil shales of Germany. Bats, now represented by some 853 species in 168 genera, diverged into two major lineages, the Megachiroptera, the Old World fruit bats, and the Microchiroptera, which comprises all the rest. Advanced flying abilities and echolocation (among Microchiroptera and one genus of Megachiroptera) have been the hallmark of bat success. In some Neotropical localities there are more species of bats than all other types of mammals combined.

As far as we know, all of these vertebrate forms, living or extinct, parachuting, gliding, or truly flying, share one feature in common: an arboreal (or other elevated place) origin. None began to fly from a purely ground-dwelling habit. This fact alone argues strongly against the theory that birds began flight as runners and jumpers.

Ground-Up Theories

In spite of the obvious objections to the ground-up theory of flight origins, over the years many researchers have advocated the cursorial theory (from the Latin *cursus*, a running, rapid motion) to explain the origin of flight, perhaps to a large extent because of the belief that birds are descended from cursorial dinosaurs that could not have been tree-dwellers. In chapter 2, I discussed myriad problems with a dinosaurian origin of birds on purely morphological grounds; I now examine theories relating to flight origins.

First to propose a cursorial origin of flight was Samuel Wendell Williston, a foremost expert on fossil reptiles and a dinosaur collector for Othniel C. Marsh. "It is not difficult," Williston wrote in 1879, "to understand how the forelegs of a dinosaur might have been changed to wings. During the great extent of time in the Triassic, for which we have scanty records, there may have been a gradual lengthening of the outer fingers and greater development of the scales, thus aiding the animal in running. The further change to feathers would have been easy. The wings must first have been used in running, next in leaping and descending from heights, and finally in soaring" (459).

Williston's rather casual hypothesis enjoyed little support among his contemporaries. Although Marsh had in 1877 supported a dinosaurian origin of birds, the following year, in 1880, he proposed, apparently for the first time, an arboreal theory: "The power of flight probably originated among small arboreal forms of reptilian birds. How this may have commenced, we have an indication in the flight of *Galeopithecus* [*Cynocephalus*, the flying lemur], the flying squirrels (*Pteromys*), the flying lizard (*Draco*) and the flying tree-frog (*Rhacophorus*). In the early arboreal birds, which jumped from branch to branch, even rudimentary feathers on the fore limbs would be an advantage, as they would tend to lengthen a downward leap, or break the force of a fall" (1880, 189). The intuitively facile arboreal theory has recruited the majority of adherents since then and was actually proposed as early as 1859 by Charles Darwin to account for the origin of flight in bats, but the cursorial alternative for the origin of avian flight has had numerous adherents and at least three significant revivals in the last hundred years.

One of these was offered in 1907 (and 1923) by a flamboyant Hungarian nobleman and self-trained Transylvanian paleontologist, Franz Baron Nopcsa (1877–1933), whose biography includes spying in World War I, volunteering his services as heir-designate to the vacant Albanian throne, murdering his homosexual lover, and then killing himself. Nopcsa was among the leading European dinosaur paleontologists of his time, and the government of Romania is now reconstructing the Nopcsa castle to house a new dinosaur museum. Many of Nopcsa's ideas were as colorful as his life. He envisioned the predecessors of birds as long-tailed reptiles that flapped their forelimbs for increasing speed while running across the ground. The reptilian scales on their forearms became elongated in the process, and their hind margins eventually sprouted feathers. In a paper entitled "Ideas on the Origin of Flight," Nopcsa wrote:

> We may quite well suppose that birds originated from bipedal long-tailed cursorial reptiles which during running oared along in the air by flapping their free anterior extremities.
>
> By gradually increasing in size, the enlarged but perhaps horny hypothetical scales of the antibrachial [forearm area] margin would in time enable the yet carnivorous and cursorial ancestor of Birds to take long strides and leaps much in the same manner of the domesticated Goose or Storks when starting, and ultimately develop to actual feathers; this epidermic cover would also raise the temperature of the body, and thus help to increase the mental and bodily activities of these rapacious forms. (1907, 234)

Turning specifically to *Archaeopteryx*, Nopcsa reckoned that the fossil creature, though not an accomplished flier, was quite far removed from the early stages leading to flight: "The rounded contour of the *Archaeopteryx* wing, together with the feebly developed sternum, show us that *Archaeopteryx*, though perhaps not an altogether badly flying creature, can on no account have been a soaring bird, but a bird that was yet in the first stage of active flight" (1907, 16).

In Nopcsa's scenario, the transition from reptiles to birds began with a primitive, quadrupedal animal that could rise up to run on its hind legs, as do the living basilisks, *Basiliscus*, of Central America or the frilled lizards, *Chlamydosaurus*, of Australia. Nopcsa's hypothetical animals then became obligatory bipeds, such as the small theropod dinosaur *Compsognathus*, and from them the forebears of the birds arose. In reality, the basilisks and the Australian frilled lizard are agile tree-climbers, and the basilisks are capable of running across water for short distances to escape predation. What these animals do illustrate, however, is the extreme behavioral plasticity of animals and the impossibility of ascribing behavioral repertoires to fossil animals; animals are always capable of at least twice the behavior that their anatomies would suggest. The Philippine mud-skipper (*Periopthalmus*), a fairly normal looking teleost fish, can walk on land and climb trees! It is, in fact, probable that behavior has preceded anatomy in vertebrate history in the invasion of virtually every major new adaptive zone.

Nopcsa's theory was undermined by its aerodynamic absurdity. Imagine the incredible energy that his creature would have to expend to get airborne, since the cost of running is much greater than climbing and gliding (Rayner 1985a, 1991). And how would thrust be maintained with the legs while the creature was attaining height while jumping off the ground? Nopcsa held that the avian protowing developed not as a true wing but as a sort of propeller, by which the flailing forelimbs, augmented by the development of feathers, would add to the thrust of the running bipedal dinosaur and increase the running speed. Once aloft, however, where would it find the power to stay in flight? Nopcsa simply failed to take into account the fact that the animal would be fighting gravity all the time and that the development of feathers, producing drag, would in the initial stages actually impede the forward progress of the runner. The main thrust, which, in Nopcsa's conception, came from the traction of the hind feet on the ground, would have disappeared. It is highly unlikely that the flapping forearm propellers Nopcsa imagined would have prevented the bird from crashing promptly back to earth.

Decades after Nopcsa's "running terrestrial" theory for the origin of bird flight (1907, 1923), and nearly a hundred years after Williston (1879) advanced the first cursorial theory, John H. Ostrom (1974, 1976b, 1979) proposed a very different version of it, which has been termed the "insect net" theory. Unlike its predecessors, Ostrom's theory was very widely accepted, especially by paleontologists and those advocating a dinosaurian origin of birds, and his view of *Archaeopteryx* as a nonflying reptilian flyswatter is found in many current textbooks. One major general biology text (Villee, Walker, and Barnes 1984, 714), for example, puts the theory thus: "John Ostrom of Yale University presented compelling evidence that birds evolved from coelurosaurs, a group of early saurischian dinosaurs. It is possible that the ancestors of birds were becoming more active and possibly warm-blooded, and feathers may have first been of value in helping to conserve heat . . . . their enlargement, along the posterior edge of the forelimb, enabled bird ancestors to use these limbs as nets to seize insects. Further enlargement of wing and tail feathers may have conferred stability in running rapidly along the ground. . . . On reaching a certain threshold of size, they could be used for true flight."

Ostrom reasoned from features of *Archaeopteryx* anatomy and from features of dinosaurs contemporary with it, as we saw in chapter 2. Along with some other paleontologists, he argued that warm-bloodedness, or endothermy, first evolved among dinosaurs. The first feathers, in this scheme, served certain groups of dinosaurs as a thermoregulatory pelt. Accordingly, the small theropod dinosaurs that *Archaeopteryx* at least superficially resembles anatomically—the coelurosaurians—would have been warm-blooded animals, and the first feathers covered them not as aids to flight but as insulating material. The head and mouth of *Archaeopteryx*, Ostrom argues, indicate that it preyed on relatively small animals, such as insects, lizards, and small mammals. Running after such creatures on its two hind legs, *Archaeopteryx* used its forelimbs to catch them. In time, elongation of the forelimb feathers made them more efficient for trapping prey. They became a kind of butterfly net that *Archaeopteryx* used to corral its prey.

Over the years Ostrom's views have changed in detail, but he has steadfastly maintained that *Archaeopteryx* and protobirds were terrestrial cursors and that avian flight originated from the ground up:

1970. Unique preservation of the horny sheaths of the manus claws provides new evidence that may be relevant to the question of the origins of avian flight. Tentative interpretation suggests a cursorial rather than [an] arboreal origin. (537)

1974. I conclude that *Archaeopteryx* was not capable of powered flight other than that of fluttering leaps while assaulting its prey. *Archaeopteryx* was a ground-dwelling, bipedal, cursorial predator . . . probably not very different from its contemporary, *Compsognathus*. (46)

1978. *Archaeopteryx* supports two theories: warm-bloodedness in dinosaurs and dinosaurian ancestry of birds. The theropods . . . and especially the smaller kinds like *Compsognathus*, *Deinonychus*, and the struthiomimids, I suspect may have been true endotherms. And add to that the evidence that *Archaeopteryx* and other birds evolved from a small theropod dinosaur. (1978b, 168, 178)

1985. It is my conviction that *Archaeopteryx* was still learning to fly—from the ground up—and that avian flight began in a running, leaping, ground-dwelling biped. . . . this animal (*Archaeopteryx*) was a highly adapted bipedal and cursorial ground-dwelling predator. That lends powerful credence to the cursorial theory of the origin of bird flight. (1985b, 169, 174)

1986. There is no evidence—and I underline the word *evidence*—that bird flight began from the trees down. What evidence there is all points to a highly adapted bipedal cursor and a "ground-up" origin of flight. . . . the actual physical evidence . . . points very strongly toward a cursorial origin of avian flight. (1986a, 80–81)

1994. It does appear that "predatory hand movements" such as grasping action by swift-moving bipedal cursors was one of the key stages in the evolution of avian powered flight—somewhere in its pre-flight ancestry. . . . However this scenario plays, it does dictate the certainty of a bipedal pre-avian, probably predaceous organism. The best candidate (if not the only candidate) is a coelurosaurian theropod—all of which were cursors (necessarily so) after prey. (175)

As stated, in this account of protobird evolution, the feathers that served first for thermoregulation acquired a second, unrelated function when they began helping to trap prey. This was an instance of what is now being termed *exaptation* (Gould and Vrba 1982): a structure evolved for one specific function (insulation) acquired a form that made it easy to assume a second, biologically unrelated one (insect trapping). In this case, the elongated feathers, by providing lift to the animal during running and leaping, began to serve the purpose of flight, another case of exaptation.

How exaptation differs from preadaptation needs a little explaining here if we are to understand Ostrom's theory. In 1982, Stephen Jay Gould of Harvard and Elizabeth Vrba of Yale (1982; also see Lewin 1982) raised objections to the term *preadaptation*, which they claim overemphasizes the power of natural selection and incorporates an adaptationist notion that all features are specifically adapted for a particular function. They have suggested that *adaptation* be restricted to features that have been built by selection for their current role and that *exaptation* be used for "features that now enhance fitness but were not built by natural selection for their current role" (4). Such features would have performed a different function

Ostrom's cursorial predator theory for *Archaeopteryx* envisions it as a small terrestrial theropod dinosaur using its wings as an insect trap and has been accepted by the paleontological community, to the extent that the group of theropods ancestral to birds, the dromaeosaurs and *Archaeopteryx,* are placed in the Maniraptora (hand-raptors). (*Upper left,* modified after Ostrom 1979; *upper right,* modified after Briggs 1991; *lower left,* modified after Bakker 1975; *lower right,* modified after Sereno from Monastersky 1990)

The double-crested basilisk (*Basiliscus plumbifrons*), a tropical American lizard that can rear up and run on its hind legs in a semierect manner. Nopcsa thought that an animal much like the basilisk was the first step in the transition from reptile to bird. In reality, basilisks and the Australian frilled lizards are adept tree-climbers, and basilisks have special adaptations of the feet that permit them to run over the surface of water for short distances. It is almost impossible to determine an animal's behavioral repertoire by a fossil specimen; behavior probably always precedes anatomical change in the invasion of new adaptive zones. (Photo by New York Zoological Society)

Mark Hallett's painting graphically illustrating the dinosaurian, cursorial origin of avian flight. (From Funston 1992; © 1984 Mark Hallett, all rights reserved; photo by Veronica Tagland)

in ancestors (this is classical preadaptation) or would represent a nonfunctional structure that was later available for some new function. Gould and Vrba have thus argued that one simply cannot infer the historical genesis of a feature from its current utility. Their primary example, however, is the evolution of feathers, the basic design of which they claim "is an adaptation for thermoregulation and, later, an exaptation for catching insects" (8). They thus accept Ostrom's view that hot-blooded dinosaurs became insulated with feathers that were available later for novel functions.

The evolution of insect wings in essence presents the same problem: how to explain the micro-elongation of feathers or wings as being adaptive at each successive stage—that is, without having to invoke a saltatory jump in the process of evolution. In the case of insect wings, there appears to be reasonable evidence that the incipient ancestral wings functioned primarily for thermoregulation (Kingsolver and Koehl 1985; Lewin 1985) and that insects, by simply increasing in size, may have created aerodynamic wings without any change in body shape or relative wing length. Accordingly, the modern insect wing, used for flight, would be an example of Gould and Vrba's exaptation.

John Ostrom's view is that not only feathers but also the avian wing evolved first for a function other than its current use. He has reasoned that the wing stroke involved in trapping insects was similar to the forearm movement of such theropods as *Deinonychus*, which was, according to Padian (1985), similar to a bird's flight stroke. Ostrom summarizes:

> It is my contention that *Archaeopteryx* was not especially arboreal in its habits, but rather was a very active, fleet-footed, bipedal, cursorial predator in which the hands, arms, and pectoral arch were primarily for seizing and holding small prey, as was almost certainly the situation in *Ornitholestes*, *Velociraptor*, *Deinonychus* and other small theropods. . . . I suggest that it was the prior release of the forelimb from normal terrestrial locomotion (probably for purposes of predation) and its modification into an elongated, predatory, grasping appendage with *strong powers of abduction* that pre-adapted the forelimb as a "proto-wing." (1974, 34)

Ostrom continues the argument as follows:

> If vigorous flapping of the feathered forelimbs played a part at any stage in the business of catching prey, the increased surface area of the enlarged contour feathers would undoubtedly have produced some lift during such assaults. From this point, it is a small evolutionary step for selection to improve those features that were important for flapping, leaping attacks on prey—perhaps to "fly up" after escaping insects: e.g., enlargement of the primaries and secondaries and their firm, rigid attachment to the forelimb skeleton; elongation and specialization of the bones of the forelimb and hand; retention of the theropodlike scapula and coracoids; enlargement of the pectoral abductor muscles; and stabilization of the shoulder joints by fusion of the clavicle. . . . Thus, selection would tend to improve not only the "flight power," but also the "flight controls"—the associated sensory and motor neural components. (1974, 35)

Ostrom discounts certain features of *Archaeopteryx*'s anatomy that suggest that the animal was arboreal, not ground-dwelling, for instance, the hind toe, or hallux. But, as we have seen, and shall continue to see, *Archaeopteryx* was much more birdlike in many features than has been previously thought. In *Archaeopteryx* the hallux is turned backward, as in all modern birds. This is an adaptation that characterizes animals that make their way through trees; it helps them grasp the branches. Ostrom, however, interprets the feature differently (1974). Noting that the backward-directed toe "apparently was present in all carnivorous theropods, but never existed in equally bipedal but herbivorous ornithopods," he concludes that the original function of the reversed toe was related not to climbing but to diet. The foot with the backward-directed toe was useful in capturing prey. However, the view that any theropods had a hallux has since been seriously challenged (Martin 1991; Tarsitano 1991; Feduccia 1993a), and there is no current acceptable evidence for it. As Walter Bock of Columbia University, the foremost proponent of the arboreal theory of avian flight, has noted, "Without a doubt the hallux in *Archaeopteryx* is reversed, and its foot functioned as a grasping one. Specialized ground-dwelling birds tend to reduce and lose the reversed hallux, and . . . it is difficult to conceive of aspects of terrestrial-dwelling that would have provided the selective agents needed for the evolution of a reversed hallux" (1986, 2).

Although Nopcsa's "running terrestrial" and Ostrom's "insect net" theories have enjoyed support at various times, a new twist on the cursorial theory came with the publication by Gerald Caple, Russell Balda, and William Willis in 1983 (see also 1984 and Balda et al. 1985) of what may be termed the "terrestrial leaping" (or fluttering) theory for the origin of avian flight. So popular was this version of the cursorial theory that in 1983 Ostrom abandoned his insect net idea in favor of this new scenario (Lewin 1983). These authors argue emphatically that a gliding stage cannot evolve from parachuting and that flapping flight cannot evolve from gliding flight, therefore, that avian flight could not have evolved from the trees down. However, they do not really explain the large evolutionary step from leaping, assisted by flapping, to fully developed flapping flight. In this theory, the avian feathered wing and concomitant flapping evolved in creatures with a bipedal cursorial habit to control body orientation during leaping and in landing. Feathered wings and flapping, argue the authors, would increase the distance of the leap and therefore the running creature's ability to capture insect prey. Because selection pressure was directed toward enhancing the ability to orient and stabilize the body, the evolution of the wing and flight feathers would have begun from the outer, or distal, end, and so the primary feathers must have evolved before the secondaries. Caple, Balda, and Willis's scenario assumes a small cursor with a rigid lightweight body. Their position is that "the result of these collective changes will lead to powered flight" (1983, 473). For them, the most appropriate selection pressure for the enhancement of wings is stability while running at high speed, perhaps to escape predation.

Objections to this terrestrial leaping or fluttering model have emerged primarily from Jeremy Rayner of the University of Bristol and Walter Bock. Rayner has calculated (1985a, 1991) that for a bipedal, cursorial theropod with a mass of 0.2 kilos (4.4 lbs.), running at speeds up to 2 meters (6.6 ft.) per second, the drop in running speed associated with the jump would be on the order of 30–40 percent, presenting very serious problems in terms of attaining any type of flight. According to Rayner, "The first 'flights' of a fluttering proto-flapper would have been at low speeds, where the energetic demands of flight are at their most extreme (Clark, 1977), and the wingbeat cycle is at its most complex. The fluttering model fails because it takes no account of the extreme morphological, physiological and behavioural specializations required for flight" (1988, 278). Rayner further argues that although the strategy of running, jumping, and gliding is attractive from the standpoint of energy use, it is apparently too slow to favor flying; "it has insufficient energy to reach speeds at which flapping is mechanically straightforward, and the costs of flight at these low speeds are so high that the demands on the forelimb musculature become extreme. . . . Given the simultaneous problems of attaining sufficient speed and sufficiently long jumps, it is unlikely that a cursorial runner could have begun to fly" (1988, 280–281).

Bock (1986) likewise has criticized the model, but primarily because it fails to provide a suitable pseudophylogeny, or chronological sequence of intermediate forms, using analogous known organisms for the various stages in the flight evolution scenario and because of the large, unexplained evolutionary jump between leaping and powered flight. As Bock writes, "I know of no small tetrapods about the size of *Archaeopteryx* that are primarily terrestrial (e.g., not flying-running forms, or secondarily flightless or degenerate flying forms) and use their forelimbs for balance during fast running or during a leap. And I know of none using the forelimbs as flapping structures to provide forward thrust to increase the length of its leap" (1986, 68).

A central argument against the arboreal theory for the origin of avian flight has been the insistence of certain authors (for example, Caple et al. 1983; Balda et al. 1985) that lift cannot be achieved when a gliding animal starts to flap its proto-wings. However, Ulla Norberg (1985, 1986), using an elegant aerodynamic model, has shown that the transition to active, powered flight from gliding is me-

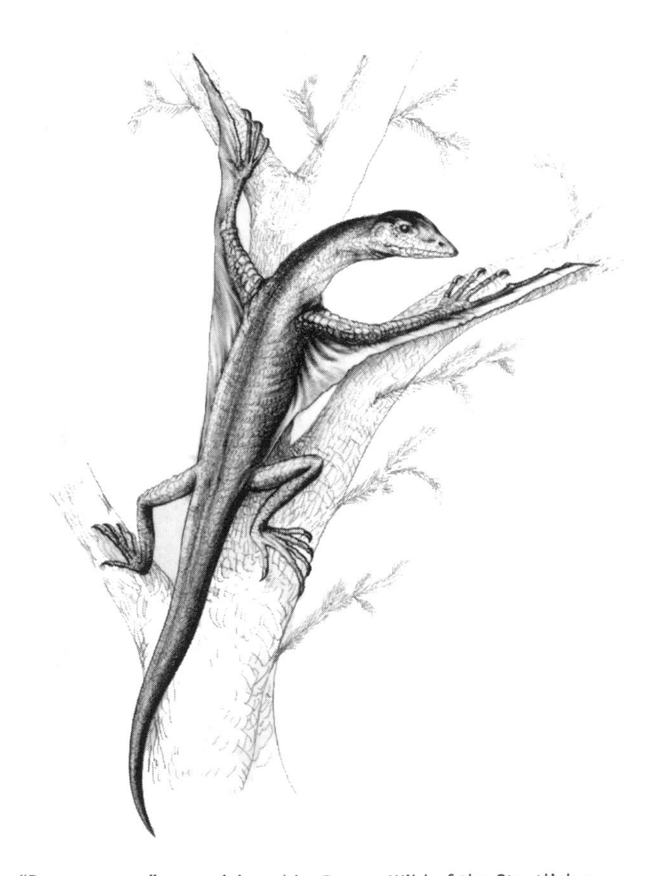

"Propterosaur," as envisioned by Rupert Wild of the Staatliches Museum für Naturkunde, Stuttgart, illustrates the ancestor of pterosaurs as a small, arboreal reptile that lived in the late Permian or early Triassic. As a glider, it had developed as another "arboreal experiment" lateral skin folds that stretched between the forelimbs and hindlimbs and were attached to the elongated fourth finger. The fifth finger had been lost, but the first three fingers were free-moving and could be used in climbing. To add to the surface area of the "wing," additional skin folds may have developed in front of the forelimbs (analogous to the avian propatagium), as well as between the tail and the hindlimbs. (From Wild 1984; courtesy Rupert Wild and Springer-Verlag)

chanically and aerodynamically very feasible. In fact, "for every step along the hypothetical route from gliding, through stages of incipient flapping, to fully powered flight, there would have been an advantage over previous stages in terms of length and control of the flight path" (U. M. Norberg 1990, 260). Gliders the size of *Archaeopteryx* could clearly have begun flapping flight and thus acquire forward thrust without any disadvantages claimed by advocates of the cursorial theory.

In sum, all of these cursorial theories share an assumption that protobirds and *Archaeopteryx* are specialized terrestrial, bipedal organisms, and they make much of the fact that *Archaeopteryx* is little more than a feathered dinosaur. These assumptions are probably wrong, as we shall see.

An interesting aside relates to the origin of pterosaurs, which descended from eosuchian or early archosaurian stock, perhaps during the Upper Permian or Lower Triassic. Pterosaurs have variously been thought to be batlike or birdlike in their terrestrial locomotion—that is, quadrupedal or bipedal—and the arguments have raged since the 1870s (Wellnhofer 1991). However, recent evidence from biomechanics indicates that they could only have been quadrupedal (U. M. Norberg 1990). Yet Kevin Padian (1983, 1984) has argued that the pterosaurs were bipedal and that their flight originated from the ground up. Many cladists have adopted Padian's view, including Paul Sereno (1991), who, apparently having adopted the ground-up scenario for bird origins, states that "bipedalism . . . may be a functional prerequisite for achieving powered flight from the ground up; the forelimbs of a biped are freed from the functional constraints of quadrupedal locomotion" (1991, 45). Pterosaur expert Rupert Wild (1983, 1984), however, has argued that a small, arboreal, climbing reptile is a perfectly suitable ancestor for pterosaurs. In the hypothetical "protopterosaur" he envisions, body surface area increased by expanding skin folds to reduce the rate of sinking while gliding. A lateral skin fold extended from the rear of the forelimbs and the fourth digit of the hand and eventually encompassed the flanks and upper hindlimbs. The fourth finger grew much longer, the fifth digit regressed, and the other fingers remained much the same length, while acquiring sharp claws adapted for climbing. The flight membrane was immediately advantageous in initially checking falls and promoting safe landings. As the wing expanded, the flight membrane increased in size and digit 4 lengthened to some twenty times the length of the others; the advent of gliding allowed the protopterosaur to exploit new space in its search for food. Ultimately, with the evolution of an ossified, keeled sternum and a strengthened pectoral girdle, powered flight emerged.

The Birdlike Features of *Archaeopteryx*

As described in chapter 2, most recent studies have shown *Archaeopteryx* to be much more birdlike than previously thought. The head, skull, and brain are quite birdlike, less reptilelike (Whetstone 1983; Bühler 1985; Walker 1985; Haubitz et al. 1988), while the reversed hallux is a feature found only in birds and known in no dinosaur. As noted in chapter 1, the hallux is a feature associated with life in the trees: digit 1, the hind toe, is opposed to the front three, making an effective grasping mechanism for perching in trees. Ostrom has maintained that the small, ground-dwelling, chickenlike dinosaur *Compsognathus* had a reversed hallux, but the thrust of his argument was based on his maintaining that this dinosaur's hallux is *not* preserved in the correct anatomical position. This special pleading is not convincing and has been seriously questioned by Sam Tarsitano (1991) and Larry Martin (1991). It is extremely difficult to imagine a situation in a terrestrial dinosaur in which a reversed, opposable rear toe would be selectively advantageous. Indeed, a reversed hallux would be a hindrance to creatures that ran; this is apparent in cursorial ground-dwelling birds, which have repeatedly reduced or lost the hallux. As Walter Bock has pointed out, "The reversed hallux seen in *Archaeopteryx* and presumably present in protobirds can be best interpreted as a feature used for grasping branches during arboreal locomotion; hence the reversed hallux is associated with life in trees" (1986, 62).

There is also the question of obligate bipedality, which John Ostrom has maintained all along is characteristic not only of birds and *Archaeopteryx* but of their supposed close relatives, the theropods (Ostrom 1991b). However, Bock (1986) has questioned whether the bipedal features of birds are all that specialized in *Archaeopteryx*, noting that the pelvis of *Archaeopteryx* is not as large as in modern birds and that the bones are not highly fused. The synsacrum is also smaller and less fused, and the pubis is not fully reversed. These features suggest to Bock that bipedal locomotion in *Archaeopteryx* was less specialized than in modern birds. He further hypothesized that *Archaeopteryx*'s lack of a rigid trunk skeleton, along with smaller and less-fused pectoral and pelvic girdles, may indicate that climbing in trees—that is, quadrupedal locomotion—may have been an important part of *Archaeopteryx*'s locomotion. Others hold slightly differing views. Considering the lateral compression of the pelvis and other features, Larry Martin (1991) has suggested a slightly sprawled hindlimb arrangement for *Archaeopteryx*, and Sam Tarsitano (1985, 1991) holds that bipedality evolved concurrently with flight and not before. No matter how *Archaeopteryx* moved, there is little doubt that both *Ar-*

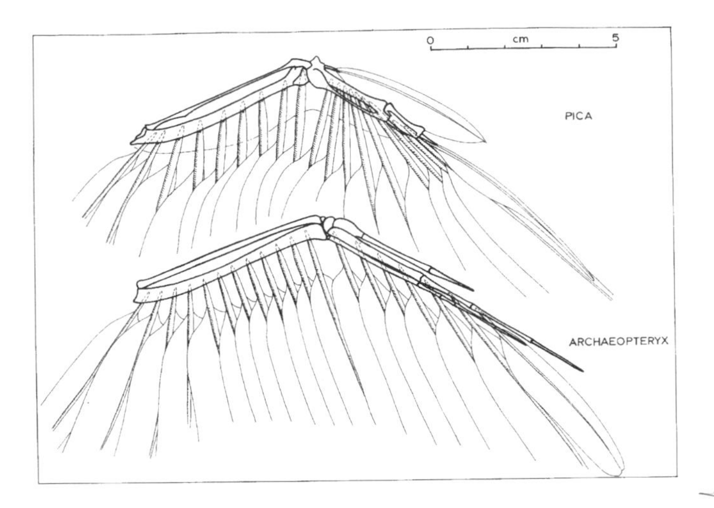

Scale drawings, ventral view, of the forearms of a magpie (*Pica*) and *Archaeopteryx,* showing the insertion of the remiges. The remiges, which insert dorsally on the bones of the hand in *Pica,* and presumably did also in *Archaeopteryx,* are shown in dotted lines where they pass behind the bones. The number of primaries ranges in publications from six to twelve; Rietschel (1985) states that there are eleven primaries, with a twelfth one suspected. Yalden (1985) recognizes, like a number of previous authors, that there are seven primaries in the Berlin specimen. Given the near-modern arrangement of flight feathers in *Archaeopteryx,* it would stretch credulity to assume that *Archaeopteryx* could not fold its wings. (From Yalden 1985; courtesy G. Viohl and Jura-Museum)

chaeopteryx and protobirds could have climbed tree trunks whether they were bipedal or quadrupedal. Bipedal climbers exist today—witness the tree kangaroos (*Dendrolagus*) of New Guinea and Queensland (Bock 1986)—and bipedal trunk-climbers are common among such birds as woodpeckers and creepers.

Also disputed is *Archaeopteryx*'s relative ability to fold the wings. In modern birds the wrist joint is highly modified to permit hyperflexion of the manus against the forearm so that the wings can be folded against the body during rest and during the recovery phase of the flight stroke. Paul Sereno and Chenggang Rao have maintained that *Archaeopteryx* was little more than a "maniraptoran theropod" and that it was incapable of flexing the manus and forearm at an angle less than 90°: "We maintain that *Archaeopteryx* could not fold its wing in a fashion similar to that in volant living birds" (1992, 847). They based this conclusion partly on biomechanical reasoning and partly on the fact that the manus and forearm are preserved in the articulated *Archaeopteryx* specimens at an angle of flexion less than 90°. However, the primaries and secondaries of *Archaeopteryx* and other birds are designed so that the wing can be folded in the manner of an old-fashioned hand fan; as Burkhard Stephan noted, "The rather long primaries and secondaries presuppose an efficient manner of closing the wings and of holding them tightly to the body" (1985, 263). And Larry Martin (1983b, 1985) has shown that a modern avian osteological system for wing folding is present in *Archaeopteryx*. Although the *Archaeopteryx* wing appears to be incapable of flexion beyond the point in which the wings are preserved, one must remember that after death the cartilage and tendinous connections contract, leaving a distorted impression of the original structure; indeed, many Tertiary bird fossils are preserved with the wing opened in *Archaeopteryx* fashion. As the lesson of the basilisk has taught, it is extremely difficult to place biomechanical constraints on fossils without having seen the living creature in action. Most living beings are capable of a magnitude of behaviors that could not be predicted by anatomy alone.

Claws Provide Important Clues. A critical aspect of the anatomy of *Archaeopteryx*, one that should yield significant information on its paleobiology, is its claw structure. Although other authors, including Burkhard Stephan of Berlin's Humboldt Museum für Naturkunde (1987), Stefan Peters and Ernst Görgner of Frankfurt's Senckenberg Museum (1992), and Derek Yalden of the University of Manchester (1985), have studied the claws of *Archaeopteryx* from various points of view, more recently I conducted a study of the pes claws of modern birds and *Archaeopteryx* (Feduccia 1993a) to ascertain if ground birds were separable from tree dwellers and to see just where *Archaeopteryx* fit into the hierarchy. My study covered more than five hundred species of birds from a great variety of ecological types, and the middle toe claw of each was photographed and measured for an average claw arc.

The degree of claw arc curvature is a general indication of habit: ground dwellers have flat, straight claws, perching birds have moderately curved claws, and trunk climbers have the more highly curved claws. The major division of birds on the basis of claw arch curvature, however, is between ground dwellers and all others. Perching birds and trunk climbers are generally separable, with some overlapping. The pes claws of the middle digit (digit 3) of the measurable specimens of *Archaeopteryx* (from London, Berlin, and Eichstätt), conservatively reconstructed, are 125°, 120°, and 115°, for a composite mean of 120°. The curvatures of the middle claws of the manus of the measurable *Archaeopteryx* fossils (from Teyler, Berlin, and Solnhofen) measured 155°, 142°, and 145°, giving a composite mean of 147.3°. Clearly, the pes claw curvatures of *Archaeopteryx* fall within the range of perching birds and completely outside the range of ground dwellers, and the manus claws fall within the range of trunk climbers.

Earlier, Derek Yalden (1985) had discovered that the manus claws of *Archaeopteryx* closely resembled those of such trunk-climbing birds as woodpeckers and such trunk-climbing or clinging mammals as the flying lemur

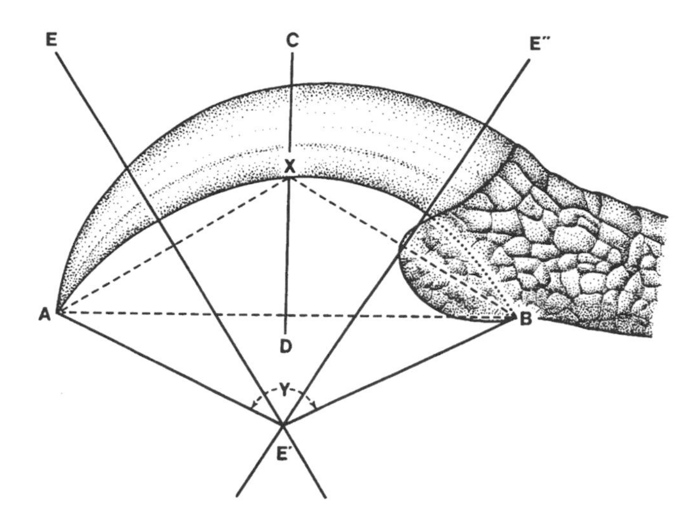

Claws are very complicated geometrically, but for comparative purposes they can be measured as simple arcs. The claw is measured here in terms of relative degrees of a circle. The angle (Y) is a measure of the degrees of arc. (From Feduccia 1993a; copyright 1993 AAAS)

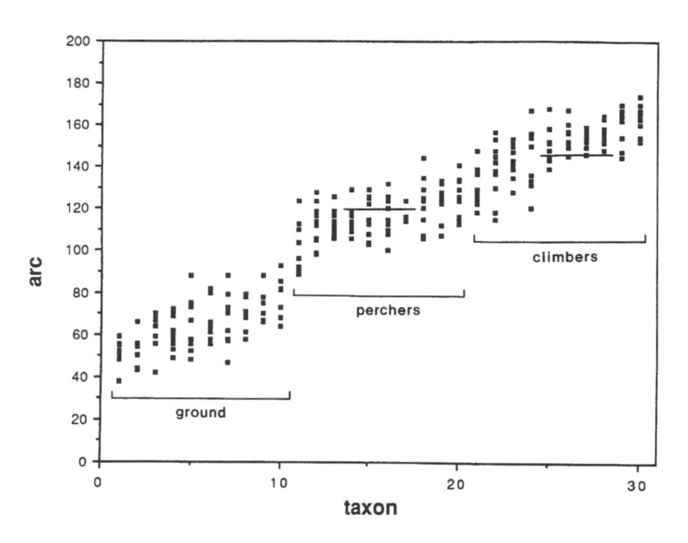

Scatter diagram of 30 species showing individual data points of claw arcs in degrees of curvature. Note the almost complete segregation of the ground dwellers. The mean value for *Archaeopteryx* pes claws is indicated by the line in the center of the perchers, and the mean for manus claws, by the line in the middle of the climbers (to the right of the diagram). (From Feduccia 1993a; copyright 1993 AAAS)

Left, external morphology of the foot of the lyrebird *Menura novaehollandiae*, a predominantly ground-dwelling bird, which shows the distinctive straight claws of a ground dweller; compared with (*right*) the foot of a bowerbird (*Chlamydera nuchalis*), a predominantly perching bird, which shows curved claws like those of *Archaeopteryx*. Not to scale. (From Feduccia 1993a; copyright 1993 AAAS)

Claw (including the bony ungual and the horny sheath) of the third finger (digit 4 of ornithologists), right manus of Teyler *Archaeopteryx*. Note the extreme curvature, lateral compression, and needlelike point. Actual straight length of claw, 10.5 mm (0.4 in.). (Photo courtesy John Ostrom)

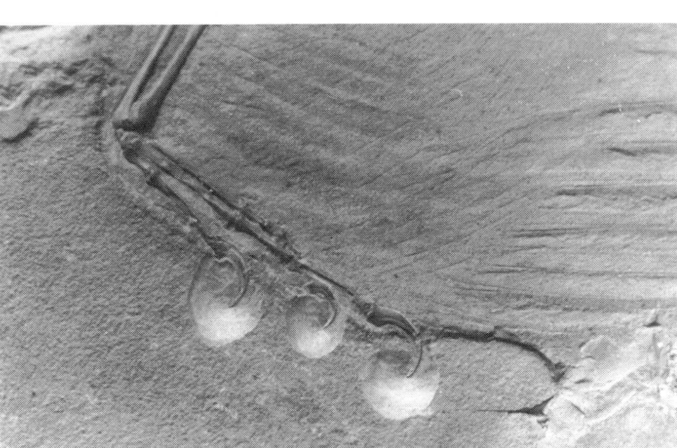

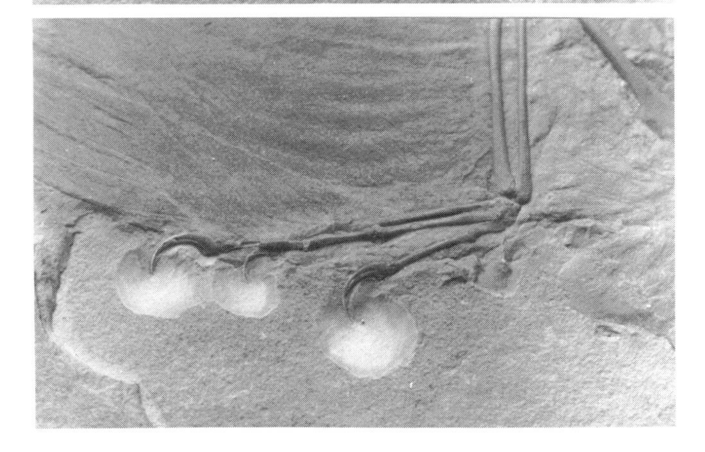

Above, medial view of the left pes of the London *Archaeopteryx,* which shows the strong claw curvature of perching birds. *Below,* dorsal views of the left (*upper*) and right (*lower*) wings and manus claws of the Berlin specimen, which exhibit claws closest in morphology to those of trunk-climbing birds, especially Neotropical woodcreepers (Dendrocolaptidae). (*Above,* courtesy S. Tarsitano; *below,* courtesy B. Stephan and Humboldt Museum für Naturkunde)

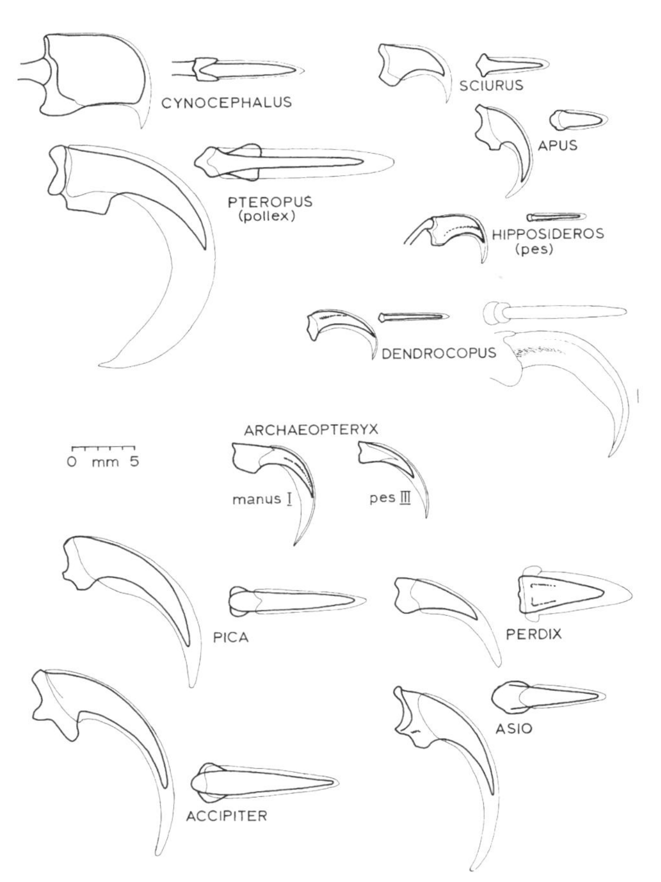

Scale drawings of the ungual phalanges (heavy line) and horny claw (thin line) of a range of birds and mammals, in lateral and extensor views. *Archaeopteryx* claws are shown for comparison. The selection includes a perching bird that forages on the ground (magpie, *Pica pica*); two avian predators that use their feet to grasp prey (long-eared owl, *Asio otus;* sparrowhawk, *Accipiter nisus*); a cliff-nesting bird (alpine swift, *Apus melba*); two trunk-climbing woodpeckers (lesser spotted woodpecker, *Dendrocopus minor;* white-backed woodpecker, *D. leucotus*); two trunk-climbing mammals (gray squirrel, *Sciurus carolinensis;* cobego, *Cynocephalus volans*); two bats (fruit bat, *Pteropus* sp.; leaf-nosed bat, *Hipposideros* sp.); and a ground-dwelling galliform (gray partridge, *Perdix perdix*). The extreme laterally compressed, strongly curved claws with needle-sharp points are characteristic of trunk climbers and do not fit the model of an earthbound predator. (From Yalden 1985; courtesy G. Viohl and Jura-Museum)

(*Cynocephalus*), various gliding opossums of Australia, flying squirrels, and fruit bats. The claws of the hand are morphologically almost identical to the foot claws of the Neotropical woodcreepers (Dendrocolaptidae), and they differ from those of predatory dinosaurs in that they exhibit extreme lateral compression and have needlelike points. The claws of predators, including theropod dinosaurs, tend to be more conical and smoothly tapered, while predatory birds have wide foot claws with lateral cutting edges. The conclusion seems inescapable: the manus claws of *Archaeopteryx* were used in climbing and clinging to trunks, and even more perhaps in clinging to branches, because *Archaeopteryx* had not yet achieved the balance that is characteristic of modern birds and its center of gravity was more reptilian.

Yalden (1985) has greatly clarified the orientation of the digits and hand claws as they appear in the preserved specimens of *Archaeopteryx*. These claws have toppled or been twisted to lie on their sides, and in most specimens digit 3 is overlain by digit 2, which is clearly not a natural condition but the result of this toppling, as the Maxberg *Archaeopteryx* confirms. The orientation preserved in the Maxberg specimen would indicate that in life the claws pointed ventrally from below the wing surface. Although this would not preclude a predatory function, no other orientation would be suitable for trunk climbing.

There can be no doubt that the pes claws of *Archaeopteryx* were those of a perching bird; the claw arcs are completely outside the range of ground-dwelling birds, and any other interpretation would stretch credulity. An interesting comparison has frequently been made between the morphological proportions of *Archaeopteryx* and such living birds as magpies, chachalacas, and coucals of the genus *Centropus*, particularly the pheasant coucal (*Centropus phasianinus*) of Australia and New Guinea. In profile this coucal, which inhabits thickets and is primarily terrestrial, is a dead ringer for the Berlin urvogel, with nearly identical size and proportions. Unlike *Archaeopteryx*, however, and like other ground-dwelling birds, the pheasant coucal has a loosely constructed tail, frayed at the tip, and a loosely constructed wing, with less ellipticity than *Archaeopteryx*. In addition, its claw arc measurements average 85.7°, well within the range of ground dwellers and outside the range of *Archaeopteryx*.

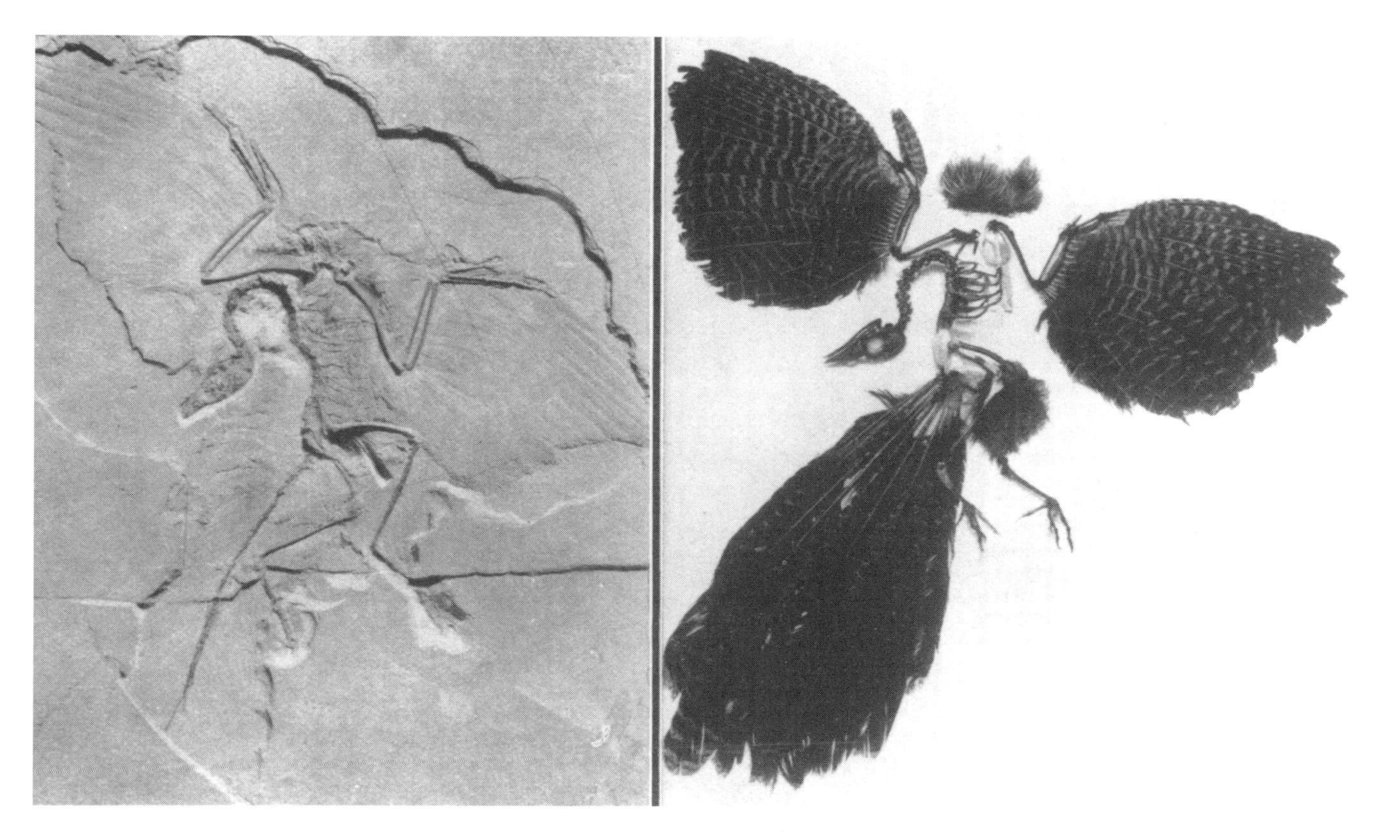

Archaeopteryx look-alikes. The Berlin *Archaeopteryx, left,* compared with its "look-alike," the pheasant coucal, *Centropus phasianinus,* of Australia, *right,* prepared in the same posture, showing remarkable superficial similarity in overall form and proportions. But the pheasant coucal is predominantly terrestrial. Therefore, unlike *Archaeopteryx,* it has a loosely constructed tail that is normally frayed at the tip and a loosely constructed, less elliptical wing. Its claw arc measurements give a mean of 85.7° (N=10), typical of ground dwellers (Feduccia 1993a). (From Heinroth 1923; courtesy *Journal für Ornithologie*)

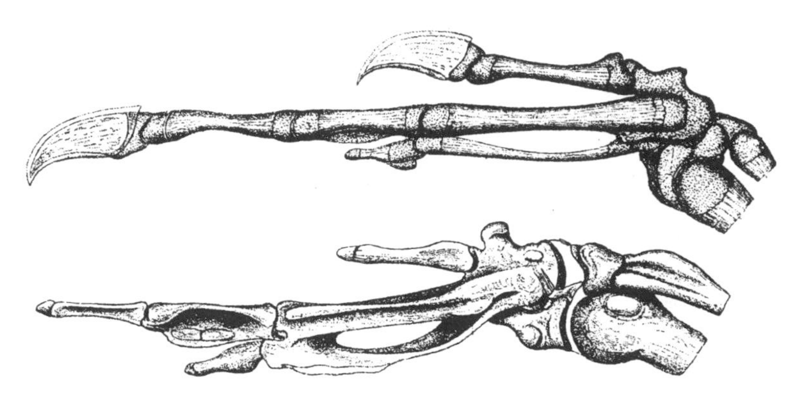

The left hand of a hoatzin embryo, *above,* compared with that of the adult, *below*. (From Heilmann 1926)

Above, a baby hoatzin (*Opisthocomus hoazin*) using its claws to climb back up to the nest; *below,* the adult hoatzin. The two claws on the wings of the baby hoatzin enable it to clamber about the limbs around the nest. At any sign of danger, the baby plops into the water and then uses its claws to return to the nest. (Photo by Stuart Strahl)

Below, left to right, dorsal views of the left hand of *Archaeopteryx,* a nestling hoatzin, and a young pigeon. *Above,* Heilmann's reconstruction of the outstretched wing of *Archaeopteryx* in dorsal and ventral views to illustrate the external anatomical relation of the claws to the wing. The insert illustrates the position of the claws. (From Heilmann 1926)

In 1985, Günter Viohl ruled out "trunk climbing as a mode of life of *Archaeopteryx*" (1985a, 31) because of the lack of evidence of forests in the Solnhofen deposits: no preserved tree trunks have been discovered there. Not surprisingly, this view was readily embraced by advocates of a dinosaurian origin of birds and a cursorial origin of avian flight, and it led Jeremy Rayner to propose a model of flight origin with the protobird running and gliding into a wind (!), because: (1) "the avian ancestors were cursorial"; (2) "*Archaeopteryx* did not climb"; and (3) "there was no suitable habitat for a climbing proto-bird in the Solenhofen vicinity" (1991, 206). Since all three assumptions are wrong, how could this model for the origin of flight be correct? Rayner further argued that *Archaeopteryx* could not have been a climber because, as he put it, "the recently discovered 'Solnhofen' specimen . . . shows exceptionally preserved keratinous claw sheaths, which show no sign of the abrasion to be expected from climbing (J. M. V. Rayner and K. Padian, personal observations)" (191). This criticism is easily discarded: that lack of wear has little bearing on claw use; in fact, trunk-climbing birds and mammals exhibit little wear on their claws, because the keratin is being constantly redeposited to conform to the active claw phenotype of the species. Indeed, it is the lack of wear that leads to claw deformity—cats, to take but one example, continually sharpen their claws on bark to keep the ends sharp. This phenomenon is easily demonstrated by examining the claws of birds and mammals that are confined in zoos, where lack of normal use frequently produces grossly deformed and elongated claws.

Trees at Solnhofen. The lack of evidence for trees at Solnhofen led Stefan Peters and Ernst Görgner (1992) to interpret the curved manus claws of *Archaeopteryx* as an adaptation for climbing cliffs, strengthening the hypothesis that avian flight began from protobirds launching from cliffs or ridges.

The problem in Solnhofen paleoecology is that the absence of tree trunks, as in so many fossil sites, cannot be used to prove the absence of trees. Dead foliage, trees, parts of trunks, and other remnants would not be found in the limestone deposits if no streams entered the ancient lagoon, as is often the case today with lagoons in drier regions. The lack of sources to bring flora into the lagoon for preservation probably means that we have grossly underestimated the plant life in the region. In this semiarid bushland setting the major vegetative areas were probably around springs and watercourses, not near the sea edge, where preservation would have been possible. We do know that the flora did not change appreciably during the Jurassic period (Vakhrameev 1991), much less during the late Jurassic, and there was forest inland from arid lagoons. If the flora from the late Jurassic but slightly older Nusplingen deposits, some 250 kilometers (150 mi.) southwest of Solnhofen, is combined with that of the Solnhofen deposit, a reasonably accurate picture of a moderately large flora emerges. The flora includes many types of arborescent, tree-like plants, including numerous seed ferns (pteridosperms, including *Cycadopteris*), conifers (coniferopsids, including *Brachyphyllum* and *Palaeocyparis*), Bennettitales (*Zamites*), and ginkgoes (*Ginkgo* and *Baiera*), any of which could have provided suitable places to perch and trunks to climb. Certain conifers were bushlike, such as the coastal halophytes *Brachyphyllum* and *Palaeocyparis* (Mutschler 1927; Jung 1974; Viohl 1985a; Barthel, Swinburne, and Morris 1990), and probably did not exceed 3 meters (10 ft.) in height. These are among the most frequently recovered land plants, and Jung (1974) has shown that they were stem succulents, ecologically adapted like cactus, with a woody central cylinder surrounded by water-storing tissue. Plants such as *Cycadopteris* grew perhaps 10 or so meters (33 ft.), and forms of *Zamites* were substantial, arborescent plants. Unfortunately, the best ginkgo material from Solnhofen was lost during World War II, according to Barthel and colleagues (1990, 107). However, ginkgos (including the genus *Ginkgo*) are well known from Europe throughout the Jurassic, and there can be no doubt that *Ginkgo* occurred at Solnhofen and Nusplingen during the late Jurassic. Ginkgoes, typified by today's *Ginkgo biloba*, a large, ornamental, gymnospermous tree native to China and Japan, may grow up to some 20 meters (66 ft.), and ancient ginkgoes could have provided not only suitable perches but myriad branches that an urvogel could have clung to.

More than a century ago, the long clawed fingers of *Archaeopteryx* intrigued C. H. Hurst, a paleontologist interested in fossil reptiles; he proposed in 1893 that they could have been used for climbing in trees. A year later, William Plane Pycraft (1894), a distinguished English ornithologist, suggested that claws would be useful only to nestlings, which would need them for climbing back to the nest in the event of an unlucky fall. Pycraft drew an analogy with the hoatzin (*Opisthocomus hoazin*), a strange South American bird that nests gregariously in bushes overhanging streams of the Amazon basin. At hatching, the young of this species have well-developed claws on both the second and third digits (Parker 1891), and Pycraft (1903) reported a third claw on an early prehatchling chick. Soon after hatching, the nestlings use their claws to clamber about the branches near the nest. When threatened with danger, the young birds fling themselves into the water, swim back to the riverbank, and then use their claws to climb back up the tree to the nest. However, as the young birds develop, the claws regress and the wings develop like other modern birds.

At the time Pycraft wrote his paper (1894), the hoatzin's claws were thought to be a primitive feature retained from *Archaeopteryx*. Now, however, it is generally believed that the hoatzin's claws evolved secondarily. Whatever the case, the young hoatzin's claws may well be analogous to some stage in evolution preceding *Archaeopteryx*, and the same genetic background that produced the claws of *Archaeopteryx* has produced a clawed hand that closely resembles the primitive condition, perhaps best termed atavistic. Interestingly, vestigial wing claws are not uncommon among modern birds, especially in hatchlings. Claws are particularly well known in the adult ratites (Parker 1888b; Fisher 1940; Stephan 1992): *Apteryx* (kiwis) with a claw on digit 3 (2 and 4 are absent), *Struthio* (ostriches) with two and sometimes three claws, *Rhea* (rheas) with a claw on digit 2, *Casuarius* (cassowaries) with claws on digits 3 and 4, and *Dromaius* (emus) with a claw on digit 3 (2 and 4 are absent). And wing claws are found in the adults of many other species, including the well-known case of screamers (Anhimidae), and such forms as New World vultures, eagles, hawks, ducks, geese, swans, and jacanas, among others. Claws are in fact widespread in nestlings; Harvey Fisher (1940) was unable to find claws in only six orders.

Although *Archaeopteryx* had well-developed flight feathers, as we shall see, with its wings partly folded, it could have used its claws as an aid in climbing, and it is not difficult to imagine the young *Archaeopteryx* or other primitive fledgling urvogels using their claws as a hoatzin does. It is also possible that the three-clawed hand of *Archaeopteryx* aided flight as well as climbing, and Bock has noted that "the flight feathers, being attached to the posterior edge of the forelimb, would not interfere with a climbing role of the grasping hand" (1985, 204). The fingers might also have acted as a primitive midwing slot, comparable to the alula, or bastard wing, located at the same position in modern birds.

The Arboreal Theory for the Origin of Flight

A theory that contradicts none of the evidence from either *Archaeopteryx* or other fossil finds is the arboreal hypothesis that Othniel C. Marsh first sketched out in 1880. Marsh's views were rather widely accepted by other students of avian evolution and were subsequently elaborated (Steiner 1917; Reichel 1941). The arboreal theory was tremendously bolstered in 1926 by Gerhard Heilmann in *The Origin of Birds*. There Heilmann reconstructed in convincing detail the hypothetical stages of evolution from terrestrial to tree-dwelling to flying animal. But he did not pay much attention to the adaptive advantage of each microevolutionary step that eventually led to the macroevolutionary change from reptile to bird. More recently, Walter Bock (1965, 1985, 1986) has carefully analyzed the arboreal theory and identified the adaptive purpose of each intermediate stage; he has shown vividly that the arboreal theory is compatible with the principle of microevolution. Bock's model is in harmony with the preference of modern theorists for evolutionary pathways that follow simple, direct routes without elaborate intermediate steps.

In Bock's version of the arboreal theory, the ancestral form is a reasonably small, ground-dwelling reptile with either bipedal or quadrupedal locomotion. It might well have been facultatively warm-blooded, using behavioral mechanisms to regulate its body temperature. A critical point in Bock's arboreal theory is the animal's invasion of the trees, which might have occurred for hiding, sleeping overnight safe from predation, or nesting. He views this invasion of trees and the concomitant intrusion into a cooler microclimate as the main impetus for selection favoring the evolution of endothermy and of feathers for insulation. The protoavis took to climbing and the arboreal life, then began leaping from tree to tree. Tree-dwelling animals often fall and are faced with the problem of impact forces when they strike the ground. The initial stages might be as simple as selection favoring any adaptation that might decrease the rate of descent or lessen the impact. Such simple changes as involving the proper orientation of the body—that is, flattening the body and spreading the limbs horizontally—would increase body surface area and lessen the impact. The lengthening of feathers would also improve parachuting ability, and from there the protoavis would expand its repertoire to include not only parachuting but gliding and, finally, active, powered flight. This sequence of events, fully adaptive at each stage, led not only from quadrupedal ambulation to powered flight but from reptile to bird.

As for *Archaeopteryx*, it and other hypothetical protobirds have usually been thought of as either entirely ground-dwelling or entirely arboreal, but there is no reason why they could not have been both, at least in part. In living reptiles, such as the Central American iguanid basilisk (genus *Basiliscus*), and in living birds we find species capable of both ground-dwelling and arboreal activity. And in its skeletal proportions, *Archaeopteryx* is very similar to the living turacos (Musophagidae) and chachalacas (Cracidae), birds that are both arboreal and terrestrial and capable of a myriad of behaviors. *Archaeopteryx*, then, was primarily an arboreal animal, but it may well have been at home both on the ground and in trees and could well have been capable of trunk-climbing. Perhaps Pierce Brodkorb of the University of Florida has best summarized its probable behavioral repertoire:

> Without question *Archaeopteryx* had arboreal habits similar to those of the Cracidae and Musophagidae. It probably used its clawed fingers to cling to twigs in the manner of young hoatzins. It may have come down to drink, as do the cracids (guans) . . . but the long flat tail was not that of a ground dweller or a water bird. Its soft, wide rectrices soon would become frayed on the ground or sodden in the water. . . . Like the touracos . . . *Archaeopteryx* probably ran agilely along the branches, leaping from perch to perch, and swooping from tree to tree. Its flight must have been uncertain and probably consisted of a combination of gliding and flapping, with the tail used both for support and for steering. (1971, 34)

The Urvogel Has Modern Feathers and Wings. The wings that sustained *Archaeopteryx*'s combination of flapping and gliding closely resemble, in their basic design, those of many arboreal, perching birds. Almost perfectly preserved in the Berlin specimen, the wings have the outline of an aerodynamic, elliptical structure. As Josselyn Van Tyne and Andrew J. Berger point out, the elliptical wing is "characterized by a low aspect ratio and only a slight amount of wing-tip vortex. This type [of wing] is found in birds which 'must move easily through restricted openings in vegetation: gallinaceous birds, doves, woodcocks, woodpeckers, and most passerine birds'" (1976, 383). The aerodynamic wing of birds, a design that modern airplanes improve on only in detail, provides good lift and offers great maneuverability. The only feature of modern birds missing from the wing of *Archaeopteryx* is the midwing slot.

The basic pattern of the wing feathers in *Archaeopteryx* is also essentially that of modern birds and has been described by Siegfried Rietschel (1985) and Burkhard Stephan (1985). One problem with interpreting the remains is that feather impressions are found only some distance from the actual feather insertions, and the decay of soft tissue near the insertions may have caused an unfavorable preservational environment (Martin 1991, 493). The number of primaries has been determined to be either eleven or twelve (modern birds have nine to twelve), according to Rietschel, with the first three reduced in length progressively inward, as in many modern birds. Helms (1982) and Stephan have suggested the presence of a small outer primary called a remicle, or little remex. Likewise, *Archaeopteryx* has fourteen secondaries, well within the range of seven to thirty-two that characterizes modern birds. Larry Martin (1991) has made a silastic cast of one of the feather impressions of the Berlin specimen and has discovered that the fabric of the feather was essentially that of a modern bird, scanning electron microscopy revealing regular spacing of barbs throughout their length and faint but clear indications of barbules.

Were the feathers themselves designed for flight? In any feather of modern birds, the long, tapered central shaft, called the rachis, supports the interlocking microscopic barbs that form a sheet on either side, called the vane. The contour, or body, feathers have symmetrical or nearly symmetrical vanes. But strongly asymmetric vanes appear in the flight feathers; the asymmetry is most marked in the primaries, somewhat less pronounced in the secondaries, and present in all the tail feathers except the central pair. In all asymmetric feathers the rachis lies toward the leading edge of the feather, which is thicker, stiffer, and narrower than the trailing vane. As a result of their asymmetry, flight feathers have an airfoil cross-section. In most birds the outer primaries function as individual airfoils to produce lift during flapping flight. In the strongest fliers, such as hawks, swifts, and hummingbirds, the asymmetry is most pronounced. The short, stiffened vanes on the leading edge of feathers in the inner wing and tail contribute stability to the overall aerodynamic design.

In short, the presence of asymmetry in primary flight feathers, "within the range of modern birds using flapping flight" (R. Å. Norberg 1995, 221; see also Speakman and Thompson 1994, 1995), argues that they are used for flight, a generalization confirmed by evidence from birds that have given up flying. The flightless ostriches (*Struthio*) and South American rheas (*Rhea* and *Pterocnemia*) descended from flying birds, but their wing and flight feathers remain only for display, perhaps for thermoregulation, and for balance in running. In these birds the vanes of the primary feathers are symmetrical. In birds that have become flightless more recently, such as the flightless rails on islands of the South Pacific (especially the genera *Atlantisia* and *Gallirallus*), the symmetry is strikingly perfect (Feduccia and Tordoff 1979).

But what of *Archaeopteryx?* Its outer primaries are clearly asymmetric, and the outer vanes, which would form the leading edge in flight, are reduced. Indeed, the very first specimen, a secondary feather discovered in 1861, is asymmetric: evidence that *Archaeopteryx* could fly was present from the very beginning. The fact that vane asymmetry in *Archaeopteryx* extended into the secondary feathers is remarkable and can only indicate strong aerodynamic function.

As an aside, the presence of feather impressions on the leg of the Berlin specimen led William Beebe (1915) to propose that a four-wing "tetrapteryx" stage had preceded *Archaeopteryx* in the evolution of flapping flight. Beebe's hypothetical tetrapteryx bird possessed not only feathers extending backward from the wings but also "hindlimb

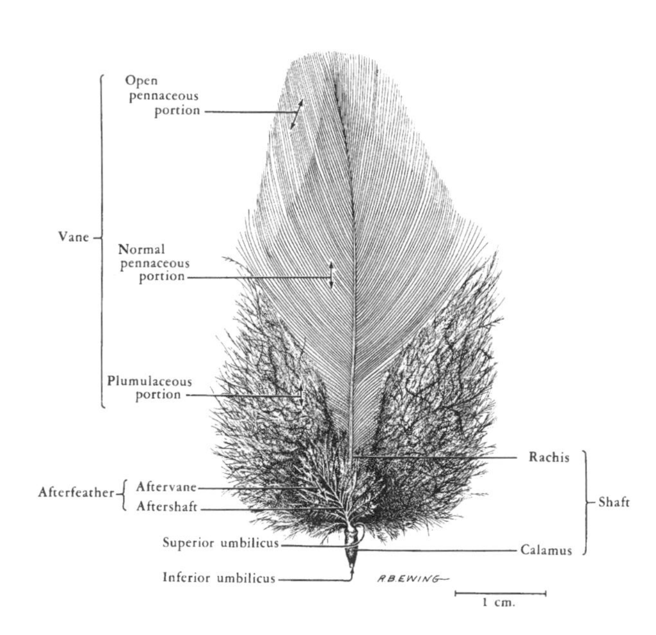

Main parts of a typical contour feather, exemplified by a feather from the middle of the back of a domestic chicken. (From Lucas and Stettenheim 1972; courtesy Peter Stettenheim)

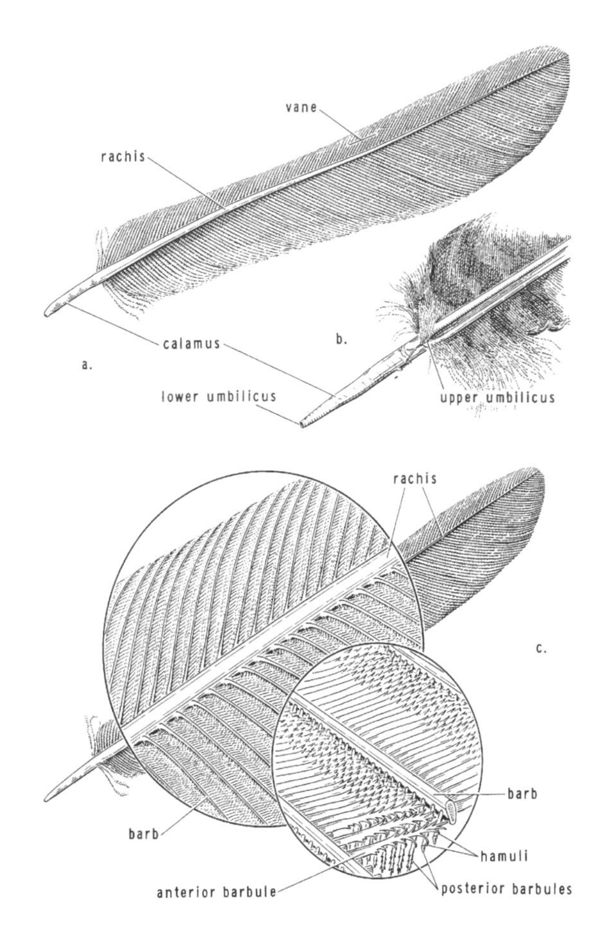

A typical flight feather and the nomenclature of its parts: *a,* general view; *b,* detail of the base of the feather; *c,* detail of the vane. (From Van Tyne and Berger, *Fundamentals of ornithology,* copyright © 1959; reprinted by permission of John Wiley and Sons)

Feathers of the domestic chicken: *right,* contour feather with small aftershaft from the breast region; *left,* primary wing feather. Note the symmetry of the vanes in the breast feather and the asymmetric vanes in the flight feather. The leading edge of the flight feather, the smaller vane, is in direct contact with the air in flight. (From Lucas and Stettenheim 1972; courtesy Peter Stettenheim)

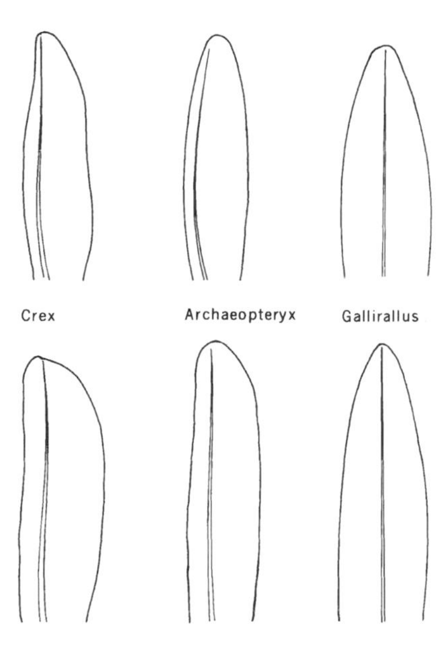

Flight feathers from the left wings of *Crex crex* (a flying rail), *Archaeopteryx,* and *Gallirallus australis* (a flightless rail): *above,* distal ends of the second primaries (counting inward); *below,* distal ends of the sixth primaries. All to scale. (From Feduccia and Tordoff 1979; drawing by Ellen Paige; copyright 1979 AAAS)

wings" that served as passive parachutes. In reality, the impressions of feathers on the hindlimb of the Berlin specimen are simply typical contour feathers. More extensive evidence for contour feathers in *Archaeopteryx* is present in a tracing done by an unknown artist sometime before 1879 that was part of failed negotiations to sell the Berlin specimen to Yale's Peabody Museum of Natural History (Ostrom 1985b). This treasure (p. 31), "first image ever made of the famous Berlin specimen" (366), shows body feathers on the neck, back, breast, and legs, all long since prepared away.

Åke Norberg (1985) has shown that the primary flight feathers of *Archaeopteryx* exhibit curvature as an expression of their complex aerodynamic design, and Rietschel (1985) has shown that they possess a reinforcing furrow along the ventral shaft, a feature unique to modern birds. As Ulla Norberg has summarized about birds in general, "The inherent aeroelastic stability of the feather is achieved passively by an appropriate combination of the three main feather characteristics. . . : (1) vane asymmetry, (2) the curved shaft, and (3) the greater flexural stiffness dorsoventrally . . . achieved partly by a ventral furrow of the feather shaft" (1990, 225). All of these features are characteristic of the feathers of *Archaeopteryx*.

Regardless of the degree to which *Archaeopteryx's* skeleton was reptilian, there can be no doubt that its feathers were indistinguishable in any important ways from those of living birds. Its wings had the basic pattern and proportions of the modern bird's wing; indeed, there has been no essential change in this aerodynamic structure for about 150 million years. So, by the late Jurassic there existed a reptile-bird that had developed the firm, bladelike pennaceous feather characteristic of modern birds and that was, as D. B. O. Savile observes, "appreciably advanced aerodynamically" (1957, 222).

Establishing that the feathers of *Archaeopteryx* permitted it to fly does little to resolve the question of how feathers came about in the first place. Feathers are unique to birds, and no known structure intermediate between scales and feathers has been identified. Nevertheless, it has generally been accepted that feathers are directly derived from reptilian scales, with models provided by Paul Maderson (1972) and Philip Regal (1975). Although feathers are broadly similar to reptilian scales at the embryonic level, there is no strong evidence to support the hypothesis that avian ø-keratins are derived directly from a reptilian epidermal protein. Feathers are widely appreciated to be unique at the molecular level, so it should be evident that, in an evolutionary sense, it follows that an important event occurred at the molecular level (Brush 1993, 129). Thus, although evolutionary experimentation with scale morphology may have been extensive during the archosaurian radiation, with some scales tending toward feathers, as we saw in *Longisquama,* in all likelihood feathers were produced by new mutations that resulted in a brand new epidermal appendage, the feather, and a complete reorganization of the uniquely thin avian integument, nearly devoid of glands. However, the early evolution of feathers, whether in the form of highly modified featherlike scales or the novel avian feather, can still be addressed in the context of selection for the gradual elongation of epidermal appendages in protobirds.

Feather Function and the Context of Early Feather Evolution. We know that feathers in modern birds have two physiological functions: thermoregulation and flight. Presumably the very first feathers served one or the other purpose, but not both, because initially feathers would have evolved under selection pressure for only one function. There is no fossil evidence to support either theory, yet, by the same token, neither is precluded. Contexts for the early evolution of feathers other than flight and thermoregulation have been suggested from time to time. Ernst Mayr in 1960 suggested that feathers evolved as a device for communication between the sexes. To be sure, feathers are important in avian courtship displays and thus in choosing mates. But the evolution of myriad structures can be explained as due to sexual selection, and given the anatomical complexity of feathers, I agree with Philip Regal that "sexual selection as a general evolutionary force, used to explain miscellaneous conditions that are difficult to account for, is unsatisfying" (1975, 43). Richard Cowen and Jere Lipps proposed in 1982 that feathers evolved in the context of such behaviors as display and threat along with thermoregulation. Still other models have feathers evolving for water repellence (Dyck 1985) or in the context of heronlike aquatic foraging, using wings to shade the water in pursuit of prey (Thulborn and Hamley 1985). Although these views are interesting to ponder, they all require that feathers, extraordinarily complex integumentary appendages, evolved first in a context other than their current primary function—that is, as aerodynamic structures. Thus, all of these exaptation models require complex explanations and do not account for the distinctive, almost magical aerodynamic structure of feathers.

Biothermal Semantics. The hypothesis that feathers originated as insulating cover takes one of two forms: that they kept heat in or that they kept it out. Currently, the more familiar and popular theory is that they helped to keep the animal warm. Among the living vertebrates, only birds and mammals are warm-blooded, or endothermic (meaning "heat from within"); that is, they maintain high and constant metabolic rates and body temperatures by relying mainly on heat generated endogenously, from within (Bennett and Ruben 1979). More precisely, endothermy is a state of constant elevated aerobic heat production (breakdown of glucose in the presence of oxygen) and metabolism (about five- to tenfold that of reptiles), generally combined with an insulation to prevent heat loss. As a result, birds and mammals sustain activity levels over a wide range of ambient temperatures and well beyond the capacity of their cold-blooded reptilian counterparts. In an ecological context, endotherms are able to sustain relatively high levels of activity for extended periods, enabling them to forage widely and to migrate (Bennett and Ruben 1979; Bennett 1991; Ruben 1996). The cost of endothermy in terms of caloric expenditure is measured in a relentless pursuit and intake of food, orders of magnitude greater than that of similar-size reptiles. In contrast, most of the reptiles, fishes, and amphibians are cold-blooded, or ectothermic ("heat from outside")—their main source of body heat is the ambient environment and their body temperature fluctuates accordingly.

Ectothermic reptiles typically rely on nonsustainable anaerobic metabolism (breakdown of glucose in the absence of oxygen) for their activities, and although these animals are capable of spectacular bursts of intense activity, they generally fatigue as a result of lactic acid accumulation (Bennett 1991; Ruben 1995). Many texts use the older, traditional thermoregulatory terms *poikilotherm* and *homeotherm*, which refer to relative constancy of daily body temperature rather than the source of the heat and are therefore ambiguous. Poikilotherm (*poikilos*, "various") refers to organisms whose body temperature varies with ambient temperature throughout the day, and homeotherm (also *homoiotherm*, "same heat") refers to organisms that maintain a fairly constant body temperature despite ambient fluctuations. By these definitions, bats and hummingbirds, whose body temperatures drop during the day and night, respectively, might not be considered homeotherms, so the source of body heat presents a better framework for definition. In addition, the stark contrast of cold-blooded and warm-blooded is grossly oversimplified and inappropriate as a classification because there are many shades of gray in temperature regulation, animals produce differing levels of endogenous heat, and endothermy, not directly equivalent to that of birds and mammals, is known in such fish as tuna and billfishes (Block et al. 1993), lamnid sharks, and some sea turtles. Further, elevated temperatures resulting from internal thermogenesis are known in some monitor lizards and in brooding pythons. Birds and mammals are perhaps most appropriately characterized as endothermic homeotherms.

Hot-blooded Dinosaurs: Notes and Comments

Dinosaurs Become Endothermic. It has usually been assumed that endothermy originated twice, once each in mammals and in birds, but a currently popular theory holds that dinosaurs were also endotherms and achieved metabolic rates, hence activity levels, comparable to those of birds and mammals. If dinosaurs were endothermic, as this theory goes, feathers might have been an effective insulating mechanism for them and, once present, could have been a preadaptation for flight, the view of those who advocate the dinosaurian origin of birds and the cursorial origin of flight.

The foremost original proponents of the theory that dinosaurs were endothermic or "hot-blooded" are the popular dinosaur "spokesman" Robert Bakker and John Ostrom, both of whom, working independently, concluded that dinosaurs, with their bipedal, erect posture, did not fit the picture of the sluggish, cold-blooded reptile (Ostrom 1969a; Bakker 1975; Desmond 1976). Of the two, Ostrom tends to be more conservative, now maintaining that dinosaurian sophistication has been greatly exaggerated and that only some of the smaller theropod dinosaurs were warm-blooded (Ostrom 1987b), whereas Bakker (1975, 1986) is fanatical, viewing all dinosaurs as fleet-footed, "hot-blooded" miracles of the Mesozoic, a view that many have since adopted, including especially Gregory Paul (1988). According to Bakker and Paul, all dinosaurs were hot-blooded endotherms, with high body temperature and high metabolic rates, and had the same level of activity as living birds and mammals; they were energetic and intelligent. In Bakker's book *The Dinosaur Heresies* (1986; see the excellent but critical review by Ostrom, 1987b), the chapter on the "flying dinosaur" features a feathered *Deinonychus* dramatically leaping about.

The main arguments for "hot-blooded" dinosaurs can be summarized as follows: (1) dinosaurs achieved erect posture, a stance found today only in living endotherms, the birds and mammals; (2) the microscopic structure of certain dinosaur bones is like that of certain living mammals and can be interpreted to reflect warm-bloodedness; (3) the ratio of predators to prey in fossil deposits of dinosaurs is similar to the ratio found in living communities of mammals and, in Bakker's view, reflects the high metabolic requirement of endothermic predators; and (4) dinosaurs were intelligent, with the large brain size associated with endothermy. All four arguments are seriously flawed as regards either the actual data or their interpretation, and the case has not been proven. The first rebuttal, "Dinosaurs as Reptiles," was published in 1973 (Feduccia 1973a), along with another by Albert Bennett and Bonnie Dalzell (1973), and a constant stream of objections has appeared over the years (to mention a few, Marx 1978; Benton 1979; Thomas and Olson 1980; Ostrom 1987b; Carroll 1988; McGowan 1991; Ruben 1996, 1995; Chinsamy and Dodson 1995; Farlow, Dodson, and Chinsamy 1995).

Posture and Bone Histology. There is certainly no causal relation between erect posture and endothermy. The first mammals were semierect endotherms, much like the egg-laying monotremes of today. Erect posture, combined with a complex structural system that included extensive vertebral excavation, was necessary to support the tremendous body weight that characterized the dinosaurs. And the same complex microscopic bone Haversian canal structures (first described by British anatomist Havers) Bakker cites as an indication of endothermy are now known to be present in such ectothermic vertebrates as turtles and absent in such endotherms as a number of small mammals and passerine birds (Bouvier 1977), and secondary Haversian canals are present in such large endotherms as crocodiles and sea turtles. These bone structures also tend to be associated with large size; thus, their presence in dinosaur bones cannot be viewed as proof of endothermy. More problematic is that the dinosaurs whose bone histology is closest to the pattern seen in mammals are the giant sauropods, forms that on other anatomical grounds would appear the least like mammals in physiological parameters and activity levels. Among theropods, a small specimen of *Allosaurus* has a much more typical reptilian bone histology, although the skeletal anatomy and "relatively" larger brain size would suggest higher levels of activity (Madsen 1976). Further complicating the matter, a group of highly advanced mammal-like reptiles, the tritylodonts, which must have approached mammals in their metabolic parameters, have a bone histology similar to that of typical reptiles (Carroll 1988, 321).

Haversian bone, however, is but one histological type of mammalian bone, and another type, known as fibro-lamellar bone, is typically found in large mammals and is associated with rapid growth. Cattle, for example, which attain adult size in about three years, exhibit this histological type of bone. Yet juvenile crocodiles raised on farms under optimum conditions are known to form fibro-lamellar bone (Chinsamy and Dodson 1995, 177), and fibro-lamellar bone is in a labyrinthodont amphibian and a number of clearly ectothermic cotylosaurs and pelycosaurs. Armand de Ricqlès of the University of Paris (1969, 1974, 1980), an expert on bone histology and the first to note that dinosaurs have both Haversian and fibro-lamellar bone, concluded that dinosaurs had rapid rates of growth and hence high metabolic rates. Later studies of sauropods, however, have shown that although sauropods have fibro-lamellar bone, the bone exhibits distinct zones of cyclical growth (Reid 1981), similar to the growth rings of trees. The inescapable conclusion seemed to be that sauropods were not endotherms, although by their size alone, they would have maintained a reasonably constant body temperature. In contrast, David Varricchio (1993) discovered that bone microstructure of the late Cretaceous theropod *Troödon* (*Stenonychosaurus*) showed several ontogenetic bone types, including fibro-lamellar bone, and he estimated that *Troödon* would have attained adult size in three to five years. Varricchio concluded that "abundant vasculature together with dense Haversian bone suggests [that] *T. formosus* had a relatively high metabolism" (99). This, combined with other morphological features, indicated to him that *Troödon* was "very active and possibly endothermic" (99). Yet *Troödon* had growth rings in the bones, suggesting that growth may have halted at times, and this would be more in line with ectothermy than endothermy or would be somewhere in between. Further, John Ruben showed that "alligator growth rates are virtually indistinguishable from estimated growth rates for the bipedal theropod *Troödon*" and notes that "fibro-lamellar bone is absent in many small, rapidly growing endotherms, and its presence in a labyrinthodont amphibian and in some clearly ectothermic cotylosaurs, pelycosaurs, and dicynodont therapsids is particularly puzzling" (1995, 88, 87).

All in all, the evidence from bone histology is not firmly interpretable. Michael Benton (1979, 985) has noted that because dinosaur hatchlings were merely some 1–2 percent of the weight of their parents, being limited in size by constraints on egg size, the vulnerable young dinosaurs may have been forced to grow rapidly, and this would have involved continuous remodeling of the bones. Whatever the case, bone histology is clearly not an invariable correlate of endothermy, and by itself cannot provide definitive evidence for temperature regulation and levels of activity.

In 1993, Claudia Barreto and colleagues studied the histology and ultrastructure of the ends of long bones of the juvenile *Maiasaura*, an ornithischian dinosaur (hadrosaur) from the late Cretaceous of Montana. These fossils preserve the growth plates, discs of cartilage near the ends of growing long bones; comparison of these plates with those of living animals indicated that they were homologous with living birds, but not with mammals or lizards. But similar growth patterns are in fact found in some living reptiles; crocodilians, the only living remnants of Mesozoic archosaurs, were not examined, and no saurischian dinosaurs were studied (Fischman 1993). Further, even if the study were substantiated, one could easily conclude from this evidence that it supported the common-ancestry hypothesis of birds and dinosaurs from a thecodont ancestor.

The dust had hardly settled from the Barreto study when another interesting paper was published by Anusuya Chinsamy, Luis Chiappe, and Peter Dodson (1994), who determined that the microscopic study of the leg bones of Cretaceous enantiornithine birds (chapter 4) and a late Cretaceous hen-sized flightless bird named *Patagopteryx* (chapter 6) revealed growth rings or lines of arrested development. These lines of arrested development, typical of ectotherms, have been observed in isolated bones (Reid 1981) and in the growth series of limbs of dinosaurs (Varricchio 1993), as mentioned above. To Chinsamy, Chiappe, and Dodson, the presence of growth rings in the Cretaceous birds "suggests that these birds differed physiologically from their living relatives and were not fully homeothermic" (197). Histological studies of extant birds as well as the extinct hesperornithiforms have never revealed these lines of arrested development. Yet, like most studies of bone histology, tremendous variation exists, and although periods of arrested development may have occurred in these birds, it cannot automatically be assumed that they were not endothermic, and debate on this paper has already begun (Browne 1994). Nevertheless, it is altogether possible that these early birds were not fully endothermic and that they may have been intermediate between their ectothermic ancestors and modern birds in thermoregulatory capacity.

The Chinsamy paper raises some interesting points. We know very little about the physiology of fossil birds, and we cannot automatically assume that feathered Cretaceous birds were fully endothermic. Even some modern birds permit a glimpse, albeit a hazy one, into the thermoregulatory past in both avian embryos and in a variety of birds that are not fully homeothermic. For example, although most precocial hatchlings (well developed, down covered) have developed a high degree of homeothermy, altricial hatchlings (naked, helpless) are poikilotherms and normally take a number of days to develop full endothermy (Ricklefs 1983; Whittow and Tazawa 1991). In addition, many living birds are not absolute and comprehensive homeotherms (Gill 1995). For example, the turkey vulture (*Cathartes aura*) normally lowers its body temperature about 6°C (11°F) at night until its body temperature is 34°C (93°F); it becomes mildly hypothermic. A number of small birds, notably hummingbirds, can enter a state of torpor, or profound hypothermia, in which they are unresponsive to most stimuli, and their oxygen consumption can drop by 75 percent when the body temperature drops by 10°C (18°F) (Calder and King 1974). The most dramatic example of torpor is found in the common poorwill (*Phalaenoptilus nuttallii*) of the southwestern United States, which hibernates at a body temperature of 6°C (11°F) for two to three months in winter. These birds are capable of spontaneous arousal but normally require about seven hours to warm up (Ligon 1970). One must wonder, given the studies of Chinsamy and colleagues, just what the thermoregulatory parameters of Mesozoic birds were. They could have ranged from near ectothermy through the level of the poorwill to that seen in most modern birds, but we will never know.

There is now evidence for contour body feathers in a late Jurassic, beaked bird from China as well as in *Archaeopteryx*, but a fully feathered urvogel could have been either ectothermic or endothermic, and a coating of feathers might have indicated flight refinement rather than endothermy. Some living birds use behavioral thermoregulation to absorb ambient heat. For example, on cold nights, body temperature in the greater roadrunner (*Geococcyx californianus*) may decline by about 4°C (7°F); but after sunrise the roadrunner will bask in solar radiation to warm ectothermally to its normal body temperature (Ruben 1995, 86). Likewise, an ectothermic *Archaeopteryx*, because it lived in a warm, tropical setting, could easily have thermoregulated itself with similar behaviors. As we have seen, *Archaeopteryx* was almost certainly fully clothed in body contour feathers, and so presumably were the later descendants of the avian lineage, but the original body contour feathers may have served first to produce smooth contours and hence laminar flow in an aerodynamic sense, with less emphasis on thermoregulatory insulation.

John Ruben (1991, 1) argued on the basis of data from muscle physiology of extant reptiles that *Archaeopteryx* was a flying ectotherm. He estimated the pectoral muscles and deltoids (flight muscles) together to have been about 9 percent of its 200-gram (7-oz.) body mass, as compared to an average of some 25 percent in modern birds (ranging from 11.5 percent to 43.9 percent; Marden 1987). Ectothermic muscle, twice as "capable" as modern avian flight muscle, would have provided *Archaeopteryx* with the necessary equipment for powered flight. Ruben's argument has been challenged by Speakman (1993; see also reply by Ruben 1993), who found a much closer metabolic capability of avian and reptilian muscle than Ruben had suggested. Speakman concluded that "*Archaeopteryx*, with a flight muscle comprising only 9% of body mass, could not take off from the ground from a stationary position" (339). Whatever the case, these data and arguments have little negative bearing on the "trees-down" theory for the origin of avian flight, because protobirds, if grounded, could simply climb up the trunk of the next tree with their hand claws. With the discovery of an ossified sternum in the latest specimen of *Archaeopteryx*, the estimates of flight muscle mass will no doubt have to be reexamined and revised upward.

One implication of the Ruben study, if correct, would be that feathers originated in some context other than insulation for hot-blooded dinosaurs. In addition, an ectothermic *Archaeopteryx* would have had indeterminate growth, which might best explain the broad range of size variation among the seven specimens (Chinsamy and Dodson 1995). And if *Archaeopteryx* was ectothermic, and it is assumed that it derives from theropods, then birds certainly did not originate from endothermic dinosaurs! This is particularly interesting in that Paul Sereno, among the more vocal advocates of a theropod origin of birds, following Ruben, states that "*Archaeopteryx* may have been a 'flying ectotherm' capable of flight over short distances, powered by lightweight anaerobic flight muscles and an ectothermic physiology as suggested recently by Ruben" (Sereno and Rao 1992, 847). According to this statement, Sereno apparently does not believe that feathers originated in the context of insulation for hot-blooded dinosaurs, but he does firmly believe that birds evolved from ground-dwelling dinosaurs, which would have also been small ectotherms.

Aggregations, Social Behavior, and Living Reptiles. Jack Horner of the Museum of the Rockies in Bozeman, Montana, has been excavating and studying Montana nests sites of the hadrosaur *Maiasaura* ("the good mother lizard") since the mid-1970s and has concluded, among other things, that fibro-lamellar bone is present, growth rates must have been high, and therefore that metabolic rates were high. Horner's studies, like Varricchio's work, however, are tempered by Reid (1981; 1984), who pointed out that rapid growth does not necessarily imply high metabolism and endothermy. Horner's discovery of nests of the "good mother lizard" created quite a sensation, but the evidence appears to have been grossly overinterpreted. Horner, Bakker, and others see dinosaurs as active endotherms, like living birds, which they consider to have descended from dinosaurs with only minor modification (Horner 1984; Horner and Makela 1979; Horner and Weishampel 1989; Horner and Gorman 1990).

The nests Horner has uncovered provide evidence of colonial nesting in dinosaurs: the nests are appropriately spaced, as in a nesting colony. In fact, advanced juveniles have been preserved in some nests, which may indicate extended parental care and feeding at the nest site. Even more interesting, one remarkable fossil site preserves the remains of a huge herd of maiasaurs that was overwhelmed by a massive fall of volcanic ash. This herd was composed of animals that fall into four rough size classes, of about 3, 4, 5.2, and 7 meters (9.8, 13, 17, and 23 ft.). These size differences have been interpreted as indicating four years of age classes; maiasaurs, Horner thus infers, must have matured in four years and so had determinate growth, like birds and mammals. If we can extrapolate from the growth rates of living vertebrates, and if maiasaurs matured in four years, doubling their weights each year until adulthood, this means that they would have grown about ten times faster than modern alligators, or about two-thirds as fast as similar-sized mammals.

The massive kill from the ash fall covered an area of some 1.5 square kilometers and contained skeletal elements from an estimated 135,000 individuals. Because no modern environment or paleoenvironment could support 135,000 "hot-blooded" herbivores, Horner concluded that the massive ash kill represented a herd of dinosaurs migrating like the great mammals of Africa's Great Rift Valley. One can easily see how unproven assumptions, layered one on top of each other, can lead to an overall picture of fossil organisms that appears to be rock solid but hangs by a mere spider's web. It is interesting here to note that literally hundreds of the late Triassic *Coelophysis* skeletons have been discovered close together (Colbert

1989, 148), having perished in some cataclysm, and this is a dinosaur with few features that would hint at endothermy.

Modern ectothermic reptiles, especially snakes and lizards (Galápagos marine iguanas, for example), may aggregate in huge assemblages for many reasons, including prehibernation and mating, and, in the case of marine iguanas, as part of their everyday behavior. Various crocodilians may also occur in large concentrations—even overcrowded conditions, in the common caiman of Venezuela, particularly during the dry season, when riverine populations tend to congregate in river pools (Ross 1989, 141). If such concentrations as these were to be preserved, they would be interpreted by modern paleontologists as evidence of herding in crocodilians!

Colonial nesting and social behavior are known in Panama's green iguanas (*Iguana iguana*) (Burghardt, Greene, and Rand 1977). The hatchlings of this species emerge from the ground in small groups and engage in complex social interactions. They often join with hatchlings from other clutches, and they depart the nest site, move about the islet, and migrate from the islet to the larger adjacent landmass in social groups. And crocodilians—the only true living archosaurs—exhibit complicated colonial nesting and fairly involved parental care, including guarding their nests, helping their young hatch by digging down to the eggs and carefully breaking the eggs open, transporting their chicks to the water using their jaws, and then guarding their chicks for some time (Ross 1989). Hatchlings frequently remain near the nest site in nurseries, where they form social groups, or pods (Ross 1989, 110), and American alligators (*Alligator mississipiensis*) stay in family groups for years, with females remaining at their nests surrounded by young from clutches from several previous seasons. Crocodilians are also very agile beasts, capable of "tail walking," a maneuver in which the entire animal emerges above the water, and even of bounding "galloping," in which speeds of some 17 kilometers (10 mi.) an hour can be achieved (Ross 1989, 46). Crocodilians may also hunt in groups and participate in group feeding to dismember such large mammals as wildebeest.

In part because of their large size, crocodilians maintain considerable thermal stability, and American and Chinese alligators can extend their ranges northward into the temperate zone where several hard freezes may occur each year. Research into the habits of these animals has brought forth surprising facts about their ability to endure the cold. According to Lehr Brisbin, "Radiotelemetric studies of free-ranging adult alligators faced with brief periods of cold weather, with air temperatures sometimes falling as low as −3.1°C (26.5°F), have found, surprisingly, that in no case did any of these animals ever seek shelter in either underwater or subterranean dens as had been previously assumed. Rather, with the approach of the cold spell, these alligators consistently sought out shallow backwaters where they could position their nostrils in such a way as to keep a small round hole open in the ice that was forming above them. . . . Temperature sensing probes implanted in radio-collared wild animals have recorded core body temperatures as low as 5°C (41°F) during cold spells, with full recovery later" (quoted in Ross 1989, 50).

Dinosaurs in the Cold and Dark? Apropos of the above discussion, much has been made of the discovery of dinosaurs on Alaska's North Slope (Brouwers et al. 1987; Clemens and Nelms 1993), in high latitudes in Australia (Rich et al. 1988; Vickers-Rich and Rich 1993), and in Antarctica (Hammer and Hickerson 1994), the general conclusion being that these animals spent part of the year in cold and darkness. The implications of these discoveries is that dinosaurs living at such latitudes must have had special physiological or behavioral capabilities to cope with this seasonally harsh environment. However, these faunas include many other fossils that raise serious questions as to whether these animals lived in the cold and dark. Because the continents have changed positions through the geologic past, fossil sites had different latitudes during their deposition; further, the science of determining paleolatitudes is inexact and in constant need of revision. For example, fossils discovered in the Australian early Cretaceous high-latitude site include a lungfish, a labyrinthodont amphibian, lepidosaurs, and pterosaurs. Aside from dinosaurs, turtles were the most common terrestrial fauna; all this points to a moderate environment. It seems almost inconceivable that the mean annual temperature was "less that 5°C" and that "the inhabiting biota had to cope with 1 to 2 months of continuous darkness" (Rich et al. 1988).

The late Cretaceous Arctic dinosaurs, it is argued, lived in a climatic zone equivalent to modern-day Anchorage, with a mean annual temperature of 2–8°C (34–47°F), with three months of annual darkness. But there is really nothing to preclude dinosaurs having slowly migrated south during severe times of the year (Currie 1989), and perhaps they only spent several months in the far north, enjoying the breeding season on the Northern Slope (now famous as a massive breeding grounds for shorebirds), then meandering south with the food supply. To the south, late Cretaceous faunas are known from Dinosaur Provincial Park in the southern Alberta plains, with some thirty types of dinosaurs that would be the dinosaurian equivalents of a combined African plains and Amazonian fauna of today, inhabiting lush tropical, bayou-type habitats, with vegetation ranging from bald cypress to cycads, tree ferns, and herbaceous lilies. Dale Russell of the National Museum of Natural Sciences, Ottawa, estimates that during the late Cretaceous period worldwide mean annual temperatures were about 10°C (18°F) higher; he proclaims that there are no modern counterparts "for the warm subhumid and polar deciduous forests" where polar dinosaurs have been found (Russell 1989, 149). "In Alaska and Yukon," notes Russell, "a large part of each year passed beneath the polar night, but there was no frost. The mean annual temperature remained near 10° Celsius, and in many respects the humid climate resembled that of Vancouver" (149).

We know that Mesozoic climates were extremely equable compared to those of today, which may be among the harshest the earth has experienced. We also know that the continents have moved away from the equator since the Mesozoic. Data must therefore be extremely convincing to argue that, for example, the Australian fossils, including lungfish and amphibians, lived for long periods each year in cold and darkness. In fact, data from a polar forest from the Permian of Antarctica indicate that "the temperature rarely went below freezing" (Taylor, Taylor, and Cuneo 1992), and recently discovered early Jurassic fossils from Antarctica (before the breakup of Pangaea) include a crested theropod (*Cryolophosaurus*), two other theropods, a large prosauropod, a pterosaur, a synapsid tritylodont, and a flora that includes tree trunks, indicating forested areas. William Hammer and William Hickerson propose the unlikely scenario that "the large prosauropods and theropods and the pterosaur may have migrated away from harsher winter temperatures; smaller animals such as the tritylodont may have hibernated through a cold season" (1994, 830). They then state, however, that "the existence of this fauna suggests that conditions were at least seasonally mild at high latitudes during the early part of the Jurassic" (830). None of these faunas, from Antarctic Jurassic to Arctic Cretaceous, would appear to have suffered through extremes in climate or prolonged darkness.

Predator to Prey Ratios Are Flawed. Going back to Bakker's proposals, perhaps the most faulty of his arguments concerns the predator to prey ratio of dinosaurs (Tracy 1976; Farlow 1976; Hotton 1980; Carroll 1988). To begin with, statistics based on fossil deposits are notoriously unreliable, because such deposits do not accurately reflect either the number of species in the fauna or the number of animals in each species. For example, during the late Cretaceous, the preservation of large numbers of hadrosaurs or duckbills reflects the taxonomic or preservational environment; these animals were aquatic and were therefore easily preserved after sinking to the bottom of lakes and bogs. Contrarily, tyrannosaurids and other theropods are relatively rare as fossils because they were terrestrial, and after death their remains were scavenged, scattered, and otherwise destroyed. Preservation occurred only through some bigger accident.

Even more important, Bakker's calculations are based on the unwarranted assumption that large carnivorous dinosaurs were primary predators—feeding only on herbivorous dinosaurs. There is no reason to believe that the carnivores did not feed on a great diversity of dinosaurs, including the young of all species (which are noticeably absent from the fossil record), or that they did not also feed on carrion. In the American alligator, for example, cannibalism accounts for about 50 percent of hatchling mortality and some 64 percent of total mortality in alligators of age eleven months or older (Rootes and Chabreck 1993), and large living reptilian carnivores, especially crocodilians and monitors, are quite fond of carrion at virtually any stage of decay. Many adult herbivorous dinosaurs, which Bakker has argued constituted a large part of the carnivores' diet, were so large that, like adult elephants today, they were probably almost immune to predation once they had reached adulthood. In reality, the relative abundance of fossil species in a given deposit tells very little about the relative abundance of species in the area (Farlow 1976).

Medium-Sized Vertebrates Are Rare in the Cretaceous. An overlooked phenomenon, and a remarkable feature of the Cretaceous fossil record, is the rarity of medium-sized vertebrates, especially those weighing less than 10 kilos (22 lbs.). Equally absent are juvenile specimens of larger animals, especially dinosaurs (Elzanowski 1983; Richmond 1965). These facts are particular striking when one considers that the microvertebrates, such as small mammals and lizards, are well represented in Cretaceous deposits. Andrzej Elzanowski has suggested that a major cause for the lack of medium-sized vertebrates is scavenging and predation by carnivorous reptiles, mainly theropod dinosaurs. The feeding method of large reptilian carnivores is essentially different from that of mammalian predators: crocodiles and large monitor lizards swallow prey items or carrion whole and will swallow items as large as can pass through their gullets. Cannibalism may have also been quite common; as H. B. Cott stated in 1961 in his study of the Nile crocodile (*Crocodylus niloticus*), "Crocodiles are much addicted to cannibalism" (296), and Walter Auffenberg, who has conducted extensive studies of the Komodo dragons, or oras (*Varanus komodensis*), noted in 1981 that "intraspecific predation is quite common. . . .

a female dug into an ora nest . . . and she attempted to eat the young and had to be driven away. They also kill and eat one another at later life stages" (152). In the late Triassic we have evidence of cannibalism in the early theropod *Coelophysis*, in which two large individuals are preserved with the remains of young individuals within the rib cage (Colbert 1989, 144). Finally, Komodo dragons will swallow bones even if cleaned of flesh, and skeletal remains of dragons are rare because of scavenging by other dragons.

In contrast, large carnivorous mammals tend to disarticulate larger prey and eat only selected parts, frequently caching prey remains, and skeletal remains are very common in the wild areas of open country of Asia and Africa. Dinosaurs certainly followed the eating habits of large reptilian carnivores. Elzanowski suggests that although scavenging behavior is largely nonselective with respect to size, medium-sized vertebrates are in fact selected over smaller ones because scavenging is sight-directed and very small vertebrates, lying in vegetation or natural crevices, are harder to find. This means that only carrion that is both large enough to be detected and small enough to be swallowed whole is in the size range to be eaten. Central to this hypothesis is the ability of reptiles to digest bone: crocodiles are known to completely decalcify bones and teeth (Fisher 1981), whereas mammalian digestion of bone is largely incomplete because food passes so rapidly through the alimentary canal. If dinosaurs are considered not as endotherms but as having some degree of homeothermy, this would not imply a mammalian level of body temperature and metabolic rate, and the rate of food passage would be the same as that of living reptiles. In this respect, W. E. Swinton (1970) noted the absence of any solid materials in dinosaur coprolites and suggested that the digestion of bones was as effective as that of crocodiles.

No Differentiated, Heterodont Dentition. The teeth of theropod dinosaurs, uniformly shaped, curved, serrated, and laterally compressed, are not the complex heterodont mastication factories of the mammalian jaw, with its complex, differentiated tooth row. The serrated teeth of tyrannosaurid dinosaurs, typical of theropods, were analyzed in 1992 by William Abler, who showed by experiments that the cutting action of tyrannosaurid teeth resembles that of a dull, smooth blade and that the spaces between the serrations act as minute frictional vises that grip and hold meat fibers. Abler argued that the chambers between neighboring serrations would receive and retain small fragments of meat and would therefore become havens for stored bacteria. Wounds inflicted by such teeth would cause infections. These infections would be analogous to those inflicted by the living Komodo dragon, the largest living monitor lizard, whose bite is infectious. Komodo dragons apparently subdue previously bitten, weakened prey often enough that the storage of bacteria in their teeth is thought to have coevolved with their hunting behavior. Abler's idea is that tyrannosaurids may have preyed like Komodo dragons, "hunting from ambush, occasional carrion feeding to renew bacterial growth, and feeding by ripping and tearing off chunks of meat" (1992, 180).

The complex heterodont dentition of mammals is totally different. As Abler writes, "Mammals have precisely fitting, complex, differentiated teeth that work only when their action is precisely controlled. It is tempting to link the neural sophistication of mammalian brains to the sophistication of their teeth and the jaw movements required to operate them efficiently" (1992, 181). As we shall see, almost all dinosaurs had small reptilian brains, matching their masticatory machinery.

Did Theropods Ambush Their Prey? As pointed out in chapter 2, not enough has been made of the "pubic boot" characteristic of theropod dinosaurs and also found in such advanced thecodonts as the late Triassic *Postosuchus*. It seems clear that this structure was more than just part of the pelvic girdle as envisioned in other vertebrates; Sankar Chatterjee has suggested that "when seated, or stretched on the ground, the pubic 'foot' would receive some of the weight of the body" (1985, 439). The late Cretaceous *Tyrannosaurus rex* has a well-developed pubic foot; it has a pitted appearance on the distal end and in life was probably covered by a thick skin-pad, which contacted the ground. Robert Bakker (1986) has asserted that six-ton *Tyrannosaurus* could move at 75 kilometers (45 mi.) an hour. As John Ostrom notes, "A six-ton African elephant might reach eighteen miles an hour running full speed, downhill, and a half-ton Thoroughbred might—just might—hit forty miles an hour on a level straightaway. Bearing this in mind, the notion of six tons of dinosaur flesh being routinely propelled at forty-five miles an hour is preposterous" (1987b, 63). Could it be that these theropods normally rested by squatting on their haunches, with legs folded, balanced in front by pubic foot and in back by their huge tails, and sallied out only on occasion to ambush prey, some already weakened by previously inflicted wounds? Large chunks of flesh were torn off and swallowed, bones were decalcified, and little trace of the carnage remained.

Most Dinosaurs Had Indeterminate Growth. To add to the mound of evidence against the concept of endothermy in dinosaurs, the vast majority of dinosaurs, unlike modern birds and mammals, had indeterminate growth, continuing to increase in size as they grew. Chinsamy's work on dinosaur bone histology has yielded a steadily confusing array of growth types, including *Syntarsus*, which showed the growth rings characteristic of ectotherms but apparently reached maximum body size in seven to eight years; and the late Cretaceous *Troödon* is interpreted as having reached maturity within three to five years and stopped growing occasionally during its formative years. However, in the vast majority of dinosaurs studied, including *Tyrannosaurus rex*, growth rings are visible, specimens differ dramatically in size, and there is little question that they exhibited indeterminate growth, like living ectothermic reptiles (Chinsamy and Dodson 1995; Chinsamy 1995). Chinsamy notes that "bone histology cannot provide any definite answer as to whether dinosaurs were endotherms or ectotherms[;] its real value lies in the morphological and descriptive analysis which pertains to its patterns (notably rates) of accretion" (1995b, 96). In addition to long-term histology, joint surfaces in dinosaurs are poorly defined, and there no doubt was a thick cartilaginous layer interfacing between the joints, not the complex separate terminal ossifications of long bones typical of birds and mammals.

Another interesting and tell-tale observation came in 1979 from Paul A. Johnston, who studied dinosaur teeth and determined that they exhibit the growth rings characteristic of ectotherms (see also Johnston 1980; Meinke, Padian, and Koppelman 1980). Such growth rings are produced by variation in the growth rate of living ectotherms, and their presence in dinosaur specimens suggests that in those animals, too, the growth rate fluctuated with the seasons.

Respiratory Turbinates. Most of the histological and anatomical features used to illustrate endothermy in dinosaurs are related only indirectly to metabolism and are therefore unable to be used as indicators of whether an animal is definitively an endotherm. As Albert Bennett and John Ruben have noted, in order for a structure to be diagnostic of endothermy, it must be clearly and exclusively functional with high metabolic rates, and it must not be present in ectotherms. Claims of the discovery of such a feature have been made by Willem Hillenius and John Ruben.

The turbinates, or conchae, of birds and mammals are complex, thin, folded bony scrolls in the nasal passages that bear, in addition to the well-known olfactory epithelium, used for smell, an exclusively respiratory maxilloturbinal that has a clear functional relation with endothermy (Hillenius 1992, 1994, 207). Through a series of experiments Hillenius showed that the maxilloturbinals in live rats, squirrels, ferrets, rabits, and opossums conserve water, recovering moisture from the breath of these mammals; in other words the respiratory turbinates work as a humidifying and dehumidifying system, reducing water losses associated with rapid, continuous pulmonary ventilation. The conchae of reptiles, in contrast, are relatively simple, recurved structures, bearing the olfactory epithelium but lacking the complex maxilloturbinals, the respiratory turbinates.

Nasal respiratory turbinates are a rigorous indicator of metabolic status and are found only in mammals and all extant birds (Bang 1971); they should therefore be excellent indicators of endothermy in extinct vertebrates. John Ruben has recently used computed axial tomography, or CAT-scans (CT-scans), to examine the fine details of the nasal region of dinosaurs. He has determined, for example, that a small theropod, the tyrannosaurid *Nanotyrannus*, exhibited well-developed olfactory turbinates but had no respiratory turbinates, a condition reminiscent of the nasal region of crocodilians; to Ruben, this is strong evidence of ectothermy (1996). After preliminary evaluation of other specimens, Ruben has concluded that respiratory turbinates were also absent in the dromaeosaurian theropods, as well as *Archaeopteryx* itself, suggesting that not only were the convergently birdlike theropods ectotherms, but that fully developed endothermy may have not appeared particularly early in avian evolution, and certainly well after the appearance of a body covering of contour feathers.

Dinosaurs Were "Pea-Brains." In spite of numerous recent claims of "intelligent dinosaurs," the vast majority of dinosaurs had tiny brains: the brain of *Stegosaurus* was the size of a walnut and some twenty times smaller than its "sacral brain"—an enlargement of the spinal cord to accommodate the motor activity of the massive hind limbs. Most of the excitement about dinosaur brains was generated by James Hopson (1977, 1980), who argued for a general correspondence between the brain size of dinosaurs and their metabolic rates. Within mammals, however, relative brain size and metabolic rate have very little correlation. Dinosaurs have historically been aptly characterized by the extremely low ratio of their brain to their body weight (appropriately scaled). According to Hopson's revisitation of the problem, the sauropods have the lowest encephalization quotient (EQ), as expected, and when compared to crocodiles, only ornithopods and theropods score substantially higher, with the highest EQ going to the late Cretaceous two-meter-long (6.5 ft.) "coelurosaur" *Troödon* (*Stenonychosaurus*)—also the sole theropod to overlap with birds and mammals. Dale Russell (1969, 1972) originally compared the brain of this late Cretaceous "coyote," an opportunistic carnivore

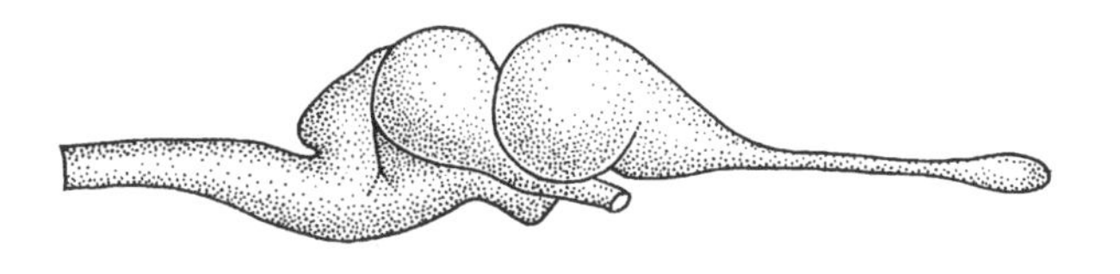

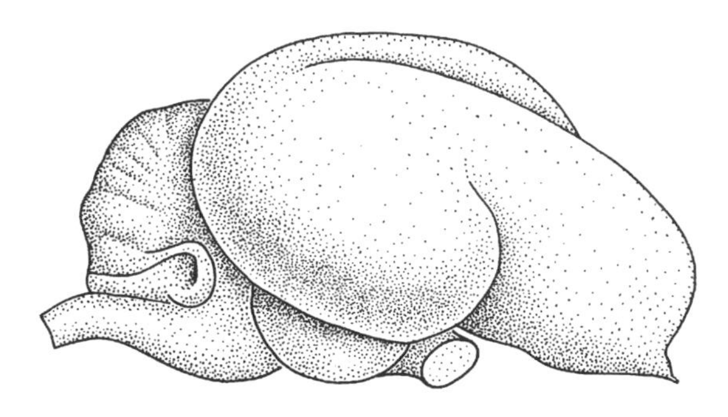

Above left, brains of a monitor lizard (*Varanus*), a typical reptile brain; *right,* a macaw (*Ara*), a typical avian brain, drawn to the same scale. Note the mass development of the integration centers, especially of the well-developed cerebral hemispheres and cerebellum characteristic of the avian brain. This type of brain development was approached within reptiles *only* within some pterosaurs. (Modified after Portmann and Stingelin 1961)

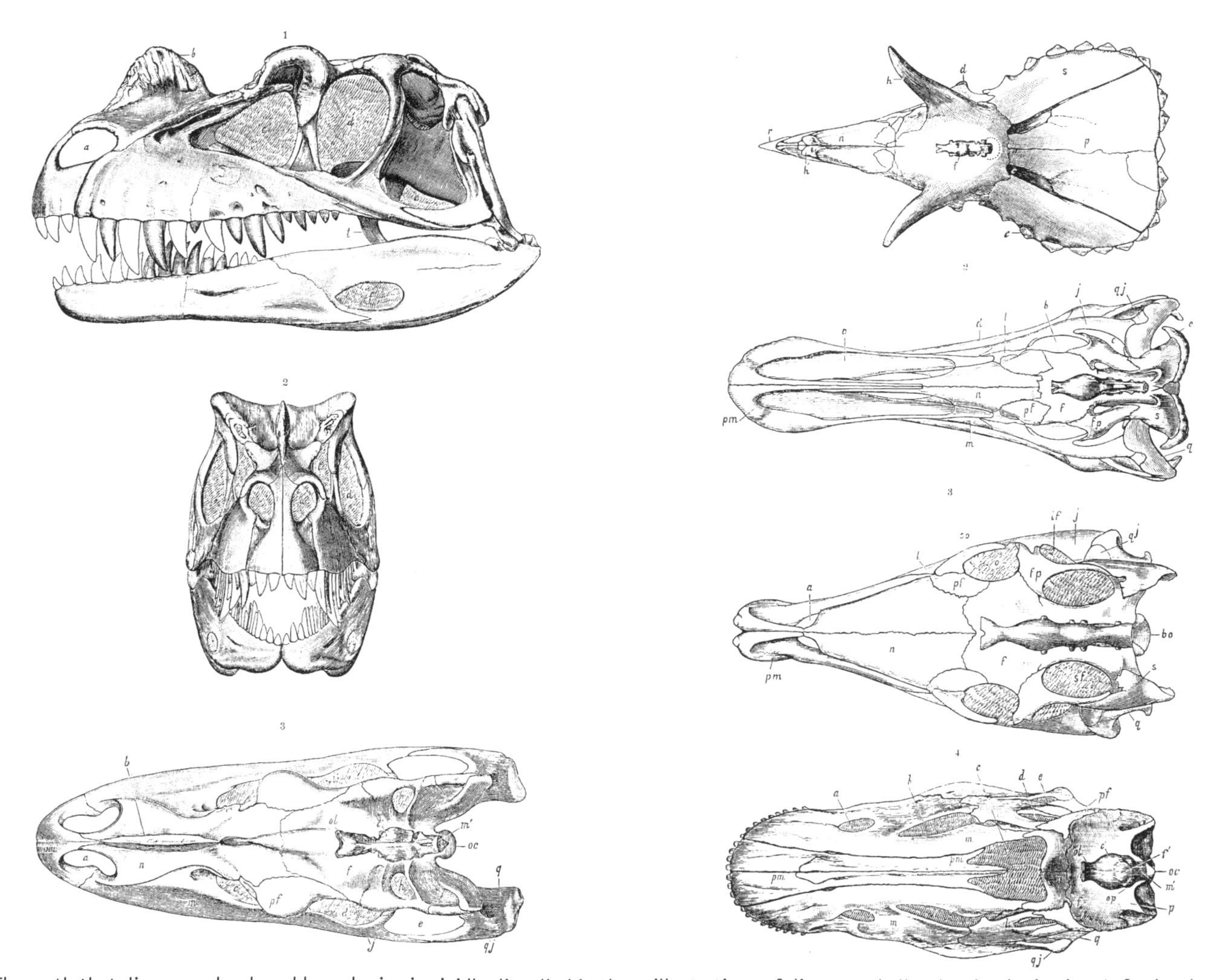

The myth that dinosaurs developed large brains is vividly dispelled in these illustrations of dinosaur skulls, showing brain size. *Left,* the theropod *Ceratosaurus* (Upper Jurassic), in lateral frontal and dorsal view (with brain cast); *right,* dorsal views of the skulls with brain casts of, *from upper to lower,* the ornithischians, *Triceratops* (Upper Cretaceous ceratopsian), *Claosaurus* (Upper Cretaceous primitive hadrosaur), *Camptosaurus* (Upper Jurassic ornithopod), and the sauropod *Diplodocus* (Upper Jurassic to Upper Cretaceous). (From Marsh 1896)

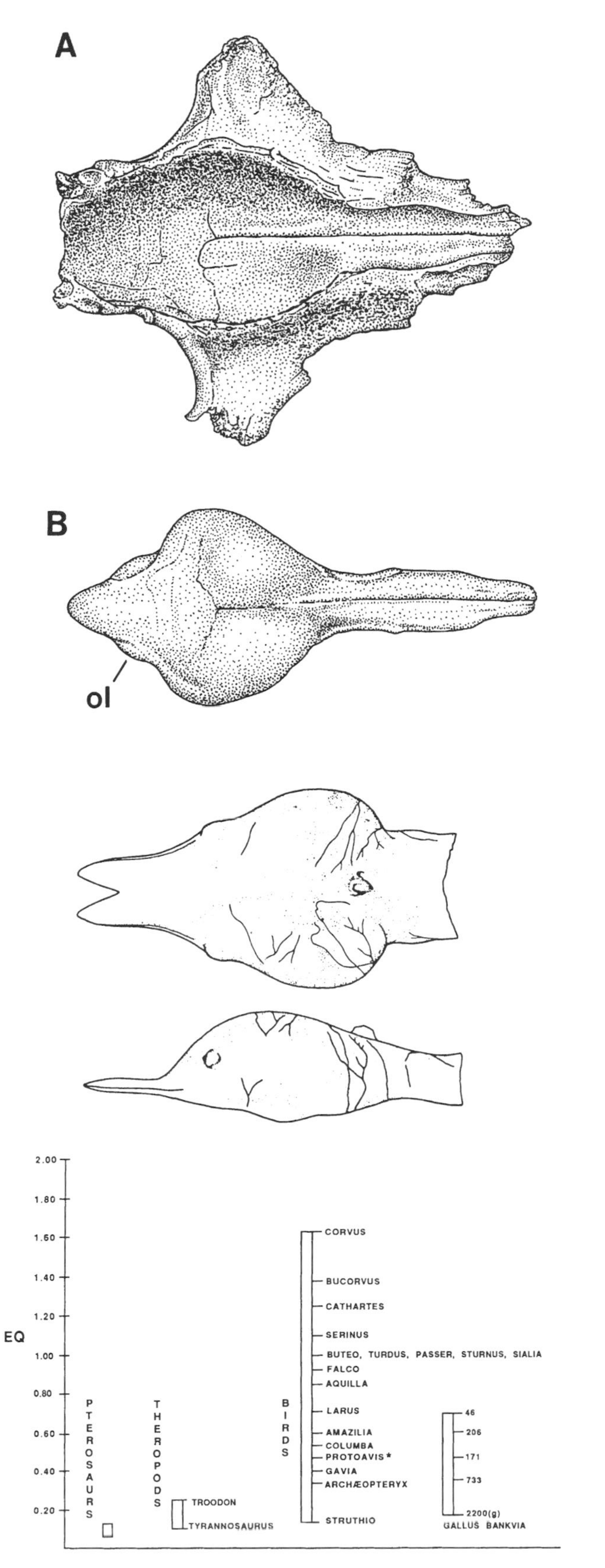

Large-brained dinosaurs (small by bird standards) are restricted to the Upper Cretaceous *Troödon* (*Stenonychosaurus inequalis*) and a few other ornithomimids, such as *Dromiceiomimus*. *Left,* the largest-brained dinosaur, the Upper Cretaceous *Troödon: A,* posterodorsal region of skull, in ventral aspect (× 0.4); *B,* endocranial mold, *ol* = optic lobe. *Below,* the emu-mimic *Dromiceiomimus brevitertius: right,* skull (× 1/4), in dorsal and lateral views; *left,* endocranial mold (× 1/3), in dorsal and lateral views. Although the general relation of brain to body weight may have approached that of an ostrich (*Struthio*) of similar size, the ostrich is derived from flying ancestors and has evolved a disproportionately large body size with respect to its brain size. The general morphology of the brain of ornithomimid dinosaurs differs dramatically from that of birds; they lack, for example, an expanded cerebellum, among many other features. Aside from these latest Cretaceous ornithomimids, dinosaurs had brains that were typically reptilian in size. (*Right,* from Russell 1969; *left,* from Russell 1972; both courtesy Dale Russell)

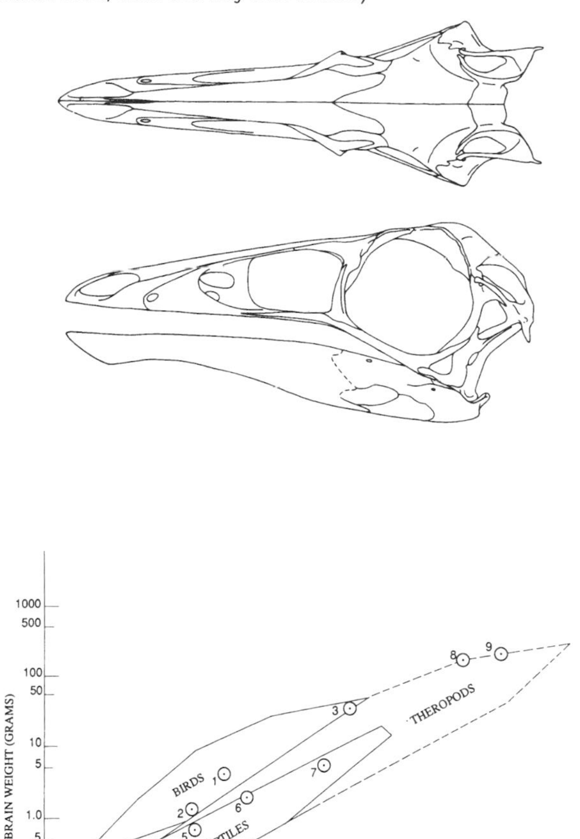

Above, brain to body size relations in some theropods, pterosaurs, and birds: *1, Protoavis; 2, Archaeopteryx; 3, Troödon; 4, Pterodactulus; 5, Rhamphorhynchus; 6, Scaphygnathus; 7, Pteranodon; 8, Allosaurus; 9, Tyrannosaurus*. (From Chatterjee 1991, fig. 18 [modified from Jerison 1973 and Hopson 1980]; courtesy Royal Society, London)

Encephalization Quotients (EQs) for pterosaurs, theropods, and birds calculated using the average mammal equation of Jerison (1973), where EQ = brain weight/[0.12(body weight)$^{0.66}$]. EQs for pterosaurs were calculated from Jerison (1973), for theropods from Hopson (1980), and for recent birds from Crile and Quiring (1940). Note that *Buteo* (hawk), *Turdus* (robin), *Passer* (sparrow), *Sturnus* (starling) and *Sialia* (bluebird) lie at 1.00, whereas *Corvus* (crow) is at the apex. *Archaeopteryx* and *Protoavis* lie at the lower range of EQ, whereas ratites are found at the bottom. This is because ratites evolved their large size through neoteny: their body size has increased allometrically with respect to their brain size. Body weight is inversely correlated with EQs (righthand bar): a large individual (2,200 g, 7 lbs., 12 oz.) of a leghorn fowl (*Gallus*) shows a smaller EQ than that of the small individual (46 g, 1.6 oz.). This in part explains the high position enjoyed by the small late Cretaceous dinosaurs *Troödon* (1.75 m, 6 ft; 45 kg, 100± pounds) and *Dromiceiomimus* (2.9 m, 10 ft.; and 80 kg, 175 lbs.). (From Chatterjee 1991, fig. 19; courtesy Royal Society, London) with enormous eyes, favorably with that of certain ratites: "It is . . . possible that *Stenonychosaurus* [=*Troödon*] had a brain . . . comparable to, if not larger than those of living ratites," he wrote in 1969 (599). In 1972 he described *Dromiceiomimus* (emu-mimic), an ornithomimid, or ostrich dinosaur: "The general relation of brain size to body weight . . . appears to be similar to that of an ostrich of comparable size" (390).

These dinosaurs were no doubt large-brained, but they have been compared to birds that have the smallest brains of all birds, the living ratites, and the ratites are birds that have evolved their large size through neoteny, or arrested development, whereby the body size has increased dramatically and allometrically with respect to brain size. At the other end of the spectrum, comparisons are made with crocodilians, another inappropriate choice, because these reptiles, being semiaquatic, have an unusually high specific gravity, and therefore by these measurements their relative brain size appears diminished. The overall average EQ of dinosaurs, by Hopson's calculations, is similar to that of modern reptiles, and certainly does not support the contention that dinosaurs were endotherms. In addition, it is in the latter half of the Cretaceous—especially in the late Cretaceous—that the "large-brained" coelurosaurian dinosaurs occur, as do the superficially birdlike dinosaurs the ornithomimids (bird-mimics) and dinosaurs with features that are interpreted to indicate trends toward endothermy, such as the hadrosaurs. This is some 80 or so million years after *Archaeopteryx* and so could have little bearing on bird origins. Any adaptations that occurred in dinosaurs with any bearing on bird origins should be present in the earliest dinosaurs, such as *Eoraptor, Herrerasaurus,* and *Coelophysis,* which occur in the late Triassic. Yet here one sees little in the way of birdlike features.

Experimentation in Thermoregulation. One clue indicating that dinosaurs as a group were not endotherms is that numerous unrelated dinosaurs developed various heat-exchanging devices on their backs. To mention a few spectacular examples, the Cretaceous ornithopod *Ouranosaurus* had a dorsal "sail," and the Cretaceous carnosaur *Spinosaurus* had a sail similar to that of the Permian pelycosaur *Dimetrodon* (which was also probably experimenting with thermoregulation in the line that led to the endothermic mammals). The popular North American museum *Stegosaurus* and its African counterpart *Kentrosaurus* had dorsal plates that were highly vascularized for heat exchange (Farlow, Thompson, and Rosner, 1976). Dorsal sails and plates helped warm the body in the morning sun, facing sideways and shunting blood into the "sails," which by mid-day could have been devoid of blood and faced directly into the sun's rays.

Large Size and Inertial Homeothermy. That dinosaurs could have maintained some degree of thermal stability, hence reasonable levels of activity—that is, that they were homeothermic—has been suggested many times as far back as G. S. Wieland in 1942. The hallmark of dinosaurs as a group is their large size. Only a few species weighed less than 5 kilos (11 lbs.), and numerous species weighed more than a ton. Nicholas Hotton III (1980) of the Smithsonian Institution has estimated that more than half of all species were as large as, or larger than, the largest living terrestrial mammals—among them, the African elephant, the hippopotamus, and the white rhinoceros. Many paleontologists believe that size was, in fact, the secret to the dinosaurs' success, both behaviorally and thermally. As far back as 1946, the preeminent paleontologist Edwin Colbert, along with physiologist Raymond Cowles and herpetologist Charles Bogert, showed that for alligators, the larger the basking animal, the slower the rates of heat absorption and loss. They concluded that by their size, ectothermic dinosaurs would have been reasonably stable thermally and in essence "homeotherms." Dinosaurs would thus have been able to maintain a relatively constant level of activity without having to eat huge amounts of food to generate endogenously produced body heat.

The larger dinosaurs in the warm, equable climates of the Mesozoic would have maintained nearly uniform body temperatures without the incredible physiological and ecological burden of endothermy. Spotila and coworkers (Spotila et al. 1973; Spotila 1980) demonstrated that, based on heat contributions from external heat only, reptiles with body diameters of 1 meter (3.3 ft.) living in a subtropical climate would have had a mean body temperature of approximately 34°C (95°F), with a daily fluctuation of less than ± 1°C. Contrarily, endothermic dinosaurs with these dimensions would require special structures or behavioral patterns to dissipate body heat, as such large mammals as elephants do today. It seems fair to say that ectotherms the size of dinosaurs could have achieved near mammalian temperature control with a typical reptilian metabolism, without burdening themselves with enormous food consumption and the other problems associated with true warm-bloodedness. In 1980, Hotton introduced the term *inertial homeothermy,* now sometimes called gigantothermy, to describe this condition.

Inertial homeothermy, then, makes sense for the larger dinosaurs. But many of the smaller theropod dinosaurs of the late Cretaceous, the ornithomimids (bird-mimics), were too small to have been inertial homeotherms. They were lightly built and showed considerable cursorial adaptations; in many features they converged on such ground-dwelling birds as chickens (*Gallimimus*) and ostriches (*Struthiomimus*). Whether these small dinosaurs became active only after acquiring heat from the morning sun or produced endogenously generated heat we shall never know, but as McNeill Alexander aptly points out, "Endothermy is a matter of degree" (1989, 94), and these small Cretaceous dinosaurs may have employed a variety of thermal strategies.

How these thermal strategies of the late Cretaceous can be linked into the Triassic or early Jurassic origin of birds, however, is an even more enigmatic problem, especially considering recent evidence for an ectothermic *Archaeopteryx* (Ruben 1991; Chinsamy and Dodson 1995). Small dinosaurs, if endothermic, would have had to have an insulatory covering to conserve the heat, once generated, because their large surface areas would quickly dissipate what heat had been produced. Indeed, no feathered dinosaur that would support Robert Bakker's theory of dinosaur endothermy has ever been discovered. The one dinosaur that should have been feathered and might have some bearing on bird origins, the chicken-sized *Compsognathus*, is preserved devoid of feathers in the fine-grained Solnhofen limestone, and if feathers had been present in *Compsognathus* they would definitely have been preserved (Ostrom 1978a).

Endothermy Is Costly. As we have seen, dinosaurs probably had a wide range of physiological strategies and activity levels, and in most cases their size would have given them a relatively constant and high body temperature. In fact, endothermy would not necessarily have been a good thing, especially in the warm climates of the Mesozoic. For one thing, it is expensive. Lions consume their body weight every nine days, the Komodo dragon does so every ninety days (McGowan 1991, 155). And the amount of food needed for endothermic activity, especially for the large sauropods, would have been astounding, with an endothermic "Supersaurus" needing to consume some 2,267 kilos (5,000 lbs.) daily, compared to some 227 to 454 kilos (500–1,000 lbs.) for an ectothermic counterpart (Ostrom 1978b, 172). In his review of Robert Bakker's *Dinosaur Heresies*, John Ostrom noted, "With a huge head (often making up fifteen percent of its body weight) and large grinding molars, a six-ton African elephant spends as much as fifteen hours a day finding, picking, and eating its daily food requirement—about three hundred pounds of leaves and limbs. At twenty-five to thirty tons, an endothermic brontosaur would have had to consume at least two thousand pounds of fodder a day, a feat that seems inconceivable, given that the creature's head was about the size of a horse's but lacked a horse's grinders" (1987b, 61). Small heads, mouths, and teeth were characteristic of all of the herbivorous dinosaurs, with the exception of most ornithopods and the ceratopsians. Compounding the difficulties, for most of the Mesozoic era high-nourishment plants were absent; flowering plants (angiosperms), including grasses, fruits, grains, and tubers, had not yet appeared. Available were plants equivalent in protein and caloric value to that of the modern gymnosperms, with which few living animals bother. Such plants as ginkgoes, cycads, pines, and firs were the rule, and most herbivorous dinosaurs, except duckbills and ceratopsians, died out long before high-energy plant sources began to appear in the early Cretaceous. With the appearance of Cretaceous flowering plants there is a broad pattern of dinosaur "gut food-processors" declining in favor of "oral food-processors" (Norman 1992, 181).

With their relatively minute heads, small mouths, and tiny teeth, the large sauropods would have had to crash through the landscape, eating constantly, to have been able to meet the nutritional demands of endothermy. Most mammals use about 90 percent of daily food energy consumed simply to maintain a constant body temperature! Modern reptiles, in contrast, need far less energy to survive. When Bennett and Nagy studied free-ranging fence lizards (*Sceloporus occidentalis*) in 1977, they found that daily field metabolism was just 3 to 4 percent that of a bird or mammal of equal size. We must not view ectothermy as an inferior way of life but merely as an alternative evolutionary way of life. One need only look at the highly active, large tropical reptiles, such as Nile crocodiles and Amazonian anacondas, to see how ectotherms have surpassed endotherms in the warm climates of the tropical zone. If the present is any key to the past, it should be evident that, owing to the physiological restraints of surface-volume relations, endotherms would tend to become large in cold climates, and ectothermic dinosaurs would have become large in warm, moderate climates, such as those of the Jurassic and Cretaceous.

Endothermy versus Ectothermy in Early Modern Birds. Although most authors have considered feathers to have originated in the context of insulation and heat regulation for endothermic birds or their predecessors, a number of workers see no evidence to indicate that the urvogel was an early endotherm but, rather, that it was possibly an ectotherm (Pennycuick 1986; Ruben 1991; Martin 1992; Ruben 1995, 1996; Chinsamy and Dodson 1995). Ruben (1991, 12) has noted that an ectothermic *Archaeopteryx* would have still been capable of anaerobically powered, flapping flight of a relatively short distance and duration but that it could have made such flights frequently. Likewise, "some aspects of modern avian physiology, including homeothermy, were not well developed in *Archaeopteryx*, and . . . the organism was not well suited for sustained flight" (Martin 1991, 532). Even more recent circumstantial and indirect evidence from respiratory turbinates (Ruben 1995, 1996), noted earlier, and from growth rings in long bones of archaic Cretaceous opposite birds allied with *Archaeopteryx* (Chinsamy, Chiappe, and Dodson 1994; Chinsamy and Dodson 1995), suggesting that these birds grew like modern reptiles in a cyclic pattern, has provided additional suggestive evidence that *Archaeopteryx* may have indeed been ectothermic. And, as Ruben has noted, a fully feathered ectothermic *Archaeopteryx* could easily have achieved homeothermy.

Hillenius (1994, 225) noted that there are two main theories for the evolution of endothermy in birds and mammals. In the first scenario, high resting metabolic rates were developed for endothermic thermoregulation associated with the maintenance of constant body temperature independent of temperature fluctuations in the environment. This was the model preferred by Bock (1985, 1986), who proposed that obligate endothermy evolved in association with the environmental demands of protobirds in trees, where the ambient temperature is considerably cooler than the ground and there is greater shade and wind. Thus, protobirds would have had greater problems maintaining a constant elevated body temperature. Likewise, many authors have imagined endothermy in the mammalian lineage having evolved to cope with the fluctuating temperatures of Mesozoic nights, reasoning that the subdominant mammals had been forced into the nocturnal habitat, partitioning them ecologically from the dominant diurnal Mesozoic reptiles. This is supported in part because although mammals evolved at about the same time as dinosaurs, they did not begin their modern explosive evolution until the beginning of the Tertiary, following the dinosaurs' extinction.

In the second scenario, termed the "aerobic capacity" model, endothermy evolved in the context of expanded capacity for sustained, aerobically supported activity; in other words, it amplified the organism's ability to maintain strenuous activity (Bennett and Ruben 1979). The findings by Hillenius (1994) of slow temporal development of mammalian metabolic rates are more consistent with the second model, and Ruben (1995, 1996) has proposed that the same model applies to birds, pointing out that the capacity of birds to sustain long-distance powered flight is far beyond that of equivalent ectotherms. This model is concordant with the fact that most features of the anatomy as well as reproduction of modern birds strongly argues that flight preceded endothermy (Randolph 1994) and points to the possibility that selection may have favored increased levels of daily sustainable activity supported by a dramatic increase in metabolically expensive, high-endurance, oxidative-type skeletal muscle fibers (Ruben 1996).

By the above model, then, avian endothermy would not have evolved directly in association with flight itself but rather in association with increased capacity for long-distance flight, which ectothermic flying birds could not have attained (Ruben 1996). Ruben has suggested further that avian endothermy may not have evolved until the mid-Cretaceous, some 50 million years after the appearance of *Archaeopteryx*, perhaps in association with selection for long-distance migration associated with widespread Cretaceous aridity. However, the birds that are circumstantially thought to be ectothermic are *Archaeopteryx* and its close allies, the enantiornithines (opposite birds), the archaic birds of the Mesozoic. However, at about the same time, Lower Cretaceous, as the appearance of the earliest opposite birds, we have evidence, as we shall see in chapter 4, of the first modern birds in the form of *Ambiortus* and *Gansus*, from Mongolia and China, respectively. *Ambiortus*, for example, is essentially a modern, fully volant carinate, with a pectoral flight architecture virtually identical to that of modern birds. And although growth rings in the enantiornithines provided evidence for reptilian ectothermy, the late Cretaceous hesperornithiforms, ichthyornithiforms, and the "transitional shorebird" *Cimolopteryx* exhibit no such growth rings (Houde 1987a; Peter Houde and Anusuya Chinsamy, pers. comm.), suggesting that these birds had attained determinate growth and presumable endothermy. This is further evidenced by the fact that both *Hesperornis* and especially the ternlike *Ichthyornis* are found hundreds of miles out to sea, the equivalent of long-distance migration. It is tempting to suggest that the early ornithurine lineage of birds had attained aerobically based sustained activity and endothermy in the early Cretaceous, whereas the archaic opposite birds enjoyed the less demanding role of inertial homeotherms in the tropical Mesozoic. Looking ahead to chapter 4, could this be the main reason that only the endothermic ornithurines made it through the Cretaceous-Tertiary bottleneck?

Finally, if *Archaeopteryx*, clothed with body contour feathers, was ectothermic, then the evolution of feathers could only be explained in a context other than insulation for endothermy.

Debate Continues. The debate over whether dinosaurs were hot-blooded will of course continue. In a new line of questioning, Reese Barrick, a paleontologist at the University of Southern California, and William Showers, a geochemist at North Carolina State University, have attempted to measure body temperature in extinct animals by measuring the ratio of certain isotopes of oxygen in *Tyrannosaurus rex* specimens (Barrick and Showers 1994). According to this scheme, because oxygen-16 is taken up in higher proportions at higher body temperatures, different parts of the body should show differences if the body temperature varied along a gradient, as in modern ectotherms. This study and the theory behind it are premature, however, and such speculations on the physiology of extinct animals are extraordinarily hasty. In fact, geochemist Yehosua Kolodny, who pioneered these isotope studies, has questioned whether fossil bone actually preserves the original "signal" from the animal, pointing out that signals can be altered during burial, when bone is mineralized fossil, or from groundwater (Morell 1994). Perhaps more important, moreover, if dinosaurs were somewhat warm-blooded and homeothermic because they were big—that is, they were inertial homeotherms—then how could such an analysis distinguish between homeothermy, which gives no clues about metabolic rate and physiological energetics, and endothermy? Additional objections to the study by Barrick and Showers have been raised (Millard 1995; Chinsamy and Dodson 1995; Ruben 1995, 1996); and the oxygen isotope study suffers perhaps mostly because its underlying assumptions are not in accord with the physiological literature, as "extremity vs core body temperatures in extant birds and mammals are often as variable as those of ectotherms" (Ruben 1995, 88). Yet, as one might guess, Barrick and Showers are already discovering (Folger 1993) that the dinosaurs were, as Bakker surmised nearly two decades ago, hot-blooded!

In reviewing "dinosaur energetics and thermal biology," James Farlow lamented the entire hot-blooded dinosaur debate:

> Unfortunately, the strongest impression gained from reading the literature of the dinosaur physiology controversy is that some of the participants have behaved more like politicians or attorneys than scientists, passionately coming to dogmatic conclusions via arguments based on questionable assumptions and/or data subject to other interpretations. Many of the arguments have been published only in popular or at best semi-technical works, accompanied by rather disdainful comments about the stodgy "orthodoxy" of those holding contrary views; what began as a fresh way of considering paleontological problems has degenerated into an exercise in name-calling. All of this has made the whole field of dinosaur studies suspect in the minds of many scientists. (1990, 43)

Feather Origins

Feathered, hot-blooded dinosaurs aside, we can now return to the central topic—namely, the origin of feathers. As mentioned earlier, in 1975 Philip Regal proposed another thermoregulatory theory, that feathers may have evolved from reptilian scales to serve as shields against hot sunlight. Regal showed that scales in several genera of modern lizards are relatively more elongated in warm climates than in cooler ones, and he argued from his experiments that the longer scales, by improving microcirculation patterns and by casting shadows, act as a barrier against solar radiation. In Regal's model, the scales of the reptilian ancestors of birds progressed toward featherlike structures because they permitted the animals to be more active in the hot midday sun; in other words, elongated scales or protofeathers were heat shields. The subdivision of scales would also have added flexibility to an otherwise cumbersome, rigid structure, and subdivisions, like modern feathers, would have enhanced the scales' reflective capabilities. Although it is speculative, especially regarding the context for the early evolution of feathers, Regal's model is certainly plausible, and it works equally well with feathers evolving for thermoregulation or for an early aerodynamic function. Regal's elongated scales would have first barred the influx of heat. After endothermy evolved, the same scales or their modified epidermal descendants would thus have been preadapted for aerodynamic functions. There's one problem, however: the hypothesis fails to account for the complex, aerodynamically adapted microarchitecture of the avian feather, which would represent gross morphological overkill for evolving

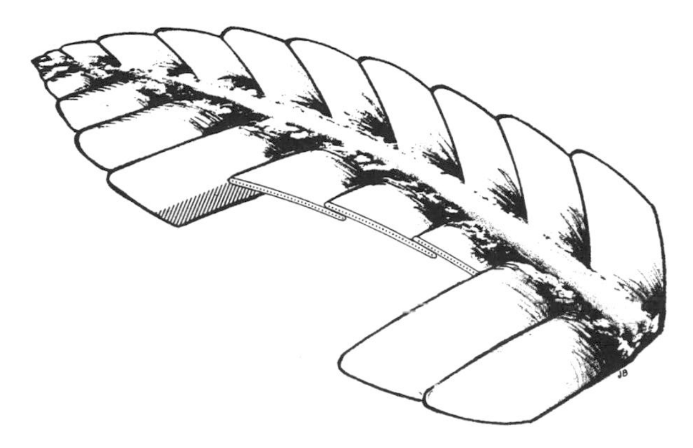

A hypothetical intermediate stage between an enlarged scale and a feather as envisioned by Philip Regal. (From Regal 1975; courtesy *Quarterly Review of Biology*)

such complex integumentary appendages for any function other than flight.

General models for the invention of feathers, at least in an embryological sense, seem plausible. The earliest processes—cellular organization of the epidermis, the formation of special epidermal thickenings called placodes, invagination (or in-pocketing)—are features common to feathers and reptilian scales. And in the embryos of various breeds of scale-footed domestic fowl, feather primordia appear as conical projections during development at the margins of foot scales, and in some breeds, down feathers may persist on the feet for weeks after hatching (Rawles 1960, 208–209). But feathers are not simply modified scales; they are the products of a genetic and developmental reorganization of both the individual feathers and of the feathers' arrangement in thin tracts over the body surface. Even birds' most scalelike features—the leg scutes (scales), claws, and the epidermally derived beak—are formed from a single category of protein, the ø-keratins. As Alan Brush has written regarding feather development, "The genes that direct synthesis of the avian ø-keratins represent a significant divergence from those of their reptilian ancestor. . . . Regulation of morphological details of feathers is, to a great degree, the result of follicular timing and differential growth rates. . . . Thus a plasticity of shape emerged based on control of timing, while the constructional materials and processing remained relatively unchanged" (1993, 152). As regards reptiles, of course, their "scales" are anything but uniform, ranging from the tiny scales of small anolid lizards to the large scutes of crocodilians, and reptile scales are even shed in a variety of ways, ranging from complete ecdysis, or molting, of the outer layer, the stratum coreum, exemplified by the well-known dried "snake skins," to the sloughing of individual epidermal scutes in turtles. In each case, new mutations reorganized the integumentary developmental ontogeny and adult characteristics, or phenotype.

All of this, however, does not prevent us from trying to reconstruct the selective context in which the initial feather evolved. The appearance of feathers no doubt followed mutations and some developmental reorganization. Some genetic material was probably duplicated, enabling a rapid change from scales to feathers.

Feathers in Narrow Tracts. Although Regal's theory that feathers evolved first as heat shields has gained considerable acceptance, it shares similarities with the theory that feathers evolved to insulate endothermic dinosaurs. That is, both theories envision feathers as having evolved in a context other than aerodynamic, their current primary function. Both theories would initially require that the body be covered evenly and entirely by modified protofeathers, because reptiles have a uniform, complete covering of scales. A critical stumbling block, therefore, is that feathers do not, from the most primitive to the most advanced state in birds, cover the entire body. In fact, just the opposite: feathers are arranged in definite rows, or pterylae, and between these feather tracts are apteria, areas devoid of any integumentary structures. Pterylae are often quite narrow and apteria are very extensive; as Storrs Olson has put it, "It is remarkable just how few feathers are actually needed to cover a bird completely. . . . this may be best appreciated in herons (Ardeidae), in which two rows of two feathers serve to cover most of the dorsal and the ventral surfaces of the body" (1985a, 86).

These pterylae are deeply ingrained in the genetic and developmental program of birds: the tracts are well defined between the ninth and tenth days of development. Each tract is composed of a specific number of primoidea that occupy definitive positions in regard to each other and to the tract as a whole (Rawles 1960), and the number of feathers of the adult bird corresponds closely with the number of embryonic papillae in the tracts. And these developmental facts support the theory that the avian integument was completely reorganized. As an aside, little has been made of the fact that even avian skin, though it is composed of, as in other vertebrates, an outer epidermis and an underlying dermis, is also different from the skin of any other vertebrates.The skin of flying birds is paper-thin, perhaps as an adaptation for reducing weight for flight, and it is virtually devoid of glands.

Even birds with feathers uniformly distributed over the body have vestigial or embryonic apteria (Olson 1985a; Clench 1970). Such "primitive" birds as the ostrich and penguins, do have uninterrupted feather follicles. In the ostrich embryo, however, the feather follicles are arranged in definite feather tracts, proving that continuous feathering is secondary (Beddard 1898; personal observation). And the plumage of penguins has been completely reorganized to deal with propulsion through water and thermoregulation in extreme environments (Taylor 1986).

The arrangement of feather follicles in pterylae, or feather tracts, is as much a characteristic of birds as the possession of feathers itself, and as Storrs Olson notes, "There is no reason to regard this as being anything other than the primitive state in the class Aves" (1985a, 86). This lack of a continuous covering of feather follicles in birds would appear to render unlikely any theory to account for early feather evolution in a thermoregulatory context, whether as heat shields or as insulation for endothermy, in that feathers evolving in any thermoregulatory context would have to be evenly distributed over the body to be effective.

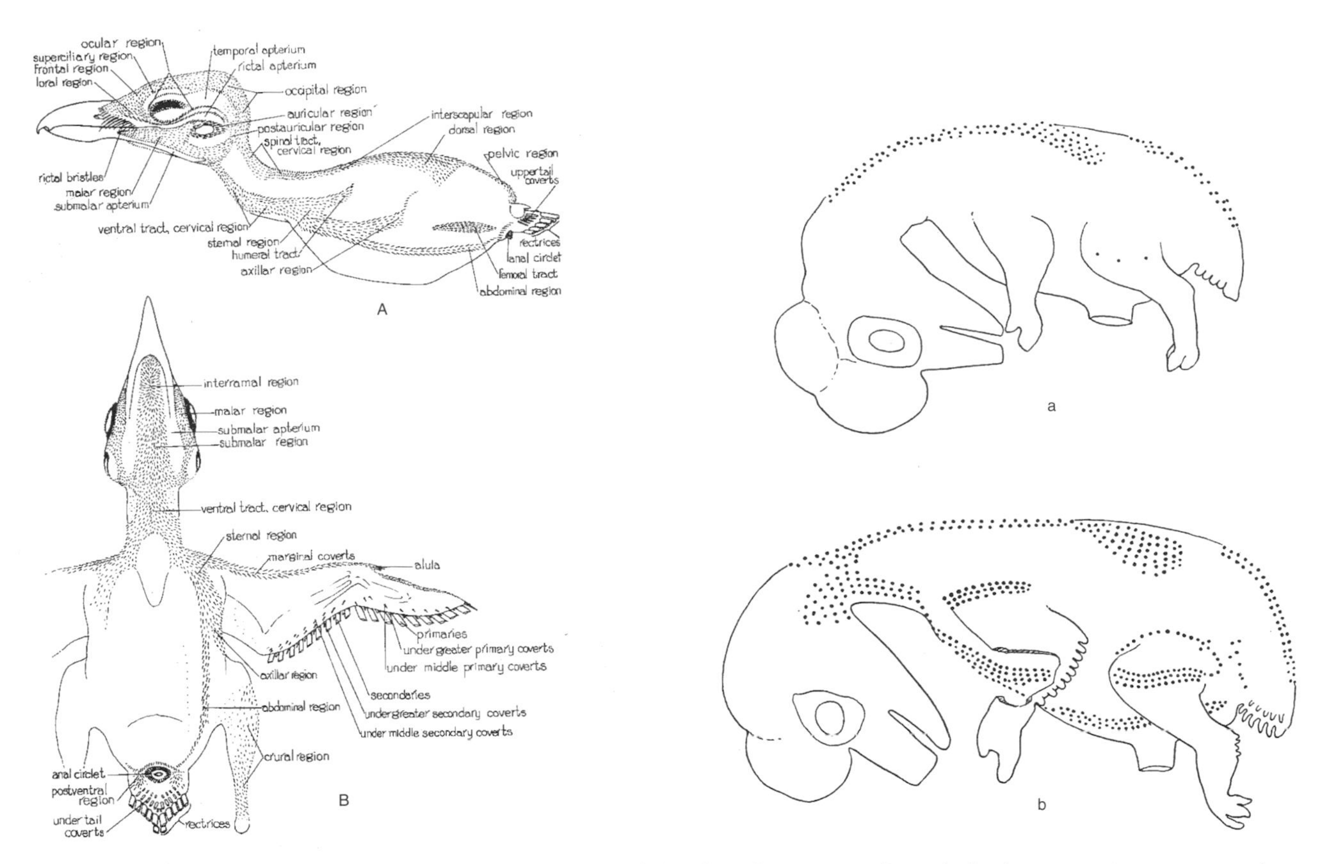

Right, embryos of the northern flicker (*Colaptes auratus*), showing relative time of emergence of certain feather tracts: *A,* a 13.5-mm embryo showing spinal tract, three papillae on crural tract, and first five rectrices; *B,* a 17-mm embryo showing, in addition to first stage, the outermost tail feathers, first five upper tail coverts, ventral tract humeral tract, secondaries, crural tract, femoral tract, and posterior margin of capital tract. *Left,* adult feather tracts, or pterylae, of the loggerhead shrike (*Lanius ludovicianus*). (*Right,* from Burt 1929; *left,* from A. H. Miller 1928)

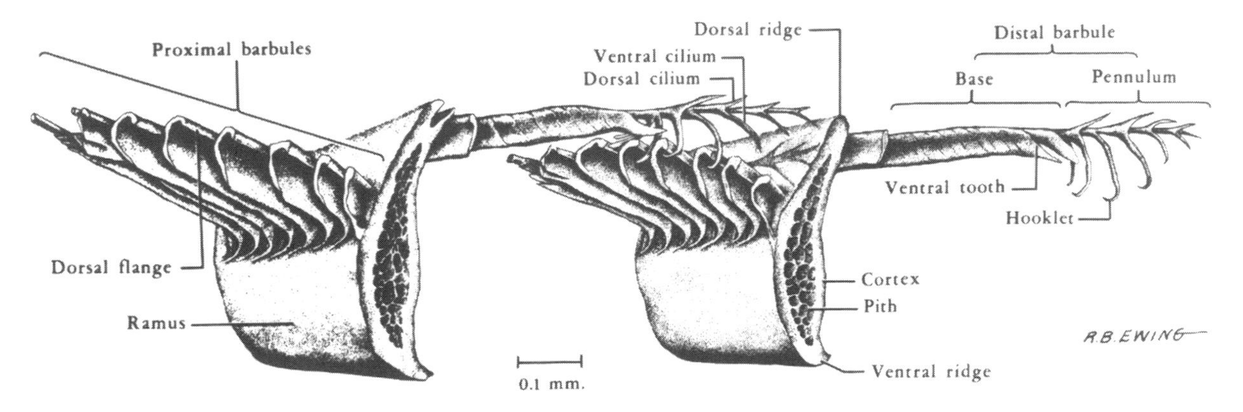

Segments of two pennaceous barbs from a contour feather of a chicken (*Gallus*). The barbs are seen obliquely from the distal end to show the interlocking parts. The two barbules, in side view, are distally armed with minute hooklets, or hamuli, to form an interlocking mechanism between the adjacent barbs. Each distal barbule has some four to six hooklets, which catch on to the flanges of the proximal barbules, somewhat similar to the way hooked and looped parts of Velcro fasteners come together. It is not difficult to visualize that feathers are the most complicated integumentary structures, both morphologically and embryologically, ever produced by the vertebrate integument. (From Lucas and Stettenheim 1972; courtesy Peter Stettenheim)

Why Evolve Feathers? Indeed, the simplest and most profound objection to all thermoregulatory theories is: why feather? Feathers are extremely complex both structurally and embryologically; they are "morphogenetically the most complex appendages produced by the epidermis in any vertebrate" (Spearman and Hardy 1985, 17). Almost every structural detail of feathers has aerodynamic significance. Feathers are a near-perfect aerodynamic design. They are lightweight (an adaptation for flight), possess an unusually high strength-to-weight ratio, have graded flexibility, possess a ventral reinforcing furrow on the ventral aspect of the shaft, and have great resilience—when broken, they come back together in original aerodynamic form, and their modular construction limits damage. Feathers create individual airfoil cross-sections, slotted wings, and wings with overall airfoil cross-sections. Their smooth, aerodynamic contours produce laminar flow on the body, thus reducing drag. Last, they provide waterproofing by the fine structure of the vane (Dyck 1985), which is enhanced by preening with oil from the uropygial gland. Especially important is the feather's ability to be repaired immediately upon injury. When a feather is struck a blow, the structure will momentarily split apart, but microscopic hooklets on the tiny barbules connecting the barbs will quickly reattach themselves and reestablish the feather's smooth aerodynamic contour, the vanes coming together like Velcro. As D. B. O. Savile has put it, feathers "allow a mechanical and aerodynamic refinement never achieved by other means" (1962, 161).

Another curious point is that body, or contour, feathers are virtually identical, down to the tiniest detail, to the aerodynamically designed flight feathers. They are indeed miniature flight feathers and are thought to derive from them (Parkes 1966, 83). This fact alone becomes almost paradoxical if feathers evolved originally for insulation. If feathers evolved initially as thermoregulatory structures, then the contour feathers of modern birds should be essentially similar to the original feather—that is, flight feathers should be secondary. They also should be uniform in structure in both flying and flightless birds, which is not the case.

For purposes of thermoregulation, why not a simpler structure, such as mammalian hair? In the flightless ratites—the ostrich, rhea, cassowary, and kiwi—for which feathers have only thermoregulatory value, they are very loosely constructed and superficially resemble hair. Gavin de Beer (1956) saw the degeneration of feathers in ratites as evidence for neoteny, or arrested development, in the group, and the fluffy, hairlike "juvenile" feathers of many flightless species, such as the Inaccessible Island rail (*Atlantisia rogersi*), actually produce a mouselike appearance (Lowe 1928; Stresemann 1932; see also chapter 6). Speaking of the flightless rail *Atlantisia*, Percy Lowe noted in 1928 that its body had a "distinct appearance of hairiness" and noted that in many flightless rails "the barbules of the wing-feathers . . . exhibit a failure to develop along the lines of the normal volant barbule"; he even suggested that flightless rails could be divided according to the presence or absence of "a juvenile or undeveloped condition of these barbules" (1928, 104, 131).

This difference in feather structure is more than superficial; scanning electron micrographs show that the feathers of flightless forms lack many of the hooklets and barbules typical of flight feathers. Explained in evolutionary terms, the release from strong selection pressure to maintain aerodynamic feathers results in a loss of feathers' distinctive characteristics. Although McGowan (1989) has challenged this notion, his arguments are seriously flawed; as an example, among the birds he studied were penguins, which are in fact perhaps the strongest of all "flying birds," having to propel themselves through the constraining, dense medium of water. And penguin feathers, as is well known, have undergone complete reorganization as adaptations to extremes in cold and underwater propulsion (Stettenheim 1976; Taylor 1986). There can be no doubt that when birds become secondarily flightless, their feathers degenerate; the amount of degeneration varies in different types of birds and may reach the feather microstructure. Although this in itself does not prove that feathers evolved in an aerodynamic context, it is exactly the opposite of what one would expect if feathers evolved originally as insulatory devices, heat shields, for sexual selection, or in any context other than an aerodynamic one (Feduccia 1985a, 1995a).

Feathered Dinosaurs? In spite of the detailed aerodynamic structure of feathers, various authors have insisted that feathers evolved for insulating hot-blooded dinosaurs, as we have seen, and many dinosaurs have been portrayed with a coating of aerodynamic contour feathers with absolutely no documentation (Bakker 1986; Paul 1988). A shocking portrayal on the cover of *Time* magazine in April 1993 of the late Cretaceous flightless, turkey-sized theropod *Mononykus* (Perle, Norell, Chiappe, and Clark 1993), incorrectly thought to be a bird (Feduccia 1994a; Ostrom 1994), shows *Mononykus* with the fully developed contour feathers of a flying bird. What's more, in a commentary on the description and phylogenetic significance of *Mononykus* later that year, Angela Milner asserted that "feathers may, therefore, have been widespread among bipedal carnivorous dinosaurs as an insulating outer layer" and that "down feathers must have arisen, initially as a means of insulation, in the lineage of manirap-

The flightless Inaccessible Island rail (*Atlantisia rogersi*); its feathers are decomposed and superficially hairlike. (From Lowe 1928)

A feathered "gracile *Coelophysis rhodesiensis* (=*Syntarsus*) in a fast run," according to Gregory Paul (1988, 264). There is no evidence that any dinosaur possessed feathers; feathers are absolutely unique to birds. (Modified from Paul 1988)

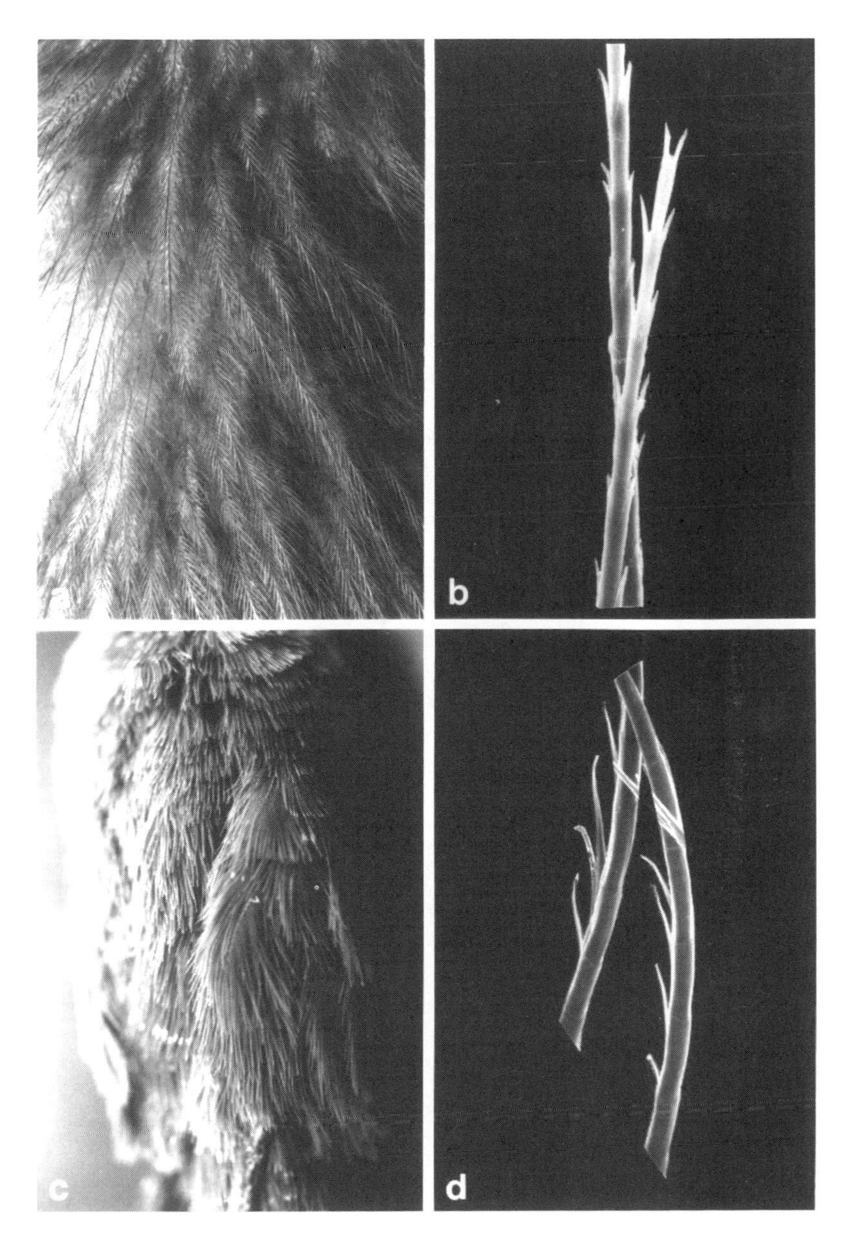

Degenerate feathers of flightless birds. With the evolution of flightlessness, feathers lose their complex structure, hooklets become vestigial, and birds gain a hairy appearance. *Above, a,* feathers of the brown kiwi (*Apteryx australis*), in superficial view, and *b,* a scanning electron micrograph (× 400) of a barbule from the vane of a contour breast feather showing vestigial hooklets. *Below, c,* feathers of the Inaccessible Island rail (*Atlantisia rogersi*), in superficial view, and *d,* a scanning electron micrograph (× 500) of a barbule from the vane of a breast contour feather showing degenerate hooklets. (Photos by author; SEM of *A. rogersi* courtesy Roxie Laybourne; from Feduccia 1995a; courtesy Forschungsinstitut Senckenberg)

toran theropod dinosaurs from which birds arose" (589). Those who advocate the view that feathers arose in earthbound dinosaurs as insulatory structures need to explain a number of issues: (1) why one would expect dinosaurs to be insulated with down feathers that would be poorly suited in that they would become mucky and soggy in rain and therefore useless and maladaptive; young downy ostriches, if not protected under their parent's wings in rain, become soaked and chilled, and frequently die; (2) why dinosaurs would have down feathers since they are thought to be secondary in modern birds (see arguments by Parkes 1966 and Dyck 1985); (3) why feathers, the most complex of all integumentary structures, would be used to insulate dinosaurs when a simple hairlike structure would be just as efficient an insulator; (4) why feathers degenerate in secondarily flightless birds, and flightless birds from distinctive origins, from kiwis to rails, become "hairlike" in their general appearance and lose the structural integrity of their feathers; (5) why contour feathers, thought to be derived from flight feathers, also degenerate in secondarily flightless birds; (6) why pterylae in even the most primitive living birds are arranged in sparse rows, with broad, bare areas devoid of feathers; in its rudimentary state, such an arrangement would be useless for insulating dinosaurs; and (7) why virtually every structural detail of feather structure relates to aerodynamic function.

Finally, no feathered dinosaur has ever been found, although many dinosaur mummies with well-preserved skin are known from diverse localities. In addition, the search for feathers in the Solnhofen dinosaur *Compsognathus* produced nothing. John Ostrom stated that "if the speculative question about feathered 'coelurosaurs' can ever be answered, the Munich specimen of *Compsognathus* is the critical specimen to examine. . . . The reader can be sure that I made an exhaustive examination, . . . but to no avail. If feathers had been present in *Compsognathus*, it is inconceivable to me that no evidence of them would be preserved. . . . But the fine-grained matrix shows nothing. Thus, I conclude that *Compsognathus* almost certainly was not feathered" (1978a, 115–116).

Certainly, feathers, with their almost magical structural complexity, from both the embryological and adult morphological point of view, would be gross overkill for insulatory structures. As I've said elsewhere, "To clothe a ground-dwelling, endothermic reptile with feathers for insulation is tantamount to insulating an ice truck with heat shields from the space shuttle" (1985a, 78). One must also ask how feathers would evolve in the ground-up scenario of avian flight, when any extension of feathers either on the forearms or on the body would increase the frictional drag on a running biped and so would impede forward progress and would be selected against.

First Feathers as Aerodynamic Structures. Having covered many aspects of the flawed cursorial theory, and associated problems with clothing an earthbound reptile with feathers, let's now look at a much simpler scenario, that feathers evolved directly in an aerodynamic context. The theory that feathers evolved directly for flight was promoted by Heilmann (1926) and Savile (1962), was reviewed by Kenneth C. Parkes (1966), curator of birds at the Carnegie Museum of Natural History in Pittsburgh, and has been bolstered by Nachtigall (1977), Larry Martin (1983), Sam Tarsitano (1985), and myself (1985a, 1993b, 1995a). This model pictures the predecessors of birds as highly active, arboreal creatures that took to jumping from branch to branch, as many modern animals do. The initial stages might involve the evolution of adaptations that lessen the impact from accidental falls, which often occur in this setting. Along with flattening the body and holding the appendages in the horizontal plane, the slightest fringe of elongated scale or protofeather along the trailing edge of the forelimb would confer an immediate advantage in parachuting or jumping; at this stage, balance would be required, but little or no ability to steer. As Savile has pointed out, such elongation "is clearly adaptive in protecting from injury in accidental falls, as an escape mechanism and in pursuit of prey" (1962, 161). If any elongation of these scales was immediately adaptive, then the emergence of feathers as an aid to airborne travel can be explained without recourse to preadaptations (or exaptations) of one sort or another. The problem of accounting for the adaptiveness of stages intermediate between scales and feathers disappears.

Consider an analogy: if a common gray or red squirrel accidentally falls from a branch, or is shaken from it, the fall at first appears almost vertical. But instead of falling to the ground willy-nilly, the animal spreads its limbs and assumes the attitude typical of the flying (more accurately, gliding) squirrels. This maneuver allows the squirrel to swerve at an angle of as much as 60° and land relatively lightly. Thus, in arboreal animals, jumping and parachuting are immediately adaptive, and it is a short step to the evolution of a true patagium, or flight membrane, to permit gliding.

As Walter Bock (1986) has pointed out, in the evolution of novel features each step must be adaptive in terms of the selective agents acting on the organism. Natural selection cannot operate on features that are functionally inoperative, and, what is paramount, each stage in a pseudophylogeny, or chronological sequence of intermediate forms, should be represented by an analogous known organism.

In small, arboreal Triassic basal archosaurs (thecodonts) somewhat similar to *Megalancosaurus*, *Scleromochlus*, and *Longisquama*, extension of reptilian scales or their replacements, primitive featherlike structures, along the posterior aspect of the forearm, perhaps in conjunction with a "gliding" membrane, would convey an immediate adaptive aerodynamic advantage. Transformation of these protofeathers to light, sophisticated, aerodynamically designed feathers (Maderson 1972) would produce the refined flight mechanism necessary for the evolution of flight. Feathers have the advantage of being able to lengthen to produce stiff wing feathers for flight. In the evolution of a proto-wing, each stage would be fully adaptive in early reptilelike birds with short, rearward-projecting scales or scalelike protofeathers from the posterior antebrachium, or forearm. By this model any projection of these protofeathers would immediately produce drag and thus reduce the rate of descent. Each stage of elongation would likewise be aerodynamically advantageous.

A model for feather evolution from archosaurian scales advanced by Pennycuick (1972) shows that the lifting force acting on protofeathers must be transmitted to the limbs and body to act as a buoyant force. The evolution of a rachis, the functional strut of the feather, was inevitable for this reason, and it would also reduce the weight of the protofeather. Interestingly, the elongated scales of the small, arboreal Triassic thecodontian *Longisquama* (mentioned in chapter 2) show some of the characteristics one would expect in early protofeathers. In the tiny specimen one can clearly discern elongated scales that closely resemble what one might predict for integumentary structures intermediate between reptilian scales and feathers. Under low, raking illumination, the dorsal, featherlike scales of *Longisquama* exhibit transverse thickenings in series, perhaps as a result of repeated hypertrophy of germinal tissue, as well as a central thickening along the main axis of the elongated scales that closely resembles a feather rachis and functionally adds support to these featherlike scales (pers. obs.; pers. comm. P. Stettenheim). The extreme length of the scales relative to the specimen, as well as the transverse thickenings and crossbands, are almost exactly what one might expect to see in a protofeather. Interestingly, A. G. Sharov distinguished *Longisquama* from other Pseudosuchia "by the presence of long, featherlike appendages along the dorsum and concrescent clavicles [furcula]" and noted that "the structure of the dorsal appendages shows that they functioned as a kind of parachute, breaking the animal's fall as it jumped from branch to branch, or from the trees to the ground" (1970, 112). No doubt, as we have noted, the scales of *Longisquama* were not transmuted into feathers, but the specimen does show that there was tremendous experimentation in featherlike scales in the basal archosaurs before the advent of feathers. No Mesozoic reptiles or reptilelike organisms but urvogels and *Longisquama* exhibit feathers or featherlike scales as integumentary structures.

Early Bird Flight

A particularly important feature in the evolution of flight that usually receives scant attention involves size, and Jeremy Rayner emphasizes that "there are strong mechanical reasons . . . for expecting flight to have evolved in relatively small animals, certainly much smaller than the closest known non-feathered dinosaur *Deinonychus*" (1991, 192; see also Rayner 1985b and Pennycuick 1986). This, of course, could be construed to be a critical argument against dinosaurs having given rise to birds, because they are basically large animals with short forelimbs, exactly what one would not like to see in an avian ancestor, whether cursorial or arboreal.

Small size is absolutely essential for arboreal life and arboreal flight origin. Sam Tarsitano points out that "parachuting requires that the proavis be small, lightly built, and able to extend its limbs to present as much surface area as possible to the airflow" (1985, 321). Likewise, Walter Bock notes that a decrease in size is a correlate of specialization for arboreal life and that small size would confer a favorable mass–surface area relation for vertical climbing as well as for lessening the effect of impact should the animal fall or jump to the ground: "Specialization for arboreal life would result in decreased size of the tetrapod to a small size, possibly less than 500 gm or less, if the ancestral form was not already that small" (1986, 62).

Arboreal reptiles may fall quite frequently, as Schlesinger and co-workers have shown. In a thirty-one-month study of the litterfall in a California oak woodland they noticed that the 15–23-centimeter (6–9 in.) western fence lizard (*Sceloporus occidentalis*) frequently fell from perches in the trees. As for numbers, an average of five thousand arboreal western fence lizards crash from an acre (0.4 ha.) of California oak trees per year, or an average lizard falls three to six times per year. The authors attribute the falls, which are more frequent in summer, to "lack of surefootedness during the enthusiastic pursuit of canopy insects, or escape from predators" (1993, 2467). The number of males taking the dive also increases dramatically in spring, when male lizards perform an active sexual display that involves pumping their legs up and down.

Smaller size also confers on an animal the ability to move through the air at lower speed, with smaller wings. In contrast, a terrestrial, cursorial way of life favors an in-

Specimens of the tiny arboreal thecodont *Longisquama insignis* (late Triassic of Kirghizia), in the collections of the Palaeontological Institute, Moscow. *Above left,* holotype showing the insertion of the dorsal appendages along the vertebral column; *left,* isolated distal end of a dorsal appendage, showing the expanded extremity; *above right,* group of close-set dorsal appendages, indicating the original outline of the wing. Scale bars, 1 cm, diameter of coin, 20 mm. (From Haubold and Buffetaut 1987; courtesy L'Académie des Sciences, Paris)

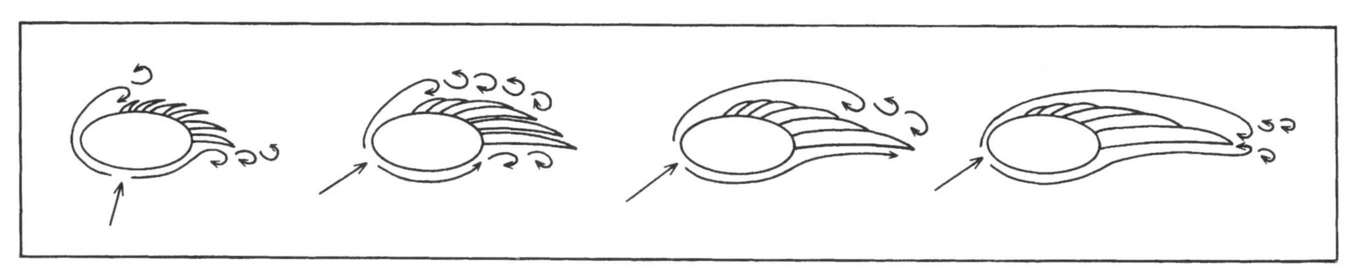

Above, left to right, hypothetical stages in the evolution of scales or feathers to increase the component of drag eventually developing an airfoil. Arrows indicate direction of air flow (modified after Tarsitano 1985). (From Feduccia 1993b; courtesy Academic Press)

Right, top to bottom, a, energy-saving mode of locomotion. It costs less for an animal to climb a tree and then glide to the next tree than *b,* to climb up and down in a tree and then run to the next (as modeled by R. Å. Norberg 1981, 1983); *c,* the run-jump-fly scenario in birds; *d,* the same as *c,* but with a gliding stage before the flying one. In these running scenarios the origin of feathers would produce drag, which would tend to slow down a running animal. Then, once the animal was aloft, where would the energy come from to maintain flight? (From U. M. Norberg 1990; courtesy Springer-Verlag)

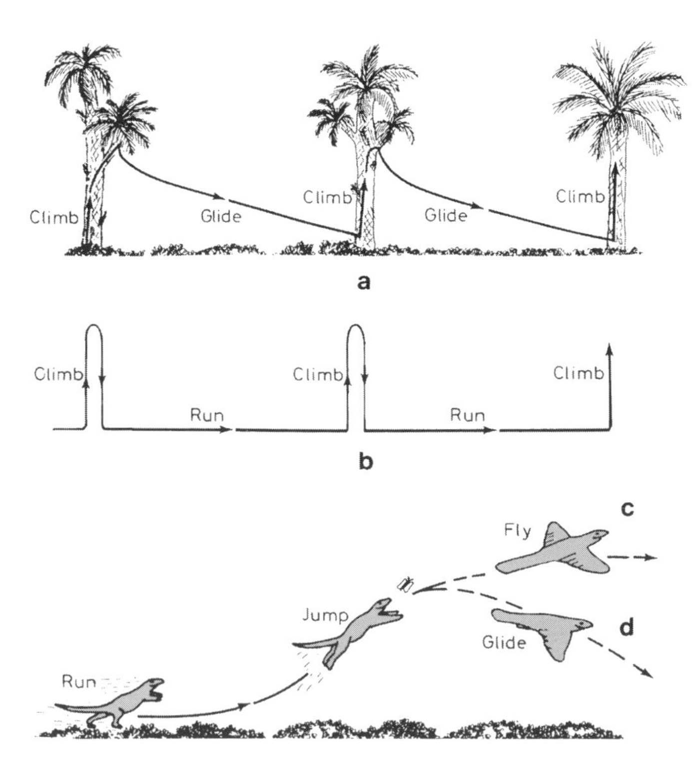

crease in size, as seen in theropod dinosaurs, because this favors increased running speed and defense against predators (Bock 1986).

Aerodynamic constraints clearly do not favor a theropod origin of birds yet do favor an avian origin from small, quadrupedal, arboreal Triassic thecodonts. As noted earlier, we do know of a radiation of small arboreal, Triassic archosaurs, represented by such disparate forms as *Longisquama*, *Scleromochlus*, and *Megalancosaurus*, and the last of these is small and particularly birdlike in many features (Feduccia and Wild 1993). Sam Tarsitano (1985) proposed appropriately that these forms be known as *avimorph archosaurs*. If the ancestors of birds were these small, arboreal archosaurs, there would be little need to consider a bipedal ancestral form, as in Bock's model; instead, one would proceed directly from a terrestrial quadrupedal to an arboreal quadrupedal archosaur, and bipedality in birds would have evolved as a result of releasing the forelimbs to fly, and not as a result of having evolved from terrestrial cursors with foreshortened forelimbs.

As Ulla Norberg (1990, 58) has suggested, maximization of net energy gain from foraging in trees might have augmented selection pressure for increased gliding performance. And gliding from one tree to another and then climbing back up the next tree during foraging activity has been shown to be a mechanism to maximize net energy gain by modern birds, as we have seen (R. Å. Norberg 1981, 1983). Simply stated, it costs less for an animal to climb a tree and then glide to the next tree than for the same animal to climb up and down in a tree and then run to the next. Then, once the glide surface had evolved, the animal's foraging efficiency would be dramatically increased and the time demands for locomotion during foraging activity would be drastically reduced. Parachuting or gliding to escape from enemies may also have played an important part in this evolutionary pathway. Ulla Norberg has correctly pointed out that "gliding must have been used only for commuting and not for insect-catching, which would require high maneuverability, which did not evolve until true flight was well established" (1990, 259).

To do more than glide, *Archaeopteryx* needed not only its aerodynamic feathers but a skeletal architecture that would allow its muscles to produce the powerful flapping required to maintain it in the air. Many scientists, notably Gavin de Beer (1954), former director of the British Museum (Natural History), have pointed out that the coracoid and other structures of the pectoral girdle in *Archaeopteryx* are weakly developed compared with those of modern flying birds and that the sternum is not preserved in any of the fossil specimens. The absence of a fossilized sternum, to which the powerful flight muscles would have been attached, does not mean that it was not present in the living creature. It could well have been cartilaginous, inasmuch as the structure was an innovation in birds, and cartilage would not have been preserved in the fossil. "That birds with largely cartilaginous sterna," as Kenneth Parkes has pointed out, "are capable of adequate flapping flight is well illustrated by the precociously flying chicks of modern gallinaceous birds. I regard the sternal structure of *Archaeopteryx*, therefore, as an open question and would not use it as either evidence for or against the ability to fly" (1966, 79). And Storrs Olson has shown that "at least in rails, even after ossification of the sternum has begun, the cartilaginous outline of the carina has still not reached its fullest development. In its early stages, the developing sternum of flying rails goes through stages resembling the ossified sternum of various flightless forms" (1973, 35). The discovery of the newest specimen of *Archaeopteryx*, with sternal plates preserved, would appear to lay this question to rest. The well-developed sternal structure of this specimen implies that the other specimens of *Archaeopteryx* likely had a cartilaginous sternum and were capable of powered flight.

The one feature of *Archaeopteryx*, other than feathers, that is unique to birds is the furcula. The London specimen has a furcula that would be considered very large in a modern bird, even hypertrophied relative to the furculas of modern birds of the same size. Some workers, such as Gavin de Beer and John Ostrom, have paid scant attention to this important structure, Ostrom merely asking, "Did it function as a transverse spacer between the shoulder sockets?" (1976a, 11). But Ostrom has also conceded that it is the one feature of *Archaeopteryx*'s skeleton that is strictly avian. Storrs Olson and I (1979) have argued that the best interpretation of this robust furcula is that it served as a major site of origin of a well-developed pectoralis muscle, which would have been responsible for the power stroke of the wing in flight. It is a widespread misconception of the pectoral girdle in birds to assume that the keel of the sternum is the only site for attachment of the massive pectoralis. In reality, this muscle originates from a number of structures, including the furcula and the coraco-clavicular membrane, which extends from the furcula to the coracoid. The posterior fibers of the pectoralis originate on the sternum only where the supracoracoideus, the muscle largely responsible for effecting the wing's recovery stroke in modern birds, is *not* present—typically only the periphery of the keel and the lateral and posterior margins. The main function of the keel, therefore, is to serve as the site of origin of the supracoracoideus and the tendon it sends to the humerus through the triosseal canal formed by the junction of the coracoid, scapula, and furcula.

It has often been thought that avian flight would be

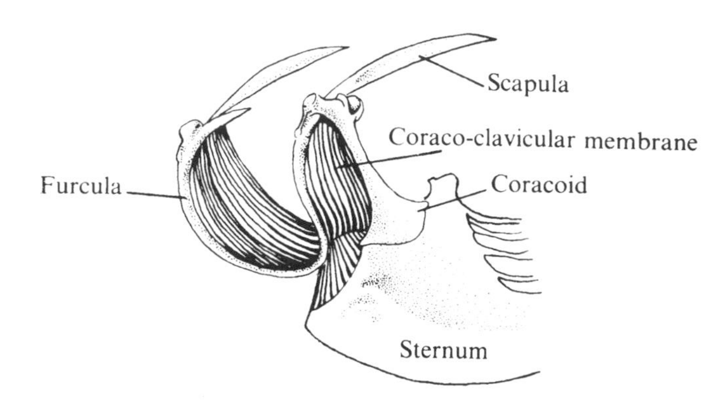

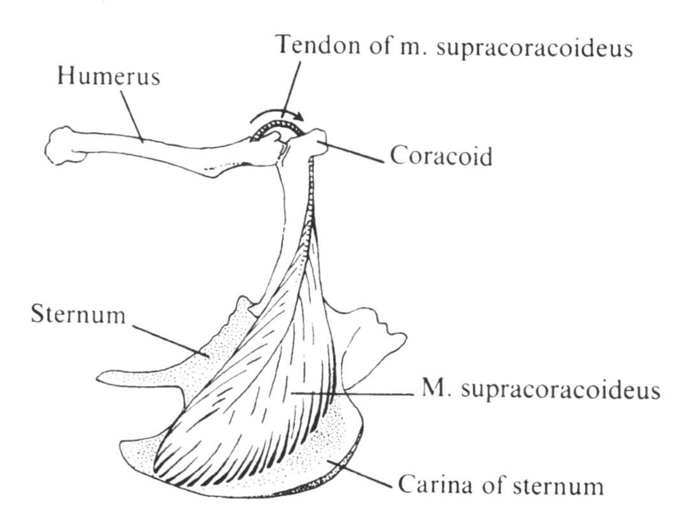

Above, left side of the pectoral girdle of the common goldeneye (*Bucephala clangula*), showing the furcula and the extensive coraco-clavicular membrane, where the pectoralis muscle in part originates. *Below,* right side of the pectoral girdle of the common pigeon (*Columba livia*), showing the action of the supracoracoideus muscle and the extensive area of the sternum it occupies. The pectoralis muscle attaches to the sternum only on those areas that are stippled. (From Olson and Feduccia 1979; drawing by Jaquin Schultz, after George and Berger 1966; reprinted with permission from *Nature* 278:247–248, © 1979, Macmillan Journals Ltd.)

impossible without a well-developed supracoracoideus muscle with its circuitous tendon. But, as we saw in chapter 1, this is not true. The dorsal elevators, principally the deltoideus major, can effect the recovery stroke by themselves, as they did in *Archaeopteryx*. The German anatomist Maxheinz J. Sy proved this when he cut the tendons of the supracoracoideus in living crows and pigeons (1936). Sy found that pigeons were capable of normal, sustained flight; the only capacity they lost was the ability to take off from level ground.

Another skeletal feature of *Archaeopteryx* that allies it with the flying birds is the angle of the scapula with the coracoid. In both, the angle is acute; as a result, the distance through which the dorsal elevators must act is shortened and they are thus more powerful. By contrast, the living flightless birds have lost the acute angle; in them the

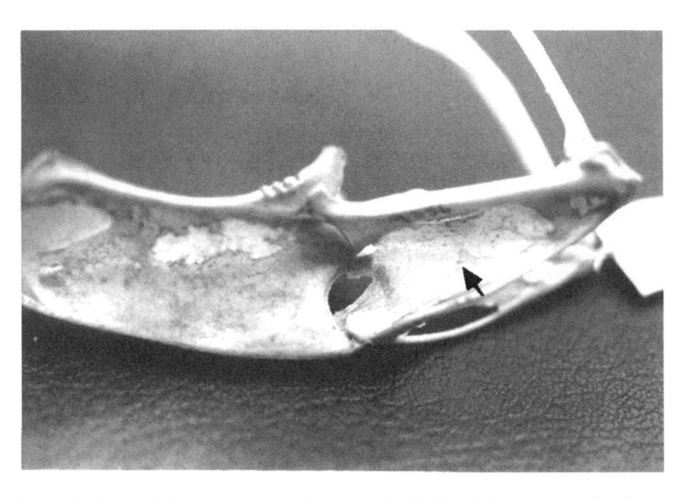

Lateral view of the sternum and pectoral girdle of *Centropus phasianius,* showing the preserved coraco-clavicular membrane (arrow), here extending from the coracoid to the furcula and serving for the origin of the pectoralis muscle. (Photo by author; U.S. National Museum specimen #9356)

scapula is more nearly perpendicular to the coracoid, and their clavicles are reduced or lost.

In *Archaeopteryx,* then, we see an early stage in the evolution of powered flight. The pectoral muscles accomplish the power stroke, but, in the probable absence of the supracoracoideus muscle, the dorsal elevators still effect the recovery stroke. Modern birds have an ossified sternum and carina, as well as the supracoracoideus with its tendon attached to the humerus, but this is a single functional complex that is not necessary for flight. No doubt it was superimposed in later birds on a pectoral architecture that was already capable of full flight.

The evidence is overwhelming. Birds and therefore avian flight originated in the trees, taking advantage of the cheap energy provided by high places in the form of gravity. The ancestors of birds, then, must be represented by small, arboreal archosaurs, and we now know that such an adaptive radiation, with divergent morphological representatives, did take place. The entire issue of flight origins has been sidetracked by *Archaeopteryx,* the first known bird from the late Jurassic, which was, despite some popular belief, already in the modern sense a bird, with fully developed flight adaptations, and can therefore tell us very little about the early evolution of flight. The issue has been further confused by paleontologists who see in *Archaeopteryx* a small, feathered coelurosaurian dinosaur. As we have seen in this and the preceding chapter, the resemblances are superficial and the homologies of virtually every element with equivalent dinosaur anatomy have been seriously challenged. The advocates of the dinosaurian origin of birds, especially those who demand that birds be derived from the deinonychosaurian-type coelurosaurs, are strapped with the burden of several untenable proposals: a late origin of birds from large, bipedal, earthbound ancestors with shortened forelimbs; the cursorial origin of avian flight; and the origin of feathers as insulation for hot-blooded dinosaurs. Before the proposal of a dinosaurian origin of birds can be taken seriously, a biophysically convincing model for the cursorial origin of avian flight—which would be unique among flying vertebrates, extant and extinct—must be modeled. Otherwise, we can assume that late Cretaceous birdlike dinosaurs were convergent on, rather than related to, birds and that avian flight originated according to the so-called arboreal theory.

Provisional schematic hypothetical model for the evolutionary changes leading to the origin of modern avian flight, *left, bottom to top,* along with temporal changes in the lineage of theropods, *right*.

Modern endothermic carinate birds, keeled sternum,
triosseal canal, fully developed flight architecture, pygostyle.
Extinction of archaic ectothermic "opposite birds" at close of Cretaceous.
early Cretaceous

Primitive powered flight (*Archaeopteryx*), incipient
obligate bipedality, reversed hallux, perching ability,
wing claws to aid balance, trunk-climbing ability, full
body contour feathers, quasi-ectothermy,
primitive powered flight.
mid- to late Jurassic

Dromaeosaurids and ornithomimids:
Ornithomimus, Struthiomimus,
Dromiceiomimus, Gallimimus—
smaller, superficially birdlike dinosaurs,
converging on large, flightless ratites and Galliformes.
early to late Cretaceous

Gliding proavis, early feathers, eventual release of
arms, incipient bipedal perching, incremental
growth of cerebellum and cerebral hemispheres.
early Jurassic

Advanced theropod dinosaurs. Large,
ground-dwelling, obligate bipedality.
Advanced cursorial adaptations. Small brain size.
Jurassic

Parachuting stage, quadrupedal pre-proavis with primoideal
featherlike scales or arm feathers used in feather-
assisted jumping; beginning digital reduction.
late Triassic

Late thecodonts (*Postosuchus*) and early
theropods (*Herrerasaurus*) with facultative
and later obligate bipedality, beginning
digital reduction, dramatic forelimb
shortening. Curved, laterally compressed,
serrated teeth. Large, balancing
tail. Release of forearms
for bipedal running.
late Triassic

Arboreal life, small size, increase in brain
and eye size, quadrupedal locomotion,
elongate forearms, leaping between
branches and trees, patagial membranes.
late Triassic

Small, ancestral basal archosaur (thecodont),
quadrupedal terrestrial locomotion,
mesotarsal joint. "Pre-lagosuchid."
early to mid-Triassic

"Opposite birds" were the predominant land birds of the Mesozoic. Here *Iberomesornis* from the early Cretaceous of Spain is shown near a shoreline. The small, nuthatch-sized opposite birds of the early Cretaceous had strongly curved foot claws, equivalent in curvature to those of modern trunk-climbers. (Drawing by John P. O'Neill)

4

The Cretaceous—Divers and Seabirds, Old and New

During the Cretaceous period, 146 million to 65 million years ago, the continents as we know them had not yet drifted into their present positions, and many of the landmasses of the modern world were covered with vast seaways. One of these, the Western Interior Seaway, divided North America, extending at its maximum expanse from the Gulf of Mexico to north of the Arctic Circle, spreading out a thousand miles (1,609 km.) in width from eastern Colorado and Wyoming across to Missouri and Iowa, and then extending as the Hudson Seaway northeastward through most of Manitoba and what is now Hudson Bay. The land itself, tropical or subtropical in climate, was presided over by dinosaurs such as *Triceratops* and *Tyrannosaurus*. Land birds, having conquered the air with powered flight, had gained cosmopolitan distribution. Feather impressions have been recovered from a lake deposit in the Lower Cretaceous Koonawarra claystones of southeastern Victoria, Australia (Talent, Duncan, and Handley 1966; Waldman 1970; Rich 1976; Molnar 1980; Rogers 1987; Vickers-Rich 1991), and from the Lower Cretaceous Santana Formation of Brazil (Martins-Neto and Kellner 1988), proving that birds were well established in the Southern Hemisphere by the beginning of the Cretaceous. Worldwide distribution of birds is indicated by discoveries of feathers from numerous localities of the Lower Cretaceous of Mongolia and from Bajsa in Transbaikalia (Kurochkin 1985a, 1988), from Lowermost Cretaceous ambers in Lebanon (Schlee 1973), Lower Cretaceous ambers in Siberia (Kurochkin 1985b), and Lower Cretaceous sediments in Spain (Ferrer-Condal 1954; Lacasa 1985), and footprints attributed to birds are known from the Lower Cretaceous of British Columbia (Currie 1982). In addition, the late Jurassic Chinese *Confuciusornis* is preserved with feather impressions (Hou, Zhou, Martin, and Feduccia 1995). These finds indicate an impressive adaptive radiation of birds, as we shall see.

Most of the fossil record of the Cretaceous, however, is from the ancient continental seas, and until recently our early knowledge of Cretaceous birds was confined largely to the toothed, principally flightless forms that inhabited the marine environment. As in the case of *Archaeopteryx*, the medium of fossil preservation was limestone, chalky white limestone best exemplified by the White Cliffs of Dover, laid down in Cretaceous oceans. It is this stone, extensive in Cretaceous formations, that gives the period its name, from the Latin *creta*, chalk (a more or less porous type of limestone). In many locations these chalks are formed primarily from shells or calcareous tests (external shells) of foraminifera (marine protozoans) and shells of similar invertebrates. The classic formation of a Cretaceous ocean in North America is the celebrated Niobrara Formation of the Great Plains, exposed as chalky white cliffs along ravines and deep washes especially in western Kansas and Nebraska.

Many fossils of the nonavian denizens of these oceans have also been recovered, and they alone could have inspired all subsequent images of "sea monsters." We can imagine huge marine turtles, such as the 4.5-meter-long (15 ft.) *Archelon*, basking near the surface while large pterosaurs such as *Pteranodon*, with wingspans of up to 7 meters (22 ft.), glided overhead. Even more bizarre were the plesiosaurs, such as the 9.8-meter-long (32 ft.) *Elasmosaurus*, long-necked creatures with fat bodies and wide, paddlelike limbs, often described as extremely large turtles with a snake strung through them, but in reality related to neither turtles nor snakes. Species of the late Cretaceous ranged from 6 to 12 meters (20–40 ft.) or more in length, and one short-necked Australian form, *Kronosaurus*, had a skull 3.7 meters (12 ft.) long. Large, voracious, seagoing monitor lizards called mosasaurs were exemplified by the 9-meter-long (30 ft.) *Tylosaurus*. There were also myriads of fish, representing the evolutionary mainsprings of our modern bony fishes. The largest of these, known as *Xiphactinus*, was over 5 meters (16 ft.) in length.

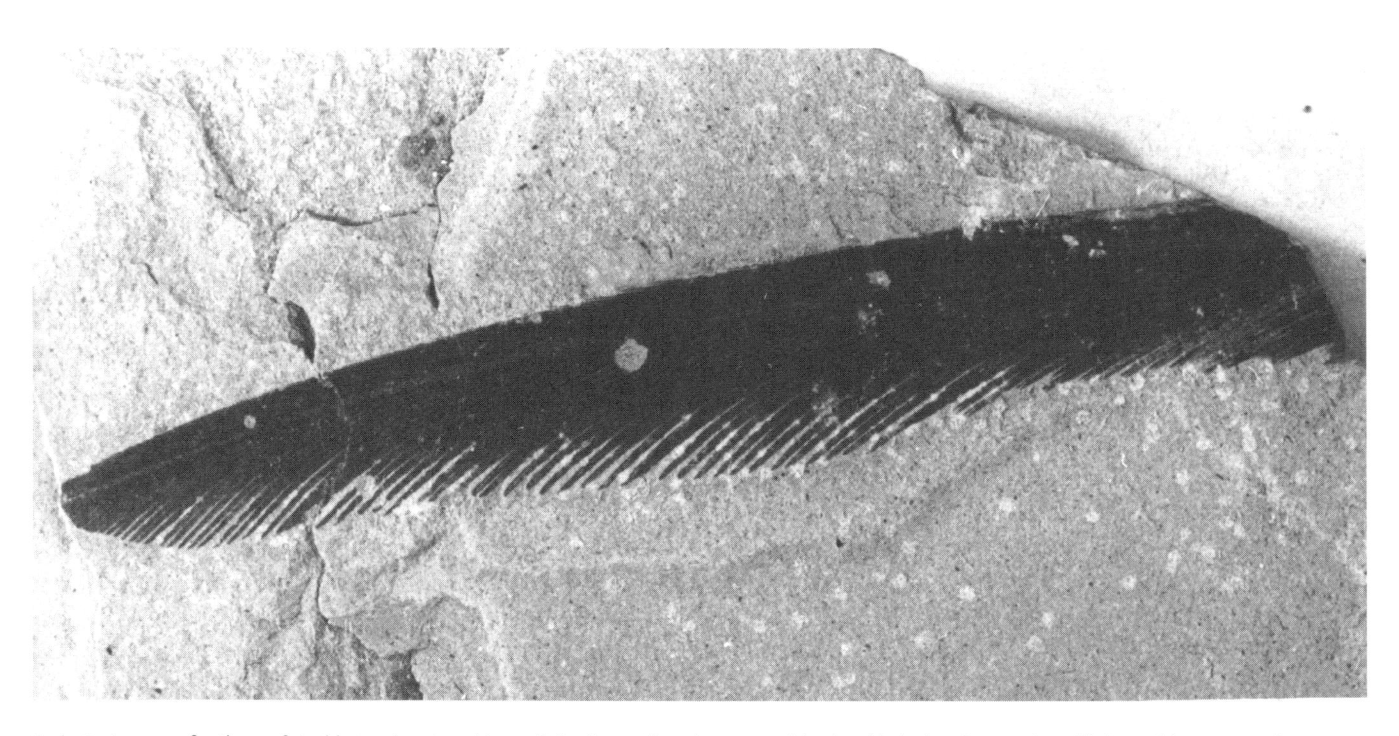

Early Cretaceous feathers of *Ambiortus* (western Mongolia), the earliest known ornithurine bird, showing modern flight architecture and asymmetric vanes characteristic of the primary feathers of modern flying birds. *Above right,* fossil feathers as preserved, × 0.74. *Above left,* enlargement of presumed tail feather, upper left of small photo; *below,* enlargement of primary feather. (Courtesy E. N. Kurochkin, Moscow)

North American epicontinental seaways from the early Jurassic period (*above*) to the late Cretaceous period (*below*). In the seaways of the early Jurassic, some 200 million years ago, two types of aquatic reptiles were abundant, the long-necked plesiosaurs and the porpoiselike ichthyosaurs. Abundant fishes were preyed on by such ichthyosaurs as *Stenopterygius* (*above right*). The most specialized of all aquatic reptiles, *Ichthyosaurs* evolved in the Triassic but became extinct well before the late Cretaceous mass extinctions. Their extinction is linked to competition from other marine reptiles and the growing dominance of advanced sharks in the late Mesozoic. Plesiosaurs such as *Plesiosaurus* (*above left*) ranged from approximately 6 to 12 m (20–40 ft.) in length and survived into the late Cretaceous. These long-necked fish-eaters propelled themselves with strong, broad flippers. Cretaceous continental seaways covered much of western North America (*below*). Toothed hesperornithiform birds abounded in these waterways, inhabited also by the giant fish-eating mosasaur *Tylosaurus* (*below center*), a creature that grew to 9 m (30 ft.) in length. The sea turtle *Protostega* had a shell as much as 1.8 m (6 ft.) long, and in the air, the flying reptile *Pteranodon* soared on wings that spanned up to 7 m (23 ft.). (Murals by Charles R. Knight; courtesy Field Museum of Natural History, neg. nos. 72379 [*above*], 66186 [*below*])

Although the Cretaceous seaways preserved near-complete faunas, much less is known of terrestrial habitats and their biotas during this period. Many dinosaurs are known, to be sure, but early collectors were interested only in the spectacular fossils, so many of the small vertebrates were neglected and not collected. Many fragments of bird or birdlike fossils have also been recovered, named, and given great significance in the literature of avian phylogeny. Unfortunately, these bone scraps are exactly that, and carry no useful information. Thus, *Gallornis straeleni* (Lambrecht 1931b), from the early Cretaceous of France, often termed the oldest Cretaceous bird, was diagnosed on the basis of a fragmentary end of a femur as belonging to the Anseriformes (ducks and allies) and was later placed near the flamingos. In reality, it simply represents an unidentifiable scrap of bone, period, as does a worn humerus named *Wyelyia* from the early Cretaceous of England (Harrison and Walker 1973). To continue, the lower jaw described as the bird *Caenagnathus* and new avian order Caenagnathiformes (Cracraft 1971b), from the Upper Cretaceous of Canada, is a coelurosaurian dinosaur related to *Oviraptor* (Martin 1983a, 301), and the Cretaceous owl family Bradycnemidae (Harrison and Walker 1975b) is likewise of dinosaurian affinity. More recently, there are unconfirmed reports that isolated bones of *Archaeopteryx* have been recovered in association with bones of true birds in rocks of the Romanian earliest Cretaceous (Jurcsák and Kessler 1984; Kessler and Jurcsák 1984, 1986; Kurochkin 1995), including the named forms *Palaeocursornis* and *Eurolimnornis* (Kessler and Jurcsák 1984, 1986). Isolated bone pieces from *Archaeopteryx* would be almost impossible to identify with certainty, as is, for example, the distal piece of femur named *Palaeocursornis!* A number of additional fragments of Cretaceous bird fossils have been described from time to time, but they invariably end up in the fossil bird trash pile.

To summarize, unless there is a uniquely adapted bone that can be unequivocally associated with a specific avian type, as in the case of the highly specialized foot-propelled divers, the hesperornithiforms, single bones or fragments from the Mesozoic usually cannot be diagnosed and certainly should not be given names; they have been around for millions of years, let them rest in peace! Such studies historically have cluttered the literature and have retarded advances in the field of avian paleontology and evolution.

Nevertheless, many spectacular bird fossils have recently come to light, and we are just now beginning to understand the land birds of this interesting interval, spanning some 85 million years, between the time of *Archaeopteryx*, the late Jurassic, and the beginning of the Cenozoic era, 65 million years ago.

The Opposite Birds

In recent decades a literal explosion in our knowledge of Cretaceous land birds has occurred with the discovery of birds in Europe, Asia, Australia, and North and South America (Chiappe 1995; Elzanowski 1995; Kurochkin 1995; Martin 1995b; Zhou, 1995a). Much of the revelation of this great avian radiation owes a considerable debt to José Bonaparte of the Museo Argentino de Ciencias Naturales in Buenos Aires, who during the 1970s conducted field explorations and excavations of the continental deposits of the Upper Cretaceous Lecho Formation of northwestern Argentina (Bonaparte et al. 1977; Bonaparte and Powell 1980). A rich assemblage of bird bones recovered from these deposits in 1975 was the first of many discoveries of a completely new group of primitive land birds not far removed from *Archaeopteryx*. The first collection included some sixty or so bones, some of them articulated, and were studied by Cyril A. Walker (1981) of the British Museum of Natural History in London. After preliminary study of the specimens, Walker concluded that they represented a new subclass of primitive flying birds, which he called the Enantiornithes, or opposite birds. The birds are so named because, among many distinctive features, there is a unique formation of the triosseal canal and the metatarsals are fused proximally to distally, the opposite of that of modern birds. Diagnosis of the opposite birds was aided by the fact that the tarsometatarsus is commonly preserved in birds.

Walker grouped the bones into four or five genera and species in three size classes, but he named just one genus and species. Almost all of the major skeletal elements, except for the all-important skull, as well as the pubis and tail, were present. Although hindlimbs differed among the birds, their wings and pectoral girdles were uniform. These opposite birds were powered fliers with a precocious flight architecture that was more advanced than *Archaeopteryx*, and they exhibited a fused carpometacarpus and an ossified, keeled—though primitive—sternum.

Walker announced the discovery in a letter to *Nature* in 1981, in which he described *Enantiornis leali* as the first member of the new suborder Enantiornithes. His findings were not seen as terribly significant at first, given the incompleteness of the material and the lack of skulls. Today, however, Walker's work is recognized as a momentous discovery in the history of avian paleontology.

The Creation of the Subclass Sauriurae. While studying Bonaparte's enantiornithine birds, Cyril Walker noted similarities with the late Cretaceous, sparrow-sized land bird *Alexornis antecedens*, from Baja California,

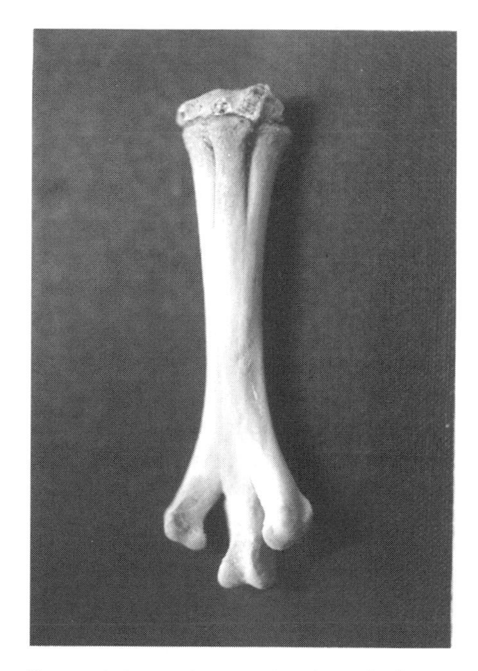

Tarsometatarsus, in posterior view, of a juvenile kiwi (*Apteryx*), showing the tarsal cap and the disto-proximal lines of fusion characteristic of modern ornithurine birds. (Specimen #18277, U.S. National Museum; photo by author)

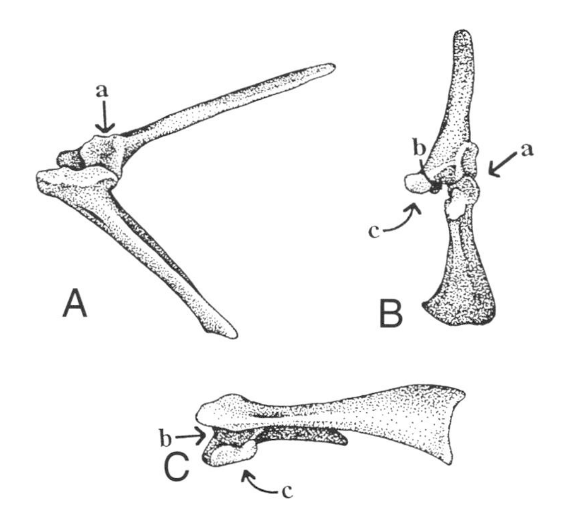

Articulated scapula and coracoid of *Enantiornis* showing the distinctive formation of the triosseal canal by a broad, rectangular process of the scapula, in contrast to modern birds, where the coracoid forms most of the canal: *A,* lateral view; *B,* anterior view; *C,* ventral view. Abbreviations: *a,* glenoid; *b,* triosseal canal; *c,* ventral facet on scapula. (From Martin 1995b; courtesy L. D. Martin and Forschungsinstitut Senckenberg)

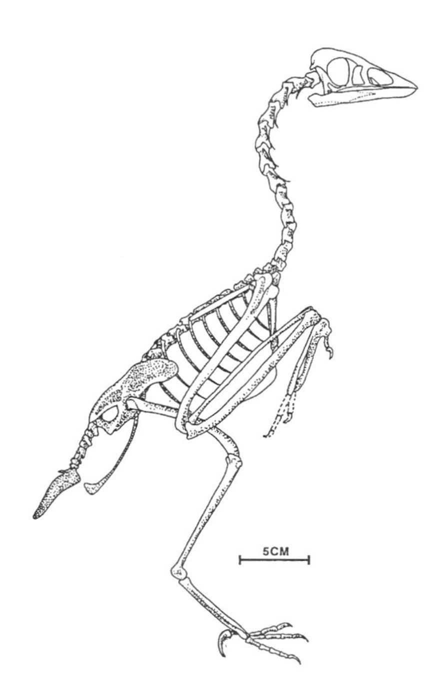

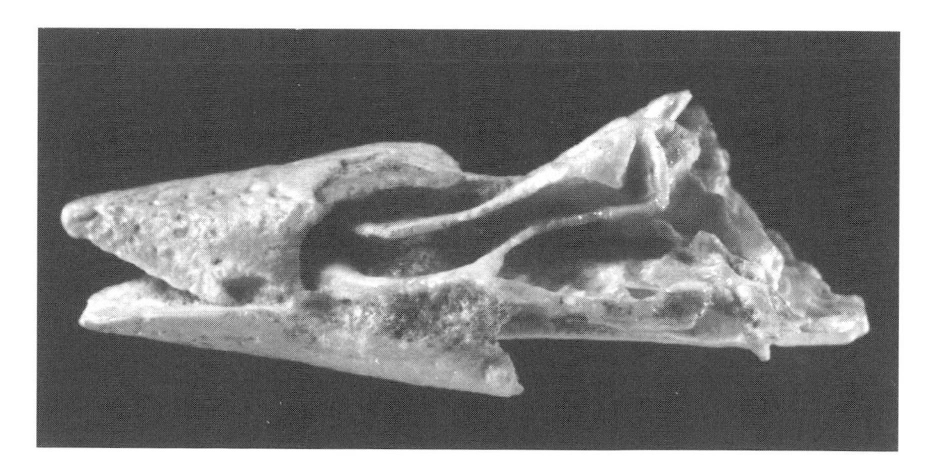

Left, reconstruction of the skeleton of *Enantiornis,* with a skull based on *Gobipteryx,* a pygostyle on *Iberomesornis,* and a pubis on *Archaeopteryx*. *Above,* skull of the embryo of *Gobipteryx,* from the Upper Cretaceous of Mongolia. Elzanowski estimated the skull of *Gobipteryx* to be 45 mm long, "roughly the size of a partridge." (Skeletal reconstruction courtesy L. D. Martin; courtesy L. D. Martin and Forschungsinstitut Senckenberg; photo courtesy A. Elzanowski)

which had been described by Pierce Brodkorb of the University of Florida. In 1976, Brodkorb had placed *Alexornis* in a new order, Alexornithiformes, which he thought could have been ancestral to the modern orders Coraciiformes (rollers and kingfishers) and Piciformes (barbets, toucans, and woodpeckers). Several years later, in 1987, Larry Martin also noted similarities between the *Alexornis* and enantiornithine material. He suggested placing *Alexornis*, together with *Archaeopteryx* and the Cretaceous Mongolian material of the genus *Gobipteryx* (described below), in the subclass Sauriurae, first used by Ernst Haeckel in 1886. This arrangement united *Alexornis*, *Gobipteryx*, and enantiornithines into a group that Martin termed the Enantiornithes and that he postulated was the sister-group of Archaeornithes, the sole member of which was the late Jurassic genus *Archaeopteryx*.

***Gobipteryx* Was Precociously Developed at Hatching.** *Gobipteryx* had first been described by the Polish avian paleontologist Andrzej Elzanowski (1974, 1976, 1977, 1981), from a collection gathered by the Polish Mongolian Expedition, to the late Cretaceous of Mongolia. The collection consisted of several skulls and fossil eggs containing well-preserved skeletons. Elzanowski considered *Gobipteryx* to be a member of the paleognathous assemblage the Palaeognathae (ostrich and allies), but the fragmentary nature of the material made the assignment somewhat uncertain, and Martin (1983b, 1987, 1995) has since shown that these are indeed enantiornithine birds. More recently, Elzanowski (1995) has noted similarities between the *Gobipteryx* material and both newly discovered fossils from the late Cretaceous of Uzbekistan, described by Lev Nessov and L. J. Borkin in 1983, and an enantiornithine late embryonic specimen from the late Cretaceous Rio Colorado Formation of Argentina, described by Luis Chiappe (1991a, 1991b, 1992), which Chiappe has assigned to the Avisauridae (a family that may prove identical with the Gobipterygidae; Elzanowski 1995).

Gobipteryx is unusual among Mesozoic birds in lacking teeth, but it is also of great interest in that the unhatched fossilized skeletons reveal many features, indicating that the chicks were well formed and precocious at hatching (Martin 1983b). As Elzanowski (1995) has pointed out, the late embryonic skeletons of both *Gobipteryx* and the Rio Colorado specimen demonstrate a well-developed shoulder girdle and wing bones in these birds at hatching, whereas the hindlimbs are poorly developed. This superprecocial development of the flight apparatus and the concomitant ability to fly immediately after hatching, as Elzanowski (1981) initially proposed, rather than advanced development of the ambulatory, pelvic architecture, amounts to high energy costs of development, perhaps somewhat comparable to the precocious living galliforms the moundbuilders, or megapodes (Megapodiidae). This differential development of the flight apparatus over the pelvic architecture may reflect the key feature of the enantiornithine radiation, as well as the morphological uniformity of the enantiornithine pectoral architecture, as opposed to the tremendous diversity and adaptive radiation in their hindlimb morphology.

The Avisauridae. In 1985, Michael K. Brett-Surman and Gregory Paul described a new family of supposed small theropod dinosaurs, the Avisauridae, from the late Cretaceous Hell Creek Formation of Montana and the Lecho Formation of Argentina based only on the metatarsus, which exhibited partially fused tarsal elements reminiscent of some Cretaceous dinosaurs. They interpreted this discovery as evidence of a faunal link between the two supercontinents Laurasia and Gondwanaland at the end of the Cretaceous, a time when the two supercontinents had been thought to be separate. However, *Avisaurus archibaldi* Brett-Surman and Paul (1985) has since been shown to be not a small dinosaur but an enantiornithine bird and to occur not only in Argentina but also in the latest Cretaceous, Maastrichtian, of Montana, as the latest record of the enantiornithines (Martin 1991; Chiappe 1992, 1993). The proper systematic allocation of *Avisaurus* has been confirmed by Howard Hutchinson (1993), who is describing a new species based on a partial headless skeleton recovered from the Upper Cretaceous Campanian of the Kaiparowits Formation of southern Utah, and still another, *Avisaurus gloriae*, has been added to the list from a lake deposit of the Upper Cretaceous Two Medicine Formation of Montana (Varricchio and Chiappe 1995); it is less than half the size of *Avisaurus archibaldi*.

In 1994, Luis Chiappe and Jorge Calvo described another new species, the crow-sized *Neuquenornis volans*, from the Upper Cretaceous of Argentina. Chiappe (1993) added three new genera and species of enantiornithines based on tarsometatarsi from the Upper Cretaceous Maastrichtian Lecho Formation of northwestern Argentina, including the avisaurid *Soroavisaurus australis*, and two additional genera of less certain affinity, *Yungavolucris brevipedalis* and *Lectavis bretincola*, which at least illustrate the tremendous morphological diversity in the late Cretaceous South American opposite birds. Although enantiornithines were once considered to be cursorial (Brett-Surman and Paul 1985), the new avisaurid material clearly illustrates perching, arboreal habits as well as strong flying ability. Above all, the extensive Argentine col-

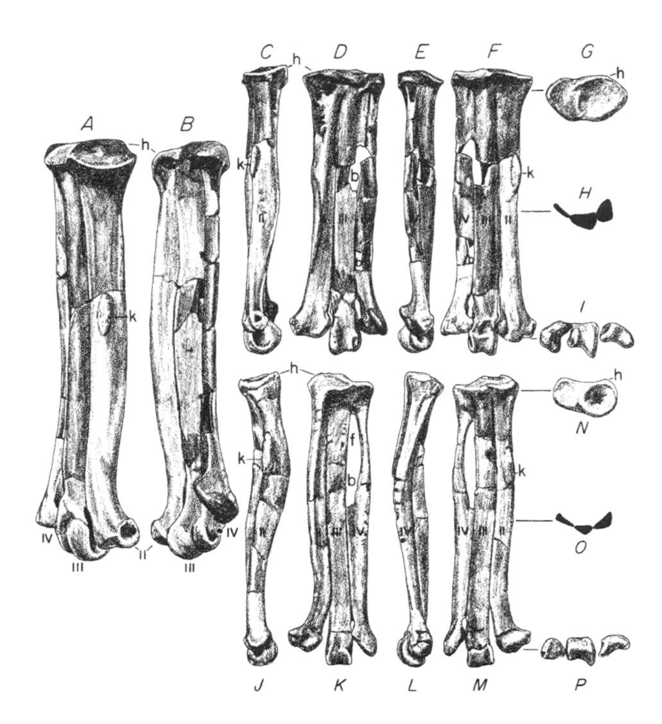

A–I, right metatarsus of *Avisaurus archibaldi,* holotype: *A,* anterior view; *B,* posterior view; *C,* medial view; *D,* posterior view; *E,* lateral view; *F,* anterior view; *G,* proximal view; *H,* cross-section; *I,* distal view. *J–P,* left (reversed) metatarsus of *Avisaurus* species from the Lecho Formation of Argentina: *J,* medial view; *K,* posterior view; *L,* lateral view; *M,* anterior view; *N,* proximal view; *O,* cross-section; *P,* distal view. Abbreviations: *h,* hypotarsus; *k,* knob for proposed origin of m. tibialis anticus; *f,* possible fenestra; *b,* postmortem breakage. Maximum length: *A,* 75 mm; *J,* 47.5 mm. (From Brett-Surman and Paul 1985; courtesy Society of Vertebrate Paleontology)

lections dramatically depict an impressive adaptive radiation of enantiornithines, showing that these were the dominant land birds of the Cretaceous.

Opposite Birds in Spain. Although feathers have been known for many years from the Lower Cretaceous of Spain (thought earlier to be late Jurassic; Beer 1954), in 1986, Antonio Lacasa first reported a partial skeleton from the Cretaceous locality near Barcelona known as Sierra del Montsec, and in 1989 he described it in more detail, naming it *Noguerornis gonzalezi*. The incomplete specimen consisted of most of one articulated wing and portions of the other, the furcula, ischium, vertebrae, and tibia fragments, as well as feathers. In 1988, in a letter to *Nature*, José Sanz, José Bonaparte, and Antonio Lacasa reported another unusual early Cretaceous (some 130–120 million years old) bird from Spain, this one a fairly complete articulated skeleton minus the skull, from a locality known as Las Hoyas in the province of Cuenca. The Las Hoyas bird created quite a sensation, combining as it did a strange assemblage of advanced avian characters with many primitive features (Cracraft 1988a); Sanz, Bona-

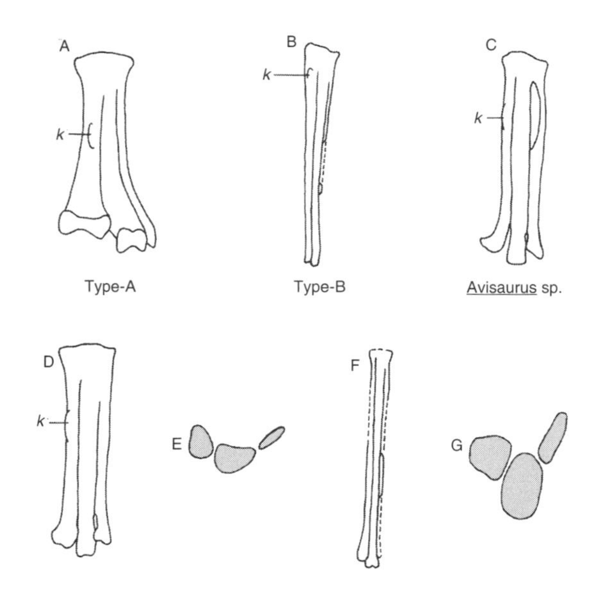

Left enantiornithine tarsometatarsi: *A–C,* El Brete tarsometatarsi in anterior view. *A,* Type A; *B,* Type B; *C, Avisaurus* species; *D, E,* anterior view and midshaft section of *Avisaurus archibaldi,* based on right tarsometatarsus; *F, G,* anterior view and midshaft section of the Patagonian specimen. Note the prominent knob (*k*) on metatarsal 2 and the reduction of metatarsal 4, which show conclusively that *Avisaurus* was a bird and not a theropod. Note also that metatarsals are fused from proximal end to distal; in ornithurine birds (including the modern avifauna), the fusion is from distal end to proximal. Note also the absence of a tarsal cap. (From Chiappe 1992; courtesy Society of Vertebrate Paleontology)

parte, and Lacasa described it as being "intermediate between *Archaeopteryx* and later birds" (1988a, 434). It was similar to modern birds in having strutlike coracoids, a furcula with a hypocleidium, and a well-developed py-

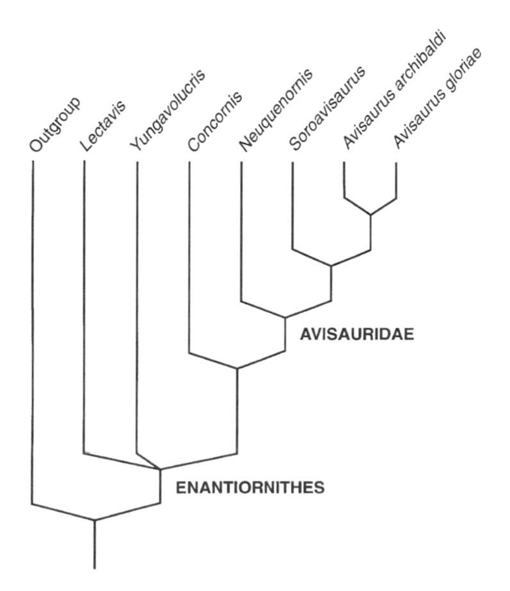

Cladograms illustrating the phylogenetic relations of *Concornis* to other enantiornithines, particularly the avisaurids. (Modified after Sanz, Chiappe, and Buscalioni 1995)

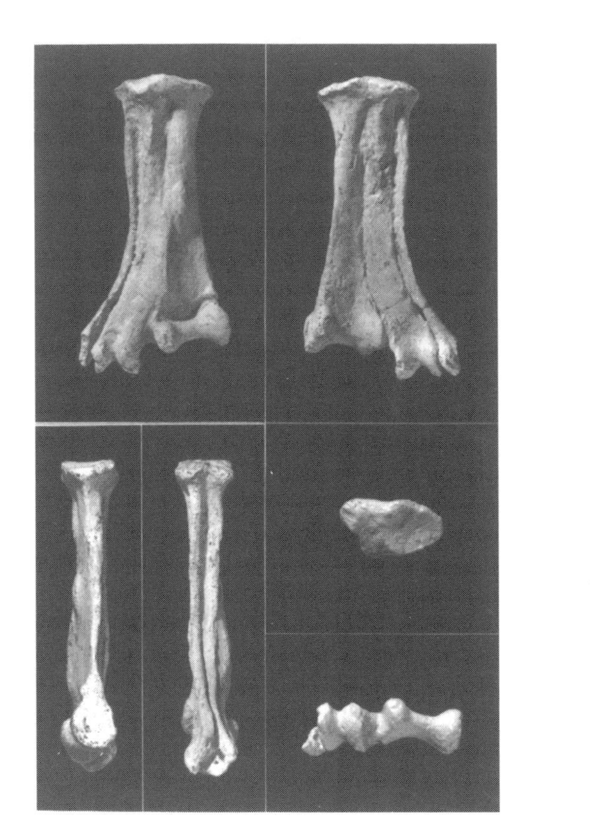

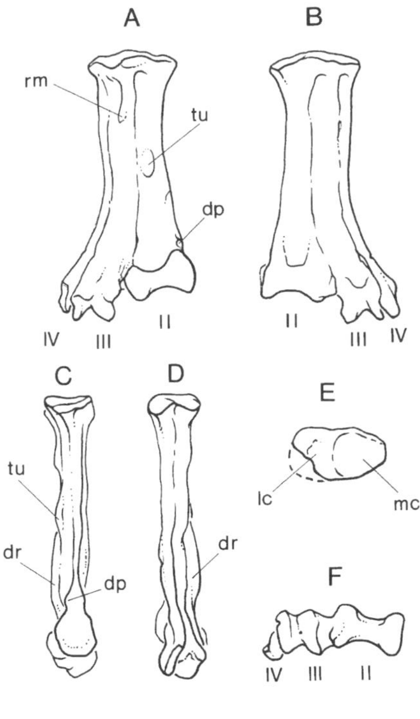

Yungavolucris brevipedalis, a new species of enantiornithine bird from the Upper Cretaceous Lecho Formation of Argentina. Right tarsometatarsus in dorsal (*A*), plantar (*B*), medial (*C*), lateral (*D*), proximal (*E*), and distal (*F*) views. Abbreviations: *dp*, dorsomedial projection of the distal end of metatarsal 2; *dr*, distal ridge between the contact of metatarsals 2 and 3; *tu*, tubercle for the attachment of *m. tibialis cranialis; lc*, lateral cotyla; *mc*, medial cotyla; *rm*, elongate proximal ridge for muscle attachment; *II–IV*, metatarsals 2–4. (From Chiappe 1993; courtesy American Museum of Natural History)

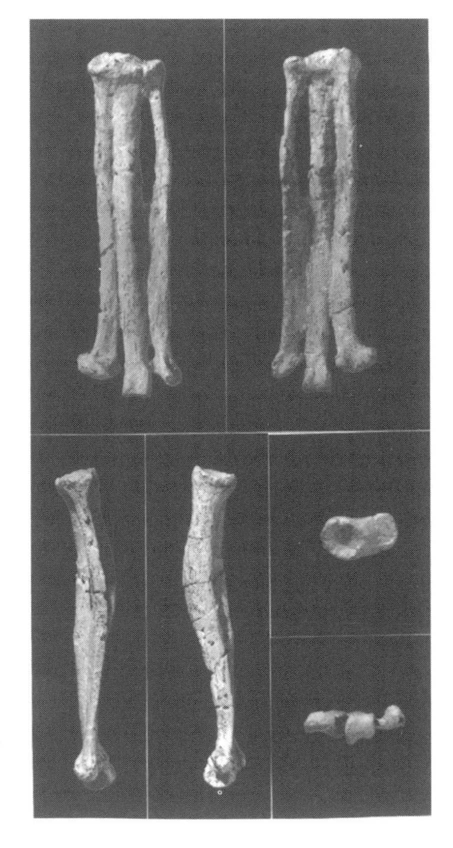

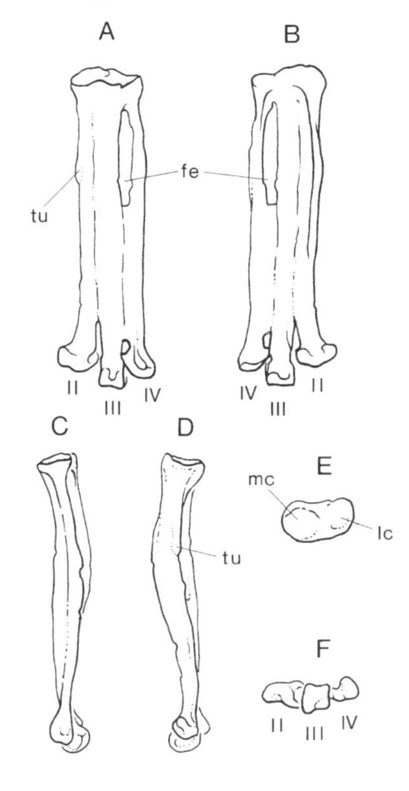

Soroavisaurus australis, another new species from the Lecho Formation of Argentina, illustrating another distinctive morphological type of enantiornithine bird. Left tarsometatarsus in dorsal (*A*), plantar (*B*), lateral (*C*), medial (*D*), proximal (*E*), and distal (*F*) views. Abbreviations: *fe*, fenestra between metatarsals 3 and 4; *tu*, tubercle for the attachment of *m. tibialis cranialis; lc*, lateral cotyla; *mc*, medial cotyla; *II–IV*, metatarsals 2–4. (From Chiappe 1993; courtesy American Museum of Natural History)

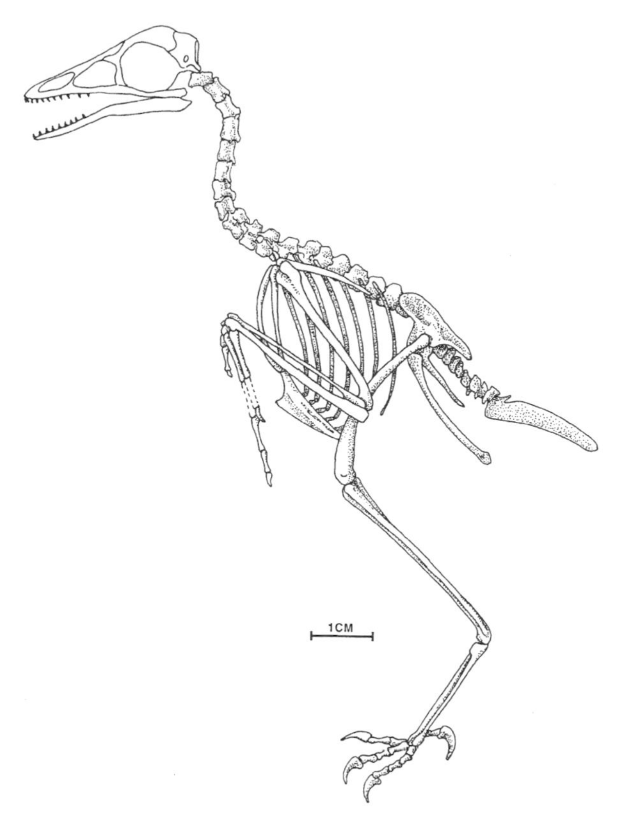

Skeletal reconstruction of *Iberomesornis romerali,* new order of birds, the Iberomesornithiformes, from the Lower Cretaceous of Las Hoyas, in Cuenca, Spain. *Iberomesornis,* like other enantiornithine birds, had a well-developed pectoral flight apparatus, but the pelvic region was still quite primitive. Note the extremely long pygostyle; anteriorly, three or four fused vertebrae are distinct, and the total number of pygostyle vertebrae is estimated at about 10–15. Note also the highly curved foot claws. (From Martin 1995b; courtesy L. D. Martin and Forschungsinstitut Senckenberg)

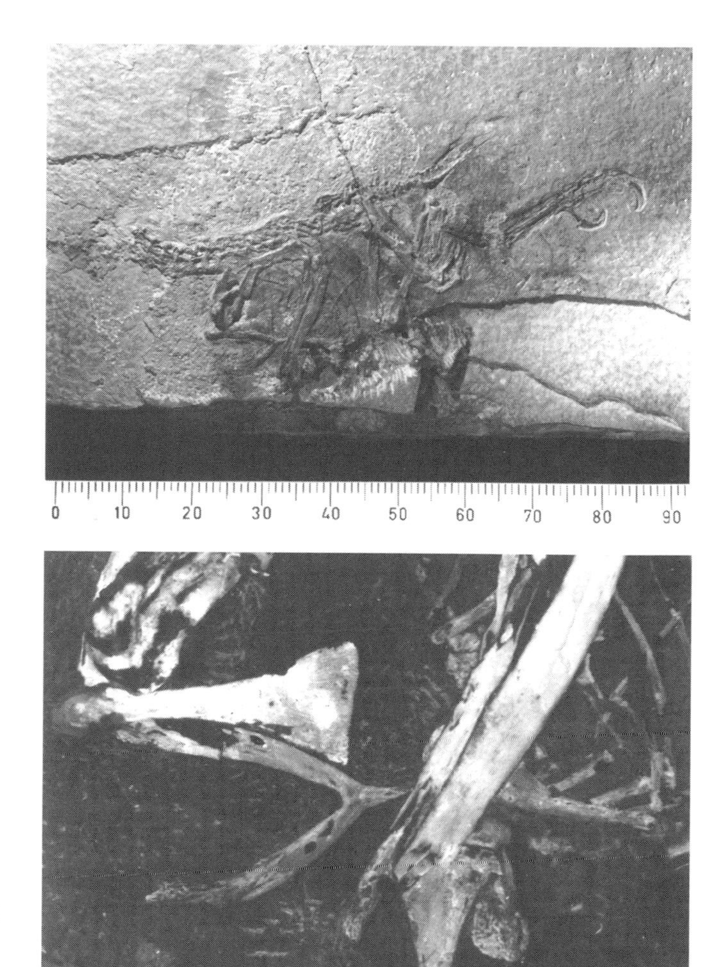

Iberomesornis romerali (LH022R). *Above,* type specimen; *below,* close-up, ultraviolet-induced photograph, showing furcula. (From Sanz and Bonaparte 1992; courtesy *Los Angeles County Museum of Natural History Science Series* 36; photos by G. F. Kurtz)

The small enantiornithine bird *Concornis lacustris,* from the early Cretaceous of Las Hoyas, Spain. Scale bar in cm. Note particularly the furcula (*upper center*) with long hypocleidium and pubes (*right center*) joined distally in a pubic apron. (From Sanz and Buscalioni 1992; courtesy J. L. Sanz)

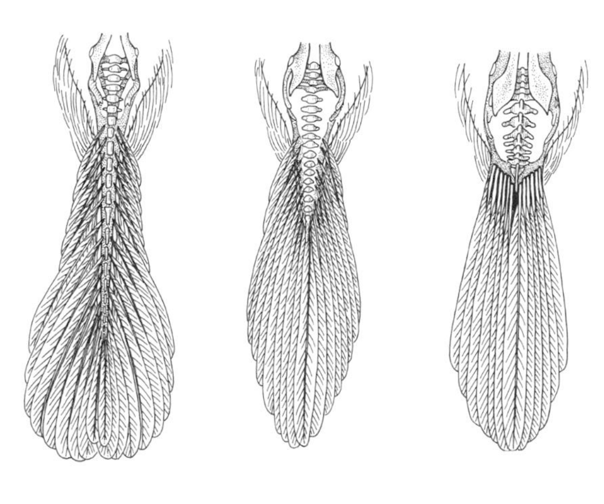

Hypothetical stages in the evolution of the avian tail and pygostyle, illustrated by Steiner in 1938, correspond roughly to recent finds of opposite birds. (Modified after Steiner 1938)

gostyle (unusually elongate), but it was primitive in features of the pelvic girdle, sacrum, and hindlimb morphology. Although the tarsometatarsus may have exhibited some proximal fusion, it was unfused distally for most of its length. A fully reversed hallux and strongly curved claws showed the Las Hoyas specimen to be a perching bird, a denizen of the trees. The locality of Las Hoyas was itself of interest, in that it represented an outcrop of lithographic-grade limestone that permitted the preservation of such fine structures as bird feathers and insect wings; it was comparable to other localities representing lacustrian or lagoonal depositional environments that were situated in the European area of the ancient Tethys Sea from the late Jurassic to the early Cretaceous (Sanz and Bonaparte 1992).

Sanz and Bonaparte fully described the Las Hoyas bird in 1992, naming it *Iberomesornis romerali*, the sole member of a new order of birds, the Iberomesornithiformes. This tiny bird, under a hundred millimeters in length, like *Gobipteryx* illustrates a stage in avian evolution in which suites of primitive *Archaeopteryx*-like characters in the pelvic girdle and hindlimb were combined with an advanced morphology, typical of the Ornithurae, or modern birds, in the pectoral girdle and flight apparatus. Sanz and Bonaparte suggested that because of this differential of primitive and derived characters, "the early evolution of birds from *Archaeopteryx* was first concentrated in the pectoral girdle, forelimb, and pygostyle to achieve the structural requirements for active flight" (39). It is interesting to note that in 1938, Hans Steiner attempted to reconstruct, using avian embryology, the sequence of anatomical transformations in the avian tail skeleton and its relation to the tail feathers, as illustrated here, proceeding from *Archaeopteryx* through some hypothetical intermediate to modern ornithurines. Steiner's intermediate form is remarkably close to the tail seen in the enantiornithines.

A second Lower Cretaceous bird from Las Hoyas was described by Sanz and Buscalioni in 1992 as *Concornis lacustris* and represented a bird approximately twice the size of the sparrow-sized *Iberomesornis;* its phylogenetic position has recently been reevaluated (Sanz et al. 1995). A more derived bird than *Iberomesornis*, *Concornis* has a true tibiotarsus and a fanlike distal metatarsal zone with distinct trochlear structures, and only the proximal metatarsal region is fused. It lacks a true tarsometatarsus characteristic of modern birds. The sternum has a small but conspicuous posterior median keel, and the posterior border is highly emarginate, excavated with two deep lateral sulci on each side of the keel. It is possible that the location of the keel only in the posterior part of the sternum may account for the elongate hypocleidium characteristic of the furcula of these birds. Although the phalangeal structure of the hand is like that of modern birds, the metacarpus is not fused distally. As in *Iberomesornis*, there is a fully reversed hallux with a hypertrophied hallical claw. Sanz and Buscalioni estimated the mass of *Concornis* to be some 76.5 grams (2.7 oz.), some three to four times the estimated weight of *Iberomesornis*. Although *Iberomesornis* was not considered to be a juvenile, Larry Martin, in examining the specimen, stated that "there . . . seems little doubt that *Iberomesornis* is a juvenile bird but proportional differences in the skeleton may still distinguish *Concornis* from it" (1995, 27).

The sparrow-sized bird *Sinornis santensis* from the Lower Cretaceous of Liaoning Province, China (Jiufotang Formation, Valanginian age). The anatomy of *Sinornis* and other enantiornithines suggests that birds had a fully developed pectoral flight apparatus some 135 million years ago, just 15 or so million years after *Archaeopteryx*. *Sinornis* exhibits strutlike coracoids, the first avian carpometacarpus, and such primitive features as gastralia, ribs without uncinate processes, and an incompletely retroverted pubis with pubic apron and unfused metatarsals. *Above,* reconstruction of *Sinornis santensis* (*A*); cross-hatching indicates missing parts. *Below,* holotype skeleton as originally preserved (the figure combines information from the slab and counterslab). Abbreviations: *I–IV,* digits 1–4; *dc,* distal carpal; *f,* frontal; *fe,* femur; *fi,* fibula; *fu,* furcula; *il,* ilium; *m,* maxilla; *pf,* pubic foot; *py,* pygostyle; *sc,* scapula; *st,* sternum; *ti,* tibia; *ul,* ulna; *ule,* ulnare. Scale bar = 1 cm. (From Sereno and Rao 1992; copyright 1992 AAAS)

Opposite Birds in China. In addition to the Lower Cretaceous birds of Spain, a nearly complete avian fossil

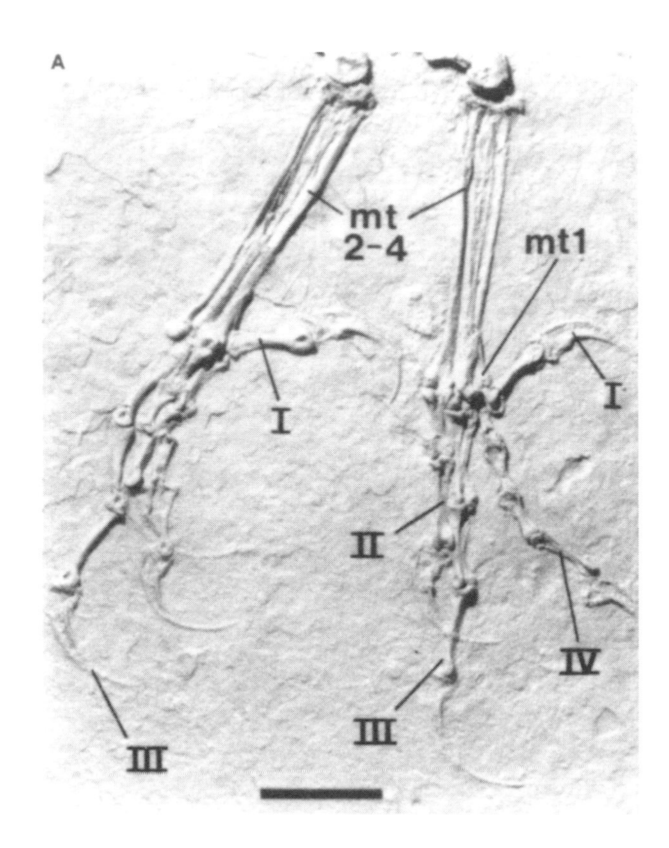

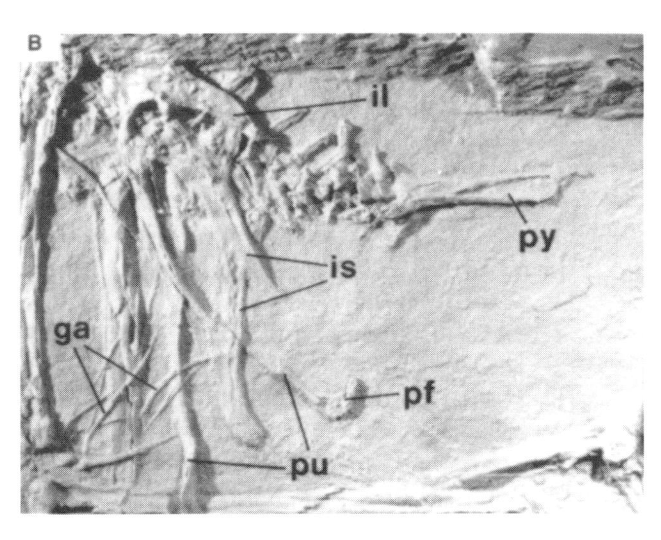

Sinornis santensis postcranium (epoxy cast from natural mold). *A,* feet showing unfused metatarsals (*mt*), retroverted digit 1 (*I*), and highly recurved unguals; scale bar = 5 mm; *B,* pelvis and tail showing the erect ilium (*il*), blade-shaped ischium (*is*), pubis (*pu*) with pubic foot (*pf*), large pygostyle (*py*), and gastralia (*ga*). (From Sereno and Rao 1992; copyright 1992 AAAS)

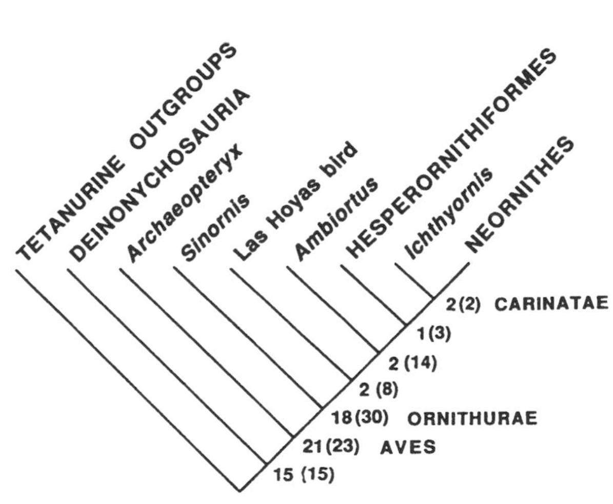

Cladistic hypothesis of early avian relations based on a numeral cladistic analysis of ninety-three osteological characters with PAUP computer program. Numbers at nodes indicate the number of supporting unequivocal synapomorphies (total number of synapomorphies under accelerated-transformation optimization shown in parentheses; consistency index equals 0.91). (From Sereno and Rao 1992; copyright 1992 AAAS)

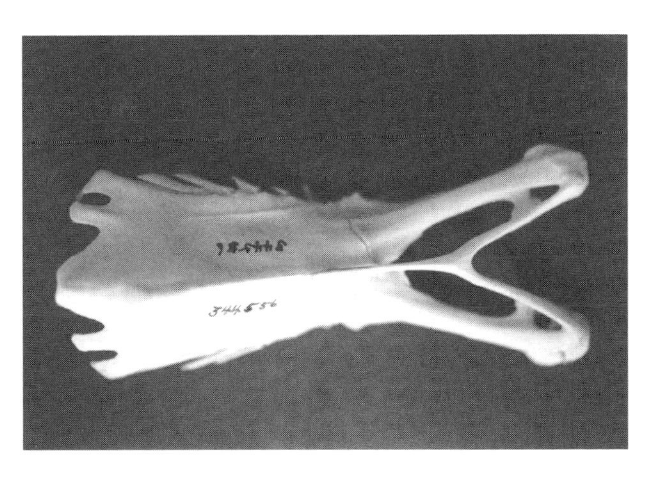

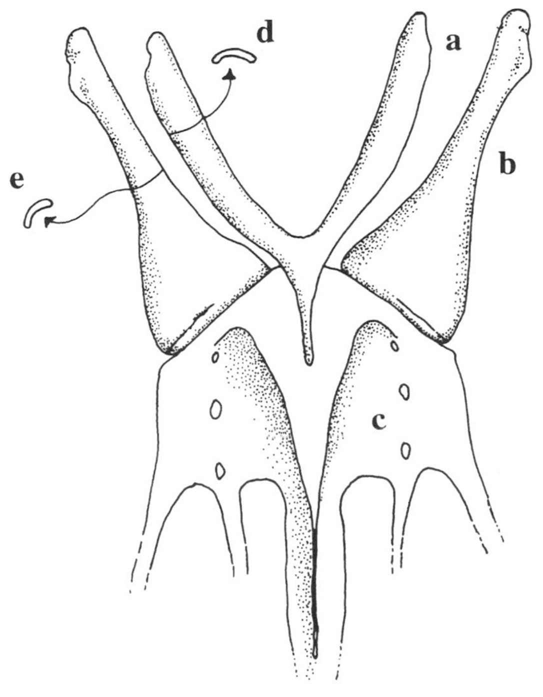

Right, sternum and pectoral girdle of the primitive living land bird the hoatzin (*Opisthocomus*), *above,* compared to that of *Iberomesornis, below*. Abbreviations: *a,* furcula, *b,* coracoid, and *c,* anterior sternum; *d* and *e,* cross-sections of furcula and coracoid. Note the broad furcula, grooved posteriorly as in *Archaeopteryx* (p. 153). The hoatzin's sternum is somewhat convergent on that of enantiornithines, with a posterior median keel that accommodates a large crop, and a long hypocleidium. Two gut structures, a crop and an esophagus, serve hoatzin folivores much as the rumens serve cows, providing a site for symbiotic bacteria to ferment fibrous plant materials and produce volatile fatty acids (Crajal et al. 1989). One wonders what digestive adaptations were present in opposite birds. (Photo by author, specimen #344586 U.S. National Museum; drawing from Martin 1995b; courtesy Forschungsinstitut Senckenberg)

of the approximate same age was discovered in 1987 by a farmer in Liaoning Province in northeastern China. The announcement in *Science* of the discovery of *Sinornis santensis* by Paul Sereno and Rao Chenggang (1992) created a sensation, and the fossil was described as belonging to a new order, the Sinornithiformes. Here at last was a small Lower Cretaceous bird, some 10 to 15 million years younger than *Archaeopteryx*, that was fully articulated with a skull exhibiting a short, toothed snout. The specimen was recovered from fine freshwater lake sediments of the Jiufotang Formation and was associated with a rich fish and insect fauna and an abundance of plant material. Aside from its *Archaeopteryx*-like skull, the sparrow-sized *Sinornis* was highly enantiornithine in many features, with a primitive pelvic girdle and hindlimb, yet it had a fully reversed hallux and strong, curved claws, similar to those of the modern trunk-climbing nuthatches of the family Sittidae. Like the enantiornithine birds, the pectoral girdle of *Sinornis* was advanced and showed full capability of powered flight. The carpus and manus were separate rather than fused into the carpometacarpus, and the hand was composed of freely articulating metacarpals, with phalanges and clawed unguals on the first and second digits. *Sinornis* exhibits some very primitive features, such as ribs without uncinate processes and gastralia, or stomach ribs (it is the only bird other than *Archaeopteryx* with this primitive structure). The correct orientation of the pubis has been disputed; Sereno and Rao (1992) have reconstructed the pelvis of *Sinornis* directed ventrally for a more dinosaurlike appearance. The preserved material indicates to me, and to Larry Martin (1995b), who has examined most of the enantiornithine material worldwide, that the pubes were more retroverted than their reconstruction presents.

The extreme curvature of *Sinornis*'s claws raises at least the possibility that these birds, members of the new avian adaptive radiation, were, perhaps for the first time, exploiting the ecological zone of the tree trunk. And the general—at least superficial—resemblance of *Sinornis*'s furcula and anterior sternal region to that of the primitive South American bird the hoatzin (*Opisthocomus*)—with a long, posteriorly directed rodlike hypocleidium and the keel of the sternum restricted to the posterior sternal region to accommodate a large crop as part of the digestive apparatus of this folivore—may suggest that *Sinornis* was exploiting a similar ecological resource, although plants of that period were far less appealing as a food source than in later times.

Actually, the first early Cretaceous bird discovered from China was not an enantiornithine but a modern ornithurine. *Gansus yumenensis* was recovered in Gansu Province, northwestern China, in 1981, and described by Hou and Liu in 1984. Though it consisted only of an incomplete left hindlimb, *Gansus* is of particular interest in illustrating the early adaptive radiation of birds into shoreline habitat. With its long toes and elongate fourth digit, it was probably the morphological equivalent of a modern shorebird, though it may have had an *Ichthyornis*-like toothed skull.

Lacustrine locality of the Lower Cretaceous Jiufotang Formation in Liaoning Province, China, where more than twenty specimens of nearly complete skeletons of small enantiornithine birds, notably *Cathayornis yandica,* have been recovered by Zhou Zhonghe of the Academia Sinica, Beijing. Three distinctive types of birds are represented in the avifauna, and the locality has produced abundant plant material as well as a rich fish assemblage, including teleosts, bowfins, sturgeons, and paddlefishes. (Photos courtesy Zhou Zhonghe)

A new era in our understanding of these Mesozoic birds was ushered in during September 1990, when Zhou Zhonghe, a graduate student of Academia Sinica's Institute of Vertebrate Paleontology and Paleoanthropology, discovered two small, articulated bird fossils while attempting to excavate paddlefish (*Polyodon*) fossils from lacustrine shales of the Jiufotang Formation. This rich lacustrine locality, which produces abundant plant material as well as invertebrate remains and fossils of teleosts,

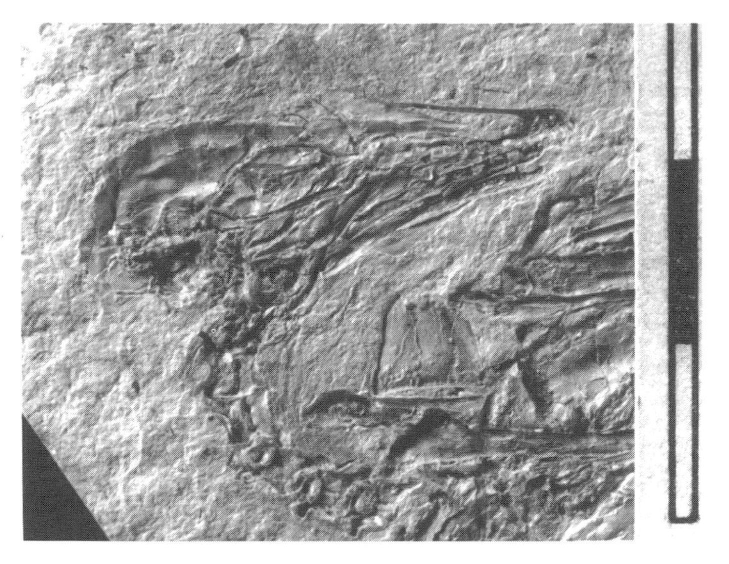

Archaeopteryx-like skull of *Cathayornis yandica,* length approximately 29 mm. The anterior part of the premaxilla is wide, with four teeth on its ventral margin, and the maxilla exhibits some fragmentary teeth. Two teeth are found on one separate thin-walled dentary. The conical teeth are slightly constricted at the bases of the crowns. Scale in cm. (Photo courtesy Zhou Zhonghe)

Slab (*top*) and counterslab (*above*) of the type specimen of the nutchatch-size enantiornithine *Cathayornis yandica,* discovered in September 1990, from the Lower Cretaceous locality shown in nearby photos. Scale bar at top = 3 cm. Skull, 29 mm. The synsacrum is composed of eight free vertebrae, posterior to which are eight free caudal vertebrae followed by the pygostyle, which is 15 mm long and tapers posteriorly. (Photos courtesy Zhou Zhonghe)

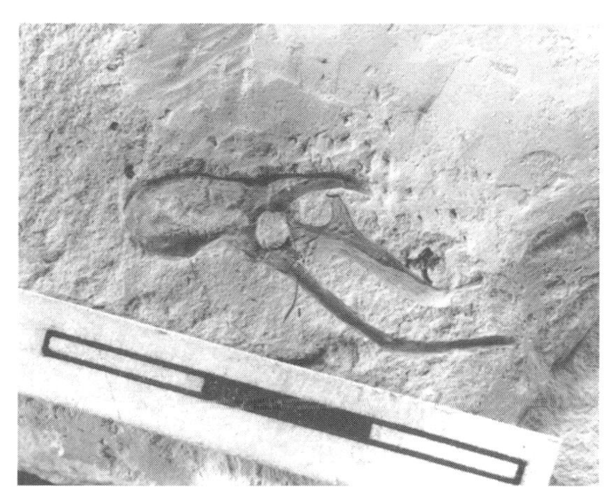

Above left, type specimen of *Chaoyangia beishanensis,* an ornithurine from the Lower Cretaceous of China. It illustrates the *"Archaeopteryx"* type of pelvic girdle and pubic apron but has uncinate processes on the ribs and a tibial crest. *Left,* unpublished pelvic girdle from the same locality preserved in mudstone (making preservational displacement highly unlikely), illustrating the avian opisthopubic orientation of the pubes. Scale in cm. (Photos courtesy Zhou Zhonghe)

bowfins, sturgeons, and paddlefishes, has since yielded more than twenty individuals representing three adaptive types of birds, based on hindlimb morphology (Zhou 1995a, b).

The first bird described from Jiufotang was *Cathayornis yandica* (Zhou, Jin, and Zhang 1992), a tiny bird about the size of a nuthatch. *Cathayornis* has a skull somewhat like that of *Archaeopteryx*, with four teeth emanating from the premaxilla, each of which inclines anteriorly and is constricted at the base of the crown. Two teeth are preserved on the dentary, one near the anterior end and the other close to the middle of the bone. Like other enantiornithine birds, *Cathayornis* was a fully powered flier, with a derived pectoral girdle and forelimb, but it had a primitive pelvic girdle and hindlimb morphology. Its sternum is very like that of the early Cretaceous Spanish birds and that of *Sinornis;* in fact, not only in morphology but in size *Cathayornis* and *Sinornis* are so similar to the Spanish opposite birds that many paleontologists would claim near congeneric status for many. Had they occurred side by side, all would be accepted as very close allies.

Many additional specimens representing numerous adaptive types have been discovered and are rapidly being described, including *Boluochia zhengi* from the Jiufotang Formation (Zhou 1995c), which was slightly larger than *Cathayornis* but had a hooked beak and may have had raptorially adapted claws. In addition, Dong (1993) reported an unnamed enantiornithine from the Ordos Basin of Inner Mongolia, and Hou Lian-Hai (1994) described *Otogornis genghisi*, preserved in graygreen mudstones from the Yijinhuoluo Formation of Inner Mongolia, as the earliest enantiornithine from China. The list of Chinese enantiornithines will greatly expand in ensuing decades.

The Wide Distribution of Enantiornithines. The Russian paleontologist Lev Nessov has discovered a number of enantiornithines in the Lower Cretaceous of former Soviet middle Asia. He named one *Horezmavis eocretacea* (Nessov and Borkin 1983; Nessov 1992). *Horezmavis* was the size of a modern common moorhen (*Gallinula chloropus*) and apparently lived on the edge of a large bay. Although the Russian paleornithologist Evgeny Kurochkin (1995) considers it a modern neognathus bird, the identification is based on a tarsometatarsus, which, as in other enantiornithines, is fused proximally but not distally. Other enantiornithines include, from the late Cretaceous of Asia, *Sazavis prisca*, described on the basis of a distal tibiotarsus (Nessov and Jarkov 1989), and from the Coniaian of the late Cretaceous, also in Asia, *Kizylkumavis cretacea*, based on a partial humerus (Nessov 1984). In addition, two new species of *Enantiornis* have been described from Uzbekistan (Nessov and Pantheleev 1993). Eggs thought to be enantiornithine found in Mongolia exhibit a distinctive elongate shape and shell structure (Kurochkin 1994). And from the island continent of Australia comes the description, based on a tibiotarsus from the Lower Cretaceous, of the sparrow-sized bird *Nanantius* (Molnar 1986), which now combines with an almost complete three-dimensional skeleton of *Nanantius* from the late Cretaceous of Mongolia (Kurochkin 1995).

Adding to the excitement, a marine enantiornithine has been discovered from the Upper Cretaceous Mooreville Chalk of Alabama (Lamb, Chiappe, and Ericson 1993), a deposit that previously had only yielded the avian *Ichthyornis*. This find, along with the earlier Spanish fossils, provides evidence that these birds may have played a more important role in marine shoreline environments than was previously thought. Because the preponderance of enantiornithines are known from nonmarine, usually lacustrine or lake, deposits, it is tempting to speculate that they were less mobile and incapable of extended periods of aerobic powered flight, possibly as a result of a quasi-ectothermic thermoregulation, and that only the fully endothermic ornithurine birds, such as *Ichthyornis* and *Hesperornis*, could venture far out to sea. Given the near absence of ornithurine bird fossils in late Cretaceous continental deposits and their near exclusive occurrence as "transitional shorebirds," ichthyornithiforms and hesperornithiforms, it is also tempting to speculate that ornithurines may have been more or less restricted to shoreline and marine environments during this time.

The list of new enantiornithines continues to expand, and opposite birds are now known nearly worldwide from throughout the entire span of the Cretaceous. They occur in an impressive array of sizes and adaptive types from a great variety of habitats, everything from tiny arboreal creepers to long-legged shore-dwelling and fowl-like species to the large, vulture-sized Argentine *Enantiornis*, whose wingspan reached some 1.2 meters (4 ft.). There can no longer be any doubt that enantiornithine birds represent the major adaptive radiation of land birds during the Cretaceous.

This singularly amazing upheaval in our knowledge of Mesozoic birds occurred so fast that until recently the pieces were difficult to assemble. Longtime student of Mesozoic birds Larry Martin (1995) has shown convincingly that all of these birds are not only related but fit nicely into the subclass Enantiornithes, which would now include the Lower Cretaceous birds from Spain and China, the Iberomesornithiformes and possibly synonymous Sinornithiformes, as well as *Nanantius* from Australia, *Gobipteryx* from Mongolia, *Alexornis* from Mexico, and all of the original enantiornithine material that was described from South and North America and Asia. As noted, these birds are characterized primarily by the development of a highly derived pectoral flight apparatus and a primitive pelvic girdle and hindlimb morphology. Enantiornithines have a pectoral architecture quite distinctive from that of modern birds, and the roof of the triosseal canal is formed by a broad, rectangular process of the scapula, as opposed to the modern condition in which the coracoid forms most of the canal.

Although these birds have a keeled sternum, strutlike coracoids, and a pygostyle, as in modern ornithurine birds, the pygostyle is quite elongate and the metatarsals are fused proximally but not distally, are fused in a row, and lack a hypotarsal cap. In addition, these birds lack uncinate processes on the ribs. These and more are all features of *Archaeopteryx*. One of the most provocative questions, therefore, concerns the relation of enantiornithines to the first bird, *Archaeopteryx*. Most of the differences between the two do in fact point to the improved flight capabilities of enantiornithes, and in profile their skulls are remarkably alike. The small Chinese and Spanish enantiornithines also share with *Archaeopteryx* a broad furcula that is grooved posteriorly. This furcula, unlike that of modern birds, cannot be compressed laterally, therefore could not have served as the carinate spring-spacer, and probably was mostly a site for the attachment of pectoralis flight muscles (Olson and Feduccia 1979; Martin 1995b). At a minimum, it is clear that *Archaeopteryx* is much more closely related to the Enantiornithes than it is to modern birds and that "the progenitors of modern birds must have been contemporary with [*Archaeopteryx*] even though we have no record of them" (Martin 1995b, 27).

Other Early Cretaceous Ornithurine Birds. The enantiornithine birds from Spain and China are only some 10 to 15 million years younger than *Archaeopteryx*, but they are not much older, if at all, than the Chinese bird *Gansus*, mentioned earlier, which would appear to be an ornithurine (Hou and Liu 1984)—that is, a bird related to the modern avian radiation. In addition, in 1993, Hou Lian-Hai and Zhang Jiangyong described another interesting bird from the same *Cathayornis* locality, which they named *Chaoyangia beishanensis*, on the basis of a well-preserved but incomplete skeleton that included the pelvic girdle and hindlimbs and part of the vertebral column and ribs, though no skull or forelimbs. This specimen is of particular interest because its pubic apron is preserved in almost the exact position as the London specimen of *Archaeopteryx* and the structures appear to be very similar. As mentioned, the orientation of the pubes has been much disputed in Mesozoic birds because the pubes appear to have been loosely attached to the pelvis, and in normal preservation, with displacement and other disturbances, it is often difficult to ascertain the original orientation. In *Chaoyangia*, however, Hou and Zhang note that "the pubis and ischium are posteriorly inclined" (224). Additional characters of interest in *Chaoyangia* include its unfused pelvic bones, non-heterocoelous (modern bird) vertebrae, unfused sacral vertebrae, and such advanced characters as uncinate processes on the ribs, ossified sternal ribs, a femoral caput with the fourth trochanter absent, and a tibial crest. Because uncinate processes are unknown in any enantiornithine birds and are almost universal in ornithurines, restudy of the fossil by Hou, Zhou, Martin, and myself led us tentatively to consider *Chaoyangia* an ornithurine. Several other ornithurine birds from the Lower Cretaceous of China remain undescribed, and a synsacral fragment of a large cassowary-size, possibly ornithurine bird has been reported from continental deposits of the Upper Cretaceous of southeastern France (Buffetaut et al. 1995), but it is far too fragmentary to be named or to shed much light on avian radiation. If avian, it is little more than a tantalizing relic to stoke the imagination. Another late Cretaceous avian of uncertain affinity, *Patagopteryx*, is a flightless, fowl-size Argentine fossil that, like the enantiornithines, exhibits ectothermic-style growth rings in its long bones (see chapter 6).

The most famous ornithurine land bird was described in 1982 and 1985 by Evgeny Kurochkin (1985a) as *Ambiortus dementjevi*, from lacustrine shales deposited from a shallow lake during the early Cretaceous of central Mongolia. This remarkable species represents the oldest verifiable modern-type bird. *Ambiortus* had a normal modernized furcula, a fused carpometacarpus, and a keeled sternum and flight architecture that are nearly identical to that of modern carinates—birds with a keeled sternum and modern flight apparatus. Kurochkin's specimen of *Ambiortus* also preserves body contour feathers and asymmetric primaries,which indicate that it flew (Kurochkin 1985a, 275). Similarities between *Ambiortus* and the ornithurine bird *Apatornis*, of the Niobrara Chalk of Kansas, have also been noted (Martin 1987).

Marine Cretaceous Birds

The radiation of Cretaceous land birds is just now coming to light, but the Cretaceous birds known to have shared the marine habitat with the ichthyosaurs, plesiosaurs, and mosasaurs have been well known for more than a hundred years. These marine birds constitute two principal orders: the loonlike Hesperornithiformes and the unrelated and ternlike Ichthyornithiformes, both members of the modern ornithurine avian clade. The oldest known specimen of either order is the hesperornithiform *Enaliornis* from the early Cretaceous of England, almost

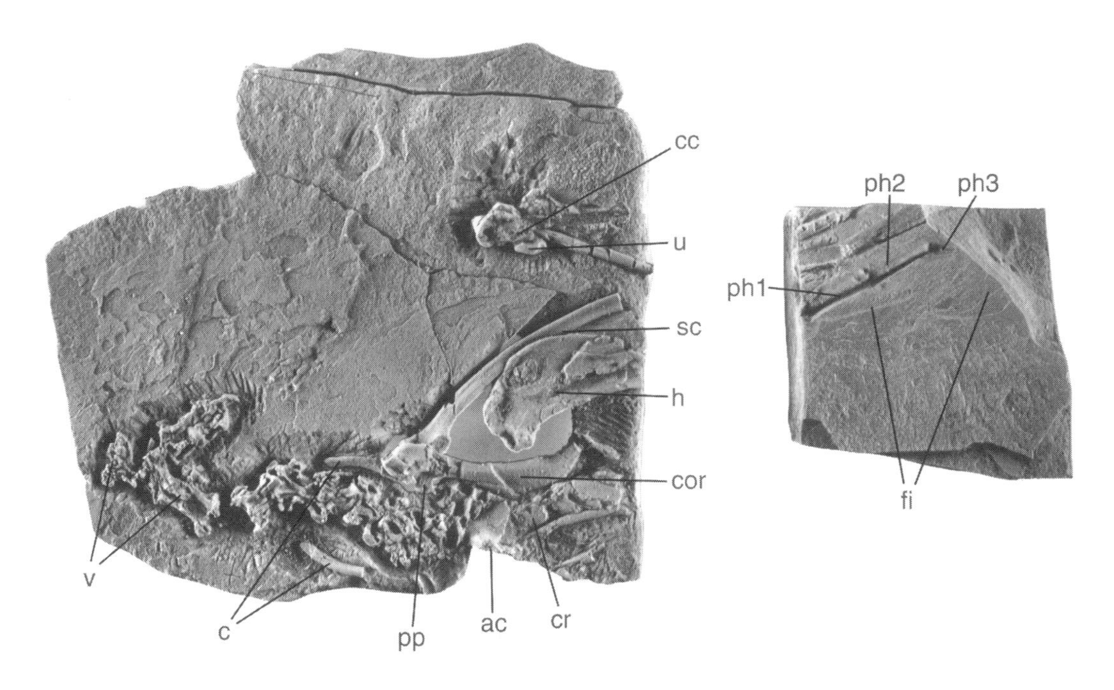

The Lower Cretaceous ornithurine bird *Ambiortus* proved that the two major lineages of birds, the enantiornithines and the ornithurines, diverged by early Cretaceous times. The bird was discovered in 1977 in shales of a shallow lacustrine deposit in central Mongolia. Such deposits are common in the region and produce numerous fish, insect, and plant fossils. *Ambiortus* was a true flying carinate bird, with a well-developed modern-type keeled sternum and a typically avian shoulder girdle. Also preserved are nice fossil feathers, including some asymmetric primaries. Abbreviations: *ac,* apex of carina; *c,* clavicle; *cc,* carpometacarpus; *cor,* coracoid; *cr,* carina of sternum; *fi,* feather impressions; *h,* humerus; *ph1–ph3,* phalanges of major digit of wing; *pp,* procoracoid process; *sc,* scapula; *u,* ulnare (cuneiform); *v,* vertebrae. × 0.78. (Photos courtesy E. N. Kurochkin)

Gansus, fossil foot, *right,* and tentative restoration, *above,* based on *Ichthyornis* skull. Scale in cm. (Photo courtesy Hou Lian-Hai; drawing by John P. O'Neill)

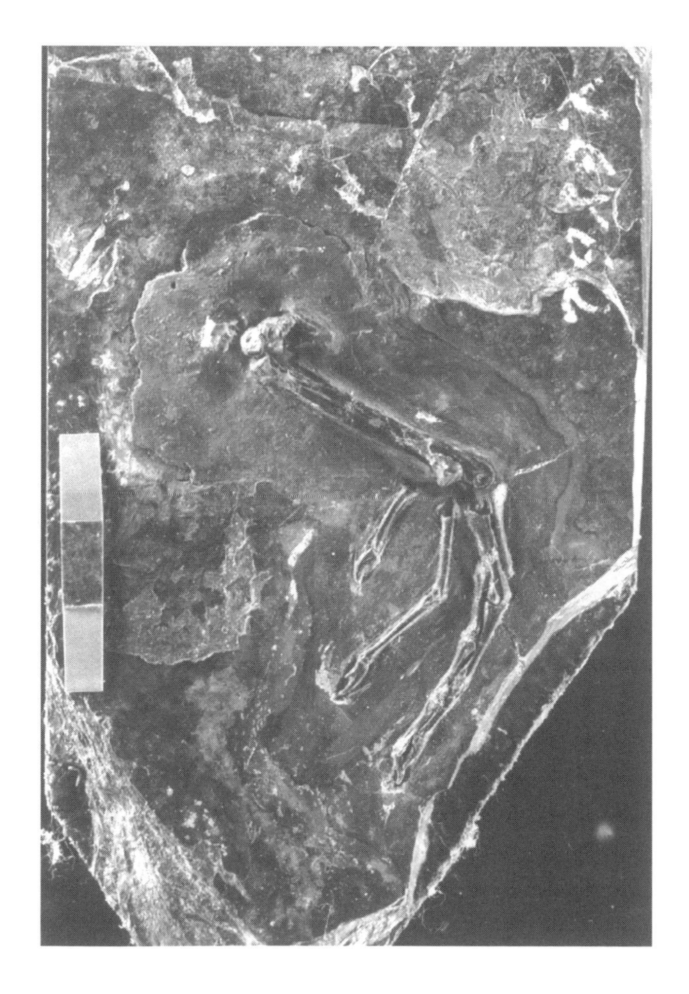

coeval with *Archaeopteryx* and predating the next well-preserved fossils of foot-propelled divers by as much as 50 million years. Pierce Brodkorb (1963a) had included *Enaliornis* within the order of loons, the Gaviiformes, but Larry Martin and James Tate (1976) showed convincingly that *Enaliornis* was a hesperornithiform bird. Andrzej Elzanowski and Peter Galton (1991) studied the braincase of *Enaliornis* and concluded that it had a primitive avian brain similar to those of *Hesperornis* and to such living birds as the primitive pelecaniform *Phaethon* (tropicbirds). In both hesperornithiform and ichthyornithiform birds, unlike the enantiornithine birds and *Archaeopteryx,* and like all modern birds and the Lower Cretaceous genera *Gansus* and *Ambiortus,* the distal end of the tarsometatarsus fuses before the proximal end, there is a wedge-shaped middle metatarsal, and there is a well-formed tarsal cap (Martin 1984). Other indications that hesperornithiform birds are members of the modern avian radiation are the presence of ossified uncinate processes on the ribs and a bone histology that is remarkably similar to that of modern birds (Houde 1987a), showing no growth rings as in the enantiornithines (Houde, Chinsamy, pers. comm.).

Hesperornithiforms Are Distinctive from Ichthyornithiforms. *Enaliornis* is known only from poorly preserved material, but there is no doubt that it, like all hesperornithiforms, was a foot-propelled diver. Clearly, birds were entering the adaptive zone of foot-propelled diving at about the same time, late Jurassic or early Cretaceous, that they were developing powered flight. Another toothed bird of the Cretaceous marine environment is the stout, superficially ternlike ichthyornithiform *Ichthyornis* ("fish bird"), a small-footed bird with large wings. Although *Ichthyornis* was undoubtedly a sideline of avian evolution (its link with the terns is merely a descriptive convenience), it was, at the time of its discovery, the oldest known bird with a keeled sternum similar to that of modern carinate birds. Both the hesperornithiforms and the ichthyornithiforms, along with the enantiornithines, died out at the same time the dinosaurs disappeared, at the close of the Cretaceous, the hesperornithiforms after having attained a high degree of adaptation for diving.

The hesperornithiforms are very distinctive from all other known birds and have been placed in their own subclass, the Odontognathae (*odon,* tooth + *gnathos,* jaw; sometimes Infraclass Odontoholcae), whereas the ichthyornithiform birds, though neornithine in general structure, are presently of unknown affinity, though similarities of *Ichthyornis* with *Ambiortus* have been noted (Olson 1985a; Elzanowski 1995). *Ichthyornis* was about the size of modern small gulls and terns and inhabited a similar habitat, which may have led various authors (see Lambrecht 1933; Brodkorb 1967, 1971) to place the Ichthyornithiformes near the modern-day Charadriiformes (shorebirds, gulls, and terns); but their humerus is unique (Harrison 1973), and they are almost certainly representative of a long-extinct earlier radiation of ornithurine carinate birds.

Both groups, like many enantiornithine birds, exhibit a feature reminiscent of *Archaeopteryx* that has generated controversy from the time of these birds' discovery until the present: the reptilian carry-over of teeth on the dentary and maxilla. Larry Martin and J. D. Stewart (1977) showed that the teeth of *Ichthyornis* (*I. dispar*) are similar to those of the hesperornithiforms and suggested that in both forms the young had teeth set in an open groove that developed discrete sockets with age. Complicating matters, the dentary bones of *Hesperornis, Parahesperornis,* and *Ichthyornis* differ from all other birds in being blunt anteriorly and bearing a terminal facet for the articulation of a predentary bone (an intersymphysial structure). The only other vertebrates known to possess such a structure are the ornithischian dinosaurs, also known as the predentates.

Another distinctive feature of both *Ichthyornis* and *Hesperornis* is an intramandibular articulation; Philip Gingerich has suggested that this is a link to theropod dinosaurs, which also had a distinctive intramandibular joint—one that "allowed the lower jaws to flex, with the anterior toothed half capable of rotating approximately 15° against the posterior half in *Herrerasaurus*" (Sereno and Novas 1992, 1138). This joint is a prime synapomorphy, or shared derived character, of theropods, but it is absent in *Archaeopteryx* and enantiornithine birds, and in fact the *Ichthyornis* and *Hesperornis* joint is closer to that of mosasaurs and living varanid lizards, which have an intramandibular joint analogous to that of theropods; it provides a flexible, grasping bite and may be an adaptation to subdue live prey (Sereno and Novas 1992, 1138). In the case of *Ichthyornis* and *Hesperornis,* these joints appear to have little phylogenetic significance in linking them to theropod dinosaurs.

The Early Discovery of Marine, Toothed Birds. In considering the Cretaceous toothed birds it is helpful to keep in mind that their discovery followed by only a few years the discovery of *Archaeopteryx* and occurred while vestiges of reptilian ancestry in birds were still confounding much of the scientific world. It was while the disputes between the likes of Huxley and Owen were still echoing through the museums of Europe that other men of science

Yale University expedition of 1870 in the field near Bridger, Wyoming. At center, O. C. Marsh stands holding a rifle. This was among the first of the "student" expeditions, and George Bird Grinnell of the class of 1870 (third from left) later wrote that probably none of the lot, except the leader, "had any motive for going other than the hope of adventure with wild animals or wild Indians." Reclining on the left is Eli Whitney (class of 1869), grandson of the inventor of the cotton gin. (Courtesy John H. Ostrom and Peabody Museum of Natural History, Yale University)

Upper Cretaceous Niobrara Chalk deposits, western Kansas. *Hesperornis* is known from this region, where it flourished along with mosasaurs, plesiosaurs, and a variety of fish and other marine life. (Photo by author)

Right, Restoration of *Hesperornis regalis,* the largest bird in the genus, about 1.5 m (5 ft.) long. Note the lobate webbing on the feet. (From Lucas 1901)

were making their way across the North American badlands in search of new fossils, many of which were lying exposed on the surface ready for the taking. The richest vein for this "bone rush" was the Upper Cretaceous chalk sediments of the Niobrara Formation.

Not surprisingly, the first discovery of a Cretaceous toothed bird from the Cretaceous seaways, that of *Hesperornis* ("western bird"), was made by O. C. Marsh. His own account, taken from his now classic monograph *Odontornithes* (1880), gives some flavor of the times and of the conditions under which this work was carried out:

> The first Bird fossil discovered in this region was the lower end of the tibia of *Hesperornis*, found by the writer in December, 1870, near the Smoky Hill River in Western Kansas. Specimens belonging to another genus . . . were discovered on the same expedition. The extreme cold, and danger from hostile Indians, rendered a careful exploration at that time impossible.
>
> In June of the following year, the writer again visited the same region, with a larger party, and a stronger escort of United States troops, and was rewarded by the discovery of the skeleton which forms the type of *Hesperornis regalis*, Marsh. Various other remains . . . were secured, and have since been described by the writer. Although the fossils obtained during two months of exploration were important, the results of this trip did not equal our expectations, owing in part to the extreme heat (100° to 120° Fahrenheit, in the shade) which, causing sunstroke and fever, weakened and discouraged guides and explorers alike. (1880, 2)

These first specimens of *Hesperornis* showed that it had very large feet, diminutive wings, and a flat, unkeeled sternum, indicating it could not fly. But the fossils were headless, so the prospect of debate over teeth in birds was postponed. Then in 1872, Marsh's colleague Benjamin F. Mudge recovered from the Niobrara Chalk the fossil of what was obviously a small bird with strong wing bones indicative of powered flight. This was the ternlike *Ichthyornis*, which was, unfortunately, also thought to be headless. However, a portion of the lower jaw was retained in the slab. Mudge sent the specimen to Marsh, who gave it the name *Ichthyornis dispar* and identified the jaw on the slab as coming from a new species of small reptile he called *Colonosaurus mudgei*. Soon, however, Marsh was forced to change his mind: "When the remains of this species were first described, the portion of the lower jaws found with them were regarded by the writer as reptilian; the possibility of their forming part of the same skeleton, although considered at the time, was deemed insufficiently strong to be placed on record. On subsequently removing the surrounding shale, the skull and additional portion of both jaws were brought to light, so that there can not now be a reasonable doubt that all are part of the same bird" (1873, 162).

Two years later, Marsh published his description of the teeth in *Hesperornis regalis*, the foot-propelled diver. This was 1875, a year before the discovery of the Berlin specimen of *Archaeopteryx*. The controversy over ancient toothed birds was not completely laid to rest, however. As recently as the 1950s, Joseph Gregory restudied Marsh's types and concluded that the jaw found with *Ichthyornis* was actually that of a small or juvenile marine lizard (Gregory 1951, 1952). He even went so far as to identify the jaw as coming from a specific genus of mosasaur, *Clidastes*. But no small mosasaur approaching the size of *Ichthyornis* has ever been found, and although some workers accepted Gregory's arguments for some twenty years (Brodkorb 1971), it is now certain that *Ichthyornis* did have teeth (Gingerich 1972; Martin and Stewart 1977).

***Ichthyornis* Was Abundant and Widespread.** Over the years Marsh described six species of *Ichthyornis* from the Smoky Hill Chalk of the Niobrara Formation, the best-known two being *I. dispar* and *I. victor*, and a seventh species, *I. lentus*, from the Upper Cretaceous Austin Chalk of Texas. Storrs Olson (1975a) showed that a fossil from the Upper Cretaceous Selma Chalk of Alabama that was originally described as the small ibis *Plegadornis* is really *Ichthyornis*, a finding that indicated a widespread occurrence of *Ichthyornis* over North America. Since then, numerous additional discoveries have been made. This expanded range for *Ichthyornis* was greatly amplified by the identification of a late Cretaceous vertebral centrum by Martin and Stewart (1982), which extended the range northward to southern Manitoba, and the discovery of a fossil humerus of *Ichthyornis* by Spencer Lucas and Robert Sullivan (1982), which extended the range westward to northwestern New Mexico, indicating that *Ichthyornis* was widespread along the late Cretaceous seaways of North America.

In 1985, William Zinsmeister identified *Ichthyornis* remains from the late Cretaceous of Seymour Island, Antarctica. This finding, combined with the discovery of a fossil humerus from the early late Cretaceous (Turonian) of Alberta (Fox 1984), extended the range of the genus throughout the Upper Cretaceous; the Canadian discovery was called the "oldest Cretaceous skeletal fossil of a bird from the Western Hemisphere" (258). In addition, David Parris and Joan Echols (1992) have documented four more specimens of *Ichthyornis* from Texas, two belonging to Marsh's *I. dispar* and two to *I. antecessor*, and they

Skeleton of *Hesperornis* from O. C. Marsh's classic monograph *Odontornithes* (1880). Marsh envisioned the bird as a large flightless form superficially similar to modern loons.

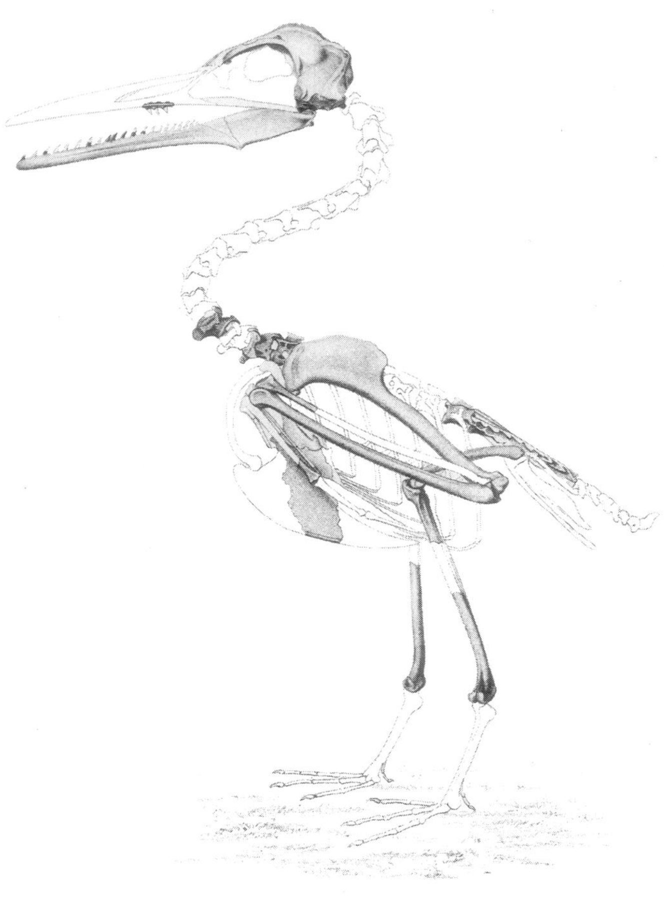

Skeleton of *Ichthyornis* from Marsh's classic monograph *Odontornithes* (1880). Note the large wing bones and keeled sternum.

Artist's reconstruction of the superficially ternlike, widespread marine Cretaceous toothed bird *Ichthyornis*. (Drawing by Sigrid K. James)

range during the late Cretaceous over a time span equivalent to that of the deposition of the Niobrara Formation.

Ichthyornis has also been recovered from the Old World. The Russian paleontologist Lev Nessov (1986, 1992a) described two new species of *Ichthyornis* from the Upper Cretaceous Kizylkum Desert of Uzbekistan, based on partial synsacra. The main Uzbekistan species was some 50 percent larger than the type species of *Ichthyornis* described by Marsh in 1880. Nessov (1984, 1992a) has also described several new species of supposed ichthyornithiforms, *Zhyraornis* (Zhyraornithidae) from the same locality, but these may turn out to be enantiornithines (Nessov and Pantheleev 1993; Kurochkin 1995). Basing new definitive taxa on such fragmentary evidence is totally unwarranted; such material should simply be reported in the literature, unnamed. However, Nessov's discoveries now make it clear that ichthyornithiforms were widespread and dominant during late Cretaceous time.

Othniel C. Marsh also named a species very similar to *Ichthyornis*, *Apatornis celer*, and placed it in its own family, the Apatornithidae. Two species of *Apatornis* are currently known, though the fossil remains are too fragmentary to permit reliable reconstruction of the bird. Larry Martin, however, has noted that "*Ambiortus* is extremely similar to *Apatornis* from the Niobrara Chalk of Kansas" (1987, 12). What we can say about the habits of *Ichthyornis* and *Apatornis* in life is therefore limited to what we can infer from their structure and from the location in which the remains have been found. It does seem abundantly clear that with strong powers of flight, long jaws, and recurved teeth for capturing prey, these birds were well adapted for the life of flying carnivores in the Cretaceous seaways. Their resemblance to gulls and terns is in size and habitat only, however, and their inclusion within the Charadriiformes, the order of shorebirds, has no taxonomic reality (Olson 1985a).

The Diversity and Abundance of Hesperornithiforms. The hesperornithiform birds were quite differently adapted and show evidence of a great diversity and worldwide distribution during the Cretaceous. Larry Martin, who has been studying the hesperornithiforms for many years, thinks that as many as seven genera and thirteen species may have existed. In addition to *Hesperornis*, the distinctive genus *Baptornis*, a more primitive and less specialized diver, is now known from rather extensive material, described in detail by Martin and James Tate (1976). Another genus, *Coniornis*, is known from the late Cretaceous of Montana. *Coniornis* and *Hesperornis* are included in the family Hesperornithidae; *Baptornis* is known from the late Cretaceous of Kansas and is placed in the Baptornithidae. Until 1992, *Neogaeornis wetzeli*, described on the basis of a tarsometatarsus from the late Cretaceous of Chile by Kálmán Lambrecht in 1929, was included within the Baptornithidae. Storrs Olson's restudy of the fossil material, however, led him to conclude that *Neogaeornis* "definitely does not belong among the Hesperornithiformes. . . . Its similarities to modern loons are such that it may be placed in the order Gaviiformes and in the modern family Gaviidae" (1992a, 124). The material is quite fragmentary, however, and given the time when definitive loons appeared, this identification, as we shall see, must remain extremely tentative.

One of the most complete hesperornithiform specimens ever recovered was an associated skeleton collected in western Kansas in 1894 and called *Hesperornis gracilis* by Samuel Wendell Williston (1898) because of its small size. The specimen was so beautifully preserved that he illustrated impressions of associated feathers and tarsometatarsal scutes. On reexamining the specimen, Larry Martin concluded that a new genus should be erected for the bird, and in 1984 he named it *Parahesperornis alexi*, in honor of "the late Alexander Wetmore, a former University of Kansas student and the most distinguished avian paleontologist of his time" (143). *Parahesperornis* was some 30 percent smaller than the type of *Hesperornis gracilis*, but it was larger than *Baptornis*, with an extremely reduced wing and a narrow, elongate body. There are indications that the premaxilla had been covered by a horny bill that was slightly downturned at the tip. In addition, the tarsometatarsus and toes show that toe rotation, characteristic of *Hesperornis*, and convergently present in grebes, was fully developed; in the Baptornithidae, this advanced toe rotation was not present.

In recent years a number of important discoveries of hesperornithiforms have come to light, and they tell of a much more amplified, worldwide adaptive radiation. Canadians Tim Tokaryk, John Storer, and Stephen Cumbaa have discovered some seventeen avians from early late Cretaceous marine deposits along the Carrot River in Saskatchewan. Their collection of more than a hundred fossils constitutes the oldest and most diverse avifauna from the North American Cretaceous. There is an enantiornithine, and four species represent the oldest North American ornithurines, two identified as *Ichthyornis* (four individuals) and two representing a new genus and two new species of baptornithid (twelve individuals). The new genus, however, is more primitive than *Baptornis* and, very interestingly, is a weak diver and possibly a flying form, its humerus and femur bearing resemblances to those of flying birds (Tokaryk, pers. comm.).

In Russia, Ukraine, and middle Asia, numerous fossils of newly unearthed hesperornithiforms are being described, including the first discovered Asian hesperornithi-

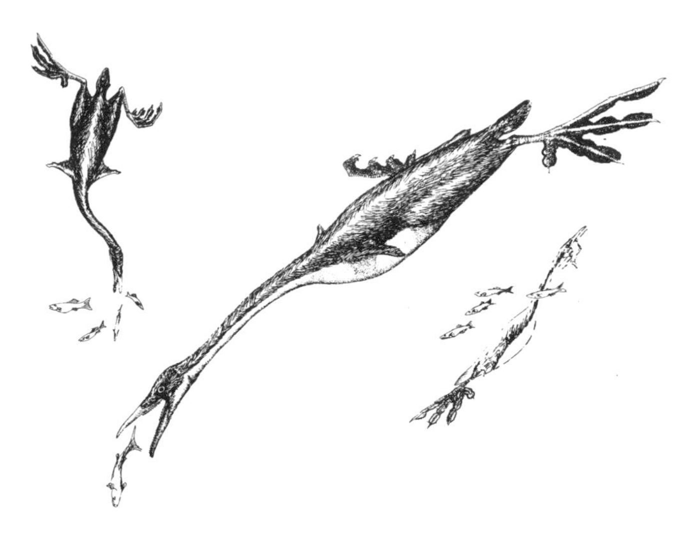

Restoration of *Baptornis advenus* ("the bird that dives under the water"). Although the feet are often depicted as having webbing across all four toes, as in pelecaniform birds, they were more probably lobed, as in *Hesperornis*. (From Martin and Tate 1976; restoration by B. Dalzell; courtesy Larry Martin)

Above, Gerhard Heilmann's life reconstruction of *Hesperornis regalis*. (From Heilmann 1926)

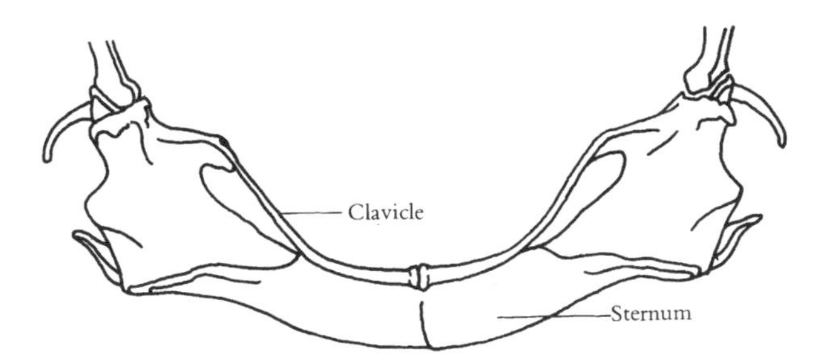

Front view of the flat, unkeeled sternum of *Hesperornis*. The lack of a carina indicates that the bird had no flight muscles and therefore could not fly. The clavicles are not fused to form the furcula typical of most modern birds, and the wing, so far as is known, consisted only of a vestigial humerus. (After Heilmann 1926)

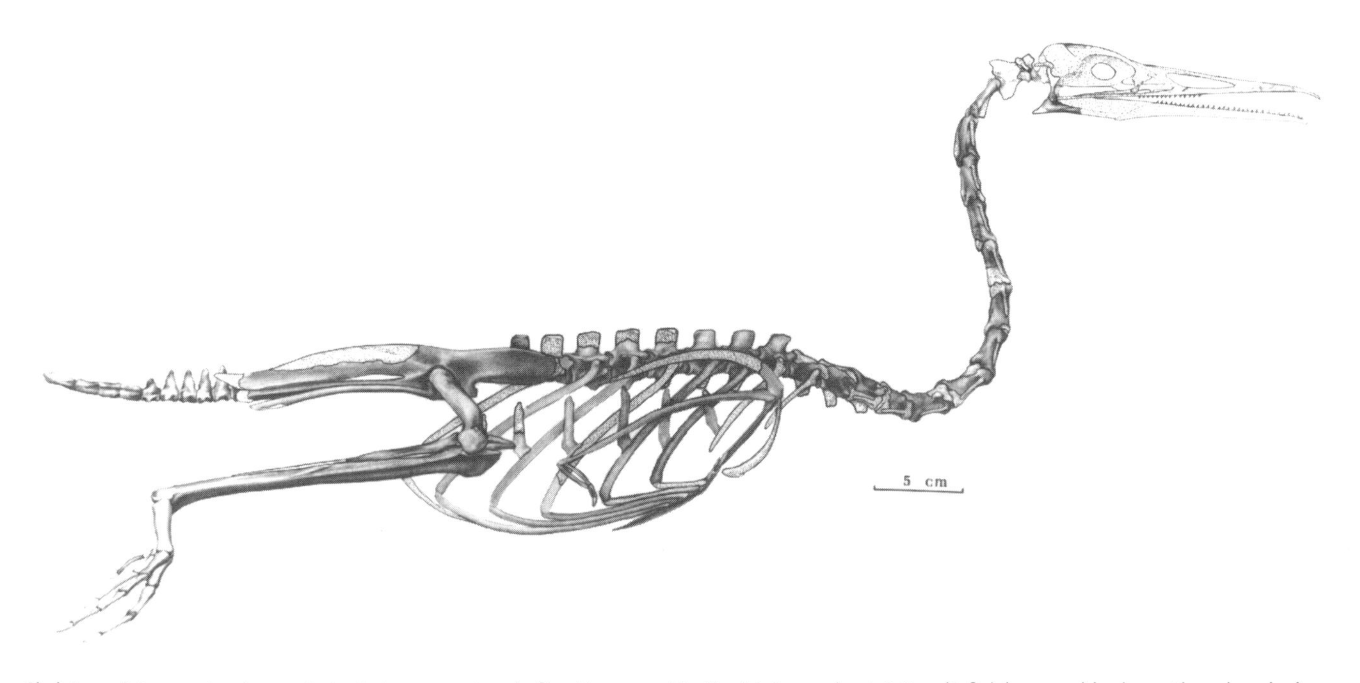

Skeleton of *Baptornis advenus* (stippled areas restored after *Hesperornis*). The bird was about 0.9 m (3 ft.) long and had greatly reduced wings. Like *Hesperornis,* it had a flat, keelless sternum. (From Martin and Tate 1976; restoration by B. Dalzell; courtesy Larry Martin)

form, a vertebra of a late Cretaceous (late Campanian–early Maastrichtian) baptornithid, *Judinornis*, from Mongolia (Nessov and Borkin 1983; Nessov 1986, 1992a and 1992b). Another fossil from another locality but the same beds was first identified as *Baptornis* but was later shown to be closely allied to *Parahesperornis* (Kurochkin 1988, 1995). In addition, other fragmentary tibiae, tarsometatarsi, and vertebrae apparently represent the flightless genera *Hesperornis*, *Asiahesperornis*, and *Parascaniornis*, from Kazakhstan, central Russia, and southern Sweden (Nessov and Prizemlin 1991; Nessov and Jarkov 1993). *Hesperornis rossica* (Nessov and Jarkov 1993), a very large advanced form, exceeding *H. regalis* in size, from the late Cretaceous of Kazakhstan and southern Sweden (Nessov 1992c), and the small *Parascaniornis* lived in the Turgai and Fenno-Scandian sea straits and adjacent areas, along with the other Eurasian hesperornithiforms (Kurochkin 1995). Much additional work must be done to understand these fragmentary fossils, but they do tell of a heretofore unknown adaptive radiation of these amazing divers and illustrate the cosmopolitan distribution of this bizarre group. Hou Lian-Hai is currently describing a hesperornithiform from the Lower Cretaceous of Antarctica.

The Habits and Diving Characteristics of Hesperornithiforms. An exciting discovery was the recovery of one specimen of *Hesperornis* from an estuarine deposit in Alberta, Canada (Fox 1974), and Larry Martin is currently describing a freshwater subfamily of hesperornithiform birds from the late Cretaceous of South Dakota. But the marine environment is by far the more commonly associated habitat. Martin has suggested that the hesperornithiforms nested in rookeries on isolated coastlines or perhaps on islands in the Cretaceous seaways. The Upper Cretaceous Niobrara Chalk in Kansas and elsewhere is a carbonate deposit that shows no evidence of any continental deposits, and the apparent absence of a nearby shoreline would imply that the hesperornithiforms, as well as the pteranodons and ichthyornithiforms whose fossils were found in these same deposits, were accustomed to venturing many hundreds of miles into the open sea (Martin and Tate 1976).

Dale A. Russell (1967) has reported remains of subadults of *Hesperornis* from the region of the Anderson River in Canada at 69° North latitude and has suggested that a nesting colony may have existed nearby. In addition to the Kansas finds, *Hesperornis* fossils are known from marine deposits in Alaska, the Northwest Territories, Manitoba, South Dakota, Montana, and Wyoming (Bryant 1983; Elzanowski 1983; Olson 1985a). Could it be that *Hesperornis* migrated to northern breeding grounds near the Arctic rim of the continent? It is also possible, as indicated by the distances the birds ventured into the open sea, that *Hesperornis*, like the earlier ichthyosaurs, gave birth to live young and never ventured onto dry land. We know that all hesperornithiforms would have been clumsy on land, because they could not fly and could propel themselves only with their large hind feet.

The first attempts to reconstruct the skeleton of *Hesperornis* placed the bird in an upright posture, like a modern auk, but it was discovered that the articulation of the head of the femur with the pelvis was such that the legs stood out to the side. Hesperornithids were strongly adapted pursuit divers of the epicontinental seas, and if they ever managed to slink onto land, they must have remained prostrate on the breast. In reconstructing *Baptornis*, Martin and Tate noted that "it seems almost certain that *Baptornis* could not walk upright on land and in fact must have pushed itself along on its stomach like a seal or a loon" (1976, 63). It is entirely possible that their habits were in some ways like those of modern diving mammals such as seals. Their advanced, almost mammal-like skeletal diving adaptations no doubt reflect the fact that as Mesozoic foot-propelled divers, they occupied the stable environment of inland seaways for the full range of the Cretaceous Period, some 80 million years.

Without doubt, the hesperornithids were fish-eaters: the one set of coprolites associated with the *Baptornis* skeleton (the only known for any Cretaceous bird) contain bones of the common Niobrara fish *Enchodus*. Hesperornithids ranged in size from that of a large grebe to the largest of living penguins, and like penguins, they were probably good indicators of abundance in marine waters. Andrzej Elzanowski has speculated that the larger species could probably reach bottom-dwelling prey—the Niobrara Sea is thought to have reached a depth of about 40 meters (131 ft.) during Smoky Hill Member deposition (Elzanowski 1983; H. W. Miller 1968). Ecologically, these birds had to coexist with isurid (mako and mackerel) and carchariid (gray) sharks, as well as with large teleost fishes, such as the huge *Xiphactinus*, and it has been suggested that for breeding they would have needed to seek protected sites in coastal estuarine areas (Elzanowski 1983; Fox 1974).

For the hesperornithiforms, selection favored adaptations that pointed more toward those of the Mesozoic marine mammals and marine reptiles than toward those associated with modern birds. In contrast to the adaptations for light weight favored in flying birds, including even the toothed *Ichthyornis*, adaptations for foot-propelled diving aimed at overcoming light weight and its resultant buoyancy. The hesperornithiforms, like the marine mammals and reptiles of their era, and unlike avian surface swimmers, had dense, heavy (pachyostosic), non-pneumatic bones, and they had a relatively high specific gravity that made diving easier. Martin has suggested that these dense bones enabled them to swim slightly beneath the surface, as do the modern snakebirds. Their tiny, vestigial wings (humerus only known from *Hesperornis*) must have been nearly useless, but may have been used, at least in *Baptornis*, for stabilizing and steering functions, much like the pectoral fins in fishes (Martin and Tate 1976). Some modern diving ducks, such as scoters (*Melanitta*), dive with the wings partly spread but with the alula, or bastard wing, partly projecting, perhaps for steering or stabilization, in a manner analogous to the suggested use of vestigial wings in *Baptornis*. And *Baptornis* had a relatively long tail with a long, laterally compressed pygostyle that may have acted as a rudder. The hesperornithiforms were indeed highly specialized birds, though they also show the salient features of the modern ornithurine radiation and belong within the subclass Ornithurae. And they were distinctive; responding to the view expressed by Gingerich (1973, 1976) that *Hesperornis* possessed the primitive "paleognathous" palate characteristic of the "ratites," Storrs Olson has rightly asserted, "The palate of *Hesperornis* is utterly different from that of other known birds." In addition, although much debate has centered around the type of cranial kinesis in hesperornithiform birds, Bühler, Martin, and Witmer (1988) firmly established that the *Hesperornis* had a prokinetic skull. This supports the suggestion (Bock, 1964; Zusi, 1984) that prokinesis is primitive at least for post-*Archaeopteryx* birds, and Bühler (1985) suggested that *Archaeopteryx* also may have had a type of prokinesis. "Recognition of a separate subclass, Odontoholcae, for the Hesperornithiformes, is still justified" (Olson, 1985a, 91).

Hesperornithiforms, Loons, and Grebes. In most skeletal features the hesperornithiforms paralleled modern grebes and loons, although there is no phylogenetic relation between them. At least one species of *Hesperornis* is similar in size to modern grebes, though *Hesperornis regalis* is almost 1.5 meters (5 ft.) long. As in all divers, the foot bones of the hesperornithiforms were laterally compressed, allowing a streamlined forward sweep through the water. These ancient birds also share with grebes the special adaptations of the toes of the hind foot that allow them to rotate sideways on the recovery stroke. It is probable that, also as in grebes, the toes were lobed, with each toe separately fringed with webbing, instead of fully webbed. Other shared adaptations include a reduced air-sac volume, a leg that is united with the body musculature nearly to the ankle joint, a long, narrow pelvis, a short femur, and a long tibiotarsus with a long bony extension to which the powerful extensor muscles of the leg are attached. This extension, known as the cnemial crest, among other things, demonstrates that the hesperornithiforms are merely convergent with, rather than related to, loons and grebes (see chapter 2). In *Hesperornis* the cnemial crest is formed by the kneecap, or patella, alone. In loons it is an exclusive extension of the tibia, and in grebes it is compounded by a fusion of the tibia and patella. This revelation was first published by Max Stolpe (1932, 1935), who concluded on the basis of hindlimb myology and osteology that any similarity between loons and grebes was due to convergent evolution. Stolpe was also able to show that the movement of the toes in swimming differed substantially between loons and grebes. In loons, the toes are flexed before the recovery stroke without being rotated, as in most swimming birds. Grebes, in contrast, are able to rotate the toes through a 90° arc while flexing them; in this manner, the longer mesial, or inner, lobes trail and the shorter lateral, or outer, lobes fold against the underside of the toe.

Still another unappreciated difference between loons and the hesperornithids is that modern loons are known to use their wings occasionally in diving, especially in "spurts or turns" (Palmer 1962, 34). According to Storrs Olson, they also probably use their wings underwater on a regular basis (1985a). He has shown that loons have become more highly adapted for wing-assisted diving since the early Miocene and that specialization for wing-propelled diving may have arisen more than once in the loon lineage.

Stolpe's observations on loon-grebe-hesperornithiform relations strongly influenced subsequent views of the nonrelation of loons and grebes. In fact, Joel Cracraft's cladistic study of 1982 stands alone in reasserting the monophyly of loons, grebes, and hesperornithiforms (see chapter 2)—it has been rejected nearly universally by subsequent workers (Olson 1985a; Sibley and Ahlquist 1990).

By the late Cretaceous, the toothed birds of the inland seaways, the hesperornithiforms and ichthyornithiforms, were probably oceanic relicts, "much as the Phaethontidae and Fregatidae are today" (Olson 1985a, 95). They were an evolutionary dead end, vanishing at the end of the Cretaceous during the demise of the great epicontinental seaways. They were soon replaced, however; highly specialized foot-propelled divers subsequently arose several times from completely independent lineages of birds.

Late Cretaceous Extinctions and "Transitional Shorebirds." It is interesting to ponder this process of late Cretaceous extinction in light of the tremendous attention that has been given to the extinction of the di-

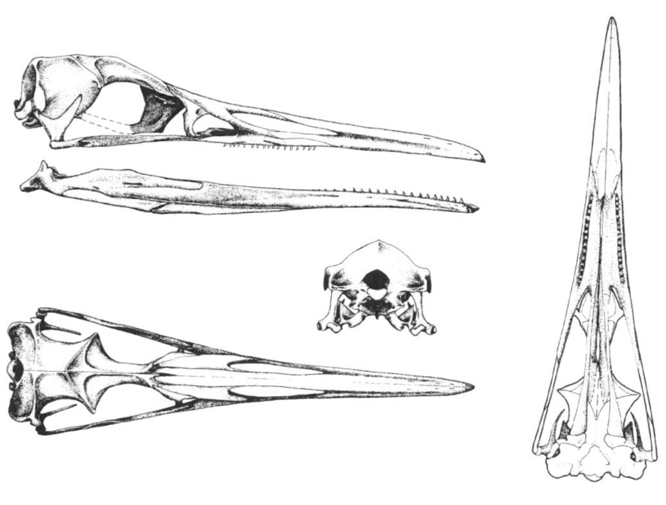

Restoration of the skull and lower jaw of *Hesperornis:* dorsal, lateral, palatal, and occipital views. As in advanced birds, teeth have been lost from the premaxilla, but they are retained on the maxilla and dentary. (Courtesy Larry Martin)

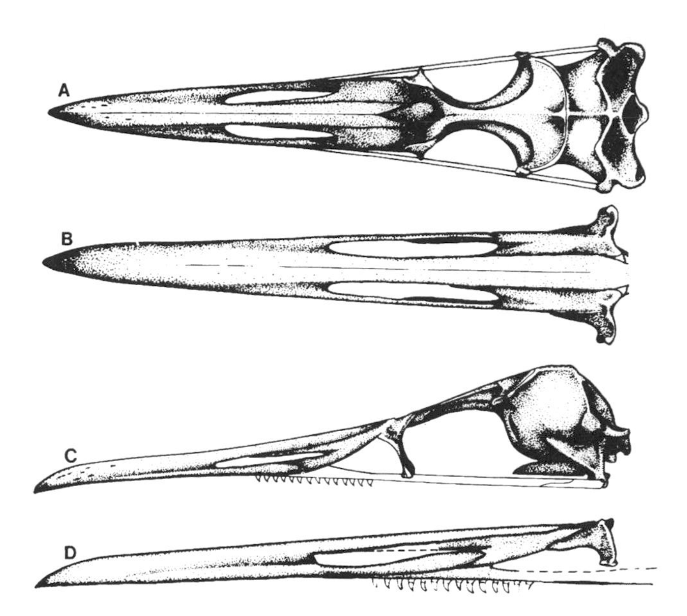

Restoration of the skull of *Parahesperornis alexi* (*A, C*) and the bill of *Hesperornis regalis* (*B, D*). (From Martin 1984; courtesy Larry Martin)

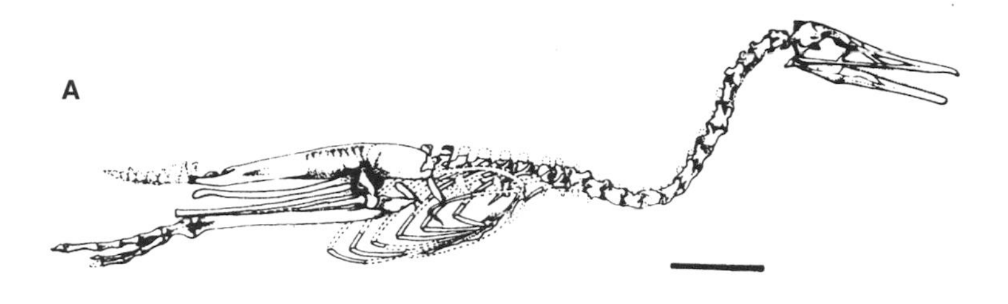

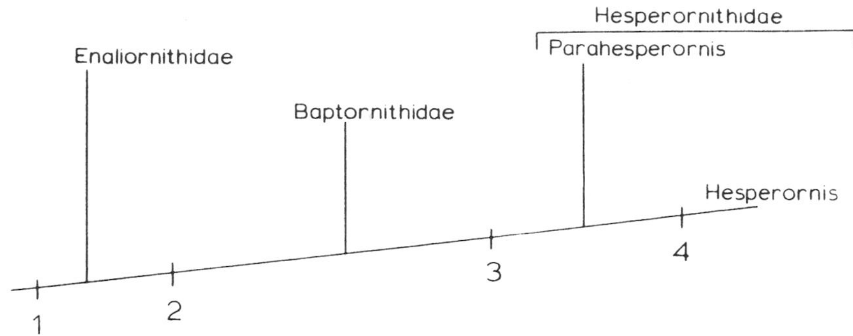

A, restored skeleton of *Parahesperornis alexi* with restored areas shown as dashed lines. Scale = 10 cm. *B,* relations within the Hesperornithiformes; characters employed: (1) characters 11–19 in Martin 1984, (2) posterior dorsal vertebrae heterocoelous, (3) advanced toe rotation, (4) fusion of the mesokinetic joint on the skull. *C,* relations of the Hesperornithiformes to other birds (characters in text). (From Martin 1984; courtesy Larry Martin)

nosaurs. Everyone is familiar with the great Cretaceous extinctions, once attributed to everything from cosmic radiation to lack of standing room in Noah's Ark, and artist Michael Ramus (Jepsen 1964) has cleverly satirized superstitions about anthropic emotions and the dinosaur extinctions. Yet the debate still rages. Although some people now theorize that the extinctions were gradual, perhaps as a result of combined climatic and geologic change, most scientists today believe the extinctions to have resulted, at least in part, or as a final blow, from a cataclysmic event such as the impact of an extraterrestrial body, perhaps an asteroid or a comet, which would have hurled thousands of tons of debris into the atmosphere, darkening and cooling the globe. Not only did no dinosaurs survive, but a myriad of other vertebrate groups, such as the plesiosaurs and mosasaurs, perished along with many invertebrates, such as the diverse and widespread ammonites. Amazingly, few have pondered avian extinctions at the close of the Cretaceous; yet birds (witness the miner's canary) would be the first to go.

Although our knowledge of birds is relatively incomplete, it seems clear that most Cretaceous birds perished along with the dinosaurs. The entire enantiornithine assemblage almost certainly succumbed, and in fact, the ornithurines *Ichthyornis* and *Apatornis* are restricted to the Upper Cretaceous. The foot-propelled hesperornithiforms are first known in the early Cretaceous and appear to have had an extensive radiation during the late Cretaceous, a time known as the Campanian, yet as one approaches the close of the Cretaceous, to deposits of the Maastrichtian age, these birds are almost absent. Hence, their extinction cannot, based on our current knowledge, be attributed to the same cause as the other terminal Cretaceous extinctions (Olson 1985a). Elzanowski suggested that their demise "may have been caused by the explosive radiation of acanthopterygian [modern percomorph] fishes" (1983, 75). Another influence may have been that by the late Cretaceous the epicontinental seaways were retreating, radically altering the geologic landscape.

Artist Michael Ramus's epitome of dinosaur history through Mesozoic time, satirizing current superstitions about anthropic emotions and the extinction of the giant reptiles. (From Glenn L. Jepsen, Terrible lizards revisited, *Princeton Alumni Weekly,* Nov. 26, 1963, 6; courtesy Princeton Alumni Publications, Inc.)

As far as we know, however, no toothed bird, enantiornithine or ornithurine, survived the extinction event of the late Cretaceous; nor is there any substantial evidence that, with the exception of shorebirds, any modern avian groups extend past the Tertiary back into the Cretaceous. An obvious question thus attends the possible presence of modern birds in the Cretaceous, especially given Joel Cracraft's use (1973a, 1974a, 1974b, and other papers) of continental vicariance biogeography to explain the distribution of many modern avian groups—that is, that they were distributed by the drifting of the continents.

At present, the only modern birds reported from the late Cretaceous are putative loons or loonlike forms from Chile, Antarctica, and Uzbekistan (all questionable), a collection of fragmentary fossils from the Lance Formation (Maastrichtian, latest Cretaceous) of Wyoming, containing at one time eight species described variously as loons, flamingos, shorebirds, and an ichthyornithiform (Brodkorb 1963b), and a procellariiform, limnofregatid, and cormorantlike bird from Mongolia (Kurochkin 1994). All of the birds reported from Mongolia are difficult to assess because of their fragmentary nature (usually an isolated bone) and because so little is known about the geological dates of the region's strata. There is also, in general, the difficult problem of dating discontinuous lake deposits. In spite of our increased sophistication in geologic dating, "biologic evolutionary history . . . has given us not only the principal means of time-correlation but the basis of the unique progressive traditional stratigraphic scale" (Harland et al. 1990, 4), and very little is known of biologic progression in the Asian region, especially for disparate lake deposits. Further, two of the pelecaniform birds (based on bony fragments), the limnofregatid *Volgavis* (Nessov and Jarkov 1989) and the cormorantlike bird (rejected by Kurochkin 1995), if correctly identified, would occur some 15 to 20 million years before the first known truly ancestral pelecaniform, the tropicbird (*Prophaethon*), from the Lower Eocene London Clay of England.

The same general objections apply to reports of loons from the Upper Cretaceous, in that the well-known and beautifully preserved primitive loon *Colymboides* is known from the Eocene. Could it be that the ancestral loon occurred millions of years after a well-developed form? And is it possible that modern loons coexisted with the ancient toothed divers, the hesperornithiforms? Yes, this is possible. However, more doubt is cast on reports of New World putative loons because of the fragmentary nature of *Neogaeornis:* a single right tarsometatarsus lacking the proximal articular surface; hesperornithiform expert Larry Martin (pers. comm.) is convinced that the femur is that of a hesperornithid. The putative Antarctic loon, first announced by Sankar Chatterjee in 1989, has yet to be published, and the Upper Cretaceous Lopez de Bartodano Formation deposits from which it was recovered have units that extend upward stratigraphically, some beyond the Cretaceous-Tertiary boundary (Zinsmeister 1982).

The problem of identifying fossil birds based on single elements is vividly illustrated by Storrs Olson's description of the Eocene frigatebirdlike form *Limnofregata,* which was, like most early Paleogene birds, a skeletal mosaic of numerous living groups and provides perhaps the greatest lesson to anyone studying avian paleontology:

> The pelvis and probably the sternum are the only elements of *Limnofregata* that by themselves would most likely be recognized as being similar to those elements in the Fregatidae. The skull and possibly the tarsometatarsus, taken in their entirety but alone, might also have been properly assigned by an alert paleontologist. The humerus, radius, ulna, carpometacarpus, coracoid, furcula, scapula, tibiotarsus and probably the femur, if found in isolation, either whole or in part, would have stood very little chance of being correctly identified as once having formed a part of a frigatebird, and almost any one of them might have been said to constitute a new family, although it is far from certain that in every instance each specimen would have been assigned to the proper order. The distal end of the tarsometatarsus might well have been referred to the Sulidae [boobies], whereas the proximal end of the tarsometatarsus would have proved impossible to assign to a modern family. The inadvisability of basing higher taxonomic categories of Paleogene birds [much less Cretaceous birds] on fragmentary limb elements . . . thus becomes evident. (1987, 31)

No real stock, therefore, can be placed in single fossil elements or bone fragments that have so frequently been described from both the Cretaceous and early Tertiary; regretfully, many must simply be ignored. As Storrs Olson has written, "The charm of these fragments has very frequently exceeded their scientific merit" (1985a, 82). Throughout the remainder of this book, therefore, numerous time-honored fossil taxa are not mentioned where the evidence is insufficient to determine their taxonomic affinity.

Of the Lance Formation birds, the ichthyornithiform has been placed in Pierce Brodkorb's charadriiform family Cimolopterygidae, in the new genus *Palintropus*. Brodkorb (1963b) included in the initial family four species of shorebirdlike fragmentary fossils in two genera, *Cimolopteryx* (*C. rara, C. minima,* and *C. maxima*) and *Ceramornis* (*C. major*), and found favorable comparisons of these latest Cretaceous shorebirds with the extant families Recurvirostridae, Burhinidae, and Glareolidae. Later, Storrs Olson and I showed that the supposed Cretaceous loons and flamingos belonged within the Charadriiformes (Olson and Feduccia 1980a), with Olson suggesting that "it is quite likely that the entire known avifauna of the Lance Formation, as well as *Apatornis,* may consist of 'transitional' charadriiforms and that the families Cimolopterygidae, Torotigidae, Lonchodytidae, and Apatornithidae may be subject to synonymization" (1985a, 173). Given the spectacular and dramatic nature of these putative Upper Cretaceous finds, much more evidence must be marshaled in their support before acceptance is possible.

In addition, fragmentary fossils of putative procellariiforms and "transitional shorebirds," or graculavids (Graculavidae), from the late Cretaceous greensands of New Jersey (Olson and Parris 1987) are now known to date from the early Paleocene (Olson 1994). The oldest reliable dates for the Procellariiformes are from the Lower Eocene London Clay (Daniels 1994) and a single *Pterodroma*-like (petrel) tibiotarsus from the Eocene of Louisiana (Feduccia and McPherson 1993), though it would not be surprising to find them somewhat earlier. Evidence from the fossil record of birds of the late Cretaceous and early Paleogene points to the transitional shorebirds (Olson 1985a, 171), which show similarity in their postcranial skeletons to such forms as the thick-knees (*Burhinus*), the shorebird-duck mosaic *Presbyornis,* and such related forms, as prime candidates for the wellsprings of most of the modern avian radiation. In addition to the transitional shorebirds from the late Cretaceous and early Paleocene of North America, several new graculavids are now known from the late Cretaceous Nemegt Formation of Mongolia (Kurochkin 1995) and from the Paleocene of France (Mourer-Chauviré 1994). Other avian types, such as primitive paleognaths, may also have made the transition through the Cretaceous-Tertiary boundary, but their fossils have not been forthcoming. Indeed, the transitional shorebirds may well be analogous to the mammalian insectivores in their evolutionary meanderings.

The basal status of the transitional shorebirds at the beginning of the Tertiary is evidenced not only by the presence of numerous latest Cretaceous and early Paleocene fossils of transitional shorebirds but by the discovery of a number of shorebird–modern order mosaics in the Eocene: *Juncitarsus, Presbyornis,* and *Rhynchaeites,* which are shorebird-flamingo, shorebird-duck, and shorebird-ibis mosaics, respectively. This is not to say that no other avian groups made it through the Cretaceous-Tertiary boundary; no evidence precludes that possibility, and one might well predict that some type of basal paleognaths made it through to give rise to the primitive paleognaths, the lithornithids and subsequent ratites. But most of the evidence now indicates that, aside from transitional shorebirds, few if any modern avian groups can be identified with certainty from the Cretaceous.

Explosive Radiation in the Early Tertiary

Fossil evidence now suggests that with the close of the Cretaceous, a new era began in avian evolution. However, the traditional view of avian evolution in the past century has been one of sluggish gradualism, in which most orders of living birds are assumed to have originated during the Middle Cretaceous or so (Sibley and Ahlquist 1990) and, passing unblemished through the Cretaceous-Tertiary boundary, diversified slowly into the present avian morphological landscape. As we shall see in chapter 6, Thomas Huxley (1867) viewed the living ratites as "waifs and strays" of the primeval avian radiation, a view most still hold, although the earliest ratite-like fossils are known only from the Paleocene epoch of the Tertiary. Because of this widespread view, however, many authors have attempted to explain current bird biogeographic patterns by examining the historic drifting of the continents, a form of vicariance biogeography. As an example, in their analysis of avian relationships based on DNA-DNA hybridization, Charles Sibley and Jon Ahlquist stated, "We have assumed that the divergence between the lineages that produced the living ostrich and rheas was caused by the opening of the Atlantic, and we assigned a date of [about] 80 million years ago to that event" (1990, 701). In contrast, recent studies by Storrs Olson and Helen James (1991) have shown that ratite-like morphology has evolved in flightless Hawaiian gooselike ducks, or moa-nalos (*Thambetochen*, chapter 6), which can be no older than the islands themselves, some 4 million years, and these large flightless birds must therefore have evolved in a much shorter time. So how could it be that ratites would have had to come, as is now often thought, from the Mesozoic?

The recent discoveries of the enantiornithines, discussed earlier in this chapter, have revolutionized our view of avian evolution, in that we now know that these archaic birds were the dominant land birds of the Mesozoic. Once this became apparent, fossils thought to represent modern bird orders in the Cretaceous have invariably proved, on reexamination, to be of enantiornithine stock. And recent discoveries and studies of early Tertiary bird fossils have shown that most of the major orders of living birds were present by the Eocene. Primary deposits that have yielded early Tertiary birds are the Lower Eocene London Clay of England, the Lower Eocene Green River and Willwood Formations of Wyoming and adjacent areas, the Middle Eocene Messel oil shales of Germany, and the Eo-Oligocene deposits at Quercy, France. In these and other Eocene fossil deposits, we find, with the exception of certain extinct groups, a pattern of the presence of modern orders (although some, as we shall see, are such mosaics that their true evolutionary affinities are difficult to ascertain). Virtually all the modern avian orders, therefore, are evidenced by Eocene time; the only structurally distinctive order conspicuous in its absence is the modern songbird order Passeriformes, which is thought to have originated, at least in its modern form, somewhat later, by the latest Eocene or early Oligocene.

One of the most remarkable collections of early Eocene birds is that amassed by Michael Daniels (1994), who in the past two decades has accumulated a collection of some six hundred individual birds from the Lower Eocene London Clay at Walton-on-the-Naze, Essex, a kilometer-long stretch of coastal intertidal exposures. The bird-rich silty marine layer generally does not exceed 3 meters (10 ft.) in thickness. Many of the specimens consist of associated skeletons in varying stages of completeness, and although a number of the birds are tentatively assigned to living orders, most represent remarkable mosaics, such that assignment to modern orders is more a convenience than a reality. Indeed, these fossils, like those of other Eocene deposits, illustrate that although avian orders are somewhat recognizable, the modern orders as we know them were in the formative stages. Nevertheless, the currently assigned ordinal affinities at least permit us a glimpse into the past, and we see that many, if not most, of the elements of the modern avifauna were present at Naze in the early Eocene, some 53 million years ago. We also see that many more types, mosaics of several orders, were present but subsequently became extinct.

A number of these London Clay birds, many from the Isle of Sheppey, were described by Harrison and Walker (1977 and many other papers; also see review by Steadman 1981). Because of the mosaic nature of these birds, however, many of the identifications cannot be assured, and the same applies to myriad bird fossils described from the British Upper Eocene (Harrison and Walker 1976a). For example, the supposed passerine *Primoscens minutus*, described by Harrison and Walker from the Lower Eocene, has been shown by Daniels, in collaboration with Storrs Olson, to be a zygodactyl bird, probably a member of the extinct family Zygodactylidae. And a cuckoo described from the London Clay is in reality a swift. Among the best-preserved birds are tiny forms the

Table 4.1 Avifauna from Naze, London Clay, Essex, England (Lower Eocene, 53 million years before present). Preliminary and tentative identifications to indicate approximate avian type, general ordinal affinity, and diversity of the early Eocene avifauna.

Common name	Tentative identification
Lithornithid paleognaths (table 6.2)	Lithornithiformes
Diver	Gaviidae
Petrel	Procellariidae
Tropicbird	Phaethontidae
Pelecaniform	Pelecaniformes
Ibislike mosaic	Threskiornithidae
Screamer	Anhimidae
Waterfowl	Anatidae
Hawk	Accipitridae
Falcon/caracara	Falconidae
Osprey	Pandionidae?
New World vulture	Vulturidae
Galliform	Galliformes
Moundbuilder	Megapodiidae
Galliform/rail/wader mosaic	Gruiformes
Rail	Gruiformes
Finfoot	Heliornithidae
Gruiform	Phorusrhacidae
Rail/wader	gruiform/charadriiform
Bustard	Otididae
Wader	Charadriiformes
Pratincole/plover	Glareolidae/Charadriidae
Oystercatcher	Haematopodidae
Jacana-like bird	Jacanidae
Thick-knee	Burhinidae
Gull	Laridae
Auk	Alcidae
Pigeon	Columbidae
Pigeon/sandgrouse	Columbiformes
Sandgrouse	Pteroclidae
Parrot	Psittaciformes
Cuckoo	Cuculiformes
Cuckoo/owl mosaic	*"Primobucco olsoni"*
Owl	Strigidae
Nightjar, etc.	Caprimulgiformes
Potoo, frogmouth, owlet-nightjar	Caprimulgiformes
Oilbird (*Prefica*)	Steatornithidae
Caprimulgiform	Archaetrogonidae
Swift	Apodidae
Crested-swift	Hemiprocnidae
Early swift	Aegialornithidae
Hummingbird-size bird	*incertae sedis*
Trogon	Trogonidae
Hoatzin, etc., mosaic	Foratidae
Mousebird	Coliiformes
Coly mosaics	Sandcoleiformes
Kingfisher	Alcedinidae
Roller-like birds	Coraciiformes
Roller	Coraciidae
Tody	Todidae
Motmot-like bird	Momotidae
Wood-hoopoe	Phoeniculidae
Barbet/piciform	Piciformes
Perching bird mosaic	*incertae sedis*
Perching bird/coly mosaic, etc., mosaics with zygodactyl feet	Zygodactylidae(?)

Source: Courtesy Michael Daniels

size of wrens and warblers that bear some passerine and mousebirdlike features. Many of the London Clay birds show considerable similarity to birds from the Lower Eocene Green River Formation of Wyoming; in both cases, because of the mosaic nature of the skeletons, identifications cannot be based on single elements.

It is nothing less than remarkable that the entire modern avian radiation is present at least in rudimentary form by the early to mid-Eocene. This can only be characterized as an extraordinarily explosive evolution, one that produced all of the living orders of nonpasserine birds within a time frame of some 5 to 10 million years (Feduccia 1995b). In a sense, the situation is somewhat like that of the famous animals of the Cambrian Burgess Shale, reviewed by Stephen J. Gould (1989), which provide a sensational evolutionary example in which striking anatomical diversity was achieved during the early stages of the group's radiation.

In timing, the avian radiation paralleled the mammalian ascendancy (Martin 1983b), whose rapidity is also just beginning to be appreciated: the evolution of fully developed marine whales from terrestrial ungulates, for example, is now thought to have occurred in less than 10 million years (Gingerich et al. 1994; Novacek 1994). The radiation of birds paralleled that of their hairy counterparts in being explosive, and to some extent, in timing: modern bird orders appear by the Paleocene and Eocene, modern families by the late Eocene and early Oligocene, and modern genera begin to be clearly discernible by the Miocene.

Evolutionary biologist George Gaylord Simpson, however, was struck by both the lack of progression of birds compared to mammals during the late Tertiary and the great similarity of Miocene and Recent birds, commenting that "Miocene birds are in general very, one might say amazingly, like Recent birds" (1946, 83). As an example, he cited the deposits of the California Miocene diatomites, which contain an oceanic bird fauna: "Of this fauna Miller (1935) has said, 'Some dozen species have been determined and other specimens have been assigned to family or genus. . . . Most of the species belong to genera that survive to present. . . . Horses had lost but two of

Table 4.2 Avifauna of western North American Green River and Willwood Formations (Lower Eocene, 50± million years before present)

Lithornithiformes (Lithornithidae—lithornithid paleognaths)
- *Paracathartes howardae* Harrison 1979
- *Lithornis promiscuus* Houde 1988
- *Lithornis plebius* Houde 1988
- *Lithornis ?nasi* (Harrison 1984)
- *Pseudocrypturus cercanaxius* Houde 1988

Pelecaniformes (Limnofregatidae, frigatebirdlike bird)
- *Limnofregata azygosteron* Olson 1977

Ciconiiformes (Ardeidae? heronlike bird)
- *Calcardea junnei* Gingerich 1987

Falconiformes (Accipitridae?)
- accipitrid-like bird (P. Houde—under study)

Falconiformes (Vulturidae?)
- cathartidlike bird (P. Houde—under study)

Galliformes (Gallinuloididae)
- *Gallinuloides wyomingensis* Eastman 1900

Gastornithiformes (Diatrymidae)
- *Diatryma gigantea* Cope 1876

Gruiformes (Phorusrhacidae?)
- phorusrhacid-like bird (P. Houde—under study)

Gruiformes (Rallidae?)
- *Palaeorallus troxelli* Wetmore 1931

Gruiformes (Messelornithidae—Messel rails)
- *Messelornis nearctica* Hesse 1991

Gruiformes (Geranoididae)
- *Geranoides jepseni* Wetmore 1933
- *Paragrus prentici* (Loomis 1906)
- *Paragrus shufeldti* Cracraft 1969
- *Eogeranoides campivagus* Cracraft 1969
- *Palaeophasianus incompletus* Shufeldt 1913
- *Palaeophasianus meleagroides* Cracraft 1969
- other unidentified gruiforms

Charadriiformes (Phoenicopteridae—flamingo-shorebird mosaics)
- *Juncitarsus gracillimus* Olson and Feduccia 1980

Charadriiformes (Presbyornithidae—duck-shorebird mosaics)
- *Presbyornis pervetus* (Wetmore 1926—possibly two species)

Charadriiformes (Anhimidae?)
- screamer (P. Houde—under study)

Cuculiformes? (Foratidae—hoatzinlike mosaic)
- *Foro panarium* Olson 1992

Strigiformes (family?)
- *"Eostrix"*—miscellaneous undescribed owls

Caprimulgiformes (Steatornithidae[?], oilbirdlike bird)
- *Prefica nivea* Olson 1987

Apodiformes
- undescribed swift

Sandcoleiformes (Sandcoleidae, mousebirdlike birds)
- *Sandcoleus copiosus* Houde and Olson 1992
- *Chascacocolius oscitans* Houde and Olson 1992
- *Anneavis anneae* Houde and Olson 1992

Coraciiformes
- kingfisherlike bird (Feduccia—under study)
- small bee-eater-like birds (P. Houde—under study)
- roller-like bird (possibly same family as bee-eater-like bird)
- *Primobucco mcgrewi* Brodkorb 1970 (Brachypteraciidae?)
- *Neanis schucherti* Shufeldt 1913 (aerial specialist—*incertae sedis*)

Piciformes–Coraciiformes (Galbulae; Primobucconidae)
- *Primobucco olsoni* Feduccia and Martin 1976
- *Neanis kistneri* (Feduccia 1973)

Table 4.3 Avifauna of Messel oil shales, Germany (Middle Eocene, 48.7 million years before present)

Struthioniformes (ostrich[?] mosaic)
- *Palaeotis weigelti* Lambrecht 1928

Ciconiiformes (Threskiornithidae? ibis-shorebird mosaic)
- *Rhynchaeites messelensis* Wittich 1898

Falconiformes (Accipitridae)
- *Messelastur gratulator* Peters 1994

Falconiformes (Falconidae—Polyborinae)
- caracaralike bird

Galliformes
- galliform bird

Gastornithiformes (Diatrymidae)
- *Diatryma geiselensis* Fischer 1978

Gruiformes (Cariamidae?)
- seriema-like birds (three undescribed species)

Gruiformes (Idiornithidae)
- *Idiornis tuberculata* Peters 1995, and second species

Gruiformes (Messelornithidae—Messel rails)
- *Messelornis cristata* Hesse 1988

Gruiformes (Phorusrhacidae—phorusrhacid)
- *Aenigmavis sapaea* Peters 1987

Charadriiformes (Phoenicopteridae—flamingo-shorebird mosaic)
- *Juncitarsus merkeli* Peters 1987

Strigiformes (Palaeoglaucidae)
- *Palaeoglaux artophoron* Peters 1991

Caprimulgiformes (Caprimulgidae)
- nightjar

Caprimulgiformes (Podargidae?)
- frogmouth-like bird

Apodiformes (Aegialornithidae—swift, size of medium-sized hummingbird)
- *Aegialornis szarskii* Peters 1985

Coraciiformes (roller-like birds)
- true roller (Coraciidae)
- true roller (bird of prey feet—Coraciidae?)
- ground-roller—(Brachypteraciidae?)
- possible hoopoelike bird (close to Upupidae?)

Piciformes
- (miscellaneous forms; one barbetlike)

Source: Peters 1991, 1992

Table 4.4 Eo-Oligocene fossil birds identified from the Phosphorites du Quercy, southwestern France

- Pelecaniformes
  - Phalacrocoracidae
    - Still undescribed genus and species. Old collections
- Ciconiiformes
  - Ardeidae
    - *Proardea amissa* (Milne-Edwards 1892). Old collections
- Falconiformes
  - Cathartidae
    - *Plesiocathartes europaeus* Gaillard 1908. Old collections
    - *Diatropornis ellioti* (Milne-Edwards 1892). Upper Eocene and old collections
  - Accipitridae
    - *Aquilavus hypogaeus* (Milne-Edwards 1892). Old collections
    - *Aquilavus corroyi* (Gaillard 1939). Old collections
  - Sagittariidae
    - Genus *Pelargopappus* Stejneger 1885 (syn. *Amphiserpentarius* Gaillard 1908)
    - *Pelargopappus schlosseri* (Gaillard 1908) (syn. *P. stehlini* Gaillard 1908 and *P. trouessarti* Gaillard 1908). Lower and Upper Oligocene
  - Horusornithidae Mourer-Chauviré 1991
    - *Horusornis vianeyliaudae* Mourer-Chauviré 1991. Upper Eocene
  - Falconidae
    - Undescribed genus and species. Upper Eocene and Lower Oligocene
- Galliformes
  - Gallinuloididae Lucas 1900
    - *Taoperdix* sp. Milne-Edwards 1869. Upper Eocene
  - Paraortygidae Mourer-Chauviré 1992
    - *Paraortyx lorteti* Gaillard 1908 (syn. *Palaeortyx cayluxensis* Milne-Edwards 1892). Upper Eocene–Lower Oligocene
    - *Paraortyx brancoi* Gaillard 1908. Upper Eocene
    - *Pirortyx major* (Gaillard 1939). Upper Oligocene
  - Quercymegapodiidae Mourer-Chauviré 1992
    - *Quercymegapodius depereti* (Gaillard 1908). Upper Eocene
    - *Quercymegapodius brodkorbi* Mourer-Chauviré 1992. Upper Eocene
- Phasianidae
  - Subfamily Phasianinae
    - *Palaeortyx brevipes* Milne-Edwards 1869 (syn. *P. ocyptera* Milne-Edwards 1892). End of Lower Oligocene–Upper Oligocene
    - *Palaeortyx gallica* Milne-Edwards 1869. Upper Oligocene
    - *Palaeortyx intermedia* Ballmann 1966. Upper Oligocene
- Gruiformes
  - Idiornithidae Brodkorb 1965
    - Genus *Elaphrocnemus* Milne-Edwards 1892 (syn. pars *Filholornis* Milne-Edwards 1892)
    - *Elaphrocnemus phasianus* Milne-Edwards 1892 (syn. *Filholornis paradoxa* Milne-Edwards 1892, *F. debilis* Milne-Edwards 1892, and *Telecrex peregrinus* Milkovsky 1989). Upper Eocene–Lower Oligocene
    - *Elaphrocnemus crex* Milne-Edwards 1892. Lower and Upper Oligocene
    - *Elaphrocnemus brodkorbi* Mourer-Chauviré 1983. Old collections
    - *Idiornis gallicus* (Milne-Edwards 1892) (syn. *Filholornis gravis* Milne-Edwards 1892). Upper Eocene
    - *Idiornis cursor* (Milne-Edwards 1892) (syn. *Orthocnemus major* Milne-Edwards 1892). Lower and Upper Oligocene
    - *Idiornis minor* (Milne-Edwards 1892). Upper Eocene
    - *Idiornis gaillardi* Cracraft 1973. Upper Eocene
    - *Idiornis gracilis* (Milne-Edwards 1892). Lower Oligocene
    - *Idiornis itardiensis* Mourer-Chauviré 1983. Lower and Upper Oligocene
    - *Propelargus cayluxensis* Lydekker 1891. Old collections
    - Genus *Occitaniavis* Mourer-Chauviré 1983 (syn. pars *Geranopsis* Lydekker 1891)
    - *Occitaniavis elatus* (Milne-Edwards 1892). Old collections
    - *Oblitavis insolitus* Mourer-Chauviré 1983. Old collections
  - Phorusrhacidae (Phororhacidae Ameghino 1889)
    - Subfamily Ameghinornithinae Mourer-Chauviré 1981
    - Genus *Ameghinornis* Mourer-Chauviré 1981 (syn. pars *Strigogyps* Gaillard 1908)
    - *Ameghinornis minor* (Gaillard 1939). Old collections and Upper Eocene?–Lower Oligocene
  - Messelornithidae Hesse 1988
    - *Itardiornis hessae* Mourer-Chauviré 1995. Upper Eocene–Lower Oligocene
  - Rallidae
    - *Quercyrallus arenarius* (Milne-Edwards 1892). Old collections

Table 4.4 Eo-Oligocene fossil birds identified from the Phosphorites du Quercy, southwestern France (*continued*)

- *Quercyrallus dasypus* (Milne-Edwards 1892). Old collections
- *Quercyrallus quercy* Cracraft 1973. Old collections
- Gruidae
  - Undescribed genus and species. Old collections
- Otididae
  - Undescribed genus and species. Old collections

- Charadriiformes
  - Recurvirostridae
    - *Recurvirostra sanctaeneboulae* Mourer-Chauviré 1978. Upper Eocene
  - Scolopacidae
    - *Totanus edwardsi* Gaillard 1908. Old collections
  - Laridae
    - Undescribed genus and species. Old collections
- Columbiformes
  - Pteroclidae
    - *Archaeoganga pinguis* Mourer-Chauviré 1992. Old collections
    - *Archaeoganga larvatus* (Milne-Edwards 1892). Old collections
    - *Archaeoganga validus* (Milne-Edwards 1892). Old collections
    - *Leptoganga sepultus* (Milne-Edwards 1869). Upper Oligocene
- Apodiformes
  - Aegialornithidae Lydekker 1891
    - Genus *Aegialornis* Lydekker 1891 (syn. *Tachyornis* Milne-Edwards 1892)
    - *Aegialornis gallicus* Lydekker 1891 (syn. *Tachyornis hirundo* Milne-Edwards 1892). Upper Eocene
    - *Aegialornis leehnardti* Gaillard 1908. Upper Eocene
    - *Aegialornis wetmorei* Collins 1976. Upper Eocene
    - *Aegialornis broweri* Collins 1976. Upper Eocene
  - Hemiprocnidae
    - *Cypselavus gallicus* Gaillard 1908. Upper Eocene–Lower Oligocene
  - Apodidae
    - *Cypseloides mourerchauvirae* Mlikovsky 1989. Old collections
  - Jungornithidae Karhu 1988
    - *Palescyvus escampensis* Karhu 1988. Upper Eocene
- Cuculiformes
  - Cuculidae
    - *Dynamopterus velox* Milne-Edwards 1892. Old collections
    - *Dynamopterus boulei* Gaillard 1939. Old collections
- Strigiformes
  - Tytonidae
    - Subfamily *Necrobyinae* Mourer-Chauviré 1987
    - *Necrobyas harpax* Milne-Edwards 1892. Lower Oligocene
    - *Necrobyas rossignoli* Milne-Edwards 1892. Upper Eocene
    - *Necrobyas edwardsi* Gaillard 1939. Upper Eocene
    - *Necrobyas medius* Mourer-Chauviré 1987. Old collections
    - *Necrobyas minimus* Mourer-Chauviré 1987. Upper Oligocene
    - *Nocturnavis incerta* (Milne-Edwards 1892). Upper Eocene
    - *Palaeobyas cracrafti* Mourer-Chauviré 1987. Old collections
    - *Palaeotyto cadurcensis* Mourer-Chauviré 1987. Old collections
    - Subfamily Selenornithinae Mourer-Chauviré 1987
    - *Selenornis henrici* (Milne-Edwards 1892). Old collections
  - Palaeoglaucidae Mourer-Chauviré 1987 (Peters, 1992)
    - *Palaeoglaux perrierensis* Mourer-Chauviré 1987. Upper Eocene
  - Sophiornithidae Mourer-Chauviré 1987
    - *Sophiornis quercynus* Mourer-Chauviré 1987. Old collections and Lower Oligocene?
    - *Strigogyps dubius* Gaillard 1908. Old collections
- Psittaciformes
  - Quercypsittidae Mourer-Chauviré 1992
    - *Quercypsitta sudrei* Mourer-Chauviré 1992. Upper Eocene
    - *Quercypsitta ivani* Mourer-Chauviré 1992. Upper Eocene
- Caprimulgiformes
  - Archaeotrogonidae Mourer-Chauviré 1980
    - *Archaeotrogon venustus* Milne-Edwards 1892. Upper Eocene–Lower Oligocene–Upper Oligocene
    - *Archaeotrogon zitteli* Gaillard 1908. Lower and Upper Oligocene

Table 4.4 Eo-Oligocene fossil birds identified from the Phosphorites du Quercy, southwestern France (*continued*)

		Archaeotrogon cayluxensis Gaillard 1908. Lower and Upper Oligocene
		Archaeotrogon hoffstetteri Mourer-Chauviré 1980. Old collections
	Caprimulgidae	
		Ventivorus ragei Mourer-Chauviré 1988. Upper Eocene
	Podargidae	
		Quercypodargus olsoni Mourer-Chauviré 1988. Upper Eocene
	Nyctibiidae	
		Euronyctibius kurochkini Mourer-Chauviré 1988. Old collections
	Aegothelidae(?)	
		Undescribed genus and species. Lower Oligocene
	Steatornithidae(?)	
		Undescribed genus and species. Upper Oligocene
Coliiformes		
	Coliidae	
		Primocolius sigei Mourer-Chauviré 1988. Upper Eocene
		Primocolius minor Mourer-Chauviré 1988. Upper Eocene
Coraciiformes		
	Sylphornithidae Mourer-Chauviré 1988	
		Sylphornis bretouensis Mourer-Chauviré 1988. Upper Eocene
	Todidae	
		Palaeotodus escampsiensis Mourer-Chauviré 1985. Upper Eocene
		Palaeotodus itardiensis Mourer-Chauviré 1985. Lower Oligocene
	Coraciidae	
		Geranopterus alatus Milne-Edwards 1892. Old collections and Upper Eocene
	Alcedinidae	
		Undescribed genus and species. Upper Eocene–Lower Oligocene
	Meropidae	
		Undescribed genus and species. Upper Oligocene
	Upupidae	
		Undescribed genus and species. Upper Eocene
Passeriformes		
	Suborder Suboscines, undetermined family. Beginning of Upper Oligocene	

Note: Material from the old collections dates from the beginning of the Upper Eocene to almost the end of the Upper Oligocene
Source: Courtesy Cécile Mourer-Chauviré

their five toes and were just showing a prophecy of cement-covered teeth at the time these quite modern-looking gannets and shearwaters were fishing in the quiet waters of Miocene California.' Wetmore and others have also emphasized this lack of progression in birds since the early Tertiary, and it is one of the most striking generalizations of paleornithology" (83).

A second phase of explosive avian radiation produced myriad passerines by the late Tertiary, during the late Oligocene and Miocene. Passerines, which now constitute 5,700 species, nearly 60 percent of the living avian species, are known from somewhat earlier in the Tertiary, but only from rare, fragmentary remains. The Miocene was their period of triumph; this is dramatically illustrated in fossil deposits in Europe, in which passerines generally are absent in the Oligocene but are then recovered in excess of all other fossil birds in certain Miocene deposits (Ballmann 1969b, 1976). Interestingly, rodents, which constitute some 40 percent (1,700 species) of mammal species, with their small size and high reproductive rates appear to have paralleled the avian passerines in their explosive evolution, though they evolved somewhat earlier in the Tertiary.

Among the problems for systematists that arise from this scenario, if it is correct, are that, with all of the avian orders coming off their phyletic nodes (points of divergence) within such a restrictive time frame, the difficulty of ascertaining higher-level relations through DNA-DNA hybridization or cladistic methodology is grossly compounded, and the resolution of many avian phylogenies may well be lost to the past unless telltale fossils are recovered, such as the various shorebird–modern order mosaics. It is within the modern avian genera, which appeared by the Miocene, following the same pattern as mammals, that successful molecular comparisons are beginning to produce highly corroborated phylogenies that agree with the fossil record. Among the more successful efforts is that of cranes, discussed in chapter 6.

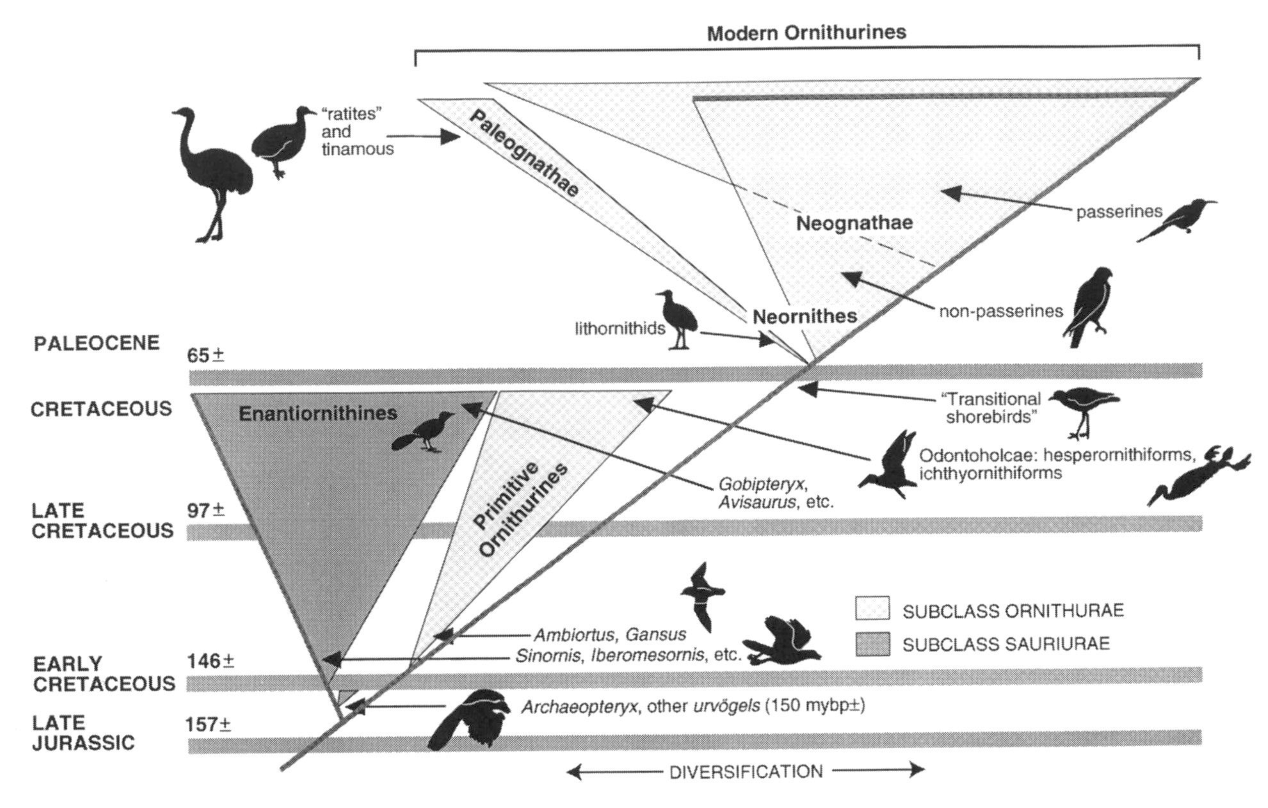

Generalized cladogram illustrating mammals' and birds' parallel evolutionary reorganization after the terminal Cretaceous extinctions. Enantiornithines (opposite birds) were the dominant land birds of the Mesozoic, but they coexisted with modern-type ornithurine birds in the early Cretaceous. *Archaeopteryx* may be closely allied with the enantiornithines; together they constitute the subclass Sauriurae. After the late Cretaceous extinctions the ornithurine birds began a modern, explosive adaptive radiation, with almost all orders appearing within 5 to 10 million years. Modern bird and mammal orders appear in the late Paleocene and Eocene, modern families by the Eocene and Oligocene, and modern genera generally by the Miocene. By the Miocene, passerines had become the dominant land birds. Silhouettes not to scale. mybp = million years before present. (From *Science* [1995]:267; courtesy AAAS)

The double-striped thick-knee, or Mexican stone curlew (*Burhinus bistriatus*), of Mexico and Central America (*left*), a member of the thick-knee family, Burhinidae. Thick-knees structurally resemble the "transitional shorebirds" found in the late Cretaceous and early Paleocene and are thought to be among the few types of birds that transcended the Cretaceous-Tertiary boundary. The nine species, so named because of their thick "knees" (actually the heel, or intertarsal joint), are crepuscular or nocturnal and feed on a variety of worms, insects, crustaceans, mollusks, frogs, seeds, and even small mammals. They occur in dry, open country, in open stony or sandy ground, and some along beaches. The great tinamou (*Tinamus major*) of southern Mexico to South America (*right*), is another candidate for an avian type that may have transcended the Cretaceous-Tertiary boundary. Although no paleognaths are known before the Tertiary, this separate lineage of modern birds must have had its wellsprings in the late Cretaceous. Tinamous are ground-dwelling birds of Central and South America that eat a variety of fruit, seeds, and insects. (Thick-knee photo by M. Marin/VIREO; tinamou photo by R. and N. Bowers/VIREO)

Naturally, many questions remain; but a general picture of bird evolution is emerging which illustrates that, like many other groups of organisms, birds underwent an initial Mesozoic adaptive radiation of archaic types, urvogels, then enantiornithines, or opposite birds, and archaic ornithurines. Birds then underwent a dramatic and cataclysmic late Cretaceous demise. Whether the extinctions resulted from gradual geologic and climatic changes or from an extraterrestrial event, it is now apparent that the Cretaceous-Tertiary extinctions were as dramatic in birds as they were in other vertebrate groups. The old model of gradualistic evolution of birds through the Cretaceous-Tertiary boundary has never made any sense. With myriad vertebrate groups undergoing massive extinctions, birds, with their high metabolism and dependence on sight, would have been the first group to go out. In a Cretaceous-Tertiary bottleneck, transitional shorebirds and perhaps a few other groups made it through to the early Tertiary. Birds then paralleled mammals in an explosive evolution that produced the modern groups in a very short time.

This model, interestingly, is concordant with speculative conclusions reached from molecular data more than a decade ago. In 1983, Jeff Wyles, Joseph Kunkel, and Allan Wilson advanced the view that "the time scale for bird evolution could be quite short," based on the "small molecular differences observed among modern birds. Prager and Wilson's comparisons [1980] of proteins from all twenty-seven orders of birds show that the accumulation of point mutations causing amino acid substitutions has been modest in birds compared with other vertebrates" (4394, 4396). They also noted that the "anatomical divergence among birds has been unusually fast in relation to both point-mutation divergence and to time," and, intriguingly, "the only other group of land vertebrates with comparable rates of anatomical evolution are the mammals," hypothesizing that "in higher vertebrates, behavior, rather than environmental change, is the major driving force for evolution at the organismal level" (4396). Time and time again their hypothesis is supported by this treatment of avian evolution, especially as it applies to the explosive evolution within the passerines.

Transitional shorebirds, in fact, are exactly the types of birds that could be expected to have survived the Cretaceous-Tertiary boundary, slipping through the keyhole into the Tertiary, especially if there had been a cataclysmic extraterrestrial impact. Daniel Janzen of the University of Pennsylvania, commenting on the new model, has posed the question of which animals would have been most likely to survive a serious nuclear winter produced by a huge earth impact:

> Those whose food in some form does not directly depend on immediate photosynthesis. That is to say, those that eat dormant seeds and insects, those that eat decaying organic matter (especially non-green plant parts), and those that eat these eaters. And especially those that are very good at finding small particulate bits of these resources, scattered and dwindling until sunlight again can penetrate the clouds in amounts sufficient for serious vegetation growth. That is to say, seed- and detritus-eating invertebrates that eat them and each other. . . . The transitional shorebirds . . . are precisely among these morphs, along with insectivores, little marsupials, small rodents, snakes, small lizards, frogs, granivorous birds, small raptors, and lots of arthropods. (Janzen 1995, 785)

In this sense, some ancient paleognaths would also fit the bill.

As Janzen notes, after slate-cleaning that ended the Cretaceous, an explosive evolution would be hard to avoid. And that appears to have been exactly what happened. With the close of the Cretaceous, a complete evolutionary reorganization of the class Aves had begun.

Modern Foot-propelled Divers

During the medial Tertiary many groups of variously adapted foot-propelled divers evolved within the modern avian radiation. Within the pelecaniform assemblage, cormorants and snakebirds (discussed later) became agile divers, and many of the Anseriformes (ducks, geese, and swans) also evolved adaptations for diving. One of the more bizarre divers was the recently extinct Pleistocene and early Holocene California goose *Chendytes*, a member of the Mergini, closely allied to the eiders (*Somateria*), which was a totally flightless foot-propelled diver. Several lines of the modern duck assemblage are so highly adapted for foot-propelled diving that they seldom venture onto land. The North American ruddy duck (*Oxyura jamaicensis*), for example, "cannot walk on land for more than a few steps without falling on its breast" (Raikow 1970, 5). But the champion among the many diving ducks is the long-tailed duck, or oldsquaw (*Clangula hyemalis*), which has been recorded at depths of about 55 meters (180 ft.), the same as for the common loon (*Gavia immer*).

Diving ducks generally fit into the category of foot-propelled divers, but two varieties of diving can be distinguished. The flightless *Chendytes* no doubt shared the strict reliance on foot propulsion typical of such ducks as buffleheads and goldeneyes (*Bucephala*), canvasbacks and scaup (*Aythya*), mergansers (*Mergus*), and stiff-tailed ducks (Oxyurini). Yet many sea ducks, such as the steamerducks (*Tachyeres*), eiders (*Somateria*), the harlequin duck (*Histrionicus*), scoters (*Melanitta*), and the long-tailed duck

(*Clangula*), have more advanced diving abilities: they dive with wings partly spread, but with the alula projecting, and they intermittently flap their wings to various degrees (Brooks 1945; Tome and Wrubleski 1988; Livezey 1993a). Documentation of underwater propulsion is scant, and great variation may exist (Schorger 1947). Richard Snell reported close-up observations of five long-tailed ducks diving: "Each diving sequence consisted of extension of partially folded wings, partial spreading of the tail, head submergence, and then steady, rapid wing-flapping. . . . Wing-flapping was maintained throughout the visible portion of descent. Feet were observed during descent, but they did not move relative to the bird." He also remarked that "in contrast to the passive ascent of steamer ducks *Tachyeres* . . . the Oldsquae actively propelled themselves upwards with their wings" (1985, 267).

The four aberrant species of steamerducks (*Tachyeres*), one flighted and the other three flightless or nearly so, are generally classified with the dipping ducks. All have more or less equally sized wings, but the flightless species have bodies nearly twice the size of the flighted one, *T. patachonicus*. These masterful divers occur in extreme southern South America and the Falkland Islands. Their name comes from their habit of "steaming," running across the water's surface, propelled by wings and feet in a manner reminiscent of the old paddle-wheeled steamers.

Avian behavior further complicates diving techniques. Wintering flocks of surf scoters (*Melanitta perspicillata*) and Barrow's goldeneyes (*Bucephala islandica*), for example, tend to dive and surface in a highly synchronous fashion; this may help these large groups maintain cohesion while foraging (Beauchamp 1992).

Loons and Grebes—Convergent Foot-propelled Divers. Loons and grebes, once classified together, undoubtedly represent independently evolved lines of foot-propelled divers. The phylogenetic relations of the grebes (order Podicipediformes) remain enigmatic, but they show some similarities to gruiform birds in their neck musculature and skull (Olson 1985a, 168), and evidence from DNA-DNA hybridization offers no solution, except that "the loons are not closely related to the grebes and . . . the grebes have no close living relatives" (Sibley and Ahlquist 1990, 490). In the fossil record grebes have been found as far back as the early Miocene, about 25 million years ago, but the similarity between ancient and modern species prevents our gaining much insight into evolutionary relations, and by the Pliocene all fossil grebes are inseparable from living genera. Ignoring these, and *Podiceps oligocaenus* from the early Miocene of Oregon, there are only two valid records for the group, a nearly complete Miocene grebe from Spain, *Thiornis sociata*, recently redescribed by Storrs Olson (1995), and *Miobaptus walteri* from the Lower Miocene (Aquitanian) of the former Czechoslovakia.

There are twenty-two living species of grebes divided among six genera, distributed worldwide, on all continents and numerous islands (such as Madagascar and New Zealand) in freshwater lakes and ponds; some species migrate great distances to winter near coastal waters. Like hesperornithiform birds, these medium-sized diving birds have legs set back on the body and lobate webbing on their feet. Among living birds, lobate webbing is known only in the phylogenetically distant phalaropes (order Charadriiformes, family Phalaropodidae), coots (Gruiformes, Rallidae), and finfoots (Gruiformes, Heliornithidae) and thus represents a remarkable example of convergent evolution. Grebes are generally weak flyers with poorly developed sternal keels and have shown a tendency to become nearly flightless, as in the Atitlán grebe of Guatemala (*Podilymbus gigas*), the nearly keelless short-winged grebe (*Rollandia microptera*) of Lake Titicaca, on the border between Peru and Bolivia, and the Puna grebe (*Podiceps taczanowskii*) of Lake Junín in Peru—each endemic to a high-altitude Neotropical lake or lake system (Livezey 1989a). Although the short-winged grebe is termed flightless, I have observed it fluttering across the water's surface, and the flight feathers show little sign of losing their strong asymmetry and flight characteristics.

The five species of modern loons, or divers (Gaviiformes), by contrast, have a large keel, are strong flyers, and occur only in the Northern Hemisphere. Robert Storer (1956) of the University of Michigan studied the primitive teal-sized loon *Colymboides minutus*, which is well represented in Upper Oligocene and Lower Miocene fossil deposits, and concluded that it was less highly adapted for diving than modern loons. The oldest fossil to be described as a loon is *Lonchodytes* from the late Cretaceous of North America, but it has now been shown to be a shorebird (Olson and Feduccia 1980a). *Neogaeornis wetzeli*, mentioned earlier, from the Upper Cretaceous of Chile, was originally assigned on the basis of a poorly preserved tarsometatarsus to an order "Colymbo-Podicipediformes" along with loons and grebes (Lambrecht 1933) and was later included within the Baptornithidae (Martin and Tate 1976); it has since been placed in the Gaviiformes (Olson 1992a). Despite the fossil's modern appearance, however, the fragmentary nature of the material, its temporal occurrence, in rocks of late Cretaceous age, and its location in the Southern Hemisphere must render any identification somewhat tentative, and it is best left in the Hesperornithiformes (Martin, pers. comm.). The oldest truly reliable records for loons are the two species of *Colymboides*, *C. anglicus* from the Eocene of England and

Mancalla (left) and *Chendytes* (right)

Mancalla

Chendytes

California's extinct flightless Lucas, or mancalline, auk, *Mancalla,* and the extinct flightless goose *Chendytes,* two flightless wing- and foot-propelled divers. (From Howard 1947; illustrations by Arminta Neal; courtesy Los Angeles County Museum of Natural History)

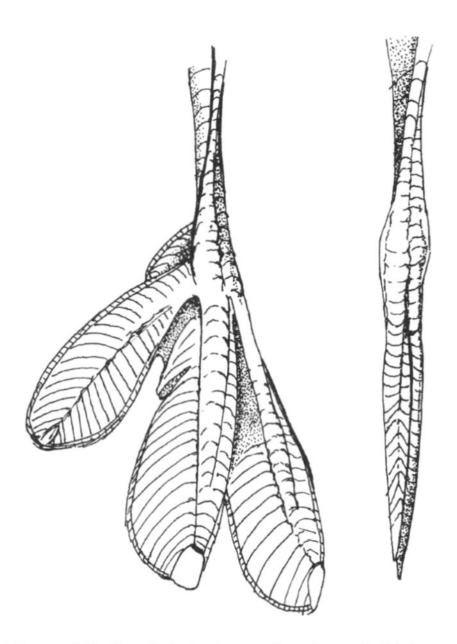

The paddlelike, lobate toes of grebes: (*left*) in position for the backward power stroke; (*right*) in position for the recovery stroke, in which the toes fold sideways by rotation to minimize friction. Hesperornithiform birds rotated the toes in similar fashion and therefore apparently had lobate webbing. (Adapted from Peterson 1963)

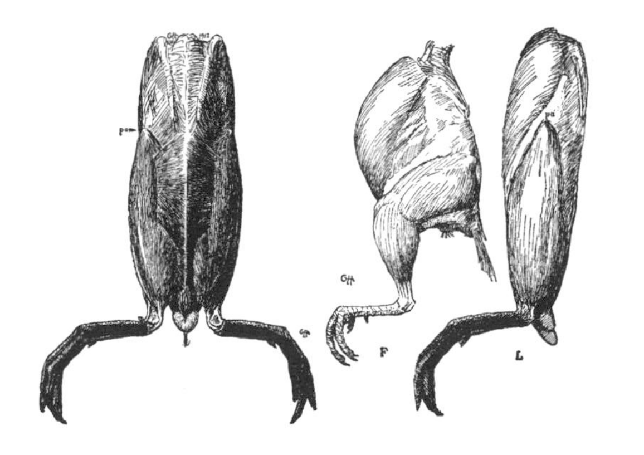

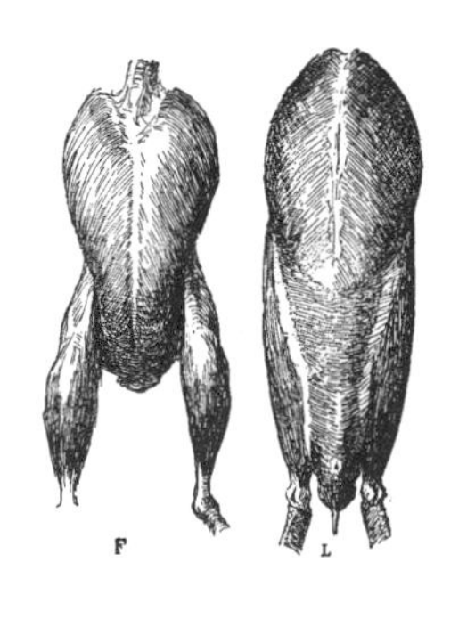

Comparison of skinned bodies, *from left to right,* of a loon in dorsal view, and a pheasant (*F*) and a loon (*L*) in lateral and ventral views with the forelimbs cut away, illustrating how the loon's leg musculature has, as a diving adaptation, become incorporated into the body mass. (From Heilmann 1926)

Grebes (Podicipedidae, worldwide, 6 genera, 22 species): *left,* pied-billed grebe (*Podilymbus podiceps*). Loons (Gaviidae, Holarctic, 1 genus, 5 species): *right,* Arctic loon (*Gavia arctica*). Grebes are medium-sized foot-propelled divers with lobate webbing on the feet, whereas loons are large foot-propelled divers with the first three toes fully webbed; in both families, the legs are positioned far back on the body for diving. (Drawings by George Miksch Sutton)

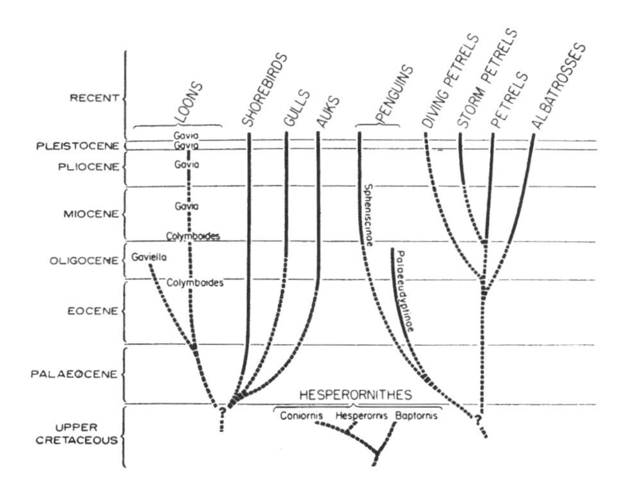

Phylogenetic trees of the toothed divers, loons, shorebirds, gulls, auks, penguins, and procellariiform birds. (From Storer 1960; courtesy Robert W. Storer)

C. minutus from the early Miocene of France and the former Czechoslovakia (Svec 1980; Olson 1985a). The modern genus *Gavia* first appears in the Lower Miocene of the former Czechoslovakia as *Gavia egeriana* from deposits that have also yielded the primitive loon *Colymboides minutus*, thus illustrating that the two lived contemporaneously (Svec 1982).

On the basis of his extensive study of the fossil loon *Colymboides minutus*, Storer (1956, 1960) suggested that loons may have been derived from a shorebird (charadriiform) ancestry. However, the feather structure and loonlike bill and vertebrae of some Eocene penguins may point to a "close common ancestry between the Gaviiformes and Sphenisciformes" (Olson 1985a, 216), a conclusion that is concordant with DNA comparisons, which "support a relationship among loons, penguins, and procellariids, but not a close relationship between loons and grebes" (Sibley and Ahlquist 1990, 550).

Biochemical Systematics. Various methodologies of systematics were described in chapter 2, but because the origin of birds can be analyzed only through the study of fossils, the realm of biochemical systematics was omitted from that discussion. Knowledge of DNA has increased dramatically in recent decades, and the field of biochemical systematics has grown accordingly. Many early DNA comparisons were based primarily on protein differences as revealed by electrophoretic separation of different morphological forms of enzymes in a weak electric current. Some of the more interesting studies involved surveys of some thirty or more genetic loci (sites on a chromosome), which is thought to be a reasonable estimate of the genetic divergence between species. In one study involving wood warblers (Parulidae), George Barrowclough and Kendall Corbin (1978) showed that protein differences clustered around three genera among nine species (*Vermivora*, *Dendroica*, and *Seiurus*), findings that corroborated previous morphological studies. Beginning in the late 1950s and increasing during the 1960s, exhaustive electrophoretic studies of avian egg-white proteins by Charles Sibley and Jon Ahlquist of Yale University produced a number of very interesting finds (Sibley 1970; Sibley and Ahlquist 1972), but comparisons of the higher categories of birds was less than satisfactory because the results were in the form of uninterpretable protein bands on gels; the only indicator of relations was the overall similarity of the patterns. More recent studies have used biochemical systematics to study geographic variation, speciation, and biogeography as well as phylogenetic relations, and include mitochondrial DNA restriction endonuclease cleavage analysis, analysis of repeated DNA segments, immunological comparisons using microcomplement fixation, polymerase chain reaction (PCR), and direct sequencing of DNA (Zink and Baverstock 1991). The field of avian molecular systematics has been reviewed by Frederick Sheldon and Anthony Bledsoe (1993).

Pioneering studies in DNA-DNA hybridization as a tool in avian systematics have been conducted by Charles Sibley and Jon Ahlquist and published in an exhaustive book entitled *Phylogeny and Classification of Birds* (1990). This encyclopedic book culminates in a finale, commonly known as the "tapestry," that is a thirty-figure representation of the phylogeny of birds. First displayed as a fifty-foot-long poster at the 1986 International Ornithological Congress in Ottawa, the "tapestry" not only shows Sibley and Ahlquist's DNA-based phylogeny but forms the basis for a new classification of birds presented in another huge volume, *Distribution and Taxonomy of Birds of the World* (1990), which Sibley wrote with Burt L. Monroe, Jr. *A World Checklist of Birds*, also by Monroe and Sibley and based on the same taxonomy, appeared in 1993. These works have had and continue to have a tremendous influence on the field of avian systematics, but they have also been severely criticized, and the entire field is now undergoing intense evaluation and debate (Cracraft 1987; Houde 1987b; Lewin 1988a, 1988b; Sarich et al. 1989; Bledsoe and Raikow 1990; Gill and Sheldon 1991; Lanyon 1992; Mindell 1992; Harshman 1994; Mooers and Cotgreave 1994).

Nevertheless, an arsenal of modern molecular techniques now exists for studying questions at various levels of taxonomic ascendancy, and scientists can now directly sequence the actual hereditary code, DNA. The coming decades will be an exciting time in this new arena of systematic inquiry, and numerous papers involving varying types of DNA analyses are already appearing. For example, Bledsoe looked at single-copy nuclear-DNA sequences for thirteen species representing the major groups of oscines with nine primary feathers (New World warblers, tanagers, and others) and discovered that "convergence in feeding specializations among lineages is more extensive than traditional arrangements of the assemblage would suggest" (1988a, 504). In another well-done study, Sheldon and co-workers (1992) used DNA-DNA hybridization to study the phylogenetic relations among the major lineages of *Parus*, the titmice and chickadees. These studies, however, were aimed at family, generic, and species levels; the transition to the ordinal level is fraught with problems, not the least of which are the broad dimension of time spans involved and the rapid early Tertiary divergence of major groups.

Given the swirl of controversy that surrounds the still untested DNA-DNA hybridization analyses of Sibley and Ahlquist, their revolutionary new classification, except for the major groupings of passerines, is not adopted here. A further consideration for the conservative approach adopted here (Mayr and Bock 1994) is the scores of books, including field guides, that are based on more traditional classifications, which have changed very little since the landmark work of Hans Friedrich Gadow in 1893. Even if we assume that the DNA hybridization studies do give a certain percentage of correct phylogenetic answers, the question then becomes *which* results are the correct ones. The hybridization studies have confirmed much of the classical morphological classification of birds, and these results are often cited in the remainder of this book, because they corroborate well-known relations and testify to the validity of known avian phylogeny. The DNA studies have also revealed numerous previously unsuspected possible relations as well as the tremendous influence of convergence and adaptive radiation among the world's ecological avian equivalents. Examples include the revelation of the paraphyly (multiple origin) of barbets (Capitonidae), now supported by other molecular studies (Lanyon and Hall 1994), and the radiation of a major corvine assemblage, including the birds-of-paradise, bowerbirds, shrikes, drongos, monarch flycatchers, and nuthatchlike Australian sitellas. Also newly understood is that many Australian songbirds that had previously been thought to be related to similar-looking Asian and European groups are in fact related to each other—meaning, in essence, that the adaptive radiation of Australian passerines, in isolation on the island continent, paralleled that of the marsupials (Sibley and Ahlquist 1985).

Yet many of the results are highly suspect. For example, the DNA hybridization data support limpkins (*Aramus*) and sungrebes (*Heliornis*) as sister-taxa and place the limpkin in the family Heliornithidae, in stark contrast to all of the morphological data, which show the limpkin to be near the base of the crane lineage. This result is so incongruous as to be completely implausible (Houde 1994). Another example of radical departure from morphology is Sibley and Ahlquist's order Ciconiiformes, which incorporates shorebirds, hawks and falcons, grebes, loons, penguins, and others. These results, as well as other highly controversial ones, may show that, although DNA hybridization is now a powerful new tool for avian systematics, its power may reside primarily below the ordinal level (Prager and Wilson 1980; Houde 1987b).

The DNA results are replete with additional problems, among them the fact that the evolution of DNA is not uniform across the various lineages of organisms, yet this fixed "molecular clock" is a pervasive basis of the entire study. Most modern bird orders are assumed to have diverged deep within the Cretaceous; the fossil record does not substantiate this view. In addition, those who advocate the strict use of cladistic methodology generally reject on first principles the idea that distance data generated from DNA-DNA hybridization can be clustered into hierarchies that reflect a phylogeny. We must keep in mind, as emphasized in chapter 2, that only derived similarities carry phylogenetic information. In essence, two relations of similarity are subsumed under the term *homology*. One involves primitive, or ancestral, homology, which comprises most of the genome, while the other involves derived homology, which carries phylogenetic information. What, then, are we measuring when we hybridize DNA strands? Finally, even if we assume that the results of the DNA hybridization studies are valid—that is, that DNA hybridization works—Sibley and Ahlquist's methods of interpreting and measuring the data have come under substantial criticism (see Gill and Sheldon 1991). The Sibley and Ahlquist studies are a remarkable accomplishment, a watershed in the progress of avian systematics; few workers could have accomplished this feat. It is undoubtedly the launching pad for a new age of phylogenetic inquiry, brimming over with the excitement of anticipated discoveries, but it is premature to adopt a radical new classification of avian relations that are based on such tenuous conclusions that cannot be substantiated from studies of morphology and the fossil record.

Nevertheless, if, as I postulate here, there was an explosive evolution within the class Aves, in which all the modern avian orders emerged in the early Tertiary—say, within a 10-million-or-so-year time frame—then the branching nodes would be clustered in an extremely tight fashion. Such a mode of evolution would present a formidable barrier indeed for DNA hybridization, much less cladistic methodology.

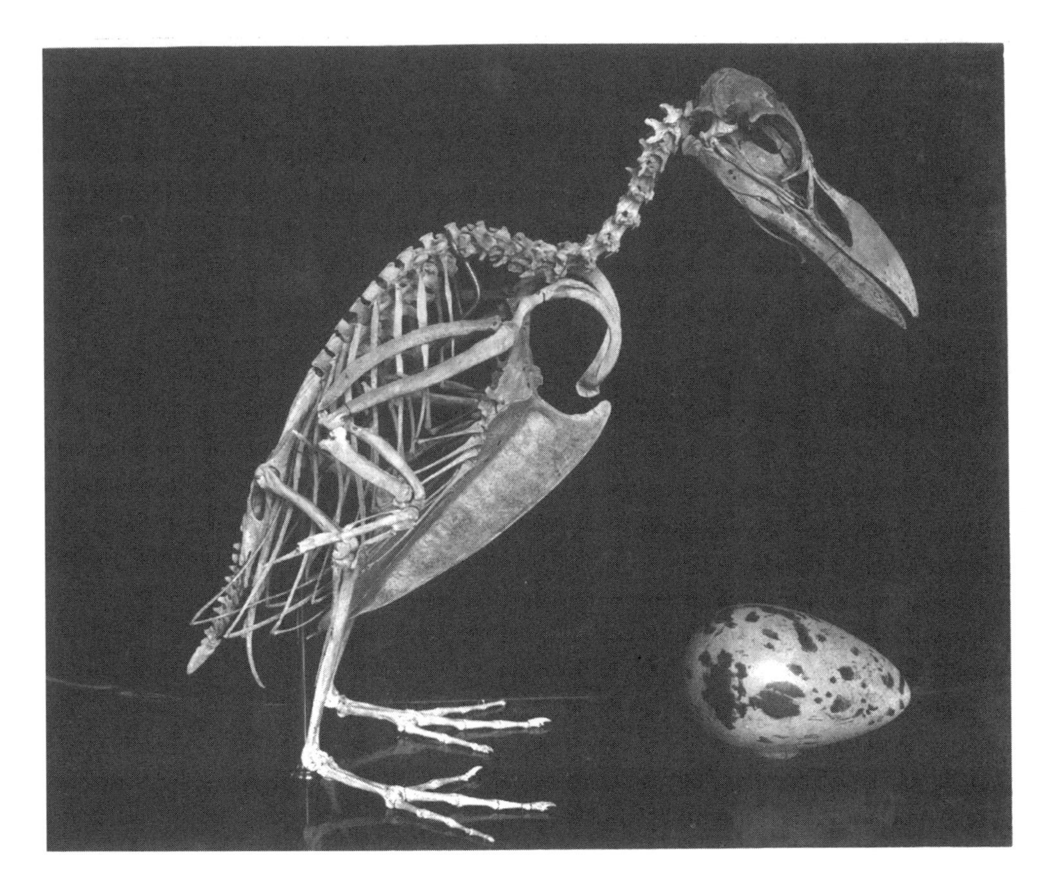

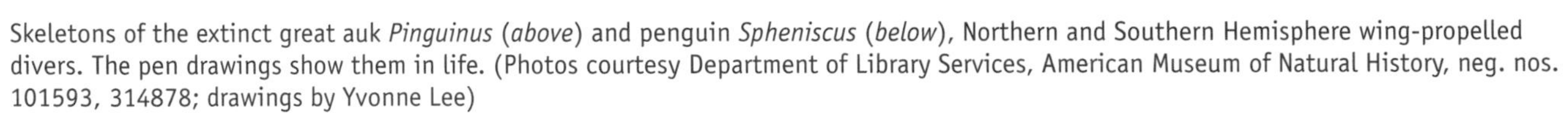

Skeletons of the extinct great auk *Pinguinus* (*above*) and penguin *Spheniscus* (*below*), Northern and Southern Hemisphere wing-propelled divers. The pen drawings show them in life. (Photos courtesy Department of Library Services, American Museum of Natural History, neg. nos. 101593, 314878; drawings by Yvonne Lee)

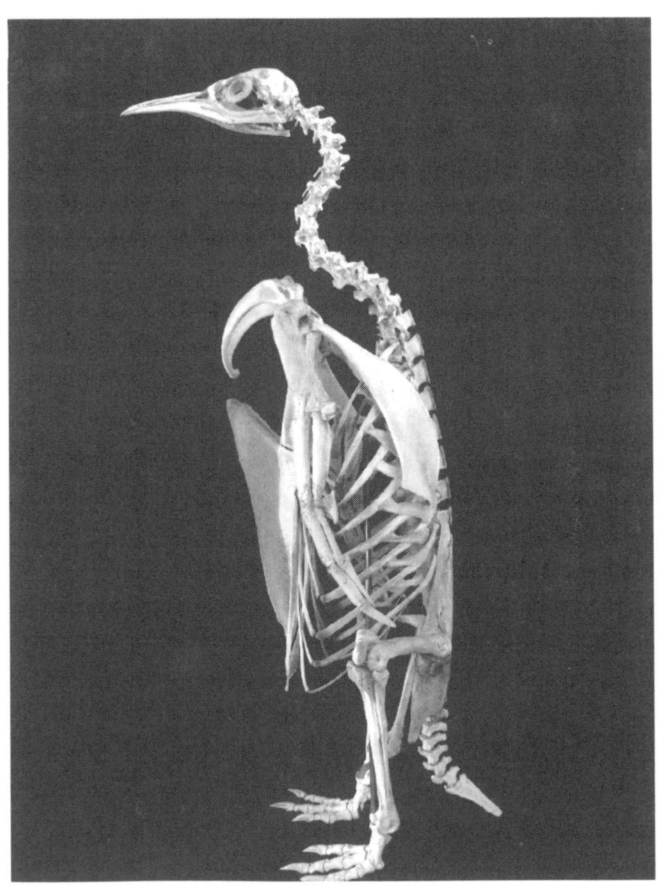

Diverse Wing-propelled Divers

Wing-propelled divers have arisen from at least four independent evolutionary lineages, but not until after the close of the Cretaceous, in Cenozoic time. From the petrels (Procellariiformes), diving-petrels (Pelecanoididae) evolved in the Southern Hemisphere. The Southern Hemisphere penguins (Sphenisciformes) are also thought to be derived from procellariiform ancestors. In the Northern Hemisphere, wing-propelled divers are known among the auks and allies (Alcidae), both living and extinct, forms derived from shorebird ancestry. Referred to collectively by ornithologists as alcids, these birds comprise the auks, auklets, murres, murrelets, guillemots, dovekies, and puffins. Finally, there are the plotopterids, an extinct Northern Hemisphere radiation of wing-propelled divers derived from pelecaniform stock that converged on penguins.

Penguins. Perhaps the birds most highly adapted for wing-propelled diving, which might be described as "flying" under water, are the penguins. These masters of the art are well known for their diving abilities. King penguins (*Aptenodytes patagonica*), for example, frequently dive deeper than 100 meters (328 ft.) and have been recorded as deep as 240 meters (787 ft.) (Kooyman et al. 1982), and emperor penguins (*A. forsteri*) have been recorded at an even deeper 534 meters (1,752 ft.) (Kooyman and Kooyman 1995). The birds studied by the Kooymans averaged 213 dives per day, and the modal depth of foraging divers was 21–40 meters (69–131 ft.); the modal duration was between four and five minutes, and the longest dive was almost sixteen minutes. Ancient birds whose existence has been documented by abundant remains of at least eighteen fossil species dating back to the late Eocene (Seymour Island, Antarctica, Australia, and New Zealand; Jenkins 1974; Olson 1985a; Livezey 1989b), penguins have always been confined to the oceans of the Southern Hemisphere, and their northernmost range is now represented by one species living in the Galápagos in the cold Humboldt Current. There are no fossil finds of penguins outside their current geographical range. Their most obvious counterparts in the Northern Hemisphere are wing-propelled diving auks, but the extinct giant penguins, which first appeared in the Eocene and died out by the early Miocene, had, as we shall see, counterparts in the Northern Hemisphere among the pelecaniform birds, also extinct.

The seventeen species of living penguins are so distinctive from all other living birds that the course of their evolution was debated for years. Most modern analyses date back to Max Fürbringer (1888), who solidified their proximity to the Procellariiformes, although he did not claim that they were particularly closely related. Fürbringer considered penguins' resemblances to loons to be superficial. In 1933 one ornithologist, Percy Lowe, even proposed that penguins were derived not from flying ancestors but from primitive flightless birds that took to the water. His arguments rested primarily on several unusual qualities of penguins: that they have a proliferation of feathers over the entire body and appear to lack apteria; that they lack pneumaticity in the bones; that Miocene penguins are similar to modern forms and show no signs of intermediacy to a presumed flying ancestor; and that the tarsometatarsus was similar to such bipedal dinosaurs as *Ceratosaurus*.

In fact, however, in penguins the three tarsal elements, though fused, do show the distinctive lines separating the individual elements. As for the lack of apteria, this is an adaptation for insulation—vestigial pterylae are present in embryonic stages (Gadow 1896; Clench 1970). Penguin feathers are also structured with a long aftershaft of down, and the feather tips overlap to form an oily, waterproof garment. Combined with a thick underlayer of fat, the penguin integument is unique among birds in adapting them to the vagaries of Antarctica. Further, buoyancy, which would be increased by pneumatic bones, is an obvious disadvantage for diving birds; loons and other diving groups all exhibit a reduction in the size of air sacs and reduced pneumaticity in the bones, if it is present at all. The unusual tarsus is no doubt due to its broadening to produce an effective "walking" foot for these large animals.

George Gaylord Simpson (1946), who saw no reason why penguins could not reasonably have evolved from such aerial oceanic birds as the diving-petrels, which adopted submarine as well as retained aerial flight, was quick to criticize Percy Lowe. As he stated, "Excepting only the wing and the tarsometatarsus, the recent penguin skeleton is remarkably like that of many flying carinates and particularly of the Procellariiformes, as has been repeatedly noticed and can be confirmed by comparisons of almost any two genera of the two groups" (84). To add to this proposed affinity, F. C. Kinsky (1960) observed that the nostrils of the little penguin chick (*Eudyptula minor*), considered to be the most primitive living penguin, are tubular, with large, nearly round apertures. He noted that the tubes start to recede and flatten during the sixth week; by the age of forty-three days, slit-like adult nostrils have formed. Kinsky viewed this as evidence for affinity of the penguins with the tube-nosed seabirds or procellariiform birds. However, Alberto Simonetta (1963) postulated that the cranial morphology and kinesis of penguins were most similar to those of loons and grebes, and Storrs Olson

Penguins (Spheniscidae, southern oceans, 6 genera, 17 species): king penguin (*Aptenodytes patagonicus*). Penguins are highly adapted, flightless, wing-propelled diving birds. (Drawing by George Miksch Sutton)

Sizes of fossil and recent penguins compared with a six-foot man. The outlines in *B, C,* and *E* are meant to suggest size only and are not meant to be true-life restorations of particular genera or species. *A, Homo sapiens; B,* the largest Miocene penguins, largest species of *Pachydyptes* and *Anthropornis; C,* other "giant" Miocene penguins, approximately *Palaeoeudyptes antarcticus* or *"Arthrodytes" andrewsi; D,* the largest living penguin, the emperor penguin (*Aptenodytes forsteri*); *E,* Miocene penguins of moderate size, such as *Palaeospheniscus patagonicus; F,* one of the smaller living penguins, the Galápagos penguin (*Spheniscus mendiculus*). Scale at right in feet. (From Simpson 1946; courtesy American Museum of Natural History)

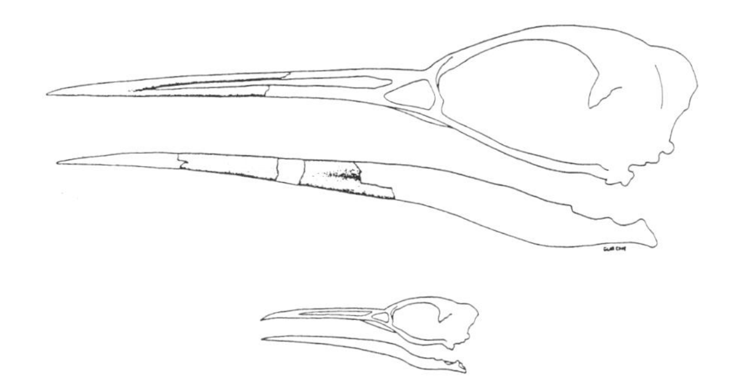

Above, reconstructed skull of a giant late Eocene penguin (?*Palaeoeudyptes* sp.) from Seymour Island, Antarctica, based on specimens of rostrum and mandible (stippled) in the Smithsonian Institution collected by William Zinsmeister. This is contrasted, *below,* with the skull of the king penguin (*Aptenodytes patagonicus*), which has the longest skull of any living penguin. (Olson 1985a; courtesy Storrs Olson)

(1985a) has noted that a giant late Eocene penguin from Seymour Island had a bill that was long and daggerlike, unlike any modern penguin but quite loonlike in overall morphology. Olson suggested that the earlier penguins may have been adapted for spearing large prey and that "a close common ancestry between the Gaviiformes and Sphenisciformes is highly likely" (216). DNA hybridization data (Sibley and Ahlquist 1990) and earlier biochemical comparisons (Ho et al. 1976) indicate that the penguin lineage branched from a "loon-tubenose clade."

As far as the penguin's wing is concerned, Simpson reasoned in 1946 that the origin of the penguin flipper would be inconceivable "except on the theory that it never ceased to be a functional wing, but only changed the medium in which it functioned" (86). His idea was that one could in essence see a pseudophylogeny in the living birds that "fly both in the air and in the water," such as some of the Northern Hemisphere auks and the diving-petrels (*Pelecanoides*). The four small species of diving-petrels bear a striking convergent resemblance to the unrelated auklets of northern seas. Diving-petrels fly close to the water with rapid, whirring wingbeats, and suddenly dive into a wave and emerge from the other side, still flying. Simpson was particularly intrigued with these birds, which he felt were not only related to penguins but showed an adaptive stage through which the penguins passed in achieving submarine flight. The diving-petrels, he wrote, "have carried aquatic adaptation about as far as possible without ceasing to fly in the air, but the . . . existence of aerial flight sets limits to the possible degree of aquatic adaptation" (86).

Robert Cushman Murphy illustrated the point in describing the Peruvian diving-petrel (*Pelecanoides garnotii*):

> How little the Potoyuncos are dependent upon the power of flight is shown by the fact that they moult their wing quills all together, and thus for a time each year become exclusively aquatic. . . . The stomachs of the naturally "crippled," temporarily penguin-like Diving Petrels, which have lost all their flight feathers, prove to be as well filled with crustaceans and small fishes as those of their flying contemporaries. . . . Beneath the surface they literally fly with their wings, whether or not the remiges are full-sized. . . . Thus, so far as feeding is concerned, they might just as well be flightless birds. The only indispensable use of full-grown primaries would seem to be to bear the Potoyuncos to and from their nesting burrows on the islands. (1936, 776–777)

The Ecological Partitioning of Seabirds. Developing Simpson's ideas further, Robert Storer (1960) argued that ecological equivalents for the various stages of wing-propelled diving evolved in different groups in the Northern and Southern Hemispheres. In the earliest stage, represented by the modern gulls in the north and the true petrels in the south, wings were used only for flight in the air. Then came a compromise stage in which wings were used both for submarine and for aerial flight; this stage is represented today by the Southern Hemisphere diving-petrels and the Northern Hemisphere razor-billed auks. Ultimately, the wings came to be used for submarine flight only, a stage represented in the Southern Hemisphere by the penguins and in the Northern Hemisphere by the now-extinct mancalline, or Lucas, auks and the great auks.

Storer (1960) also used five examples from different avian groups to illustrate the probable course of evolution that produced the wings of penguins. The completely aerial stage might be represented by a gull, followed by the compromise adaptation represented by the razor-billed auk. The now-extinct great auk would represent a stage shortly after the loss of aerial flight; then would come the Pliocene auklike genus *Mancalla* (the mancalline auk), which has a still more penguinlike morphology; and finally the penguins themselves. One can see as a sort of pseudophylogeny the progressive specialization in the wing skeleton from the slim, rounded wing bones of a gull through shorter and heavier bones to the broad, flat wings of a penguin's flipper.

A word of caution should be injected into this rather simplistic approach to the world's ecological partitioning of seabirds into such neat categories. Ecological partitioning of the world is not so neat, and this exercise is intended as a gross oversimplification to emphasize special adaptations. In reality, the gulls (Laridae) and the petrels (Procellariidae) are very distinctive groups, and, for example, the shearwaters (*Puffinus*) may use their wings and, to some extent, their feet for underwater propulsion (Kuroda 1954). They may also dive from the surface in pursuit diving or plunge from the air. Further, digestive adaptations render gulls and petrels totally incomparable. Gulls and their charadriiform allies are restricted in their range, particularly in breeding season, because they must capture food and return to the nesting site to feed the young. In contrast, petrels are able to catch and digest food at sea, eliminating excess water, and only later return to the nest to feed the young on special stomach oils plus the partially digested remains of the food most recently acquired (Ashmole 1971; Place et al. 1989). This ability to use stomach oils enables procellariiform birds to range widely over

Diving-petrels (*Pelecanoides*) in their feeding analogize a stage through which penguin evolution must have proceeded, flying into the waves, "flying underwater," and then flying out of the wave into the air. Penguins are thought to have proceeded through such a stage and are thought to have evolved from procellariiform stock. Diving-petrels feed primarily on planktonic invertebrates, which are collected in a distensile pouch beneath the tongue. (From Harrison 1978; drawing by Ad Camerion; courtesy Andromeda Oxford Ltd.)

SOUTHERN HEMISPHERE Petrel-Penguin Stock	Adaptive stage	NORTHERN HEMISPHERE Gull-Auk Stock
Penguins	Wings used for submarine flight only STAGE C	Great Auk
Diving Petrels	Wings used for both submarine and aerial flight STAGE B	Razor-Bill
Petrels	Wings used for aerial flight only STAGE A	Gulls

Parallel adaptive stages in the evolution of two stocks of wing-propelled divers, one in the Southern Hemisphere and one in the Northern Hemisphere. (From Storer 1960; courtesy Robert W. Storer)

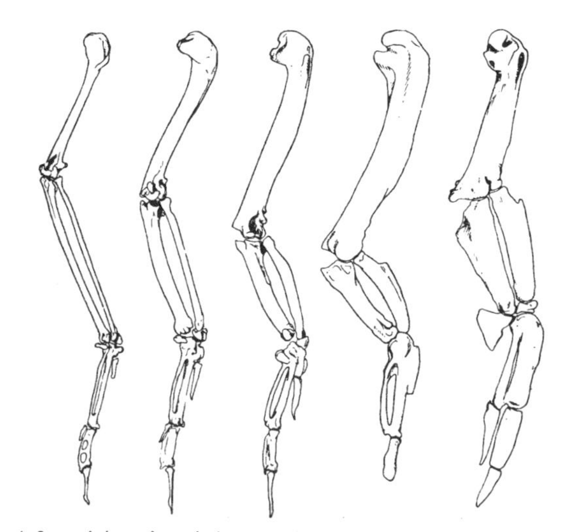

From left to right, wing skeletons of a gull (*Larus*), razorbill auk (*Alca*), great auk (*Pinguinus*), Lucas auk (*Mancalla*), and penguin (*Spheniscus*), showing adaptive stages similar to those through which the penguin flipper evolved. (From Storer 1960; courtesy Robert W. Storer)

The tube-nosed seabirds, Procellariiformes, consist of four families of pelagic birds that make a living harvesting the ocean surface. They are birds of the southern oceans, though many are also found in northern waters. *Left to right and top to bottom,* Diving-petrels (Pelecanoididae, southern oceans, 1 genus, 4 species): Peruvian diving-petrel (*Pelecanoides garnotii*). Shearwaters, petrels, and fulmars (Procellariidae, all oceans, 12 genera, 75 species): northern fulmar (*Fulmarus glacialis*). Albatrosses (Diomedeidae, southern oceans, North Pacific, 2 genera, 14 species): black-footed albatross (*Diomedea nigripes*). Storm-petrels (Hydrobatidae, all oceans, 7 genera, 21 species): Wilson's storm-petrel (*Oceanites oceanicus*). (Drawings by George Miksch Sutton)

vast oceans and to collect food for their chicks over a period of several days far from the colony.

The Fossil Record of Tube-nose Seabirds. The fossil record of tube-nose seabirds is austere, and although these birds must have appeared much earlier, the first completely reliable record for the order is that of petrel-like birds from the Lower Eocene London Clay and Eocene of Louisiana (Feduccia and McPherson 1993). Storrs Olson and David Parris (1987) proposed a new family, genus, and species (Tytthostonychidae, *Tyttostonyx glauconiticus*) for a humerus recovered from the early Paleocene greensands of New Jersey that shows similarities to the Pelecaniformes and Procellariiformes and tentatively assigned it to the Procellariiformes.

The remaining record is equally sparse. Albatrosses (Diomedeidae) are known from the late Oligocene, and we know little of their history. Interestingly, they are common in the early Pliocene of North Carolina (and as many as four species may be present); these records and finds from England and Florida led Olson to state that "albatrosses were common in the North Atlantic in the late Tertiary and . . . their subsequent disappearance from that ocean was a Quaternary event" (1985a, 210). Just as sparse are fossils of the shearwater family, Procellariidae, with scattered Oligocene fossils; the first remains of the Procellariidae in the subantarctic region are from the early Pliocene of South Africa (Olson 1985a, 1985b, 1985c). Storm-petrels (Hydrobatidae) are not known before the late Miocene (California), and the only Tertiary record of the diving-petrels is a species from the early Pliocene of South Africa (Olson 1985a).

Penguins and Auks. The association of penguins and auks has a long tradition. The word *penguin* was, in fact, originally used to designate the extinct flightless great auk (*Pinguinus impennis*) of the North Atlantic. These birds, which stood over 60 centimeters (2 ft.) tall, bred on some of the islands off Newfoundland, Iceland, and even Great Britain. Vast hordes were exterminated by sailors, the last auks having been killed in 1844; Funk Island, off Newfoundland's northeast coast, alone contained as many as 200,000 of these birds (Montevecchi 1994). The species wintered south to New England in the western Atlantic and to Spain in the eastern Atlantic, and fossils found in locations from Florida to Italy indicate a much more extensive distribution in the past. Storrs Olson (1977a) described the ancestor of the great auk from Lower Pliocene deposits in North Carolina. This fossil species, *Pinguinus alfrednewtoni*, was quite similar in size and proportions to the Recent great auk, but its skeleton was slightly less specialized for wing-propelled, submarine locomotion. In these auks flightlessness was a consequence of extreme specialization for pursuit diving, at a level convergent with that seen in the penguins, and involved primarily increase in size, shortening of the wings, and dorsoventrally flattening the wing elements (Livezey 1988). The small extent of evolutionary change between the Recent great auk and its five-million-year-old ancestor is extraordinary.

Alcids. The twenty-three living species of alcids come in a variety of forms, but they can be characterized as wing-propelled diving birds that propel themselves and maneuver underwater almost exclusively using strokes of partly folded wings. Alcids are master divers: one study of

Gulls, alcids, and allies are, in contrast to tube-nosed seabirds, predominantly birds of the northern oceans, and alcids are restricted to the Northern Hemisphere, but gulls and terns are found worldwide, though they occur in lesser numbers in southern lakes and oceans. *Left to right and top to bottom,* Gulls and terns (Laridae, worldwide, 13 genera, 94 species): great black-backed gull (*Larus marinus*). Skuas and jaegers (Stercorariidae, polar regions of the Northern and Southern Hemispheres, often included within Laridae, 2 genera, 8 species): long-tailed jaeger (*Stercorarius longicaudus*). Skimmers (Rynchopidae, Americas, Africa, Southeast Asia, 1 genus, 3 species): black skimmer (*Rynchops niger*). Auks, murres, and puffins (Alcidae, northern oceans, 12 genera, 23 species): black guillemot (*Cepphus grylle*). (Drawings by George Miksch Sutton)

alcids caught incidentally in stationary gill nets off the coast of Newfoundland recorded common murres (*Uria aalge*), razorbills (*Alca torda*), Atlantic puffins (*Fratercula arctica*), and black guillemots (*Cepphus grylle*) at maximum depths of 180, 120, 60, and 50 meters (591, 349, 197, 164 ft.), respectively (Piatt and Nettleship 1985). The depths attained by the various species correlated nicely with body size, and it was postulated that these similar, sympatric species of alcids might exploit different parts of the water column. In addition, Alan Berger and Mark Simpson (1986) used depth gauges to record Atlantic puffins foraging mostly at depths of less than 60 meters (197 ft.), whereas common murres dived to at least 138 meters (453 ft.). It was their impression that "the underwater swimming abilities of alcids are comparable to those of the penguins" (829), thus lending support to Storer's concept of Northern and Southern Hemisphere ecological equivalents.

A phylogenetic analysis of the Alcidae by Joseph Strauch (1985) found the puffins to be a sister-group to all other alcids, and four other clusters or groupings are represented by successive dichotomies (bifurcations), with the auklets as a sister-group to the remaining species.

Intermediate between the great auks and penguins in structural specialization for wing-propelled diving were the flightless Lucas, or mancalline, auks. They were initially named for Frederick Lucas, who described the first species, *Mancalla californiensis*. These auks are known from hundreds of elements assigned to eight species in three genera from Upper Miocene to Pleistocene deposits along the California coast. A less specialized auk is known from the Miocene of Orange County, California: *Premancalla* is considered a probable ancestor of the largely Pliocene *Mancalla*, but neither is thought to be closely related to the great auks. These fossils have been described in detail by the eminent avian paleontologist Hildegarde Howard (1966a, 1970, 1976), chief curator emeritus of the Natural History Museum of Los Angeles County. Howard also described a poorly known mancalline genus, *Alcodes*, from the late Miocene of California, as "progressing towards flightlessness" (1968, 19).

The earliest alcid is *Hydrotherikornis oregonus* from the late Eocene of Oregon (A. H. Miller 1931), but the fossil record of the Alcidae is poor until the late Miocene, when we find essentially modern forms. Murres and guillemots of the genera *Uria* and *Cepphus* are commonly found in the eastern Pacific from late Miocene to the present and have no doubt had a long association with the puffins and small auklets of the region. But the most abundant seabird fossils of the eastern Pacific are the mancalline alcids. Kenneth Warheit (1992) reviewed the fossil seabirds of the North Pacific Tertiary, noting that of the ninety-four species described during the past century, most are from the medial Miocene through the Pliocene (16–1.6 million years) sediments of southern California. However, species from the Eocene to the Miocene (52–22 million years) are found in Japan, British Columbia, Washington, and Oregon. Warheit correlates most of the changes in North Pacific seabird distribution patterns with oceanographic events and tectonic changes that led to the thermal isolation and refrigeration of Antarctica. Intensification of coastal upwelling in the California Current appeared to have had the most dramatic impact on seabirds in southern California. In general, medial to late Miocene and Pliocene seabird faunas from California resemble Recent communities, with the exceptions of the great abundance and subsequent extinctions of gannets and flightless alcids, or mancallids; the low abundance of cormorants and shags until the late Pliocene; and the absence of marine Laridae until the late Pliocene.

In the Atlantic, the earliest alcid is *Miocepphus mcclungi*, from the Middle Miocene of Maryland and North Carolina, and by the medial Miocene there were only a few rather small *Miocepphus* species in the mid-Atlantic. By the early Pliocene, however, a diverse alcid radiation occurred in the same region that included, as Olson has listed, "*Alca* [razorbills], also from the Pliocene of Italy, *Australca* [extinct razorbill-like auk, Pliocene of Florida], and *Pinguinus* [great auks], as well as certain undescribed genera. *Alle* [dovekies], too, may be part of this radiation. . . . Another interesting phenomenon is the occurrence of two species of *Fratercula* [puffins] in the early Pliocene of the Atlantic" (1985a, 186).

A recent discovery of a spectacular late Pliocene (about 2 million years ago) fossil deposit of seabirds in Sarasota County, Florida (Emslie and Morgan 1994; Emslie 1995), has added to a growing body of evidence for cold-water upwelling in the Gulf of Mexico during the Pliocene, correlated with periods of high sea level and the submergence of the Panamanian isthmus. A single species of cormorant (*Phalacrocorax*) is represented by 137 skeletons and is closest phylogenetically to the living Brandt's cormorant (*P. penicillatus*), which is currently restricted to a cold-water upwelling system in the eastern Pacific. The fauna includes a walrus, three species of seals, five species of dolphins, and nine species of whales. The avifauna contains, among other types of birds, an albatross, two additional cormorants, three boobies, a gannet, and five species of auks—three as yet unidentified species, *Pinguinus* sp. (great auk), and *Australca grandis*, also known from Florida's Lower Pliocene Bone Valley fauna. This impressive concentration of fossil alcids illustrates the dramatically changing world of seabird distribution.

Reconstruction of the largest species of plotopterid, pelicaniform allies, shown to scale with the outline of the largest living penguin. Plotopterids thrived in the northern Pacific about 30 million years ago. (From Olson and Hasegawa 1979 [cover illustration]; drawing by B. Dalzell; copyright 1979 AAAS)

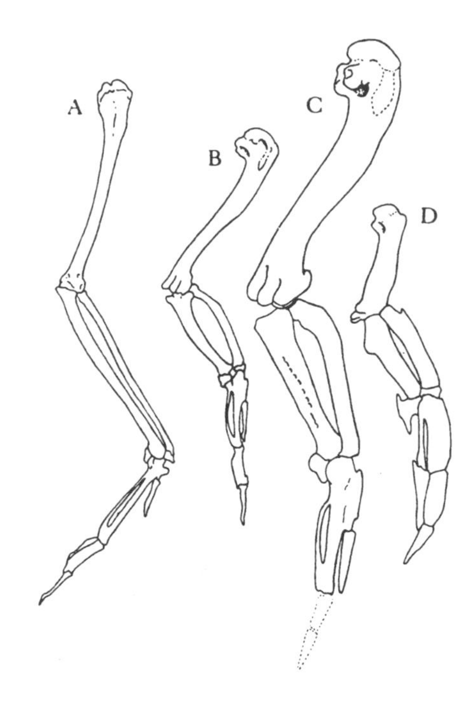

Dorsal view of right wing skeletons. *A,* Anhinga (Pelecaniformes); *B,* great auk (Charadriiformes); *C,* plotopterid (Pelecaniformes; largest Japanese species); *D,* penguin (Sphenisciformes). The similarities between the three wings on the right are due to convergence. The plotopterid evolved from an ancestor with a wing like that of the anhinga. Drawn to scale. (From Olson and Hasegawa 1979; copyright 1979 AAAS)

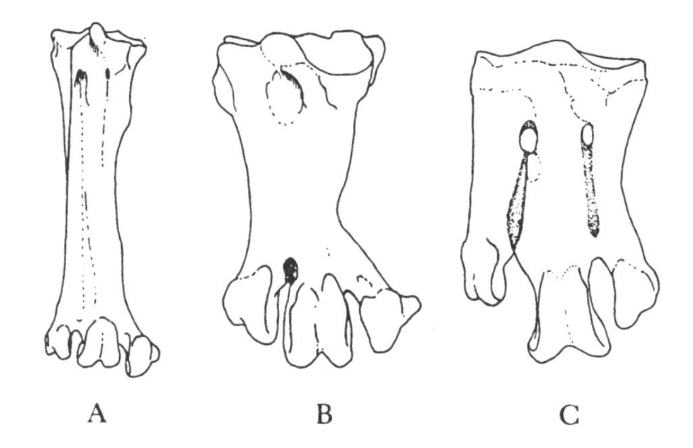

Anterior view of right tarsometatarsi. *A,* anhinga (Pelecaniformes); *B,* plotopterid (Pelecaniformes; largest Japanese species); and *C,* penguin (Sphenisciformes). Not to scale. (From Olson and Hasegawa 1979; copyright 1979 AAAS)

Plotopterids: Wing-propelled Pelecaniform Divers. Discoveries of mid-Tertiary fossils by Storrs Olson and Yoshikazu Hasegawa (1979) indicate that the Northern Hemisphere, specifically the northern Pacific, was once inhabited by yet another group of penguinlike birds that converged on the giant penguins of the Southern Hemisphere in their pectoral architecture. This is the fossil family Plotopteridae, the only wing-propelled divers in the order Pelecaniformes (pelicans and allies), derived from the suborder Sulae among the pelecaniform birds, which includes the Sulidae (boobies), Anhingidae (snakebirds), and Phalacrocoracidae (cormorants, also called shags). The plotopterids offer a strange mosaic of characters, combining a rather typical pelecaniform sternum with a wing convergent on penguins and flightless auks and hindlimb elements that most strongly resemble the Anhingidae (Olson 1985a). This family contains the largest diving birds known; one species is estimated to have measured 1.8 meters (6 ft.) from the bill to the tip of its tail. Amazingly, the family was originally diagnosed from a fragment of a coracoid from the early Miocene of California by Hildegard Howard (1969). She described it as *Plotopterum joaquinensis*, a bird about the size of a medium-sized cormorant. Howard's diagnosis has since been confirmed, and plotopterid specimens have been found in rocks of late Oligocene and early Miocene age in Japan (Hasegawa et al. 1979), Washington (Olson 1980), and California, and *Phocavis maritimus* was described from the late Eocene deep water sediments (500–1,000 m; 1,640–3280 ft.) of the Keasey Formation in Oregon (Goedert 1988), providing not only the earliest record of plotopterids but evidence that these birds ventured far off shore. Plotopterids probably became extinct at the same time as the giant penguins. George Gaylord Simpson's hypothesis for why the giant penguins became extinct by the end of the early Miocene is that the niches for pelagic endotherms of that size were occupied by seals and porpoises, both of which were undergoing adaptive radiation at that time. The same hypothesis would apply to the plotopterids, adding credence to Simpson's theory.

Pelecaniforms. The spectacular birds of the order Pelecaniformes are united by a totipalmate foot—one in which all four toes are connected by a web—and a gular, or throat, pouch, which is less developed in the primitive tropicbirds, Phaethontidae. The order itself is not obviously related to any other group of birds, living or extinct, although certain features hint at affinities with the storks, the Ciconiiformes, and oceanic birds, the Procellariiformes, and their affinities have been sought primarily in these two orders. Traditional treatments of the Pelecaniformes, such as that of Alexander Wetmore (1960), and that of the *A.O.U. Check-list of North American Birds* (1983), have recognized the tropicbirds (*Phaethon*) as being primitive within the order but have placed the frigatebirds (*Fregata*) as the most derived, terminal group. In 1977, however, Storrs Olson described a nearly complete skeleton of an Eocene frigatebird, *Limnofregata*, from the Green River Formation of Wyoming, and his extensive monographic study (1977b) strongly supported the phylogeny of Urless Lanham (1947), "who treated the Phaethontes and Fregatae as being primitive within the order, to be followed by the more specialized families of the Pelecani" (Olson 1977b, 29). In 1985, Joel Cracraft conducted an extensive "phylogenetic analysis" of the order, and he, too, concluded that the tropicbirds were primitive within the order and were to be followed first by the frigatebirds and then by the more specialized pelicans, boobies, cormorants, and anhingas. Cracraft's analysis supported a close, sister-group relation between the pelecaniforms and the procellariiforms. Sibley and Ahlquist's DNA hybridization data (1990) from the Pelecaniformes, bolstered by DNA sequence data (Hedges and Sibley 1994), are completely at odds with all of the morphological and other evidence, placing *Pelecanus* as the sister-group of the strange shoebill (*Balaeniceps*), the frigatebirds as members of the Procellarioidea (penguins, loons, petrels, shearwaters, and albatrosses), and making the tropicbirds out to be descendants of an ancient divergence. This, obviously, is a case in which the DNA hybridization data cannot provide a reasonable assessment of phylogenetic affinities and must be viewed with great caution.

A number of fragmentary Cretaceous bird fossils have been assigned to the Pelecaniformes, but they have been shown to be either misidentified or too fragmentary to be identifiable as to order (Olson 1985a, 192–193). The earliest traces of the order are the early Eocene remains of a primitive extinct tropicbird called *Prophaethon*, an extinct genus known from the London Clay, Isle of Sheppy, England, that because of substantial morphological distinctiveness was placed in its own family, Prophaethontidae, by Colin Harrison and Cyril Walker (1976a). However, *Prophaethon*'s skull is quite similar to that of modern tropicbirds, and the sternum has salient pelecaniform features. Of additional interest are the nostrils of *Prophaethon*: very long and open, similar to that seen in embryos of modern pelecaniforms and in the early Eocene frigatebird *Limnofregata*, helping to confirm the basal position of these two groups. Modern tropicbirds, *Phaethon*, are regarded as primitive within the order, and *Prophaethon* is considered by Harrison and Walker (1976a) to show some affin-

The Pelecaniformes comprises a variety of aquatic birds that eat fish or squid and have a totipalmate foot. Most have a more or less distensible gular pouch between the rami of the mandibles. *Left to right and top to bottom,* Tropicbirds, primitive pelicaniforms (Phaethontidae, pantropical oceans, 1 genus, 3 species): white-tailed tropicbird (*Phaethon lepturus*). Frigatebirds (Fregatidae, pantropical oceans, 1 genus, 5 species): magnificent frigatebird (*Fregata magnificens*). Gannets and boobies (Sulidae, all oceans, 3 genera, 9 species): blue-faced booby (*Sula dactylatra*). Cormorants (Phalacrocoracidae, worldwide, 1 genus, 38 species): great cormorant (*Phalacrocorax carbo*). Snakebirds, or anhingas (Anhingidae, pantropical, 1 genus, 4 species): anhinga (*Anhinga anhinga*). Pelicans (Pelecanidae, all continents, 1 genus, 8 species): brown pelican (*Pelecanus occidentalis*). (Drawings by George Miksch Sutton)

ity, not surprisingly, with both the Charadriiformes and the Procellariiformes. Modern tropicbirds are adapted for obtaining food by plunge-diving into tropical and subtropical seas in pursuit of fish and squid, whereas modern frigatebirds are adapted for catching fish from the surface of the water but are also noted for their well-known "pirate" habit of pursuing boobies and gulls and forcing them to disgorge and drop their food.

Pseudodontorns: Bony-toothed Pelecaniforms. One of the most unusual of all avian groups is the now-extinct pelecaniform group known as the pseudodontorns, or bony-toothed pelecaniforms of the family Pelagornithidae. These were sensational, gigantic marine gliders with immense bills that bore numerous bony toothlike projections as extensions of the rostrum and mandible. The pseudodontorns also had excessively thin-boned and lightweight skeletons as an adaptation for flight—they probably surpassed the frigatebirds in this feature. They were true giants among flying birds: Howard (1957) estimated the wingspan of the well-known *Osteodontornis orri,* from the late Miocene of California, at more than 4.8 meters (16 ft.), and Olson (1985a), who studied additional material, estimated the same bird at 5.5 to 6 meters (18–20 ft.). The earliest European finds are those of *Odontopteryx,* from the Lower Eocene deposits on the Isle of Sheppy, England, where these birds occur in association with a flora and fauna normally found in Malaysia today, with sharks, turtles, crocodiles, and *Nipa* palms (Harrison and Walker 1976b). Although some authors place these unusual birds in two distinctive families based on pseudotooth structure, *Odontopteryx* being placed in the Odontopterygidae and *Osteodontornis* and *Pseudontornis* in the Pseudodontornithidae, the group is in need of a total revision, and a more conservative approach is to term them all *pseudodontorns* and place them all in the family Pelagornithidae.

Richard Zusi and Kenneth Warheit (1992) discovered that the pseudodontorns differ from extant birds in lacking a bony symphysis of the mandible. They also discovered that extant pelecaniforms have evolved intraramal mandibular joints (joints on each lower jaw) that aid in widening the lower jaw in feeding by flexing the jaws outward (lateral bowing) and then back inward, and that these joints are best developed in the pseudodontorns. Joints in the lower jaws have developed convergently in *Hesperornis* and *Ichthyornis* and in mosasaurs and are a synapomorphy of theropod dinosaurs, as discussed in chapter 4. However, in each case the joints were evolved independently, as is the case in pseudodontorns, where the joints are formed by a different configuration of bones.

Because of their highly specialized gliding wings, pseudodontorns probably did not plunge into the water in the fashion of boobies or tropicbirds, but they may have grasped prey from the water surface while in flight by a downward nod of the head, frigatebird style, or while swimming (Zusi and Warheit 1992, 359). Storrs Olson (1985a) has noted that the pseudoteeth of the pseudodontorns were irregularly sized and not particularly reinforced, indicating that their prey was likely such soft-bodied creatures as squid and other unshelled cephalopods. "Bowing of the mandible to an extraordinary degree admitted large prey into the throat; pseudoteeth and hooked maxilla served to secure live prey in jaws that were weakened for grasping by lack of a bony symphysis" (Zusi and Warheit 1992, 351).

Pseudodontorns must have been nearly cosmopolitan. Certainly they were widespread throughout the Tertiary; remains have been recovered from the Eocene of England, Nigeria, and Antarctica (Tonni 1980a), the Oligocene of South Carolina and the Caucasus, and the Miocene of France, Maryland, Virginia, North Carolina, California, Oregon, British Columbia, and New Zealand (Olson 1985a, 199).

Today's Pelicans and Allies. Pelicans themselves are rather spectacular birds, and they come in two varieties. Brown pelicans dive from heights of up to 23 meters (75 ft.) into the water to 0.6 meters (2 ft.) or so beneath the surface, where they scoop up fish with a mouthful of some two and a half gallons (9.5 l) of water. After forcing the water out, the pelican eats its prey. Extensive air sacs along the brown pelican's chest cushion the plunge. White pelicans, in contrast, swim in pursuit of prey and, instead of diving, plunge their heads underwater to catch fish. The oldest record of a pelican is *Pelecanus gracilis* Milne-Edwards, from the Lower Miocene of France, and there is a fairly abundant Middle Miocene form, *P. intermedius,* from Germany. Several small Tertiary species have also been named. Pelicans are particularly abundant in the Miocene and younger deposits of Australia (Vickers-Rich 1991).

The order Pelecaniformes, as mentioned, has produced several lines of foot-propelled divers in addition to the more spectacular wing-propelled plotopterids. The snakebirds (Anhingidae) are strong divers and swimmers adapted for spearing fish with their sharp beaks. The earliest fossil referred to the Anhingidae is *Protoplotus beauforti* (Lambrecht 1931a), a nearly complete skeleton from the late Eocene of Sumatra, but this form may be referable to a new family (Olson 1985a). The fossil record for this family is spotty throughout the Tertiary, but considerable diversity is indicated by the discovery of a giant anhinga, *Anhinga grandis,* from the latest Miocene of Nebraska, that was some 25 percent larger than the modern North Amer-

Limnofregata azygosternon, a primitive frigatebirdlike pelecaniform from freshwater lake deposits in the Lower Eocene Green River Formation of Wyoming, is represented by a nearly complete skeleton with feather impressions (retouched photograph). It is placed in a separate subfamily, the Limnofregatinae, and represents the first known Tertiary record of the pelecaniform family Fregatidae. Scale = 40 mm. (From Olson 1977)

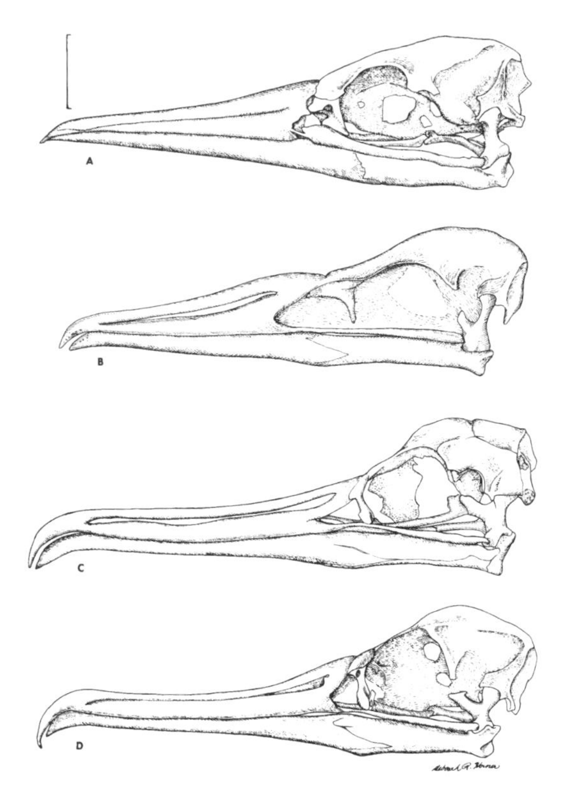

Skulls and mandibles in lateral view. *A,* red-footed booby (*Sula sula*); *B, Limnofregata azygosternon* (generalized reconstruction); *C,* juvenile great frigatebird (*Fregata minor*) showing the very long, open nostril and reduced ossification; *D,* adult lesser frigatebird (*Fregata ariel*). Scale = 20 mm. (From Olson 1977)

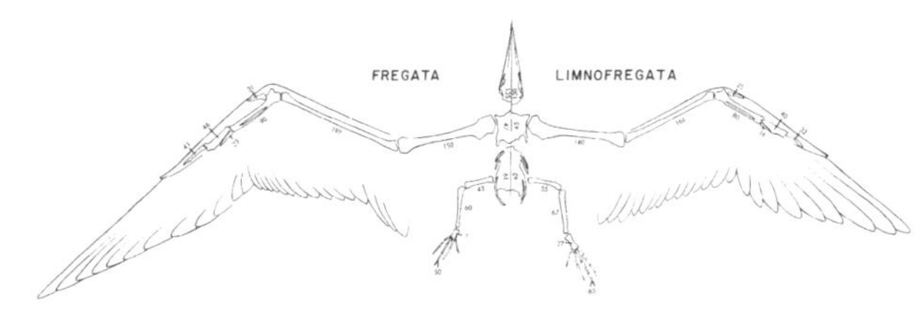

Outline comparing the overall skeletal proportions of modern and fossil frigatebirds: *left half, Fregata; right half, Limnofregata*. Numbers refer to the actual length (mm) of individual bones in the holotype of *Limnofregata azygosternon* and in a small male lesser frigatebird (*Fregata ariel*). Note that the body size of the two species is virtually the same. Scale = 100 mm. (From Olson 1977)

Reconstruction of the early Eocene frigatebirdlike *Limnofregata azygosternon*. (From Olson 1977)

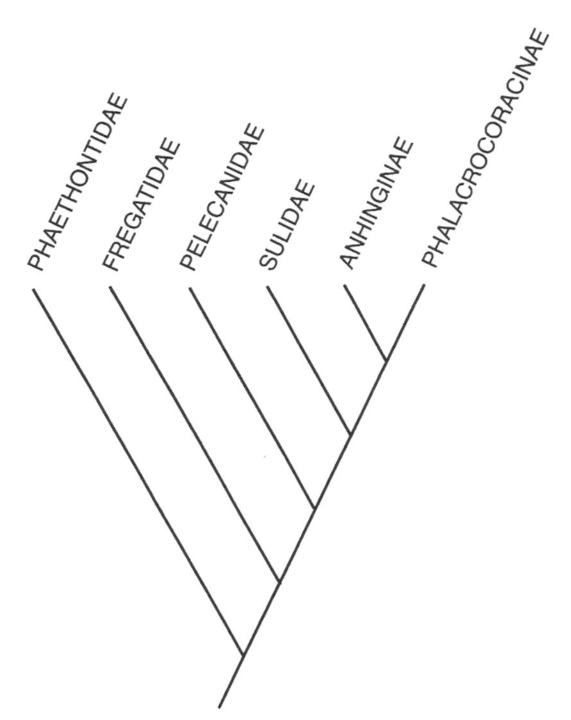

Phylogenetic analysis of the Pelecaniformes by Joel Cracraft (1985), based on a 52-character numerical cladistic analysis. Pelecaniform monophyly was highly corroborated, and within the Pelecaniformes, the tropicbirds, thought to be primitive, are the sister-group of the other pelecaniform families, which are divided into two lineages: the fregatids and the pelecanids, sulids, and phalacrocoracids (including anhingids). Within the second clade, according to this scheme, the sulids and phalacrocoracids are each other's closest relatives. This study also provides corroborating evidence that pelecaniforms and procellariiforms are sister-taxa. (Modified after Cracraft 1985)

Storrs L. Olson of the Smithsonian Institution's Natural History Museum compares a modern albatross's wing bone with that of a fossil of the largest flying seabird ever discovered. With a wingspan of more than 5.5 m (18 ft.), pseudodontorns, or "pseudo-toothed" birds, were once widespread but became extinct about 5 million years ago. (Photo courtesy Storrs L. Olson)

Feeding methods of seabirds. In most cases the genus depicted is only one of several that perform the action. The genera shown, *from left to right: top row, Stercorarius* pursuing *Phalaropus, Catharcta* harrying *Larus; second row, Fregata, Anous, Larus, Rynchops, Oceanites, Pachyptila, Daption, Macronectes; third row, Sterna, Pelecanus, Phaethon, Morus, Diomedea, Phalaropus; under water, Puffinus, Uria, Pelecanoides, Pygoscelis, Phalacrocorax, Melanitta*. (Drawing by Jon Ahlquist; from Ashmole 1971; courtesy Academic Press)

A nondiving pelecaniform: the California Miocene pseudodontorn *Osteodontornis orri,* with bony toothlike projections along the sides of the bill rather than true reptilian teeth. Pseudodontorns, which became extinct at the end of the Pliocene, were among the largest flying birds, well adapted for soaring, with wingspans as large as 6 m (20 ft.). They are known first from the early Eocene of England and occurred all over the world throughout the Tertiary, including well-known sites from the late Miocene of both coasts of North America. Depicted here is the fossil slab containing *Osteodontornis orri* as discovered, with reconstructions of its skeleton, musculature, and total form, scaled by a large gull. The life reconstruction (*at bottom*) shows the bird eating a squid. (Reconstructions by Mark Hallet; courtesy Los Angeles County Museum of Natural History)

ican anhinga (*A. anhinga*) (Martin and Mengel 1975). *Anhinga grandis* has also been found in the late Miocene of Florida (Becker 1987a), and the smaller late Miocene *A. subvolans*, also of Florida, from approximately 18 million years ago, is the oldest reliable record for the family (Becker 1986a).

Other giant anhingas are known to have occurred in the Miocene. An anhinga similar in size to *A. grandis*, from the Middle Miocene of Colombia's Upper Magdalena Valley, has been described but not named (Rasmussen and Kay 1992), and Herculano Alvarenga (1995) has described *Meganhinga chilensis*, a gigantic, possibly flightless anhinga, from the Miocene of Chile. Gigantism in these South American anhingas has been attributed to extensive development of huge inland freshwater swamps during the Miocene of both North and South America; however, recent evidence indicates that during the medial to late Miocene (8–12 million years ago) the sea level was higher and portions of interior South America were flooded, forming a huge Amazon seaway (Räsänen et al. 1995; Webb 1995). As a consequence, marine species were introduced into the Amazon River system, which exceeds all other river systems in marine-derived species.

Cormorants (Phalacrocoracidae) are also very powerful foot-propelled divers that seize fish with their hooked bills. One species, the Galápagos cormorant (*Phalacrocorax harrisi*), has become completely flightless (Livezey 1992). With this exception, there is little more throughout the Tertiary in the way of bizarre cormorants, and the family, like so many others, appears in the late Eocene of Europe and is widely distributed throughout the Old and New Worlds throughout the Tertiary. Boobies and gannets (Sulidae) are also strongly adapted and are perhaps the grandest aerial divers of them all, dropping from 9 to 30 meters (30–100 ft.) through the air to descend to great depths under the water using their powerful feet and half-opened wings. The earliest known sulid is *Sula ronzoni*, from the early Oligocene of France, and more than twenty paleospecies are named throughout the Tertiary of both hemispheres, including a very large gannet, *Morus magnus*, from the late Miocene of California (Howard 1978).

The Realm of Seabirds. The almost bewildering variety of seabirds produced by the Mesozoic and Tertiary adaptive radiation of birds indicates the wealth of evolutionary solutions to the problem of obtaining food (in competition with other birds, marine reptiles and mammals, and large fish) from the sea, an environment not suitable either for avian reproduction or for year-round living. Certainly the abundance and variety of the highly adapted flightless, foot-propelled divers of the tropical, epicontinental Mesozoic seaways stands in stark contrast to the seabird life of Cenozoic seas, where the great radiation of highly adapted, flightless (or nearly flightless), wing-propelled seabirds was restricted to temperate, indeed sub-Arctic and sub-Antarctic, seas. In modern seabird communities the tropical and subtropical seabirds are found almost exclusively in the larids, pelecaniforms, and some procellariiforms, all good fliers, and all birds incapable of deep pursuit diving. As the Ashmoles (1967) have pointed out, most of these birds, and particularly the smaller species, depend partly or entirely on schools of predaceous fish, particularly tunas, to drive small prey close to the water surface. In this sense it is interesting to ponder (as pointed out by Elzanowski 1983) the ecological consequences that the rise of such large perciform, or perchlike, predators as scombroids (including the endothermic tunas and mackerels; Block et al. 1993), sphyraenids (including barracudas), and large percoids must have had not only in terms of competition with seabirds but indeed in terms of the evolution of seabird communities. This is particularly striking when one considers that the only acanthopterygian fish, or modern-type bony fish, in the Niobrara Formation is the 7-centimeter-long (2.8 in.) holocentrid berycoid *Kansius sternbergi* (Patterson 1964), a primitive percomorph.

The astonishing variety of morphological forms in modern seabird communities reflects the effects of multifarious evolutionary interactions. It is one version of the diversity of form that evolved in different avian groups at different times in the past, especially during the Tertiary, and produced some of the most spectacular examples of clearly demonstrable convergent evolution seen in all the vertebrates.

Flightless cormorant (*Phalacrocorax harrisi*) of the Galápagos Islands, sunning what remains of once powerful cormorant wings. (Photo courtesy Mike Tove)

Reconstruction of the common Paleocene and Eocene *Presbyornis*. Highly colonial in habit, these shorebird-duck mosaics filtered algae and other small biota from saline lakes and, like flamingos, often formed huge nesting and feeding aggregations. (Drawing by John P. O'Neill)

5

Flamingos, Ducks, and Long-legged Waders

One might, at a glance, conclude that this chapter was designed as a dumping ground for all the avian groups not covered elsewhere in this book. To speak of ducks, or anseriforms, and long-legged waders, or ciconiiforms, in the same breath would appear to be pure heresy, but in truth, although many authors might argue that our knowledge of these groups is exceptionally well established, we actually know little of how they are related to other groups. Storrs Olson (1979), for example, has suggested that the order Ciconiiformes may be an entirely artificial collection of unrelated families, and few advocates of a monophyletic wading-bird assemblage have suggested ways these birds could have derived from other groups.

As far back as the eighteenth century, the Swedish taxonomist Linnaeus (1758) created an order Grallae that included not only the long-legged waders but the shorebirds and cranes, his assumption being that long necks and long legs have had some phylogenetic significance. At the end of the nineteenth century, Alfred Newton, considering this question, was exasperated by the fact that the heron family had, "through the neglect and ignorance of ornithologists . . . been for many years encumbered by a considerable number of alien forms, belonging truly to the Gruidae (Crane) and Ciconiidae (stork), whose structure and characteristics are wholly distinct" (1896, 416). Yet most classifications until recently followed, at least roughly, those of Hans Friedrich Gadow (1893), Ernst Mayr and Dean Amadon (1951), and Alexander Wetmore (1960). In these classifications, the number of forms contained within the Ciconiiformes varied but generally included the herons, storks, and ibises, plus a miscellany of other wading birds, with the flamingos squeezed between the ciconiiforms and the ducks, the anseriforms.

The classification was radically revised when Joel Cracraft published "a first attempt at a phylogenetic classification of Recent birds" (1981, 681), but, as Olson noted, Cracraft's classification "is not constructed according to cladistic principles" (1982a, 733). Its Division 3 contains not only the traditional ciconiiform birds, with the flamingos and storks in the same superfamily, but also the New World vultures, the secretarybird, and owls, followed by the traditional falconiform groups. Charles Sibley and Jon Ahlquist remarked of Cracraft's classification "that it does not consistently separate homology from analogy, as indicated by his placing the loons and grebes together and the owls with the Falconiformes" (1990, 244). For their part, Sibley and Ahlquist, following their DNA-DNA hybridization data, proposed an order Ciconiiformes that includes the shorebirds and sandgrouse, falconiforms, grebes, pelecaniforms, traditional ciconiiforms, New World vultures and storks, and a superfamily Procellarioidea, which includes the formerly pelecaniform frigatebirds, penguins, and loons, along with the traditional tube-nosed swimmers.

The flamingos, the most spectacular of the long-legged waders, have been placed everywhere in the taxonomy, and determining their phylogenetic position has been one of the most perplexing problems in avian systematics. Thomas Huxley argued in 1867 that "the genus *Phoenicopterus* is so completely intermediate between the anserine birds [ducks] on the one side, and the storks and herons on the other, that it can be ranged with neither of these groups, but must stand as the type of a division by itself" (460). In 1950, J. Berlioz included the flamingos in his order Anseriformes, and a year later, Mayr and Amadon, noting the conflicting evidence, placed the flamingos in an order of their own, Phoenicopteriformes, because "they may be related to both" (1951, 7). As we consider the odd assemblage of birds that make up this chapter, the flamingos make a fine place to begin.

Today's Flamingos and the World of Filter-feeding. Time and again during his long voyage on H.M.S. *Beagle* (1831–1836), Charles Darwin was fascinated by the flamingos he encountered in Argentina, Chile, and the

Galápagos, wondering how these strange creatures could exist on lakes that for half the year became covered with stark white salt, at times more than a foot thick (Darwin 1934). He did not fully understand that flamingos are tied to saline lakes with large algal blooms, where these master strainers make their livelihood by filtering algae and other small aquatic creatures from the water. Little did Darwin know that in these briny inland lakes he was observing the highly specialized relicts of a strange lineage of evolution that originated, in all probability, from ancient filter-feeding, stiltlike shorebirds. Flamingos, today confined to several isolated areas of the earth, once enjoyed worldwide distribution and were represented by two evolutionary lines.

The surviving six (or five) species are all quite similar in anatomy. The largest, the greater flamingo (*Phoenicopterus ruber roseus*), and the smaller lesser flamingo (*Phoeniconaias minor*) occur widely in the Old World and are especially prevalent in East Africa's Great Rift Valley. The lesser flamingo is the most numerous species in Africa, while the greater flamingo is more widespread, nesting not only in Kenya but also from the shores of the Mediterranean east to western India and Sri Lanka. The other species are confined to the New World, with one, the American flamingo (*Phoenicopterus ruber ruber*), ranging from the Bahamas through the Yucatán and northwestern South America to the Galápagos Islands. The remaining three, the Chilean, Andean, and Puna flamingos (respectively, *Phoenicopterus chilensis*, *Phoenicoparrus andinus*, and *Phoenicoparrus jamesi*), live primarily in the soda lakes of the high Andean plateaus.

Lake Nakuru in Kenya is a world-famous haunt of flamingos. Located in the Great Rift Valley, Nakuru is a shallow, alkaline lake, about 38 square kilometers (24 sq. mi.) in extent, that at times may have concentrations of more than a million flamingos. One such swarm inspired Roger Tory Peterson to label it "the most fabulous bird spectacle in the world." What is it about the seemingly sterile lakes of the Great Rift Valley that attracts such huge aggregations of birds? Although apparently barren, these lakes actually produce a great abundance of small animal and plant life at certain times of the year. As Philip Kahl has estimated, "When a million or more lesser flamingos and thousands of greater flamingos congregate on relatively small bodies of water, such as Kenya's Lake Nakuru, they may consume 200 tons of food a day" (1970, 279). Flamingos flock in such vast numbers because of the unpredictability of the food supply. As water levels in different lakes vary, so do salinity and food levels; the flamingos wander from lake to lake, converging on those that contain an abundance of food.

These filter-feeding flamingos exert an important influence on the lake ecosystem. In one experiment, Stuart Hurlbert and Cecily Chang (1983) excluded the Andean flamingo from a shallow saline lake in the Bolivian Andes. Within seventeen days, this exclusion had resulted in a dramatic increase in the biomass of microscopic organisms inhabiting the lake sediments—the food of flamingos. The large size of flamingo feeding flocks and the flamingos' high feeding rates clearly have a considerable impact on aquatic ecosystems.

On a much larger scale, the Bering Sea floor has been altered by filter-feeding gray whales (*Eschrichtius robustus*). As gray whales feed on the bottom they roll on one side with their mouth paralleling the sea floor, sucking in fauna-rich sediment through one side. In the search for food they excavate huge patches of sediment, introducing more sediment into the water of the northeastern Bering Sea than does the Yukon River (Nelson and Johnson 1987). Filter-feeding is an ancient occupation for vertebrates, and sievelike filtering mechanisms evolved early on in groups ranging from sharks and seagoing reptiles to pterosaurs.

The most characteristic feature of modern flamingos, one that is unique among birds, is their strange expanded, bent bill, superbly designed for filtering algae and other small plants and animals from water. Modern flamingos feed with their bill inverted, the upper bill below the lower; the lower bill moves against the upper to sieve water (containing algae and other small plants and animals) and mud through slits on the upper bill and toothlike protuberances on the tongue, the pumping organ. The tongue is so big that it prevents large organisms from entering the mouth. Though algae are a very important food, flamingos may feed on a great variety of organisms depending on the habitat, including diatoms, protozoans, small worms, insect larvae, small mollusks and crustaceans, and occasionally very small fish.

The various species may feed quite differently, and where two species occur side by side, the ecological niche may be partitioned very finely for harmonious coexistence. On East Africa's flamingo lakes where both the greater and lesser flamingos occur, the lesser is primarily an algal filter-feeder, straining algae from the upper 5 to 8 centimeters (2–3 in.) of the lake. On calm days lesser flamingos swim out into the center of the large lakes and continue to filter the upper layer of water for its rich algal supply. In contrast, the much less numerous greater flamingos, with coarser toothlike projections on the bill and tongue, feed principally on larger items, such as mollusks and crustaceans, that the lesser flamingo cannot strain. The greater flamingos also stir up the bottom mud with their feet (a behavior known in a number of modern birds), thus scooping up microor-

Once widespread, the six living species of flamingos survive in inhospitable corners of the earth. Of the six species, the largest, the greater flamingo (*Phoenicopterus ruber roseus*), and the smallest, the lesser flamingo (*Phoeniconaias minor*), occur in the Old World, primarily in East Africa. The remaining four inhabit the New World. The American flamingo (*Phoenicopterus ruber ruber*), *above,* ranges from the Bahamas to the Galápagos; the other three are found in Andean South America. Charles Darwin wrote in August 1833 after a visit to a large alkaline lake near the mouth of the River Negro in Argentina, "I met with these birds wherever there were lakes of brine." (Drawing by George Miksch Sutton)

Andean flamingo (*Phoenicoparrus andinus*) feeding downy chick. (Photo by M. P. Kahl/VIREO)

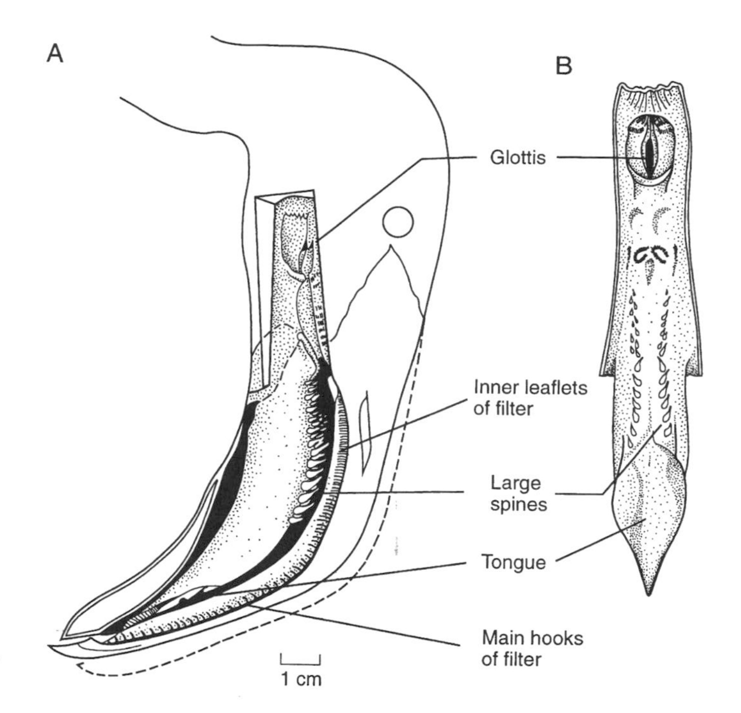

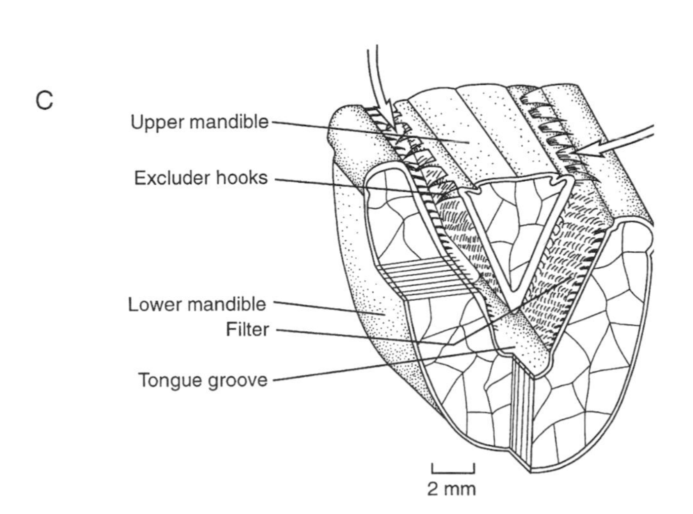

A, head of a modern flamingo in the normal upside-down feeding position; *B,* the tongue; *C,* cross-section of the bill of the lesser flamingo, showing flow of water currents caused by pumping the tongue. When the tongue is depressed in the tongue groove, inflowing currents (arrows) are drawn through the large, hooklike lamellae of the upper mandible, which exclude large objects. When the tongue is elevated, outflowing currents pass through the succession of filters on both jaws formed by the smaller, velvet-textured lamellae, called fringed inner platelets. (Modified after Jenkin 1957)

ganisms in a souplike suspension that it strains along with algae (Allen 1956).

Flamingo nests are unique in the world of birds, and all flamingos build more or less the same type. Made of mud, the nests are, depending on the species, 12 to 46 centimeters (5–18 in.) tall, about 30.5 centimeters (12 in.) in diameter at the top, and perhaps 38 to 76 centimeters (15–30 in.) across the base. In the concave top the female deposits a single egg, which both sexes incubate. Young flamingos are hatched as downy chicks, which are white and unpatterned, have two coats of nestling down, and are highly precocious. Several weeks after hatching the young are capable of swimming and leave the nest to herd with other youngsters in a large aggregation called a crèche, tended by several adult "nursemaids." Both parents feed the young by regurgitation for up to seven weeks, unerringly locating their own offspring in the huge crèche.

The History of Flamingo Relations. The relations of flamingos has been one of the most debated issues in avian evolution (Sibley and Ahlquist 1972, 1990). At least fifteen separate studies list the flamingos somewhere near either the ducks, geese, and swans (Anseriformes) or the storks, ibises, and herons (Ciconiiformes), long-legged wading birds adapted for life in shallow waters or marshes. Evolutionary kinship with the ducks and geese has been based primarily on the structure of the bill and feet, the development of the young (young flamingos have a straight bill that is superficially reminiscent of that of ducks, as well as a plumage of fluffy down like ducklings), and the similarity of duck and flamingo feather parasites. Adult flamingos also make a loud honking sound very like that of geese. The main reason for believing that the flamingos share kinship with the storks, ibises, and herons has been superficial similarity in external appearance. But historically the question has been whether flamingos are most closely related to the herons and storks and merely convergent to the anseriform birds or whether they were derived from the ducks and geese and only later converged toward the stork and heron body form because of their wading habits. And a further possibility, advanced by myself and Storrs Olson (Feduccia 1976a, 1977a, 1977b, 1978a, 1978b; Olson and Feduccia 1980a, 1980b), is that flamingos derived from some other group of birds—namely, the shorebirds, primitive Charadriiformes, and have become similar to both ducks and herons through convergent evolution.

As we noted, Joel Cracraft (1981) placed the Ciconiiformes and Falconiformes in his "Division 3," and in the Ciconiiformes he defined two lineages: flamingos, storks, and ibises; and herons and *Balaeniceps* (the shoebill). Almost a decade later, Charles Sibley and Jon Ahlquist concluded from DNA-DNA hybridization evidence that "the herons, Hamerkop, flamingos, ibises, Shoebill, pelicans, New World vultures, and storks are most closely related to one another than any of them is to another group" (1990, 527). What any of these approaches fails to explain, however, is how either storklike birds or ducklike species could have evolved the complex filter-feeding mechanism of flamingos. In the case of storks, the transmutation of a hard, pointed stork bill into the soft, lamellate flamingo filtering apparatus defies imagination; likewise, the basic structural adaptation for filtering in the duck lineage is completely dissimilar and must have evolved independently. In contrast, one can, by looking at a variety of modern shorebirds, generate a fairly complete pseudophylogeny from the very primitive beginnings of filter-feeding as seen in phalaropes leading to more complex and distinct filtering mechanisms as seen in flamingos on one hand and in ducks on the other.

Whether flamingos are actually allied with living stilts is another question, and we might do well to focus on whether they are derived from ancestral "transitional shorebirds" that may have resembled modern avocets and stilts. If this were the case, and if the similarities to modern recurvirostrid shorebirds were a parallel development, it would certainly explain the biochemical dissimilarities, the DNA evidence, and the fact that the modern shorebird assemblage is linked by a unique form of malate dehydrogenase (Kitto and Wilson 1966). Could it be that all modern shorebirds likewise represent a complete reorganization from Cretaceous "transitional shorebirds" and are therefore a monophyletic group apart from their Cretaceous forebears? In either case, however, a shorebird derivation of both flamingos and ducks is the only model that can generate a pseudophylogeny in which each level sets the stage for the next evolutionary stage, each step is small and fully adaptive, the organism at each stage constitutes a functional and adaptive whole, and each stage is represented by an analogous known organism (Bock 1986).

Similarity to Avocets and Stilts. Flamingos are almost identical to recurvirostrid shorebirds (avocets and stilts) in much of their postcranial anatomy, and in life history, ecology, and behavior they match up remarkably well with the highly colonial Australian banded stilt (*Cladorhynchus leucocephalus*). One need only examine the tarsometatarsus of a stilt and a flamingo to be convinced of their remarkable similarity. Another link is that they share a remarkable number of host-specific cestode parasites. As Jean Baer has stated: "One finds in flamingos four monotypic, endemic genera (*Amabilia, Cladogynia, Gynandrotaenia, Leptotaenia*). The last two, very specialized overall, are derived from a family of cestodes characteristic of the Charadrii" (1957, 274; my translation).

The life history of flamingos and the banded stilt of Australia are astonishingly alike. Like flamingos, banded stilts are colonial, breed in great, densely packed colonies, nest in depressions in the ground and are regularly spaced about 30 centimeters (12 in.) apart, have white, unpatterned downy young with two coats of nestling down (unknown in other shorebirds but found in flamingos), and frequent saline temporary lakes (in this case, in southern Australia) where they feed on enormous quantities of small crustaceans (for details, see McGilp and Morgan 1931; Jones 1945; McNamara 1976; and Olson and Feduccia 1980a). Also, as Olson and I have written, "McNamara quotes Mr. Tom Spence, Director, Perth Zoological Gardens: 'the display of the Banded Stilt is quite spectacular and very reminiscent of that of the Lesser Flamingo. They parade in a tight little pack with their long mantle feathers raised (just like a Flamingo) and synchronise their movements pretty closely, I have also seen this in the wild' " (1980a, 12). And Philip Kahl (1975) noted that this type of behavior may be the exact equivalent of the flamingo marching display. Interesting, too, is that flamingos, while swimming, may feed by tipping up in a manner reminiscent of dabbling ducks, which many workers have noted in trying to relate the two groups. What is often unremarked, however, is that avocets and the banded stilt also feed while swimming and tip up in order to obtain food.

Yet another similarity is nesting behavior. Flamingos may lay their eggs in a depression in the ground, in shorebird fashion, but they usually construct a truncated cone of mud to accommodate their single egg. According to J. Rooth, "the function of these high constructions is clear at high water. . . . It is typical that the wader *Himantopus himantopus* [black-winged stilt, Recurvirostridae] which breeds along the shores of the salinas [on Bonaire] also makes such high nests!" (1965, 99–100). As Storrs Olson and I have written, "Not only can all the supposed ciconiiform and anseriform behavioral traits of flamingos be accounted for by having flamingos derived from the Charadrii, but also the seemingly unique characteristics of the Phoenicopteridae that are irreconcilable with either storks or ducks can be explained as well" (1980a, 15).

The preponderance of morphological, behavioral, parasitic, and other complementary evidence for shorebird affinities of the flamingos would appear to lay the case to rest, but like most problems in systematics some nagging contradiction remains, this time coming from the cladistic and biochemical arenas. Cladistic analyses of the ciconiiform birds and flamingos (Cracraft 1981, 1986) have not supported a shorebird-flamingo link, and Cracraft has asserted, "I believe that the similarities between flamingos and *Cladorhynchus* will eventually be interpreted as convergences" (1981, 691). But his contention that the hypotarsus of the fossil flamingo *Juncitarsus* "is totally unlike all known phoenicopterid genera, both fossil and Recent," is hard to comprehend when one views the figures of these birds as seen in these pages. Further, because the postcranial anatomy of these birds is of the primitive, "transitional shorebird" type, it is extremely challenging to identify "derived" characters except in the highly specialized filtering apparatus. I must concur with Storrs Olson, who has noted, "Cracraft has completely misrepresented the evidence, and not one of the characters he cites in any way refutes the charadriiform relationships for flamingos" (1982a, 736). However, a number of biochemical studies (for example, Kitto and Wilson 1966), including Sibley and Ahlquist's much-maligned DNA hybridization studies, have not supported the shorebird-flamingo connection, but this may well be because improper comparisons were made: flamingos may well derive from long extinct ancestral "stiltlike" shorebirds, and, the modern shorebird radiation is very likely post-Cretaceous. Sibley and Ahlquist noted that their DNA comparisons "support a phylogeny that is incompatible with the conclusions of Olson and Feduccia" (1990, 521). Earlier, however, a study of the uropygial waxy secretions of flamingos and other birds indicated "relationships of the flamingos to wading birds (Charadriiformes)," with the authors concluding, "It is, therefore, supposed that all of them are of common origin" (Jacob and Hoerschelmann 1985, 49).

This is clearly another case where the morphological evidence is unequivocally in support of a shorebird derivation of flamingos, and it is in conflict with the biochemical comparisons to date. Focal to the debate is the Eocene avian fossil *Juncitarsus*, the oldest flamingolike bird, which was given its generic name from the Latin *juncus*, a reed or rush, and *tarsus*, in reference to the slender gracile tarsometatarsus that closely resembles that of recurvirostrid shorebirds as well as flamingos. The fossil's derived features, however, are those of flamingos, and it is clearly an early form of flamingo. As Stefan Peters has noted, "The ancestors of the flamingos must have already been waders before they evolved the specialized beak. *Juncitarsus* correlates entirely with this stage of evolution. The body proportions and, most of all, the legs, are already those of a flamingo-like wader; the beak, however, is still not flamingo-like. But it is also not stork- or goose-like, and suggests a charadriiform relationship. The opinion of Olson and Feduccia is there supported by virtually irrefutable evidence" (1992a, 144). The resolution of this conflict would be an exciting venture in biochemical systematics.

Banded stilt (*Cladorhynchus leucocephalus*), family Recurvirostridae; *left,* individual, *above,* flock. It is the closest analogous living link between shorebirds and flamingos. Like flamingos, these stilts are highly colonial, filter-feeding denizens of alkaline lakes. Their downy young, like those of flamingos, soon leave the nest to congregate in crèches. (Photo by M. P. Kahl/VIREO)

Downy chick of the banded stilt, showing the dense, unpatterned white down that is very similar to flamingos. (From Olson and Feduccia 1980a)

The Flamingo Fossil Record. In 1931, the Hungarian paleontologist Kálmán Lambrecht (1931b) described what was then the oldest Cretaceous bird, *Gallornis straeleni*, as a representative of the duck line on the basis of a portion of a femur and a scrap of humerus, and then it was placed near the flamingos by Pierce Brodkorb (1963a). It has since been shown again and again that it is impossible to determine the affinities of Mesozoic or early Tertiary birds on the basis of such fragmentary evidence (Olson and Feduccia 1980a)—indeed, in this case *Gallornis*'s status as a bird cannot be verified. In 1933, Lambrecht erected a family of Cretaceous flamingos known as the Scaniornithidae, but the entire Mesozoic fossil record of flamingos has proved to be a myth, in all probability represented by "transitional shorebirds," and the family Scaniornithidae simply cannot be diagnosed.

The further one looks back in time, the more flamingo bones resemble those of ancient shorebirds, but the oldest certain record is the primitive flamingo already mentioned, *Juncitarsus*, from the Middle Eocene of Wyoming and Germany. This fossil is critical to the record in that it is a clear shorebird-flamingo mosaic, a stiltlike flamingo, intermediate in size between the largest of the extant shorebirds and the smallest of the living flamingos. Important corroboration of this evolutionary path was the discovery and subsequent description of another species of *Juncitarsus* (*J. merkeli*), from the Middle Eocene Messel deposits of Germany, by Stefan Peters (1987a), which he claimed "provided important arguments supporting the view that flamingos are charadriiform birds" (1991, 574). Had it not been for associated skeletal elements, both fossils would undoubtedly have been classified as recurvirostrid shorebirds because of their similarities to the avocets and stilts.

The presence of additional specimens of young individuals, too young to fly, indicated that *Juncitarsus gracillimus* nested in colonies. The frontal bone of one of these juvenile specimens exhibits depressions for salt glands, which indicate that *Juncitarsus* inhabited saline environments, common in Wyoming during the Eocene. Both features are, of course, characteristic of flamingos, and *Juncitarsus*'s presence in ancient saline lakes at least places it in an appropriate paleoenvironment for filter-feeding.

After *Juncitarsus* there is a gap in the fossil record, and the next flamingo fossil is that of *Phoenicopterus croizeti*, which occurs in the late Oligocene to early Miocene (Aquitanian) deposits of France. This mid-Tertiary flamingo, characterized by a straighter bill than modern forms, is also known from deposits of the same age in Germany and the former Czechoslovakia (Olson 1985a). The modern genera of flamingos are distinguished on the basis of differences in bill morphology that in turn reflect differences in their filter-feeding mechanisms. It was therefore not surprising that a flamingo named *Phoenicopterus aethiopicus* (Harrison and Walker 1976b), discovered in the early Miocene of Kenya, would have a bill morphology intermediate between that of *Phoenicopterus* and *Phoeniconarrus* (including *Phoeniconaias*), and Pat Rich and Cyril Walker (1983) have erected a new genus, *Leakeyornis*, for this new flamingo. In my opinion, it would be much less confusing to place all living and modern-type fossil flamingos in the same genus *Phoenicopterus*, as Monroe and Sibley (1993) have done, but nevertheless, *Leakeyornis aethiopicus* does illustrate that by the early Miocene the divergence in bill morphology seen in modern flamingos was under way. Extinct species of flamingos of the modern type are commonly found in the Pliocene and Pleistocene of North America and Mexico, from deposits as old as late Oligocene in Europe, as noted above, and from the Miocene to Pleistocene of Australia, where flamingos were once extremely common and diverse in the fossil record, although they no longer exist there (Rich and Van Tets 1982).

Aside from modern types of fossil flamingos, there is only one other valid group, represented by abundant fossils from the early Miocene to the early Pliocene of the Northern Hemisphere. These are the swimming flamingos of the family Palaelodidae. According to Jacques Cheneval and François Escuillié, who have studied all the palaelodids, "They are closely related to the family Phoenicopteridae, with possible affinities with the families Anatidae and Recurvirostridae" (1992, 209). Palaelodids represent one of the truly remarkable and abundant groups of Tertiary fossil birds. They are most common in the Lower Miocene deposits of Saint-Gérand-le-Puy, in the Department of Allier, France, where literally thousands of bones representing all the species have been recovered, and are also known in the fossil record from early Miocene to early Pliocene deposits in North America (*Megapaloelodus*), in the Miocene to the medial Pleistocene of Australia, and as far back as the early Oligocene of Egypt (Olson and Rasmussen 1986; Olson 1985a). Cheneval and Escuillié's studies of all the palaelodid material from Saint-Gérand-le-Puy included "for *P. ambiguus*, 8,333 bones, at least 477 minimum number of individuals (MNI); for *P. gracilipes*, 690 bones, 55 MNI; for *P. crassipes*, 320 bones, 32 MNI; and for *Megapaloelodus goliath*, 63 bones, 13 MNI" (1992, 209). The new material they studied included a nearly complete articulated skeleton with skull, the bill of which, they concluded, "looks very much like that of the recently described primitive flamingo-like bird *Juncitarsus merkeli*" (216) but has more similarities "in comparison with the skull of Recurvirostridae" (218).

The palaelodids were swimmers and perhaps divers,

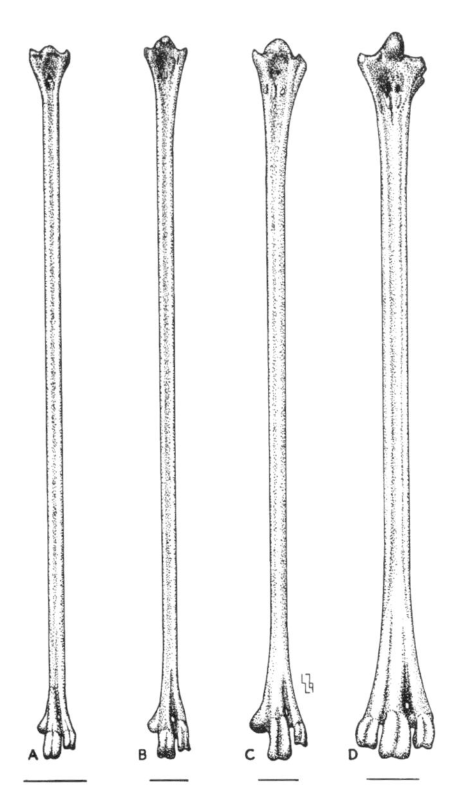

Anterior views of the left tarsometarsi: *A,* black-winged stilt (*Himantopus mexicanus,* Recurvirostridae); *B, Juncitarsus gracillimus* (Phoenicopteridae); *C,* lesser flamingo *(Phoeniconaias minor,* Phoenicopteridae); *D,* Abdim's stork, *Ciconia abdimii* (Ciconiidae), illustrating the extreme morphological similarity between shorebird and flamingo and how different both are from typical ciconiiforms, the storks and allies. (From Olson and Feduccia 1980a)

Holotype tarsometatarsus of *Juncitarsus gracillimus* in *A,* anterior view; *B,* posterior view; *C,* lateral view; and *D,* medial view. Total length, 182 mm (7 in.). (From Olson and Feduccia 1980a)

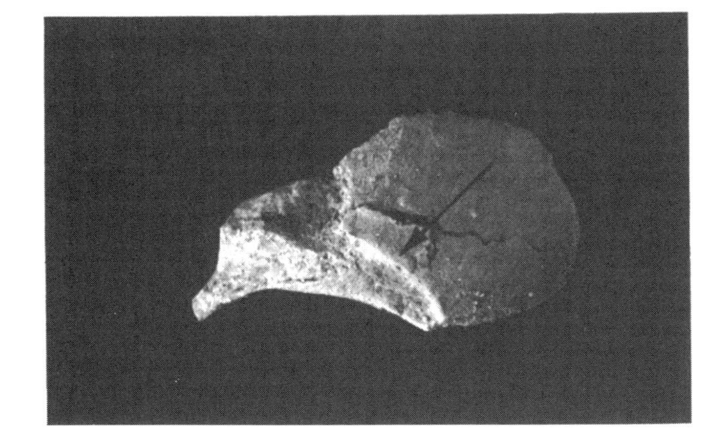

Unfused frontal bone of juvenile specimen of *Juncitarsus gracillimus* showing depression for salt gland (arrow). Note the constriction in nasofrontal region characteristic of flamingos. × 3. (From Olson and Feduccia 1980a)

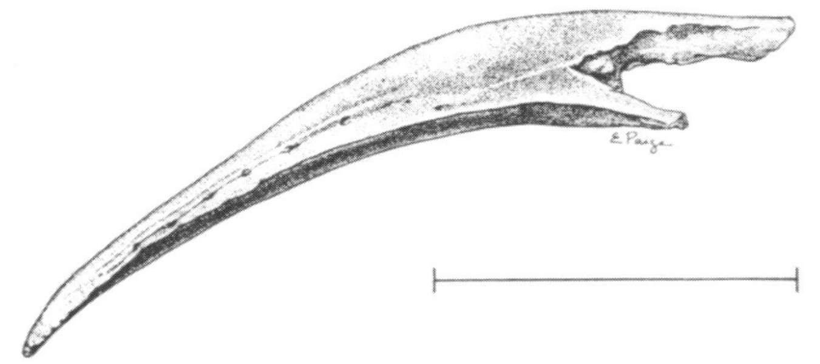

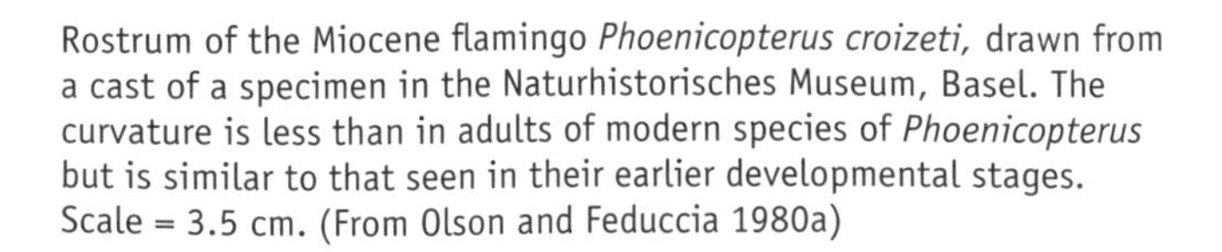

Rostrum of the Miocene flamingo *Phoenicopterus croizeti,* drawn from a cast of a specimen in the Naturhistorisches Museum, Basel. The curvature is less than in adults of modern species of *Phoenicopterus* but is similar to that seen in their earlier developmental stages. Scale = 3.5 cm. (From Olson and Feduccia 1980a)

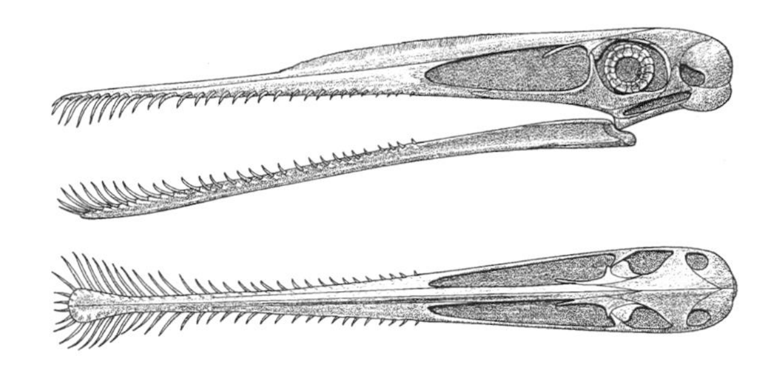

Gnathosaurus ("jaw-reptile") skull restoration by Peter Wellnhofer: *above,* lateral view; *below,* ventral view. This Solnhofen pterosaur (late Jurassic of Bavaria) exhibits a low bony crest along the midline of the skull roof. The dense arrangement of teeth points to a filter-feeding mode of existence and a diet of small marine organisms. Skull length 28 cm (11 in.). (Courtesy Peter Wellnhofer)

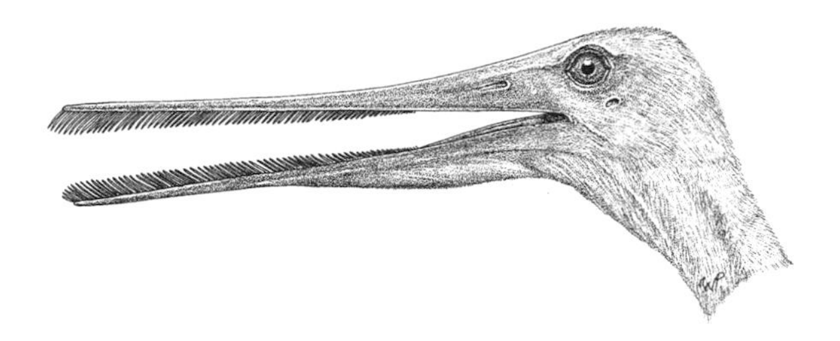

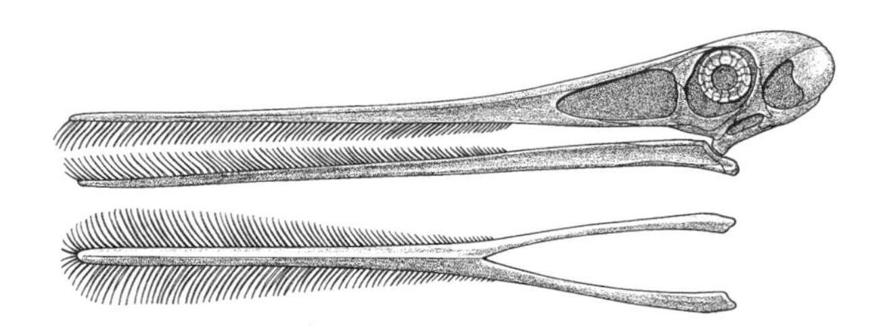

Ctenochasma ("comb-jaw") head and skull (with lower jaw in ventral view) restoration by Peter Wellnhofer. The jaws of this pterosaur (late Jurassic of Solnhofen, northern Germany, and France) housed some 250 tightly packed teeth, indicating a fairly advanced stage in filter-feeding. *Ctenochasma* and *Gnathosaurus* are placed in their own family, the Ctenochasmatidae. Skull length 10.4 cm (4 in.). (Courtesy Peter Wellnhofer)

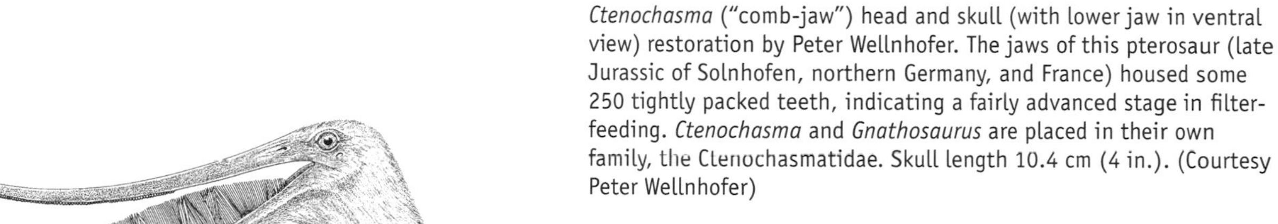

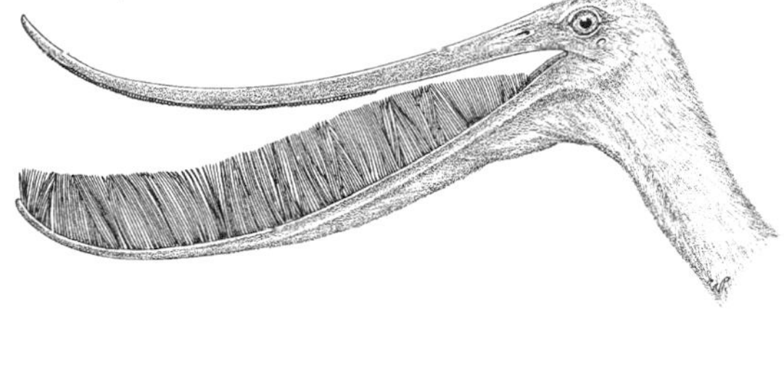

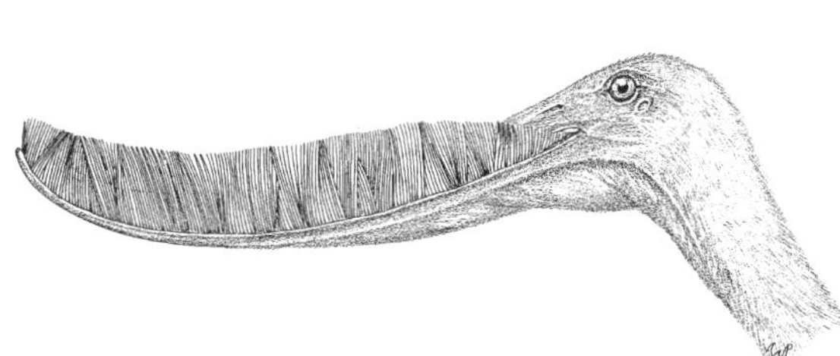

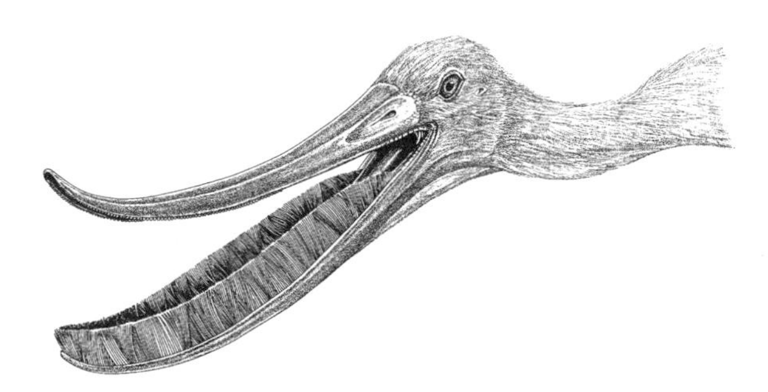

Pterodaustro ("south-wing") life portraits by Peter Wellnhofer in three views. The *Pterodaustro* (early Cretaceous of Argentina), dubbed the "flamingo pterosaur," is the most highly evolved pterosaur filter-feeder. It had a comblike array of long teeth in the lower jaw for sieving small creatures from the water. There were 24 "bristle teeth" per centimeter, or about 500 along each side of the jaw. Its upper jaw had a series of short, blunt teeth for chopping up food items into smaller pieces to be swallowed. Skull length 23.5 cm (9.25 in.). (Courtesy Peter Wellnhofer)

and they "may have already developed a filter-feeding apparatus, which was probably more primitive than in modern flamingos, but perhaps comparable to plankton-feeding auks and penguins" (Cheneval and Escuillié 1992, 221). It is also interesting to note their possession of salt glands, suggesting that they lived near saline lakes, and that the lakes of Saint-Gérand-le-Puy were brackish water expanses inhabited by other saline aquatic birds, such as a more or less modern flamingo (*Phoenicopterus croizeti*), an ibis (*Plegadis paganus*), and a cormorant (*Phalacrocorax littoralis*), whose modern representatives preferentially occur in brackish water (Cheneval 1983a, 1989). It is also noteworthy that the paelaelodids occur in the same deposit (as well as in Germany and the former Czechoslovakia) as a modern flamingo, *P. croizeti*, indicating that the two groups must have diverged from a common ancestor somewhat earlier.

To summarize, in the fossil record of the Phoenicopteridae there are three basic groups: *Juncitarsus*, ancestral shorebirdlike flamingos from the medial Eocene of Wyoming and Germany; *Palaelodus*, short-legged swimming flamingos, widespread from the late Oligocene or early Miocene to early Pliocene; and modern *Phoenicopterus* flamingos, fairly widespread from the late Oligocene or early Miocene to the present.

The Evolution of Filter-Feeding

Filter-feeding, the benchmark of flamingos, is a common form of invertebrate and early chordate feeding and is also found in amphibian tadpoles and a number of fish, particularly in their young. The filter-feeding selachians, basking sharks and whale sharks, are the largest living fish, and among the batoideans, the prize for the largest goes to another filter-feeder, the manta ray. Unlike their terrestrial counterparts, these marine creatures have little pressure to restrict their size, and these giants have made a profession of quietly harvesting the ocean's plankton, freed by their large size from the ravages of most ocean predators. Among the higher vertebrates, the development of sieving and filtering mechanisms has been relatively restricted. The ancient mesosaurs (*Mesosaurus*) of the Middle Permian of South America and South Africa (incidentally, this range is commonly used as evidence that the two southern continents were once together) were meter-long (3.3 ft.) aquatic reptiles that developed a greatly elongated snout with very long, slender teeth that must have formed a straining device to feed on the small crustaceans that abound in the same deposits (Carroll 1988, 206). Later, a number of pterosaur genera—*Ctenochasma* and *Gnathosaurus*, from the late Jurassic Solnhofen deposits of Germany, northern Germany, and France, and *Pterodaustro*, from the Lower Cretaceous of Argentina—evolved sieving and filtering devices to varying degrees. In the present, filter-feeding is well known in the baleen whales, and well-developed filtering mechanisms are known in three groups of birds—flamingos, ducks, and a single genus of petrels, the prions (*Pachyptila*)—and other groups filter-feed to varying degrees.

The striking similarities in the heads of flamingos and baleen whales was first noted by the naturalist Leonhard Stejneger (1885, 153), but only in passing. Storrs Olson and I (1980a) have shown that there is an almost uncanny similarity between the head of a flamingo and that of the right and bowhead whales (*Eubalaena* and *Balaena*). Right and bowhead whales have a very narrow upper jaw with a ventral keel that bears long plates of baleen. This narrow upper jaw is quite visible and is perhaps the most outstanding feature of these whales as they swim near the surface. Their lower jaws are very deep and swollen, accommodating a huge tongue and rising up to cover the baleen plates of the upper jaw. No baleen whales have any filtering devices associated with the lower jaw, but otherwise the general similarity in appearance of the head of right and bowhead whales to that of flamingos is an outstanding example of convergent evolution in the animal kingdom. This overall similarity of heads is a good indication that the "bent bill" structure of the flamingos arose primarily in accordance with the constraints of filter-feeding alone and that the "upside-down" feeding posture evolved secondarily.

Variations on straining or modified filter-feeding mechanisms are seen in the primitive apparatus of certain plankton-eating species of penguins and auks, which have evolved features that enable them to feed effectively on small organisms (Bédard 1969; Zusi 1974; Olson and Feduccia 1980a; Zweers, Berkhardt, and Vanden Berge 1994; Zweers et al. 1995), but in these forms, lamellate strainers are not fully developed. Instead, plankton-feeding penguins have an enlarged tongue, a broadened rostrum, and cornified papillae on the roof of the mouth, inner surface of the lower jaw, and even the tongue and larynx. The penguin genus *Eudyptes*, which feeds largely on small crustaceans, takes its food items individually, holding and manipulating them toward the throat by action of the tongue, which has deepened the bones of the lower jaw. In similarly adapted alcids, the enlarged tongue is accommodated by a distensible gular pouch (Bédard 1969; see Olson and Feduccia 1980a for a detailed discussion of the evolution of filter-feeding in birds).

If one views this style of tongue manipulation as a first stage in a pseudophylogeny of filter-feeding, then perhaps the prions (*Pachyptila*), also whalebirds because of their baleenlike feeding apparatus, may illustrate a second

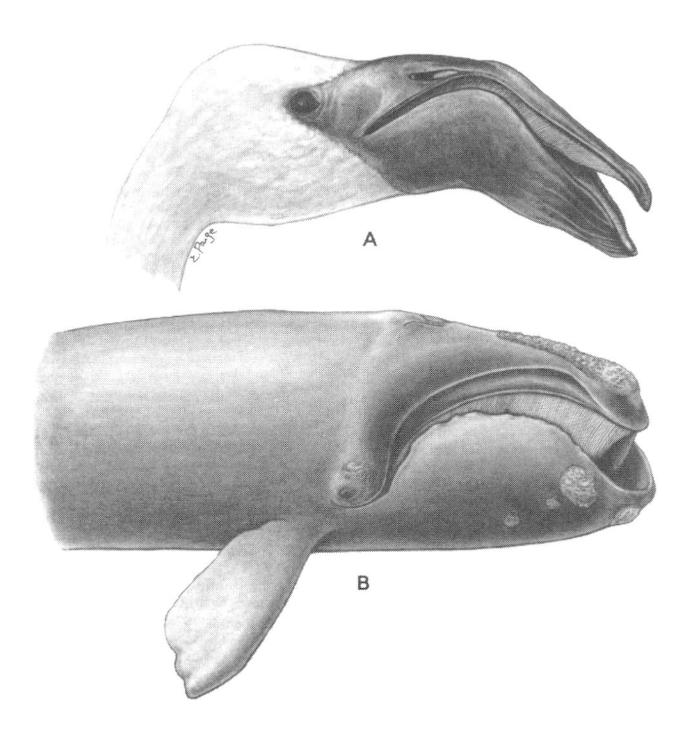

A, head of the lesser flamingo (*Phoeniconaias minor*), compared with that of *B,* the right whale (*Eubalaena glacialis*), showing convergent similarities in the filter-feeding apparatus. Both heads have a feeding apparatus that includes a narrow upper jaw, a large, fleshy tongue accommodated by a deep lower jaw, and a bend in the jaw that provides a greater surface area for the filters, either lamellae or baleen. (From Olson and Feduccia 1980a; drawing by E. Paige)

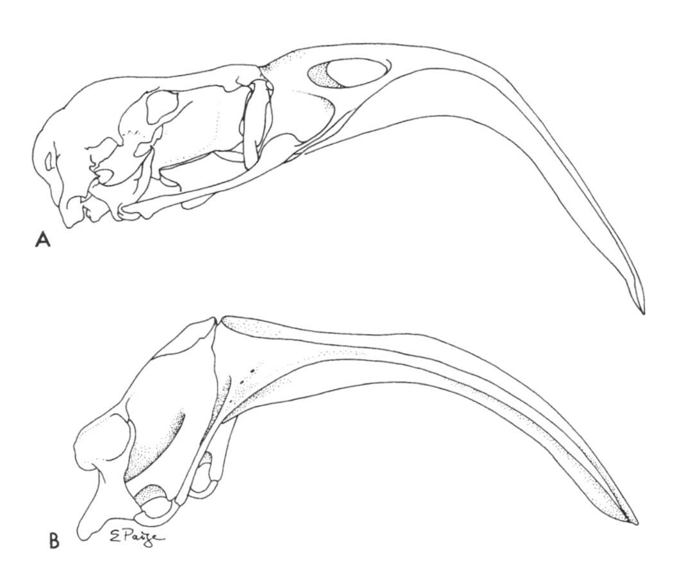

A, skull of the lesser flamingo, compared with that of *B,* the right whale, showing the similarity in the shape of the long, decurved bony structure of the rostrum. (From Olson and Feduccia 1980a; drawing by E. Paige)

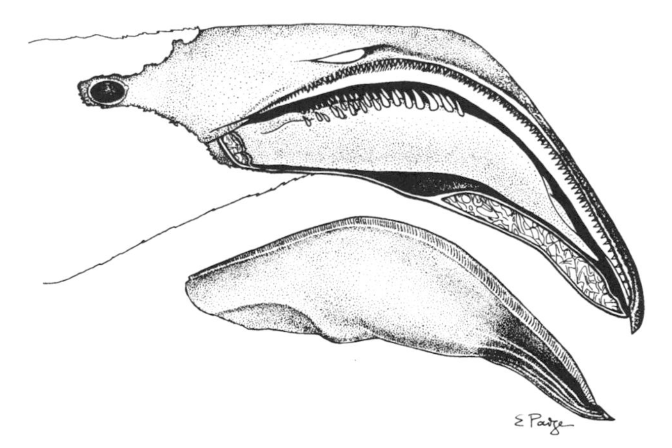

Head of the greater flamingo (*Phoenicopterus ruber*), with the lower jaw removed and displaced below. The tongue is housed in the lower jaw, and neither the lower jaw nor the keel on the upper jaw are as deep as in the more specialized *Phoeniconaias*. (From Olson and Feduccia 1980a; drawing by E. Paige)

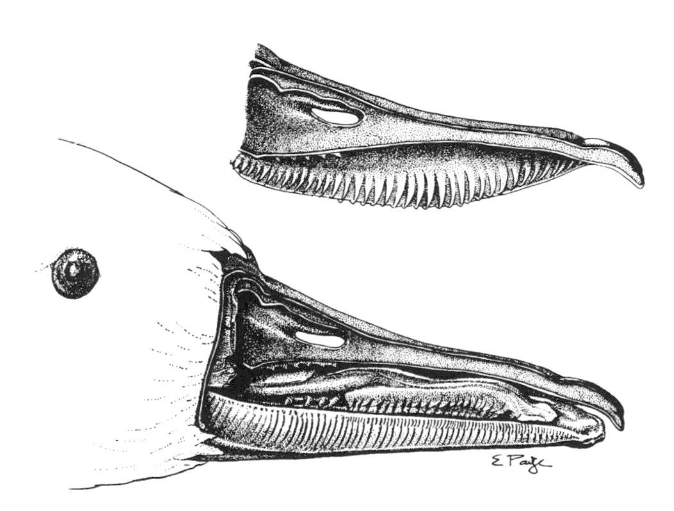

Head of the mallard (*Anas platyrhynchos*) with the upper jaw removed and displaced above. In ducks the tongue is accommodated by the upper jaw, and there are two bulges in the tongue that form a double-piston suction pump, quite unlike flamingos' filter-feeding mechanism. (From Olson and Feduccia 1980a; drawing by E. Paige)

stage. In these birds the bills vary from relatively narrow to greatly expanded, and those with narrow bills tend to have only faint striations, the very beginnings of lamellae. The rudimentary lamellae provide small gaps for water expulsion, indicating that the initial function of lamellae may have been to expel water rather than to sieve prey. As feeding specialization increased, the lamellae have enlarged to take on the role of food retention, while the tongue functions to force out water and manipulate food toward the throat. Interestingly, increasing specialization of the feeding apparatus in *Pachyptila* is correlated with a decrease in the prions' size of prey (Watson 1975). This also holds true in penguins (Zusi 1974) and flamingos (Jenkin 1957) and is in all probability a generality that could be extended to all filtering-feeding organisms.

As Storrs Olson and I have pointed out elsewhere: "The filter-feeding petrels . . . illustrate several important points. First, the initial adaptation for filter-feeding is enlargement of the tongue. This is accompanied or followed by widening and deepening of the bill and the development of lamellae. Second, the initial function of lamellae may be to permit the expulsion of water rather than to retain prey. Third, the species of *Pachyptila* show progression from a primitive, nearly unmodified species, to a highly evolved lamellate filter feeder within a single extant genus, suggesting that the great broadening of the bill and the development of a fine strainer may occur rapidly in birds" (Olson and Feduccia 1980a, 59).

Flamingos, the most specialized of the filter-feeding vertebrates, use their fine lamellae to filter out microorganisms. The large flamingo lower jaw is in essence a troughlike cylinder that houses a large, fleshy tongue equipped with a series of spiny protuberances. The dorsal surface of the lower jaw is lined with a series of fine ridges. The upper jaw is small and lidlike, and lined to varying degrees, depending on the species, with a series of lamellae. The most highly derived living flamingos are the genera *Phoenicoparrus* and *Phoeniconaias*, whereas the least specialized is *Phoenicopterus;* as might be expected, *Phoenicopterus* feeds on the largest food particles (Jenkin 1957). The upper jaw of *Phoenicoparrus* and *Phoeniconaias* is much narrower and is equipped with long rows of featherlike lamellae, or platelets. In feeding the mouth is opened and water enters along the entire length of the gape; when the gape is closed, the tongue forces water back through the filtering device. The American and greater flamingos (*Phoenicopterus*) feed mainly on invertebrates (small mollusks and crustaceans) in the bottom mud while wading in shallow soda lakes but may feed while swimming. The smaller lesser flamingos (*Phoeniconaias minor*), with their finer filter-feeding apparatus, feed primarily on blue-green algae but may also feed on the surface vegetation of alkaline lakes while swimming.

The filtering apparatus of ducks is quite different from that of flamingos—in ducks, the enlarged tongue is accommodated by the upper rather than lower jaw—but they, too, have the essential elements for filter-feeding: enlarged tongue, broad bill, and straining lamellae (see Zweers 1974; Kooloos et al. 1989; Kooloos and Zweers 1991; and Zweers et al. 1994 for a complete discussion of filter-feeding in ducks). Typical filter-feeding ducks have two rows of lamellae that lie along the inner surfaces of both the upper and lower jaw, but these lamellae vary considerably, even within genera. Mallards (*Anas platyrhynchos*), for example, have coarse lamellae, while the northern shoveler (*Anas clypeata*), known for its exceptional filter-feeding bill, has lamellae that extend into fine hairlike fringes associated with the size of prey taken. The enlarged tongue of ducks also differs from that of flamingos in having two bulges along its upper surface. The duck's bill and tongue act in concert as a suction pump consisting of two pistons in a cylinder, with water entering at the tip of the bill and being expelled posteriorly (Zweers et al. 1977). Ducks and flamingos have thus evolved distinctive filter-feeding mechanisms along multiple evolutionary pathways to achieve similar functional solutions.

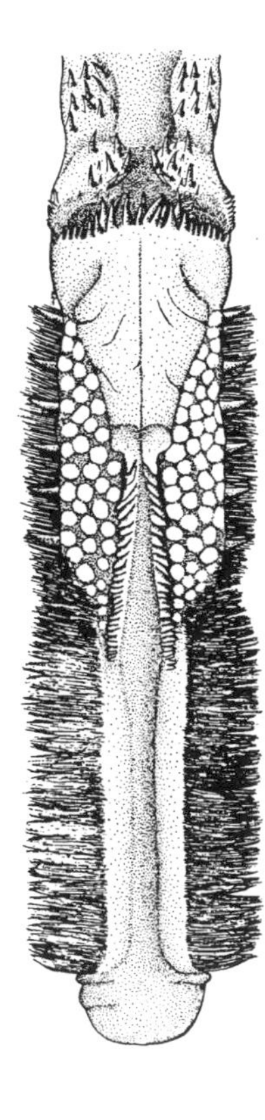

Tongue of the northern shoveler (*Anas clypeata*), showing a highly advanced duck-type of piston tongue fringed with filtering lamellae. (From Gardner 1925)

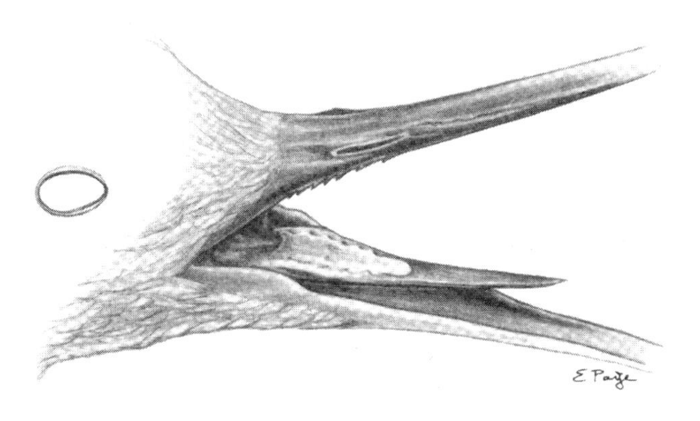

Mouth of the American avocet (*Recurvirostra americana*), showing its broad, fleshy tongue. The acquisition of such a tongue is the first stage in the evolution of filter-feeding. (From Olson and Feduccia 1980a; drawing by E. Paige)

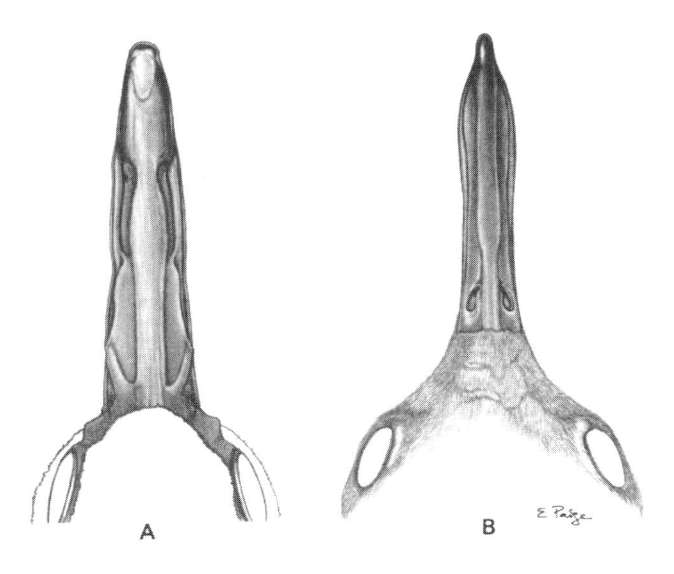

A, dorsal view of the bill of a downy chick of the greater flamingo (*Phoenicopterus ruber*), compared with that of *B,* an adult red phalarope (*Phalaropus fulicarius*). (From Olson and Feduccia 1980a; drawing by E. Paige)

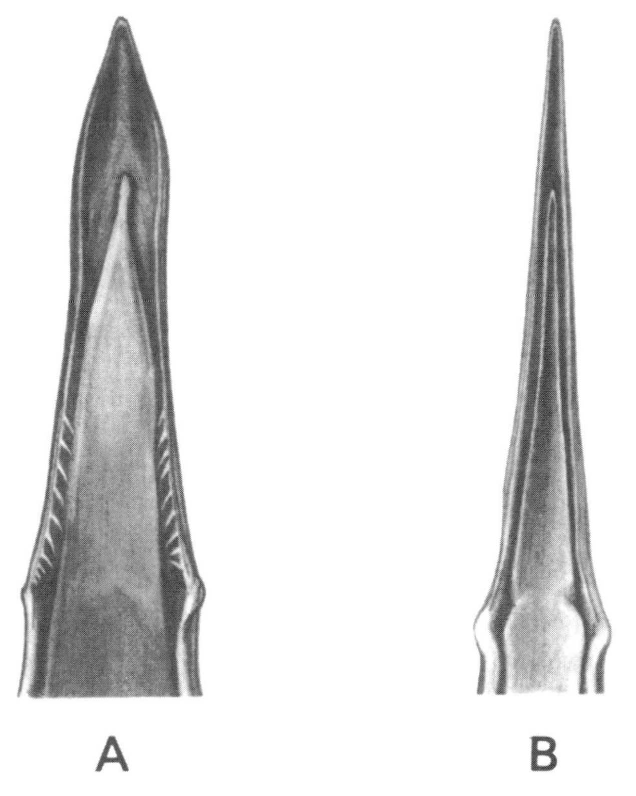

A, dorsal view of the lower jaw and tongue of the red phalarope (*Phalaropus fulicarius*), compared with that of *B,* the northern phalarope (*Lobipes lobatus*). The red phalarope's broad bill, enlarged tongue, and papillae probably indicate a capacity for primitive filter-feeding. (From Olson and Feduccia 1980a; drawing by E. Paige)

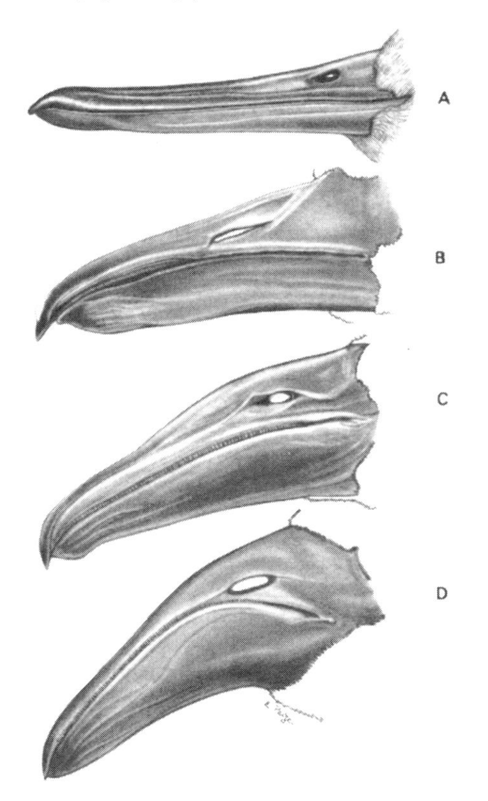

Lateral view of bills. *A,* red phalarope (*Phalaropus fulicarius*). *B–D,* progressive developmental stages of flamingos: *B,* downy chick of the greater flamingo (*Phoenicopterus ruber*) at less than a week; *C,* chick of *P. ruber* at about 30 days; *D,* chick of the lesser flamingo (*Phoeniconaias minor*) at 7 weeks (adapted from Kear and Duplaix-Hall 1975, pl. 48). Note the progressive deepening of the mandible and bending of the bill with increasing age. (From Olson and Feduccia 1980a; drawing by E. Paige)

A Pseudophylogeny of Filter-feeding. In proposing an independent shorebird ancestry for both ducks and flamingos, Storrs Olson and I (1980a, 1980b) were forced to construct a pseudophylogeny of the groups involving shorebirds as ancestors, and the results were surprising. As I have pointed out, the primary requisites for filter-feeding are: an enlarged tongue, a broad bill, and straining lamellae. However, in groups previously thought to be allied with flamingos, such as the storks, the tongue is rudimentary (Gardner 1925). And although it is difficult to envision the chicken beak transmuting into a soft, lamellate duck bill, the chicken tongue, likewise, is little more than a bony structure covered with cornified flesh. By contrast, among the shorebirds, the recurvirostrids, both *Recurvirostra* and *Cladorhynchus* are highly adapted for feeding rapidly on small prey items, and the tongue, the first stage in the development of any filter-feeding apparatus, is enlarged and fleshy to facilitate movement of small prey backward to the throat.

The red phalarope (*Phalaropus fulicaria*) is an example of a shorebird in the beginning stages of the development of a filter-feeding apparatus. The other two species of phalarope have a diminutive, slender bill, but the red phalarope's bill is broad and deep, with wide, flexible margins, and narrow papillae that must serve as strainers project from the lower jaw. The tongue is correspondingly enlarged to fit in the expanded bill.

Red phalaropes feed very rapidly on small crustaceans and insects and have been observed to extract chironomid midge larvae from floating vegetation and to have strands of vegetation hanging from the mouth. Their stomachs almost always contain bits of vegetation apparently taken accidentally while feeding. Red phalaropes appear to be engaging in a type of primitive, straining type of filter-feeding, making them an excellent living example of the type of morphology that would have been found at an early stage in the evolution of filter-feeding in birds. Charadriiform birds typically have relatively weak and extremely flexible bills, and the adaptive radiation of the order is testimony to the morphological diversity that can derive from the shorebird feeding apparatus. By taking the shorebird bill a step further in evolution we can easily envision its transformation into the bills of either flamingos or ducks.

It is commonly observed that the bills of young flamingos are straight, not unlike the narrow, somewhat flexible bills found in such shorebirds as phalaropes. The ontogenetic changes in the flamingo jaw involve increasing the depth of the mandible and bending both jaws. In fact, if one considers the jaws of *Juncitarsus* and *Palaelodus* with the early Miocene flamingo *Phoenicopterus croizeti*, one can reconstruct an imaginary and quasi-phylogeny

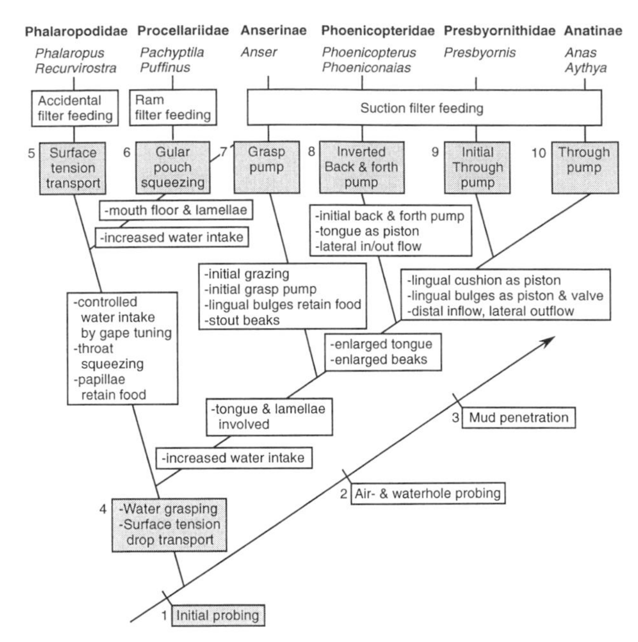

Array of myriad filter-feeding mechanisms and their probable evolutionary pathways, beginning with initial probing (*1*), illustrated by a wader that forages along shores. Levels *2* and *3* illustrate more specialized levels of probing. At *4*, water is grasped while the bird pecks at submerged food, and accidental filter-feeding can occur (*5*). Maximizing the water intake volume leads to two parallel lines of filter-feeding, one of gulping, in which mouth floor motions occur (*6*), and one of suction mechanisms, in which lingual motions are involved (*7–10*). In addition to the suction mechanisms, the filter capacity is maximized by the addition of lamellae along the tongue or the edges of the mandibles (*7–10*). (Modified after Zweers et al. 1995; courtesy Gart Zweers)

that is reflected, in general form, in the ontogeny of the modern flamingo bill. That the flamingo bill could derive from that of a stork seems completely improbable, whereas the transformation of a shorebird bill to a highly modified filter-feeding apparatus is highly probable, with an easily constructed pseudophylogeny. These same arguments apply equally to attempts to reconstruct a pseudophylogeny of the duck filter-feeding apparatus from that of galliform birds.

The bend in the bill of flamingos is certainly an interesting morphological feature, and it varies considerably within both living and fossil flamingos. In the palaelolids the bill is more or less straight, and the extinct flamingo *Phoenicopterus croizeti* from the late Oligocene to early Miocene of France and Germany has a bill that is straighter than in modern forms of the genus. Within modern flamingos, the lesser flamingo exhibits a strongly bent rostrum with a deep keel, whereas *Phoenicopterus* has a smaller bend and a shallow keel. The bent rostrum in flamingos provides space for the greatly enlarged tongue while allowing the tips of the jaws to come together, a necessity for flamingos to be able to pick up individual food

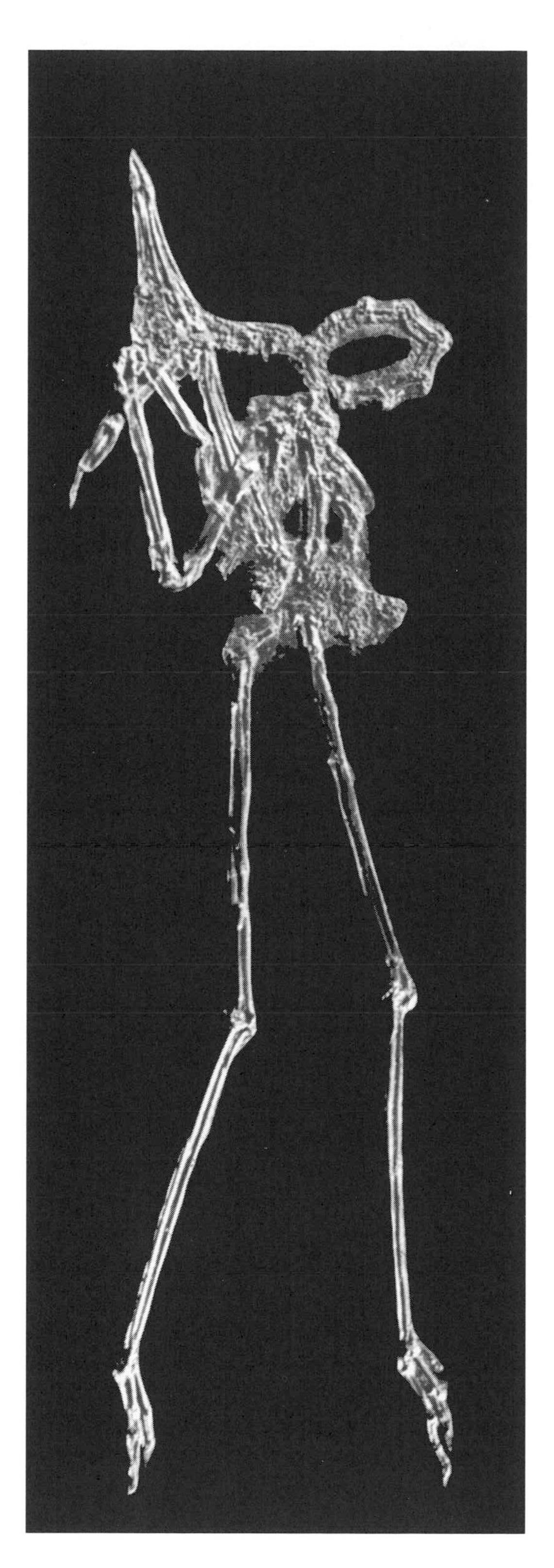

Juncitarsus merkeli, from the Middle Eocene of Germany, was described by Stefan Peters in 1987, who confirmed Storrs Olson and my observations (1980) that *Juncitarsus* was a shorebird-flamingo mosaic. Peters posited three valid subfamilies of flamingos: Juncitarsinae (stilt-flamingo mosaics), Palaelodinae (swimming flamingos), and the living Phoenicopterinae. × 0.33. (Courtesy S. Peters and Forschungsinstitut Senckenberg; private collection Kessler)

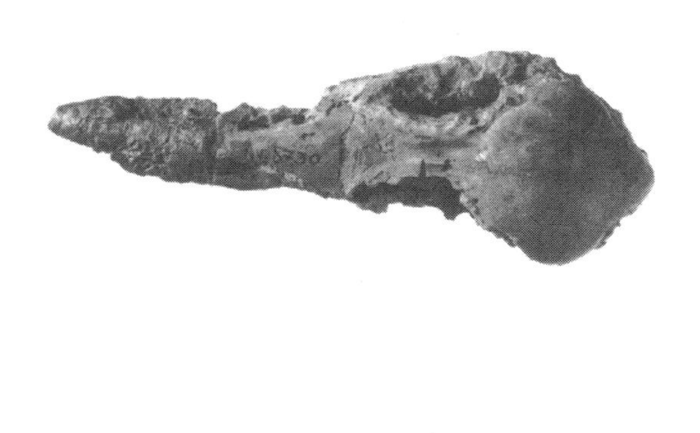

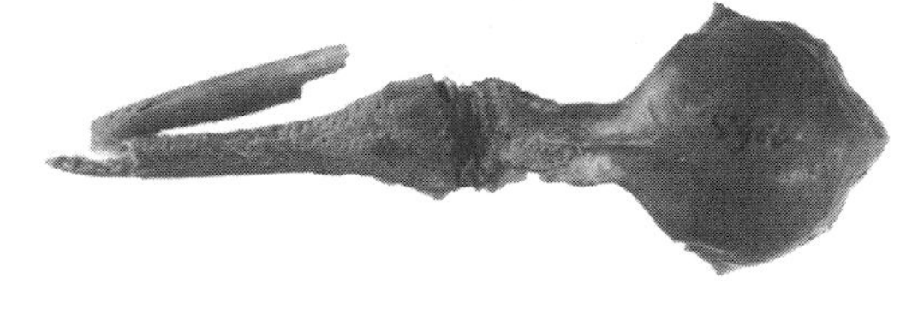

Skulls of two specimens of *Palaelodus ambiguus,* the Miocene swimming flamingo, in dorsal view, showing the straight bill of these widespread mid-Tertiary flamingos of the subfamily Palaelodinae. Jacques Cheneval and François Escuillié note that "In its general aspect, the bill looks very much like that of the recently described primitive flamingo-like bird *Juncitarsus merkeli*" (1992, 216). × 0.75. (Courtesy J. Cheneval)

items (Rooth 1965) and manipulate nesting materials. Perhaps more important, the bend in the bill allows the entire length of the gape to be opened with a minimum of jaw movement (Jenkin 1957). One would imagine that the bend must have arisen first for such functions and was modified only later into the more advanced filter-feeding apparatus found in such forms as the lesser flamingo.

All modern flamingos feed in highly saline waters, and as we have seen, all of the major finds of fossil flamingos for which paleoecological information is known would appear to have resided in the same type of habitat. This is true, for example, for the earliest fossil flamingo, *Juncitarsus*, and the swimming flamingos, *Palaelodus*. The same is certainly true for the highly colonial *Presbyornis*, a filter-feeding charadriiform near the ancestry of ducks, discussed below. These highly saline inland lakes, such as those seen today in Africa's Great Rift Valley, provide superabundant food in the form of blue-green and other algae blooms, as well as swarms of brine-adapted arthropods. But such an environment would favor selection of feeding devices that would prevent the ingestion of salt water, thus avoiding the energy costs associated with excreting salt. This stark environment would favor selection for the development of anatomical solutions to the problem of obtaining small prey from saline waters, and the specialized filtering mechanisms of today's ducks and flamingos might never have evolved under less difficult conditions. The ephemeral nature of the food supply in these stark and rapidly transforming environments would also inevitably have led the birds to adopt colonial behavior, and this is observed not only in the fossil flamingos *Juncitarsus* and *Palaelodus* and in *Presbyornis* but also in the banded stilt (*Cladorhynchus*) of Australia.

Presbyornis: A Shorebird-Duck Mosaic

The fossil record of most groups of vertebrates is replete with "missing links" that tell of interrelationships, but until recently there were no fossils that truly linked any orders of modern birds. Then in 1971, students and field parties under the direction of Paul O. McGrew of the University of Wyoming began to discover huge concentrations of associated skeletons of *Presbyornis*, a strange evolutionary mosaic with characteristics of several living orders of birds. This versatile filter-feeding bird, which probably had worldwide distribution, has allowed us to completely reevaluate the relations of several groups of modern birds, but particularly modern ducks. The initial *Presbyornis* fossils came from Lower Eocene deposits in the Green River Formation of southwestern Wyoming and southeastern Utah, and date back roughly 50 million years. The study of *Presbyornis* planted the idea that shorebirds are the basic ancestral stock for both flamingolike birds and the anseriforms, ducks and their allies, and that the long-legged waders, the traditional ciconiiforms, have little, if anything, to do with the flamingos or ducks (Feduccia 1976a, 1977b, 1978a; Olson and Feduccia 1980a, 1980b; Olson 1985a; Peters 1991). Their resemblance to flamingos, with their long legs for wading, is strictly superficial and the result of convergent evolution. In fact, even the various groups within the Ciconiiformes appear not to be closely related, and it is highly unlikely that herons have any close phylogenetic affinity with either the storks or the ibises (Olson 1979).

Dissociated skeletons of *Presbyornis* had been discovered long before McGrew's efforts, and were first described, in 1926, by Alexander Wetmore of the Smithsonian Institution as a new family of shorebirds close to the Recurvirostridae. It is an example of Wetmore's perspicacity that he properly identified the fossils as being of shorebird affinity. Even now, had the recently discovered skeletons not been associated, different elements of *Presbyornis* could be described as a duck, because of the skull and bill; a flamingo, because of features in the region of the frontal and nasal bones; or a shorebird, on the basis of certain bones, such as the humerus, tibiotarsus, and especially the tarsometatarsus.

When I first began to study *Presbyornis* in 1971, I tentatively classified it as a flamingolike wader because the postcranial bones were quite similar to those of flamingos. But I had wondered why Wetmore placed *Presbyornis* near the shorebird family Recurvirostridae, and on making the proper comparisons I was quick to discern that the postcranial anatomy of flamingos and recurvirostrid shorebirds is very similar, and distinctive from that of all other ciconiiform birds. This discovery led me to conclude that flamingos were related to and derived from ancient shorebird stock (Feduccia 1976a).

The grayish-green mudstone of McGrew's original *Presbyornis* quarry had, however, supplied more than postcranial bones. Many bills were unearthed, and, of equal interest, the braincases of a number of individuals were preserved with nasal and frontal bones intact. Both the bills and the braincases were quite ducklike, but the nasal and frontal bones were arranged in a V-shaped conformation found elsewhere only in modern flamingos, in modified form in flamingo adults, but almost exactly similar in their young. In fact, this region of the skull of *Presbyornis*, in size and morphology, is nearly identical to that of a thirty-day-old chick of the American flamingo. Thus the initial identification of *Presbyornis* as a flamingolike wader with ducklike features of the skull was unavoidable (Feduccia and McGrew 1974).

During the summer of 1975, McGrew and I col-

The original *Presbyornis* quarry explored by Paul McGrew and associates, north of Rock Springs, Wyoming, in Sweetwater County. This view shows the Laney escarpment above Parnell Creek in the background. The quarry itself was once part of a vast saline lake of the Eocene Period. The discovery of logs encrusted with algae indicates that this briny lake produced large algal blooms like those found today in the East African alkaline lakes. (Courtesy Paul O. McGrew)

Braincase (*center*) and nasofrontal region of *Presbyornis,* as discovered at the Canyon Creek locality. The constricted region in front of the hinge accommodated large nasal salt glands for excreting salt; salt glands are found in marine ducks and all living flamingos and occur in almost all other birds that frequent saline lakes or salt water. (Photo by author)

View of the Canyon Creek butte, about 91 m (300 ft.) high, in the southern part of Sweetwater County, Wyoming. This butte preserves part of the shoreline of a very large Eocene lake, perhaps the same lake that covered the original *Presbyornis* quarry, 152 km (95 mi.) to the north. Part of the Green River Formation, the butte is composed of a number of materials: mottled mudstone interbedded with varicolored sandstone resistant to erosion, some detrital material derived from the Unita Mountains to the southwest, and a few thin layers of tan algal limestone. Fossil bones of *Presbyornis* project in vast numbers from a thin layer below the rim of the butte. The author is shown examining the *Presbyornis* zone in one of the few places where it can be reached, in the summer of 1975. (Courtesy Paul O. McGrew)

lected at another Eocene locality about 150 kilometers (95 mi.) south of the original quarry, a resistant steamboat-shaped butte near the Colorado-Wyoming border called the Canyon Creek butte. This enormous butte preserves part of the shoreline of an Eocene lake, a shoreline that was apparently teeming with the highly colonial *Presbyornis*. In this locality the bills were much better preserved, and we began to see that *Presbyornis* was even more duck-like than we had thought.

Since that time I have collaborated with Storrs Olson on the relationships and evolution of ducks and flamingos (Olson and Feduccia 1980a, 1980b). More recently, a complete monographic phylogenetic analysis of *Presbyornis* and allies has been undertaken by Per Ericson, ornithologist at the Stockholm Museum of National Antiquities. This work has made it clear that the flamingolike features of *Presbyornis* are in fact primitive shorebird features and therefore that flamingos derived from ancient shorebirds. The discovery in 1977 of a completely articulated skull of *Presbyornis* at Canyon Creek butte made the affinity between *Presbyornis* and ducks even clearer. Collected by Olson and Robert J. Emry of the Smithsonian Institution, this skull is so similar in overall architecture to duck skulls, with only a few flamingo characteristics, most of them found in certain primitive ducks, that it erases all doubts that *Presbyornis* was on the line leading to ducks and not to modern flamingos.

The evolutionary choice in the ancient highly colonial filter-feeding shorebird ancestors was apparently whether the tongue would be accommodated by the lower jaw, as in flamingos, or whether it would be accommodated by the upper jaw, as in ducks. The choice of the lower jaw taken by flamingos (and by baleen whales convergently) led to a highly specialized feeding apparatus that allowed very little adaptive radiation, restricting flamingos largely to alkaline lakes with algal blooms. By contrast, the anseriform bill is much more versatile, being used not only for filter-feeding but in the case of geese for cropping or grazing and in some ducks for seizing fish.

The Paleoenvironment of *Presbyornis*. What we know about *Presbyornis* and its paleoenvironment allows us to reconstruct hypothetical stages that may have occurred in the evolution of modern ducks. The habitat of *Presbyornis*, the Green River Lake System of Wyoming and Utah, underwent dramatic changes over geological time that presented highly variable selective forces. This lake system came into existence in the early Eocene and occupied thousands of square kilometers. During a period known as the Wilkins Peak regression, the water in the area became very saline, and vast amounts of trona, or natron, a white or yellowish-white hydrated form of sodium carbonate, were deposited. This suggests that the environment was very much like that of the lakes of the Great Rift Valley, the only place where trona is being deposited in large quantities today. Like the most primitive living ducks, *Presbyornis* filter-fed primarily on the large algal blooms and other vegetable matter in alkaline lakes. Early Paleogene flamingos were like the more primitive living greater flamingo and fed primarily on small animal matter, thus permitting the nearly simultaneous evolution of two different filter-feeders in the same restricted, ephemeral environment.

Presbyornis was highly colonial; literally thousands of bones protrude from the coarse sandstones of the enormous Canyon Creek butte, showing that *Presbyornis* flocked in numbers comparable to those seen today in the flamingo lakes of East Africa. In another locality north of the Canyon Creek butte, large numbers of *Presbyornis* fossils have been found along with an abundance of eggshells, indicating that a large nesting colony lived near the shoreline of a lake. The fossil of a probable predator on *Presbyornis* chicks has also been found, the primitive frigatebirdlike *Limnofregata azygosternon*, a Green River bird that, like *Presbyornis*, could not have been properly identified without the discovery of an associated skeleton (Olson 1977b).

Fossils also provide evidence that a tropical climate characterized most of North America in the Eocene. Abundant remains of tropical plants, including palm trees, litter the lake deposits, as do the remains of crocodiles and soft-shelled turtles. Some other important finds are the more than thirty-five species of fossil fish that have been recovered from the Laney Shale and Tipton Tongue of the Green River Formation. Many of these are marine derivatives, and they include a stingray (*Heliobatis*), herrings (especially *Knightia*), and a number of primitive perchlike species. However, primary freshwater species also have been found, fish that have no tolerance for any salinity, including three species of strictly freshwater catfish, the chondrostean paddlefish (*Crossopholis*), the holostean bowfin (*Amia*), and the mooneye (*Eohiodon*), which probably gained access to the Eocene lakes by freshwater connections (Grande 1980). We know from geological evidence that the Green River Lakes underwent major changes in size and in the position of shorelines and deltas; obviously, they must also have varied from being entirely freshwater.

Stages in Duck Filter-Feeding. From what we know of the ancient Green River System and of the modern Great Rift Valley, it is not difficult to imagine the scenario in which the filter-feeding mechanisms of either

The first articulated skull of *Presbyornis,* recovered in 1977 from the Canyon Creek butte by Storrs Olson and Robert J. Emry. The slightly upturned bill is not an artifact of preservation; all of the many bills now recovered exhibit the same recurved characteristic. Skull length, 90 mm (3.54 in.). (From Olson and Feduccia 1980a)

Slab as removed from the *Presbyornis* quarry at Canyon Creek Butte, Wyoming, showing ducklike bills in association with long bones. No other kinds of bills and no ducklike postcranial elements occur at this site, and there can be no question that the shorebirdlike body of *Presbyornis* is correctly associated with a ducklike head. Scale = 4 cm. (Courtesy Storrs Olson and Per Ericson)

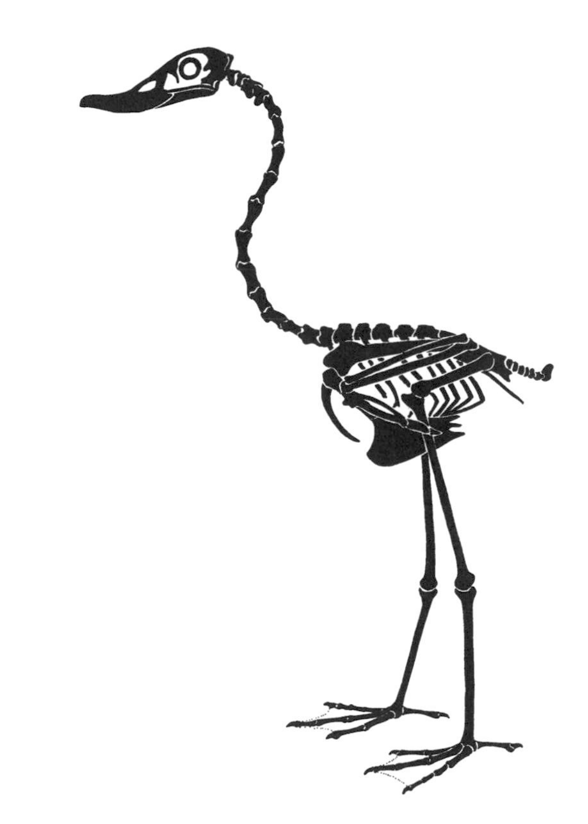

Provisional reconstruction of the skeleton of *Presbyornis* by Storrs Olson, showing overall proportions and the ducklike head on a shorebird body. (From Olson and Feduccia 1980a)

Fossils from the Green River Formation equivalent in quality of preservation to the Solnhofen limestone. *Upper left to lower right,* stingray, undescribed, about 26 cm (10 in.); frog (*Eopelobates* sp.), about 6 cm (2.4 in.); perchlike *Mioplosus labracoides,* about 12 cm (4.7 in.) swallowing the herringlike *Knightia humilis;* dragonfly (Libellulidae), about 5 cm (2 in.); crayfish (*Procambarus primaerus*), about 5.5 cm (2.2 in.); prawn (*Bechleja rostrata*), without antennae, about 8.3 cm (3.3 in.). (Photos by Ed Gerken; courtesy Black Hills Institute of Geological Research, Inc.)

ducks or flamingos could have evolved. The essential elements are highly colonial stiltlike shorebirds occurring on ancient lakes with the unstable salinity of those of the Green River System. Strong selective forces would have favored the evolution of long legs modified for wading into deeper water and, eventually, along with a myriad of other adaptations, the perfection of a suction filter-feeding mechanism that required great modification of either the lower or upper jaw to accommodate a large tongue.

Obviously, though, the great variety of feeding zones provided by the fluctuating lakes of the Eocene led also to different diversifications of the feeding apparatus. Adaptations leading toward ducks were surely favored in many circumstances. *Presbyornis* already had a ducklike bill, quite similar to the primitive Australian freckled duck (*Stictonetta naevosa*), and with only minor modifications it could have diverged to give rise to the beak of modern anseriforms. We can easily imagine also that some lakes had very narrow shorelines, and in such situations it would have been advantageous for *Presbyornis* or a *Presbyornis*-like bird to begin to perfect swimming adaptations leading to a duck prototype. Alteration from *Presbyornis* to modern anseriform structure would have involved mainly shortening of the leg bones to yield a limb with a more efficient swimming stroke.

The skull of *Presbyornis* is unmistakably ducklike; yet it has a peculiar upturned bill, with a very small rostral nail, a long, slender mandibular symphysis, and a deep groove in the ventral surface of the anterior portion of the mandibular rami. Among living ducks, these features are characteristic of two monotypic Australian genera, the freckled duck (*Stictonetta naevosa*) and the pink-eared duck (*Malacorhynchus membranaceus*). Of all living ducks, only *Stictonetta* has the peculiar upturned bill with a very small rostral nail and a long, narrow mandibular symphysis. Although highly specialized in its feeding apparatus, only *Malacorhynchus* among living ducks has the same grooves on the ventral surface of the mandibular rami. All in all, the skull of *Stictonetta* is remarkably similar to that of *Presbyornis*, and not surprisingly, this little-known duck is now thought to be a very primitive member of the Anseriformes. Indeed, Australian waterfowl expert Harry Frith remarked of *Stictonetta* that "few would deny that the waterfowl family of today passed, in evolution, through duck-like ancestors before differentiating into swans, geese, and others. The Freckled Duck is probably the closest living waterfowl to that ancestor" (1967, 114).

With this background, and because *Stictonetta* is the only living duck with a bill substantially similar to that of *Presbyornis*, it seems logical to examine the feeding behavior of the little-known freckled duck to gain insight into the original anseriform feeding adaptation. Here are some of Frith's observations of the freckled duck:

> Bottom filtering is by far the most common method of feeding; the birds wade slowly in shallow water, seldom more than two inches deep. The bill is immersed and is held immediately above the soil surface and a rapid filtering action is set up so that the fine particles of mud on the surface swirl up. . . . The bill has not yet been seen to enter the mud itself, but is maintained immediately above it—a true filtering, rather than a dabbling, action.
>
> Filtering of surface water is the least frequently seen. . . . The birds swim slowly, filtering and nibbling at surface particles. Captive birds spend a great deal of time running their bills along the edges of logs and posts and concrete walls in the water that have become encrusted with algae. The bill action is nibbling and they have been clearly seen to be feeding extensively on the algae. One of the most constant sources of food was algae, nearly every stomach examined contained some and it accounted for 30 percent of the total volume; seeds of smartweeds and docks were also found in most stomachs and provided 22 percent of the food. Various aquatic grasses . . . accounted for 16 percent of the food. . . . The whole bulk of the animal food accounted for 11 percent of the total food. (1967, 118–119)

That these ducks generally feed by wading rather than swimming and that they use the bill in a true filter-feeding capacity, with their principal food being algae, are of course interesting pieces of evidence. In conjunction with these findings, Paul McGrew and I noted of the original *Presbyornis* locality that an "interesting feature of this fossil site is the occurrence of many logs and branches that are heavily encrusted, presumably by algae. . . . The encrustations demonstrate the abundance of algae in the waters. . . . it seems reasonable that [*Presbyornis* was] feeding on algae and microorganisms" (1973, 164).

Using this model, the anseriform filter-feeding mechanism may originally have arisen as an adaptation for feeding on vegetable matter, particularly algae. On the other hand, the primitive living flamingos are adapted to feed on small invertebrates and only later became obligate algal strainers. Perhaps the adaptation to two different food sources was the factor that permitted the nearly simultaneous evolution of two groups of highly derived filter-feeding shorebirds in the same limited, ephemeral environment.

Screamers: Enigmatic Anseriforms. The relations of the strange birds known as screamers, family Anhimidae, endemic to South America, have long been debated, and they have been thought to represent a primitive form

Freckled duck (*Stictonetta naevosa*), a primitive anseriform of Australia that shares characteristics with dabbling ducks and geese. Interestingly, the structure of the bill of *Stictonetta* is the closest match for that of *Presbyornis* among the living anseriforms. (Photo by B. Gadsby/VIREO)

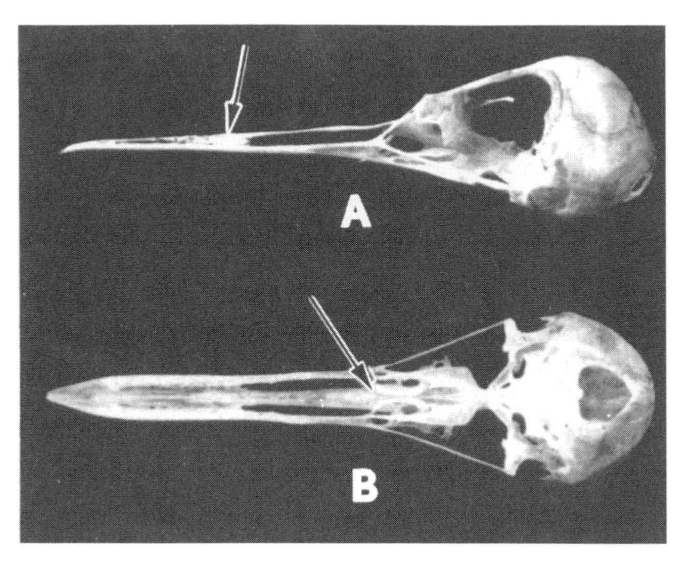

Lateral (*A*) and ventral (*B*) views of the skull of a red phalarope (*Phalaropus fulicarius*), a charadriiform with rudimentary filter-feeding adaptations. By ossification of the nostrils and the maxillopalatine area (arrows), a ducklike aspect could be derived from such a bill. The nostrils of *Phalaropus* have in fact already begun such ossification. (From Olson and Feduccia 1980a)

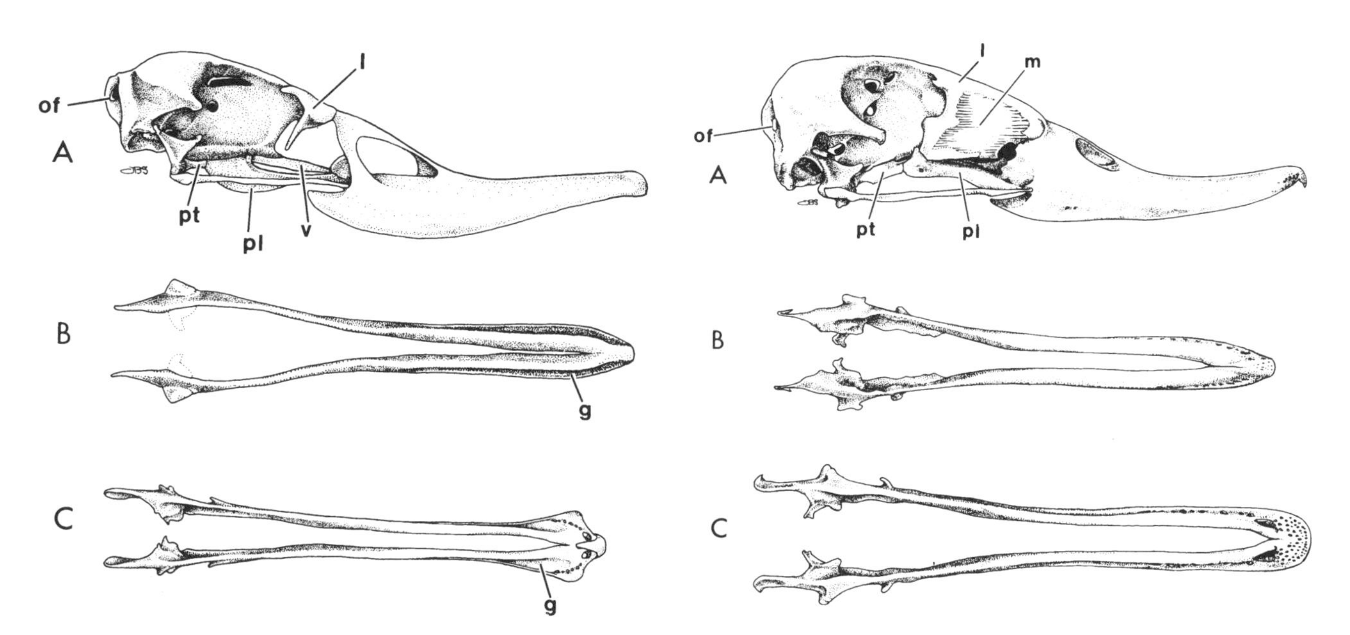

Left, lateral view of the skull (*A*) and ventral view of mandible (*B*) of *Presbyornis*. Note the characteristic groove (*g*) in the mandibular ramus. This is also present in (*C*) the extant pink-eared duck (*Malacorhynchus membranaceus*). *Right,* lateral view of skull (*A*) and ventral view of mandible (*B*) of the freckled duck (*Stictonetta naevosa*). Note the upturned bill and the long, narrow mandibular symphysis, as in *Presbyornis,* contrasted with the mandible of a typical duck (*C*), the mallard (*Anas platyrhynchos*). Abbreviations: *l,* lacrimal; *m,* lacrimal membrane; *of,* occipital fontanelle; *pl,* palatine; *pt,* pterygoid; *v,* vomer. (From Olson and Feduccia 1980a)

that perhaps links the anseriforms and galliforms. These turkey-sized marsh-dwellers occur in tropical and subtropical lowlands. The screamer's bill in superficial silhouette resembles that of a chicken, but that is as far as the similarity extends. Storrs Olson and I (1980b) have pointed out that there is considerable evidence to link the screamers to the ducks because of their morphological and behavioral similarities to the Australian magpie goose (*Anseranas semipalmata*), which has long been recognized as distinctive (Woolfenden 1961). We pointed out that although *Anseranas* differs from the Anhimidae in many respects, "it certainly represents a morphological stage similar to that through which the Anhimidae must have passed in becoming increasingly terrestrial (as shown by the reduction of webbing in the feet) and in relinquishing filtering as a means of feeding (as shown by the strong nail and reduced lamellae of the bill)" (1980b, 7–8). Indeed, although the modern screamers are not filter-feeders, they retain vestigial lamellae in the inside of the jaw, thus proving that they are highly derived descendants of some initially filter-feeding anseriform that probably went through a stage similar in morphology to the magpie goose, which has a long, unelevated hallux, very long toes with greatly reduced webbing, and an anhimid-like bill with the lamellae much reduced. The screamers share virtually no similarities with the Galliformes and certainly do not support a "duck-chicken" relationship.

Fossil Anseriforms. The fossil record of anseriforms begins, as we have seen, with the duck-shorebird mosaic *Presbyornis pervetus* and a slightly smaller species from the Paleocene of Utah and Mongolia (Olson 1985a, 171). The similar but distinctive contemporaneous genus *Telmabates*, from Patagonia, apparently differs sufficiently in limb proportions to be retained as a separate genus (Ericson, pers. comm.); the type species *Telmabates antiquus* is considerably larger than *Presbyornis pervetus*. In addition, Storrs Olson (1994) has described a much larger species, *P. isoni*, from the Upper Paleocene Aquia Formation of Maryland, dating somewhere between 61 and 62 million years ago, but there is no skull and it could be quite distinctive from the ducklike *Presbyornis* we know.

Peter Houde of New Mexico State University has discovered a fossil "screamer" from the Eocene Green River Formation and Michael Daniels has found them in the Lower Eocene London Clay, proving not only that these birds are early anseriform derivatives but also that they were once much more widespread. Like so many other birds, today they are relictually confined to the South American continent. According to Houde (pers. comm.), the fossil is represented by a nearly complete skeleton, but the distal tibiotarsus and proximal tarsometatarsus are missing so the length of its hindlimb is unknown. Houde writes:

> It exhibits cranial characters uniquely derived in Anseriformes. I refer to the fossil as a "screamer" because it is an anseriform with a fowl-like bill, but it is truly intermediate between Anatidae and Anhimidae, and it could be a member of either or neither. Parsimony analysis . . . suggests that the "screamer" is actually sister to the Anatidae, exclusive of *Anseranas* [the magpie goose]. Postcranially the fossil, like *Presbyornis*, is similar to Charadriiformes. Since *Presbyornis* and the new fossil have bill morphologies characteristic of the two main clades of Anseriformes, then their shared postcranial morphology must be the ancestral state for Anseriformes. The skeleton differs markedly from that of modern Anhimidae in that it is non-pneumatic, with the notable exception of a series of gargantuan pneumatic foramina on the sides of the thoracic and first sacral vertebrae. The vertebral foramina may be the only derived characters uniting the fossil with modern Anhimidae. (Pers. comm.)

As for the fossil record of anatids, the first supposed duck, *Eonessa*, was identified on the basis of unreliable elements, two fragmentary wing bones from the Eocene of North America, and Michael Davis includes waterfowl in his listing of Lower Eocene London Clay birds. However, the first diagnosed duck fossils are from the Oligocene, about 10 million years later in the Middle Tertiary, where the genera *Romainvillia* and *Cygnopterus* have been found in early Oligocene deposits in France and Belgium. Jacques Cheneval (1983a, 1983b, 1984) has recorded three species of whistling ducks (*Dendrochen*) and a new species of *Cygnopterus*, a genus closely related to swans (Anserinae), from the Aquitanian of France, from the deposits of Saint-Gérand-le-Puy. In reviewing all the Anatidae from Miocene French sites, Cheneval concluded, "It is possible to say that there was a change of avifauna during the Middle Miocene. During the Oligocene and the Lower Miocene only species of Dendrocygninae and Anserinae are described, then species of Anatinae appear during the Middle Miocene. So the hypothesis according to which two successive adaptive radiations existed can be proposed; during the Middle Miocene, the Anatinae extend as the Dendrocygninae disappear except in tropical regions where it is still possible to find them now" (1987, 138).

Bradley Livezey and Larry Martin (1988) have examined the systematic position of the French Miocene anatid *Anas blanchardi* and have placed it, along with its close relatives *A. consobrina* and *A. natator*, in a new

View of the inside of the mouth of the southern screamer (*Chauna torquata,* family Anhimidae), of southern South America, showing the row of vestigial lamellae (arrow) on the inside of the upper jaw. (From Olson and Feduccia 1980a)

The northern screamer (*Chauna charvaria*) of northern Columbia and Venezuela, one of two species of *Chauna*. The screamers represent a very aberrant and highly derived offshoot of the anseriforms, with only a trace of webbing in the feet. Thought by some to be a link between the anseriforms and galliforms because of their superficially chickenlike bill, the screamers are no doubt derived from ancient anseriforms resembling the magpie goose (*Anseranas semipalmata*) of Australia. Screamers are strong fliers, and their wings are "armed" with two sharp and very large spurs. Living in marshes and wet grasslands, screamers feed exclusively on water plants. (Photo by author)

The magpie goose (*Anseranas semipalmata*) resembles a quasi-morphological stage through which the screamers are thought to have passed in transition to their highly derived structure. (Courtesy Serendip Sanctuary, Victoria, and Rory O'Brien)

The horned screamer (*Anhima cornuta)* represents the second genus of the "screamers." *Anhima* is a monotypic genus. Note the long toes and semi-webbed feet, adapted for walking in marshes and on mats of floating vegetation; horned screamers occasionally swim. Note also the large wing spur on the forward edge of the manus. (Drawing by George Miksch Sutton)

Black-bellied whistling-duck (*Dendrocygna autumnalis*), family Dendrocygnidae, thought to represent a primitive stage in the evolution of modern waterfowl. (Photo by A. and E. Morris/VIREO)

genus, *Mionetta*, with *Dendrochen* as a sister-genus. More generally they have concluded that "*Mionetta blanchardi* evidently was a small 'duck-like' member of the relatively primitive grade of anatids that includes the whistling ducks (*Dendrocygna*), geese and swans (Anserinae), white-backed duck (*Thalassornis*), and freckled duck (*Stictonetta*), and it probably represents an extinct branch that diverged after *Dendrocygna* but before *Stictonetta*" (208).

The North American fossil anseriforms include a duck, *Paranyroca magna*, from the early Miocene of South Dakota (Miller and Compton 1939). Ducks are noticeably rare in deposits of Paleogene age, however, though they

Ducks, geese, and swans (Anatidae, worldwide, 43 genera, 148 species): snow goose (*Anser caerulescens*). The bill of most waterfowl is a broad, flat, lamellate structure designed for straining microscopic food, but there is much variety: mergansers have serrated bills for capturing fish, and geese have highly modified bills adapted for terrestrial grazing and pulling aquatic plants from shallow water. (Drawing by George Miksch Sutton)

become extremely common in the Neogene. In fact, in the relatively recent North American freshwater deposits of the Pliocene and Pleistocene ducks are often the dominant group of birds, demonstrating not only that they were common but also that they are easily preserved, much like the duck-billed dinosaurs of the Cretaceous that may make up to 90 percent of dinosaurs in certain localities. The lack of early Tertiary anseriform fossils therefore could well indicate that ducks and allies may have originated in the medial Tertiary, after the Eocene, for the proposal that anatids arose as early as the Cretaceous would render the absence of duck fossils throughout the Paleogene almost incomprehensible. *Presbyornis*, as we have seen, embodies many of the morphological features that one would associate with the ancestry of modern ducks, and it occurs in the Paleocene and Eocene, probably before actual transitions occurred. Even though *Presbyornis* was somewhat specialized for a particular mode of life, there is really nothing in morphology or time to preclude it from the actual ancestry of modern anseriform birds.

Storrs Olson has proposed that the "principal radiation of modern tribes and genera of Anatidae took place by the Miocene" (1985a, 187). Ducks from the early Miocene of France have been assigned to the whistling-ducks of the genus *Dendrochen*, subfamily Dendrocygninae, and by the medial Miocene the modern fish-eating ducks of the genus *Mergus* are present (Alvarez and Olson 1978). Another interesting find is that the modern shelducks of the tribe Tadornini, genus *Tadorna*, distributed more or less worldwide today, are known from the Middle Miocene of Germany, as well as from the North American Pleistocene (Olson 1985a). The relations of modern dabbling ducks of the tribe Anatini has been a particularly thorny issue, but Bradley Livezey (1991) has attempted to clarify many of the more persistent problems in an extensive phylogenetic analysis of the group based on a comparative morphology.

The main phylogenetic proposals to date for the evolution of both flamingos and ducks do not mention shorebirds as possible ancestors. The primary candidates for the ancestry of flamingos have been either ducks or the ciconiiforms, whatever they are in terms of relationships. In the case of ducks, one can simply ask, how could their

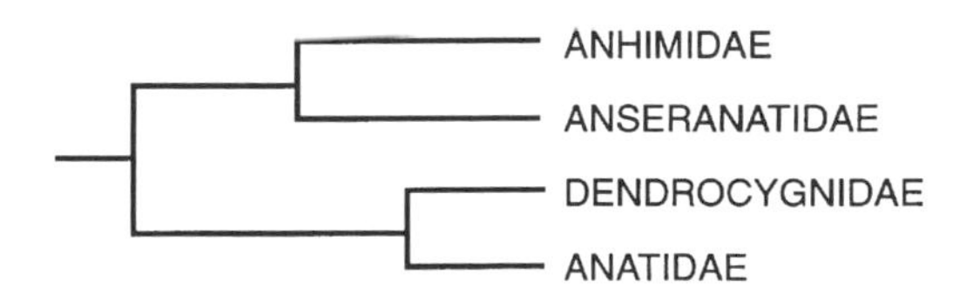

Phylogenetic tree indicating general relations of the major families of anseriform birds. (Modified after Mourer-Chauviré 1992)

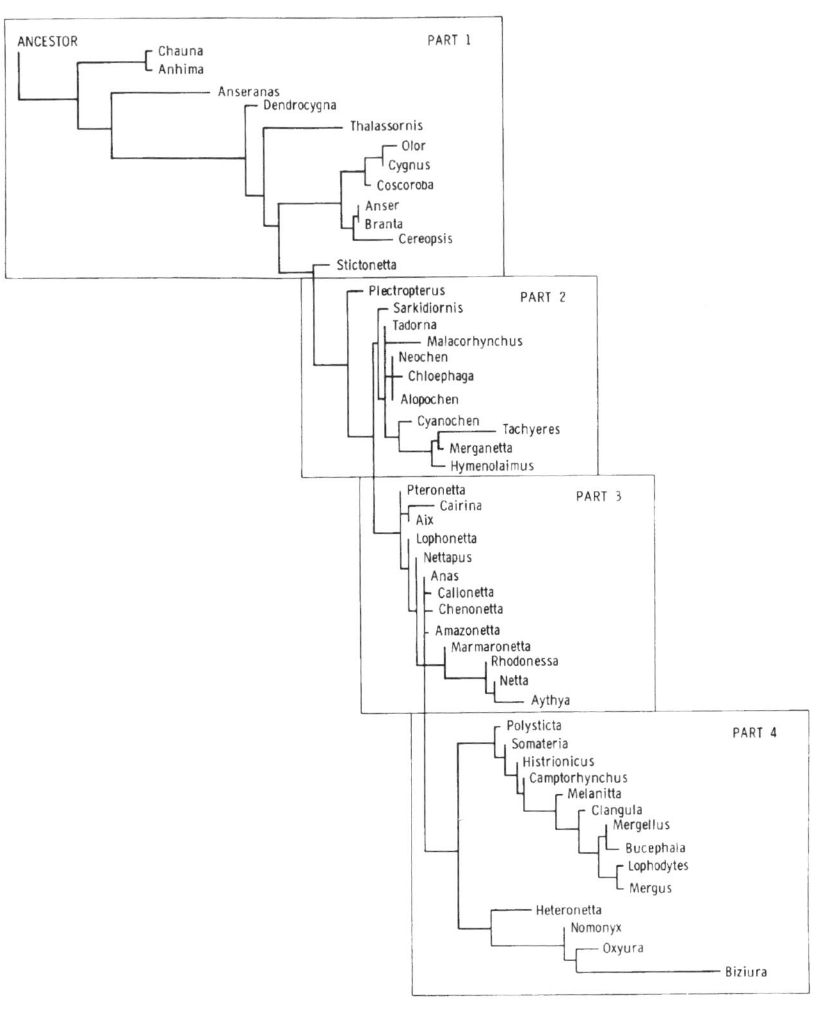

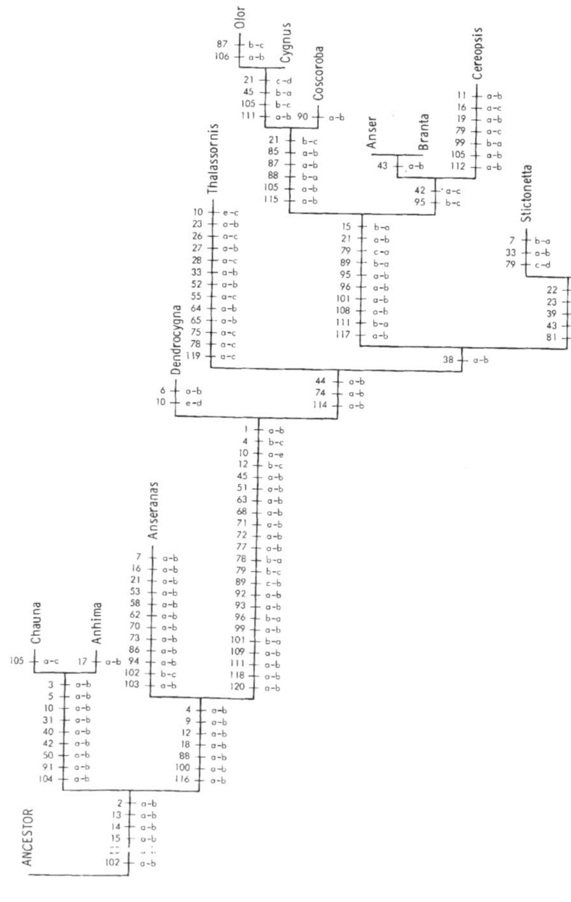

Above, phylogenetic tree of Recent anseriform genera and selected subgenera based on 120 morphological characters. Lengths of horizontal lines correspond to the number of character changes (apomorphies) in the lineages. *Below,* detailed diagram of Part 1 of the phylogenetic tree, highlighting the position of the more primitive living anseriforms, the screamers (*Chauna* and *Anhima*), magpie goose (*Anseranas*), whistling-ducks (*Dendrocygna*), and freckled duck (*Stictonetta*). (From Livezey 1986; courtesy *Auk*)

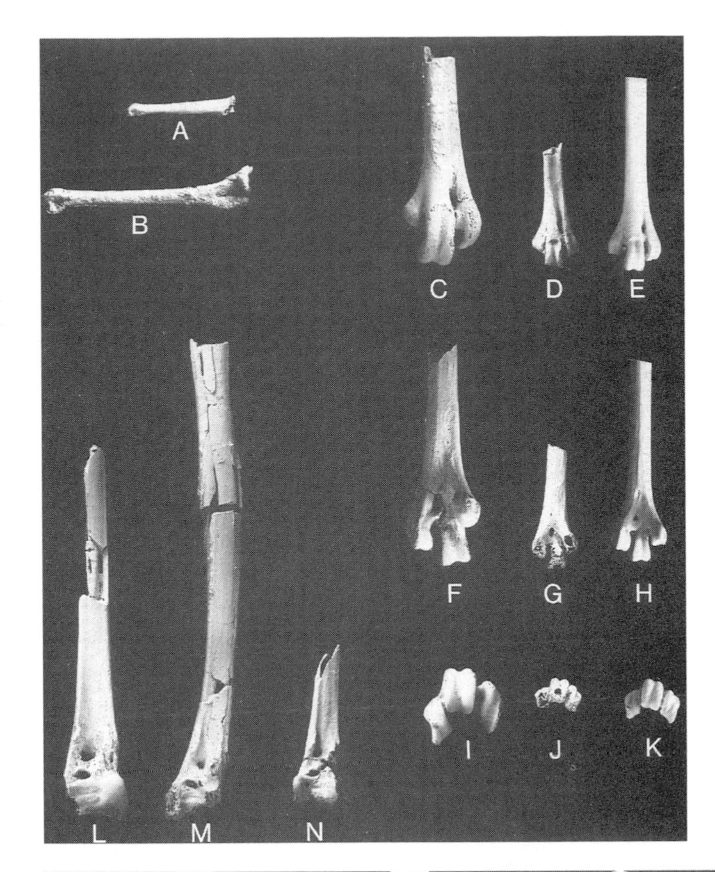

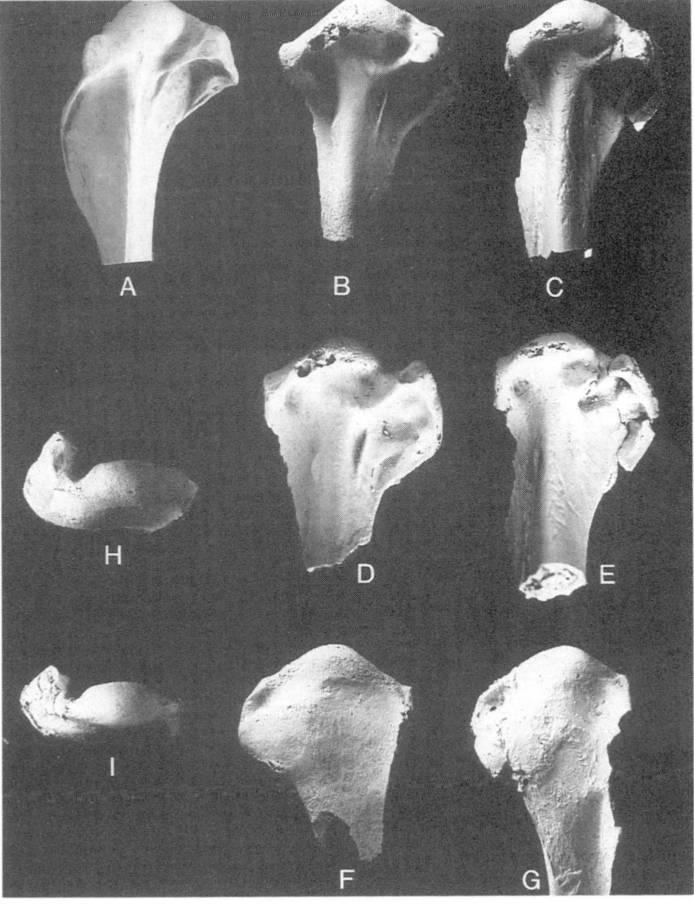

Transitional charadriiforms, represented here by the scant evidence provided by hindlimb elements from the early Paleocene (Danian) greensands of southern New Jersey (earlier thought to be late Cretaceous). All are tentatively referred to the "form family" Graculavidae, and most species belong to a group of primitive Charadriiformes ("transitional shorebirds") resembling in limb morphology the fossil family Presbyornithidae and the living Burhinidae. *Above: A, B,* right pedal phalanx of *Telmatornis* and *Presbyornis,* respectively; *C–K,* views of tarsometatarsus of *Presbyornis* (*C, F, I*), *Telmatornis* (*D, G, J*), and *Burhinus* (*E, H, K*); *L–N,* distal portions of tibiotarsi of *Palaeotringa* (*L, M*), and *Telmatornis* (*N*). *Below,* proximal ends of left humeri of *Graculavus* and related birds: *A, Esacus* (Burhinidae); *B, D, F, H, Graculavus*; *C, E, G, I, Presbyornis*. (From Olson and Parris 1987)

complicated, highly derived filter-feeding apparatus be transformed into the most derived skull among birds—namely, that of flamingos? In the case of storks, it is almost inconceivable that an intricate filter-feeding apparatus could have evolved from the stiff pointed bill of storks or storklike ancestors. The same holds true for anatid ancestry, where the most commonly held view is they evolved from some galliformlike ancestor.

Cort Madsen, Kevin McHugh, and Siwo de Kloet attempted a partial classification of waterfowl based on single-copy DNA of thirteen species of anseriforms. They concluded interestingly that "the Magpie Goose (*Anseranas*) lineage diverged very early . . . and the whistling-ducks (Dendrocygnini) . . . diverged somewhat more recently from the main lineage," as well as that "*Stictonetta* is only distantly related to the other Anatidae and that its lineage must have diverged early from the main lineage leading to other waterfowl." They found the screamers to be "distantly related to the Anatidae" (1988, 457).

The DNA-DNA hybridization studies by Sibley and Ahlquist (1990) produce a branching pattern that agrees fairly well with that of Madsen and co-workers, except for the position of *Anseranas*, the magpie goose, which comes out, concordant with the morphological picture, close to the screamers, Anhimidae. Here again, however, when we look at the broader picture of ordinal relations, the biochemical data and DNA comparisons do not fit the morphology of *Presbyornis*—namely, the head of a duck on the postcranial skeleton of a shorebird. Sibley and Ahlquist have supported the commonly held view that the Anseriformes and Galliformes shared a common ancestor "but that the divergence between them occurred in the late Cretaceous or early Tertiary" (1990, 311). From the standpoint of morphology very different conclusions have emerged, but many workers see a close alliance of galliforms and anseriforms, the latest Evgeny Kurochkin (1995), who segregates all modern birds into the subgroupings Galloanserae and Neoaves.

The Shorebird Radiation

Although difficult to trace, the various lineages of modern birds descended from ancient shorebird stock that transcended the Cretaceous-Tertiary bottleneck are probably legion. Storrs Olson (1985a) has suggested that two divisions can be recognized within the shorebird assemblage, or order Charadriiformes. The ancient groups, the "transitional shorebirds," include the late Cretaceous and early Paleogene shorebirds (collectively under the name Graculavidae), as well as the Presbyornithidae. However, these early presbyornithids were no doubt attired with the head of a shorebird rather than the duck head of the Eocene *Presbyornis pervetus* discussed in this chapter. Many of the late Cretaceous and early Paleogene shorebirds show some similarity to the living thick-knees, the Burhinidae, but it is not known whether these ancient forms are closely related to modern thick-knees. The transitional shorebirds also include some ibis-like birds that share many features of both shorebirds and gruiforms. These transitional shorebirds (as opposed to "higher," modern charadriiforms) form "a disparate amalgam of archaic birds that lack certain typical charadriiform features such as the expanded ectepicondylar spur of the humerus found in most of the 'higher' families (except Jacanidae), yet possess other characters . . . that are seldom or never met with in the Gruiformes" (Olson 1985a, 169).

As mentioned, the thick-knees are strange birds that belong within the higher Charadriiformes but share many postcranial features with the ancient shorebirds, the Graculavidae, and thus with the Presbyornithidae. Unfortunately, the fossil record of thick-knees is sparse, and the oldest fossil and only Tertiary species is *Burhinus lucorum*, from the Miocene of Nebraska (Bickart 1982).

The oldest described ibis, *Rhynchaeites messelensis*, is from the Middle Eocene Messel oil shales of Germany; interestingly, it was twice identified as a member of the Charadriiformes until, with the study of new material, Stefan Peters (1983) was able to show that it was an ibis. Peters considered that the shorebird features of *Rhynchaeites* provided additional evidence corroborating the hypothesis that ibises "are close to the groups from which Gruiformes and Charadriiformes arose" (26). Michael Davis has also discovered an ibis in the Lower Eocene London Clay.

Also included within the "transitional" Charadriiformes are the ancient shorebirds of the family Presbyornithidae, which, as we have seen, gave rise to ducks and show many similarities to the ancient Cretaceous and early Paleogene shorebirds of the family Graculavidae. The graculavid, shorebird genera, previously identified as shorebirds, rails, and cormorants, include *Cimolopteryx*, *Graculavus*, *Telmatornis*, and *Laornis*.

Most of the graculavid late Cretaceous or early Paleogene fossil birds represented by single elements that do not belong to the hesperornithiform birds are ancient shorebirds, and some, such as *Telmatornis*, are strikingly similar to the living shorebirds of the genus *Burhinus* and postcranial bones of the fossil *Presbyornis* (Cracraft 1972; Olson and Parris 1987). The several species of *Telmatornis*, all from the Paleocene of New Jersey (Olson 1994b), coexisted with a fairly diverse avifauna, indicating an early Tertiary divergence of many orders of birds.

The shorebirds, or "higher" Charadriiformes (suborder Charadrii), are known in abundance from the Lower Eocene London Clay and quickly diversified. Today they

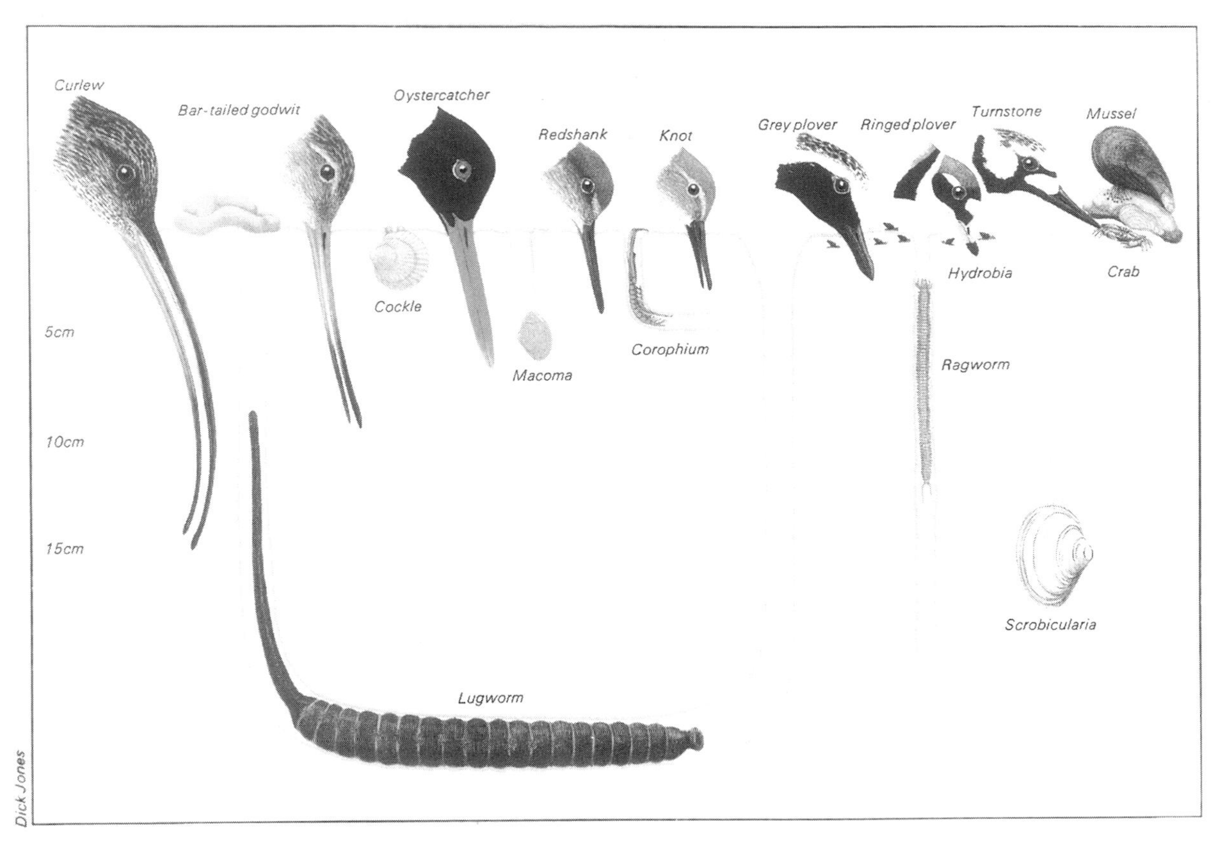

Modern shorebird adaptive radiation has resulted in varied bill lengths with differing types and degrees of rhynchokinesis, whereby the dorsal ridge of the bill has double-hinged axes, bending at the base and near the tip, enabling long-billed waders to grasp worms, insects, and crustaceans without opening the whole beak when they probe in soft sediments (Zusi 1984). For example, the woodcock (*Scolopax*), an earthworm specialist, can open just the tip of its bill to grasp a worm deep in the soft substrate. Having bills of varied lengths enable shorebird species to probe in various depths of soft sediments and sand for food. Plovers feed on small invertebrates, mainly by pecking at the surface with their short bills, whereas turnstones flip tidal stones. Other waders, such as knots and redshanks, with moderate-length bills probe the top 4 cm (1.6 in.) of the shoreline substratum, which contains worms, bivalves, and crustacea. Oystercatchers are specialists in prying open limpets and other shelled invertebrates, and such long-billed shorebirds as curlews and godwits can reach deep-burrowing prey, such as lugworms and ragworms. (From Goss-Custard 1975; courtesy Royal Society for the Protection of Birds)

Common snipe (*Gallinago gallinago*) showing how the tip of the upper jaw can be lifted separately (distal rhynchokinesis). (Photo by Johann Waskala, in Bühler 1980)

The shorebirds or waders of the suborder Charadrii of the order Charadriiformes consist of some 14 families (depending on the authority) containing about 222 species in 55 genera. Members of the Charadrii feed primarily by wading in shallow water or along the edge of the water and probing for small fauna in mud and soft sediments. They are distributed worldwide. *From left to right and top to bottom* (*row 1*), Jacanas (Jacanidae, pantropical, 6 genera, 8 species): northern jacana (*Jacana spinosa*). Painted-snipes (Rostratulidae, nearly pantropical, 1 genus, 2 species): greater painted-snipe (*Rostratula benghalensis*). Sandpipers—woodcocks, snipes, sandpipers, phalaropes, turnstones, etc. (Scolopacidae, worldwide, 21 genera, 88 species): whimbrel (*Numenius phaeopus*) and (*row 2*) red-necked phalarope (*Phalaropus labatus*). Crab plover (Dromadidae, Indian Ocean, 1 genus, 1 species): crab plover (*Dromas ardeola*). Sheathbills (Chionididae, subantarctic, 1 genus, 2 species): snowy sheathbill (*Chionis alba*). (*Row 3*) Plains-wanderer (Pedionomidae, Australia, 1 genus, 1 species): plains-wanderer (*Pedionomus torquatus*). Seedsnipes (Thinocoridae, temperate South America, 2 genera, 4 species): grey-breasted seedsnipe (*Thinocorus orbignyianus*). Oystercatchers (Haematopodidae, worldwide, 1 genus, 11 species): American oystercatcher (*Haematopus palliatus*). (*Row 4*) Avocets and stilts (Recurvirostridae, all continents, 3 genera, 10 species): black-necked stilt (*Himantopus mexicanus*). Thick-knees (Burhinidae, warm climates worldwide, except North America, 1 genus, 9 species): Eurasian thick-knee (*Burhinus oedicnemus*). Pratincoles and coursers (Glareolidae, warm climates Old World, 5 genera, 17 species): collared pratincole (*Glareola pratincola*). Plovers and lapwings (Charadriidae, worldwide, 11 genera, 66 species): dotterel (*Eudromias morinellus*). Not illustrated: Magellenic plover (Pluvianellidae, Patagonia, 1 genus, 1 species); Ibisbill (Ibidorhynchidae, Asia, 1 genus, 1 species). (Drawings by George Miksch Sutton)

Silhouettes, *left,* sandgrouse (Columbiformes, Pteroclidae), and *right,* cursor (Charadriiformes, Glareolidae). The sandgrouse, in a quasi-evolutionary sequence proposed by Jon Fjeldså (1976), lead by way of the cursors back to shorebird stock. (Drawings by Sigrid K. James)

Sandgrouse (Pteroclidae, Old World, 2 genera, 16 species): Tibetan sandgrouse (*Syrrhaptes tibetanus*), 41 cm (16 in.). Sandgrouse are thought to be intermediate between Charadriiformes and Columbiformes, and recent DNA evidence (Sibley and Ahlquist 1990) indicates that they may be shorebirds. They occur in arid regions: their feet are feathered for walking on sand, and their breast feathers are adapted for carrying water back to the nest. (Drawing by George Miksch Sutton)

are represented by some 349 species in some eighteen families falling roughly into three groups: the true shorebirds, such as sandpipers, plovers, and their wading allies; the skuas, gulls, and terns (larids); and the auks, murres, and puffins (alcids), the latter two groups discussed with seabirds in chapter 4. The adaptive radiation of modern shorebirds has been characterized primarily by varying adaptations associated with their feeding apparatus (Burton 1974a). Their bony nostrils are schizorhinal (see chapter 1), and they have a form of cranial kinesis known as rhynchokinesis that allows precise movement of the tip of the bill for probing into mud of soft sediments in search of small food items (Zusi 1984; Bühler 1987). Joseph Strauch (1978; see also Chu 1995) performed a phylogenetic analysis of shorebirds based on morphological characters and was able to discern three distinctive suborders; Joseph Jehl's (1968a) earlier study of relations in the Charadrii based on color patterns of the downy young agreed roughly with more recent DNA analyses. Sibley and Ahlquist's (1990) DNA-DNA hybridization studies led them to divide the shorebirds into two parvorders, Scolopacida and Charadriida.

The order Gruiformes (cranes, rails, and similar marsh-dwelling waders) is clearly an ancient order that is closest to the Charadriiformes, and the two orders probably shared a common ancestor derived from ancient shorebird stock. Because they have given rise to so many flightless derivatives they are considered in chapter 6. In addition to the 214 or so living species of gruiforms, classified in eleven or more extremely diverse families, ancient gruiforms gave rise to many other birds, both living and extinct. Possible additional shorebird derivatives include the sandgrouse (Fjeldså 1976), doves, and pigeons (Columbiformes) (Stegmann 1969; Fjeldså 1976; Hoch 1980), and, through the toothed-billed pigeons, the parrots (Psittaciformes) (Bock 1969), but these transitions are far more speculative.

Ella Hoch saw in bones of a shorebird-ibis mosaic, *Plumumida* (= *Rhynchaeites*, p. 229), she redescribed from the Messel oil shales of Germany, "a form 'on the line' from early shorebirds to doves" (1980, 47), and although debate continues on the exact nature of a shorebird-dove connection, there is now little doubt that sandgrouse are shorebird derivatives. The xerophilous sandgrouse (Pteroclidae) are adapted to life in the arid and semiarid conditions of Africa, Madagascar, southern Europe, and Asia. Sibley and Ahlquist's (1990) DNA comparisons indicate that the sandgrouse are more closely related to the charadriiforms than to the columbiforms.

Cécile Mourer-Chauviré (1992a) restudied the fossil sandgrouse material from the Paleogene of Quercy (Upper Eocene to Middle Oligocene), originally described by Milne-Edwards (1892) as two new species of the Recent genus *Pterocles,* and concluded that they are primitive members of the group; she assigned them to a new genus, *Archaeoganga.* She also described a very large form, *Archaeoganga pinguis,* a species estimated to have been some three times larger than the largest Recent sandgrouse. In a more extensive review of all the Paleogene to Lower Miocene French sandgrouse, Mourer-Chauviré (1993) concluded that the form from Saint-Gérand-le-Puy (Upper Oligocene), *Leptoganga sepultus,* persists into Lower Miocene localities. In addition, the indications from the mammal and bird faunas with which sandgrouse are found conform to an open and arid paleoenvironment. She concluded that "in the Upper Eocene, the Pteroclidae were already completely individualized with respect to the Charadriiformes" (74).

Many now consider the loons to be shorebird deriv-

Herons, storks, and allies of the heterogeneous order Ciconiiformes are long-legged, long-necked wading birds that feed in shallow water or on open ground. There is little to ally the varied families phylogenetically, and the order may be polyphyletic. Herons are distinctive, holding the neck in an S-shaped curve. Herons are thought by some to belong to the Gruiformes (Olson 1979), and storks may be allied with New World vultures (Rea 1983). *Left to right and top to bottom,* Boat-billed heron (Cochleariidae, Neotropics, 1 genus, 1 species, often included in Ardeidae): boat-billed heron (*Cochlearius cochlearia*). Herons, bitterns, and egrets (Ardeidae, worldwide, 21 genera, 65 species, including boat-billed heron): great blue heron (*Ardea herodias*). Ibises and spoonbills (Threskiornithidae, pantropical, some temperate, 14 genera, 33 species): white ibis (*Eudocimus albus*). Storks (Ciconiidae, pantropical, temperate Eurasia, 6 genera, 19 species): white stork (*Ciconia ciconia*). Hamerkop (Scopidae, sub-Saharan Africa, Madagascar, 1 genus, 1 species): hamerkop (*Scopus umbretta*). Shoebill (Balaenicipitidae, sub-Saharan Africa, 1 genus, 1 species): shoebill (*Balaeniceps rex*). (Drawings by George Miksch Sutton)

atives, and the grebes, too, may well be of gruiform extraction, but their true ancestry is masked by extremes in swimming adaptations. Although the thought is highly conjectural, there is also really nothing to preclude a shorebird ancestry for the petrels and allies (Procellariiformes) and perhaps the pelicans and allies (Pelecaniformes). As we saw in chapter 4, the early Eocene putative, primitive tropicbird *Prophaethon* shows a number of shorebird as well as procellariiform features and, like *Limnofregata* and *Presbyornis*, fulfills the expectation of early Tertiary avian mosaics. Indeed, in many respects the shorebirds may well be comparable to the evolutionarily prolific mammalian insectivores in their evolutionary meanderings.

Ciconiiformes: Long-legged Waders

The only large group of water birds, except for the ibises, that shows no evidence of shorebird beginnings is the order Ciconiiformes, with which some avian paleontologists once connected the flamingos. A heterogeneous group of long-legged waders living in marshlands or shallow waters, these birds include the storks (Ciconiidae), the herons and bitterns (Ardeidae), the ibises and spoonbills (Threskiornithidae or Plataleidae), and two aberrant African forms, the hamerkop (Scopidae) and the shoebill (Balaenicipitidae). Although these groups have few specific characteristics in common, they are all long-necked, long-legged birds with somewhat rounded wings and comparatively short tails; most have long, spreading toes that are sometimes slightly webbed. The sexes are usually identical or quite similar in plumage. One of the few morphological characteristics that unites these birds is a particular conformation of the bones of the palate, in which the palatine bones are fused to each other along the midline of the skull; but this arrangement is also found in many other birds.

The family of storks is represented by nineteen species that occur throughout most of the warmer regions of the world. Those that nest in the temperate zone migrate great distances to reach their wintering grounds. Storks differ in many ways from other ciconiiform birds and have been shown by David Ligon (1967) to share many features with the New World vultures (Cathartidae), including the habit, when overheated, of cooling the body by excreting liquid feces on the legs, a process termed urohidrosis, which gives stork legs their whitish appearance. Philip Kahl (1963) has suggested that storks accomplish heat loss by rapid panting and evaporation of this liquid excreta. Storks have rather short toes that are partially webbed at the base, and because they lack muscles in the voice box, storks are mute and communicate by rattling their bills.

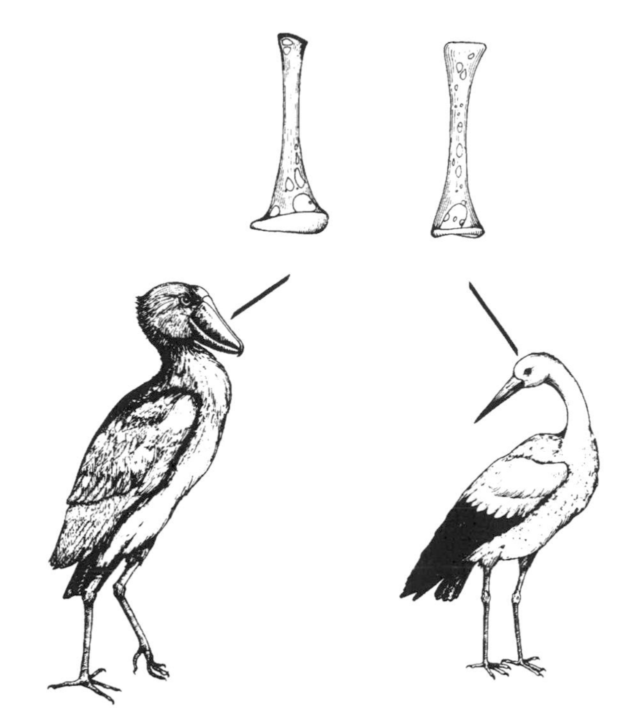

The tubular middle-ear bone, or stapes (proximal ends lower), shared by storks (here, the white stork, *Ciconia ciconia*), the shoebill (*Balaeniceps rex*), and, in modified form, the hamerkop (not shown). Stapes drawn to same scale. (From Feduccia 1977c; reprinted with permission from *Nature* 266:719, © 1977 Macmillan Journals Ltd.)

Painted stork (*Ibis leucocephalus*) facing the early morning sun in the delta-wing position, Bharatpur, India. Note the white legs, the result of expulsing liquid feces on the legs to cause cooling. This physiological trait is shared with New World vultures and is used to ally the two otherwise disparate groups (see also p. 302). (Photo courtesy M. P. Kahl/VIREO)

Shoebill (*Balaeniceps rex*). The large bill is adapted for feeding on catfish and lungfish in the murky waters of African marshes. (Photo by author, Berlin Zoo)

The earliest stork of reliable identity is *Palaeoephippiorhynchus dietrichi*, which is known from a skull and mandible from the early Oligocene of the Fayum series in Egypt; it is also the longest generic name in the class Aves (Olson 1985a). There are other fossil storks from the Tertiary of Europe, Asia, North America, and Africa, but for the most part, they are based on isolated bone fragments and tell us nothing of the relations of the Ciconiidae. Storks do, however, possess a distinctive tubular ear bone (Feduccia 1977c) that is also found in the shoebill and, in modified form, in the hamerkop, but this bone does not differ greatly from the ear bone of the pelecaniform birds.

Balaeniceps. The curious African shoebill (*Balaeniceps rex*), also called the whale-headed stork or shoebill, is now restricted to freshwater marshes in east-central Africa, where it is highly adapted for capturing large prey in heavy vegetation. Shoebills have large feet for walking on marshy vegetation, and their bills are adapted for feeding on catfish, and particularly the lungfish, which inhabit these mucky environments. Aside from having a storklike ear bone (Feduccia 1977c), the shoebill is also a bill-rattler and has historically been considered an aberrant stork. It was regarded by its original describer, John Gould (1852), as having pelecaniform affinities, and in a detailed paper in 1957, Patricia A. Cottam described several important features of the shoebill that indicate links to pelecaniforms. Storrs Olson has suggested that the shoebill, hamerkop, and storks "may be loosely interrelated and have affinities with the Pelecaniformes" (1979, 169).

Regardless of the ultimate answer to the ancestry of these birds, it is clear that storks and the shoebill share a number of characteristics with the pelecaniform birds. Fossil shoebills are known from the late Miocene of Tunisia (Rich 1972) and the early Oligocene of Egypt (Rasmussen, Olson, and Simons 1987), in the latter case, *Goliathia andrewsi*, in combination with an avifauna that included, among other species, jacanas and flamingos, which indicated "a tropical, swampy, vegetation-choked, fresh-water environment at the time of deposition" (Rasmussen, Olson, and Simons 1987, ii). From this same deposit these authors describe a new family of extinct ciconiiform birds, the Xenerodiopidae, based on a heavy decurved rostrum and a humerus (12–13).

Scopus. The hamerkop (*Scopus umbretta*), also known as the hammerhead, anvil-head, lightningbird, and hammer-headed stork, is a most unusual, medium-sized bird that has traditionally been placed in a monotypic family, Scopidae, within the Ciconiiformes. These storklike birds, which range widely in the Ethiopian and Malagasy regions, are generally held in awe by native peoples, who feel that harm will befall them if they molest the birds, and many African legends are built around hamerkops. They are particularly known for building gigantic nests of sticks, plastered with mud, with a side entrance. The nests are adorned with all sorts of decorative objects, from feathers to bones. There is but a single fossil species from the early Pliocene of South Africa, *Scopus xenopus* (Olson 1984a), which was slightly larger than the living species.

Herons and Ibises. The affinities of the herons, Ardeidae, and ibises, Threskiornithidae or Plataleidae, are another matter. Although there is a supposed ibis fossil from the Paleocene of North Dakota, it cannot be identified, and neither it nor the later fossil ibises that have been recovered tell us much about either the evolution of ibises or their relations to other avian groups. As noted above, the *Rhynchaeites messelensis* specimen from the Eocene of Germany was twice misidentified as a shorebird before being assigned to the ibis family and has given considerable credence to an ibis derivation from the Charadriiformes. Storrs Olson (1982b) reviewed the fossil ibises and concluded that except for the Messel bird, the next oldest fossil ibis is *Plegadis paganus* from the early Miocene of France. In addition, aside from a number of questionable species and some later ibises, based on rather fragmentary material, there are some Quaternary flightless insular ibises that are of great interest and are discussed in chapter 6.

Although superficially similar to other ciconiiforms,

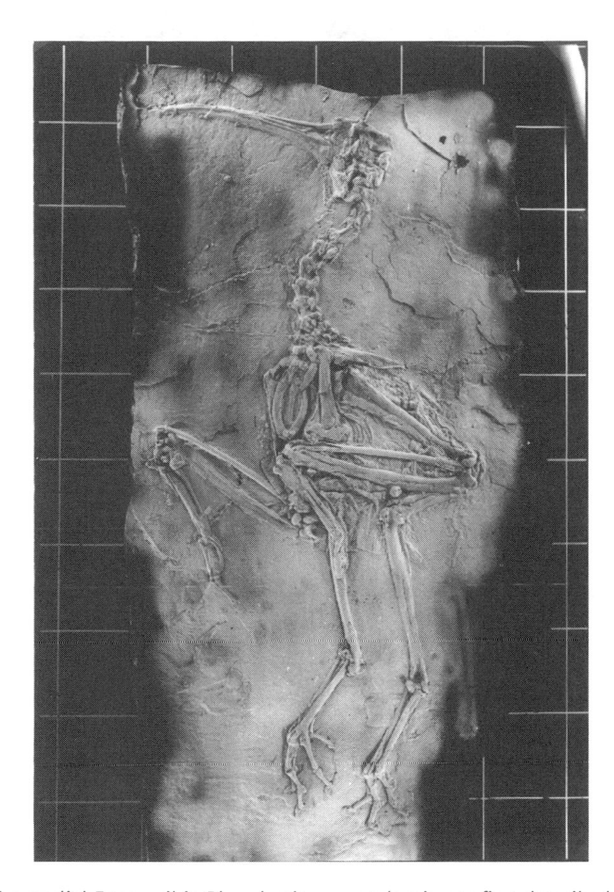

The medial Eocene ibis *Rhynchaeites messelensis* was first described in 1898 as intermediate between the shorebirds and gruiform birds. It was redescribed in 1980 under the name *Plumumida lutetialis,* and again similarities with plovers were emphasized. In 1983, Stefan Peters described the fossil a third time, this time indicating its true identity: an ibis. *Rhynchaeites,* however, does confirm the affinity of ibises to gruiform and charadriiform birds. × 0.40. (Photo courtesy S. Peters and Forschungsinstitut Senckenberg; private collection Bastelberger)

ibises differ from all other birds in the order in many features. Perhaps their most dramatic departure is their elongated schizorhinal nostrils, characteristic of shorebirds, in which the posterior margin forms a slit. The long, decurved bill of ibises is equipped with sensory pits at the tip, somewhat like those of curlews. As we have seen, there is extensive evidence, both from comparative morphology of extant birds and from the fossil record, to lend support to a shorebird derivation of the ibises.

Even more distinctive than ibises are the herons, which show almost no anatomical similarity with the other members of the Ciconiiformes. They superficially resemble the other families that are traditionally placed in the Ciconiiformes in being generally aquatic, long-legged, wading birds with long necks; and they differ from them in having, in most members, a spearlike beak, a pectinate middle toe claw, an S-shaped neck in flight, and often spectacular display feathers that are used in elaborate mating exhibitions. In addition to having very distinctive bones, herons have a highly unusual pattern of feather arrangement and powderdown patches, like those of some of the primitive gruiform birds. Olson has recently suggested that herons and bitterns "are the only currently successful group in an early radiation of primitive Gruiformes" (1979, 169). However, Frederick Sheldon (1987) of Louisiana State University performed extensive DNA-DNA hybridization analyses and concluded that "the sister groups of the herons appears to comprise other ciconiiform birds, and perhaps some Pelecaniformes as well" (107).

The evolutionary history within the herons was addressed by Robert Payne and Christopher Risley (1976), who extensively studied the morphology and other features of modern herons to provide a comparison and phenetic relations and phylogenetics within the family. Frederick Sheldon's study (1987) of the phylogeny of herons from DNA-DNA hybridization data supported Payne and Risley's linear arrangement. The strangest and most aberrant heron, the boat-billed heron (*Cochlearius cochlearius*), of the American tropics, has been treated as a separate family, Cochleariidae, but Walter Bock (1956) considered it simply an aberrant night heron. Yet Payne and Risley concluded that "the many unique behavioral and morphological features of *Cochlearius* indicate that it should be recognized as a distinctive tribe, Cochlearini, within the night herons, Nycticoracinae" (1976, 97).

A Fertile Field for Avian Systematists. Clearly, we are a long way from understanding the relations of the avian orders Gruiformes, Charadriiformes, and Ciconiiformes, as well as Galliformes (discussed in the next chapter), which have traditionally been considered to be closely allied with the ducks in a group often termed the "gallo-anseres." However, as we have seen, the postcranial morphology of the primitive ducks is that of a primitive shorebird, and as we shall see, evidence increasingly points to a sister-group relation of the Galliformes and the ratites, especially the tinamous. Another important question, as yet unaddressed, is whether the myriad Paleogene shorebird mosaics, such as *Juncitarsus* (shorebird-flamingo), *Presbyornis* (shorebird-duck), and *Rhynchaeites* (shorebird-ibis), are more closely allied with the Cretaceous transitional shorebirds and whether the modern shorebirds represent a distinctive clade from these late Cretaceous forms. If this turns out to be the case, it would explain much of the confusion concerning the biochemical evidence for relations. The phylogenies of the birds contained within the traditional orders Ciconiiformes, Gruiformes, Charadriiformes, and Galliformes remain a major challenge to evolutionary ornithologists.

The extinct New Zealand owlet-nightjar (*Aegotheles novaezealandiae*), largest member of the family Aegothelidae, was flightless or nearly so, exhibiting wing bones as big as modern species of owlet-nightjars, but with much larger legs. Known from numerous sites on both North and South Islands, New Zealand owlet-nightjars probably came out at night to feed on terrestrial insects and similar prey. The one genus and eight species of modern owlet-nightjars are distributed today in Australasia, with most forms occurring in Australia and New Guinea; there are no living New Zealand species. (Drawing by John P. O'Neill)

6

The Evolution of Flightlessness

Flight is an extremely taxing ability to maintain in terms of energy, both metabolic and embryogenic. If there is no strong and continual selection for the maintenance of the flight apparatus, it tends to disappear. In each case the reasons for the loss of flight may be quite distinctive, but the evolution of flightlessness is a pervasive avian phenomenon, occurring in both aquatic and terrestrial birds as diverse as geese and hoopoes, on continents, and especially on islands. Flightlessness occurred as early as the foot-propelled Lower Cretaceous divers of chapter 4, not too far removed temporally from *Archaeopteryx* itself, and as late as island rails that have adapted to flightlessness within the past few thousand years. Some flightless birds retain characteristics that tell of their ancestry from a specific avian group; others have become so greatly modified that we are left with no clue as to their origin or their possible interrelatedness. Yet the fact remains that flightlessness as a specific characteristic did evolve, and, as we shall see, the evidence is pervasive that all flightless birds have been derived from flying predecessors.

An Early Tertiary Experiment. One of the most sensational discoveries in the history of avian paleontology was the discovery by Edward Drinker Cope in 1876 of the giant terrestrial bird *Diatryma*, which stood slightly over 2 meters (6.5 ft.) tall, had a head nearly half a meter (1.5 ft.) long, and is estimated to have weighed some 175 kilos (385 lbs.) (Andors 1991, 1992). The name *Diatryma*, from the Greek, meaning "through a hole," refers to the large foramina (perforations) that penetrate some of the foot bones. A new era in the study of this enigmatic bird was issued in with the description in 1917 by paleontologist William D. Matthew and collector Walter Granger of a nearly complete skeleton, unveiling this "magnificent and quite unexpected bird skeleton," from 50-million-year-old Eocene deposits of Wyoming, in the *Bulletin of the American Museum of Natural History*. For Matthew and Granger, this fossil skeleton represented a giant predaceous carnivore: "*Diatryma* was a gigantic bird, ground-living and with vestigial wings. In bulk of body and limbs it equalled all but the largest of the moas and surpassed any living bird. . . . The height of the reconstructed skeleton is nearly seven feet [2.14 m]. The neck and head were totally unlike any living bird, the neck short and very massive, the head of enormous size with a huge compressed beak" (319).

For decades the prevailing belief was that at the beginning of the Age of Birds and Mammals, the Cenozoic era, about 65 million years ago, the ecological niche for a bipedal carnivore had been left vacant by the late Cretaceous extinction of the flesh-eating theropod dinosaurs. In the resulting absence of competition, giant, flightless birds adapted for a predatory existence were able to get the jump on mammals, which were just then coming out from under the domination of the dinosaurs and would not evolve into successful, advanced predators for another geologic epoch. For Matthew and Granger, the gigantic head and short, powerful neck of *Diatryma* identified it as a fierce, bipedal predator, perhaps like a diminutive version of *Tyrannosaurus*, with small, functionless forelimbs but powerful hindlimbs. As one author commented not too long ago, following the prevailing view of the past seventy years, "*Diatryma* must have kicked, clawed, and bitten its prey into submission" (Gould 1986b, 25).

Indeed, the late dean of modern vertebrate paleontology, Alfred Sherwood Romer of Harvard University's Museum of Comparative Zoology, wrote in his preeminent textbook on vertebrate paleontology in 1966: "The presence of this great bird at a time when mammals were, for the most part, of very small size (the contemporary horse was the size of a fox terrier) suggests some interesting possibilities—which never materialized. The great reptiles had died off, and the surface of the earth was open for conquest. As possible successors there were the mammals and the birds. The former succeeded in the conquest, but

Restoration of *Diatryma gigantea*. (From Heilmann 1926)

Skeleton of *Diatryma gigantea* from the early Eocene of New Mexico. (Courtesy Department of Library Services, American Museum of Natural History, neg. no. 321726)

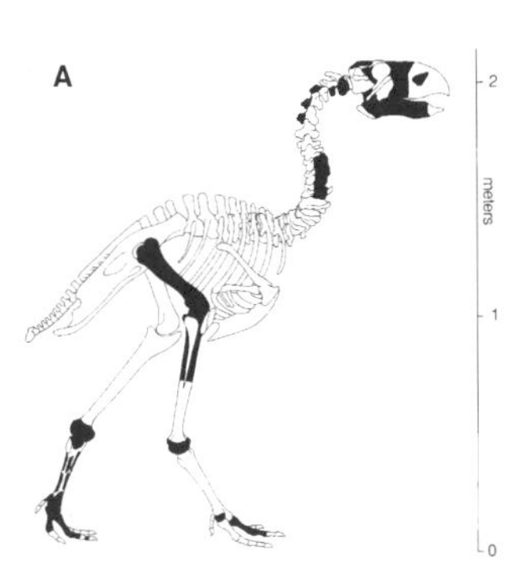

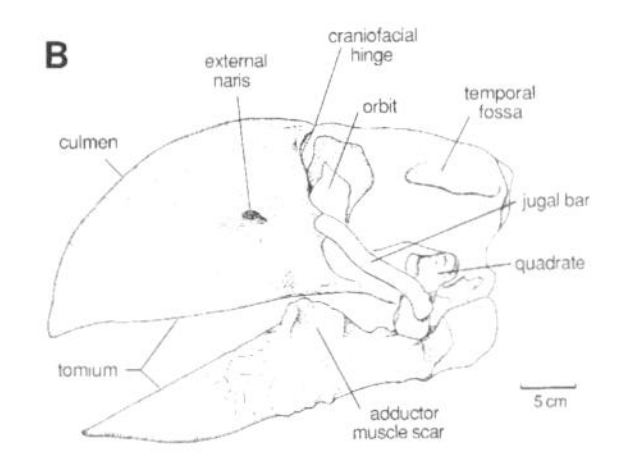

A, restoration of the skeleton of *Diatryma gigantea* (redrawn from Matthew and Granger 1917). Shaded elements are bones preserved in the new specimens studied by Witmer and Rose. *B,* restoration of the skull of *D. gigantea* in left lateral view based on the addition of new material. (From Witmer and Rose 1991; courtesy Witmer and *Paleobiology*)

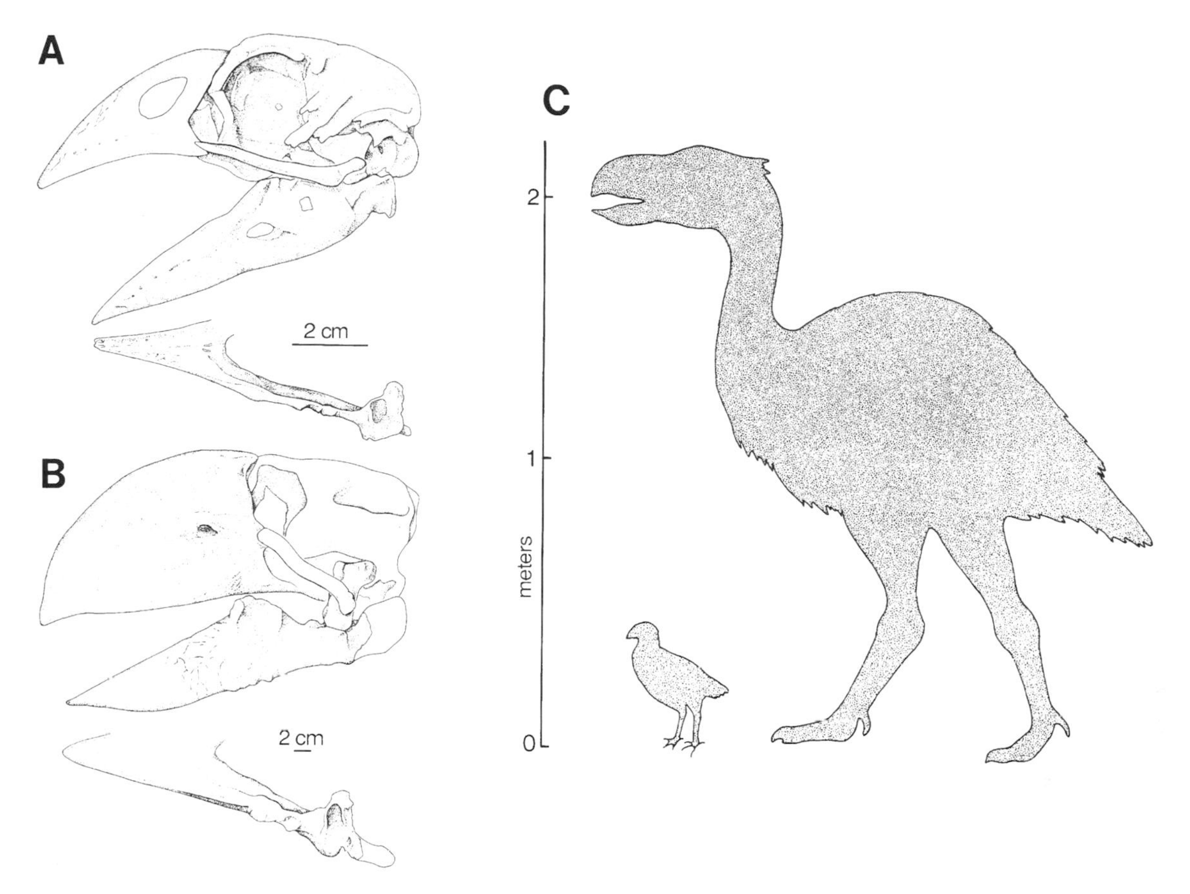

Comparison of *Diatryma gigantea* with a takahe (*Porphyrio mantelli*), a folivorous rail that has been advanced as a modern analogue for *Diatryma* by Watson (1976) and Andors (1988). Skulls in left lateral view and lower jaws in dorsal view. *A,* takahe; *B, Diatryma gigantea.* Although these skulls show numerous similarities when enlarged to unit length, the *Diatryma* skull is about five times as large as a takahe's. *C,* silhouettes of body shapes of *Diatryma* and a takahe to the same scale. (From Witmer and Rose 1991; courtesy Witmer and *Paleobiology*)

Skulls (*above*) and body silhouettes (*below*) of several Recent, subfossil, and fossil large ground birds. Each row is to the same scale. Elephant-birds (*Aepyornis*), moas (*Dinornis*), ostriches (*Struthio*), cassowaries (*Casuarius*), and rheas (*Rhea*) are ratites. All are mostly to completely herbivorous, and all have relatively small, lightly built skulls. *Diatryma* and phorusrhacids have decidedly larger, more robust skulls. Phorusrhacids are widely considered to have been carnivorous and predatory in habits. Scale marks = 1 m. (From Witmer and Rose 1991; courtesy Witmer and *Paleobiology*)

Skeletons of *Diatryma gigantea* (redrawn from Matthew and Granger 1917) and potential mammalian prey items from the Willwood fauna where *Diatryma* was recovered, drawn to same scale. (From Witmer and Rose 1991; courtesy Witmer and *Paleobiology*)

Life restoration of the *Diatryma gigantea* feeding on a carcass of the early horse *Hyracotherium* as an oxyaenid creodont (carnivorous mammal) looks on. According to Lawrence Witmer and Kenneth Rose, *Diatryma* was probably capable of both active predation and scavenging. (From Witmer and Rose 1991; courtesy Witmer and *Paleobiology*)

the appearance of such a form as *Diatryma* shows that the birds were, at the beginning, rivals of the mammals" (171). This view of *Diatryma* as a fierce carnivore continued to hold sway, Kurtén (1971, 48) labeling it a "terror crane," although several popular accounts portrayed it as a carrion-eater (Lanyon 1963) and as a herbivore (Watson 1976). George Watson was puzzled by the bird's lack of a hawk-like hook at the end of its bill. "I think *Diatryma* was a peaceful grazer, using its massive bill to scythe coarse vegetation—somewhat like *Nortornis*, largest of living rails and a voracious cropper of New Zealand's remote grasslands" (105).

A wealth of new information and analyses of these fantastic birds recently emerged. Allison Andors (1991, 1992) began an extensive study of all of the known specimens of *Diatryma* and its functional morphology and presented a completely different picture of *Diatryma*. In his view, *Diatryma* was a browsing herbivore and its habitat, deduced from depositional environments, was coastal lowlands and alluvial floodplains. Andors's biomechanical studies show *Diatryma* to have a more or less modern avian prokinetic skull with "rostral and mandibular tomia differentiated into an anterior seizing/cropping region and a posterior slicing/crushing region . . . interpreted as adaptations for folivory and paralleled in part by Recent avian folivores (including the largest rail, *Porphyrio mantelli*, and the largest parrot, *Strigops habroptilus* . . .), which have likewise lost or reduced their capacity to fly and have undergone correlated increase in body size" (1992, 117). Andors argued against the view of diatrymids as running predators (most recently revived by Witmer and Rose 1991) that "the weakly developed rostral hook, reduced flexor tubercles on the claws of the toes, and shortened tarsus seriously impeded [carnivory] and may have precluded it altogether" (117). He further pointed out that *Diatryma* had hindlimb proportions that approached those of various graviportal birds, which were adapted for supporting great body weight, such as the New Zealand moas and the Malagasy elephantbirds, and this, combined with the retention of a functional hallux and short, heavy toes, indicate that *Diatryma* moved slowly, in the manner of a bustard, and seldom ran.

By contrast, Witmer and Rose (1991) conclude from their functional analysis that *Diatryma* fits the classical picture of the past century: a somewhat slow but effective early Cenozoic carnivore. "*Diatryma* probably could run fast enough to catch most of the contemporary mammals, especially young, old, or sick individuals. . . . We suggest that *Diatryma* was carnivorous. It probably could pursue and kill live prey, could have scavenged carcasses, and may have been specialized as a bone crusher" (117).

Andors has reported many formerly unknown specimens; some fifty *Diatryma* fossils are now known from the Lower Eocene of New Mexico, Colorado, Wyoming, New Jersey, Ellesmere Island, and France and from the Middle Eocene Messel oil shales of Germany. Although there are comparatively few specimens from the Old World, the temporal range there is greater, from Lower to Middle Eocene, whereas in the New World the occurrences are broadly contemporaneous. The most productive *Diatryma* fossil deposits in North America are those of the Willwood Formation of northeastern Wyoming. As Andors points out, "The presence of flightless diatrymids on both sides of the present-day North Atlantic is *prima facie* evidence that Europe and North America were once connected by dry land" (1992, 119). Other Eocene birds and mammals of similar age have similar disjunct distributions, and "the geological and geophysical data are in basic accord with biologic evidence for the former existence of a land connection spanning the northern end of the Atlantic Ocean during part of the Early Eocene" (119).

In 1881 the French paleontologist V. Lemoine published a restored skeleton of another large bird from the Paleocene of France, which he named *Gastornis* and which many have thought belonged possibly with *Diatryma*. The two were placed in separate families, the Gastornithidae and Diatrymidae (order Gastornithiformes), however, and although most workers considered both groups to be related to the Gruiformes, possibly through the seriemas, in recent years Olson (1985a) and Andors (1988) have cast doubt on this assumption, and in 1992 Andors viewed the diatrymids as a sister-group of the Anseriformes. Larry Martin (1992) reexamined the old *Gastornis* fossils along with new fossil material and concluded that although *Gastornis* is very similar to *Diatryma*, it does not belong in the same genus. *Gastornis* fossils are known from the Upper Paleocene of France, Germany, Belgium, and England, and Hou (1980) has described a new genus of gastornithid, *Zhongyuanus*, from the early Eocene of Henan Province, China. As Andors notes, "The geochronologic ranges of the Gastornithidae and Diatrymidae overlap in the Early Eocene, with Paleocene-Eocene gastornithids ultimately being 'replaced' by Eocene diatrymids" (1992, 111).

The "Splendid Isolation" of South America. Approximately 100 million years ago, during the late Cretaceous, South America began to migrate westward, separating from Africa during the breakup of the continent Gondwana. Indeed, South America remained in perfect isolation until about 2.5 million years ago, when uplift in the northern Andes and a drop in sea level produced the Panamanian land bridge, which connected North America to South America and set in motion perhaps the most dra-

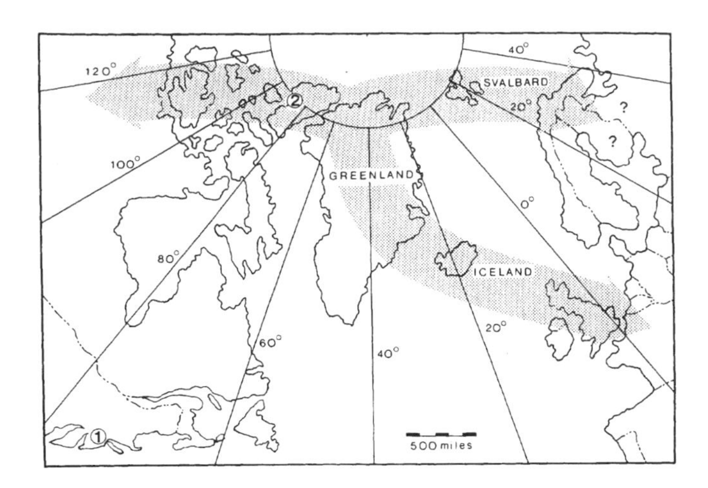

Diatryma-bearing localities and potential early Eocene terrestrial dispersal routes in the North Atlantic region. Continents and islands are shown in their present positions to facilitate recognition of geography: (1) Manasquan or Shark River Formation, Atlantic Coastal Plain, eastern New Jersey; (2) Eureka Sound Group, eastern Sverdrup Basin, west-central Ellesmere Island, Northwest Territories. (From Andors 1992; courtesy *Los Angeles County Museum of Natural History, Science Series* 36)

Epochs		N.A. Mammal Ages	Ma	Ranges of Gastornithiformes	Standard Ages
Eocene	Late	Duchesnean	40		Priabonian
	Middle	Uintan	45		Bartonian
		Bridgerian	50		Lutetian
	Early	Wasatchian	55	*Diatryma* [N.A.]; *Diatryma* [Europe]; *Gastornis* [Europe]; *Zhongyuanus* [Asia]	Ypresian
Paleocene	Late	Clarkforkian			Selandian / Thanetian
		Tiffanian	60		
		----?----			
	E.	Torrejonian			Danian

Geochronologic ranges of the genera of Gastornithiformes. Ranges indicated by broken lines are questionable, inferred, or imprecisely known. (From Andors 1992; courtesy *Los Angeles County Museum of Natural History, Science Series* 36)

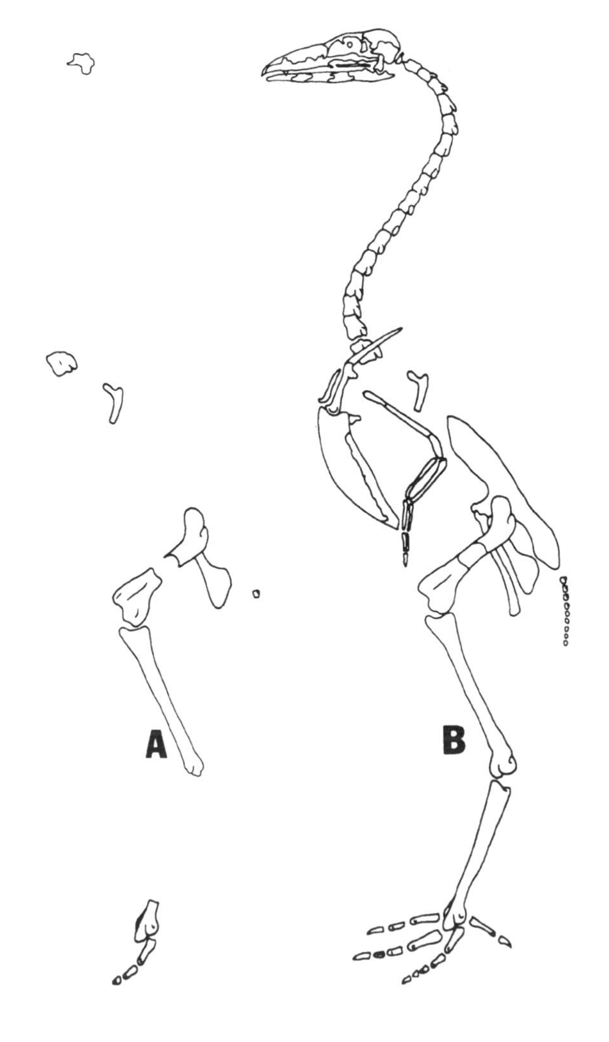

Gastornis edwardsii. A, correctly identified avian material in V. Lemoine's restoration of 1881; *B,* Lemoine's restoration. (From Martin 1992; courtesy *Los Angeles County Museum of Natural History, Science Series* 36)

Representatives of the families of land mammals that participated in the Great American Interchange. Animals shown in North America migrated northward from South America (and vice versa), beginning about 2.5 million years ago, after the appearance of the Panamanian land bridge. The flightless phorusrhacids likewise ventured northward from their splendid isolation in South America during the Tertiary. (From Marshall et al. 1982; copyright 1982 AAAS; drawing by M. H. Werner, courtesy AAAS)

matic biotic interchange of the earth's recent history. This extraordinary event, resulting from a buildup in the polar ice caps and concomitant drop in sea level by about 50 meters (164 ft.), is known as the Great American Interchange (Webb 1976, 1991; Marshall et al. 1979, 1982; Marshall 1988, 1994). The interchange was first recognized by Darwin's contemporary and competitor in the race for the theory of evolution by natural selection, Alfred Russell Wallace (1876), but not until recently was the extent of the phenomenon understood fully. The most dramatic events occurred in the mammalian exchange, South American archaic mammals, primarily marsupials, having evolved in isolation from advanced mammalian competitors in a manner quite analogous to those of Australia. With the coming of the land bridge, entire groups of South American mammals became extinct, and today some half of the families and genera of South America belong to mammal groups that emigrated from North America during the interchange.

South America's geographic isolation from advanced mammalian carnivores during most of the Cenozoic probably led to the evolution of a group of carnivorous flightless birds, the phorusrhacoids, within the suborder Cariamae, which today comprises two species of seriema (Cariamidae), a component of early land bird radiation surviving today as relicts in South America. These ancient carnivores are known from the late Paleocene (Alvarenga 1985a) and early Eocene, but primarily from the Oligocene (Alvarenga 1982; Tonni 1980b) to the close of the Pliocene. Their latest occurrence coincides with the Great American Interchange, and their extinction was presumably due to the invasion into South America of North American flesh-eating eutherian (placental) mammals (Marshall 1988, 1994).

The birds are typified by the genus *Phorusrhacos*, and this name has an odd meaning and history. The birds were originally described by the Argentincan paleontologist Florentino Ameghino in 1887, who believed he was looking at the jaw of an edentate, a ground sloth; he gave the creature the rather burdensome name of *Phorusrhacos*, which probably means "branch." In 1891, however, Ameghino realized that the jaw was avian, and for reasons unknown, he simply amended the name to *Phororhacos*, which stuck for some time (see Ameghino 1895; Andrews 1899). (Ameghino was responsible for naming most of the phorusrhacids of South America.) Today, however, following nomenclatorial rules of priority, we use the generic name *Phorusrhacos*.

In 1960, Bryan Patterson of Harvard University's Museum of Comparative Zoology and Jorge L. Kraglievich of the Municipal Museum of Mar del Plata, Argentina, revised and classified the various phorusrhacids. They placed them into three families that included forms of medium, large, and gigantic sizes, phorusrhacids ranging in height from approximately 1 to 3 meters (3.3–9.8 ft.). Larry Marshall reviewed the discovery and radiation of the phorusrhacoids in 1978 and 1994, terming them the "terror birds" (a previous name for them was "thunderbirds"; Kurtén 1971). In light of many recent discoveries, these birds are in need of revision and are under study by Herculano Alvarenga of Taubaté, Brazil.

The dozen or so species of phorusrhacids, now all placed in the family Phorusrhacidae by French paleornithologist Cécile Mourer-Chauviré (1981, 1982, 1983), were rather lightly built but very tall, averaging some 1.5 to 2.75 meters (5–9 ft.) in height, the largest with skulls as long as 48 centimeters (19 in.). They are known primarily

Various South American Tertiary phorusrhacids shown with their living relative, a seriema (*Cariama*), and a 1.8-m (6-ft.) man. (From Marshall 1978; adapted from drawings by Jean and Rudolph F. Zallinger, courtesy Field Museum of Natural History)

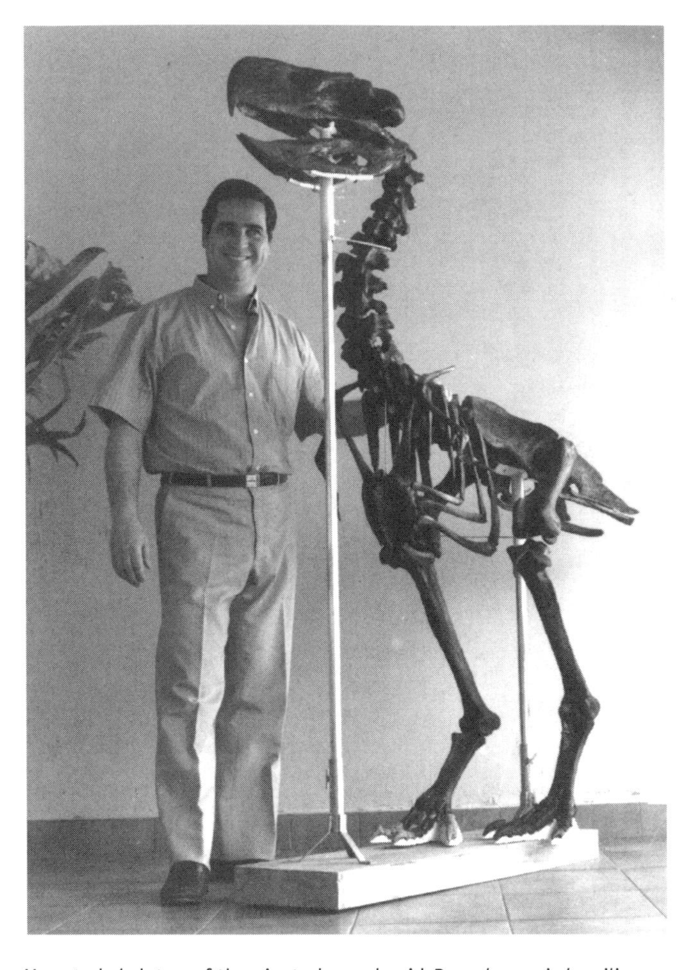

Mounted skeleton of the giant phorusrhacid *Paraphysornis brasiliensis,* a cursorial predator from the Lower Miocene of southeastern Brazil, shown next to Herculano Alvarenga, Brazilian authority on these birds. (Courtesy H. M. F. Alvarenga)

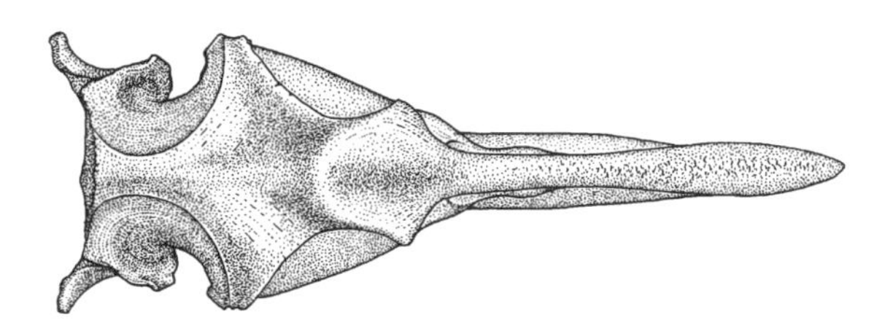

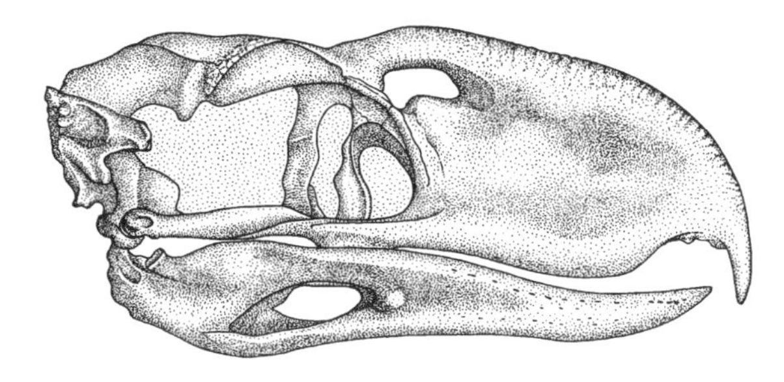

Skull of *Phorusrhacos* (*Phororhacos*) *inflatus,* in dorsal and lateral views, illustrating the powerful, laterally compressed and strongly hooked predatory bill of these terrestrial predators. Length, 35 cm (14 in.). (Redrawn from Ameghino 1891)

A restoration of the phorusrhacid *Phorusrhacos,* a giant flightless bird from the Miocene of Patagonia. (Painting by Charles R. Knight; courtesy Department of Library Services, American Museum of Natural History, neg. no. 39443)

from Argentina and Brazil, and many were flightless or nearly so. Phorusrhacids were apparently denizens of grassland ecosystems, and perhaps open areas in dry forests. They had a powerful hooked beak that was highly adapted for tearing flesh, and unlike the diatrymas, many of these presumed cariamid gruiform derivatives were undoubtedly swift cursorial predators, agile enough to run down fast-moving mammals. Some species may have specialized in eating carrion, or may at least have fed on carcasses opportunistically. Argentinean paleontologist Eduardo Tonni has reviewed the temporal distribution of phorusrhacoids in South America (1980b) and grouped them into several adaptive types, which he places in separate families: the cursorial predators (Phorusrhacidae), the graviportal scavengers (Brontornithidae), and the cursorial predators capable of limited flight (Psilopteridae). It was during the Lower Miocene (Santacruzian) that the large phorusrhacoids reached their greatest diversity, and they then began to decline, lending credence to the theories of paleontologist George Gaylord Simpson, who doubted in his book *Splendid Isolation* that invading advanced mammalian carnivores had precipitated the demise of these great birds: "It has sometimes been said that these and other flightless South American birds . . . survived because there were long no placental carnivores on that continent. That speculation is far from convincing. . . . Most of the phororhacids became extinct before, only a straggler or two after, placental carnivores reached South America. Many of the borhyaenids [doglike marsupials] that lived among these birds for many millions of years were highly predaceous. . . . The phororhacids . . . were more likely to kill than to be killed by mammals" (1980, 147, 150).

Two isolated footprints some 18 centimeters (7 in.) in length, belonging to a phorusrhacid or perhaps to an early ratite, have been found in rocks that are about 55 million years old on the Fildes Peninsula of King George Island in West Antarctica (Marshall 1994), and the anterior part of a beak, presumably belonging to a phorusrhacid, was collected from rocks that are some 40 million years old from the La Meseta Formation on Seymour Island, on the south side of the Antarctic Peninsula. The proportions of the beak indicate a bird about 2 meters (6.6 ft.) tall (Case, Woodburne, and Chaney 1987).

Stunning evidence that the phorusrhacoids must have survived in South America until the isthmian gap was closed in the late Pliocene came with the discovery of a giant North American phorusrhacoid in 1963 by Pierce Brodkorb of the University of Florida. In Florida, during the late Pliocene (latest Blancan), there existed a giant phorusrhacoid, named *Titanus walleri* and described by Brodkorb as "larger than the African ostrich and more than twice the size of the South American rhea" (1963c, 111). Additional material of *Titanus*, including cranial pieces, vertebrae, and portions of wings and legs, has recently been discovered (Chandler 1994) and shows that, unlike the greatly reduced forelimbs of ratites, this giant had robust wings, possibly armed with a claw, that may have been used to keep struggling prey under submission. At least this one species, therefore, spread up from South America through Middle America and across the Gulf Coast to Florida after the land bridge was created.

Titanis walleri, at approximately 3 m (10 ft.), was a giant, flightless phorusrhacid that made its way to Florida from the south during the Great American Interchange and represents the only large terrestrial carnivore to take part in the interchange. It probably preyed on small to medium-sized land mammals such as capybaras. (After Marshall 1988; reprinted by permission of *American Scientist,* journal of Sigma Xi, the Scientific Research Society)

The Tertiary Gruiform Radiation

The Cariamids. Although the South American phorusrhacids evolved in isolation from cariamidlike birds, other members of the suborder Cariamae are known from the Tertiary of the Northern Hemisphere and are included in two families: the Bathornithidae of the late Eocene to early Oligocene of North America and their European relatives, the Idiornithidae of the Middle Eocene (Peters 1995) to early Oligocene (Mourer-Chauviré 1983). These were moderate to large-sized, long-legged, terrestrial predators or perhaps scavengers. Among the better known of the bathornithids are *Bathornis veredus,* from the Lower Oligocene of Colorado, Nebraska, and South Dakota, *B.*

Skull and neck vertebrae of an as-yet-undescribed Middle Eocene species belonging to the Idiornithinae from the Messel oil shales of Germany. The specimen is under study by Stefan Peters of the Senckenberg Museum, Frankfurt. × 0.83. (Courtesy S. Peters and Forschungsinstitut Senckenberg; specimen in Landessammlungen für Naturkunde, Karlsruhe)

Above, skull of *Bathornis* (formerly *Neocathartes*) *grallator,* from the late Eocene of Wyoming, in dorsal (*A*), ventral (*B*), and lateral (*C*) views. *Below,* life reconstruction of *Bathornis grallator* by Walter Weber. *Neocathartes* was first described as *Eocathartes* and was characterized as a primitive terrestrial vulture with limited or no ability to fly. On reexamining the specimen, Storrs Olson (1985a, 150) concluded that it belonged to *Bathornis*. The North American bathornithids and their European relatives, the idiornithids, were long-legged terrestrial, gruiform birds, closely allied with other Cariamae, and belonging in that group. (From Wetmore 1944; courtesy Carnegie Museum of Natural History)

Aenigmavis sapea (the generic name means enigmatic bird) was described by Stefan Peters from the Middle Eocene Messel oil shales of Germany as a small member of the Phorusrhacidae. It was either a weak flyer or flightless and shares features with the Cariamidae, Rhynochetidae, and Eurypygidae. Phorusrhacids are also known from the early Tertiary of France and a skull from the London Clay, but the European forms may well represent a paraphyletic lineage of these types of birds. Scale = 8 cm. (Photo courtesy S. Peters and Forschungsinstitut Senckenberg)

celeripes, from the Middle to Upper Oligocene of South Dakota, Nebraska, and Wyoming, and *B. cursor*, from the Upper Oligocene of South Dakota and Wyoming. The bathornithid *Paracrax gigantea*, of the late Oligocene of South Dakota (Cracraft 1968), was a very large terrestrial bird. One of the bathornithids from the Eocene of Wyoming was initially identified, by Alexander Wetmore in 1944, as a "terrestrial vulture," *Eocathartes (Neocathartes) grallator*, and it became, through its reconstruction by Walter Weber, perhaps the best known Tertiary fossil bird for decades. In the 1980s, however, Storrs Olson (1985a) showed that it is indeed a bathornithid, and the fossil bird, which had a reduced coracoid and therefore reduced powers of flight, is now properly known as *Bathornis grallator*.

Large terrestrial gruiforms were abundant during the Tertiary of the Northern Hemisphere and formed an integral part of the avifauna (Cracraft 1968, 1971a). They were closely allied with the Cariamidae, as indicated by Mourer-Chauviré (1981, 1983), who studied the abundant fossils of idiornithids from the French Phosphorites du Quercy and reduced the families Bathornithidae and Idiornithidae to subfamilies of the Cariamidae. She also judged the idiornithids to be "equally close to the genus *Opisthocomus*" (39), a conclusion supported by Olson, who has speculated that "the suborder Cariamae appears to have been derived from 'basal' land birds related to *Opisthocomus*" (1985a, 152).

Knowledge that the cariamid radiation was not restricted to the New World took on a particular significance with the discovery in both France and Germany of phorusrhacid birds. In 1981, Cécile Mourer-Chauviré, while excavating the Phosphorites du Quercy, announced a remarkable discovery: a phorusrhacid from the Eo-Oligocene. She had discovered that the humerus of the supposed owl *Strigogyps minor*, described in 1939 by C. Gaillard, was in fact a primitive phorusrhacid of medium size with somewhat reduced flight abilities. She placed the fossil in a new genus, *Ameghinornis* (hence *A. minor*), in honor of the Argentinean paleontologist Ameghino.

Another phorusrhacoid, *Aenigmavis sapea*, from the Middle Eocene Messel oil shales of Germany, was described in 1987 by Stefan Peters (1987b) of Frankfurt's famed Senckenberg Museum. This specimen consisted of a nearly completely articulated skeleton, though it lacked the skull. According to Peters, *Aenigmavis* was flightless or at best a weak flier, some 50 or so centimeters (20 in.) in length, "the size of a large rooster" (1989, 2061). As he later noted, "The Phorusrhacidae seem to be part of an early radiation of gruiform birds. The modern species of Rhynochetidae, Eurypygidae, and Cariamidae are apparently the last surviving representatives of this radiation" (1989, 2061). The distribution of these terrestrial birds in the early Tertiary of Europe is enigmatic, given the extreme unlikelihood of a connection between South America and Europe during that time. However, we must remember that the phorusrhacids are large flightless de-

Red-legged seriema (*Cariama cristata*). The seriemas are long-legged, largely cursorial gruiform birds confined to Brazil and Argentina. The two monotypic genera of the family Cariamidae represent relicts of the ancient gruiform radiation that produced the phorusrhacids, bathornithids, and idiornithids. (Drawing by George Miksch Sutton)

Pale-winged trumpeter (*Psophia leucoptera*). The trumpeters (Psophiidae, 1 genus, 3 species) are another relict gruiform family of largely terrestrial birds that fly when forced. They are restricted to the humid forest of the Amazon Basin. (Drawing by George Miksch Sutton)

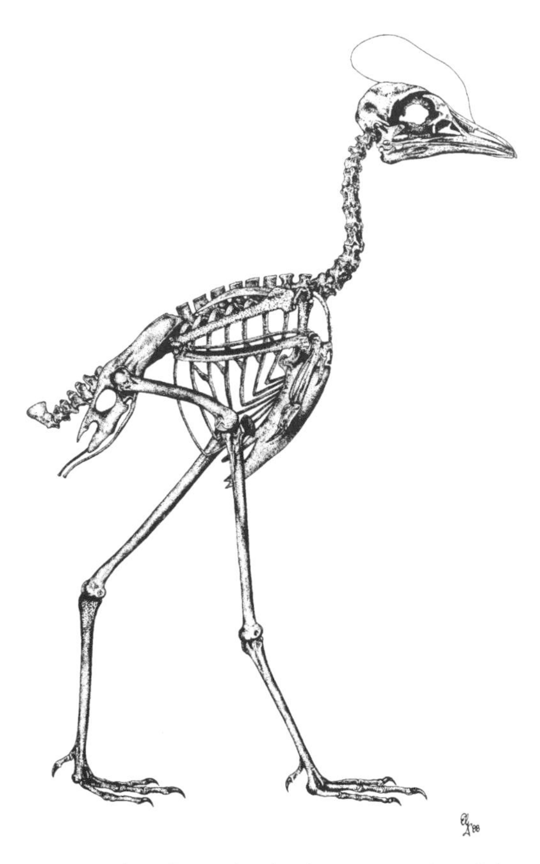

Above, reconstruction of *Messelornis cristata,* a member of the well-known Messel rails (Messelornithidae), which are the most frequently recovered birds of the Messel oil shales; *Messelornis* is also known from the Green River Formation of Wyoming and from other parts of Europe. In spite of their common name, these birds are not rails at all and are most closely allied with the living relict sunbitterns (Eurypygidae) of the Neotropics. *Messelornis cristata* had a helmetlike ornament on the head that consisted of a fleshy or horny but boneless formation, and its tail feathers were quite long. × 0.39. (From A. Hesse 1990; drawing by Elke Groening; courtesy S. Peters and Forschungsinstitut Senckenberg)

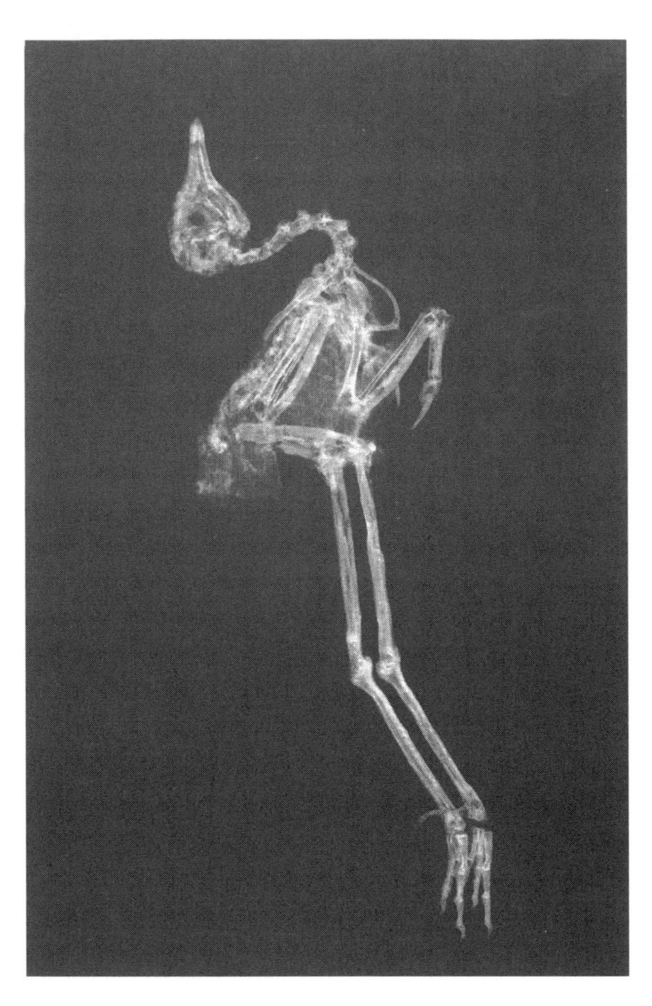

X-ray photograph of *Messelornis cristata*. × 0.47. (Courtesy A. Hesse, S. Peters, and Forschungsinstitut Senckenberg; X-ray, J. Habersetzer)

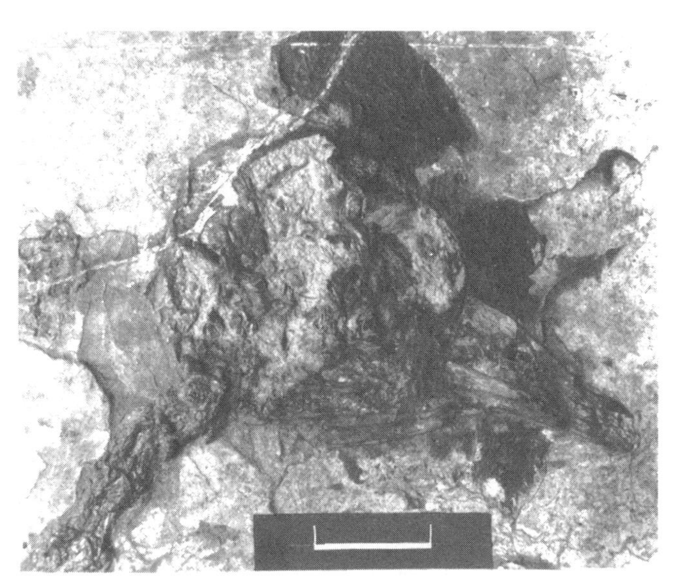

Messelornis cristata, head with vestiges of a fleshy or perhaps horny crest for which it is named. × 1.2. (Courtesy A. Hesse, S. Peters, and Forschungsinstitut Senckenberg)

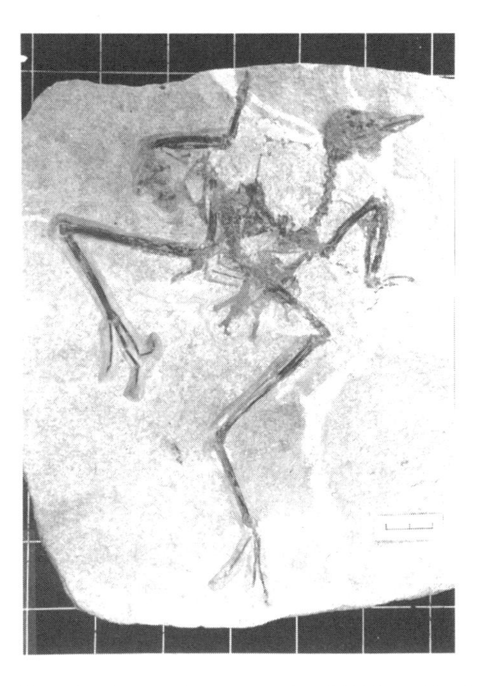

Messelornis neactica, a messelornithid from the Eocene Green River Formation of Wyoming, illustrating the widespread occurrence of Messel rails. Scale in cm. (Courtesy S. Peters and Forschungsinstitut Senckenberg; *Los Angeles County Museum of Natural History, Science Series* 36)

rivatives of the Cariamae, which contains the North American Bathornithidae and the European Idiornithidae, families that gave rise to many terrestrial, near flightless species. As Storrs Olson has noted of the French fossil phorusrhacoid, "There is a possibility that . . . *Ameghinornis* evolved . . . similarities to South American phorusrhacids in parallel, and that their similarities are due in part to degenerative reduction. Nevertheless, *Ameghinornis* was a large flightless derivative of the Cariamae, which is what phorusrhacids are" (1985a, 146). Still other phorusrhacid material (including the bill) of a chicken-sized bird from the Lower Eocene London Clay has been recovered by Michael Daniels.

The Cariamidae are known as fossils from numerous specimens from the French Upper Eocene or Oligocene Phosphorites du Quercy (Mourer-Chauviré 1982), as well as from the Upper Paleocene (Mourer-Chauviré 1994), and at least three undescribed species are known from the German Middle Eocene Messel oil shales (Peters 1991). The two living cariamids, the red-legged, or crested, seriema (*Cariama cristata*) and the black-legged, or Burmeister's, seriema (*Chunga bermeisteri*), which range from the tablelands of central Brazil to Paraguay and northern Argentina, thus provide us with a remarkable glimpse into the evolutionary past of the great Paleogene avian radiation that produced the phorusrhacids. The red-legged seriema inhabits the pampas, while the black-legged is a bird of sparse brushy forest. Both are terrestrial stalkers, usually occurring in pairs or small flocks, and they run rapidly when disturbed. They fly weakly, and not often, and roost in trees. The red-legged seriema nests on the ground, its sibling species in low bushes and small trees. Food of these omnivores consist of insects, especially ants, as well as fruit and other miscellaneous items, including small mammals, snakes, and lizards. The only living gruiforms thought to be fairly closely allied with the seriamas are the trumpeters (Psophiidae) (Stegmann 1978), a relict Amazonian family of largely terrestrial birds that now consists of three species in a single genus.

Messel Rails. But what about the other extinct gruiform birds? Perhaps the best preserved evidence comes from the Messel oil shales, where the abundant fossil birds known as Messel rails (*Messelornis*) once lived. These remarkable terrestrial, moorhen-sized birds are, in fact, not rails at all but part of a Lower Tertiary radiation of birds in the family Messelornithidae (Hesse 1988). They are most closely allied with the superfamily Eurypygoidea, especially the living relict river bird the sunbittern (*Eurypyga helias*) of South America and the flightless kagu (*Rhynochetos jubatus*) of New Caledonia, which are unknown in the fossil record. Messelornithids, best characterized and beautifully described by Angelika Hesse in her monograph on the type species, *Messelornis cristata*, are known from a huge amount of material, particularly from Messel, where several hundred specimens have been recovered, but also from the Eocene to the Oligocene of Germany and France, with *M. russelli* from the Upper Paleocene of Mont Berru, France, and *Itardiornis hessae*, from the Eo-Oligocene Phosphorites of Quercy (Mourer-Chauviré 1995). A new species, *M. nearctica*, is known from the Eocene Green River Formation of Wyoming (Hesse 1992). *Messelornis cristata*, the well-known Messel form, had a fleshy or horny helmetlike ornament on the head, and its tail feathers were quite long. Interestingly, only adult specimens are known, suggesting that the birds either visited the lake at Messel only outside the breeding season or, perhaps, that they nested in trees, as the sunbittern often does today.

Eogruids and Ergilornithids. We know that an extensive adaptive radiation of gruiform birds took place in the Eocene and Oligocene, as we have seen, memorialized by a number of flightless and nearly flightless fossil forms, such as *Bathornis* and *Idiornis*. But less well known are certain cranelike Paleogene fossil gruiforms from North America and Asia; though less well preserved, they are certainly an important aspect of the avian radiation. The family Geranoididae, known primarily from leg bones, consists of a number of large, superficially cranelike birds from the early and medial Eocene of North America (Cracraft 1969, 1973b). The other families are the Eogruidae, large cranelike birds from the late Eocene and

The sunbittern (*Eurypyga helias*) is a denizen of streams and ponds in tropical American forests. This monotypic family represents a relict of a once-great radiation of gruiform birds. (Drawing by George Miksch Sutton)

The kagu (*Rhynochetos jubatus*) is the sole member of another relictual, monotypic gruiform family now confined to the forest habitat of the island of New Caledonia. The kagu is apparently flightless. (Drawing by George Miksch Sutton)

early Oligocene of Asia, and the Ergilornithidae, large, fast-running birds, convergent on ostriches, from the early Oligocene of Asia and the late Miocene and Pliocene of Asia and Europe. Ergilornithids were didactylous, or two-toed, with ostrichlike short, flattened phalanges; like modern ostriches, the inner and hind toes were lost as an adaptation for speed. Similar morphological trends can be seen in the evolution of such mammals as horses, which have taken this principle to the extreme of retaining only one toe. As Olson has pointed out of the "geranoidid-eogruid-ergilornithid lineage," as he calls it, "it is quite possible that most, if not all, of these fossil forms were flightless" (1985a, 157).

The fossil record of the Eogruidae and Ergilornithidae, which during the Oligocene are often difficult to distinguish (Kurochkin 1976), is summarized by Evgeny Kurochkin (1981, 1982), who has studied these fossil birds extensively, and by Storrs Olson (1985a), who suggested that it was among the Ergilornithidae that the ancestors of ostriches might be sought. The proposal had been advanced earlier by Pierce Brodkorb, who, in his "Catalogue of Fossil Birds," placed the highly specialized two-toed runner *Urmiornis* in the Ergilornithidae (Gruiformes) and commented that the family was "possibly related to the Struthionidae [ostriches]" (1967, 154, following Burchak-Abramovich 1951).

Among the more interesting of these fossils are *Proergilornis minor*, from the early Oligocene of Inner Mongolia, in which the inner trochlea of the tarsometatarsus is reduced to a mere stub, indicating an early trend toward more cursorial habits than those of its probable flightless or near flightless ancestor, *Eogrus*, from the late Eocene of Mongolia. *Ergilornis rapidus*, also from the early or medial Oligocene of Inner Mongolia, has a tarsometatarsus in which the inner trochlea is largely absent, with only a faint sign of a stub. The continued reduction in the size of the inner trochlea indicates an accelerating trend toward a gracile, cursorial animal. Another fossil of great interest is *Amphipelargus* (=*Urmiornis*) *maraghanus* (see Harrison 1981), from the late Miocene to early Pliocene of western Asia. *Amphipelargus* was contemporaneous with and probably coexisted with the Pliocene ostrich, and like it, completely lacked the inner trochlea, having attained a degree of cursorial adaptation seen only in the living ostrich, *Struthio*.

All of these birds evolved in the expansive open high plains of Mongolia and western Asia, and the astonishing fact is that no ostrich fossils older than Miocene age have been found there (Mikhailov and Kurochkin 1988), though ergilornithid fossils occur back to the Oligocene. This interesting group of cursorial gruiform birds appears to have originated in North America, crossed into the Old World, like the diatrymids, via the North Atlantic land connection during or before the early Eocene, and adaptively radiated with ever-increasing cursorial adaptations in the Mongolian plains throughout the Tertiary. They extended into the Neogene in the two-toed runner that lacked any sign of the inner trochlea, *Amphipelargus*, from the late Miocene and Pliocene of Eurasia, and may have been descended from *Ergilornis rapidus*. Until evidence on ostrich origins recently surfaced (discussed later in this chapter), these cranelike but highly cursorial ergilornithids had to be considered prime candidates for ostrich ancestry. We now know, however, that instead they represent a remarkable example of avian convergent evolution.

The Cranes. True cranes, family Gruidae, are derived from deep within the Paleogene gruiform radiation, and a variety of fossils referable to the Gruidae exist from the Eocene of Europe, where *Geranopsis*, from the late Eocene of England, represents a crane thought by Joel Cracraft (1973b) to be closest to the primitive *Balearica*, the crowned-cranes of Africa. Crane fossils are known throughout the Tertiary of Europe and Asia, and from North America there are numerous fossils of small to medium-sized cranes from the early Oligocene to the late Miocene that appear to be crowned-cranes; similarly, crowned-cranes are known from the Miocene and Pliocene of Asia. These cranes have been placed variously in the genera *Probalearica* and *Aramornis* but clearly belong in *Balearica* (Olson 1985a; Feduccia and Voorhies 1992). From a late Miocene volcanic ash locality in Nebraska, we now have complete skeletons of a small crowned-crane in

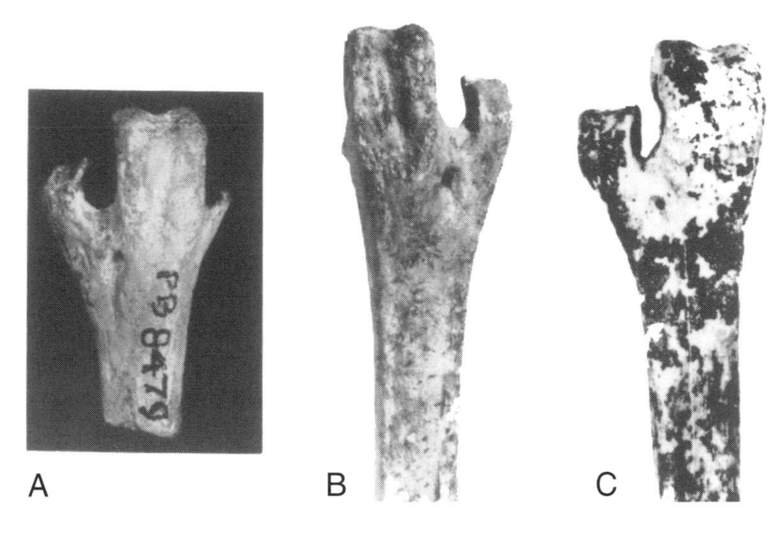

Didactylous ergilornithids. *A, Proergilornis minor* (early or medial Oligocene of Inner Mongolia), anterior view of left tarsometatarsus; cast of holotype × 1.1. *B, Ergilornis rapidus* (early or medial Oligocene of Inner Mongolia), anterior view of right tarsometatarsus; × 1.0. *C, Urmiornis maraghanus* (late Miocene and early Pliocene of Ukraine, Iran, and Kazakhstan), anterior view of left tarsometatarsus; × 1.2. The only living didactylous bird is the ostrich (*Struthio*), apparently converged on the highly cursorial ergilornithids. (From Cracraft 1973; courtesy American Museum of Natural History)

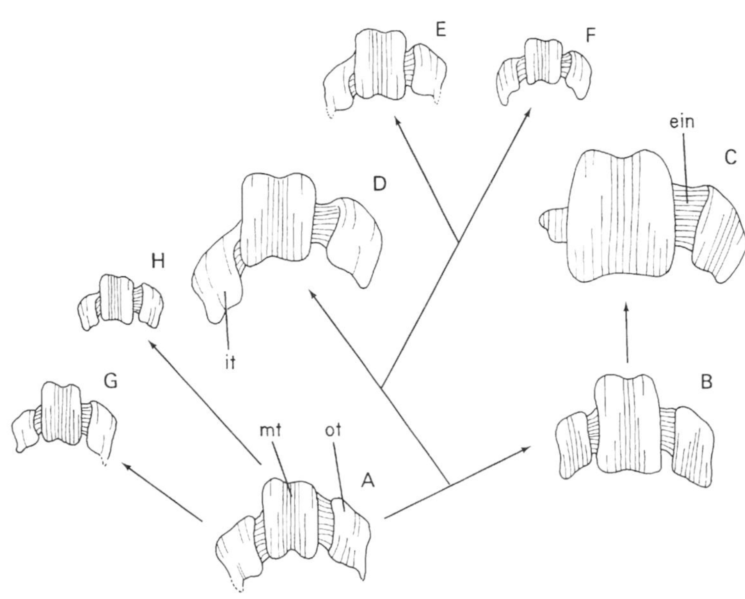

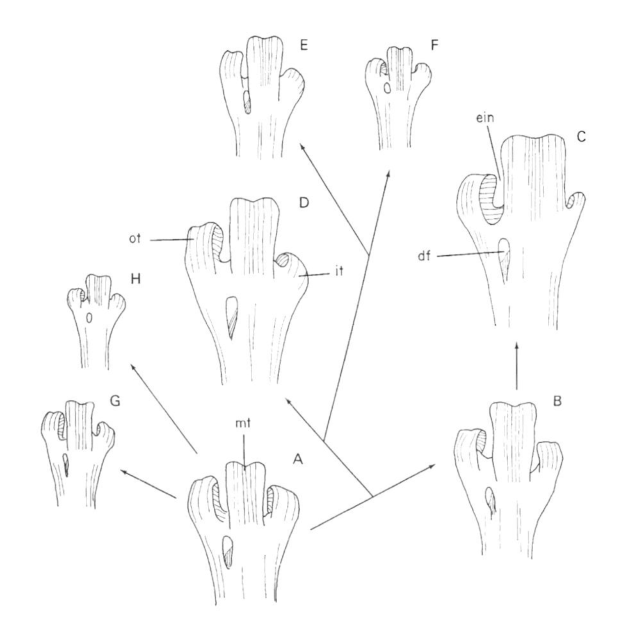

Possible evolutionary pathways of some gruiform tarsometatarsi as seen in distal view (*above*) and views of left distal ends (*below*). This is not a phylogeny. *A, Paragrus shufeldti* (Geranoididae); *B, Eogrus aeola* (Eogruidae); *C, Proergilornis minor* (Ergilornithidae); *D, Pliogrus pentici* (Gruidae); *E, Anisolornis excavatus* (Aramidae); *F, Psophia crepitans* (Psophiidae); *G, Bathornis celeripes* (Bathornithidae); *H, Elaphrocnemus phasianus* (Idiornithidae). Abbreviations: *df,* distal foramen; *ein,* external intertrochlear notch; *it,* inner trochlea; *mt,* middle trochlea; *ot,* outer trochlea. (From Cracraft 1973; courtesy American Museum of Natural History)

Demoiselle crane (*Grus virgo*). Cranes are long-legged birds with an elevated hallux; they fly with the neck extended. Successful remnants of a diverse adaptive radiation that began in the Eocene, the spectacular birds of the family Gruidae, which contains fifteen species in four genera, occur on all continents except South America. (Drawing by George Miksch Sutton)

The two species of crowned-cranes (depicted here is the black crowned-crane, *Balearica pavonina*), now confined to Africa but once widespread, are thought to be the primitive members of the crane assemblage and are frequently separated off in their own subfamily, the Balearicinae. (Photo by author)

association with the short-limbed rhinoceros *Teleoceras major*, horses, camels, and other large vertebrates (Voorhies 1981; Feduccia and Voorhies 1992). Indeed, evidence of fossil grass from *Teleoceras* specimens indicate a diet of grasses of the genus *Berriochloa* (Voorhies and Thomasson 1979) and, in combination with other paleoecological evidence, show that the crowned-crane inhabited wet grasslands, similar to crowned-cranes' current habitat in Africa; in fact, they also show that crowned-cranes have remained in essentially the same habitat for approximately 10 million years.

Pierce Brodkorb (1967) was probably correct in using a separate subfamily, Balearicinae, to separate these primitive cranes of the genus *Balearica* from the Gruinae, which appears somewhat later in geologic time and includes the genera *Grus* and *Baeopteryx*. Even more primitive is the cranelike limpkin (*Aramus*), which morphologically is clearly little more than a primitive crane (Allen 1956) and is close in structure to the crowned-cranes (*Balearica*). Many early workers considered the limpkin to be intermediate between rails and cranes (Fürbringer 1888), often with *Balearica* as a primitive crane leading into the Gruidae, whereas Olson (1985a) considers *Aramus* to be a primitive member of the Gruidae. Joel Cracraft questioned the intermediacy of the limpkin between rails and cranes but stated that "the similarity of *Aramus* and the Gruidae [is] more apparent when comparison is made to a crane with a more primitive skeleton such as that of *Balearica*" (1973b, 115). Indeed, virtually all studies, including recent DNA analyses (Ingold et al. 1989; Krajewski 1989; Sibley and Ahlquist 1990; Krajewski and Fetzner 1994), conclude that cranes of the genus *Balearica* are distinctive and primitive within the Gruidae. The early occurrence of crowned-cranes in the fossil record of cranes, therefore, is not surprising.

Given the primitive position of crowned-cranes, it is

Limpkin (*Aramus guarauna*). The monotypic gruiform family Aramidae, ranging from the Neotropics north to subtropical North America, is thought by many to represent a primitive cranelike form, not too distantly removed structurally from the ancestral cranes. (Drawing by George Miksch Sutton)

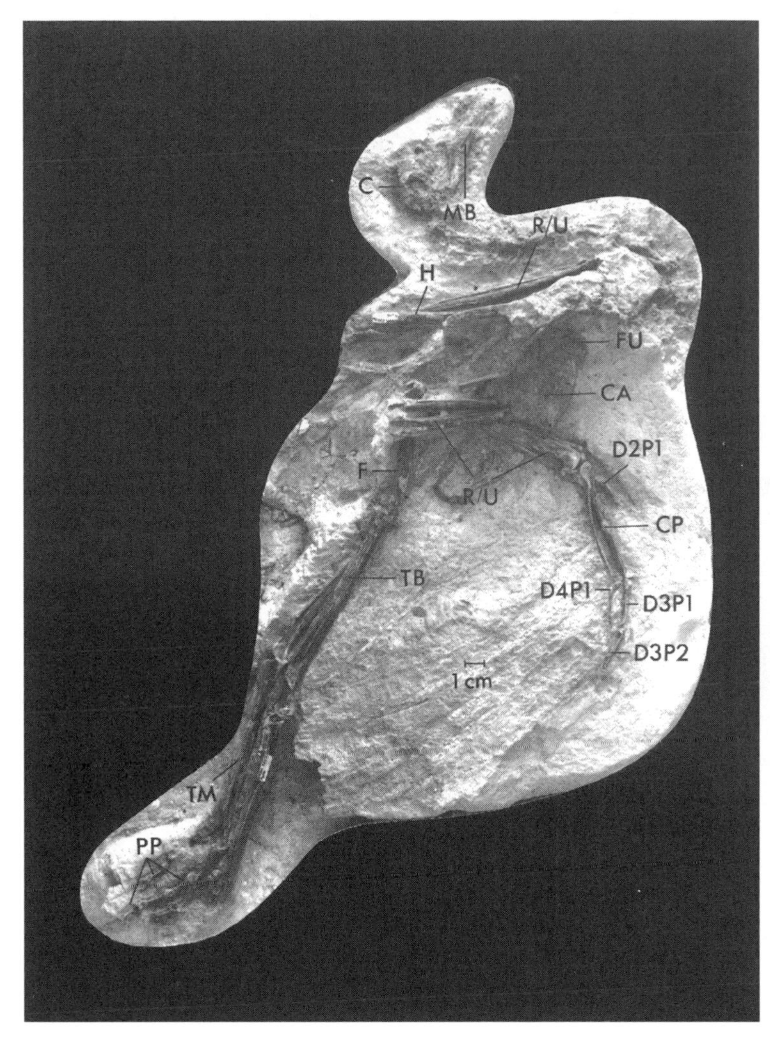

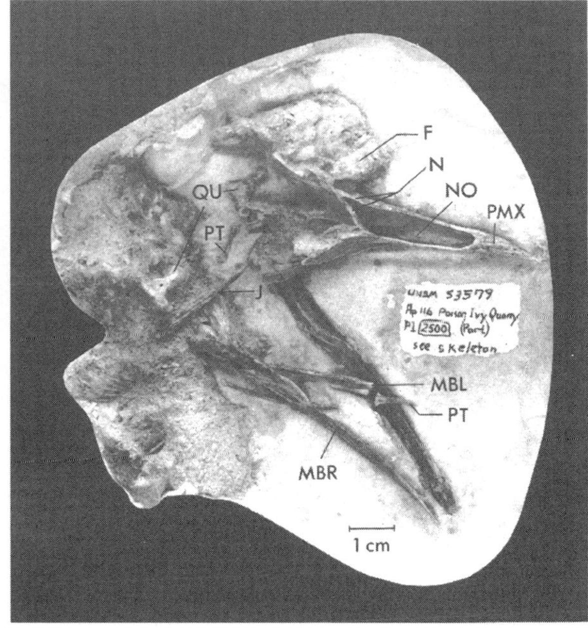

A late Miocene volcanic eruption produced a remarkable death assemblage in Nebraska some 10 million years ago, including hundreds of well-preserved fossils of rhinoceroses, horses, camels, and a small species of crowned-crane, *Balearica exigua*. Complete skeleton and skull. (From Feduccia and Voorhies 1992; courtesy *Los Angeles County Museum of Natural History, Science Series* 36)

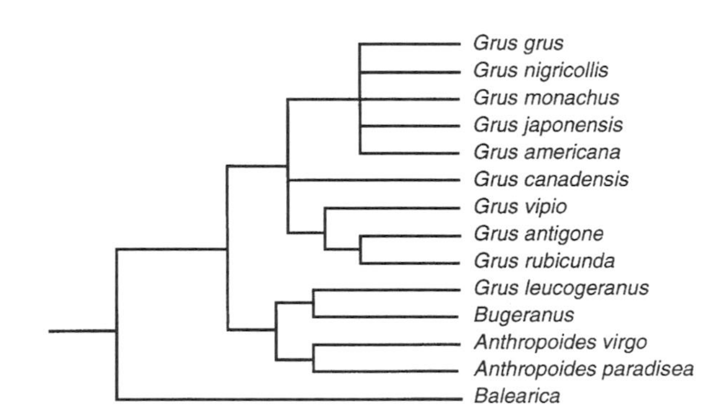

Crane relations as indicated by overall similarity of their unison calls (Archibald 1975; figure redrawn from Wood 1979). The horizontal axis is arbitrary, showing only Archibald's view of nested similarity levels. (Modified after Krajewski 1989)

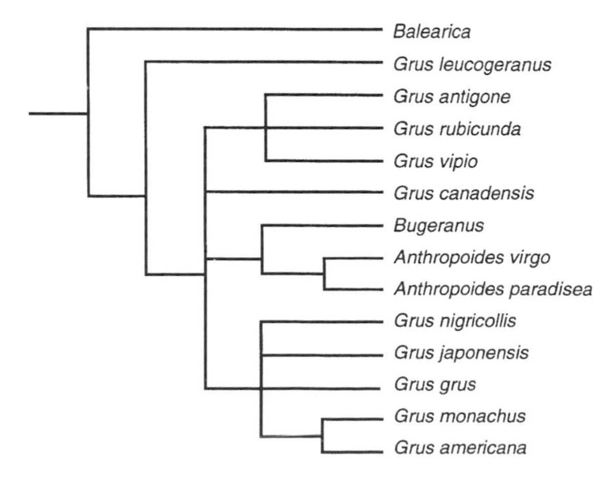

Jackknife strict consensus tree for the square matrix of distances from data on DNA hybridization by Carey Krajewski. Note that all data, including those from paleontology, point to the primitive nature of *Balearica*. The enigmatic Siberian crane (*Grus leucogeranus*) appears, from DNA hybridization, to be the sister-group of the remaining species, which fall into four closely related groups. (Modified after Krajewski 1989)

Miscellaneous distinctive gruiform families. *Left to right and top to bottom,* mesite, or roatelo, family (Mesoenatidae, Madagascar, flightless birds with powderdown and reduced clavicles, 2 genera, 3 species): brown mesite (*Mesitornis unicolor*); buttonquail family (Turnicidae, tropical and warm temperate parts of Old World, 2 genera, 17 species): spotted buttonquail (*Turnix ocellata*); finfoot, or sungrebe, family (Heliornithidae, pantropical, 3 genera, 3 species): African finfoot (*Podica senegalensis*); bustard family (Otididae, Old World, 6 genera, 25 species, sometimes classified as a charadriiform family): black-bellied bustard (*Eupodotis melanogaster*). (Drawings by George Miksch Sutton)

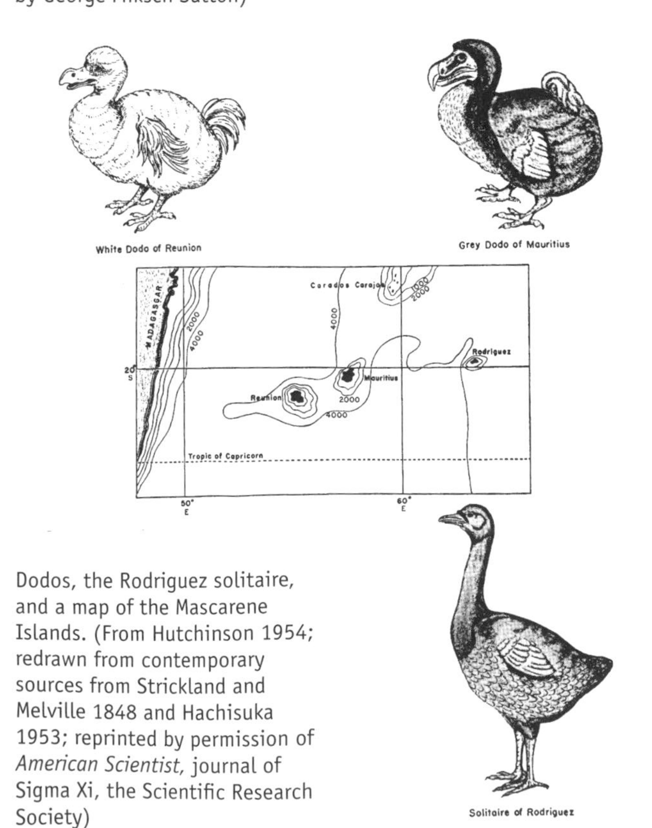

Dodos, the Rodriguez solitaire, and a map of the Mascarene Islands. (From Hutchinson 1954; redrawn from contemporary sources from Strickland and Melville 1848 and Hachisuka 1953; reprinted by permission of *American Scientist,* journal of Sigma Xi, the Scientific Research Society)

The Samoan tooth-billed pigeon (*Didunculus*), the sole member of its subfamily, Didunculinae, is thought to be the type of volant, semiterrestrial columbiform bird that flew to the Mascarene Islands and gave rise to the "didine" birds (from the generic name), the flightless dodos and solitaires. (Drawing by John P. O'Neill)

of more than passing interest that these cranes exhibit a slow rate of growth relative to other cranes. D. E. Pomeroy (1980) documented this slow growth rate and subsequent attainment of adult morphology in seven crowned-cranes raised in captivity to twenty-three months, where weight increase followed a sigmoid curve, with full adult size attained at about two years (breeding age is normally three years). In sharp contrast to sandhill cranes (*Grus canadensis*), which at four to five months had attained 91 percent of adult weight (Miller and Hatfield 1974), crowned-cranes of that age had attained less than 60 percent of adult weight (Pomeroy 1980). In addition, individual crowned-cranes vary tremendously in size, making documentation of fossil species extremely difficult.

A Miscellany of Living Gruiforms. Except for the rails, family Rallidae, the remaining groups of gruiforms—the nearly flightless Malagasy mesites, Mesitornithidae, the terrestrial South American forest birds known as trumpeters, Psophiidae, and the pantropical aquatic finfoots, Heliornithidae—have produced no fossil record, but these families no doubt represent ancient relicts of the great Tertiary gruiform radiation. Two enigmatic families of small terrestrial birds, the Pedionomidae, or plains-wanderers of Australia, have been placed in the Charadriiformes by Storrs Olson and David Steadman (1981), and the Old World and Australian Turnicidae, or buttonquails (hemipodes), are of uncertain but probable gruiform affinities.

Last, the Old World bustards, Otididae, large terrestrial birds that seldom fly and which in many respects superficially resemble such ancient gruiforms as the bathornithids, may have affinity with the Charadriiformes. Olson claims (1985a, 179) that their present taxonomic position in the Gruiformes was caused by Hans Friedrich Gadow's (1893, 188) insistence that the bustards were "steppe rails," a name that is no longer used, yet most authors treat them taxonomically as part of the gruiform assemblage. Fossil remains of bustards are known from the French late Eocene or Oligocene Phosphorites du Quercy (Mourer-Chauviré 1982; Olson 1985a) and from the Pliocene of Asia.

Big Chicks and the Origins of Flightlessness

Columbiforms and Their Derivatives. Another distinctive avian order that has given rise to flightless species is the Columbiformes, the pigeons and doves. The most familiar flightless columbiform derivative is the famous extinct dodo, *Raphus cucullatus,* of the island of Mauritius in the Indian Ocean (Hachisuka 1953; Livezey 1993b; Maddox 1993). The Mascarene Islands, notably Mauritius, Réunion, and Rodrigues, lie west of southern Madagascar. Once forested, the islands were devoid of land mammals or other large predators and were home to giant land tortoises as well as a variety of unusual birds, including the spectacular flightless Mauritius broad-billed parrot (*Lophopsittacus mauritianus*) and the recently extinct and volant Rodrigues and Réunion parrots, as well as some flightless rails of the genus *Aphanapteryx*. The broad-billed parrot was big, some 70 centimeters (28 in.) long, was either gray or blue, and was thought to be nocturnal; it had an enormous bill, a reduced sternal keel, very short wings, and was probably flightless. It may well have been a Mauritius analogue to the large nocturnal kakapo (*Strigops habroptilus*) of New Zealand.

The "didine" birds, the dodo and the kindred solitaire of Rodrigues, also extinct, were derived from columbiforms thought to be close to the ground-dwelling tooth-billed pigeon (*Didunculus strigirostris,* family Dididae), of Samoa, which has a heavy, hooked bill and feeds on large, hard fallen fruit (Burton 1974b). In my opinion these pigeons are so close to the dodos that the dodos' segregation from the Columbidae, in the monotypic family Rhaphidae, is unwarranted. Although rafting of a flightless ancestor between Mauritius and Réunion has been proposed, the dodo and solitaires appear most likely to have been large, flightless forms of the tooth-billed pigeons whose ancestors flew to the Mascarenes. A deep trench separates Rodrigues from Mauritius, which precludes any prior connection between the two.

Finally, with respect to the Mascarene dodos and solitaires, G. Evelyn Hutchinson notes that "some authorities, notably Rothschild, have . . . suspected the presence of two white forms on Reunion, one dodo-like, the other solitaire-like. Others, notably Oudemans, suppose only one species to have existed, and this a white dodo" (1954, 304). What had been thought for years to be the extinct solitaire of Réunion has been shown to be an extinct ibis with a relatively short, straight bill and is now known as *Threskiornis solitarius* and is thought to be closely related to the sacred ibis (*T. aethiopicus*) and the straw-necked ibis (*T. spincollis*) (Mourer-Chauviré, Bour, and Ribes 1995a, 1995b).

The dodo was about the size of a large turkey, and males and females weighed approximately 21 and 17 kilos (lbs.), respectively, exhibiting unusually great sexual dimorphism (Livezey 1993b). Dodos were exterminated at the close of the seventeenth century; Portuguese Mascarenhas discovered the island in 1507, and it is said that 174 years later no living dodo remained. Introduced pigs and monkeys ate their eggs, and European sailors provisioned their merchant ships with the large birds and land tortoises. The scale of the devastation is indicated in an ac-

The extinct crested, or broad-billed, parrot (*Lophopsittacus mauritianus*) of Mauritius was a very large gray or blue nocturnal flightless parrot with an enormous bill, a reduced sternal keel, and very short wings; it was perhaps similar to the flightless kakapo (*Strigops*) of New Zealand. (From Newton 1896; adapted from a tracing by A. A. Milne-Edwards of the original drawing in the MS journal kept during Wolphart Harmanszoon's voyage to Mauritius in 1601–1602)

count of small ships sent from Réunion to Rodrigues to collect tortoises and birds; thirty thousand land tortoises were collected in less than eighteen months (Halliday 1978, 124). Dutch ships provisioned in part with dodos as early as 1598 (Halliday 1978), as the Portuguese had for nearly a century, and the practice was recorded in the dairy of a Dutch ship's captain in 1602:

> They also caught birds which some named Dodaarsen, others Dronten when Jacob Van Neck was here, these birds were called Wallich-Vogels, because even long boiling would scarcely make them tender, but they remained tough and hard, with the exception of the breast and belly, which were very good. . . . on the 25th of July, Willem and his sailors brought some Dodos which were very fat; the whole crew made an ample meal from three or four of them, and a portion remained over. . . . on the 4th of August Willem's men brought 50 large birds on board the "Buyn-vis"; among them were 24 or 25 Dodos. . . . another day, Hogeneen set out from the tent with four seamen provided with sticks, nets, muskets, and other necessaries for hunting. . . . they captured another half hundred birds, including a matter of 20 Dodos, all which they brought on board and salted. (Strickland and Melville 1848, 14)

The dodos, whose name is derived from the Portuguese *doudo*, a simpleton, were ill equipped to survive such devastation and predation. Among other size and physiological disadvantages, their clutch consisted of a single egg, laid in a nest (described by one visitor as a mass of palm leaves some 46 centimeters [18 in.] high on the forest floor), and both parents spent time on the nest. The young were apparently slow to mature; the dodo's lifespan may have been about thirty years.

A fascinating note on the biology of the dodo comes from Stanley Temple of the University of Wisconsin. Temple (1977) studied the ecology of Mauritius and concluded that the nearly extinct endemic tambalacoque tree *Calvaria major* and the extinct dodo were an example of obligatory mutualism, a situation in which each of the species through evolution developed a strong dependence on the other. The elimination of one or the other would therefore cause a marked reaction in the other. *Calvaria*, once widespread, apparently became nearly extinct because its large, extremely hard seeds needed to pass through the digestive tract of the dodo in order for germination to occur. In 1973, just thirteen old, overmature, and dying trees were known to exist (there were no young trees). These trees were estimated to be about three hundred years old, the approximate time of the dodo's disappearance from Mauritius. By force-feeding fresh *Calvaria* pits to turkeys, Temple was able to overcome the seed-coat dormancy, and for the first time in three centuries *Calvaria* seedlings germinated.

With this in mind, it is of great interest that, as one writer remarked: "About 1638, Sir Hamon Lestrange tells us, as he walked London streets he saw the picture of a strange fowl hung out on a cloth canvas, and going to see it, found a great bird kept in a chamber 'somewhat bigger than the largest Turkey cock, and so legged and footed, but shorter and thicker.' The keeper called it a Dodo and shewed the visitors how his captive would swallow 'large peble stones . . . as bigge as nutmegs'" (Newton 1896, 159).

At almost exactly the same time the dodo met its demise, the slightly smaller and perhaps less modified flightless cousin of the dodo died out on neighboring Rodrigues. The Rodriguez solitaire (*Pezophaps solitaria*) received its name because it was said it was always found alone (Hutchinson 1954, 304). It was larger than the dodo and was last seen alive at the end of the seventeenth century. Like dodos, solitaires exhibited exaggerated sexual dimorphism, with females only some two-thirds as heavy as males—Bradley Livezey has estimated that male and female Rodriguez solitaires weighed 28 and 17 kilos (62 and 37 lbs.), respectively—and having much shorter bills; dimorphism in the Rodriguez solitaire may have been greater than for any known carinate bird (Livezey 1993b, 247).

Both dodos and solitaires were primarily frugivores, with an enlarged crop and gizzard stones. As part of the morphological changes associated with gigantism, these bizarre, jumbo columbiforms had an increased thermodynamic efficiency, a lower basal metabolic rate, and an improved capacity for fasting, which, combined with a great seasonal variation in the deposition of fat, may have been a tremendous advantage for these frugivores who lived on an island with trees that bore fruit sporadically (281).

Arrested Development. As we shall discover in this chapter, many large, flightless birds, including the ratites, paleognathous birds that lack a keeled sternum, are the result of the process of developmental heterochrony—relative differences in developmental schedules, either by slowing down or speeding up (McKinney 1988; McKinney and McNamara 1991). Stephen J. Gould defines heterochrony as "a phyletic change in the onset or timing of development, so that the appearance or rate of development of a feature in a descendant ontogeny is either accelerated or retarded relative to the appearance or rate of development of the same feature in an ancestor's ontogeny" (1977, 481). In the case of most flightless birds, the process is usually called paedomorphosis, or derived underdevelopment. It is also termed neoteny, a process of arrested development in which the rate of growth is reduced and the adult retains many features of the embryo of late developmental stages. The subsequent attainment of sexual maturity renders this "big chick" the adult of the species, and by this mechanism "evolutionary novelties" that may represent dramatic evolutionary change can occur (Muller and Wagner 1991).

Neoteny, as I term it here (including paedomorphosis), is thus a mechanism by which macroevolution may occur with very minor genetic change, perhaps involving only the timing of developmental events (Diamond 1981). Considerable caution must be exercised in using these terms, however, because although the initial stages in the evolution of an "avian novelty," such as flightlessness, may be initiated by heterochronous development, by neoteny, which can be attested to mainly by the lack of anatomy associated with the late stages of development—in the case of flightlessness, the pectoral flight apparatus—other novel features, such as hypertrophy of the hindlimbs and associated architecture for ambulatory locomotion may be due to subsequent selection for a specific biological role rather than heterochronous development.

Strickland and Melville's description of the dodo is the first recognition of neoteny in any bird species:

> And as these three islands form a detached cluster, as compared to other lands, so do we find in them a peculiar group of birds, specifically different in each island, yet allied together in their general characters, and remarkably isolated from any known forms in other parts of the world. These birds were of large size and grotesque proportions, the wings too short and feeble for flight, the plumage loose and decomposed, and the general aspect suggestive of gigantic immaturity. . . .
>
> We cannot form a better idea of it than by imagining a young Duck or Gosling enlarged to the dimensions of a Swan. It affords one of those cases . . . where a species, or a part of the organs in a species, remains permanently in an underdeveloped or infantine state. Such a condition has reference to peculiarities in the mode of life of the animal, which render certain organs unnecessary, and they therefore are retained through life in an imperfect state, instead of attaining that fully developed condition which marks the mature age of the generality of animals. . . . And lastly, . . . the Dodo is (or rather was) a *permanent nestling*, clothed with down instead of feathers, and with the wings and tail so short and feeble, as to be utterly unsubservient to flight. . . . Thus, if we suppose, for instance, that the abstract idea of a Mammal implied the presence of teeth, the idea of Vertebrate the presence of eyes, and the idea of a Bird the presence of wings, we may then comprehend why in the Whale, the Proteus, and the Dodo, these organs are merely suppressed, and not wholly annihilated. (1848, 4–5, 33–34)

Some of the more interesting aspects of this "gigantic immaturity" and "underdeveloped or infantine state" include a dramatic reduction in the pectoral architecture; this represents a tremendous savings in developmental expenditure as well as a reduction in the burden of having to maintain this anatomically and metabolically expensive structural complex. As for the so-called downy plumage of the adult dodo, cited as "juvenility" by Strickland and Melville, this was probably a misinterpretation of the loosely constructed pennaceous feathers that typically characterize flightless birds (see chapter 3) and to the untrained observer resemble down.

The many other morphological and physiological characteristics associated with flightlessness and gigantism (Calder 1984) include increased longevity (Calder 1983), in the case of the dodo, estimated at about thirty years by Livezey (1993b), and enhanced thermodynamic efficiency by decreasing the relative surface area while increasing volume (see chapter 3). As a rule there is a general decrease in basal metabolic rate in flightless birds from diverse lineages, from large ratites (Calder and Dawson 1978) to the tiny Inaccessible Island rail (*Atlantisia rogersi*) (Ryan, Watkins, and Siegfried 1989). As Livezey (1993b) has pointed out, large size in the dodo also conveyed an improved capacity for fasting, which could be of considerable importance for a frugivore on an island with seasonal productivity of fruit.

The Mechanism of Neoteny. It is known that removing a chick's thyroid gland soon after hatching will inhibit its growth and development (Woitkevitsch 1966), and it has been shown that in adult European starlings (*Sturnus vulgaris*), thyroidectomy can induce sexual maturation in postbreeding photorefractory (postgonadal cycle) birds (Dawson, Goldsmith, and Nicholls 1985). In a recent experiment A. Dawson and co-workers (1994) performed an interesting procedure in which four-day-old European starlings were thyroidectomized by injecting radioactive iodine. Thyroidectomy resulted in slowed body growth (sternum length), and the starlings kept the short, wide bill, comparatively large protuberant eyes, and distended abdomen characteristic of young birds. In addition, in one-year-old birds the skull sutures remained unfused, and juvenile plumage persisted. Behavior was also affected; thyroidectomized birds did not learn to feed independently until fourteen weeks, compared to only four weeks for control birds. The experimental birds also did not develop fully effective thermogenesis and had to be maintained at 25°C to prevent shivering.

Dawson and co-workers concluded that "neonatal thyroidectomy therefore caused neoteny in a passerine bird" (637), and they noted a number of similarities between the neotenic characteristics of thyroidectomized starlings and those of ratites. They suggest further that hypothyroidism may have been a factor in ratite evolution, "possibly because their evolution occurred in an area deficient in iodine," and that living ratites may even be hypothyroid. Certainly neoteny is well known within amphibians, and in these animals it is known to be thyroid-dependent (Dodd and Dodd 1976). The genetic change needed to produce hypothyroidism is undoubtedly very small and that could occur many times independently in different groups of birds.

The Radiation of Doves and Parrots. Some 313 species of pigeons and doves are found today in most parts of the world. And even today the order Columbiformes contains species that tend toward flightlessness, with such semiterrestrial forms as the so-called ground-pigeons, ground-doves, and quail-doves. The western crowned pigeon (*Goura cristata*) of New Guinea is also semiterrestrial, as is the Luzon bleeding-heart (*Gallicolumba luzonica*) of the Philippines. Giant, extinct species of imperial pigeons, genus *Dacula*, are known from a number of South Pacific islands and are thought to have been eradicated by humans (Balouet and Olson 1987).

The columbiforms' presumed close allies, the parrots (order Psittaciformes), comprise some 360 species, including a flightless species, the kakapo (*Strigops habroptilus*) of New Zealand, and a largely terrestrial form, the ground parrot (*Pezoporus wallicus*) of Australia. *Strigops* is now apparently doomed to extinction, the remaining 50 or so individuals having been transported to and confined to Stewart Island, after introduced mammalian predators were removed from the island. This large parrot has small wings and a reduced sternum; it is the only parrot that uses communal male courtship display grounds, called leks.

The fossil record of doves and parrots is meager. The earliest dove to date is *Columba calcaria*, from the early Miocene of France. Doves are fairly common in the late Tertiary, but their fossil record reveals little of their evolutionary or biogeographic history. Likewise, the fossil record of parrots tells us little of their history. There is a supposed parrot, *Palaeopsittacus*, from the Lower Eocene of England, but the first firmly identified forms come from the late Eocene of La Bouffie, France. Mourer-Chauviré (1992b) has named a new family, the Quercypsittidae, for these early parrots, which have some features that are more primitive than in the Recent family Psittacidae. The earliest modern psittacids are known from the early Miocene in the Northern Hemisphere in *Conuropsis fratercula* of Nebraska and, from at least that time on, superseded the ancient parrots of the Quercypsittidae. Mourer-Chauviré interprets this modern and fossil distribution as indicating two vicariant families, the Northern Hemisphere Quercypsittidae and the Southern Hemisphere modern, derived Psittacidae. However, as we shall see, there is little evidence for vicariant distributions in any group of birds, and it is equally possible that the Quercypsittidae represent an ancestral grade in parrot evolution.

Island Rails: Showcase of Flightless Evolution. The group of birds with the greatest tendency to become flightless is the ancient order Gruiformes, which has given rise not only to the diatrymas and phorusrhacids but also to many other flightless forms, both living and extinct. As we saw in chapter 5, the Gruiformes probably shared an ancient common ancestor with the shorebirds. We know that there was an extensive adaptive radiation of gruiform birds in the Eocene and Oligocene, memorialized by a number of flightless and nearly flightless fossil forms. Perhaps the best known of these are the six or so species of *Bathornis*, the large, cursorial ground-dwelling birds of North America, described earlier in this chapter. Other

The endangered, flightless kakapo (*Strigops habroptilus*) of New Zealand. Largely nocturnal, this large parrot (1.3–2 kg, 2.9–4.4 lbs.) emerges in the evening to feed, often climbing trees to get fruit; on ground it runs rapidly, frequently spreading its wings. Its wings are reduced and it has no sternal keel. The few remaining individuals reside on Stewart Island, and the outlook for the species is bleak. *Left,* Sinbad no. 2 (male), displaying with wing-waving activity (Milford Sound, 1975); *top right,* "Jonathon" (male) (Esperance Valley, Fiordland, 1974). (Photos by and courtesy of Don Merton.) *Right,* the kea (*Nestor notablis*), a brownish green parrot the size of a crow. A highland parrot that normally occurs above the tree line on South Island, New Zealand, the kea was once persecuted for its habit of pecking into the backs of living sheep to obtain kidney fat. Today many keas are quite tame and will feed from the hand. (Photo by author)

Pink cockatoo (*Cacatua leadbeateri*). The family Psittacidae contains 80 genera and 360 species of primarily pantropical distribution, with some temperate representatives; they are very common in Australia. Parrots are quite similar in their postcranial anatomy to the pigeons and doves and are thought to have derived from columbiform stock. (Drawing by George Miksch Sutton)

Pigeons and doves (Columbidae, worldwide, 40 genera, 313 species): common wood-pigeon (*Columba palumbus*). (Drawing by George Miksch Sutton)

fossil gruiforms we have noted include the flightless and highly cursorial ostrichlike runners *Ergilornis* and *Urmiornis*, from the Tertiary of Eurasia, once thought to be involved in the ancestry of the ostrich.

Today the order Gruiformes contains a number of extremely diverse birds, many of them already mentioned and many ancient relicts. The largest and most successful group of living gruiforms, however, is the rails (Rallidae), comprising some 143 species with a nearly cosmopolitan distribution. The earliest fossils of definite rallid affinity are known from the late Oligocene and early Miocene of Europe, and fossil rails are fairly common in younger Tertiary deposits from Europe, Asia, and North America (Olson 1985a, 162).

Some of the more spectacular flightless rails were the Mascarene rails of the genus *Aphanapteryx* (invisible wing), *Aphanapteryx bonasia* of Mauritius and *A. leguati* of Rodrigues, both extinct. In 1638, François Cauche found on Mauritius "red hens with the beak of a woodcock; to capture them one need only to present them with a piece of red cloth, they follow and let themselves be taken by hand: they are the size of our hens, excellent to eat" (translated from Strickland and Melville 1848, from Olson 1977c, 358). That same year the traveler and diarist Peter Mundy wrote that these birds were "feathered all over, butt on their wings they are soe Few and smalle that they cannot with them raise themselves From the ground" (358).

Other spectacular flightless rails have existed on islands from the Hawaiian Archipelago throughout the South Pacific, with at least one flightless species on almost every island of any size. But the most bizarre and highly modified living forms occur on New Zealand (Atkinson and Millener 1991; Weber and Hesse 1995), where another extinct flightless bird of unknown affinity, the giant woodhen, or adzebill, *Apterornis* (Apterornithidae), once shared the high-altitude tussock habitat with the large, flightless moas (Andrews 1896). One form of *Apterornis* was found on North Island, another on South Island. *Apterornis*, once thought to be a large, flightless rail has since been shown to be of unknown gruiform affinity (Olson 1977c); it was a large bird with a unique jaw mechanism. *Apterornis* is under study by Storrs Olson and Richard Zusi of the Smithsonian Institution.

Of the many island rails that are part of the species complex, and directly derived from the purple swamphen (*Porphyrio porphyrio*) of disjunct areas of the Mediterranean, sub-Saharan Africa, and southern Asia to Indonesia and Australasia, and Micronesia east to Samoa, the most highly modified are the large purplish-greenish "moorhens" of New Zealand, where two striking species reside. These are the geographically wide ranging, marsh-dwelling, fully flighted *Porphyrio porphyrio melanotus*, known by the Maori as the pukeko; and the world's largest rail, the totally flightless moorhen, known to the Maories as the takahe (*Notornis* [or, more accurately, *Porphyrio*] *mantelli*), a vegetarian, which crops snow-grass for a living in New Zealand's high-altitude tussock swamps. The takahe was first named from a skull found in North Island, and except for numerous bones of these birds that were recovered with those of moas in Pleistocene and Holocene deposits, there were only three known specimens, the first having been collected in 1849, and the last in 1855. For fifty years there were no further sightings. Then, in 1948, some hundred years later, a small population of a hundred or so takahes was discovered in a small valley high in the Murchison range in southeastern South Island between 750 and 1,200 meters (2,400–4,000 ft.), having survived the onslaught of stoats and weasels introduced by European settlers. The population declined when red deer were introduced into the region, however, and the takahe, now highly endangered, is restricted to a captive breeding program.

New Zealand's flightless wood rail, or weka (*Gallirallus australis*), is a small-chicken-sized flightless rail that is clearly a derivative of an early invasion of *Rallus* stock to New Zealand. The weka has not been particularly affected by the advent of humans, but seems to thrive in most environments. This aggressive omnivore is a formidable predator in its own right and includes introduced mice and rats in its diet.

Another rail worthy of mention is the small Inaccessible Island rail *Atlantisia rogersi* (Lowe 1928), which because of its decomposed and hairlike plumage was cited as an example of feather degeneration in flightless birds in chapter 3. This diminutive rail inhabits one of the islands in the Tristan de Cunha group in the South Atlantic, some 2,900 kilometers (1,800 mi.) west of the Cape of Good Hope and about 3,200 kilometers (2,000 mi.) east of South America. Storrs Olson (1977c) has shown that this rail and its extinct relatives on other islands are neotenic forms derived from the widespread typical rails of the genus *Rallus*. Another species of *Atlantisia*, *A. elpenor*, from Ascension Island in the South Atlantic, was a medium-sized flightless rail first recorded by Peter Mundy as "a strange kind of fowle with wings very imperfitt such as wherewith they cannot raise themselves from the ground" (from Olson 1977c, 354). Mundy and his shipmates captured and ate six of these birds on 7 June 1656 (Olson 1977c, 354), and the species was exterminated sometime thereafter by introduced mammals. As Hutchinson has noted, "The catalogue of the Rallidae is a pathetic cemetery filled

King rail (*Rallus elegans*). The rails, coots, and moorhens (Rallidae, 34 genera, 143 species) form the core of the modern Gruiformes. They also tend toward flightlessness, particularly on islands—most of the Pacific Islands at some time had at least one flightless species. (Drawing by George Miksch Sutton)

Reproduction of George Hoefnagel's painting of the extinct Mauritius Island flightless rail (*Aphanapteryx bonasia*), which was a uniform dark reddish brown, with a long decurved, pointed bill and poorly developed wings. (From Milne-Edwards 1868)

Apterornis or *Aptornis* (10–13 kg; 22–29 lbs.), commonly known as the adzebill, was a large flightless gruiform bird of New Zealand, with a distinctive species on the North and South Islands. (Drawing by John P. O'Neill)

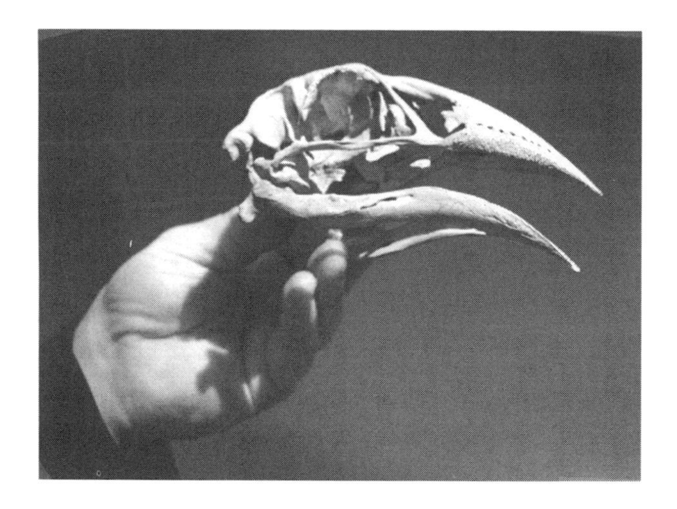

Skull of the fossil gruiform *Apterornis defossor* from Pyramid Valley, North Canterbury, which equaled in size that of the second largest moa, *Pachyornis*. The powerful beak was employed for chopping off clumps of tussock grass. (Photo courtesy Beverley McCulloch, Canterbury Museum, Christchurch)

The weka (*Gallirallus australis*) of New Zealand. Wekas are aggressive, chicken-sized, flightless rails that are formidable predators, having added introduced mice and rats to their culinary repertoire. (Photo by author)

The flightless Inaccessible Island rail (*Atlantisia rogersi*) of the remote South Atlantic island of the same name. (Photos by and courtesy of Peter Ryan, Percy Fitzpatrick Institute of African Ornithology)

New Zealand's takahe (*Porphyrio mantelli*), a large flightless gallinule weighing about 3 kg (6.6 lbs.). First described from a fossil skull in 1848 by Sir Richard Owen, and at that time thought to be extinct, it was discovered alive a few years later, and the living bird turned out to be distinctive from the fossil species. In 1948, nearly a century after the last live bird had been collected, a small population was discovered in several remote mountain valleys on the South Island. The few remaining birds are now in a captive breeding program; there is no known wild population, but some may still exist. (Photo by author)

The purple swamphen (*Porphyrio porphyrio*), a geographically widespread species with numerous races and a close relative and probable ancestral form of the takahe. (Photo by G. K. Lestrange; courtesy Gordon Maclean)

with crosses commemorating flightless forms that have developed in different directions, considered of generic value, on oceanic islands, only to become extinct with the advent of modern man and his inevitable companions" (1950, 201).

The above are but a few examples, but of all living birds, the rails are the most likely to develop flightlessness. Probably more than sixty species, more than one-fourth of all rails, living or recently extinct, have lost the ability to fly, and some ten groups have independently evolved weak-flighted or flightless derivatives (Diamond 1991). In Hawaii there were twelve species of flightless rails (Olson and James 1991), and "it is likely that all islands in the Pacific were inhabited by one or more species of flightless rail before the arrival of humans" (Steadman 1986:27). All flightless rails are island forms, although the islands may range in size from tiny, austere Laysan to the large, lush, and topographically diverse islands of New Guinea and New Zealand. Rails thus provide us with a living laboratory for examining the structural changes that occur in the evolution of flightlessness.

Why Become Flightless? "I have therefore asked the question concerning Mauritius henns and dodos, thatt seeing those could neither fly nor swymme . . . how they shold come thither? Soe now againe concerning the Ascention birds allsoe, that neither fly nor swymme. The iland being aboutt 300 leagues from the coast of Guinnea . . . , the question is, how they shold bee generated, whither created there from the beginning . . . , or whither the nature of the earth and climate have alltred the spape [shape] and nature of some other foule into this, I leave it to the learned to dispute of" (from the journal of Peter Mundy, in Temple and Anstey, 1936, 83). Peter Mundy's journal antedates Darwin's *On the Origin of Species* by some two centuries, and yet the major questions remain.

Some general selective pressure must be at work to ignite the evolutionary development of flightlessness on islands. One cause is the lack of predators: most mammals are unable to cross large bodies of water and so are generally not found on ocean islands. This removes the need for escape from predation. The ability to migrate with the seasons is likewise no longer needed. And as Darwin and many others have argued, flightlessness would be advantageous on islands because flying organisms could be blown away by storms and strong island winds and would perish in the ocean. This certainly must have some bearing on the question, because the phenomenon of flightlessness on islands is truly pervasive: it is observed also in insects and even in plants.

After reading a book on the beetles of the Madeira Islands, Darwin wrote, "There is a very curious point in the astounding proportion of the Coleoptera that are apterous [wingless]" (1855). As is now known, this is the situation on islands everywhere, with group after group of island insects taking up the flightless condition and having normal, flying relatives on the closest mainland area. In the normal pathway of morphogenesis, the wings are shortened and narrowed; this is followed by complete disappearance. Flies, moths, bees, wasps, ants, grasshoppers, crickets, and especially beetles all have flightless island forms. Among the most remarkable and best known are the New Zealand wetas, extraordinary giant crickets that have lost not only wings but wing coverts. Just why this has occurred is not understood. Because the wetas inhabit such a large island—indeed, one weta species is a denizen of the deep, quiet forest floor—fear of displacing wind does not seem a suitable explanation. Yet other grasshoppers, such as those of the Galápagos, would indeed seem to have lost the ability to fly to avoid being swept by winds out to sea.

Plants, too, lose their "flight": many oceanic island forms develop fruit without aerodynamic dispersal mechanisms—parachutes, barbs, and other devices—and are adapted for falling near the parent. Concurrently, they tend to increase in size. Good examples are the fruits of the Compositae: mainland forms have good dispersal mechanisms, but their island counterparts have much diminished abilities. The ancestors of the Hawaiian tarweed (*Wilkesia*), something similar to the mainland tarweed

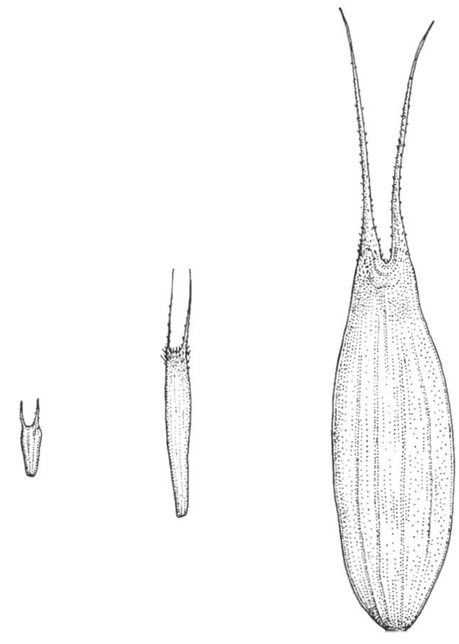

Progressive loss of dispersal ability is illustrated by fruits of closely related members of the sunflower family (Compositae). *Top,* the Spanish needle (*Bidens frondosa*), fruits of which catch easily on fur or feathers and are therefore easily dispersed. *Center, Fitchia cuneata,* from Tahaa Island, which has much larger fruits that do not attach easily to fur or feathers. *Bottom, Fitchia speciosa,* from Rarotonga Island, which has the largest fruits in the Compositae, contain heavy seeds and are virtually ineffective in dispersal much beyond the base of the tree on which they grow. (Modified from Carlquist 1965)

(*Layia*), has small, lightweight fruit that bear a fluffy "parachute," but the Hawaiian tarweed has longer, heavier fruits, with much shorter appendages (Carlquist 1965, 242). Again and again this theme is sounded when fruits of island plants are compared with their mainland relatives. Again, the answer would seem to lie, at least in part, with island winds. Being blown out to sea would be disadvantageous to an island plant; conversely, the ability to trade off this loss by an increase in size would offer advantages, such as abundant food storage to boost a seedling groping for light on the forest floor to greater heights in a shorter time.

The Architecture of Flightless Birds. When a bird becomes flightless, it undergoes dramatic morphological modification. The major change is the immediate reduction of the muscles and bones of the wing and pectoral girdle. The keel of the sternum is greatly reduced or lost, along with the loss of flight muscles. Because the body no longer needs to be lightened for flying, most flightless birds have a decided tendency to become large. So, as birds are structurally similar in possessing the ability to fly, they become structurally similar in the loss of flight. In addition, as we saw in chapter 3, the feathers, designed as aerodynamic structures, tend to degenerate and become loosely constructed, with the asymmetrical vanes of the primary feathers reverting to symmetry. In other words, if there is not continued selection to warrant the energy expenditure in the embryogenesis of these complex structures, they will tend to be lost, whether they are muscles, bones, or feathers.

Many of these adaptations, then, help the flightless bird save energy, both during embryogenesis and in adult life, by removing the burden of unwanted structures that hinder locomotion and by reducing the metabolic energy spent maintaining them. Brian McNab of the University of Florida has examined the hypothesis that the flightless condition is promoted by energy conservation by comparing factors that correlate with basal metabolic rate in flightless and flighted rails and ducks, and in kiwis (McNab 1994), and he has concluded that energy conservation contributes to flightlessness in species where the pectoral muscle mass is reduced. Energy conservation is also accomplished by the evolution of small size in flightless rails on oceanic islands, and the two factors facilitate the persistence of rails in environments with limited resources.

Among the first features to disappear are the expensive flight muscles, especially the pectoralis major and supracoracoideus. These flight muscles alone account for, on average, about 17 percent of the total body weight of most birds, ranging from a low of 7.8 percent in the white-throated rail (*Laterallus albigularis*), to a high of 36.7 percent in Cassin's dove (*Leptotila rufinucha*) (Hartman 1961; George and Berger 1966). But the flight muscles alone do not constitute the entire flight structure. There are the wing muscles and bones, and the pectoral girdle and sternum itself: the combined structure averages 20–25 percent of body weight in typical birds. The loss of this muscle and bony bulk of course decreases energy output, and further energy savings are generated by eliminating the daily maintenance of flight muscles, which rank among the body's most energy-consuming tissues. Any reduction in these high-metabolic organs immediately confers great energy savings. As Jared Diamond has put it, "The energy saved by flightlessness can be put to other purposes. A winged rail on a predator-free island is like a 60-kg backpacker forced eternally to carry 15 kg of bricks and to regurgitate half of each meal" (1981, 507).

If the flight apparatus is not used or is artificially immobilized, for example, by either pinioning or tenotomy (in which extensor tendons on the metacarpal bone are cut), as is common in zoo birds, the musculoskeletal system associated with flight degenerates. Zoo birds exhibit dramatic and bizarre bony lesions as a result of this atrophy. Birds are not anatomically suited for confinement unless they are small to medium size and kept in large aviaries (Feduccia 1991).

Neotenic Changes. Reduction or loss of the sternal and flight apparatus generally comes about through neoteny; and flightless birds, as well as those with a strong tendency to become flightless, are characterized by a number of neotenic features. To put it another way, all known birds possess the morphology of flightlessness at some stage when they are developing in the egg. They have disproportionately large pelvic regions and hind legs, greatly reduced wings, and a greatly reduced sternum with little or no keel. By the simple alteration of one or several developmental regulatory genes, those that govern the timing of hatching and therefore arresting perinatal development, the flightless state can be achieved in the adult. This evolutionary phenomenon involving heterochronous development of various structures is in fact the probable evolutionary force behind the early evolution of vertebratelike early chordates, and even the evolution of humans from apes. Whenever arrested development affects a particular part of an organism, it causes similar effects in other structures and organs. Therefore, development can only be arrested without lethal consequences after essential organs have developed sufficiently to sustain normal life processes.

Particularly interesting here is the late development of the sternum in rails, pigeons, and grebes. It has long

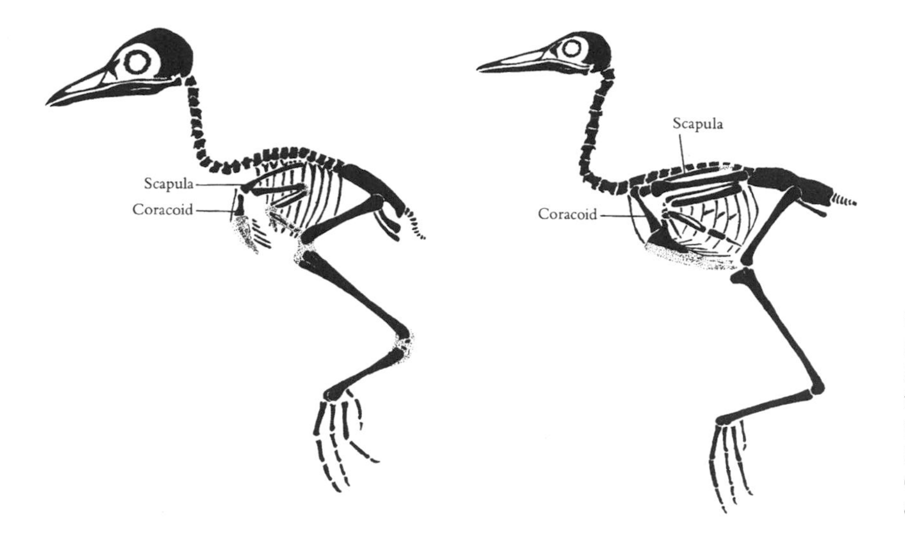

Cleared and stained skeletons of the king rail (*Rallus elegans*), a flying rail: *left,* at 17 days after hatching; *right,* at 47 days, reduced so that the femur lengths in the two drawings are equal. Stippled areas represent cartilage. Note that the articulation of the scapula and coracoid forms an obtuse angle in the younger form but an acute angle in the older form. (From Olson 1973)

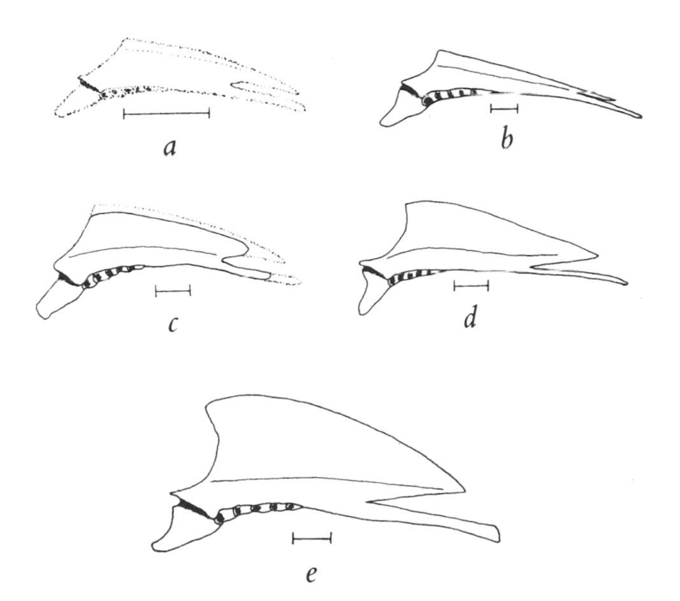

Development of the sternum of the purple gallinule (*Porphyrio martinicus*), a flying rail (*a, c, e*), showing how the carina in its early stages corresponds to the shape of the carina in two species of flightless rails (*b, d*): *a, P. martinica,* downy chick about a week old; the sternum is entirely cartilaginous but has nearly the same conformation as *b,* the weka (*Gallirallus australis*), a flightless rail, adult; *c, P. martinica,* an immature that is fully feathered but not quite volant; the sternum is still partly cartilaginous and now resembles *d,* the Guam rail (*Gallirallus owstoni*), adult; *e, P. martinicus,* adult. Scales = 5 mm. Dotted lines indicate cartilage. (From Olson 1973)

Chicks of the clapper rail (*Rallus longirostris*), *left,* and the great curassow (*Crax rubra*), a galliform, *right,* represent the two extremes in relative development of the pectoral flight apparatus of the downy young. Rails, born with essentially no development of the sternal keel and associated flight apparatus, can easily arrest development and give rise to flightless species. Galliform birds, by contrast, are endowed with nearly fully developed flight machinery at birth. For them to give rise to flightless species, development would have to be arrested far in advance of hatching, when other organs have not yet developed, and the results might well be lethal. (*Left,* adapted from Ripley 1977; *right,* adapted from Delacour and Amadon 1973)

been known that in domestic pigeons (*Columba livia*) and the great crested grebe (*Podiceps cristatus*) the sternum does not ossify until after hatching (Schinz and Zangrel 1937), and Storrs Olson (1973) has shown that the same is true for rails. As long as seventeen days after hatching, when all other major bones are relatively well ossified, the king rail (*Rallus elegans*) still has only a cartilaginous outline of the sternum, and at forty-seven days the sternum is only beginning to ossify. This postponed development of the sternal apparatus allows neoteny in rails to result in the loss of the flight apparatus without also resulting in the loss of the rail.

Galliform Birds

By contrast, birds such as the Galliformes (chickens and allies) that develop the sternal apparatus very late do not give rise to flightless forms. The one exception is the giant megapode-like flightless bird *Sylviornis neocaledoniae* from the Isle of Pines, New Caledonia, known from oral tradition as the du, which was extirpated by island dwellers (Mourer-Chauviré and Poplin 1985; Poplin and Mourer-Chauviré 1985). *Sylviornis* appears to have lost its ability to fly more through the evolution of gigantism than through any evolutionary mode of arrested development, and not unexpectedly, other giant megapodes have been discovered in recent years from the Pleistocene of Australia (Vickers-Rich 1991). Galliforms are among the strongest short-distance fliers. In the domestic fowl, a typical galliform, the major bones of the sternum begin to ossify between the eighth and twelfth days of incubation, rather than after hatching, so that the chick has almost fully developed flight capacity at birth. Galliforms could probably never survive arrested development of the sternal apparatus that would occur early enough to rid them of the keeled sternum.

We are a long way from understanding the relations of the Galliformes. The order has traditionally been considered a close ally of the ducks and has been placed in the Gallo-Anseres (Monroe and Sibley 1993; Kurochkin 1995), and in many classifications galliforms are placed after the diurnal birds of prey simply by default. It is also likely that the convergent cursorial adaptations in the pelvic architecture of chickens and theropod dinosaurs (Coombs 1978) has led to the belief that Galliformes represents a primitive lineage of birds.

The fossil record does not argue for antiquity of the group. The earliest record of a galliform bird is *Gallinuloides wyomingensis*, from the Lower Eocene Green River Formation, and the debate on its proper systematic posi-

Restoration of the giant New Caledonian megapode *Sylviornis neocaledoniae,* known as the du. The morphology of the head is based on accounts of the oral tradition surrounding the du. (From Poplin and Mourer-Chauviré 1985; courtesy Mourer-Chauviré)

Fowl-like or galliform birds are medium to large semiterrestrial birds with short, rounded wings, strong hindlimbs, and a powerful flight apparatus with a strongly keeled sternum. *Left to right and top to bottom,* Curassows, guans, and chachalacas (Cracidae, Neotropics, 11 genera, 50 species): great curassow (*Crax rubra*). Megapodes, or moundbuilders (Megapodiidae, Australasia, 6 genera, 19 species): Malleefowl (*Leipoa ocellata*). Guineafowl (Numididae, Africa, 4 genera, 6 species): vulturine guineafowl (*Acryllium vulturinum*). Pheasants, quail, grouse, and turkeys, included variously in different families (Phasianidae, nearly worldwide, 53 genera, 207 species): black grouse (*Tetrao tetrix*); silver pheasant (*Lophura nycthemera*); and ocellated turkey (*Agriocharis ocellata*). (Drawings by George Miksch Sutton)

tion has ranged from placing it in a separate monotypic family to inserting it in the Cracidae (chachalacas and allies) (Tordoff and Macdonald 1957) and to placing it with the phasianids (Cracraft 1973b). Tordoff and Macdonald (1957) described the cracid *Procrax brevipes*, from the early Oligocene of South Dakota, and found it somewhat intermediate between *Gallinuloides* and modern cracids, and there have been many cracid fossils named from the Lower Miocene to the Lower Pliocene of North America. Timothy Crowe and Lester Short described a new genus and species of gallinaceous bird, *Archaealectrornis sibleyi*, from the late Oligocene of Nebraska, and concluded that this form "and the other pre-Miocene galliform fossils studied are representatives of an extinct family of gallinaceous birds, the Gallinuloididae, which is the sister-group of the Phasianidae, and that similarities between members of this family and members of the Anseriformes are a result of character reversals to primitive states" (1992, 179).

The several genera and species of Galliformes from the late Eocene to early Oligocene of France were thought by Brodkorb (1964) to be cracids, but after careful study, Mourer-Chauviré (1982) determined that they represented an archaic group that would merit a distinct family, and fossil cracids are now known from the Upper Oligocene or the Lower Miocene of North America (Olson 1985a). Other notable galliform fossils include a small megapode from the late Eocene deposits of Quercy, France (Mourer-Chauviré 1982), but this bird has been placed in a separate family, Quercymegapodiidae (Mourer-Chauviré 1992c); the modern megapodes, or moundbuilders as they are often called, are unknown from the Tertiary. Moundbuilders are unique in their breeding habits; they bury their eggs under a huge mound of earth and vegetation, which on decomposition provides the heat for incubation. The adults regulate the temperature inside the mound by frequent digging and probing, and the highly precocious young begin running immediately on hatching and can fly within hours.

Aside from megapode fossils, there is a possible guineafowl from the late Eocene of Mongolia (Olson 1985a) and a guineafowl, *Telecrex peregrinus*, from the late Eocene of France (Mlikovsky 1989a). And there are numerous fossil New World quail ranging from the early Oligocene to the late Pliocene, Miocene grouse, phasianid-like birds in the Tertiary of Europe and Asia, and early turkeys from the early Miocene in North America; the Recent well-known history of fossil meleagrids (turkeys) is nicely reviewed by David Steadman (1980). The true fossil phasianids, placed in the genus *Palaeortyx*, appear only at the end of the early Oligocene (Mourer-Chauviré 1992c).

Although the sister-group relations of the galliform birds have been and will continue to be debated, it has long been recognized that they share a number of distinctive features with the paleognathous birds (ratites and tinamous). In the tinamous and many galliforms the posterior sternum is greatly lengthened, and the deep, open incisions are covered by a membrane that increases surface area for the origin of the flight muscles. There are also similarities in the structure of the coracoids. And the feather synapomorphies are striking; Asa Chandler found the structure of the plumaceous down of the tinamou genera *Calopezus* (*Eudromia*) and *Nothura* to be extremely similar to that of galliforms, and he interpreted this evidence as pointing to an "unmistakable relationship" (1916, 342). A sister-group relation of galliforms and paleognaths must remain a viable hypothesis.

The Anatomy of Flightlessness. As emphasized, the keel-less sternum is characteristic of birds that have become flightless and can therefore provide no evidence for evolutionary relations among avian groups. The flat-bottomed sternum has been used to ally the flightless ostriches, kiwis, emus, rheas, and cassowaries into a single assemblage called ratites (from the Latin *ratis*, meaning raft), but the keeled sternum is absent also not only in *Hesperornis* but in flightless rails, extinct flightless "geese" (*Thambetochen*), flightless ibises (*Apteribis*) from Hawaii, and a flightless gruiform (*Apterornis*) from New Zealand, to mention a few. And the keel has almost disappeared from the flightless grebe of Lake Titicaca, the flightless Galápagos cormorant, and the nearly flightless Central American passerine *Zeledonia*, as well as the flightless or nearly flightless tapaculos (Rhinocryptidae) of Central and South America.

Another anatomical feature that predictably changes with flightlessness is the angle of articulation between the coracoid and scapula. In flying birds, as we saw in chapter 3, this angle is acute; an acute angle shortens the distance through which the dorsal elevators must act, thus providing greater power. When flightlessness occurs, the angle exceeds 90°, and in the large flightless ratites the two bones meet in an extremely obtuse angle and fuse to form a single scapulocoracoid.

First recognized by T. H. Huxley as a characteristic of flightless birds, the obtuse angle of the scapula-coracoid articulation is also neotenic in birds, as shown by the development of the skeleton in rails. The angle of the articulation decreases as the sternum enlarges along with the pectoral muscle mass. Nearly hatchling embryos of rails retain the neotenic obtuse angle of the scapula-coracoid articulation, and in the recently extinct Hawaiian "goose" *Thambetochen chauliodous* the angle is so obtuse as to approach the ratite condition, although the two bones are

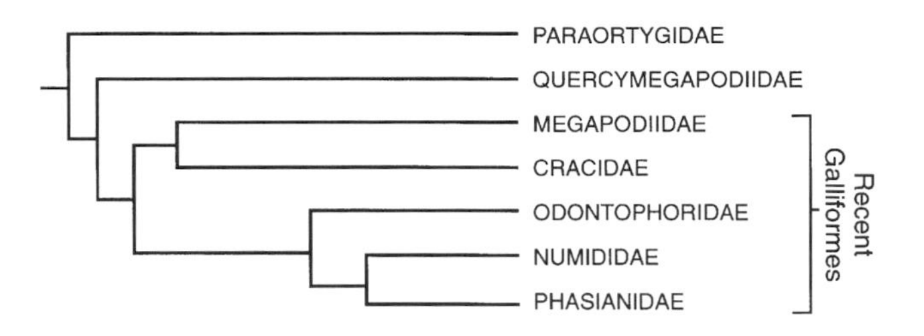

Hypothetical phylogenetic tree as reflected by the well-preserved galliforms from the Eo-Oligocene of Quercy, France. (Modified from Mourer-Chauviré 1992)

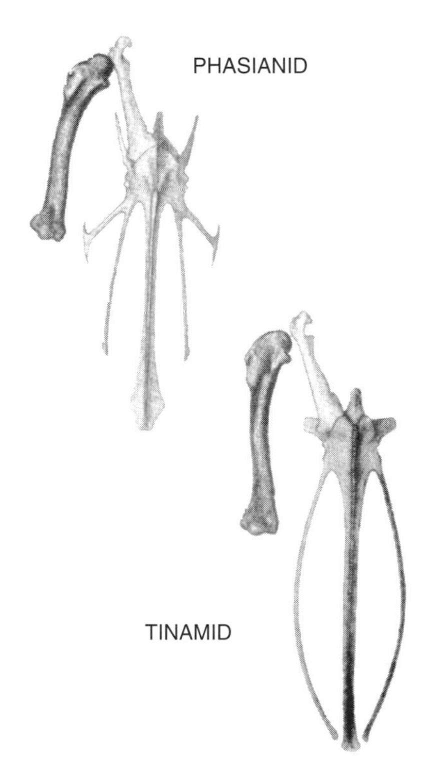

Sterna, coracoids, and humeri of a pheasant (*Phasianus*) and a paleognathous bird, a tinamou (*Tinamus*), showing the striking similarities in the deeply incised posterior region of the sternum and in the morphology of the coracoids. (From Houde 1988; courtesy Nuttall Ornithological Club)

Gallinuloides wyomingensis, the earliest well-known galliform, from the Lower Eocene Green River Formation of Wyoming. Approximately × 0.40. (Photo courtesy Museum of Comparative Zoology, Harvard University)

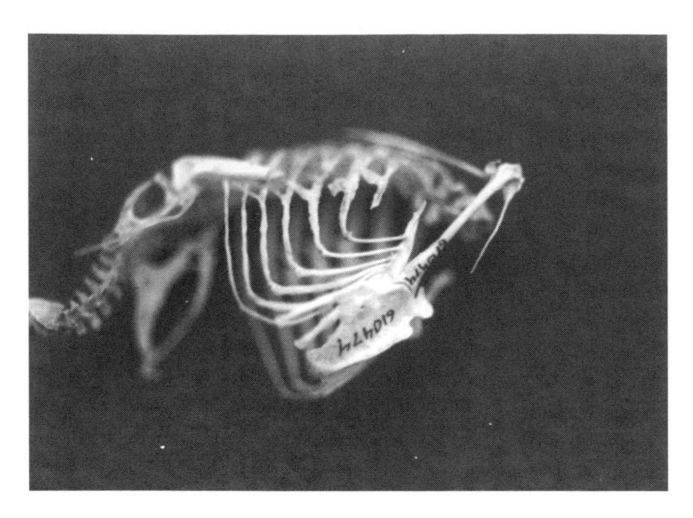

Lateral view of the skeleton of the small, flightless (or nearly so) suboscine passerine the ash-colored tapaculo (*Myornis senilis*) of Colombia, showing the absence of a keel on the sternum and the clavicular splints, all that remains of the furcula. (U.S. National Museum specimen #610474; photo by author)

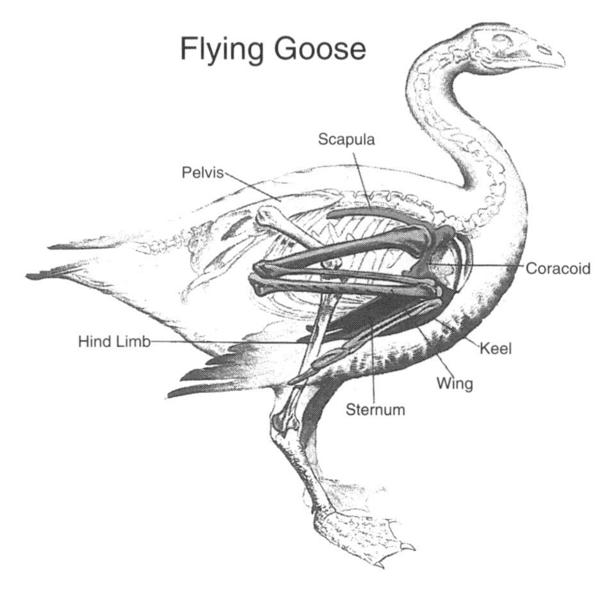

Flying birds, like this modern goose, have well-developed scapulae and coracoids that meet at a sharp angle and help brace the wing bones. They also have a sizable keel for the attachment of flight muscles. (Drawing by Douglas Cramer, from James and Olson 1983)

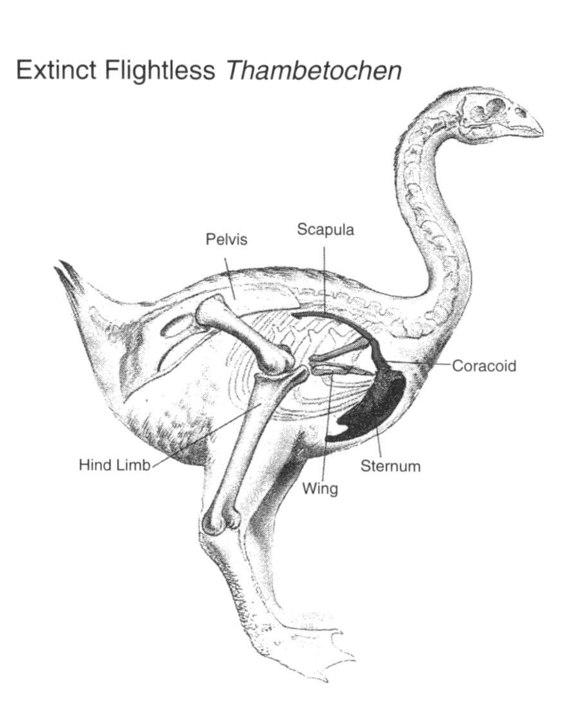

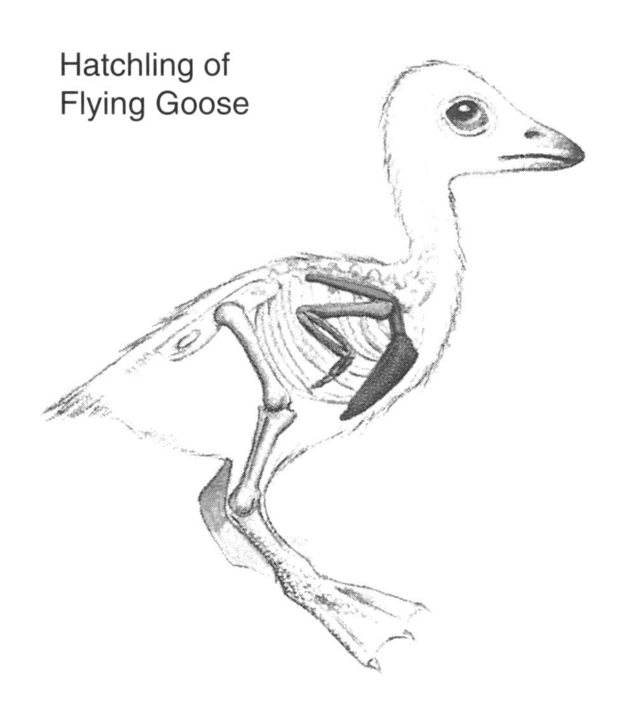

Compared with a flying goose, the extinct flightless Hawaiian "goose" *Thambetochen, left,* had stout hindlimbs, a massive pelvis, small and unsculptured wing bones, and a small sternum with no keel. The angle between the coracoid and scapula is wide, a trait common among flightless birds but never found in adult flying birds. Some of the skeletal features of adult flightless birds result from the retention of juvenile characters, a process called neoteny, or arrested development. The hatchling of a flying goose, *above,* is too young to fly. Notice the open angle between coracoid and scapula, reduced wings, and absence of a keel. (Drawing by Douglas Cramer, from James and Olson 1983)

not fused. It had a keel-less sternum and also may have lacked clavicles in some individuals (Olson and Wetmore 1976; Olson and James 1991). It was a big gosling—a striking case of arrested development.

As still another corollary of the flightless condition, the furcula (fused clavicles) become greatly reduced and even splintlike. This is true of all ratites and is also true of other flightless birds, in which the furcula is no longer needed for the attachment of the pectoralis flight muscles or as a pectoral transverse spacer, or "spring," in flight (Jenkins et al. 1988). In the case of the recently evolved moa-nalos, the furculae may be absent or "weak, with widely divergent rami" (Olson and James 1991, 38). Clavicles are splintlike in ratites, were greatly reduced or absent in *Diatryma*, and are splintlike in numerous independently evolved flightless birds (Glenny and Friedmann 1954).

The reduction or loss of the furcula raises an interesting question: If this is a pervasive characteristic of flightless birds, why would one expect to find a fully developed furcula in flightless bipedal dinosaurs? Could it be that the structure identified as a furcula in theropods is really something else, a structure not homologous with the analogous avian wishbone? Harold Bryant and Anthony Russell of the University of Calgary have studied the furcula in archosaurs and birds, and they raise the possibility that the avian furcula may be unique, a newly evolved "neomorph, or . . . the reappearance of a 'lost' structure" (Bryant and Russell 1993, 171). Their suggestion makes good sense in light of the reduced clavicles found among living flightless birds.

Still another characteristic of adult ratites that is seen in the embryos of rails and other modern birds is the broad unossified region between the ilium and ischium in the pelvic region known as the ilioischiatic fenestra. This condition, also found as a primitive state in *Archaeopteryx*, *Hesperornis*, and *Ichthyornis*, is surely neotenic in living flightless birds. Another neotenic feature seen in the embryos of today's flying birds and in the adults of some ratites, such as the ostrich, is sutured skulls rather than the extensively fused skulls characteristic of adult flying birds.

Some Flightless Birds of Hawaii. Perhaps the most remarkable study in the evolution of flightlessness comes from Storrs Olson and Helen James's monograph of 1991 on the late Holocene fossils of nonpasserine birds from the Hawaiian Islands, all thought to have been extirpated by the Polynesians shortly after they arrived around A.D. 300 (Olson and James 1982). The original avifauna probably numbered more than fifty species, outnumbering the extant Hawaiian avifauna. This work surely ranks as one of the most outstanding studies of avian biological archaeology ever undertaken.

Olson and James's discoveries include at least two flightless ibises, *Apteribis glenos* and *A. brevis*. No ornithologist would ever have dreamed of a flightless ibis, but within a year of their discovery, another flightless ibis from cave deposits in Jamaica was discovered, this time in the drawers of fossil mammals in the American Museum of Natural History; it was named *Xenicibis xympithecus* (Olson and Steadman 1977). The leg bones of these ibises were so much shorter and stouter than those of typical ibises that until additional material was discovered their identity was uncertain. The flightless ibises of Hawaii were restricted to the islands of Maui Nui and "could not have colonized these islands more than 1.8 million years ago, the age of the oldest rocks found on Molokai, the oldest of the Maui Nui group" (James and Olson 1983, 40). They are thought to have made a living probing in forest floor litter in the manner of the kiwi.

The other bizarre birds to come from these discoveries are the extinct flightless "geese," *Thambetochen* ("amazing goose"), which were in fact ducks that became gooselike in size and morphology. According to Olson and James, "The big flightless 'goose-like' birds of the Hawaiian Islands may actually have been derived from something more like a Mallard (*Anas platyrhynchos*) and certainly were not derived from true geese, although in their terrestrial, herbivorous habits they were undeniably 'goose-like'" (1991, 28). There were at least four species of these flightless moa-nalos (from the Hawaiian words *moa*, fowl, and *nalo*, lost or vanished). Olson and James believe that they were the ecological equivalents of the giant tortoises of the Galápagos and Indian Ocean islands, on which there are no large herbivorous birds, and that they evolved in response to the absence of herbivorous mammals. *Thambetochen chauliodous* (Olson and Wetmore 1976), described from Molokai, had a beak with toothlike projections on the tomia, and one genus is named *Chelychelynechen* ("turtle-jawed goose"), "so named for the decidedly chelonian aspect of the rostrum and mandible" (32).

Preying on these "geese," ibises, and other birds were a total of six hawks, goshawklike owls, and a large eagle. The moa-nalos were quite divergent from any living forms and were characterized by extremely large and robust hindlimb elements, excessively reduced wings and pectoral girdles, and a sternum that lacked a keel and was unfused along the midline of the posterior half, as well as the features noted earlier: the coracoid articulating with the scapula at an extremely obtuse angle, as in living ratites, and a weak (possibly absent, in one population) furcula, with widely divergent rami. Lacking skulls, these

Life reconstruction of the extinct Hawaiian flightless "goose" *Thambetochen chauliodous*, known from the Pleistocene. These ponderous goose-like ducks (*Thambetochen* and allies *Chelychelynechen* and *Ptaiochen*), called moa-nalos ("vanished fowl"), had bony, toothlike protuberances along the mandibles and were the major grazers of the Hawaiian ecosystem. (Painting by H. Douglas Pratt; courtesy Bernice P. Bishop Museum)

birds could have passed for ratites, so similar in morphology and proportions were they to the smaller of the moas of New Zealand. Interestingly, a large flightless goose, *Cnemiornis*, with greatly reduced wing bones and powerful legs adapted for a terrestrial existence, evolved on the islands of New Zealand. *Cnemiornis* is thought to have died out before the arrival of humans.

In addition to the truly bizarre, Olson and James have recorded no fewer than nine or ten species of flightless rails, in addition to the two known historically, that were distributed throughout the Hawaiian Archipelago, with one or more species occurring on each island. All of the extinct rails of Hawaii were flightless. The flightless characteristics of these birds therefore certainly derived independently on each island, and each developed in a similar manner, as is true for all flightless rails. One species, the Molokai rail (*Porzana menehune*), is the smallest rail yet known.

The Time Required to Evolve Flightlessness. How much time is necessary for flying birds to lose their powers of flight has been a subject of some controversy. In the past it has often been thought that vast time spans were required—tens of millions of years perhaps; but it now seems more likely that the evolution of flightlessness and the concomitant attainment of large size, as seen dramatically in the ratites, needs relatively little time, especially on islands.

Once again, the rails provide telling evidence. Storrs Olson has pointed out that "the span of time needed to evolve flightlessness in rails can probably be measured in generations rather than in millennia" (1973, 34). This generalization is bolstered by the occurrence of two nearly identical flightless moorhens some 400 kilometers (250 miles) apart in the south Atlantic, on Tristan da Cunha and on Gough Island, the moorhens *Gallinula n. nesiotis* and *G. n. comeri*, respectively. They are perhaps best treated as

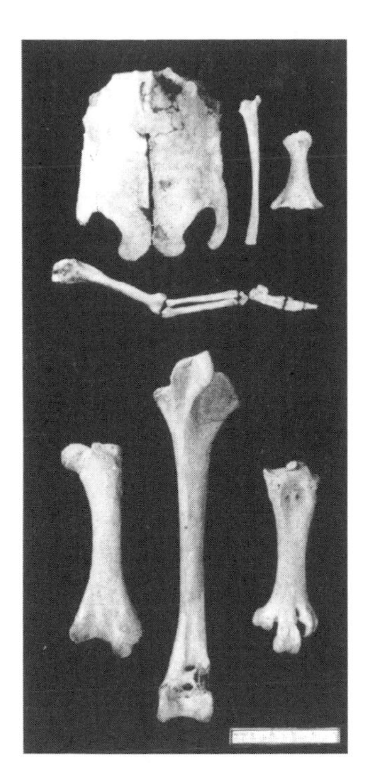

Fossil bones of *Thambetochen chauliodous*. *Left to right and top to bottom,* sternum (note lack of keel), scapula, coracoid, right wing (note drastic reduction), femur, tibiotarsus, and tarsometarsus. The pelvis and hindlimbs were extremely robust, and the wing and pectoral girdle was excessively reduced. In addition, the sternum lacked a keel and the posterior half was unfused along the midline; the furcula was weak, with widely divergent rami. In all but the head, the moa-nalos closely resembled ratites convergently. Note that the bones have come to resemble convergently those of the large ratites in loss of the sternal keel, reduction of the wing to a vestige, and the greatly increased size of the leg bones. Scale = 5 cm. (From Olson and Wetmore 1976; photo by Victor E. Krantz)

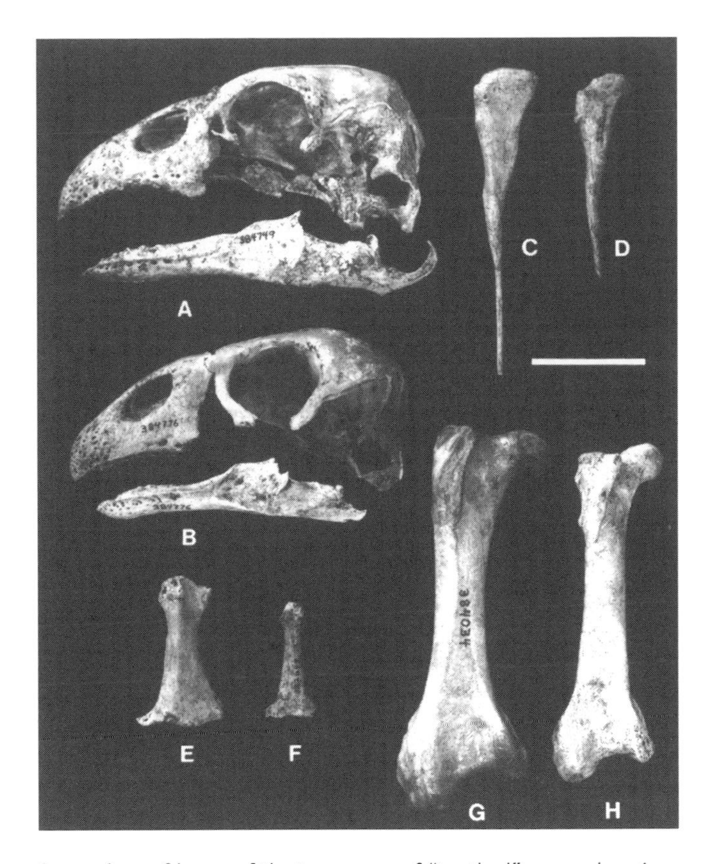

Comparison of bones of the two genera of "toothed" moa-nalos: *A,* skull and mandible of *Thambetochen chauliodous* in left lateral view; *B,* skull and mandible of *Ptaiochen pau* in left lateral view; *C,* left fibula of *T. chauliodous* in lateral view; *D,* left fibula of *P. pau* in lateral view; *E,* right coracoid of *T. chauliodous* in ventral view; *F,* right coracoid of *P. pau* in ventral view; *G,* right femur of *T. chauliodous* in anterior view; *H,* right femur of *P. pau* in anterior view. Scale = 3 cm. (From Olson and James 1991; courtesy Storrs Olson)

New Zealand's extinct South Island goose (*Cnemiornis*) was about 1 m (3.3 ft.) tall. Note the keel-less sternum, strong hindlimbs, and the open fenestrae of the pelvis. (From Owen 1879)

subspecies or as sibling species (similar populations that have become reproductively isolated) of the modern common moorhen (*Gallinula chloropus*) and are therefore of only very recent origin. *Fulica newtonii* of Mauritius and Réunion, and *Fulica chathamensis* of New Zealand and the Chatham Islands are other examples in which very similar flightless subspecies evolved independently and recently from a common ancestor on two distant islands (Diamond 1981, 508). In addition, there are flying and flightless races in a single rail species, *Dryolimnus cuvieri;* and a completely flightless rail, *Atlantisia elpenor,* is present on Ascension, an island in the south Atlantic whose oldest rocks have been dated at just 1.5 million years. Among ducks, incidentally, *Anas aucklandica* includes flying subspecies on New Zealand and Chatham Island and a well-known flightless subspecies on Auckland Island.

We do not know how long it would take for these rails to become unrecognizable as rails through the evolution of gigantism or bizarre adaptations. But we need only look at the flightless ibis *Apteribis* to see how modifications for flightlessness can disguise a bird's ancestry. "The hindlimb . . . is so modified from that of typical ibises as almost to defy identification. At first we had only the femur, tibiotarsus, tarsometatarsus to work with, and of all modern birds the proportions of these elements most closely approached those of the kiwis . . . and it was not until we received the associated material from Maui . . . that our suspicions were confirmed" (Olson and Wetmore 1976, 250–251). Had it not become extinct, what might *Apteribis* have looked like in another million years? The oldest rocks on the Hawaiian Islands date to less than 6 million years, and, as pointed out above, the ibises can be no more than 1.8 million years old, the age of the oldest rocks on Molokai. The Galápagos Islands, also of volcanic origin, with their incredible avian and chelonian adaptive radiation, have a probable maximum age of some 3 million years (Adams et al. 1976; Bailey 1976; Hall 1983), which means that the Galápagos cormorant must have developed flightlessness in a brief period of geologic time.

Huxley's "Waifs and Strays"

Of all flightless birds none has received more attention than the living ratites, which have presented avian systematists with the most intractable problem of the past century. The ratites (from the Latin *ratis*, raft, for the shape of the sternum) are generally large birds characterized by a keel-less sternum and a paleognathous palate, a distinctive arrangement of palatal bones so-named by William Plane Pycraft in 1900 to denote their presumed reptilelike configuration. The living ratites are the ostrich (*Struthio camelus*) of Africa and Arabia, two species of South American rheas, greater and lesser (*Rhea americana* and *R. pennata*), the Australian emu (*Dromaius novaehollandiae*), three species of cassowaries (*Casuarius*) of northeastern forested Australia and New Guinea, and three species of New Zealand kiwis (*Apteryx*). The long list of extinct ratites includes some eleven species of New Zealand moas (Dinornithidae), an undetermined number of Malagasy elephantbirds (Aepyornithidae), and some fifteen or more described bony fragments and eggshells dubiously assigned to one or another group of so-called ratites. Then there are some nine extinct species of Australian mihirung birds (Dromornithidae), which may not be true ratites.

Charles Darwin (1859, 106, 226) held the view that the ratites were derived from flying ancestors through "disuse" of the wings and increased use of the hindlimbs, but his adversary Richard Owen explained their development through "arrested development of the wings unfitting them for flight" (1866, 12); Owen, in other words, advocated a neotenic origin for ratites, much as Gavin de Beer (1956) would later reason. In 1867, Huxley made the first truly comprehensive attempt to unravel the evolutionary affinities of ratites by describing the dromaeognathous (now paleognathous) palate, and controversy has surrounded these enigmatic flightless birds ever since.

Huxley defined the ratites by their palatal structure, considering it primitive among birds. He envisioned these "struthious" birds (from *Struthio*, the generic name of the ostrich) as ancient relics of a once-great evolutionary radiation: "Though comparatively few genera and species of this order now exist, they differ from one another very considerably, and have a wide distribution, from Africa and Arabia over many of the islands of Malaysia and Polynesia to Australia and South America. Hence, in all probability, the existing Ratitae are but the waifs and strays of what was once a very large and important group" (1867, 419). Huxley had proposed to divide the class Aves into three orders: the Saururae, to be represented only by the fossil *Archaeopteryx;* the Ratitae, which comprised the large flightless birds and the kiwi; and the Carinatae, to accommodate all other birds. He was perplexed by the tinamous, which did not fit neatly into either the Ratitae or the Carinatae. Tinamous did possess the paleognathous palate that Huxley used to group the ratites, but they had the sternum typical of the Carinatae—indeed, it closely resembled that of the Galliformes. So, Huxley began his section on Carinatae by stating that it embraces all birds except the Ratitae and then pointed out that the tinamous are unique among the carinates in having a completely struthious palate; they are volant paleognaths.

Although many followed Huxley's proposal, others, particularly Max Fürbringer (1888, 1902), argued that the

Life reconstruction of the extinct flightless Hawaiian ibis (*Apteribis glenos*). Shown with it are two extinct flightless rails. (Painting by H. Douglas Pratt; courtesy Bernice P. Bishop Museum)

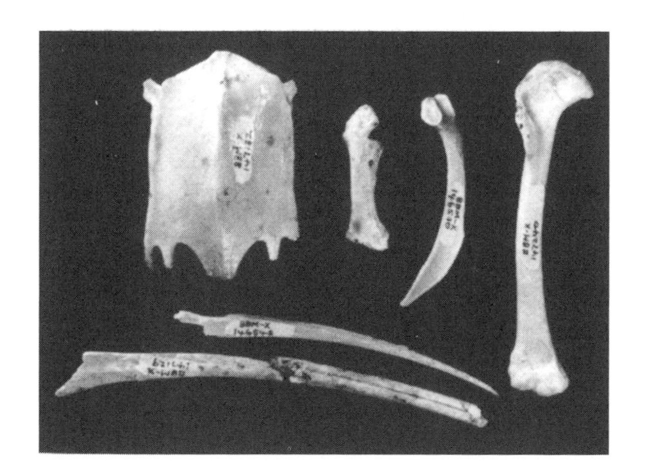

Fossil bones of *Apteribis glenos*. *Left to right and top to bottom,* sternum (note lack of keel), coracoid, scapula, humerus (note reduced size), rostrum fragment, and two mandible fragments. (From Olson and Wetmore 1976; photo by Victor E. Krantz)

The ratites are a diversified assemblage of flightless birds whose relations have remained among the most controversial within the class Aves. They are united by a number of primitive characters, such as the paleognathous palate, and neotenic characters, such as the keel-less sternum, the obtuse angle of the scapulocoracoid, and an open ilioischiatic fenestra. They are collectively characterized by a ratite type of skull, with schizorhinal nostrils and rhynchokinesis. Australian and New Zealand natives include the cassowaries, emu, and kiwi. *From left to right and top to bottom,* Cassowaries (Casuariiformes, Casuariidae, Australia and New Guinea, 1 genus, 3 species): northern cassowary (*Casuarius unappendiculatus)*. Emus (Casuariiformes, Dromiceidae, Australia, 1 genus, 1 species): emu (*Dromaius novaehollandiae)*. Kiwis (Apterygiformes, Apterygidae, New Zealand, 1 genus, 3 species): brown kiwi (*Apteryx australis)*. African ratites, ostriches (Struthioniformes, Struthionidae, southern Africa, 1 genus, 1 species): ostrich (*Struthio camelus)*. South American ratites, Rheas (Rheiformes, Rheidae, South America, 2 genera, 2 species): greater rhea (*Rhea americana)*. Tinamous are flying paleognaths related to ratites. Tinamous (Tinamiformes, Tinamidae, Neotropics, 9 genera, 47 species): red-winged tinamou (*Rhynchotus rufescens)*. (Drawings by George Miksch Sutton)

similarities among ratites were due to convergent evolution and that each group had evolved independently from unknown ancestors. In response, many systematists, including Alfred Newton (1896), called Fürbringer's arguments "hardly convincing," and the heated argument over ratite unity continued.

The antiquity of the ratites has also been argued on the basis of the similarity between the hand of the ostrich and that of such small dinosaurs as *Ornitholestes* and *Struthiomimus*. But rudimentary claws or spurs are found in an endless array of modern birds ranging from hawks to ducks and rails (Fisher 1940; Stephan 1992), and modern embryology shows the dinosaurian scapulocoracoid to be a neotenic character associated with flightlessness. As we have seen, primitive features in organisms appear earlier in embryogenesis than more specialized characters. Thus, it is not unreasonable to expect numerous primitive characteristics to appear through neoteny, especially if the ratites are, indeed, "big chicks," as proposed by de Beer (1956).

Among the many related studies that could be mentioned in this context, perhaps the most interesting is a series of papers by Percy Lowe (1928, 1935, 1944) in which he proposed that ratites and small coelurosaurian dinosaurs actually shared a common ancestor. According to Lowe, the ratites derived from creatures that had never acquired flight. Anatomical evidence, however, makes it clear that ratites, like all other flightless birds, derived from flying predecessors. We need only look at the vestigial flight quills on the wing of the cassowary to put the question to rest. Other "flight characteristics" of ratites include the presence of the same type of cerebellum (the part of the brain that controls equilibrium) found in flying birds and clearly evolved for flight; a wing skeleton built on the same general plan as that of flying birds, with the fusion of several carpal elements to form a "flight" carpometacarpus; and tail vertebrae fused to form the uniquely avian pygostyle, which lightens the skeleton for flight. And in the rheas, the feathers on the first digit are still arranged to form an alula, or midwing slot, clearly a carinate adaptation for flight. Lowe's arguments, which also posited that penguins derived from primarily terrestrial, nonvolant ancestors, were attacked by many authors, but particularly George Gaylord Simpson (1946), who, based on the existence of *Archaeopteryx* and much other hard evidence, found Lowe's views totally untenable.

The Paleognathous Palate. As for the ratites themselves, ever since Huxley's time the major arguments for or against monophyly have centered on the paleognathous palate. In 1948, Sam McDowell recognized four types of paleognathous palates (reserving a possible fifth type for the imperfectly preserved elephantbirds). These included the tinamiform type, for rheas and tinamous; the casuariiform type, for cassowaries and emus; the struthioniform type, for the ostrich; and the apterygiform type, for the kiwi. In McDowell's view, the ratite groups evolved independently and the palates could be developed through neoteny because other features of the ratite skull, such as the presence of sutures, are neotenic. This view became widely influential. In the 1951 and 1960 editions of Alexander Wetmore's widely used classification of birds, Wetmore noted: "For years I have felt that recognition of the Palaeognathae, as a separate group apart from other birds, on the basis of a supposed peculiarity in the palate, stood on flimsy ground. . . . As there is no clear-cut separation, the former Palaeognathae must be combined with the Neognathae" (1960, 4). And the influential classification of Ernst Mayr and Dean Amadon noted that "the problem of ratite phylogeny continues to receive much attention. The present consensus is that the main groups of these birds are of independent origin" (1951, 4).

Gavin de Beer (1956, 5), confronting the problem of the palate, concluded that "the palate of the ratites is not primitive but neotenous, and represents an early stage through which the palate of many Carinates passes during the development period" (1956, 5). In 1963, however, Walter Bock maintained that the paleognathous palate was indeed a definable functional unit, and indicated a common origin of all the ratites and tinamous, but that "the ratites do not appear to be primitive among birds . . . nor do they have to be any older than other typical avian orders" (1953b, 53). Kenneth Parkes and George Clark described a peculiar conformation of the rhamphotheca in ratites and tinamous not found in other birds and concluded that this was "an additional piece of evidence that resemblances among this group are to be attributed to monophyletic origin rather than to convergence" (1966, 469).

Even in Huxley's era, the paleognathous palate and other characters were disputed as panaceas for avian systematics. Writing in 1896, Alfred Newton viewed use of the palate with great skepticism:

> The present writer is inclined to think that the characters drawn thence owe more of their worth to the extraordinary perspicuity with which they were presented by Huxley than to their own intrinsic value, and that if the same power had been employed to elucidate in the same way other parts of the skeleton—say the bones of the sternal apparatus or even the pelvic girdle—either set could have been made to appear quite as instructive and perhaps more so. Adventitious value would therefore seem to have been acquired by the bones of the palate through the

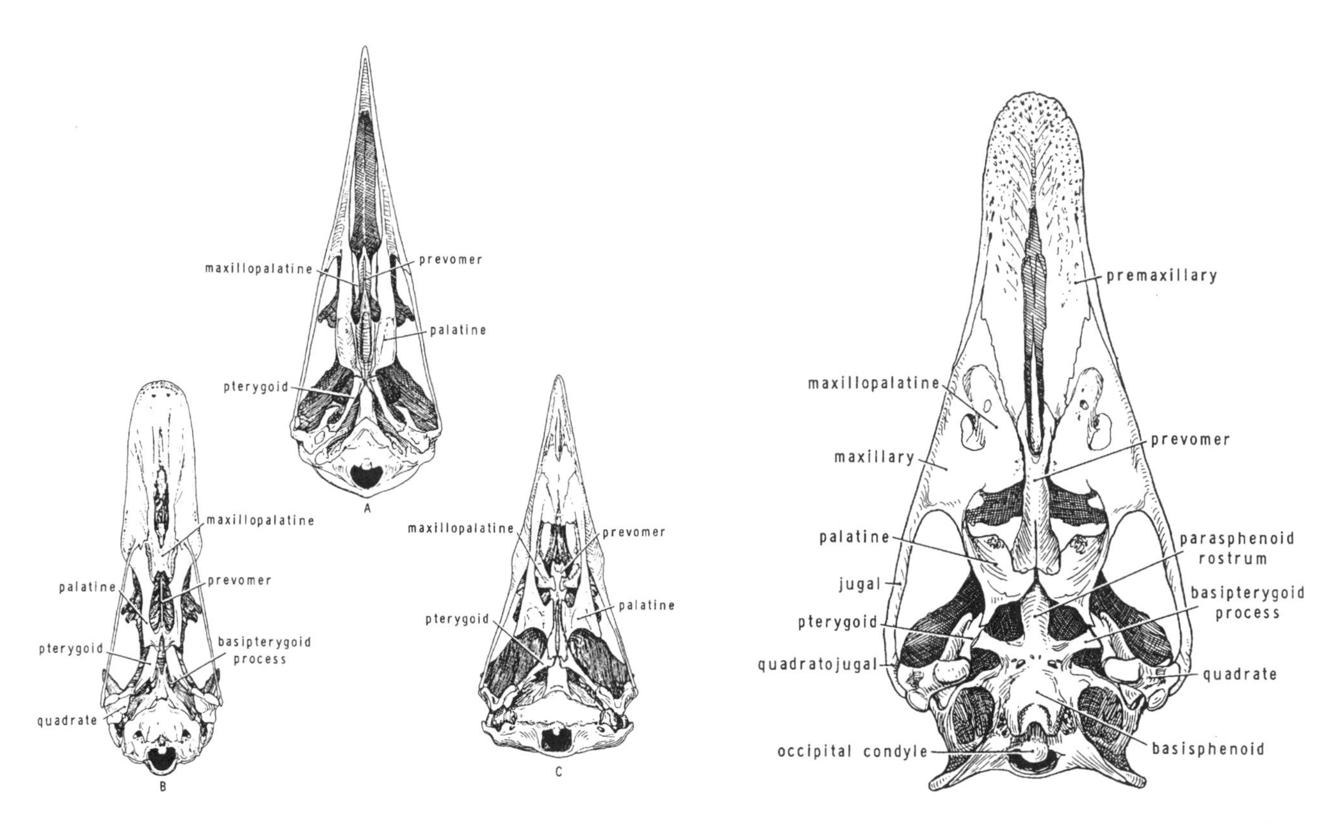

Ratites and tinamous have traditionally been defined as paleognathous birds on the belief that their palatal type indicated a relationship; it is now known to be the primitive avian palate. *Left,* the three "classical" types of palate: *A,* schizognathous, laughing gull (*Larus atricilla*); *B,* desmognathous, Canada goose (*Branta canadensis*); *C,* aegithognathous, common raven (*Corvus corax*). *Right,* paleognathous, greater rhea (*Rhea americana*). In the paleognathous palate the prevomers extend back to articulate with the palatines (at their posterior ends) and the pterygoids, thus separating both from the parasphenoid rostrum. Large basipterygoid processes are also characteristic, although they are found in a number of other birds. (From Van Tyne and Berger, *Fundamentals of ornithology,* copyright © 1959, reprinted by permission of John Wiley and Sons)

fact that so great a master of the art of exposition selected them as fitting examples upon which to exercise his skill. (1896, 84)

Recent Ratite Phylogenies and the Lithornithids. Joel Cracraft's study of the ratites using skeletal characters in 1974 was the first attempt to use cladistics on a group of birds, and it was therefore, in a sense, a test of the power of cladistic methodology. The study appears to have been undertaken to prove the group monophyletic, which would conform to Cracraft's zoogeographic hypothesis that the ratites were a very ancient group that moved to their current geographical ranges on drifting continents (Cracraft 1973a, 1974). In his reasoning, the ostrich and the rheas shared a common ancestor on the southern continent of Gondwana, and its split in the late Cretaceous made the birds' ancestors vicariant passengers on the new continents of Africa and South America. Unfortunately, Cracraft's study combined the use of primitive and neotenic characters, was unable to establish character polarities, and in general was so seriously flawed that a discussion of its shortcomings is not possible here; the reader is referred to Olson 1982b, 1985a, and Sibley and Ahlquist 1990.

Anthony Bledsoe (1988b) omitted the tinamou from the analysis, refined the general methodology, and his phylogeny, though it employed many questionable characters, generally conformed with that of the DNA-DNA hybridization comparisons made by Charles Sibley and Jon Ahlquist (1990). In sharp contrast to these studies, however, is that of Walter Bock and Paul Bühler (1988), who concluded from their study of the structure of the tongue apparatus that the paleognathous birds were divisible into two suborders: the Struthioni, containing the ostriches and elephantbirds, and the Tinami, containing all other living and recent fossil groups. They found no close relation between ostrich and rhea and therefore no evidence for direct dispersal between South America and Africa. Loss of flight, according to Bock and Bühler, could have occurred at least two and probably as many as four times. An excellent detailed study of the ratite middle ear by Matthias Starck (1995) supported the view that the paleognaths are mono-

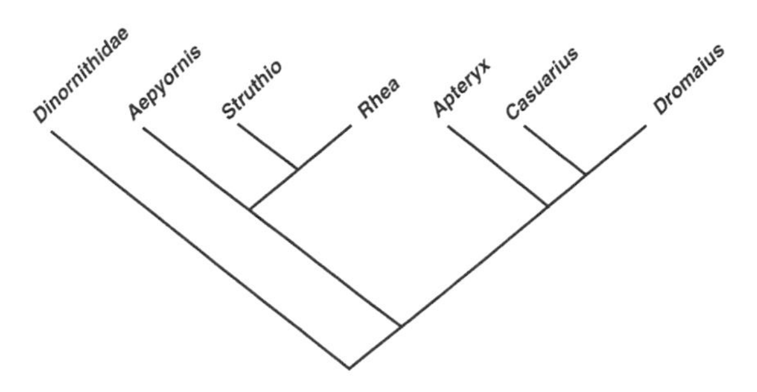

A classical view of ratite phylogeny estimated from numerical cladistic analysis of postcranial skeletal characters performed by Anthony Bledsoe. The tree is rooted to a hypothetical ancestor (not shown) with the ancestral state for all characters, which corresponds to the Tinamidae. (Modified after Bledsoe 1988)

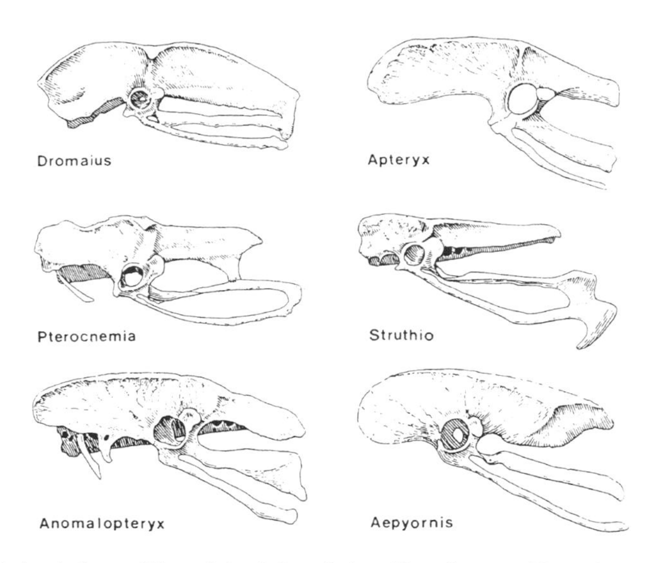

Lateral views of the pelvic girdles of six ratites: the emu (*Dromaius novaehollandiae*), the brown kiwi (*Apteryx australis*), the lesser rhea (*Rhea pennata*), the ostrich (*Struthio camelus*), a moa (*Anomalopteryx didiformis*), and an elephantbird (*Aepyornis hildebrandti*). *Anomalopteryx* is after Archey 1941; *Aepyornis* is after Andrews 1897. The diversity of ratite pelvic structure would be astounding if all derived directly from a single flightless common ancestor and is perhaps best explained by ratites having evolved through a combination of increasing body size and neoteny. (Courtesy Academic Press and R. J. Raikow)

phyletic but could not resolve any possible relation to neognathous birds. And in a review of parental care systems in birds, David Ligon of the University of New Mexico concluded that the existence of exclusive male parental care in ratites and tinamous "reflects a common evolutionary history" (1993, 7).

In 1984, Christopher McGowan examined the ontogeny of the tarsus in birds and found what he considered to be two distinctive conditions. He concluded that carinates have a pretibial bone that fuses with the calcaneum, whereas ratites and tinamous do not; instead, they have an ascending bony process that fuses with the astragalus, as in theropod dinosaurs. According to McGowan, "This implies that, since carinates possess the unique pretibial bone, they could not have given rise to the ratites, which are primitive for this feature" (733). The cleared and stained embryos of ratites McGowan studied were near hatching, however, casting doubts on the reliability of his results, and his study and conclusions have been seriously challenged (Martin and Stewart 1985).

Numerous biochemical studies (especially Ho et al. 1976; Prager et al. 1976; and Prager and Wilson 1980) using micro-complement fixation techniques in a comparison of several proteins from a variety of birds showed that "there were at least two major phases in the history of birds: the first probably involving an early adaptive radiation of paleognathous birds with limited capacity for sustained flight and the second involving a later adaptive radiation of carinate birds" (1211). Sibley and Ahlquist's extensive DNA comparisons led them to conclude the following: "The living ratites are monophyletic and . . . the tinamous are their nearest living relatives. It is clear that the Emu and cassowaries are the most closely related genera and that the kiwis are their next nearest relatives. The position of *Rhea* relative to *Struthio*, and to the kiwi-cassowary-emu group, remains uncertain" (1990, 288).

A dramatic breakthrough in this seemingly intractable problem occurred with the description in 1981 of fossil material of fully volant, medium-sized paleognathous carinate birds from the Tertiary of North America by Peter Houde and Storrs Olson. Before these fossils were discovered, no pre-Quaternary birds could be shown beyond doubt to possess the paleognathous palate. These birds, which are known from nearly all skeletal elements, including the skull, are now understood to have been widespread in the Paleocene and Eocene of western North America and Europe. They had a well-developed wing and pectoral architecture and a definite paleognathous palate. Fossils of these birds had been represented in museum collections for more than a century masquerading under the names of a number of disparate modern families of carinate birds. That these Paleogene birds, and the paleognathous palate, were primitive is indicated by the existence of at least some features of the paleognathous palate in the early ontogeny of some neognathous birds (de Beer 1956; Jollie 1958), the occurrence in these birds of a reptilian splenial bone, the overall lack of fusion of cranial elements, and the generalized nature of the postcranial elements. As Houde and Olson concluded, "The new fossil birds reported here are probably remnants of what may have been a diverse radiation of paleognathous carinates. . . . Tinamous and ratites may have descended independently from various families or orders within this radiation of paleognaths, or some of the ratites may have evolved secondarily from neognathous birds through neoteny" (1237).

Years earlier, in discussing the possible phylogeny of the ratites, Kenneth Parkes and George Clark had proposed that the living tinamous might represent a flying stage in the evolution of the flightless ratites and they postulated a more generalized ancestor that "we may thus for convenience call the proto-tinamous" (1966, 467). These Paleogene flying paleognathous birds, described in 1988 by Peter Houde, are in fact very like tinamous in many features and certainly approach the description, both in time and in morphology, of the proto-tinamou ratite ancestor postulated by Parkes and Clark.

Houde has placed these Paleogene birds in a single order, Lithornithiformes, which he describes as differing from all other known avian orders except the Tinamiformes by having a rhynchokinetic skull, a paleognathous palate, and a carinate sternum. Lithornithids differ from the tinamous in more specific characters, such as the structure of the sternum, which in living tinamous closely resembles that of galliform birds and has been the main evidence for a link between the two groups. Lithornithids had a small, weak tail, similar to that of tinamous, with a small and poorly fused pygostyle. Their claws, however, unlike the flat claws of tinamous, were large and strongly curved, adapted for perching in trees. Houde writes, "I cannot overemphasize the similarity of the Lithornithiformes and the Tinamiformes," and "analysis of the Lithornithiformes is most appropriately made . . . with the Tinamiformes, the group to which they are phenetically most similar and from which they presumably diverged" (1988, 107).

He groups the fossil taxa into a single family of three genera—*Lithornis*, *Pseudocrypturus*, and *Paracathartes*—and eight species. They had a jaw apparatus somewhat like that of kiwis, being "sensitive-billed probers," perhaps inhabiting the wooded floodplains that characterized parts of western North America during the Paleogene. Although the living tinamous are ground-dwellers capable of short bursts of powerful flight, the lithornithids were not particularly cursorial and were capable of sustained flight and gliding.

In Houde's view, the lithornithids are paraphyletic primitive birds. Because of the overall primitive nature of the lithornithids and because he believes, like many others, that DNA comparisons at this high taxonomic level may be beyond the resolution powers of the DNA hybridization technique, there still remain at least four possible phylogenies of the ratite birds and the tinamous (Feduccia 1985b; Houde 1988; Olson 1985a):

1. The group is strictly monophyletic and is derived from birds possessing the neognathous palate, as suggested by Bock (1963b) and Cracraft (1974). This would mean that flightlessness arose but once from a common flightless ancestor and that all living ratites, kiwis to ostriches, have attained their current distribution by being passengers on drifting continents; this is Cracraft's well-known vicariance biogeographic hypothesis (1973a, 1974).
2. The living paleognaths are the survivors of an ancient group of paleognathous flying birds that may be the sister-group of other living birds, but within the group, various ratites may have arisen from different groups of proto-tinamous.
3. The living ratites all arose through neoteny from neognathous ancestors on one or more occasions, so that each group or subgroup of ratites may be more closely related to a group of neognathous birds than to the other ratites.
4. Some of the ratites may be neotenic descendants from neognathous ancestors and therefore are secondarily paleognathous, and others may be ancient, having been derived from an archaic group such as the lithornithids.

Houde denies that any derived characters unite the monophyletic Palaeognathae and proposes that the living paleognathous birds are paraphyletic, with the tinamous and neognathous birds being sister-groups; this would be equivalent to figure *b* in his diagram (see p. 276). In sum, he writes, "Cladistic analysis of osteological evidence supports the primitiveness of paleognathous birds" (1988, 131). Although this is a controversial conclusion, it may well approach reality.

Certainly Houde's dramatic and provocative studies place the tinamous in a new and interesting light: living remnants of a once-great radiation of flying paleognathous birds, primitive to be sure, and possibly allied with the neognathous primitive Galliformes.

Tinamous and Kiwis. The modern tinamous are placed in their own order and family (Tinamiformes, Tinamidae) and are known from as early as only the medial Miocene of Argentina (Chandler and Chiappe, under study). The forty or so modern species range from southern Mexico to Patagonia, inhabiting regions as diverse as the deepest Amazonian jungles, the grassy slopes of the Andean plateau, and mountainous zones up to 4,270 meters (14,000 ft.). Tinamous forage entirely on the ground and run rapidly, though they can fly strongly for up to 100 meters (328 ft.) when alarmed. Their food consists of a va-

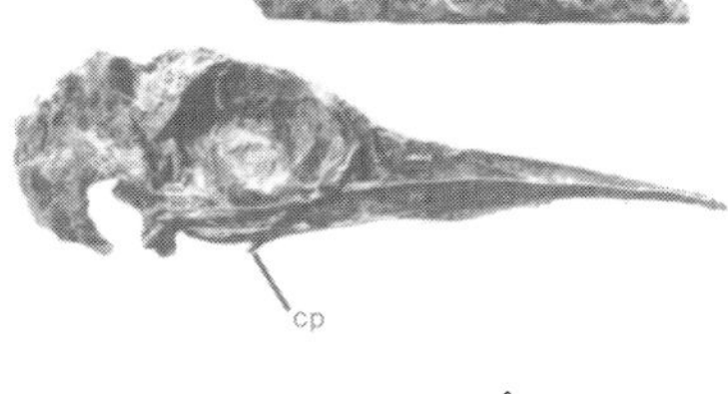

Skull of the flying paleognathous bird *Pseudocrypturus* ("false tinamou") *cercanaxius,* from the early Eocene Green River Formation of Wyoming. × 0.54. *Below,* cutout of skull to show the caudal process of the palatine (*cp*). (From Houde 1988; courtesy Nuttall Ornithological Club)

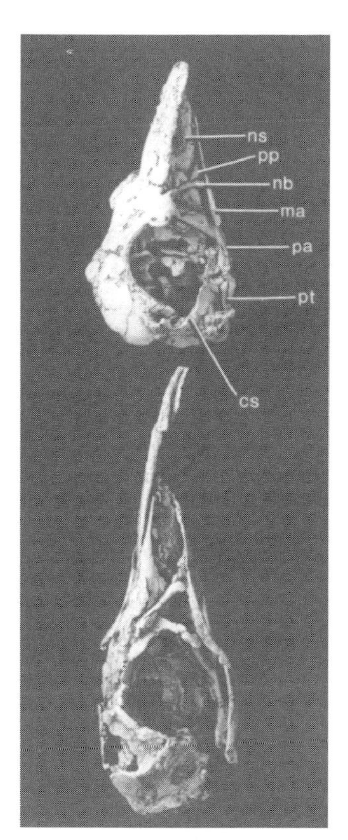

Right lateral aspect of lithornithid skulls. *Above, Lithornis celetius,* lacking rostral bill; *below, L. promiscuus*. Abbreviations: *cs,* coronal suture (disarticulated); *ma,* maxilla; *nb,* lateral nasal bar; *ns,* nasal septum; *pa,* palatine; *pp,* palatine process of maxilla; *pt,* pterygoid. (From Houde 1988; courtesy Nuttall Ornithological Club)

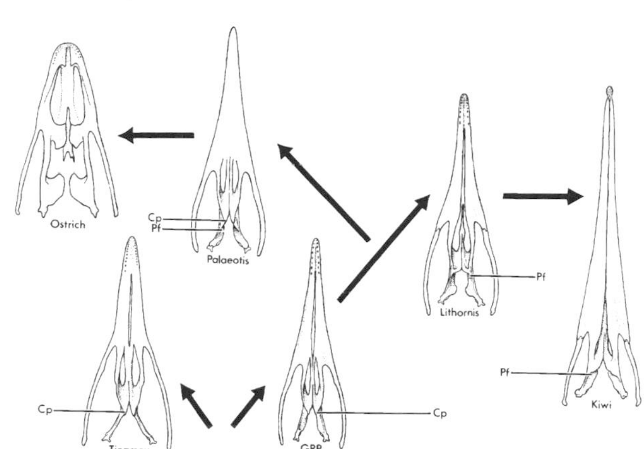

Ventral aspect of palates of lithornithids and selected other paleognathous birds (from Houde 1986). Arrows indicate hypothetical path of the evolution in grades of the palatal complex in a tinamou (*Tinamus*), *Pseudocrypturus cercanaxius* (GRP), *Lithornis promiscuus* (similar to *Paracathartes howardae*), *Palaeotis weigelti,* a brown kiwi (*Apteryx australis*), and an ostrich (*Struthio camelus*); not to scale. Abbreviations: *cp,* caudal process of palatine; *pf,* pterygoid fossa. (From Houde 1988; courtesy Nuttall Ornithological Club)

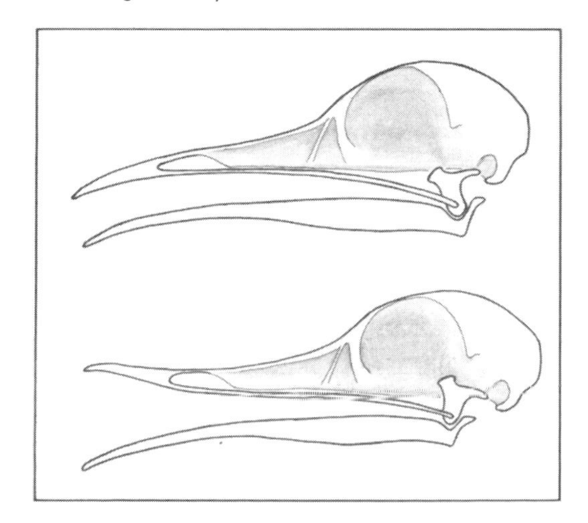

Distal rhynchokinesis in the skull of lithornithids. *Above,* bill at rest; *below,* quadrate protracted, premaxillae raised. Note the continuity of nasal and interorbital septa and the extreme rostral region of flexion of the dorsal nasal bar. (From Houde 1988; courtesy Nuttall Ornithological Club)

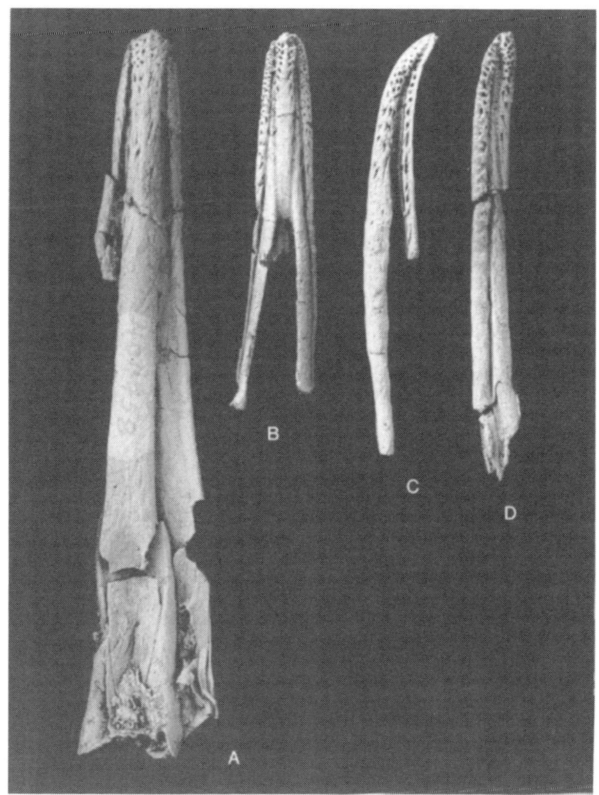

Lithornithid bills. *A,* dorsal aspect of upper bill of *Paracathartes howardae; B,* ventral aspect of rostral mandible of *P. howardae; C,* right lateral aspect of premaxilla of *Lithornis promiscuus; D,* right lateral aspect of rostral mandible of *L. promiscuus*. × 1.5. Note the foramina for rostral rami of the mandibular nerve and the "W" pattern formed by the rhamphothecal grooves, a character once thought to be a synapomorphy uniting the ratites and tinamous. (From Houde 1988; courtesy Nuttall Ornithological Club)

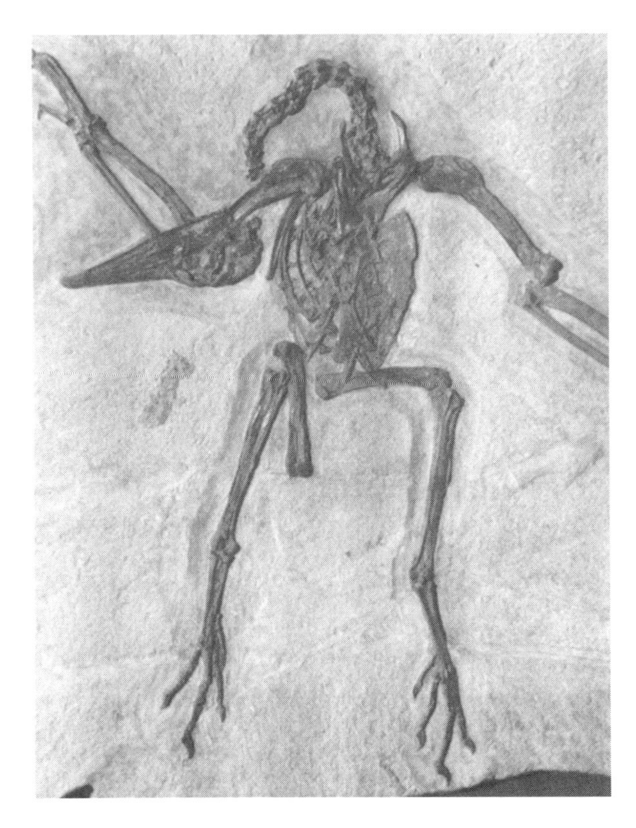

Crushed articulated skeleton of the flying paleognath *Pseudocrypturus cercanaxius.* (Privately owned by Siber and Siber; from Houde 1988; courtesy Nuttall Ornithological Club)

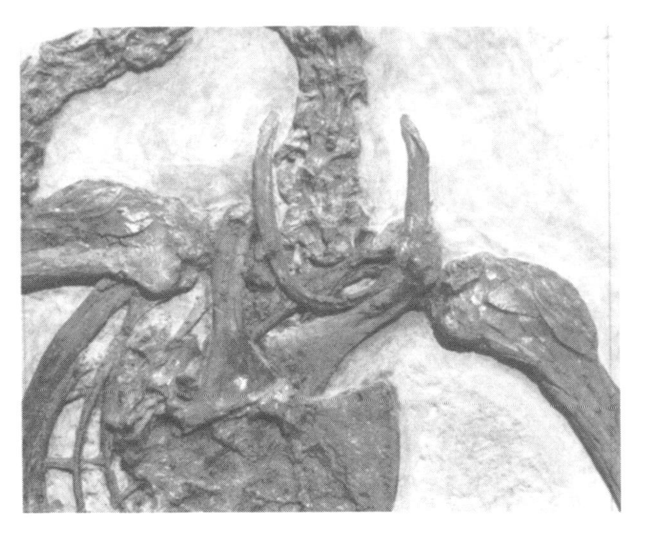

Enlargement of furcula and coracoids of crushed articulated skeleton of *Pseudocrypturus cercanaxius*. (Privately owned by Siber and Siber; from Houde 1988; courtesy Nuttall Ornithological Club)

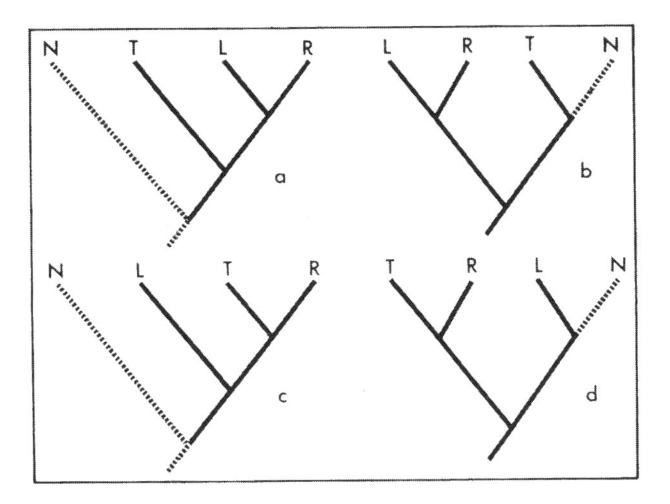

Hypothetical phylogenies of paleognathous and neognathous birds of unequal rank. Broken line = neognathous character state; solid line = paleognathous character state. Abbreviations: *L,* Lithornithiformes; *N,* neognathous birds; *R,* ratites; *T,* Tinamiformes. (From Houde 1988; courtesy Nuttall Ornithological Club)

Grades of ratite evolution as represented by the series of fossils *Lithornis celetius, L. promiscuus, Paracathartes howardae, Palaeotis weigelti,* and a modern cassowary (*Casuarius*), to scale. (From Houde 1988; courtesy Nuttall Ornithological Club)

Table 6.1 The Tertiary record of paleognathous and putatively paleognathous birds, exclusive of the Lithornithiformes

Epoch	Family	Taxon
Pliocene		
	Tinamidae	
		Eudromia olsoni, Argentina
		Nothura parvulus, Argentina
		Querandiornis romani, Argentina
	Struthionidae	
		Struthio asiaticus, India
		S. chersonensis, Greece, Ukraine, Kazakhstan
		S. wimani, China, Mongolia
		S. bradydactylus, Odessa
	Rheidae	
		Heterorhea dabbenei, Argentina
		Hinasuri nehuensis, Argentina
	Dromiceidae	
		Dromaius ocypus, Australia
Miocene		
	Tinamidae	
		Eudromia sp., Argentina
	Struthionidae	
		Struthio orlovi, Moldavia
	Opisthodactylidae (possibly Rheidae)	
		Opisthodactylus patagonicus, Argentina
Oligocene		
	Aepyornithidae	
		Stromeria fajumensis, Egypt
		Eremopezus eocaenus, Egypt
	Eleutherornithidae	
		Proceriavis martini, England
Eocene		
	Eleutherornithidae	
		Eleutherornis helveticus, Switzerland
	Struthionidae(?)	
		Palaeotis weigelti, Germany
Paleocene		
	Opisthodactylidae (possibly Rheidae)	
		Diogenornis fragilis, Brazil
	Family uncertain	
		Remiornis minor, France

Note: See table 6.2 for the Lithornithiformes. Dromornithidae not included. Familial allocations follow doctrine.

Table 6.2 Stratigraphic relations of lithornithid taxa

	North America
Eocene	
Bridgerian	Lithornithidae, genus indet.
Late Wasatchian	*Pseudocrypturus cercanaxius*
Early Wasatchian	*Paracathartes howardae*
	Lithornis cf. *nasi*
Middle Clarkforkian	*Lithornis promiscuus*
	Lithornis plebius
Paleocene	
Early Tiffanian	*Lithornis celetius*
	Europe
Eocene	
?Ypresian, Plastic Clay	*Lithornis* cf. *nasi*
Ypresian, London Clay D-E	*Lithornis vulturinus*
	Lithornis cf. *plebius*
London Clay C	*Lithornis vulturinus*
London Clay B	*Lithornis hookeri*
London Clay A	*Lithornis nasi*
	cf. *Pseudocrypturus cercanaxius*
?Ypresian, Blackheath Beds	Lithornithidae, genus indet.

Source: Houde 1988

riety of vegetable matter such as seeds, buds, and roots but also includes many insects. The nest, built on the ground, varies considerably in size, and the clutch likewise varies from one to twelve eggs. The chicks, highly precocial, are active soon after hatching.

Among the strangest of all living birds, the New Zealand kiwis are thought by Peter Houde (1988) to be derived directly from the *Paracathartes* grade of the lithornithids. The kiwis (order Apterygiformes, family Apterygidae) are represented by three living species, with two more known from Pleistocene deposits on New Zealand. Named by the native Maoris for their piercing cries, these nocturnal paleognaths live in thick, swampy tree-fern forests where they make a living probing for food in the moist, thick humus of the undergrowth with their long, nostril-tipped bills. The kiwi's principal food source, earthworms, is abundant in New Zealand—with some 178 native and 14 introduced species. Extraordinarily large olfactory lobes make the kiwi well suited to seek out these grubs and worms. The three chicken-sized species are the smallest of the ratites and have truly vestigial wings, some 5 centimeters (2 in.) long, with clawed tips that lie hidden in the coarse plumage. Kiwis spend their days quietly and nest in underground burrows, laying the largest egg relative to their body size (about 18 centimeters, 7 in., long) of any bird.

William Calder (1978, 1979) has shown that many anatomical and physiological features of the kiwi are convergent on mammals, including the kiwi's two functionally alternating ovaries. Calder also argues that the kiwi's huge egg was not reduced proportionately in size as the various species reduced in size from their presumed giant ancestors, the moas, and "could merely represent the absence of a strong selective pressure for economy in egg size, perhaps abetted by the advantages of hatching a more developed chick" (1978, 142). The problem with this theory is that there is absolutely no evidence that kiwis evolved from large ancestors. As the Taborskys note, "This explanation also fails to account for the amount of yolk and for the extraordinary fat content of kiwi eggs. We believe that large hatchling size is the ultimate goal of laying these jumbo eggs. . . . Kiwis seem to be adapted so that their population density matches the carrying capacity of the environment. These adaptations include a low rate of dispersal, longevity, slow physical growth, and a low reproductive rate. . . . natural kiwi populations were dense and stable. Kiwis that produced the biggest, healthiest chicks had an edge over their neighbors. Big eggs, big yolks, and big chicks became the reproductive ideal" (1993, 54, 56).

Although many authors have thought that kiwis are related to the moas, they are so manifestly different in morphology that a shared origin now seems highly unlikely. In work published in 1991 and 1992, Cooper and his co-workers amplified and sequenced approximately four hundred base pairs of a mitochondrial RNA gene from bones and soft tissue remains of four species of moas, as well as eight ratites and a tinamou, and the kiwis were shown to be more closely related to Australian and African ratites than to the moas. Their inescapable conclusion was that "New Zealand probably was colonized twice by ancestors of ratite birds" (1992, 8741), a finding that conforms nicely with Peter Houde's analysis of Paleogene fossils, on the basis of which he suggested, "Volant birds have been immigrating to New Zealand throughout its history, thus kiwis and moas may have arrived as volant forms as well" (1988, 133).

If the kiwis are of independent origin, then it becomes increasingly likely, in the absence of ancient fossils, that the extinct elephantbirds of Madagascar, like the moas, arrived on their island homes and evolved flightlessness independently. These extinct forms, known only from the Pleistocene and Recent sites, provide additional evidence that not only flightlessness but gigantism as well can occur in a very short time. Both moas and elephantbirds evolved to fill a grazing niche on their respective islands that was never occupied by mammalian herbivores. Unlike the ostriches, rheas, and emus, which evolved great speed in running as an adaptation for eluding predators in open continental areas, these island forms evolved great size and adaptations for a graviportal locomotion to support their ponderous weight. They had a short, stocky tarsometatarsus with short, heavy toes splayed out to form a broad and stable foot that was obviously associated with walking, not with rapid locomotion.

Moas. The moas, subject of a superb book by Atholl Anderson (1989) and an entire issue of the *New Zealand Journal of Ecology* (Rudge 1989), were a diversified group that superficially resembled large, flightless geese and were mostly confined to the North and South Islands of New Zealand. For a time, they coexisted with the Maori, who arrived in New Zealand about a millennium ago and gave these strange birds their name (*moa* is Polynesian for chicken or domestic fowl). The moas, like the now extinct birds of the Hawaiian Islands, are a dramatic symbol of the ecological devastation inflicted by the Polynesians in their trek through the South Pacific; no longer will the image of the Rousseauistic "noble savage" living in harmony with the environment be tenable. Anderson has documented that moa-hunting began about nine hundred years ago, and, although it did not involve mass-kills, as had been previously postulated, it was probably wasteful. Hunting ceased by four hundred years ago, and it is unlikely that any moas survived much longer than that.

The first published accounts of the moas are from a narrative of the early explorer J. S. Polack, who traveled in New Zealand between 1831 and 1837: "That a species of the emu, or bird of the genus *Struthio*, formerly existed in the [North] island I feel well assured, as several fossil ossifications were shewn to me when I was residing in the vicinity of the East Cape, said to have been found at the base of the inland mountain of Ikorangi. . . . The natives added that, in times long past they received the tradition that very large birds had existed, but the scarcity of animal food, as well as the easy method of entrapping them, had caused their extermination" (1838, 303). Later in his book Polack writes: "Petrifactions of the bones of large birds supposed to be wholly extinct have often been presented to me by the natives. . . . Many of the petrifactions have been the ossified parts of birds, that are at present (as far as is known) extinct in these islands, whose probable tameness, or want of volitary powers, caused them to be early extirpated by a people, driven by both hunger and superstition (either reason is quite sufficient in its way) to rid themselves of their presence" (345–346).

The first bone of a moa, a femur shaft, to be viewed by a scientist arrived in England in 1839 and was carefully studied by Richard Owen, who at first doubted that any bird could have a bone so large or that such a bird bone could have come from New Zealand. The femur that he observed, that of the largest moa, was larger than the femur of an ostrich. But eventually Owen became convinced that a giant, extinct bird had once inhabited New Zealand, and he gave it the generic name *Dinornis*. Later, when Europeans began to explore New Zealand thoroughly, they found bones of moas exposed on the surface of the ground in great profusion, particularly in caves and swamps. Bones were also found in association with old Maori cooking places, often showing traces of scorching. Where conditions were particularly favorable for preser-

X-ray photograph of a kiwi egg, ready for laying, in a brown kiwi's oviduct. The egg takes some 30 days to form and is more than 60% yolk. The male will incubate the egg for more than 80 days. (Courtesy Otorhanga Zoological Society; X-ray by Barry Rowe)

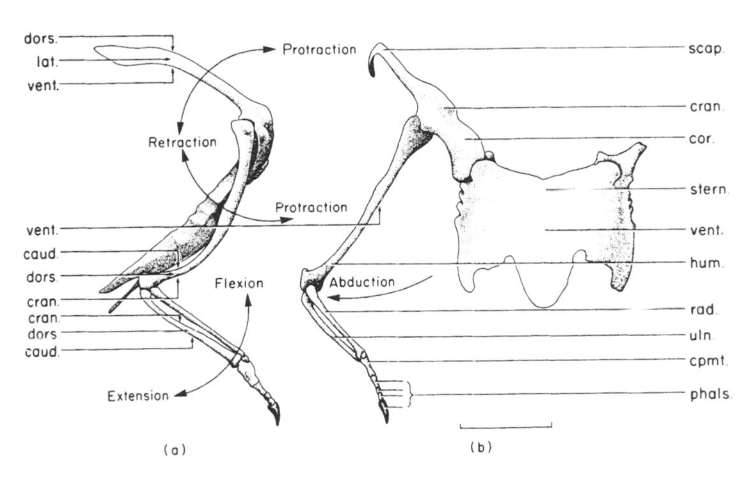

Top, an adult female specimen of the brown kiwi (*Apteryx australis*), with feathers removed from the thorax and neck, illustrating the almost bizarre body form and greatly reduced clawed wings. *Above,* right wing, scapulocoracoid and sternum, articulated as in life, to show anatomical details: *a,* lateral view; *b,* cranioventral view. Scale = 2 cm. (From McGowan 1982; courtesy Zoological Society of London)

Sir Richard Owen and *Dinornis maximus*. Owen is grasping the single bone from which came his historic announcement in 1839 that "there had existed and perhaps still exists in New Zealand a race of struthious [ostrichlike] birds of larger and more colossal size than the ostrich or any known species." Photo taken about 1877. (From *The life of Owen* [1874]; courtesy Natural History Museum, London)

Restoration of a moa (*Euryapteryx*) in assocation with other birds on the South Island of New Zealand during postglacial times (5,000 years ago). (Courtesy Department of Library Services, American Museum of Natural History, neg. no. 322337)

vation, skeletons have been found with stomach contents, ligaments, and even feathers. Small forms of the moa were common, but the very large species appear to have been relatively rare.

The many genera and species identified since Owen's original description were summarized by W. R. B. Oliver (1949; also Archey 1941), but they remained a taxonomic nightmare until Joel Cracraft (1976) studied them systematically. He concluded that approximately five genera and thirteen species existed, divisible into two groups, the greater and lesser moas. Studies since then have refined the number of species to eleven, the type elements of *Anomalopteryx oweni* having been shown to belong to different species (Millener 1982), and an additional species having been shaved off (Worthy 1991). They ranged in size from the giant moas (*Dinornis*), some possibly having stood 3.3 meters (10 ft.) high or more, to the small forms, some about the size of a large turkey, each apparently adapted for grazing or browsing in a different way.

Our extensive knowledge of moas comes from the many fossils that are preserved in the extraordinary swamps of New Zealand (Duff 1949), where moas apparently became bogged down by their great weight. In one swamp, an estimated 800 moas were packed into an area no more than 9 by 6 meters (30 by 20 ft.) across and 3 meters (10 ft.) deep. But the most remarkable of all is the Pyramid Valley Moa Swamp, discovered in North Canterbury on the South Island in 1937. Five years after its discovery, no fewer than 50 fairly complete skeletons had been excavated from the 1.4-hectare (3.5-acre) swamp, and by 1949, 140 specimens had been removed (Duff 1949). Among them were 44 specimens of the large *Dinornis*, from 3 to 3.7 meters (9–12 ft.) tall; 14 skeletons of *Pachyornis*, from 1.6 to 2.1 meters (5.5–7 ft.) tall; and 12 specimens of *Euryapteryx*, from 1.4 to 1.7 meters (4.5–5.5 ft.) tall. There were 70 specimens of *Emeus*, about the size of *Euryapteryx*. (*Pachyornis* and *Euryapteryx* are somewhat different from the other moas in having very short and relatively massive legs.) Many of these fossils have been found in upright stance with all of the bones articulated.

It has been estimated that as many as 1,700 birds per hectare (800 per acre) are preserved at Pyramid Valley (Duff 1949). Along with the moas, excavators found the giant eagle *Harpagornis*, the moa's principal predator before the arrival of the Polynesians, and a number of other flightless birds, including the kiwi (*Apteryx*), the takahe (*Porphyrio mantelli*, formerly *Notornis*), a large gruiform bird (*Apterornis*), and a large goose (*Cnemiornis*). These birds shared the wealth of New Zealand's shrubs and grasses in the absence of grazing mammals.

Aside from the usual features associated with flightlessness, including a keel-less sternum, large size, and an open pelvic region, moas shared few characters with the other ratites except the paleognathous palate. In contrast to the living ratites, the moas had no wings and no pygostyle, and they were equipped with a tendonal canal in the distal end of the tibia, a feature present in flying birds but not in other ratites. Owing to the remarkably well-pre-

Depiction of the largest South Island moa (*Dinornis maximus*) being snared and speared by New Zealand Maori. Dean Amadon (1947) estimated that *Dinornis maximus* weighed 236 kg (520 lbs.) and was 3 to 3.7 m (9–12 ft.) tall, making it the tallest bird known to have lived. The evidence from moa-hunter camps indicates that most of the moas had become extinct by about 400 years ago. (From Higham, *The Maoris*, copyright © 1981; reprinted by permission of Cambridge University Press)

Moas in two ground herbivore guilds. *Above,* North Canterbury, South Island, *from left to right: Dinornis giganteus, D. novaezealandiae, D. struthoides, Emeus crassus, Euryapteryx geranoides, Pachyornis elephantopus. Below,* Waitomo region, North Island, *from left to right: Dinornis giganteus, D. novaezealandiae, D. struthoides, Anomalopteryx didiformis, Pachyornis mappini, Euryapteryx curtus, E. geranoides.* Estimated weights (kg) are shown for each species. (From Atkinson and Millener 1991; courtesy I. A. E. Atkinson and New Zealand Ornithological Trust Board)

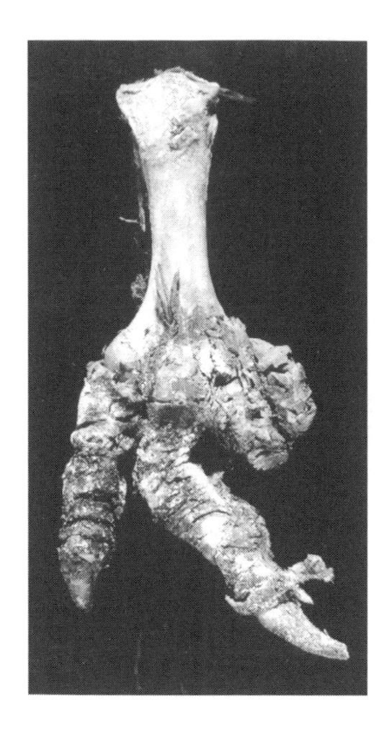

Preserved foot of a moa (*Megalapteryx*), reputedly the last genus to become extinct. These several-thousand-year-old remains were recovered from a cave in the Murchison area on the South Island in 1987. (Courtesy National Museum of New Zealand, Wellington)

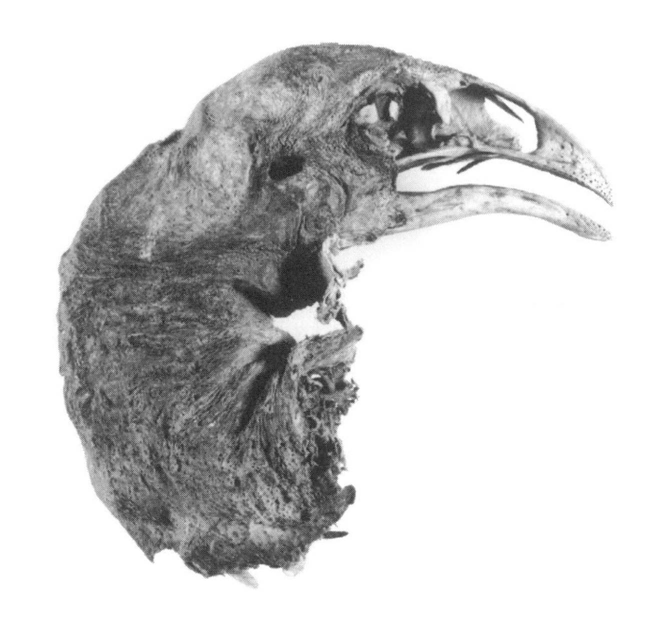

The mummified head of a moa (*Megalapteryx*), part of an almost complete skeleton found in a cave near Cromwell, on the South Island of New Zealand, in the nineteenth century. The head and upper neck are covered by skin and dried flesh, but there are no feathers. Remains such as these were once thought to be recent but are now known to be hundreds or even thousands of years old. (Courtesy National Museum of New Zealand, Wellington)

Most moa feathers have been found in caves in Central Otago, on the South Island, but usually without any indication of what species the feathers came from, as with these specimens. (Courtesy National Museum of New Zealand, Wellington)

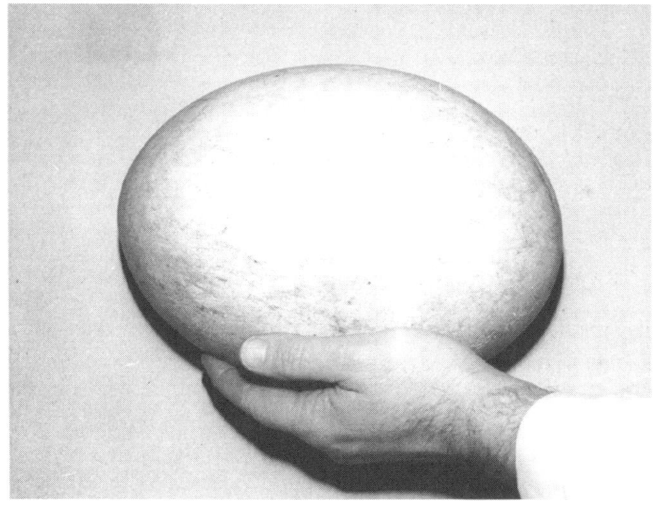

The largest known egg of a moa (species not known) from Kaikoura, on the South Island; it was found in 1857. Length 24 cm (9.5 in.); maximum width, 17.8 cm (7 in.); cubic capacity 4,302 cubic cm (or 90 average hen's eggs). (Photo courtesy Beverley McCulloch, Canterbury Museum, Christchurch)

The largest of the elephantbirds (*Aepyornis maximus*), which probably weighed close to 454 kg (1,000 lbs.). (From Wetmore 1967, drawing by Walter A. Weber; courtesy National Geographic Society)

served fossils, we know that the feathers of moas had a well-developed aftershaft, as in the living cassowaries. Large numbers of "gizzard stones" for grinding food, some as large as 5 centimeters (2 in.) across, have been found associated with many skeletons.

The moas show that herbivory can be achieved through the evolution of flightlessness. True folivory, confined to the parrots, the South American hoatzin (Grajal et al. 1989), the South American passerine plantcutters (*Phytotoma*), and several other isolated forms, is extremely rare in birds, because in a flying animal the weight of a long caecum, the intestinal appendage necessary to house bacteria to digest cellulose, cannot usually be tolerated. But once birds become flightless, the added weight of the caecum is easily accommodated.

Elephantbirds. While the Pleistocene radiation of moas in New Zealand produced the tallest bird, the heaviest, the elephantbird, evolved during the same era on the island of Madagascar. In the thirteenth century, Marco Polo recorded that the legendary roc came from Madagascar. This part of the roc tale may not have been so legendary, but if Sinbad the Sailor was borne off by huge flying birds, the elephantbird cannot be blamed; it was wholly earthbound.

The seven species of elephantbirds, some the size of a cassowary, were still present less than two millennia ago, when humans arrived on Madagascar. Radiocarbon dates of eggshell fragments (Burger et al. 1975) indicate that the birds were widespread as late as the tenth century. Their extinction probably took place gradually over many centuries because they had no major enemies—the only large predators, until humans arrived, were crocodiles and two eagles. Under these conditions, the adaptive radiation of elephantbirds produced a variety of species, though only one attained elephantine proportions, *Aepyornis maximus* ("largest tall bird"). This ponderous giant with elephant-style legs stood 2.7 to 3 meters (9–10 ft.) tall, and Dean Amadon (1947) of the American Museum of Natural History has calculated that its weight approached 450 kilos (1,000 lbs.). By contrast, a large ostrich may attain a height of 2.4 meters (8 ft.) and weigh about 135 kilos (300 lbs.). Like the moas of New Zealand, the elephantbirds were grazers, and probably, as Alexander Wetmore (1967) noted, they also cropped the lower branches of shrubs and trees, which they could easily reach with their long necks. Somewhat similar to moas in their graviportal posture, the elephantbirds differed from them in having vestigial wings and a pygostyle.

The elephantbird is first mentioned in a book by the French traveler E. de Flacourt in 1661, but the most vivid early descriptions appear in the diary of a Mr. Joliffe, surgeon of H.M.S. *Geyser,* which was cruising off Madagascar in October 1848 with a French merchant named Dumarele on board.

> M. Dumarele casually mentioned that some time previously, when in command of his own vessel trading along the coasts of Madagascar, he saw at Port Liven, on the North-west end of the island, the shell of an enormous egg, the production of an unknown bird inhabiting the wilds of the country which held the incredible quantity of *13 wine quart bottles of fluid!!!* he having himself carefully measured the quantity. It was of the colour and appearance of an ostrich egg, and the substance of the shell was about the thickness of a spanish dollar, and very hard in texture. It was brought on board by the natives . . . to be filled with rum, having a tolerably large hole at one end through which the contents of the egg had been extracted, and which served as the mouth of the vessel. M. Dumarele offered to purchase the egg from the natives, but they declined selling it, stating that it belonged to their chief, and that they could not dispose of it without his permission. The natives said the egg was found in the jungle, and observed that such eggs were *very very rarely* met with, and that the bird which produces them is still more rarely seen. (Strickland 1849, 338–339)

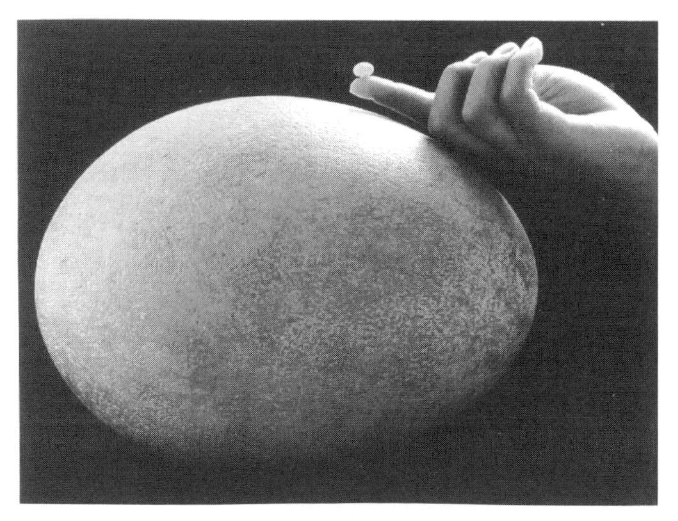

Extremes in egg size: the 8.5-l (2-gal.) capacity egg of the Malagasy elephantbird (*Aepyornis maximus*) and a hummingbird egg. (Photo by Clark Sumida; courtesy Los Angeles County Museum of Natural History)

Many eggs have now been recovered, and about thirty unbroken *Aepyornis* eggs are known to exist, some weighing more than 9 kilos (20 lbs.), measuring up to 33 centimeters (1 ft.) in length, and equal in volume to seven ostrich eggs. One actually contains an embryo, revealed by X-rays. The eggs of *Aepyornis* are the largest birds' eggs known, and as Josselyn van Tyne and Andrew J. Burger (1976) have noted, "such an egg would hold the contents of seven ostrich eggs or 183 chicken eggs or more than 12,000 hummingbird eggs" (1976, 31).

The probable fossil record of elephantbirds is confined to the Pleistocene and Recent remains known from Madagascar, but this has not stopped paleontologists from assigning fragmentary material to the elephantbirds. Two pieces of leg bones from the Fayum of Egypt, a region south of Cairo, have been identified on totally insufficient grounds as belonging to the Aepyornithiformes, the order containing elephantbirds. One of these is a small fragment of a tibiotarsus in the British Museum of Natural History named *Eremopezus,* and the other is a piece of a tarsometatarsus from the Oligocene named *Stromeria.* Even an eggshell fragment, called *Psammornis,* from the Eocene of southern Algeria has been assigned to the elephantbird. Moreover, Franz Sauer (1976) has reported "aepyornithoid" eggshells from the Miocene and Pliocene of Turkey, and Sauer and Peter Rothe (1972) have described eggshells from Miocene and Pliocene deposits on the Canary Islands as being "aepyornithoid." These attributions are based on the shells' relatively large pore size. But pore size increases with egg size, and there is no valid evidence that these specimens are from elephantbirds; they could all easily be eggshells of other large flightless birds, paleognathous, gruiform, or of other unknown affinity.

There are no confirmable Tertiary elephantbird fossils, and in the absence of a reliable fossil record extending beyond the Pleistocene, it is difficult to explain the existence of elephantbirds on Madagascar. The separation of Africa and Madagascar began in the medial Jurassic, which was about the same time as the initial breakup of Gondwanaland, and Madagascar assumed its present position in the early Cretaceous (Rabinowitz, Coffin, and Falvey 1983). To view elephantbirds as flightless passengers on the ancient drifting landmass of Madagascar can only stretch credulity to its outer limits. The ancestors of elephantbirds, like those of the moas, must have flown to Madagascar and became flightless later, perhaps as late as the Miocene or even Pliocene.

Ostriches, Rheas, and Vicariance Biogeography. The commonly accepted hypothesis to explain the current distribution of ratites is that there was a single, flightless ancestor, widely distributed in Gondwanaland during the Cretaceous period, that gave rise to the various lineages of ratites (Cracraft 1973a, 1974). As Peter Houde has put it: "Populations of this flightless pluripotent ratite ancestor already present in Gondwanaland . . . were isolated by ocean barriers as they arose. Ratites are viewed as passive passengers on rifting continents; populations isolated and diverged as daughter continents were born. The present-day distribution of ratites is due only to the distribution of this flightless ancestor in Gondwanaland and the subsequent breakup of this ancient supercontinent to form the existing South Hemisphere landmasses" (1988, 131). This has become known as the "vicariance biogeography hypothesis of ratite origins," and the dating of the DNA molecular clock used by Sibley and Ahlquist (1990) is largely based on this hypothesis; to simplify, the divergence of ratite DNA sequences was initiated by their isolation and therefore can be dated by this rift in the southern continents, the splitting of Gondwana. The nagging problem is: where are the fossils? If Cracraft's hypothesis were correct, why are there no ostrich fossils, heavy bones that are easily preserved and are commonly found, even from the early Pliocene of Africa? Fossils of the tarsometatarsus belonging to *Struthio* are recognizable at a glance. Almost all living birds have three or four toes. But the ostrich, along with the convergent gruiform ergilornithids, is didactylous—having only two toes and therefore only two trochleae, bony structures for the toe articulation, on the end of the tarsometatarsus. This is a running adaptation that converges on similar trends toward reduction in the number of digits in several mammalian lines, especially the horses.

The euphoria of continental vicariance biogeography, however, has been dampened by Peter Houde's discoveries. In addition, restudy of *Palaeotis weigelti*, previously identified as a bustard, showed that it was a crane-sized, flightless paleognathous ostrich from the Middle Eocene of Germany, and its ancestry is clearly traceable to the *Lithornis* group of Paleogene flying paleognaths. *Palaeotis* is thought to be allied with the ostrich (Houde and Haubold 1987) or with the rheas (Peters 1992b; Storch 1993). Thus, ratites were not only present in the Northern Hemisphere but almost certainly evolved there long after the breakup of Gondwanaland and subsequently emigrated from Europe to Africa. Given this startling revelation, and in the absence of evidence to the contrary, it is altogether possible that no birds owe their current distribution to continental vicariance biogeography.

The fossil record of modern-type ostriches (*Struthio*) begins in the Miocene of Moldavia with *Struthio orlovi* (Kurochkin and Lungu 1970). By the Pliocene four species ranged from China and Mongolia and India to the Ukraine, Kazakhstan, Odessa, and Greece, and there are numerous fossils through the late Pleistocene, showing a number of species widespread over the Old World and on into Africa. Though ostriches are presently confined to southern Africa, until recently they occupied North Africa as well. The last reported Arabian ostrich was killed and eaten during World War II by Saudi tribespeople. Like most other ratites, ostriches usually roam in small bands of from ten to fifty birds and eat a variety of foods, including fruits and seeds, other plant matter, insects, and small animals. Although they swim well, ostriches are particularly known for their running speed, and claims of 97 kilometers (60 mi.) per hour have been made.

The rheas or birds thought to be allied with them (Rheiformes), sometimes called South American ostriches, have left fragmentary fossils as far back as the Middle Paleocene of Argentina and Brazil, but they tell little of evolutionary relations (Tambussi 1995). The late Paleocene flightless *Diogenornis fragilis* (Opisthodactylidae, Alvarenga 1983), is the oldest unquestionable ratite bird, the next oldest being *Remiornis* from the late Paleocene of France, but its fragmentary nature renders it of little value. As Larry Martin has summarized, "*Remiornis* is an early, giant, flightless bird whose affinities lie with modern ratites" (1992, 97). As for vicariance biogeography, Houde writes: "Ancient as *Diogenornis* is, it is still about 30 million years shy of the age when South America rifted away from the remainder of Gondwanaland" (1988, 132).

There were a number of Pliocene genera of rheas, but none survived into the modern fauna (Tambussi 1995), and rhea fossils similar to modern forms occur commonly in the Pleistocene. The greater rhea (*Rhea americana*) once ranged in great flocks across the Brazilian and Argentine pampas; the smaller and still common Darwin's, or lesser, rhea (*Pterocnemia pennata*) lives in the eastern Andean foothills from Peru and Bolivia to the southern tip of South America. Small flocks of rheas, often in company with tinamous and the camellid guanacos, roam the open grasslands in search of a variety of vegetable matter, insects, and other small animals. Considerably smaller than the African ostriches, rheas stand 1.2 to 1.5 meters (4–5 ft.) tall and weigh about 23 kilos (50 lbs.). Still, they are the largest birds in the New World.

Cassowaries and Emus. Cassowaries and emus are known to be closely related and are generally placed in the same order, Casuariiformes. The oldest member of the group, *Dromaius gidju*, was originally described as a primitive emu (Patterson and Rich 1987), but additional fossils from the late Oligocene to the medial Miocene, under study by Walter Boles, show that it may be intermediate between emus and cassowaries. According to Patricia Vickers-Rich, *Dromaius gidju* "is clearly a form that lies close to the ancestry of both groups in the early Neogene or late Palaeogene" (1991, 744). Cassowary fossils are known from the late Miocene or early Pliocene of the Northern Territory of Australia, some toe bones from the Pliocene of New Guinea have been tentatively assigned to the group (Rich and van Tets 1982), and a pygmy cassowary has been uncovered at a late Quaternary site in the highlands of Papua New Guinea (Vickers-Rich 1991).

The three species of cassowaries (Casuariiformes, Casuariidae) are known as fossils only from the Pleistocene and are denizens of the forest of New Guinea, its adjacent islands, and forested northern Australia. Their habitat has favored the adaptation of a casque, or bony forehead helmet, used to deflect obstructions as they maneuver through dense rain forests. They are also strong swimmers, well adapted to crossing jungle rivers. They lack tail feathers, have very rudimentary clawed wings only 5 centimeters (2 in.) long, yet have three to five long, wiry quills that are easily seen from the side and tell of their derivation from flying predecessors. Cassowaries are known for their pugnacity and have often killed humans. They attack by leaping at their enemy feet first, slashing with their powerful, sharp claws. Somewhat nocturnal, they eat mainly berries and fruits, though they may also consume small plants and animals.

Emus (Casuariiformes, Dromiceidae) are closely related to the cassowaries, and like their cousins they swim well, lack tail plumes, and have very rudimentary wing bones with clawed tips. Now confined to Australia, they were much more diversified during the Pleistocene, when there were several species, and even in modern times three species have vanished from Tasmania and two other smaller islands off Australia's southern coast. Fossil emus are also known from as far back as the Miocene of South Australia (Olson 1985a), and *Dromaius ocypus* is known from the Lower Pliocene (A. H. Miller 1963). Among the ratites, emus are exceeded in size only by the ostrich, standing 1.5 to 1.8 meters (5–6 ft.) tall and weighing up to 54 kilos (120 lbs.). Like the other large ratites, emus live primarily on plant food, but they also eat insects and small animals.

Mihirungs. In 1979 and 1980, Patricia Rich (now Vickers-Rich) of Monash University unveiled a new group of large flightless Australian birds that she considered to be highly derived ratites. These mihirungs (Dromornithidae), as they are now known (from their aboriginal name meaning giant emus), are well represented in the Australian fossil record from the medial Miocene to the Pleistocene and are known from trackways in the late Oligocene of Tasmania, as well as from New Guinea. The dromornithids are also found in the late Paleogene and appear to have survived until at least 26,000 years ago. They were large (slightly larger than the living emu) to truly gigantic, ground-dwelling birds and have been variously allied with the ratites and galliforms, though their relations remain unclear. Storrs Olson (1985a, 104–105) argued that dromornithids were definitely not ratites based on cranial material in the Smithsonian that showed a troughlike mandible, and Vickers-Rich (1991, 742) stated that the cranial material is robust in the same manner as the Psittaciformes and the Alcedinidae, but in their detailed morphology were unique. Recently recovered cranial material shows a highly derived skull that is not clearly ratite or galliform (Vickers-Rich 1991).

At present, five genera and eight species, some quite gigantic, are known, with the greatest diversity occurring during the Miocene. Study of hindlimb morphology suggests that both graviportal (*Dromornis*) and cursorial (*Ilbandornis lawsoni*) forms existed, often sympatrically. Rich has argued that the mihirungs were herbivores, lacking both the hooked beak of such forms as the phorusrhacids and hooked ungual phalanges, and has suggested (Vickers-Rich 1991, 743) that they may have not been particularly successful at invading open grasslands, losing out to the emus and kangaroos, which were undergoing a rapid adaptive radiation as aridity, hence grassland, was overtaking Australia during the later Tertiary.

Dromornithids had a flat, unkeeled ratite-like sternum but lacked an open ilioischiatic fenestra of the pelvis. The Pleistocene genus *Genyornis* was about 2 meters (6.6 ft.) tall and was typical of the group in having a reduced medial digit of the foot and hooflike ungual phalanges. Other dromornithids, known from isolated bones, indicate a bird that may have rivaled or exceeded in weight any known fossil bird, including the elephantbird *Aepyornis maximus* (Rich 1979).

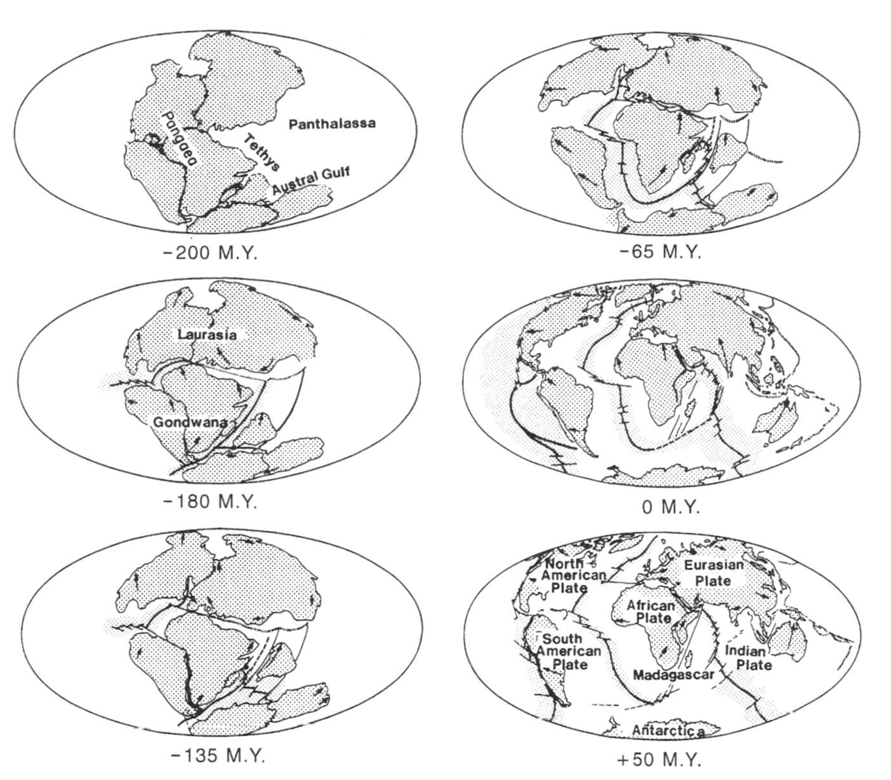

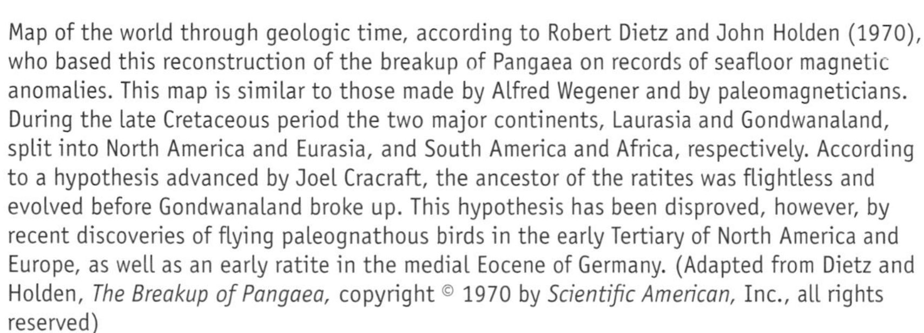
Map of the world through geologic time, according to Robert Dietz and John Holden (1970), who based this reconstruction of the breakup of Pangaea on records of seafloor magnetic anomalies. This map is similar to those made by Alfred Wegener and by paleomagneticians. During the late Cretaceous period the two major continents, Laurasia and Gondwanaland, split into North America and Eurasia, and South America and Africa, respectively. According to a hypothesis advanced by Joel Cracraft, the ancestor of the ratites was flightless and evolved before Gondwanaland broke up. This hypothesis has been disproved, however, by recent discoveries of flying paleognathous birds in the early Tertiary of North America and Europe, as well as an early ratite in the medial Eocene of Germany. (Adapted from Dietz and Holden, *The Breakup of Pangaea,* copyright © 1970 by *Scientific American,* Inc., all rights reserved)

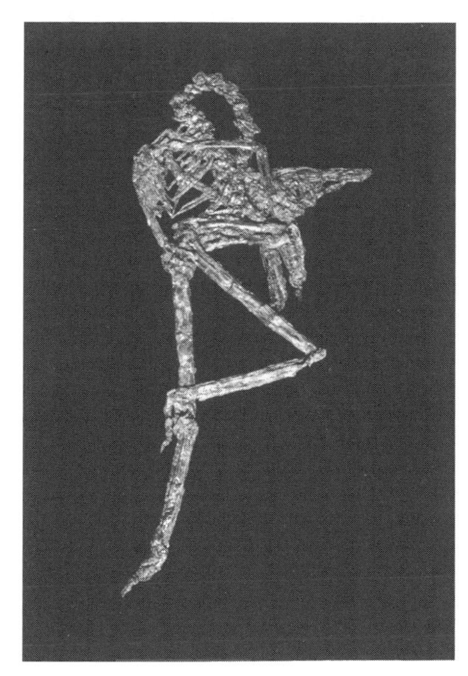
The primitive ratite *Palaeotis weigelti,* from the Messel oil shales of Germany, proves that ratites were present in the Northern Hemisphere in the early Tertiary. × 0.11. (Courtesy S. Peters and Forschungsinstitut Senckenberg)

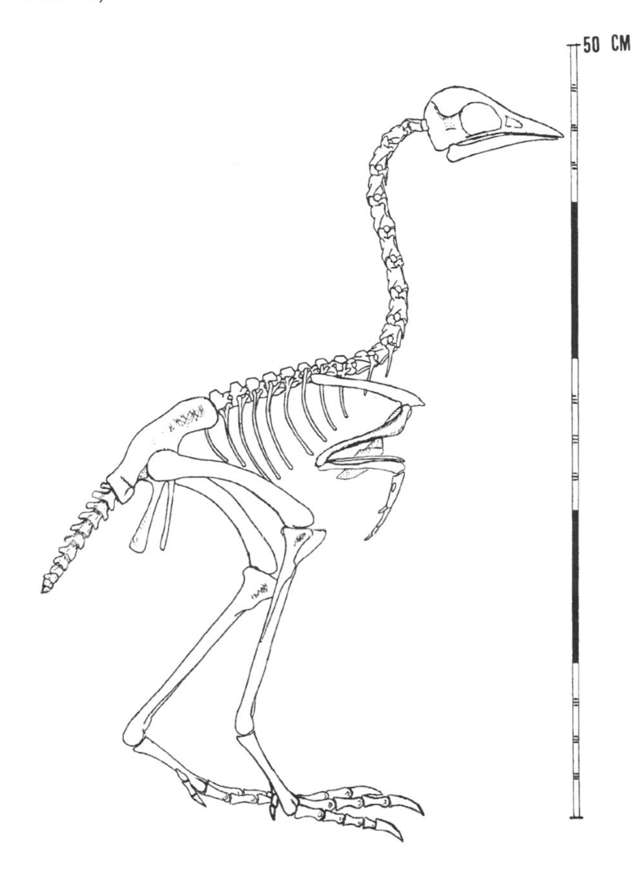

Restoration of the skeleton of the enigmatic Upper Cretaceous species *Patagopteryx deferrariisi* (Patagopterygiformes), a flightless bird the size of a chicken, of uncertain affinity. (From Alvarenga and Bonaparte 1992; courtesy Herculano Alvarenga and *Los Angeles County Museum of Natural History, Science Series* 36)

Top and middle, reconstruction of the dromornithid *Genyornis newtoni* being attacked by the giant goanna (*Megalania prisca*), a large monitor lizard, both known from the Pleistocene of Australia. *Bottom,* the family Dromornithidae, now extinct, survived until at least 26,000 years ago. Dromornithids were large terrestrial birds and are represented by a variety of fossil species extending back to the Miocene. The family included both highly cursorial and nearly graviportal species. One giant species from the late Miocene of the Northern Territory rivaled or possibly exceeded the largest known elephantbird (*Aepyornis maximus*) in size. (*Top,* from Rich 1991, courtesy P. Rich; *bottom,* modified after Rich and VonTets 1982)

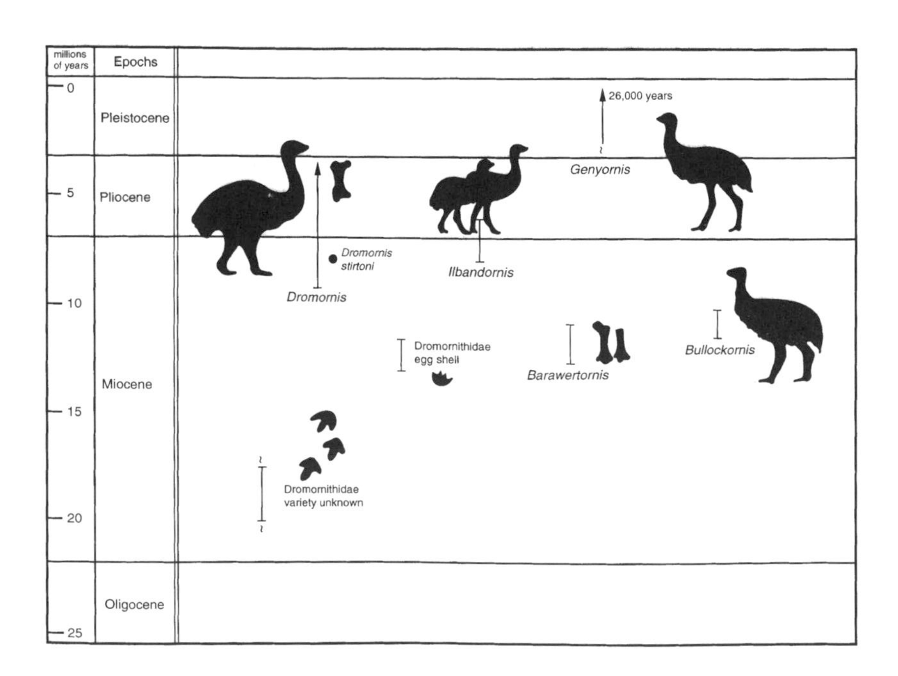

***Patagopteryx:* Flightless Bird of the Cretaceous.** Still another unique flightless bird, *Patagopteryx deferrariisi,* of unknown affinity has been recently discovered in late Cretaceous deposits from Argentina and has been placed in its own family and order, the Patagopterygiformes (Chiappe 1990; Alvarenga and Bonaparte 1992). *Patagopteryx* (named in reference to Patagonia) was the size of a domestic fowl, with robust legs and wings too small for flight. It had typical avian heterocoelous vertebrae, but the thoracic vertebrate were of the primitive procoelous type; and the ischium and pubis were slanted backward, with an open ilioischiatic fenestra. As noted in chapter 3, *Patagopteryx,* along with the enantiornithine birds of the Cretaceous, has been shown to have growth rings or lines of arrested growth in the long bones (Chinsamy, Chiappe, and Dodson 1994). Growth rings are known neither in any modern ornithurine birds nor in the Cretaceous ornithurine, hesperornithiform, and ichthyornithiform birds. This cyclical bone deposition in *Patagopteryx* and the Enantiornithes suggests at least that these birds differed physiologically from their living counterparts and may not have been fully endothermic. This finding would not be too surprising, given the delayed ontogeny of full endothermy in modern birds and the fact that a number of living birds can lower their body temperatures and undergo periods of torpor. Although different from all other birds in important characters, Herculano Alvarenga and José Bonaparte have concluded that *Patagopteryx* was an adaptive type somewhat close to the ratites, but it is so different in the morphology of its vertebrae and ilium that it must, at least for the time being, be considered a line of Cretaceous birds that, like so many others, met its demise along with the dinosaurs.

Flightlessness and Aves. The evolution of flightlessness in birds has been almost as characteristic of the class Aves since its origin as has the evolution of flight itself, the key adaptation that has molded all of the salient features of the group, including feathers and their aerodynamic design. In a sense, then, the evolution of flightlessness sheds light on the evolution of flight itself—that is, the degenerative processes can tell us a great deal about how flight came about.

Flightless birds come in a variety of types, including the large carnivorous South American phorusrhacids and possibly carnivorous Northern Hemisphere diatrymids. Open-area runners, in which cursorial adaptations were perfected for rapid escape in open landscapes, include the Eurasian ergilornithids and ostriches. Other modes of selection produced flightless birds on islands (climatically stable areas without predators) from any number of orders, including the anseriforms, the psittaciforms, the columbiforms, the ciconiiforms, and particularly the gruiforms. Of particular importance for understanding the evolution of flightlessness are the island rails—they are in essence a living laboratory for the study of the morphological and physiological changes that occur when flightlessness evolves. They also tell a story of convergent evolution, with flightless forms on distant islands acquiring the same set of morphological features. These rails prove, too, that flightlessness and bizarre adaptations can evolve very rapidly in an isolated environment. Wherever the island, whatever the group of birds, when flightlessness occurs, the same general morphological features appear time and time again, convergently, providing a veritable lesson in the rules of evolutionary design.

Above, the small accipitrid hawk *Horusornis vianeyliaudae* (Horusornithidae), from the late Eocene of France and early Oligocene of North America, seen using its intertarsal joint mechanism to extract prey from a tree cavity. Convergent evolution has produced similar intertarsal joints in two unrelated living genera of hawks, the harrier-hawks (*Polyboroides*) and the crane-hawks (*Geranospiza*). *Below,* another example of convergence in the falconiforms is seen in this small secretarybird-like accipitrid, *Apatosagittarius terrenus,* from the Miocene of Nebraska, that hunted the grassland community of the late Tertiary. The paleoenvironment indicates a Miocene setting similar to the present-day habitat of the secretarybird, with extensive grasslands and savannas. True secretarybirds are known from the late Oligocene and early Miocene of France. (Drawings by John P. O'Neill)

7

Birds of Prey

The birds of prey have traditionally consisted of two groups, the order Falconiformes, including the familiar falcons, hawks, eagles, vultures (Old and New World), and ospreys, as well as the odd secretarybirds, and the order Strigiformes, composed of the owls. Unlike most other birds, most of these raptorial forms, excluding the carrion feeders, are highly modified in the beak and claws for seizing prey. During past geologic epochs, as we have seen, the term "birds of prey" could be applied equally to such birds as the phorusrhacids and possibly the diatrymids; and even today, such birds as the passerine shrikes, or butcherbirds, and certain cuckoos and kingfishers might be considered predatory. Certainly the roadrunner is a swift, cursorial predator, effectively feeding on a variety of living prey, including snakes and lizards. The expression "birds of prey" has remained restricted to hawks and owls, however, because the two groups were once considered closely related, neatly divided into the nocturnal raptors, the owls, and the diurnal raptors, the hawks and eagles.

Falconiform Birds: Hawks and Allies

The order Falconiformes comprises a variety of families thought to be united by certain raptorial adaptations, such as the short, sharply hooked beak with a soft mass called a cere across its top, through which the nostrils pass, and powerful feet with long claws equipped with strongly curved talons and an opposable hind toe. Hawks and their allies occur everywhere in the world except Antarctica. Very strong-winged, powerful fliers, many of them are also remarkable soarers, and they all feed almost entirely on animal food, live or dead. They lay a relatively small number of eggs and thus have a low reproductive rate.

The systematics of the order Falconiformes is among the most problematic of all avian orders, and the myriad types of diurnal raptors have been grouped in a single order largely because classifications have historically depended heavily on bill and foot structure. For purposes of our discussion, the diurnal raptors include the hawks, eagles, Old World vultures, and kites, family Accipitridae; the falcons and caracaras, family Falconidae; the ospreys, family Pandionidae; and the New World vultures, family Vulturidae or Cathartidae. These diverse birds range in size from the tiny 15-centimeter (6 in.) Philippine falconet (*Microhierax erythrogenys*) to the immense 1.2-meter (4 ft.) Andean condor (*Vultur gryphus*). Unlike the nocturnal owls, falconiforms are primarily diurnal, doing most of their hunting during the daytime.

Strigiform Birds: Barn and Typical Owls

The owls, order Strigiformes, in contrast, consists of two families, the typical owls (Strigidae) and the barn owls (Tytonidae), which most ornithologists consider to be closely related to typical owls. The barn owls, divided into seventeen species by Monroe and Sibley (1993), many endemic to individual islands, are widely distributed, ranging throughout the New and Old Worlds. Fossil giant barn owls are common on islands, ranging from Cuba and the West Indies to the Mediterranean (also see Mourer-Chauviré and Marco 1988). The Tytonidae also includes two or three species of grass-owls (genus *Tyto*) that range in India, Australasia, Africa, and Madagascar and differ in having longer, almost unfeathered legs, presumably adapted to more terrestrial life in grasslands, and the bay-owls (mainly the oriental bay-owl, *Phodilus badius*), denizens of wet tropical forests, ranging from northern India to Indonesia. Interestingly, a new species of bay-owl, *Phodilus prigoginei*, was described from a single specimen collected in March 1951 in the Congo; no more is known of it. Barn owls are characterized by distinctive heart-shaped, sound-reflective facial discs, heads without ear tufts, long, slender, feathered legs, a serrated, or pectinate middle claw, and dis-

The diurnal birds of prey, Falconiformes, are raptors with short, strongly hooked bills with a fleshy cere containing the nostrils. Their legs are generally short, and the feet have sharp, strongly curved talons; the vultures are an exception. *Left to right and top to bottom,* Falcons and caracaras (Falconidae, worldwide, 10 genera, 63 species): peregrine (*Falco peregrinus*). Hawks, eagles, harriers, kites, and Old World vultures (Accipitridae, worldwide, 65 genera, 237 species): crowned hawk-eagle (*Stephanoaetus coronatus*). Osprey (Pandionidae, worldwide, 1 genus, 1 species): osprey (*Pandion haliaetus*). Secretarybird (Sagittariidae [affinities uncertain], Africa, 1 genus, 1 species): secretarybird (*Sagittarius serpentarius*). New World vultures (Vulturidae or Cathartidae [may be distantly allied with storks], North and South America, 5 genera, 7 species): king vulture (*Sarcoramphus papa*). (Drawings by George Miksch Sutton)

tinctive middle ear bones (Feduccia and Ferree 1978); they are the keenest of hunters, being capable of acoustic location of prey in total darkness (Knudsen and Konishi 1979; Konishi 1993).

The 156 or so species of typical owls, or strigids, are a reasonably closely related group, falling into some two dozen genera, but comprise a variety of types of owls. These include the eagle- and hawk-owls, screech-owls, scops-owls, pygmy-owls, fishing-owls, and burrowing owl of the subfamily Buboninae; and the snowy, tawny, short-eared, and long-eared owls of the subfamily Striginae. The strigids differ from the tytonids in having distinctive osteological characters, an unforked, rounded tail, and a round head, often with distinctive ear tufts.

Problems of Classification

In the classification proposed by Linnaeus (1758), the Accipitres, his first order, contained the hawks, eagles, kites, vultures, and falcons, but also the owls and a variety of other birds. By the early 1800s the obvious nonraptors had been excluded, and Thomas Huxley's (1867) classification included all the raptors in the Aetomorphae, though he himself doubted the validity of this grouping, stating that the raptors varied widely in many characters. Huxley recognized four main groups: Strigidae (owls), Cathartidae (New World vultures), Gypaetidae (hawks, eagles, falcons, and Old World vultures), and the secretarybird (Gypogeranidae). Many classifications followed, but most recognized the artificiality of placing the hawks and owls in the same genealogy and separated them into two separate orders. One prominent exception is Joel Cracraft (1981), who, using cladistic methodology, has revived Linnaeus's classification by placing the hawks and owls together. Most current workers, however, would concur with the view advocated by Charles Sibley and Jon Ahlquist, on the basis of their DNA comparisons (1990, 471), that the owls are instead related to the Caprimulgiformes and are similar to the hawks through convergence. Indeed, one hawk, the northern harrier, or marsh-hawk (*Circus cyaneus*), has converged with certain owls, both anatomically and physiologically, in having a curved, sound-reflecting facial ruff or disc and has been shown through experiments to be able to locate concealed prey by the use of acoustical cues (Rice 1982). The northern harrier's reliance on auditory cues explains its low foraging position as it glides back and forth over marshy fields.

Fossil Owls

The fossil record of raptorial birds is spotty, yet although it does not yield a wealth of information about their evolution, it does provide some interesting insight into certain groups. The oldest fossils described as owls have been assigned to a distinctive family, the Bradycnemidae, from the late Cretaceous of Transylvania (Harrison and Walker 1975a), but these fossils are most likely from small dinosaurs. The oldest reliable record of owls is a fossil tarsometatarsus from the Paleocene of Colorado that Pat Rich and David Bohaska (1976) described as *Ogygoptynx wetmorei*, but it yields little information on owl origins. *Ogygoptynx* was initially considered to be a mosaic, intermediate between the two extant owl families; though it was unique in many features, it shared some character-states with the Strigidae and Tytonidae. After studying the early Tertiary European owl material, Rich and Bohaska (1981) placed it in a distinctive family, the Ogygoptyngidae, which differs considerably from North American proto-strigids and all European Paleogene owls, though it exhibits a combination of characters shared with tytonid, strigid, and proto-strigid owls. The only other Paleocene owls are *Sophiornis* from France, described as a new family, Sophiornithidae, by Cécile Mourer-Chauviré in 1987, and a newly described Upper Paleocene large owl, *Berruornis*, from the same family (Mourer-Chauviré 1994); however, owls are known from the Lower Eocene London Clay and numerous Eocene localities.

It was during the Eocene or slightly earlier that the owls underwent their major adaptive radiation, producing at least four families—three extinct (Protostrigidae, Sophiornithidae, and Palaeoglaucidae) and the extant Tytonidae. Some two families comprising eleven species have now been identified from the famous Phosphorites du Quercy in France. Mourer-Chauviré (1987) studied all of the strigiform fossils from Quercy and reviewed their systematics and paleogeography. She found them to be different from the two previously described extinct owl families, the North American Paleocene Ogygoptyngidae and the North American and European Eocene Protostrigidae. Most of the fossil owls from the Phosphorites du Quercy appear, wrote Mourer-Chauviré, "nearer to Tytonidae, but the distal part of the tibiotarsus is more similar to Phodilinae. *Necrobyas* represents a kind of owl that is not very different from the recent *Tyto*, but with a shorter and heavier tarsometatarsus. In this genus, the succession of four species, *N. rossignoli*, *N. harpax*, *N. edwardsi*, and *N. arvernensis*, from the Upper Eocene to the Lower Miocene, can be observed. The general trend in this lineage is an increase in size and a lengthening of the tarsometatarsus" (1987, 90). She also transferred two previously described owls from the strigid genera *Bubo* and *Asio* to new genera, *Nocturnavis* and *Selenornis*, described three new genera, *Palaeobyas*, *Palaeotyto*, and *Palaeoglaux*, and placed all five genera, along with *Necrobyas*, in the Tytonidae. This significant study has greatly advanced knowledge of the evolution of the Strigiformes. The Quercy fossils illustrate that the family Tytonidae was highly diversified in the Paleogene of Europe, being represented at Quercy by two subfamilies, the Necrobyinae and Selenornithinae, though it is now represented by only *Tyto* and *Phodilus*.

In 1992, Stefan Peters described a new species of owl from the Middle Eocene Messel oil shales of Germany, the only owl to date recovered from that locality. He placed *Palaeoglaux artophoron* in the same family as some of the Quercy owls, the Palaeoglaucidae, noting that the fossil "provides further evidence that the genus *Palaeoglaux* displays a mixture of tytonid and strigid characters" (1992a, 169). Because the family Strigidae first appears in the Lower Miocene of North America and Europe, it is reasonable to assume that the barn owls were superseded by the typical strigid owls during that period. This general chronology conforms to the view expressed by John Burton (1973, 30) that the early to medial Tertiary diversification of many forms of small mammals may have accelerated the adaptive radiation of owls. Early and medial Oligocene owl fossils from North America that "have the stout tarsometatarsus without an ossified tendinal loop of the 'phodiline' owls" are under study by Storrs Olson (1985a, 131).

Extremely interesting fossil owls dating to the Ice Age, or Pleistocene, have been found on islands, where, in the absence of mammalian predators, different owl lineages all over the world evolved gigantism. In the Antilles these birds evolved to occupy niches usually filled by carnivorous mammals in continental areas. On Cuba, for example, as well as on Hispaniola and Puerto Rico, fossil material from Pleistocene cave deposits indicates a large number of rodents, ground sloths, and insectivores in this period. Also on Cuba, where only one mammalian carnivore is known to have existed, the canid *Cubacyon transversidens* (known from a single jaw fragment), there were four genera of very abundant rodents, four genera of ground sloths of varying size, and the two well-known insectivores *Solenodon* and *Nesophontes*. As Oscar Arrendondo (1976, 169) has noted of these caves, "It is no exaggeration to say that tens of thousands of mandibles of *Capromys pleistocenicus* [a small rodent] . . . can be extracted from a single small chamber . . . and that a small cave . . . yielded the remains of over 200 individuals of the edentate genus *Mesocnus* [a ground sloth]." The subsequent discovery that there were giant hawks and owls on these islands thus came as no great surprise, because most of the fossil rodents were derived from owl pellets! The fossil raptors included a giant eagle (*Aquila borrasi*), a vulture (*Antillovultur varonai*) similar in size to the Andean condor, and three giant species of barn owls, *Tyto alba*, *T. noeli*, and *T. riveroi*, the latter two being, respectively, much larger than modern barn owls and truly gigantic.

Interestingly, three species of barn owls that fall into the same size classes have been described from the late Miocene of the Gargano Peninsula in Italy by Peter Ballmann (1973, 1976). During the Miocene the Gargano Peninsula was apparently an island or perhaps an archipelago, and the parallel with the Cuban radiation of tytonids is remarkable. Finally, large but not gigantic barn owls are known from numerous Quaternary deposits from the Bahamas, Hispaniola, and the Plio-Pleistocene of the Mediterranean, specifically Majorca and Minorca, illustrating that gigantism has been characteristic of barn owls' insular evolution. Owls were apparently an integral part of insular avifaunas all over the world and suffered extinction like other avian taxa. A recent example is the now extinct *Mescarenotus grucheti* (Mourer-Chauviré et al. 1994) from the Mascarene Island of Réunion.

Even more spectacular than the barn owls is the gigantic owl from Cuba *Ornimegalonyx oteroi*, which probably stood more than a meter (3 ft.) tall, had a tarsometatarsus that measures over twice the length of that of the great horned owl (*Bubo virginianus*), and is the largest owl known to have existed (Arrendondo 1976, 1982). There is evidence that several species of *Ornimegalonyx* once lived on Cuba (Kurochkin and Mayo 1973), and Oscar Arrendondo (1982) has named four species. *Ornimegalonyx* was a typical strigid owl, not far removed from owls of the genus *Strix*, with a small wing, a reduced sternum, and long, robust hindlimbs. Arrendondo proposed that the Cuban owl "was little or not at all capable of flight . . . the bones of the wing are poorly developed, particularly the carpometacarpus" (183). In addition, Arrendondo and Olson (1994) have named a new species of Bubo from Cuba, *Bubo osvaldoi*, that is the only representative of the genus *Bubo* and the tribe Buconini in the Antilles and was bigger than any living owl.

Gigantism in owls was apparently not restricted to islands and is known in continental owls in both North America and Europe. In 1933, Hildegard Howard described an extinct owl from the tar pits of Rancho La Brea as *Strix brea*, larger than either of the living species of *Strix*, the barred owl (*S. varia*) and the spotted owl (*S. occidentalis*); at the time, however, the fossil was not compared with the even bigger great grey owl (*S. nebulosa*), because at the time *S. nebulosa* was regarded as belonging to the monotypic genus *Scotiaptex*. In 1984, Storrs Olson reported a mandible of a truly gigantic owl of late Pleistocene (Rancholabrean) age from Georgia, but the material was not sufficient to describe as a new taxon (Olson 1984b). The fossil material consisted only of the anterior portion of a mandible that was larger than any of the liv-

The nocturnal birds of prey, Strigiformes, are characterized by rounded heads with the eyes facing forward, and surrounded by large facial disks of feathers designed to concentrate sound and therefore greatly increase their hearing ability. Like the falconiform birds, they have raptorial bills and feet with strong, curved talons. *Left,* barn owls (Tytonidae, worldwide, 2 genera, 17 species): barn owl (*Tyto alba*). *Right,* typical owls (Strigidae, worldwide, 23 genera, 156 species): spectacled owl (*Pulsatrix perspicillata*). (Drawings by George Miksch Sutton)

Palaeoglaux artophoron is the only owl known from the Messel oil shales of Germany, and it belongs in its own family, the Palaeoglaucidae. The pinions of the left wing are to the left, and at higher magnification the body feathers reveal structures unknown in recent owls. × 0.71. (Courtesy S. Peters and Forschungsinstitut Senckenberg)

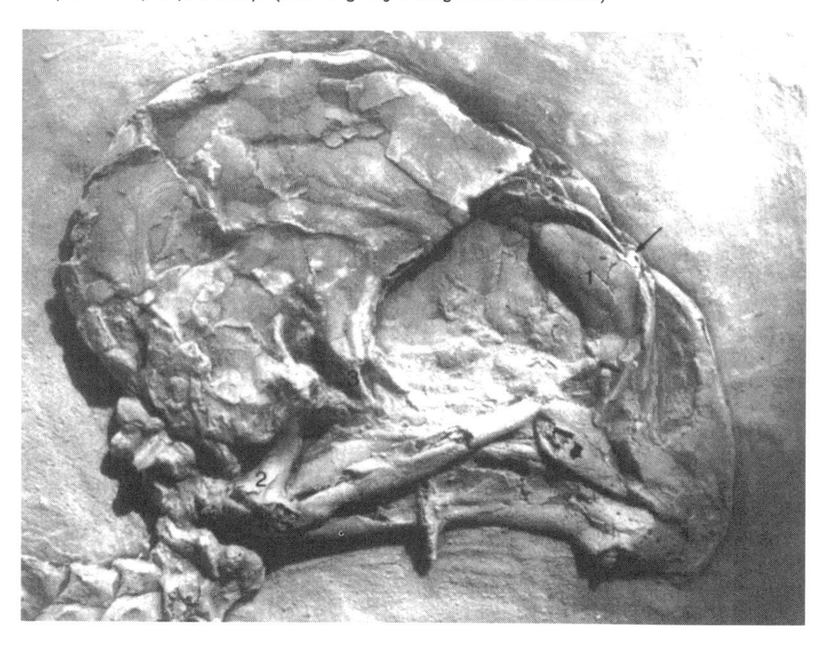

Skull of the medial Eocene acciptrid *Messelastur gratulator*. Arrow points to the nasofrontal hinge; *1,* lacrymal; *2,* quadrate; *3,* postorbital process. × 2.1. (Courtesy S. Peters and Forschungsinstitut Senckenberg)

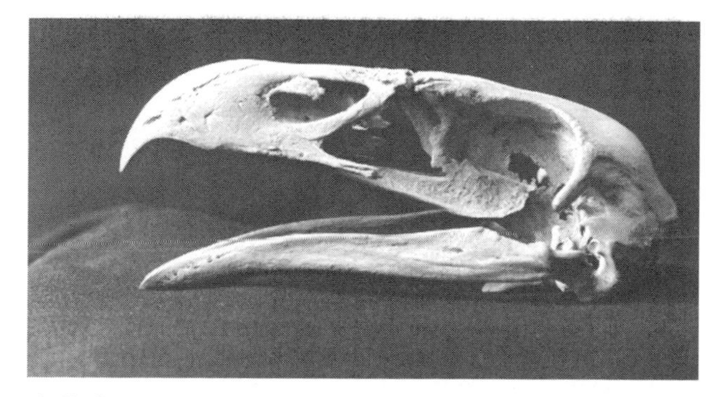

Skull of New Zealand's giant eagle (*Harpagornis morreri*), the primary predator, before the advent of the Maori, on the twelve species of moas. The skull measures 15.2 cm (6 in.) from the base to the tip of the beak. Skeletons of this giant eagle are among the rarest of all New Zealand bird remains. (Photo courtesy Beverley McCulloch and the Canterbury Museum, Christchurch)

ing species of Strigiformes and, like the giant Cuban owls, appeared in its morphology most similar to the genus *Strix*. In addition, Cécile Mourer-Chauviré and Antonio Marco (1988) report the giant barn owl (*Tyto balearica*), previously described as a barn owl endemic to the Balearic Islands, from somewhat older sites in continental localities in Spain and southern France. They relate the size of this owl to a diet that included large rodents typical of that period in Europe.

Falconiform Relations and the Fossil Record

Fossil hawks go back to the early Eocene of England, the medial Eocene of Germany, and the late Eocene or early Oligocene of France in the Old World, and in the New World they are known from the early Oligocene of South America and the medial Oligocene of North America, but they indicate no phylogenetic connections. The order Falconiformes may be polyphyletic (Jollie 1976), yet the relations among the various families of the Falconiformes, as well as the order's ties to other avian groups, are still largely unknown. Two well-preserved specimens of the order Falconiformes are known from the Middle Eocene Messel oil shales of Germany, but they appear already well developed (Peters 1992b). Storrs Olson, in reviewing the falconiform fossil record, laments its paucity: "Although extensive fossil material exists for the Accipitridae, divining its significance in our present state of systematic ignorance proves nearly impossible" (1985a, 112). For their part, Sibley and Ahlquist argue that "the Falconides . . . includes only the hawks, eagles, kites, harriers, Old World vultures, Ospreys, Secretary-bird, falcons, and caracaras. It does not include the New World vultures, which are related to the storks" (1990, 471).

Although the Falconiformes are traditionally placed next to the Galliformes, this should not be construed as having any phylogenetic implications; there is little if any evidence to connect falconiform and gallinaceous birds. Moreover, the conventional placement of the Falconiformes after the ducks, geese, and swans (Anseriformes) is unreasonable; ducks, as we saw in chapter 5, are clearly derived from an ancient shorebird stock. The question of the relations of the various families of diurnal raptors to one another and to other avian orders remains a major challenge of avian systematics.

The Accipitridae includes some 237 species in 65 genera, ranging from sea- and fish-eagles to Old World vultures, kites, harrier-hawks, goshawks, and sparrowhawks to the buzzards, or buteonine hawks. Pierce Brodkorb listed 62 paleospecies of the Accipitridae in 1964, and a number of species have been described since then, but the systematic placement of most is unclear. Although there are reports of "falconiforms" from the Lower Eocene of Europe, the earliest truly diagnosed hawks are represented by two beautifully preserved skulls of a small accipitrid from the Middle Eocene Messel oil shales of Germany. Described as *Messelastur gratulator*, it is approximately half the size of a female Eurasian sparrowhawk (*Accipiter nisus*) and is among the earliest known raptors (Peters 1994). Mourer-Chauviré (1991) has described the numerous remains of a very small accipitrid (chapter frontispiece) from the Upper Eocene locality of La Buiffie of the Phosphorites du Quercy as *Horusornis vianeyliaudae* (Horusornithidae); it is also recorded from the Lower Oligocene of the United States. *Horusornis* is the size of the extant red-footed falcon (*Falco vespertinus*) of Eurasia and Africa, and although it differs in important derived characters from other families of the Accipitridae, it is convergent with the living genera *Polyboroides* and *Geranospiza* in the shape of the articulation of the tibiotarsus and tarsometatarsus, "which probably allowed the leg to flex backwards as well as laterally and medially" (183). Mourer-Chauviré theorizes that this flexibility was related to a particular mode of feeding, perhaps involving extraction of prey from crevices in trees, in a manner similar to that employed by the living African harrier-hawk, or gymnogene (*Polyboroides*, Burton 1978), whose flexible legs can be bent up to 70° behind and 30° from side to side, enabling it to grope in tree cavities or hollows for nestlings and other small animals. Likewise, the Neotropical crane hawks (*Geranospiza*) are also characterized by great flexibility between the distal tibia and proximal tarsometatarsus, permitting them to extract prey from bark crevices, epiphytes, and other otherwise inaccessible places (Burton 1978; Jehl 1968b).

Other early species include a single small accipiter, *Messelastur gratulator*, from the Middle Eocene Messel oil shales of Germany (Peters 1994), and two forms of *Aquilavus*, from the late Eocene to early Oligocene of France. Species assigned to *Buteo* appear by the medial Oligocene, but these fossils are in need of revision. There are also such forms as *Palaeohierax gervaisii*, from the early Miocene of France, which appears most similar to the palm-nut vulture (*Gypohierax angolensis*) (Milne-Edwards 1867–1871; Rich 1980b; Olson 1985a). This fossil is of particular interest in that the Old World vultures of the subfamily Gypaetinae occur in Tertiary and Quaternary deposits in North America. For its part, the palm-nut vulture is especially interesting because it represents a living morphological intermediate between the eagles and the Old World vultures.

The largest eagle that ever lived evolved on the South Island of New Zealand. The New Zealand eagle, *Harpagornis moorei*, is estimated to have had a wingspan of up to 3 meters (9.8 ft.), with talons the size of a tiger's claws, and was able to strike down full-grown moas that weighed some 100 to 250 kilos (220–550 lbs.). According to R. N. Holdaway (1989), *Harpagornis* was a powerful flier but may have been too heavy to soar. It most likely swooped down on its prey from a high perch, such as a tall tree or a rocky bluff. There is no satisfactory explanation as to why *Harpagornis* is absent from Holocene deposits on the North Island, a time when most of the island was covered by forest (Atkinson and Millener 1991). Another large, insular eagle, similar in size and closely allied to the African mainland crowned hawk-eagle (*Stephanoaetus coronatus*, p. 292), has been described from subfossil Holocene deposits on Madagascar, which has also yielded fourteen species of lemurs, seven of them extinct (Goodman 1994b). The Malagasy crowned eagle (*Stephanoaetus mahery*, from the Malagasy adjective meaning powerful) may well have been the legendary *roc* of the tales of Sinbad and Marco Polo but probably, like its African sibling, fed on primates; the crowned hawk-eagle's diet at one locality has been recorded as 80 percent primates (Struhsaker and Leakey 1990) that can weigh up to 12 kilos (26.5 lbs.). Another Malagasy eagle of the genus *Aquila* has been discovered, and the presence of the two, combined with the probability that they ate diurnal primates, has been interpreted as a potential evolutionary force for the lemurs' development of antipredator behavior, a stereotypic response to birds of prey flying overhead even though none of the living eagles eats large primates (Goodman 1994a, 1994b).

As for falcons, Falconidae, the family has a worldwide distribution today but is concentrated in South America (Olson 1985a), with only the falconets occurring much beyond the Neotropics. Although several early Eocene fossils from England of dubious identity have been assigned to this group of falconiforms, an undescribed juvenile specimen from the Middle Eocene Messel oil shales that is clearly a falcon has been tentatively assigned to the subfamily Polyborinae (modern caracaras) of the Falconidae (Peters 1989, 1991). In South America, falconid fossils are known as far back as the early Miocene of Argentina (*Badiostes patagonicus*), and a fossil originally described as *Falco ramenta*, from the medial Miocene of Nebraska, has been removed to the genus *Pediohierax*, considered by Jonathan Becker (1987b, 270) as the primitive sister-group of the Falconidae. Cécile Mourer-Chauviré (1982) lists the Falconidae among the avian taxa identified from the Eo-Oligocene Phosphorites du Quercy in France, and Stefan Peters (1989, 1991) has identified a juvenile specimen from the Messel oil shales as a member of the falcon subfamily Polyborinae, which includes the caracaras. The caracaras, which independently evolved vulturine habits, are today restricted to the Neotropics, but one fossil caracara, *Milvago readei*, recovered from the Pleistocene (Rancholabrean) of the Itchtucknee River, Florida, is the first record of that genus from North America (Campbell 1980). Today *Milvago* is confined to South America, Panama, and Costa Rica, although a paleospecies, *M. alexandri*, is known from late Pleistocene cave deposits in Haiti (Olson 1976a). The living two species of the genus inhabit open savannas, scrub forests, and forest edges, an indication that *M. readei* extended its range into Florida from Middle America via what is known as the Gulf Coast Savanna Corridor (Webb 1974, 1978) during a glacial period in the Pleistocene. Much of Florida was covered by open savanna during dry phases of the Pleistocene, when the corridor was open for dispersal, but the open country largely disappeared at the end of the Pleistocene, remaining only in the southernmost region. As the advent into Florida of the giant phorusrhacid *Titanus* indicates (see chapter 6), numerous bird species of South and Central American origin, including a jacana (Olson 1976b), probably accompanied the large mammals into North America via this savanna corridor during glacial periods. "It would appear . . . likely that suites of savanna-adapted birds rather than just one or two species moved into Florida during the periods of lesser rainfall and more open country than now, only to retreat or become extinct when the climate returned to more humid conditions and the open country disappeared during interglacial periods" (Campbell 1980, 128). The remarkable Itchtucknee avifauna, which is primarily aquatic, with grebes and anatids accounting for 62 percent of the specimens, consists of some fifty-six species from an astonishing collection of 1,363 fossil bird bones.

The nearly cosmopolitan osprey, *Pandion haliaetus*, sole member of the family Pandionidae, is known from as far back as the Oligocene of Egypt and the Miocene of North America (Olson 1985a, 114), but the origin of these so-called fish-hawks remains elusive. Ospreys have particularly long and sharp talons, and their toes bear spiny tubercles, or nodules, on the undersurfaces for grasping fish. As in owls, the osprey's outer toe is large and can be rotated to face backward.

The Secretarybird

The most enigmatic falconiform in terms of its links with other groups is the strange secretarybird (Sagittariidae), which takes its name from the quill-like feathers behind its head. Represented in the modern fauna by the one species, *Sagittarius serpentarius*, confined to sub-Saharan Africa, the secretarybird is unique among living birds, hunting its prey by stalking and then running in zigzag fashion after snakes (its principal quarry), lizards, and small mammals. Secretarybirds usually strike their prey with one foot, usually behind the victim's head, and then batter the prey to death. However, they will also eat insects and even bird's eggs.

Because of these unusual habits, it is extremely interesting from the standpoint of both evolution and paleoecology to encounter secretarybirds in the fossil record. The only two valid paleospecies are known from France: *Pelargopappus schlosseri*, from the mid- and late Oligocene Phosphorites du Quercy, and the larger *P. magnus*, from the early Miocene of Saint Gérand-le-Puy, which was similar in size to today's secretarybird (Mourer-Chauviré and Cheneval 1983). Apparently, then, these strange terrestrial raptors were once widespread, at least in the Old World, and are now confined to a relictual distribution in Africa.

Given this past distribution, it was not surprising when what appeared to be a secretarybird was later uncovered in Miocene deposits in Nebraska, where during that period, some 10 million years ago, savannas and grasslands had become the predominant vegetation in much of the center of North America. The fossil bird (chapter frontispiece) was found in association with a rich ungulate fauna and a small version of the crowned-crane (*Balearica*), an association that is the ecological equivalent of present-day east and central Africa. Study of the fossil, named *Apatosagittarius* ("false secretarybird"), however, revealed that although the morphology of the tarsometatarsus was almost identical to that of the living secretarybird, implicating a similar method of prey capture, the morphology and proportions of the toe showed beyond doubt that *Apatosagittarius* was an accipitrid hawk, convergent on the secretarybird (Feduccia and Voorhies 1989). Thus, life in the ancient North American grassland community had produced a similar but smaller bird, deriving from accipitrid hawks. Although DNA comparisons (Sibley and Ahlquist 1990) show *Sagittarius* to be a member of the Falconiformes, many authors have questioned this relation, and some have argued for affinity with the South American seriemas (Snow 1978), which superficially resemble the secretarybirds and, not surprisingly, capture their prey in a similar fashion.

Vultures

Perhaps the most puzzling raptors are the vultures. Large, visually repugnant, bare-headed carrion-eaters, the vultures really comprise two distinct families that have little to do with each other in an evolutionary sense. The New World vultures (family Vulturidae or Cathartidae) are of unknown ancestry but are thought by some to be allied with the storks, whereas the Old World vultures (subfamily Gypaetinae) share the large family Accipitridae with a variety of forms ranging from hawks and eagles to kites.

Old World vultures differ from New World vultures in possessing strongly hooked feet, rounded nasal openings, and a voice box. Derived from eaglelike ancestors, they range in size from the gigantic griffons and eared vultures to the diminutive Egyptian vulture (*Neophron percnopterus*), which ranges throughout Africa and eastward through Arabia to India, feeding indiscriminately on any kind of garbage or carrion, even gleaning carrion from carcasses larger vultures have abandoned. Egyptian vultures are well known for their use of tools: they often hurl stones at ostrich eggs to glean the nourishing contents. This behavior was once thought to be learned within certain groups of these birds (Lawick-Goodall 1968), but experiments on hand-reared birds show no evidence of cultural transmission, and it has been suggested that the origins of aimed stone-throwing are related to another characteristic behavior, the unaimed throwing of small eggs (Thouless, Fanshawe, and Bertram 1989).

Intermediate between eagle and vulture morphology is the lammergeier, or bearded vulture (*Gypaetus barbatus*). It makes its living by dropping bones on rocks to split them open, and its white throat and breast feathers often exhibit a reddish coloration from the iron oxide in clay in which it rubs its breast (this is the only known example of cosmetic ornamentation in birds). Another bird of great interest with respect to the early evolution of Old World vultures is the palm-nut vulture (*Gypohierax angolensis*), which, as mentioned earlier, is an intermediate form in suspended animation between the eagles and the Old World vultures.

The Fossil Record of Old World Vultures. Presently confined to Africa and Eurasia, the Old World vultures are, surprisingly, represented in the Tertiary and Quaternary fossil record of North America. In 1916, Loye Miller described a fossil specimen of *Neophrontops americanus*, an Old World vulture closely related, and close in size, to the smallest of the living Old World vultures, the Egyptian vulture, along with another Old World vulture, *Neogyps errans*, from the Pleistocene Rancho La Brea tar pits in the distinctively New World setting of southern California. Miller later admitted that "announcement was withheld for two years because of the wide geographic separation from other members of the Old World vulture group" (Miller and Demay 1942, 95). Since Miller's initial discovery, many other Old World fossils, representing three diverse genera, have been discovered from the early Miocene (*Palaeoborus*), the Pliocene, and the Pleistocene of North America. But because Old World vultures also

The palm-nut vulture (*Gypohierax angolensis*) of Africa. Here a female nests in a raphia palm (*Raphia australis*) on the north coast of Natal. These birds feed on oil-palm fruit, and their distribution coincides with that of the palm *Elaeis guineensis*. Palm-nut vultures share many features with both eagles and Old World vultures and provide a valuable living intermediate. (Photo by Dave Harris)

The lammergeier, or bearded vulture (*Gypaetus barbatus*), a mountain bird of the southern Palearctic, southern Asia, and parts of Africa, is a large bird that drops bones onto rocks to split them and then eats the marrow. Lammergeiers also swallow carrion bones as big as a sheep's scapula whole. (Photo by author)

The African white-backed vulture (*Gyps africanus*) at an abandoned lion's kill of a young Cape buffalo on the Loita Plains of Kenya. Note the marabou stork (*Leptoptilos crumeniferus*) in the background of the picture above, its legs white from excreted uric acid. (Photos by author)

The American neophron (*Neophrontops americanus*) from Rancho La Brea. This common Pleistocene Old World vulture was close in morphology to the living Egyptian vulture (*Neophron percnopterus*). (Courtesy George C. Page Museum)

PERIOD/EPOCH	myBP	NEW WORLD	OLD WORLD
Pleistocene		Neogyps	Neophron, Gyps
Pliocene			
Miocene	10, 20	Neophrontops, Paleoborus, Arikarornis	Paleohierax
Oligocene	30		
Eocene	40, 50		
Paleocene	60		
Cretaceous	70, 80		

The geographic and geologic range of Old World vultures, the Gypaetinae. (Modified after Rich 1983)

occur as fossils as far back as the early Miocene of the Old World (*Palaeohierax*), the fossil record says little about their place of origin. Patricia Rich (1980b) reviewed the living and fossil Old World vultures and was unable to reach a solid conclusion, judging New World fossils of Old World vultures could represent one or more separate derivations from eaglelike birds and hawks and are unique to the New World. In contrast however, Hildegard Howard, who for years studied the extensive material of the Rancho La Brea birds, commented unequivocally that "the skeleton of *Neophrontops* is markedly like that of the Recent Old World vulture, *Neophron*" (1966b, 3), and that the differences between them "are of less note than those which exist between *Neophron* and its contemporaries among the vultures today" (Howard 1932, 70). There would seem little doubt that these are truly Old World vultures in the New World.

New World Vultures. Like storks, New World vultures cool themselves by urohidrosis, dumping their urinary liquids onto their hind legs (Kahl 1963), which accounts for the characteristic buildup of whitish uric acid on their legs. As pointed out in chapter 5, this is one line of evidence that, combined with DNA comparisons and other morphological evidence, has led to the generally accepted conclusion that the Vulturidae is closely allied with the storks (Ligon 1967; Rea 1983; Sibley and Ahlquist 1990). Experiments performed on the representatives of the two other monotypic families usually allied with the storks, the shoebill (Balaenicepitidae) and the hamerkop (Scopidae), in fact failed to induce urohidrosis; urohidrosis is thus known only in storks and New World vultures. However, as also was mentioned before, storks and the shoebill have a unique synapomorphy, a specialized tubular ear bone, or stapes, which argues for a link, whereas the New World vultures have the primitive condition of the stapes. If they are related to the storks, the New World vultures would have to be an early offshoot of the lineage, branching off before the development of the ear bone and other unique derived features that define the storks and allies.

With their powerful wings, New World vultures are excellent soarers, perhaps the world's best. There are seven living species, including the nearly extinct California condor (*Gymnogyps californianus*) and the Andean condor (*Vultur gryphus*), very large birds weighing up to 11.5 kilos (25 lbs.) and with a wingspan of almost 3 meters (10 ft.). The other five are the king vulture (*Sarcoramphus papa*), which ranges from southern Mexico to Argentina; the smaller black vulture (*Coragyps atratus*) and turkey vulture (*Cathartes aura*), more familiar to North Americans; and two species of yellow-headed vultures (*Cathartes melambrotus* and *C. burrovianus*). All have long front toes with a small web at their bases, and their feet are not adapted for grasping. Further characteristics are a longitudinal nasal opening and the absence of a voice box, which prevents them from making any sound except a hiss.

The phenomenon of urohidrosis as a mechanism of heat dissipation, in storks and particularly in the turkey vulture, has received much attention. Daniel Hatch (1970) performed a series of experiments in which he heated turkey vultures' environment, and they responded by excreting on their legs at an increased rate. These large birds were able to control deep body temperature through evaporation of a small amount of liquid on a very small portion of their body. Vultures also dissipate heat by panting and by extending their bare upper neck from the loose, densely feathered skin of the lower neck; this exposes about a couple of centimeters or so of normally covered skin. Hatch made other interesting observations about turkey vultures. He confirmed that they are well adapted to life in hot environments by keeping one in captivity in good health for one year without water. He also observed that the birds showed nasal secretions while eating; this linked with an earlier study (Cade and Greenwald 1966), in which nasal glands appeared to enhance many falconiforms' ability to conserve water through the secretion of ingested electrolytes, especially sodium.

A great debate concerning vultures in general is their relative ability to locate prey by sight or smell. In a series of field experiments K. E. Stager (1964) demonstrated that, unlike most birds, the turkey vulture has a well-developed olfactory organ, which helps it locate prey. The two other species of *Cathartes*, the greater and lesser yellow-headed vultures, have similar searching habits and are assumed also to locate food by smell. The black vulture, in contrast, has a much smaller olfactory organ (Bang 1960, 1967), and Stager's experiments indicated that it does not use smell to locate food. Although both turkey and black vultures are opportunistic scavengers that "will thrash the soil searching for invertebrates" (Rea 1983, 43), and will fish, interestingly, the black vulture is more a bird of open country, river banks, estuaries, and now garbage dumps, but it has never been a forest hunter. Experiments on captive king vultures, also forest hunters, indicate that like black vultures they cannot locate hidden food by smell (Houston 1984). David Houston suggests that king vultures, which fly high above forests, locate carcasses by watching the activities of the *Cathartes* vultures below them. Old World vultures similarly appear to lack any ability to locate food by smell, having a poorly developed olfactory apparatus, and they rarely occur in forested habitats (Stager 1964; Houston 1976; Hertel 1992). The comparative evolutionary ecology of Afrotropical and Neotropical vultures in forests indicates that a fair amount of carrion is available in Neotropical forests, which would seem capable of accommodating a high diversity of scavengers, and selection probably favored New World vultures that could locate carrion by smell (Houston 1985).

The New World vultures, now confined to the Western Hemisphere, have left a variety of fossil remains in Europe, and most of their early history is confined to the Old World. The fossil vulturids found in the Old World were considered to be first represented by *Eocathartes*, known from two species from the Middle Eocene of Germany, but Storrs Olson (1985a, 191) has claimed that *Eocathartes* and a number of other putative vulturids are misidentified. In 1972, Joel Cracraft and Patricia Rich reviewed the Old World fossils of the vulturids, confirming the systematic position of the genera *Plesiocathartes* and *Diatropornis*, from the Upper Eocene to Middle Oligocene Phosphorites du Quercy, as vulturids, as did Cécile Mourer-Chauviré in 1982. These were relatively small New World vultures, one species being only about as large as that of the living turkey vulture. According to Olson (1985a, 192, from Kurochkin, pers. comm.), a vulturid larger than *Diatropornis* or *Plesiocathartes* has been discovered in the early Oligocene of Mongolia, showing that the family was also present in Asia during the same period. There is no evidence of any cathartids in the Old World past the early Miocene, the last being *Plesiocathartes gaillardi* from the early Miocene of Spain (Crusafont and Villalta 1955).

The oldest fossil ever considered to be a New World vulture, a species mentioned in chapter 6 and firmly implanted in most ornithology textbooks, is the "running vulture" described by Alexander Wetmore in 1944 as *Neocathartes grallator*, who placed it in its own family, the Neocathartidae. Known from the late Eocene of Wyoming, this fossil was for some time thought to be a long-legged cursorial cathartid with reduced wings, but it is now known to be the cariamid gruiform *Bathornis* (Olson 1985a). Another fossil, *Palaeogyps prodromus*, from the early Oligocene of Colorado, is likewise a bathornithid. And a third form, *Eocathartes*, was emended to *Neocathartes* because the name had already been applied to a European fossil; *Neocathartes*, superficially reminiscent of storks, may serendipitously have led some systematists to postulate a relation between storks and cathartids.

The earliest valid cathartid in the New World is *Phasmagyps*, from the early Oligocene of Colorado, and *Brasilogyps*, from the early Oligocene of Brazil (Alvarenga 1985b). Most biogeographers have thought that the New World vultures originated there because of their current distribution, but the fossil record indicates that the Vulturidae have been in the Old World since the mid-Paleogene, whereas their presence in the New World clearly dates from much later. More fossil evidence, however, is needed to confirm this assertion.

Aside from the putative North American vulturid from the early Oligocene, the next chronological occurrence of a vulturid is the condorlike *Hadrogyps aigialeus*, from the medial Miocene of North America (Emslie 1988a), and New World vultures eventually appear in the form of

The marabou stork (*Leptoptilos crumeniferus*) of Africa, scavenging at a kill. (Photo by author)

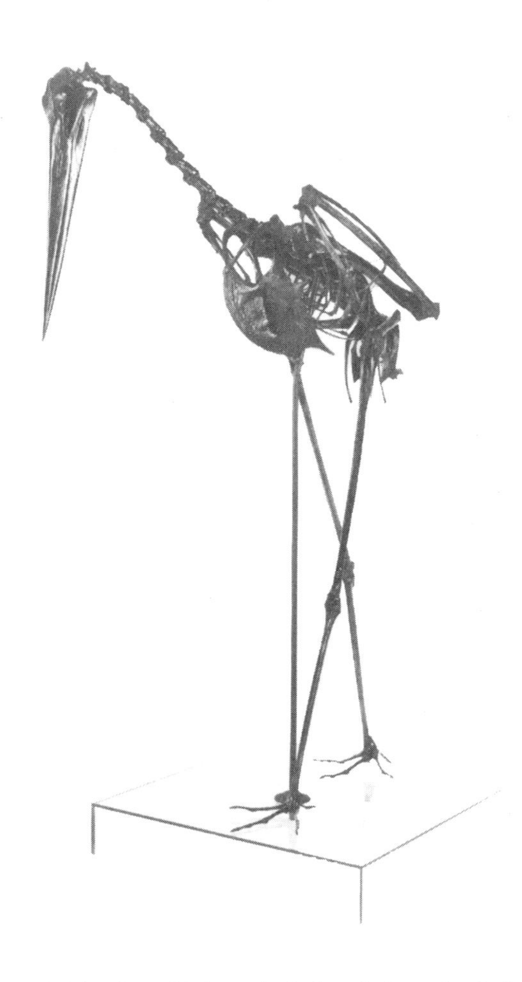

The La Brea stork (*Ciconia maltha*), ecological equivalent of today's African marabou stork. (Courtesy George C. Page Museum)

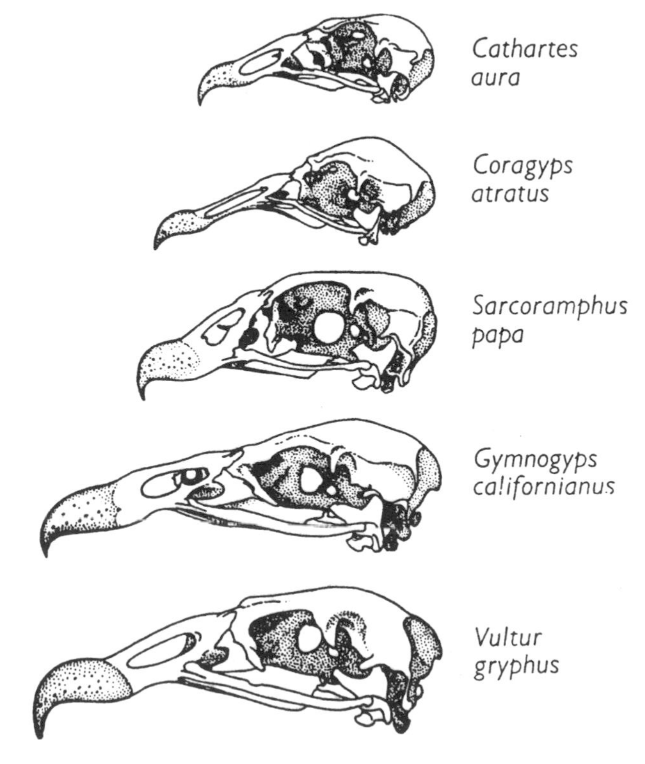

Differential size and proportions of beaks and skulls of New World vultures. (Modified after Fisher 1944; courtesy *Condor*)

The Andean condor (*Vultur gryphus*), one of the largest living flying birds. (Photo by author)

The La Brea condor (*Breagyps clarki*). (Courtesy George C. Page Museum)

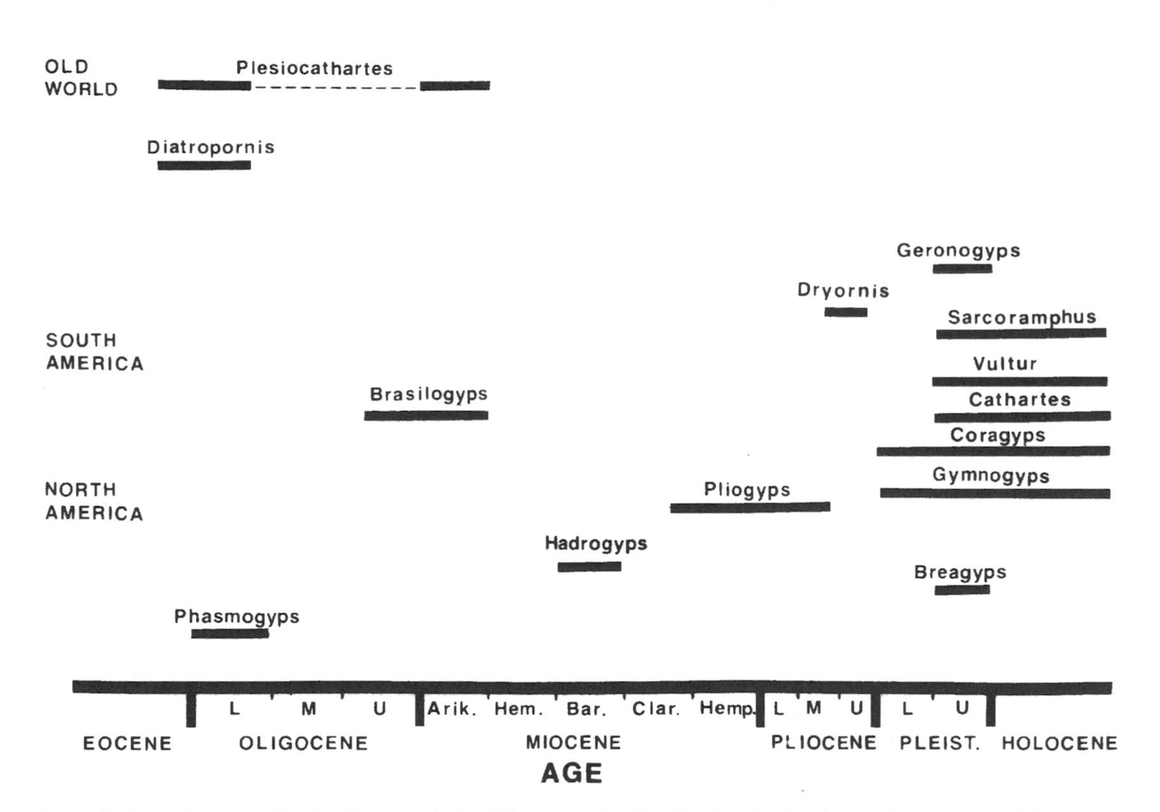

The geologic and geographic distribution of all valid genera of Vulturidae (Cathartidae). Epochs (not to scale) are divided into Lower (*L*), Middle (*M*), and Upper (*U*) periods or by land mammal ages (Miocene only). The dashed line indicates uncertain geologic distribution. Note that *Cathartes*, *Coragyps*, and *Gymnogyps* have occurred or do occur in both South and North America. (From Emslie 1988; courtesy *Auk*)

Sarcoramphus, ancestral to the king vulture, and *Vultur*, ancestral to the Andean condor. Two species of *Pliogyps* (Becker 1986b), a genus distantly related to living cathartids, are known from the late Miocene and Pliocene. Both the Miocene *Pliogyps charon* and the Pliocene *P. fisheri* appear to represent part of a radiation of short-legged, robust condorlike vulturids in North America (Becker 1986b). The only other pre-Pleistocene vulturid is the condorlike *Sarcoramphus kernense*, from the early Pliocene of North America, *Sarcoramphus* being considered somewhat intermediate between the small vultures and the condors (Emslie 1988b, 226). The earliest vulturid from South America is from the Lower Oligocene and was described by Herculano Alvarenga in 1985 as *Brasilogyps faustoi*, which is surprisingly quite similar to *Coragyps* in size and morphology and tells of an early advanced morphology of the vulturids (Alvarenga 1985b). No other vulturids are known from South America until the early to medial Pliocene, and the largest was *Dryornis pampeanus* (Tonni 1980b); all other South American condors are from the late Pleistocene. Why New World vultures should not have persisted in Europe after the early Miocene is not understood. Possibly they disappeared because they could not compete with the Old World vultures, which were then on the ascent.

During the Pleistocene of North America, the vulturids, along with the Old World vultures, multiplied in almost incredible numbers. The ancestor of the black vulture, *Coragyps*, and the ancestor of the turkey vulture, *Cathartes*, arose in this period, along with a wide variety of other forms, including the now extinct La Brea condor (*Breagyps clarki*) and the once widespread California condor (*Gymnogyps californianus*). Steven Emslie (1988a) described a new species of *Gymnogyps* (*G. kofordi*), from the early Pleistocene of Florida, concluding that the new species shared a common ancestor with the California condor, that the condors (*Gymnogyps* and *Vultur*) formed a monophyletic assemblage with *Gymnogyps* as a distinctive North American genus and *Vultur* as a distinctive South American genus, and that condors probably originated in North America and reached South America in the mid-Pliocene near the beginning of the Great American Interchange.

Vulture Paleoguilds. A guild, in ecological terms, is a group of species that use the same resource in a similar manner; paleoguilds, often difficult to study, must be delineated according to morphological comparisons (Van Valkenburg 1995). In 1992, Fritz Hertel examined character divergence in New World vultures on the basis of beak strength, body size, and body shape, finding that the Pleistocene vultures from Rancho La Brea exhibited a diversity of beak strengths and body sizes similar to that observed in extant New World vultures. Hertel's study paralleled an earlier study (Kruuk 1967) of East African vultures that showed that, despite phylogenetic differences, vultures from varied geographical areas separate in similar ways, reflecting differing abilities in processing carcasses and variations in body size. Hertel's study also reinforced Houston's (1988) conclusions that the New and Old World vulture guilds had similar feeding behaviors.

Through multivariate analysis of cranial, mandibular, and bill measurements, Fritz Hertel (1994) demonstrated conclusively a theme of convergent evolution in guilds of New and Old World vultures. In terms of food processing techniques, the two carrion guilds contained rippers, gulpers, and scrappers. Such species as the New World king vulture (*Sarcoramphus papa*) and the Old World lappet-faced vulture (*Torgos tracheliotus*), for example, are rippers in their respective guilds; the New World California condor (*Gymnogyps californianus*) and the Old World African white-backed vulture (*Gyps africanus*) are gulpers; whereas, the New World black vulture (*Coragyps atratus*) and the Egyptian vulture (*Neophron percnopterus*) are ecologically scrappers. Likewise, at the Ice Age Rancho La Brea site, rippers included the extinct *Neogyps errans*, gulpers *Breagyps clarkii*, and scrappers *Coragyps occidentalis* and *Neophrontops americanus*.

The California Condor: A Long Road to Extinction. The California condor is an example of a Pleistocene relict that once widely ranged over North America when abundant large mammal carcasses dotted the landscape. This magnificent condor has been on the wane for thousands of years, following the Ice Age extinctions, and is now on the brink of extinction. In spite of elaborate breeding programs, it is probably doomed because the mammalian fauna that evolved side by side with this great bird has long since vanished.

In an extensive study of the remains of the California condor in the Grand Canyon in Arizona, Steven Emslie (1987) concluded that the species became extinct in that region and other parts of the inland West more than ten thousand years ago. This date coincides with the extinction of the Pleistocene megafauna, including primarily the proboscidians (elephants), edentates (ground sloths and others), and perissodactyls (horses). His conclusion that these birds relied on the large mammal megafauna for food (as the recovery of food bones from a late Pleistocene nest cave suggested) was novel, but it was bolstered by the work of David Steadman and Norton Miller (1987), who discovered eleven-thousand-year-old late Pleistocene condor fossils in New York in association with spruce–jack pine woodland, and reached a similar conclusion. Stead-

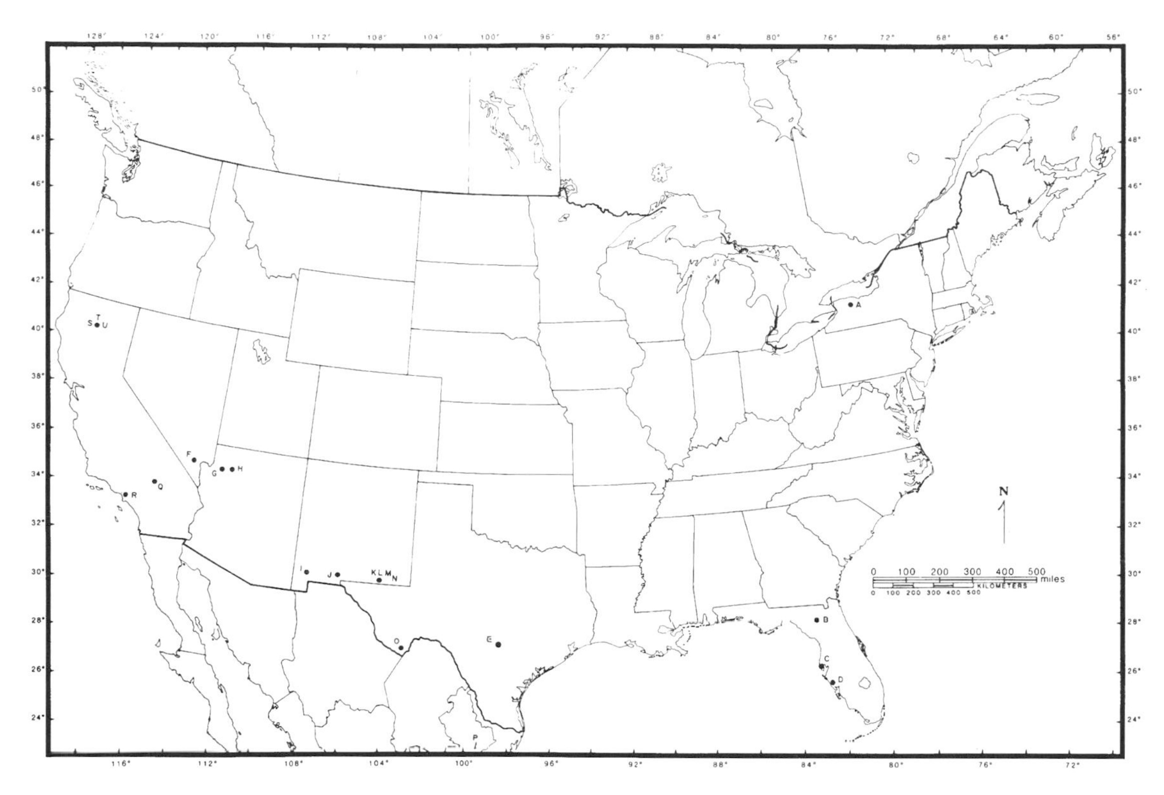

The late Pleistocene range of the California condor (*Gymnogyps californianus*). See Steadman and Miller (1987) for details. (From Steadman and Miller 1987; courtesy University of Washington Press)

man and Norton's find was of particular interest because California condor fossils were previously known only from Florida and from Nuevo León and Texas west to California. Their discovery indicated conclusively that this condor was once even more widespread and able to live in a colder boreal, coniferous setting when the North American megafauna still existed. California condors are now known to have been much more dependent on the Pleistocene megafauna than was previously thought, and they appear to be a relic of the great Ice Age radiation of large mammals and New World vultures and teratorns in North America.

New World Teratorns: A Third Group of Vultures. Perhaps the most remarkable of the Ice Age vulturine birds found in the New World were the teratorns. Teratorns belong to an extinct avian subfamily (Teratornithinae) that was long considered to be related to the New World vultures, Vulturidae (Cathartidae), because of a raptorial appearance, especially in the beak and in parts of the skeleton. However, numerous workers, including their original describer, Loye Miller, questioned this link. When Miller established the Teratornithidae, he stated that "*Teratornis* . . . shows very bold divergence in its osteology from the closely knit family of the Cathartidae, the divergence taking a number of different pathways. The degree of divergence is in excess of those osteological differences to be noted between most families of living birds classified under one order" (1925, 94). Teratorns share numerous similarities with other avian groups, including the storks and the pelecaniforms, and their systematic position is under investigation by Kenneth Campbell and Eduardo Tonni (1980, 1982, 1983). Gigantic, soaring birds, teratorns may have had vulturine habits, although Campbell and Tonni (1982, 1983) have hypothesized that teratorns were predators rather than scavengers based on their sharp, hooked beaks, which could have enabled them to catch their prey of small animals and swallow them whole. Until overwhelmingly convincing evidence is uncovered, however, their overall appearance tells of a large, soaring, open-country vulturine type of bird.

The very common and best known teratorn, *Teratornis merriami*, stood some 0.75 meters (2.5 ft.) tall, had a wingspan of 3.5 to 3.8 meters (11.5–12.5 ft.), and weighed some 15 kilos (33 lbs.); *T. incredibilis*, known from Pleistocene deposits in Nevada and California, stood about 0.75 meters (2.5 ft.) tall, had a wingspan that may have approached 5.2 to 5.9 meters (17–19.4 ft.), and was some 40 percent larger than *T. merriami* (Campbell and Tonni 1980). Two other species—*Cathartornis gracilis* and *Argentavis magnificens*—are known. Although *Teratornis*

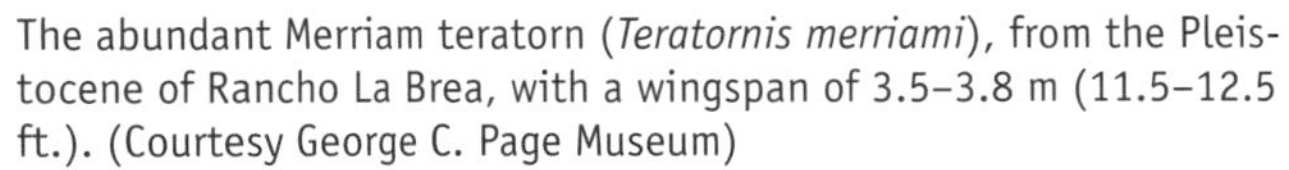

The abundant Merriam teratorn (*Teratornis merriami*), from the Pleistocene of Rancho La Brea, with a wingspan of 3.5–3.8 m (11.5–12.5 ft.). (Courtesy George C. Page Museum)

Relative sizes of the magnificent teratorn (*Argentavis*) and the living American bald eagle. The wingspan of the teratorn is 7.6 m (25 ft.), that of the bald eagle 2.4 m (8 ft.). Drawn to scale. (Drawing by John P. O'Neill; modified after Campbell 1980)

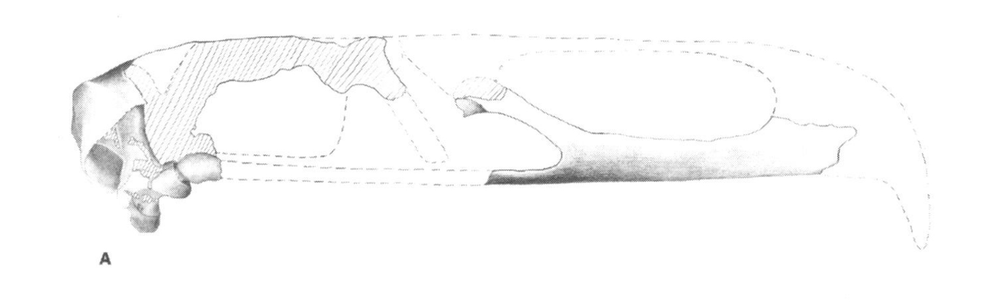

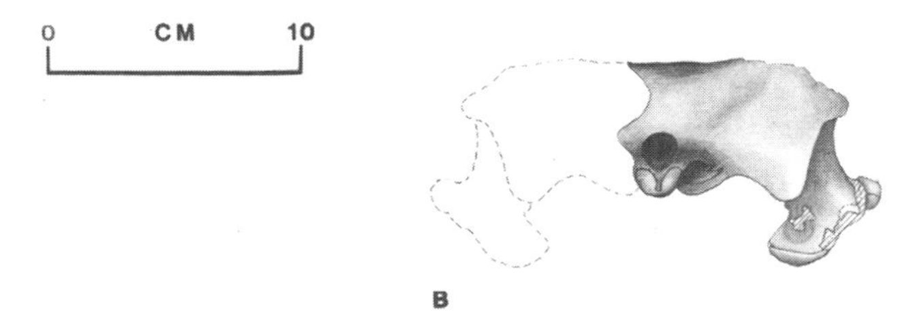

Skull of the magnificent teratorn (*Argentavis magnificens*), largest flying bird ever known, in lateral (*A*) and posterior (*B*) views. Hatch-marked areas represent portions of the specimen where the bone has flaked away but the matrix remains to show form; dotted lines show estimated outline of bone where missing, based on corresponding bones of *Teratornis merriami*. (From Campbell and Tonni 1980; courtesy *Los Angeles County Museum of Natural History, Contributions in Science* 330)

merriami is known from hundreds of fossil bones from the La Brea tar pits, *T. incredibilis* and *Cathartornis gracilis* are known only from a few bones each. Teratorns were widely distributed—fossils are known from Florida, Arizona, Nevada, Mexico, California, Peru, and Argentina—and Campbell and Tonni have suggested "that the evolution of teratorns to their large size proceeded in phase with the development of grasslands and semiarid habitat in southern South America" (1982, 271). They point out, as David Webb (1978) has explained, that grasslands and accompanying biota began to develop in the temperate latitudes of South America during the Paleocene and there was progressive drying throughout the Tertiary, producing the pampas, steppes, and semidesert regions characteristic of present-day Argentina. This habitat must have been the domain of these giant birds.

The fourth species of teratorn, *Argentavis magnificens*, is also the oldest known teratorn, dating to between 5 and 8 million years ago (late Miocene), and was the real giant of the family. This huge form, known from an Argentine fossil, was nearly twice the size of *Teratornis merriami*, standing 1.5 meters (5 feet) tall, weighing some 72 kilos (158 lbs.), and having a wingspan of about 7 to 7.5 meters (23–25 ft.); it is the largest flying bird known to science (Campbell and Tonni 1980, 1983; Campbell and Marcus 1992). Because of its large size, it must have been limited to areas of savanna or open grassland. *Argentavis* was recovered from late Miocene deposits at approximately 37° South latitude, just north of the present limits of South America's strong westerly winds, but in Miocene times the blocking action posed by the Andes on the westerlies would have not been present. As F. Prohaska has noted of Argentina's westerly winds, "In few parts of the world is the climate of the region and its life so determined by a single meteorological element, as is the climate of Patagonia by the constancy and strength of the winds" (1976, 115). Such strong and constant winds, argue Campbell and Tonni, "would have been more than sufficient to carry *Argentavis magnificens* aloft whenever it spread its huge wings" (1983, 402). The evolutionary genesis of an avian behemoth such as *Argentavis* may have been impossible without this extreme "meteorological element," and its discovery may mean that teratorns evolved in South America and spread into North America during the Great American Interchange. This would explain the temporal distribution of teratorns in North America.

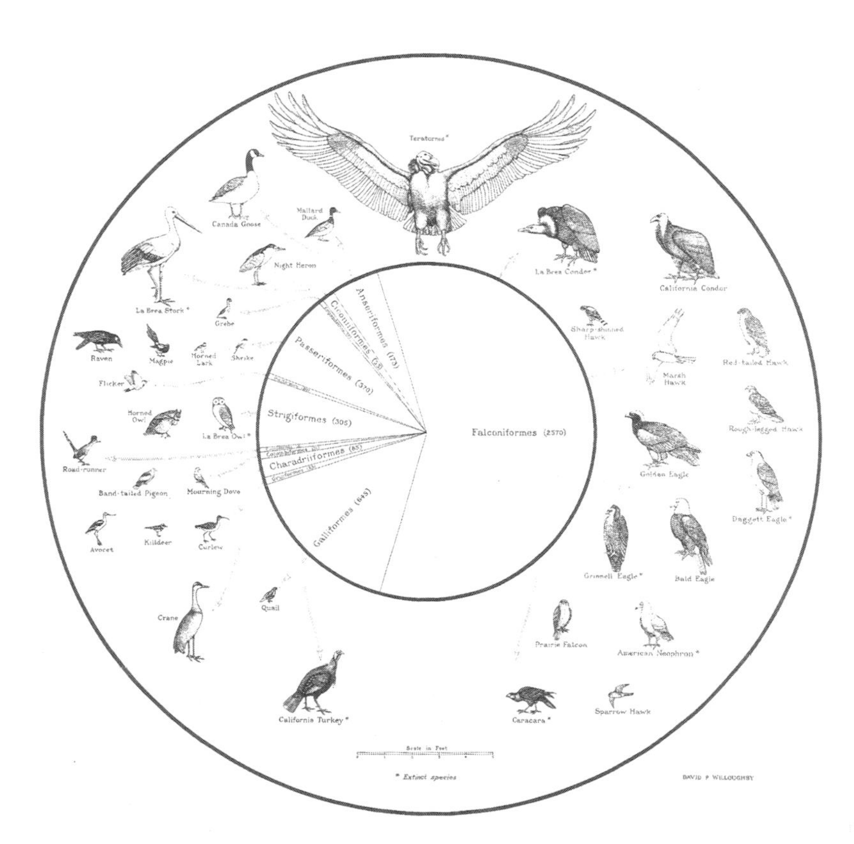

The diversity of birds in the Rancho La Brea fauna. The numbers refer to the fossils found. (Courtesy Los Angeles County Museum of Natural History)

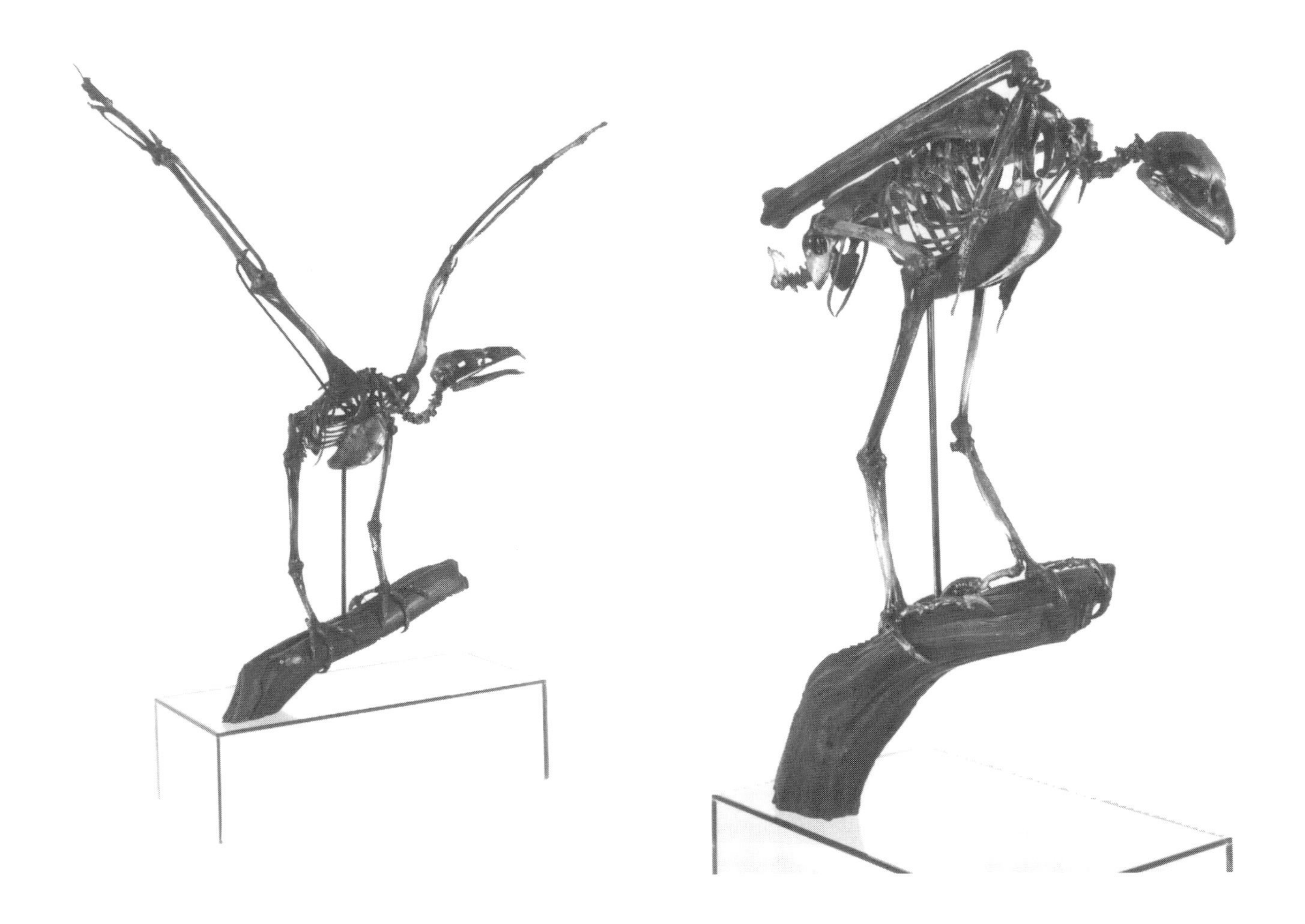

One of the more spectacular bird fossils from Rancho La Brea is the errant eagle (*Neogyps errans*), *above left,* a bird similar in size to the American golden eagle but related to Old World eagles. Other beautifully preserved birds are the Woodward eagle (*Amplibuteo woodwardi*), *above right,* the largest of the seven eagles from Rancho La Brea and known only from that locality, and the La Brea caracara (*Polyborus planchus prelutosus*), *right,* a carrion-feeding member of the falcon family. (Photos courtesy George C. Page Museum)

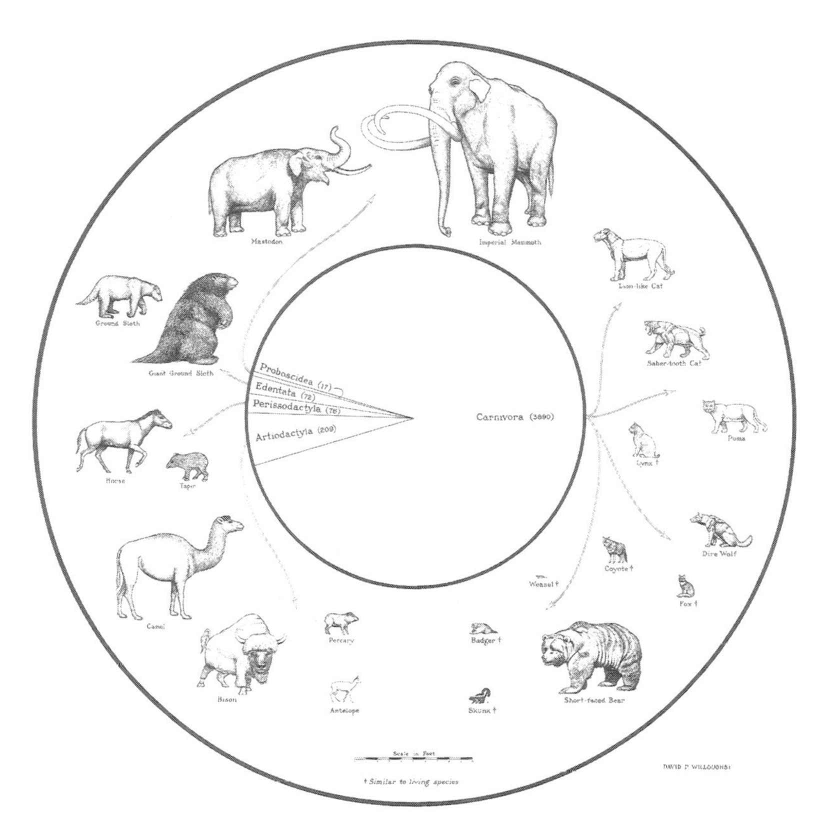

The diversity of Pleistocene mammals recovered from the Rancho La Brea tarpits, not including insectivores, bats, rodents, and rabbits. The numbers refer to the fossils found. (Courtesy Los Angeles County Museum of Natural History)

Ice Age Faunas and Extinction. Much of our knowledge of the Ice Age avifauna comes from fossils preserved in such asphalt deposits as the Rancho La Brea tar pits near Los Angeles. Lying under Pleistocene waterholes, these tarry sediments often seeped to the surface, trapping the mammals and birds that gathered there to drink. Although an estimated 85 percent of the La Brea fossils, now housed in the Page Museum of the Los Angeles County Museum of Natural History, remain unidentified, some 400,000 avian fossil specimens spanning twelve orders of birds and comprising 135 species, 15 percent of them extinct, are known. More than half of these are falconiforms, probably present in such great numbers because as carrion-eaters they flocked to waterholes to feed on mired dead or dying prey. The large vulturine population was easily sustained by the immense numbers of mammals that thrived in the North American Pleistocene. Carnivores included the saber-toothed tiger, the cheetah, and the lionlike cat. Herbivorous mammals were even more numerous, and included two common elephants—the mastodon and larger imperial mammoth—camels, horses, tapirs, giant and smaller ground sloths, bison, peccaries, and antelopes, to mention only a few. In 1979, Kenneth Campbell, of the Page Museum, described another impressive tar pit avifauna dating to almost fourteen thousand years ago from northwestern Peru. Known as the Talara Tar Seeps, this site has produced some 6,200 specimens of nonpasserine bird fossils, representing eighty-nine species, and provides a good comparison with Rancho La Brea. Interestingly, 23 percent of the nonpasserines from the Talara Tar Seeps are extinct, as compared to 17 percent of the nonpasserine species from Rancho La Brea.

By the end of the Pleistocene, most of the large mammals that had roamed the continents were extinct, and so were the Old World vultures, the teratorns, and most of the New World vultures that fed on them. What caused this massive extinction has been a matter of controversy. Sixty-seven genera of mammals, some 50 percent of large mammalian genera, had disappeared by the end of the Ice Age, but nearly half of these died out in the narrow time interval of twelve thousand to eight thousand years ago. Paul S. Martin (Martin 1973; Martin and Klein 1984; Steadman and Martin 1984) has developed a now popular theory, usually referred to as the Pleistocene "overkill hypothesis," to explain the extinction of mammals during this period. According to Martin, prehistoric hunters, crossing into North America over the Bering Strait land bridge, killed off much of the mammalian fauna, thus reducing vultures' food supply.

Martin's ideas were reinforced from a different per-

Reconstruction of a Pleistocene waterhole at Rancho La Brea approximately 100,000 years ago. The dire wolf (*Canis dirus*), on the left, and the saber-toothed tiger (*Smilodon*), on the right, came to prey on other animals and were at times trapped themselves. To the right rear are horses of an extinct species. The large teratorns waited to feed on the carcasses of mired animals. (Mural by Charles R. Knight; courtesy Field Museum of Natural History, neg. no. 72378)

Excavation of fossil bones at Rancho La Brea, showing the exposure of bones of Ice Age animals in the asphalt. (Courtesy Los Angeles County Museum of Natural History)

spective by Norman Owen-Smith (1987; Lewin 1987), who offered an explanation termed the "keystone herbivore hypothesis." In simple form, very large herbivores, or megaherbivores, make a tremendous impact on vegetation as they feed, opening up habitats in which smaller herbivores can thrive. If the megaherbivores are removed, for whatever reason, the vegetation grows thicker, eliminating the smaller herbivores and their ecological dependents. However, any overkill hypothesis, even one restricted to megaherbivores, involves a rapid spread of humans throughout the New World from about 12,000 to 10,000 years ago, and there are over fifty radiocarbon dates on paleo-Indian sites in the New World dated earlier than 12,000 years ago (L. Martin and Neuner 1978). And neither Paul Martin's nor Norman Owen-Smith's hypotheses fully account for the simultaneous extinction of at least ten avian genera (Grayson 1977) and the disappearance of such small forms as pocket gophers, and the fact that moose and bison survived, whereas every last camel, horse, and stag-moose (*Cervalces*) was extirpated.

Many authors would like to implicate humans in all major recent extinctions, extending the model of island extinction, so vividly told by the devastation of Southern Pacific fauna by the Polynesians (Steadman 1995), to continental extinction (Diamond 1991). But these sweeping generalizations are difficult to apply on a continental or global scale. Although some Ice Age species, particularly some of the larger mammals, may have been extirpated, at least in part, by human hunters, it seems more likely that a sudden change in climate caused the larger pattern of extinctions. In fact, it is now clear that a tremendous biotic reorganization took place from eleven thousand to eight thousand years ago, during the height of the Wisconsin glaciation, the last and most severe of the Pleistocene glacial advances, when almost half of North America was covered by continental glacial ice (Martin and Neuner 1978). Nowhere is this biotic reorganization more apparent, perhaps, than in southeastern North America, where not only some boreal species, but the truly tropical fauna, including capybaras, tapirs, jaguars, giant armadillos, and giant tortoises (*Geochelone*), disappeared.

Larry Martin and A. M. Neuner (1978), like others, attribute the Pleistocene extinctions to climatic change. In their scenario the extinctions were caused largely by the Pleistocene trend toward increasing seasonality, which culminated between eleven thousand and eight thousand years ago—the same period of the extinctions of most Pleistocene large mammals. The very complex communities that had been established in the mild conditions of the Pliocene and Pleistocene were suddenly exposed during the Wisconsin glaciation to severe climatic upheaval. As the huge Pleistocene herbivores were wiped out, extinction cascaded throughout the ecosystem, mandating the death of many of the avian scavengers—the Old and New World vultures and the magnificent teratorns.

The primitive Eocene land bird *Foro panarium* bears similarities to cuckoos, turacos, and the hoatzin. Its precise affinities remain obscure, and it is currently placed, more or less by default, in the Cuculiformes. (Drawing by John P. O'Neill)

8

The Rise of Land Birds

If anything is clear from the study of avian evolution it is that birds as a group are arboreal and that they originated in trees. *Archaeopteryx*, the first known bird, was arboreal, and there is no reason to think that the first "modern" birds occupied a different ecological zone. Nor does the paleontological record tell us differently. The first great radiation of birds involved at least two main lineages, the enantiornithine, or opposite, birds, which had a worldwide distribution and tremendous diversity during the Cretaceous period, and the ornithurine birds, first seen in the fully volant *Ambiortus* of the early Cretaceous, a lineage that survived the Cretaceous-Tertiary bottleneck and led to the modern post-Cretaceous avian radiation. The highly specialized lineage of foot-propelled hesperornithiform water birds from the Cretaceous were also part of the modern ornithurine radiation but were an evolutionary dead-end that concluded with the termination of the Cretaceous; and we really do not know when the radiation of modern water birds began. But it is becoming increasingly obvious that a dramatic faunal reorganization took place after the great extinctions of the late Cretaceous, and it would be extremely unlikely if the same events that led to the demise of the dinosaurs, whether cataclysmic or gradual, did not also have an equally dramatic effect on birds, the veritable "miner's canary," perhaps as a filtering of only a few adaptive types through the upheaval.

Certainly the avian fossil record is concordant with this hypothesis. No modern land birds are known from before the Cretaceous-Tertiary boundary, and, with the exception of *Neogaeornis*, a putative loon from the late Cretaceous of Chile that is based on such fragmentary material as to be suspect (the age of the fossil is also in need of verification), neither are any modern water birds known for certain from before the Cretaceous-Tertiary boundary, although some may have existed. The facts may change, but the best current hypothesis is that all modern birds, as we know them, are post-Cretaceous. Many fossils of the so-called transitional shorebirds, or graculavids, that were once thought to be late Cretaceous (Olson and Parris 1987) are now known to be Paleocene. And what was once thought to be the first modern land bird, *Alexornis*, a species described by Pierce Brodkorb (1976) as possibly intermediate between the piciform and coraciiform birds, is now known to be a member of the Cretaceous enantiornithine radiation. The real question seems to be whether modern water birds came soon after, or much later than, modern arboreal land birds. The answer may be that there was a great radiation of modern *birds* from which the modern water birds and modern land birds sprang more or less simultaneously from similar stock, perhaps involving a primitive shorebird-gruiform nexus.

Conventional wisdom has placed the arboreal land birds at the apex of the avian classifications, and the same orthodox practice has led me to end this book with them. Nevertheless, aside from the passerines, which appear somewhat later in the Tertiary and are highly derived, most orders of arboreal land birds appear at about the same time during the early Tertiary. The volant paleognathous birds, from which the ratites sprang, are in fact among the first birds to appear in the Paleogene and may have arisen separately from Cretaceous paleognaths, so there is little need to stray from past practice in placing these birds at the beginning of the classification. Further, if water birds and land birds sprang almost simultaneously from the vortex of a post-Cretaceous avian radiation, it makes little sense to stray from conventional classifications that have become ingrained through use of popular field guides and texts, especially when conclusive evidence of ordinal relations is in most cases absent or controversial (Mayr and Bock 1994). The term *land bird* itself may appear to have little meaning, in that flying as well as flightless paleognathous birds and galliforms could well be called "land birds." In this chapter, however, I restrict the discussion to the classical sense of the term, meaning the higher, more or less arboreal land birds.

Primitive Land Birds: A Miscellany

The Hoatzin. There are certainly living birds that can be considered primitive land birds: among gruiforms, the South American seriemas; among more arboreal birds, the hoatzin (*Opisthocomus*), cuckoo, and musophagid grouping; crested-swifts or treeswifts (Hemiprocnidae); mousebirds (Coliiformes); and, among the coraciiform-piciform groups, the Malagasy ground-rollers (Bracypteraciidae), cuckoo-rollers (Leptosomatide), rollers (Coraciidae), and Galbulae, the primitive basal "coraco-piciform" (roller-woodpecker) grouping. In addition, such passerines as the New Zealand wrens (Acanthisittidae) and the Australian scrub-birds (Atricornithidae) and lyrebirds (Menuridae) exhibit numerous primitive passerine features. Together with certain critical fossils that have recently come to light, these birds permit us to reconstruct a history—albeit hazy—of the rise of land birds. There are certainly morphological connections between such known primitive forms as the hoatzin and the cariamid assemblage, and it is in this focus that we might appropriately view the "basal" land bird genesis.

Suggestions that the unusual hoatzin (*Opisthocomus hoazin*) is a primitive bird are practically as old as systematic ornithology itself and have been based primarily on the presence of two well-developed reptilian claws on the wing digits of young hoatzins. In fact, some adults may retain a second phalanx on the alular digit, a feature that is also present in a number of other avian groups (Fisher 1940; Stephan 1992). The hoatzin's osteology has also been found to be "primitive," and Storrs Olson (1985a) and Cécile Mourer-Chauviré (1983) compared the hoatzin skeleton favorably with cariamid fossils of Eocene age, with Olson (1985a) emphasizing that *Opisthocomus* is one of the most primitive of living birds. Mourer-Chauviré reached this conclusion in her study of fossil elements of the gruiform idiornithids, from the Eo-Oligocene of France. Although she described the Idiornithidae as a subfamily of the Cariamidae, she stated that the idiornithids were "equally close to the genus *Opisthocomus*" (1983, 139). Most avian systematists, however, have found the hoatzin to be placed most comfortably in either the Galliformes, based on its rather primitive, nondescript postcranial skeleton, or the Cuculiformes, where DNA comparisons put it (Sibley and Ahlquist 1990). Given the apparent antiquity of both the galliform and the cuculiform group, however, the DNA results would not be surprising one way or another.

The only fossil of a hoatzin is *Hoazinoides magdalenae*, described by Alden H. Miller (1953) on the basis of a cranium from the Miocene of Columbia; unfortunately, no postcranial skeleton is known. The living hoatzin, which today inhabits flooded forests and riparian growth along the rivers and tributaries of the Amazon, is usually encountered in family groups of up to a dozen or so birds. It is a strange bird, indeed, one of few living avian folivores. It has a large, muscular crop that produces active foregut fermentation, unique among birds (Grajal et al. 1989). The hoatzin's sternum and pectoral girdle are highly modified to accommodate this large crop (p. 149) and resemble at least superficially the morphology of the early enantiornithines. This similarity is no doubt due to convergence, although it does raise the interesting question of habits in the early avian radiation.

***Foro*: Primitive Eocene Land Bird.** It has been exciting to see the unveiling of a newly acquired fossil from the Lower Eocene Green River Formation of Wyoming of a primitive land bird whose osteology is similar in many ways to the hoatzin but is also similar to the cuckoos and to the turacos (Musophagidae). For this new Green River bird, *Foro panarium* (frontispiece), Storrs Olson created a new family, Foratidae, and placed it within the Cuculiformes "by default," as he put it (1992b, 129). He found similarities between the fossil and, primarily, the cuckoos, turacos, and hoatzin. Aside from the fossil's rather nondescript postcranial skeleton and therefore lack of "derived" characters that would ally it clearly with a living order of birds, its most remarkable feature, one easily recognizable by even an amateur ornithologist, is the marked similarity of the skull with that of the hoatzin, even down to size and proportions. In summing up the discovery, Olson writes, "The mosaic of characters seen in *Foro* is not unexpected in so ancient a bird and, in the final analysis, may substantiate an origin of the modern neognath radiation among the basal land birds" (135).

Cuckoos and Turacos. It seems appropriate next to consider birds that many believe to be among the most primitive of land birds, the cuckoos and turacos. The reasons are many: the tradition of placing the hoatzin in either the Galliformes or Cuculiformes, the historical view that turacos and cuckoos are closely linked, and the interesting comparisons that have been made among the primitive Eocene fossil *Foro*, the hoatzin, and the cuckoos and turacos. The association of cuckoos and turacos goes back to some of the earliest classifications, including that of Linnaeus (1758, 111), but the two groups are distinct in many morphological features.

Fossil specimen of *Foro panarium,* from the upper Lower Eocene Green River Formation, Wyoming. (From Olson 1992; courtesy *Los Angeles County Museum of Natural History, Science Series* 36)

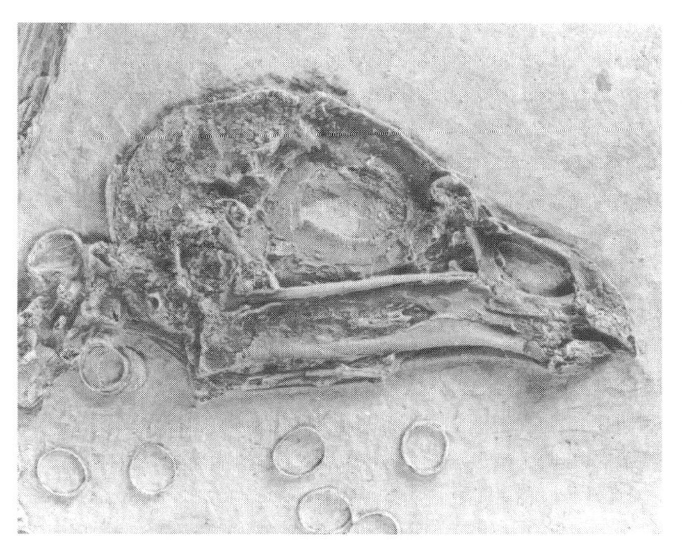

Detail of the hoatzin-like skull of *Foro panarium*. × 1.12. (From Olson 1992; courtesy *Los Angeles County Museum of Natural History, Science Series* 36)

The famed South American riverine bird the hoatzin (*Opisthocomus hoazin*). (Drawing by George Miksch Sutton)

Charles Sibley and Jon Ahlquist are emphatic in their view that evidence from DNA hybridization points to the antiquity of the cuckoos: "We conclude that the cuckoos are an ancient lineage with substantial genetic diversity among the living groups. Their next nearest living relatives are uncertain. The turacos may be their living relatives, but the evidence is not convincing. . . . It is likely that the divergence of the cuckoo lineage was so long ago that the idea of close living relative is irrelevant" (1990, 372).

Although the 142 or so species of cuckoos come in a variety of sizes and morphologies, all have zygodactyl feet, with digits 2 and 3 opposing 1 and 4 behind. More than a dozen species, such as the roadrunners, are terrestrial, and they tend to be large birds with sturdy legs, but the zygodactyl foot has adapted all cuckoos for perching in trees or bushes. Cuckoos are found more or less worldwide, with concentrations in tropical and subtropical regions and exhibiting complex zoogeographic patterns that are no doubt related to the antiquity of the group. Some 50 or more species are brood parasites, some of them producing eggs that actually mimic the eggs of the host species; on hatching, the young cuckoo usually ejects its nestmates and is reared in their place by the host parent.

Unfortunately, this group of early land birds is highly underrepresented in the fossil record. Skipping by the Paleogene fossils that have been misidentified as cuckoos, there are undescribed cuckoos from the Lower Eocene London Clay and a handful of described Tertiary forms, notably an apparent cuculid, *Dynamopteryx velox*, from the Eo-Oligocene of France. Other early Tertiary species include *Neococcyx mccorquodalei*, from the early Oligocene of Saskatchewan, and *Cursoricoccyx geraldinae*, from the early Miocene of Colorado; these fossils tell of a widespread Tertiary distribution but yield little information about the origin or evolution of this interesting group of primitive land birds.

The turacos, or plantain-eaters, of the family Musophagidae, are medium-sized inhabitants of the deep forests of Africa, although several of the twenty-three species are bushland forms. They are similar to cuckoos in their general appearance, but they have an outer toe that can be reversed at will, as opposed to a permanently reversed one, as in the cuckoos; this makes them facultatively, not truly, zygodactyl. Their diet consists mainly of fruit, with some insects as a supplement.

Turacos enjoyed a most interesting distribution in the past. The oldest identifiable fossils, inseparable from the living genus *Crinifer*, are known from the Oligocene in the Fayum of Egypt (Olson 1985a, 110). Peter Ballmann (1970) reported an unnamed turaco from the late Oligocene of Bavaria and in 1972 assigned another fossil from the early Miocene of France to the Musophagidae. The only other Tertiary fossil outside the present-day range of the family is *Musophaga meini*, from the late Miocene of France (Ballmann 1969a), which was described along with a few other unnamed musophagid fragments. Nevertheless, these fossils, meager as they are, represent indisputable evidence of a former widespread distribution of this now African group, a pattern that is repeated over and over in the zoogeography of living birds. They also indicate that most of the so-called perching birds, with the exception of the scansorial (trunk-creeping) woodpeckers, suborder Pici, and the higher passerines, are quite old, most extending back to the Eocene of the Northern Hemisphere.

Mousebirds. The mousebirds, or colies (order Coliiformes), are a drably colored, nondescript group of small birds restricted to sub-Saharan Africa. Their long tails, fluffy plumage (an effect created by long feather aftershafts), and habit of perching with their feet more or less level with their pectoral region, give them a mouselike appearance. The six species of mousebirds usually creep and crawl among bushes and small trees in small groups, often clinging upside down in titmouse fashion. Their most distinctive anatomical feature is the strange foot, termed pamprodactyl, in which the first and fourth toes are reversible and can face either forward or backward, whichever is favored by the surface being grasped. Mousebirds are more or less frugivorous but will also eat a variety of foliage as well as nectar.

North America has not yet yielded a modern mousebird fossil, but three genera of Middle Eocene fossils previously referred to the primitive piciform family Primobucconidae have been referred to a new order of birds, the Sandcoleiformes, by Peter Houde and Storrs Olson (1992). These morphologically diverse birds (particularly *Eobucco* and *Sandcolius*), well known from seven species in six genera from the Lower and Middle Eocene of Wyoming, were like mousebirds in their general appearance but exhibited very different bill specializations, ranging from gaping to thrushlike adaptations, and all appear to have been both facultatively zygodactyl and pamprodactyl. Interestingly, Stefan Peters reports two species of land birds from the Middle Eocene Messel oil shales of Germany that are "very close to *Eobucco*" but have raptorial feet that "are almost owl-like, with enormous claws" (1991, 575; 1989, 2062). These birds appear to have been generalists in their feeding habits, "eating anything from berries, fruits, and seeds to hard- and soft-bodied invertebrates" (Houde and Olson 1992, 156). Houde and Olson note that a fossil of unknown age from the Phosphorites du Quercy, France, described by Mourer-Chauviré (1988a) as *Primocolius sigei*, is actually closer to the extinct sand-

The greater roadrunner (*Geococcyx californianus*). The Cuculiformes contains 142 species, in 29 genera, of small to medium-sized birds with long tails and zygodactyl feet. A number of the larger species are terrestrial or semiterrestrial. (Drawing by George Miksch Sutton)

The yellow-billed turaco (*Tauraco macrorhynchus*). The Musophagiformes contains 23 species, in 5 genera, of medium-sized, long-tailed arboreal birds characterized by a foot with a reversible outer toe. Unlike most birds, turacos' feathers contain true red and green pigments. Turacos are today restricted to Africa. (Drawing by George Miksch Sutton)

Speckled mousebird (*Colius striatus*). The 6 species, in 2 genera, of the Coliiformes are today restricted to Africa. Mousebirds are characterized by rounded wings and a long tail and have soft plumage that gives them a mouselike appearance. Their foot is highly adaptable and may be pamprodactyl, but the first and fourth toes are reversible. Mousebirds are arboreal acrobats and, like the parrots, use the bill in climbing. (Drawing by George Miksch Sutton)

Hypothetical reconstruction of *Sandcolius copiosus,* from the early Eocene of Wyoming. The order Sandcoleiformes, though not close to any extant orders, was probably closer to the mousebirds than to other living birds. These birds resembled mousebirds but had very different bill specializations. (From Houde and Olson 1992; courtesy *Los Angeles County Museum of Natural History, Science Series* 36)

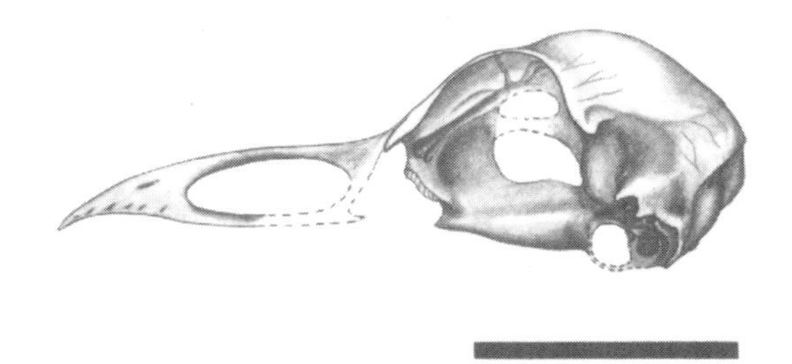

Reconstructed composite skull of *Sandcolius copiosus*. Scale bar = 2 cm. (From Houde and Olson 1992; courtesy *Los Angeles County Museum of Natural History, Science Series* 36)

coleiform genus *Chascacocolius* than to the living *Colius*, and that another fossil from the same locality is intermediate between the sandcoleiform *Anneavis* and *Colius;* the order Sandcoleiformes may therefore have been widespread during the early Tertiary of the Northern Hemisphere. According to Houde and Olson, this archaic order branched off early from the land bird radiation and as a result is not phylogenetically close to any living order of birds, though it remains closer to the living Coliiformes than to any other order and does show characters that suggest a link with the suborder Pici of the order Piciformes. With their diversified feeding adaptations and bills, the sandcoleiforms no doubt occupied many niches later filled by passerines and paralleled the temporal replacement pattern of mammals. "Replacement by modern orders of archaic early Tertiary groups, with no close living relatives, is a pattern also observed among mammals (D. E. Russell 1982), where creodonts and condylarths, for example, were replaced by carnivores and ungulates" (156).

True mousebirds that closely resemble living species occur well back in the fossil record and are known from the Upper Eocene deposits at Quercy (Mourer-Chauviré 1982). Peter Ballmann's study (1969a) of French fossils that had been previously referred to a number of odd taxa revealed two species of *Colius* from the early Miocene, one from the medial Miocene, and another from the mid- to late Miocene; Ballmann has also identified bones of *Colius* from the Miocene of Germany (Olson 1985a, 124). Mourer-Chauviré (1988a) has described two species, *Primocolius minor* and *P. sigei*, in the modern family Coliidae, from the late Eocene Phosphorites du Quercy, although *P. sigei* may belong to the Sandcoleiformes. A modern mousebird fossil, described as *Colius hendeyi*, has been described from the Miocene of South Africa (Rich and Haarhoff 1985).

Susan Berman and Robert Raikow (1982) suggested a close relation between mousebirds and parrots based on hindlimb musculature, but their study probably should have been one in convergent adaptation in musculature, because the two groups share almost no other remarkable morphological features. Interestingly, Sibley and Ahlquist concluded from DNA studies that "the mousebirds, or colies, have no close living relatives" and "that they are the only survivors of an ancient divergence" (1990, 362), which corroborates the evidence from morphological studies and the fossil record.

Fossil specimen of the sandcoleiform *Anneavis anneae*, from the Lower Eocene Green River Formation of Wyoming. A coating of ammonium chloride enhances bone detail but obscures feather impressions. (From Houde and Olson 1992; courtesy *Los Angeles County Museum of Natural History, Science Series* 36)

The Caprimulgiformes are nocturnal or crepuscular birds characterized by large, gaping mouths, short legs, and soft, cryptically colored plumage. They are particularly known for their distinctive and often eerie nocturnal vocalizations. *Left to right and top to bottom,* Frogmouths (Podargidae, southern Asia, Australasia, 2 genera, 14 species): tawny frogmouth (*Podargus strigoides*). Nightjars (Caprimulgidae, worldwide, 14 genera, 79 species): pennant-winged nightjar (*Macrodipteryx vexillarius*). Oilbirds (Steatornithidae, northern South America, 1 genus, 1 species): oilbird (*Steatornis caripensis*). Owlet-nightjars (Aegothelidae, Australasia, 1 genus, 8 species): feline owlet-nightjar (*Aegotheles insignis*). Potoos (Nyctibiidae, Neotropics, 1 genus, 7 species): grey potoo (*Nyctibius griseus*). (Drawings by George Miksch Sutton)

Goatsuckers, Swifts, and Allies

Caprimulgiform Birds. Of unknown origin but probably derived from ancient land birds are the goatsuckers and allies of the order Caprimulgiformes. Although Joel Cracraft's (1988b) cladistic analyses assigned the owls to the Falconiformes and placed the nightjars closest to the swifts, most modern systematists have concluded, as have Sibley and Ahlquist from DNA hybridization studies, that "the owls, not the swifts, are the closest living relatives of the nightjars" and that "the owls, caprimulgiforms, swifts, and hummingbirds form a monophyletic assemblage" (1990, 418, 420). (The owls were covered in chapter 7, on the birds of prey, both for convenience and as a lesson in convergence.) Active primarily in the evening and at night, the caprimulgiforms are characterized by tiny feet and enormous gaping mouths adapted for catching insects. Except for the South American oilbird (family Steatornithidae), a cave-dwelling echolocator that eats palm fruits at night, all are primarily insectivorous. The most successful family is the nightjars, the Caprimulgidae, which contains seventy-nine species found on all continents except in regions of extreme cold. The other three families are the South American potoos (Nyctibiidae), the frogmouths (Podargidae) of Australasia, and the Australian owlet-nightjars (Aegothelidae). The last two are relict families that resemble owls in both behavior and external appearance and structure. Frogmouths have strongly hooked bills and feed by fluttering down to the ground from a perch to catch beetles, centipedes, caterpillars, and even mice. The owlet-nightjars, often called moth owls by Australians, closely resemble small, long-tailed owls. They feed in the manner of frogmouths and also seize insects on the wing. Like owls, they sit crossways instead of lengthwise on branches, nest in hollow trees, and hatch young that are covered with down. Perhaps these living relicts, particularly the owlet-nightjars, indicate a link between ancient caprimulgiforms and owls.

The fossil record of the Caprimulgiformes is among the most interesting of all birds, producing a number of unexpected surprises in fossil deposits of the Northern Hemisphere. Few Caprimulgidae have been discovered, but an undescribed species is known from the Middle Eocene Messel oil shales (Peters 1991), and Mourer-Chauviré (1989) has described the caprimulgid *Ventivorus ragei*, from the Upper Eocene of Quercy, in the extinct family Archaeotrogonidae (Eo-Oligocene).

In 1980, Mourer-Chauviré published an extensive review of fossil trogons from France, creating a new family, Archaeotrogonidae, for the four species in the morphologically distinct genus *Archaeotrogon*, recovered from the late Eocene to late Oligocene Phosphorites du Quercy, described as far back as 1892 by Alphonse Milne-Edwards. Bones of *Archaeotrogon*, which did not have the characteristic trogon heterodactyl foot, are abundant at Quercy, and one species, *A. venustus*, has a temporal range of some 14 million years, from the late Eocene to the late Oligocene; the other three are found only in the Oligocene deposits.

With additional study, it was discovered, however, that the Archaeotrogonidae were in fact a group of extinct caprimulgiform birds. Even before that was determined, substantial differences between the Archaeotrogonidae and the Trogonidae had been noted: Mourer-Chauviré (1980, 1982) had even remarked on similarities between the Archaeotrogonidae and the Caprimulgiformes.

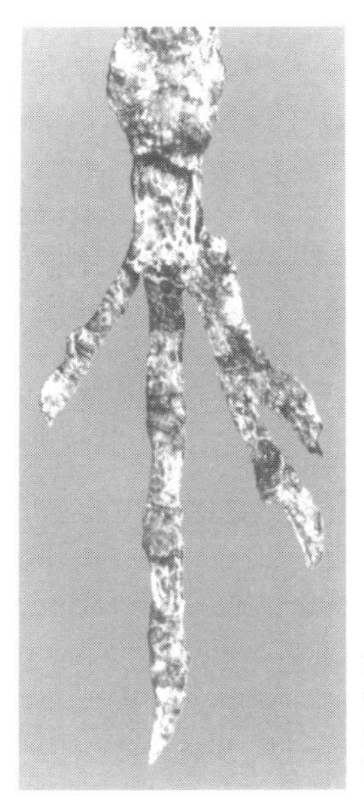

Characteristic foot of a nightjar from the Middle Eocene Messel oil shales of Germany. (Photo courtesy S. Peters and Forschungsinstitut Senckenberg)

A classic locality of the Quercy phosphorites, Mémerlin, near the small town Cajarc, Department of Lot. The site opens like a small canyon at the surface of a Jurassic limestone plateau, and the hollows have been filled with clays containing remains of terrestrial vertebrates and enriched in phosphatic concretions. There are some 120 similar localities in Quercy, and most of the clay fillings of these large hollows are late Eocene to late Oligocene. (Courtesy B. Sigé)

	Stages and Absolute Age in my	Zones of Nannoplankton after Martini	Mammal Zones	Deposits of the Phosphorites du Quercy	Species
OLIGOCENE	Chattian		Boningen	Pech du Fraysse	*A. venustus* *A. zitteli* *A. cayluxensis*
				Pech Desse	*A. venustus*
	30	NP 24			
			Antoingt		
	Rupelian Stampian		Heimersheim		
		NP 23	Montalban	Itardies	*A. venustus*
				Mounayne	*A. venustus*
			Villebramar	Mas de Got B	*A. venustus* *A. zitteli*
	32			La Plante 2	*A. venustus*
		NP 22		Roqueprune 2	*A. venustus*
			Hoogbutsel		
	"grande coupure"				
EOCENE	36	NP 21	Frohnstetten		
	Pria-bonian	NP 20	Montmartre—San Cugat	Escamps	*A. venustus*
	39		La Débruge		
		NP 19	Perrière	Perrière	*A. venustus*
		NP 18	Fons 4		
	41				
	Barto-nien s. st.	NP 17	Grisolles		

Temporal distribution of the caprimulgiform genus *Archaeotrogon* (right column) in the deposits of the Phosphorites du Quercy, France, illustrating the complexity of the geologic units. (From Mourer-Chauviré 1980; courtesy *Los Angeles County Museum of Natural History, Contributions in Science* 330)

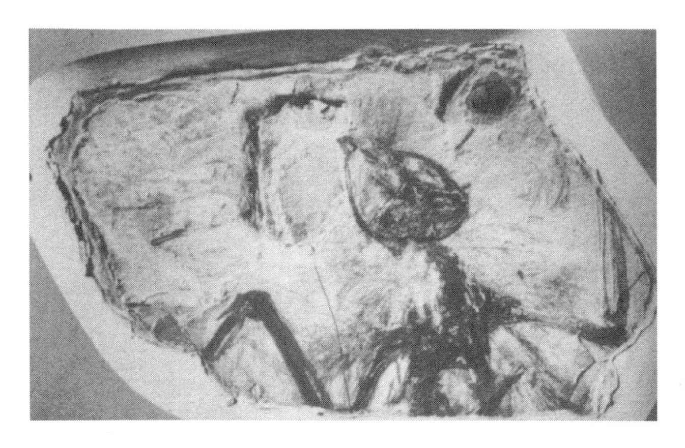

Subfossil of the extinct New Zealand owlet-nightjar *Aegotheles novaezealandiae* (chapter 6 frontispiece). These birds are thought to have been well on their way to becoming another of New Zealand's flightless birds. They were probably ground-dwelling nocturnal hunters of insects and small animals. (Photo courtesy P. Rich)

The large feline owlet-nightjar (*Aegotheles insignis*) of Australia. (Photo W. Peckover/VIREO)

Fossil skeleton of the early Eocene oilbird *Prefica nivea,* from the Green River Formation of Wyoming. (Photo by author)

Another specimen of the primitive oilbird *Prefica* (Olson, pers. comm.), first thought to be close to the living rollers *Eurystomus*. The similarity of rollers and caprimulgiforms deserves more attention. (Photo by H. Heckel; courtesy S. Rietschel, Staatliches Museum für Naturkunde Karlsruhe)

Modern (black) and fossil (X) distribution of the Steatornithidae in the New World. (From Olson 1987)

Frogmouths are known as fossils in Australia only from the Quaternary (Vickers-Rich 1991) but have been uncovered in Messel, where there is an undescribed species from the Middle Eocene (Peters 1991), and in France, where *Quercypodargus olsoni*, from the Upper Eocene of Quercy, has been described by Mourer-Chauviré (1989). This genus is quite close in morphology to the living frogmouth genus *Podargus* of Australia. Another strange occurrence in the fossil record is the discovery, in the late Eocene to Oligocene Phosphorites of Quercy, of a potoo. The family Nyctibiidae is now restricted to the Americas from southern Mexico to Brazil and the islands of Jamaica and Hispaniola and had been known in the fossil record only from the Pleistocene of Brazil and Jamaica. This species, described as *Euronyctibius kurochkini* by Mourer-Chauviré (1989), is close in morphology to the living potoo genus *Nyctibius*.

Still another surprise in the fossil record was the tentative discovery of a sternal fragment possibly belonging to an owlet-nightjar from the late Eo-Oligocene deposits at Quercy (Mourer-Chauviré 1982). The Aegothelidae, a family now restricted to Australia, New Guinea, New Caledonia, and the Indonesian islands of Halmahera and Batjan, would seem an unlikely candidate for a prior distribution in the Northern Hemisphere, but the pattern has been uncovered over and over again. In addition, a nearly complete skeleton of a primitive owlet-nightjar is known from the Middle Miocene of Australia (Rich and McEvey 1977), and a fossil of a fairly large endemic genus, *Megaegotheles* (frontispiece, chapter 6; Scarlett 1968), is known from Pleistocene to sub-Recent deposits in the North and South Islands of New Zealand (Rich and McEvey 1977; Rich and Scarlett 1977). It had a proportionately larger body with proportionately smaller wings than the living *Aegotheles* and has been described as flightless or nearly so. However, the principal characters separating *Megaegotheles* from *Aegotheles* are associated with large size and reduced flight apparatus, conditions generally associated with insular evolution. Olson, Balouet, and Fisher (1987) later considered these characters insufficient to warrant a separate genus, hence *Aegotheles novaezealandiae*.

Last of the Caprimulgiformes is perhaps the most bizarre, the strange South American guacharo, or oilbird (*Steatornis caripensis*), traditionally considered to be an avian family that evolved as a specialized South American endemic. This all changed, however, in 1982, when Mourer-Chauviré tentatively referred an unnamed fragmentary fossil sternum from the late Oligocene at Quercy to the Steatornithidae. This assertion was met with total skepticism at first, but in 1987, Storrs Olson described the first truly diagnostic fossils of an oilbirdlike caprimulgiform from the Lower Eocene Green River Formation of Wyoming (1987b). Known from two nearly complete skeletons and placed in a new subfamily, Preficinae, *Prefica nivea* was a smaller form than the living oilbird and more primitive in many ways.

Swifts. Although the origins of caprimulgiform birds remain elusive, there is a possible link, based on certain morphological similarities, between the goatsuckers and the swifts (Apodidae) in the primitive crested- or treeswifts (Hemiprocnidae) of Southeast Asia and adjacent islands. As Sibley and Ahlquist have remarked, "It seems probable that the swifts and hummingbirds are the next nearest relatives of the Strigiformes (owls, nightjars)" (1990, 420). Fossils of both caprimulgiform birds and swifts are known from the Eocene of the Northern Hemisphere, but these bones have yielded little information regarding any evolutionary affinity. Aside from the four species of crested-swifts, which may be an early evolutionary offshoot from the swifts, some ninety-nine species of modern swifts range widely from tropical to temperate regions throughout the world—almost anywhere flying insects exist.

The earliest of the swiftlike birds, classified in their own family, Aegialornithidae, are known from the early Eocene of England, where Colin Harrison and Cyril Walker (1975b) described a small bird they called *Primapus lacki*, in honor of the famed British ornithologist David Lack, but Stefan Peters (1985) later placed *Primapus* in the genus *Aegialornis*. There are four larger species of the genus *Aegialornis* from the late Eocene to Oligocene Phosphorites du Quercy (Collins 1976a) and one elsewhere in France (Mourer-Chauviré 1978, 1988b); another genus, *Cypselavus*, occurs in the same two deposits. *Aegialornis* appears to have died out at the close of the Eocene, but *Cypselavus* persisted into the Oligocene (Mourer-Chauviré 1978). *Aegialornis* was placed by Charles Collins in the Caprimulgiformes, where it remained for some time (perhaps an insight on the relations of swifts in general), but it was later removed to the swifts by Mourer-Chauviré (1978) and is now considered particularly close to the Hemiprocnidae.

As for other early swifts, there is a small swift from the Lower Eocene Green River Formation of Wyoming in private hands; for years it was on display in a small store in Farson, Wyoming. I studied the bird in 1970, and photos of the beautiful little fossil have since appeared in a number of books. I could see no reason overall to consider it anything but a swiftlike bird—its leg proportions are equivalent to those of the modern genus *Chaetura*—but careful study may prove that assumption wrong. At any rate, it is certainly a swift of some kind and in general

Swift from sediments of ancient Lake Gosiute of the Eocene Green River Formation, Sweetwater County, Wyoming, housed in a private collection in Farson, Wyoming. This Eocene fossil has leg proportions similar to a modern *Chaetura*-type swift. Total length approximately 10 cm (4 in.). (Photo by author, 1970)

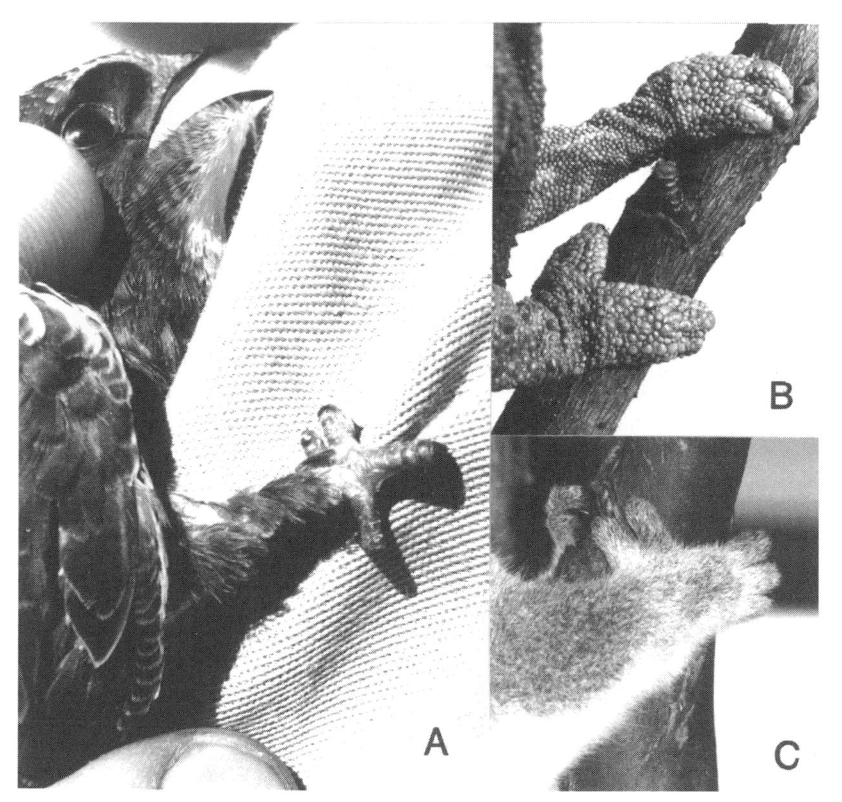

Convergent lateral grasping mechanism in climbing vertebrates: *A,* a nestling little swift (*Apus affinis*); *B,* left and right forelimbs of Jackson's chameleon (*Chamaeleo jacksoni*); *C,* right forelimb of koala (*Phascolarctos cinereus*). (From Collins 1983; courtesy Charles Collins and *Auk*)

The crested treeswift (*Hemiprocne coronata*). The crested-swift family (Hemiprocnidae), containing 4 species in 1 genus, is distributed throughout southern Asia to the Solomon Islands. Thought by some to provide a possible link between swifts and caprimulgiform birds, crested-swifts are considered to be primitive among living swifts and are known from the early Tertiary. (Drawing by George Miksch Sutton)

The white-throated swift (*Aeronautes saxatalis*). True swifts, Apodidae, 99 species in 18 genera, are distributed worldwide. (Drawing by George Miksch Sutton)

The booted racket-tail (*Ocreatus underwoodii*). The Trochilidae, hummingbirds, in the order Apodiformes, are a New World family comprising 322 species in 109 genera, and most likely derive from primitive swifts. (Drawing by George Miksch Sutton)

form qualifies as an Eocene "swift." Otherwise, the true swifts, or Apodidae, occur as fossils in the late Eocene of France as *Cypseloides mourerchauvireae* (Mlikovsky 1989b) and in the early Miocene of France as the species *C. ignotus* (Collins 1976b); both species are members of the primitive subfamily of modern swifts, the Cypseloidinae, which is now confined relictually to the New World. Medial to late Miocene swifts of the genus *Apus* are known from France and Italy.

On the basis of the morphology of the humerus, Alexandr Karkhu (1992) argued that two major lineages of swift evolution can be divined. According to this scheme, the Paleogene genera *Aegialornis, Primapus* (=*Aegialornis*), and *Cypselavus* and the Recent genus *Hemiprocne* all represent a single trend of specialization, and this provides a basis for uniting them in the family Hemiprocnidae. According to Karkhu, the Apodidae, including the fossil genera *Procypseloides* and *Scaniacypselus*, the Trochilidae (hummingbirds), and the Paleogene (early Oligocene) apodiform family Jungornithidae (Karkhu 1988) all possess a completely different structure of the humerus. He has proposed that the first grouping be placed in a suborder Hemiprocni separated from the suborder Apodi, which would contain the second grouping minus the hummingbirds.

The feet of swifts are peculiarly adapted for grasping. All of the swifts of the subfamily Chaeturinae (including the Cypseloidinae) have a typical avian anisodactyl foot, but the advanced Apodinae are generally thought to be pamprodactylous, with all four toes directed forward under normal conditions, as has been somewhat incorrectly described for mousebirds (Coliidae) and several other birds, including oilbirds and parrots of the genus *Micropsitta*. In reality, the toes of mousebirds can be altered to suit the functional demands of a particular situation (Bock and Miller 1959), and it appears that pamprodactyly is not really typical of the oilbird or *Micropsitta* (Collins 1983). The Apodinae foot has toes of nearly equal length, and they typically assume a grasping position, forming a sort of "yoke-toed" arrangement, with two toes positioned laterally on either side. In fact, the opposing toes are similar to that of the heterodactyl foot of trogons, not the zygodactyl foot of woodpeckers, so that toes 1 and 2 oppose toes 3 and 4, although the orientation is lateral instead of anterior-posterior, as in trogons. As Charles Collins (1983) has nicely documented, a similar laterally oriented grasping mechanism is found in the highly specialized "zygodactylous" grasping feet of chameleons and the forelimbs of numerous phalangeroid marsupials, such as the Australian koala, except that both chameleons and these marsupials have five toes instead of four, so that three toes always oppose two. This grasping mechanism in

The distribution and species-density of hummingbirds shows that they are South American birds and probably invaded Middle America during the Great American Interchange. (Modified after Austin 1961)

the feet of swifts, chameleons, and marsupials is a remarkable example of convergence across three classes of vertebrates.

Hummingbirds: New World Nectar-eaters. Another group often linked to the swifts, as just mentioned, is the hummingbirds (Trochilidae), a very successful New World Neotropical family of some 322 species. With a distribution centered in South America, the Trochilidae is almost certainly of mid-Tertiary origin. These small nectar-feeding birds, which also eat tiny insects, are somewhat convergent on the Old World passeriform sunbirds, the Nectariniidae. They have usually been placed with the swifts in a single order, the Apodiformes, but some authors believe that they are not related to swifts, and it is probably more reasonable to assign them to the order Trochiliformes, as some authorities have done. Some believe the hummingbirds' true alliance to be with the passerines, with which hummingbirds share a number of important anatomical characteristics (Sibley and Ahlquist 1972, 200–205). Sibley and Ahlquist have concluded, however, that "the swifts and hummingbirds diverged more recently from one another than either diverged from any other lineage" (1990, 401). Other than modern species from the Quaternary, there is no fossil record of hummingbirds.

Perching Birds

Although the coraciiforms, piciforms, and passeriforms and allies are loosely related, it has been difficult to resolve their phylogenetic relations. They are extremely uniform morphologically, and more than a century of anatomical and other study has produced few clues, either from structure or from the fossil record, that have aided the understanding of their evolution. Indeed, it is in these morphologically uniform groups that the information from DNA hybridization and other biochemical comparisons may be of particular importance, especially when they blend nicely with and corroborate morphological studies.

Most phylogenetic studies of avian structures have suffered from one major drawback: the inability to establish unequivocally the primitive nature of the particular structure or structures involved. For example, one might think, as was believed for years, that the Cretaceous toothed birds *Ichthyornis* and *Hesperornis* shared a common ancestor because they possessed very similar teeth. As we have seen, however, a plesiomorphic trait like this cannot be used as evidence for relatedness because it is a *retained* primitive reptilian characteristic. In the same way, the presence of feathers in a grebe and a loon cannot be used as evidence that they shared an immediate common ancestor because the ultimate ancestor of all birds had feathers. But if we could establish that a certain characteristic of various birds is unique and derived, or synapomorphic—evolved beyond the primitive condition—then, as we have seen, that character would convey strong evidence for evolutionary affinity. Therefore, the first step in any evolutionary study of structural characteristics is to establish whether a character is primitive or derived, that is, to establish character polarity, and in most cases of cladistic analyses, as we have seen, this has been a difficult task.

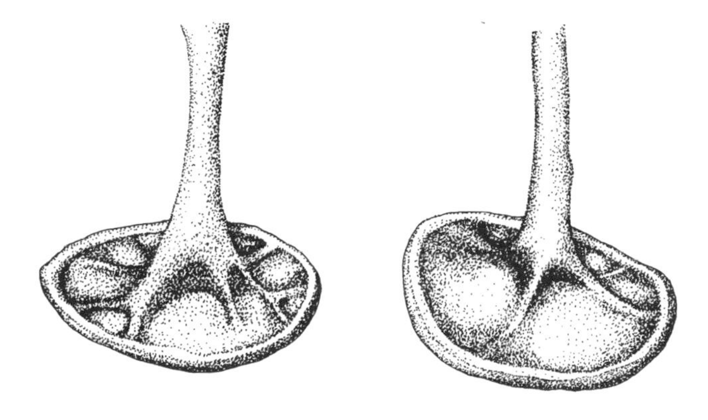

The primitive avian stapes, or columella, of the kiwi, *Apteryx* (*left*), and of a passerine, a cuckooshrike, *Coracina* (*right*). Both are drawn to similar size for easy comparison. (From Feduccia 1977c)

Much evidence has come to light through studies of the structure of the avian middle-ear ossicle, the stapes, or columella, which is homologous to the same element in reptiles. The avian stapes was one of the last remaining elements of the skeleton to be studied in detail, no doubt because of its minute size (one to several millimeters), its fragility, and its remote location in the inner recesses of the middle ear cavity. Over the past several decades I have examined more than two thousand specimens, representing nearly all of the living families of birds, and have reported these findings in various journals (see Feduccia 1975a, 1977d, and 1979 for summary). The columella has provided an exceptional opportunity for phylogenetic analysis because its primitive nature can be established beyond doubt. Unlike the mammalian middle-ear system, in which there are three bony ossicles (the malleus, incus, and stapes) that transmit sound vibrations from the tympanic membrane to the fluids of the inner ear, the avian middle-ear system, like that of reptiles, has a single stapes that performs the same function. It is a simple structure that consists of a flat footplate that fits into the oval window of the inner ear; its straight bony shaft connects via its ligaments to the tympanic membrane (Starck 1995).

The vast majority of living birds possess the primitive condition of the bony stapes that is homologous with the same element in reptiles and represents a retained primitive character. Therefore, where pockets of unique, derived morphologies occur, they must be strong evidence for evolutionary relations. Fortunately, within several groups of the enigmatic arboreal land birds we find these unique, derived middle-ear ossicles, and they have aided greatly in our understanding of interrelations. In addition, we can rely on comparisons of other skeletal and muscular features, such as those associated with the skull, syrinx, feet, and hindlimb musculature, and see how these conform to information from the fossil record and from DNA comparisons.

Coraciiforms: Primitive Perching Birds. The birds generally called "coraciiforms" are a diverse group whose best-known representatives in the Northern Hemisphere are the kingfishers. Classical systematics considered the order Coraciiformes to consist of ten distinctive families of brightly colored subtropical and tropical land birds. These forms represent an early attempt at becoming arboreal land birds, and some can now be considered relicts, occurring as well-preserved Eocene fossils. Coraciiforms were among the predominant arboreal birds of the

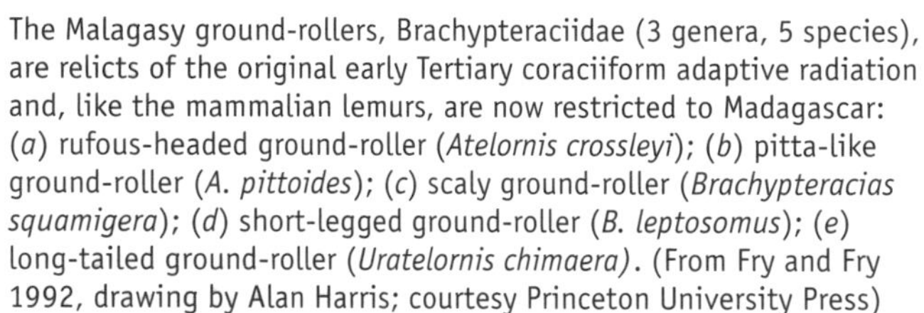
The Malagasy ground-rollers, Brachypteraciidae (3 genera, 5 species), are relicts of the original early Tertiary coraciiform adaptive radiation and, like the mammalian lemurs, are now restricted to Madagascar: (*a*) rufous-headed ground-roller (*Atelornis crossleyi*); (*b*) pitta-like ground-roller (*A. pittoides*); (*c*) scaly ground-roller (*Brachypteracias squamigera*); (*d*) short-legged ground-roller (*B. leptosomus*); (*e*) long-tailed ground-roller (*Uratelornis chimaera*). (From Fry and Fry 1992, drawing by Alan Harris; courtesy Princeton University Press)

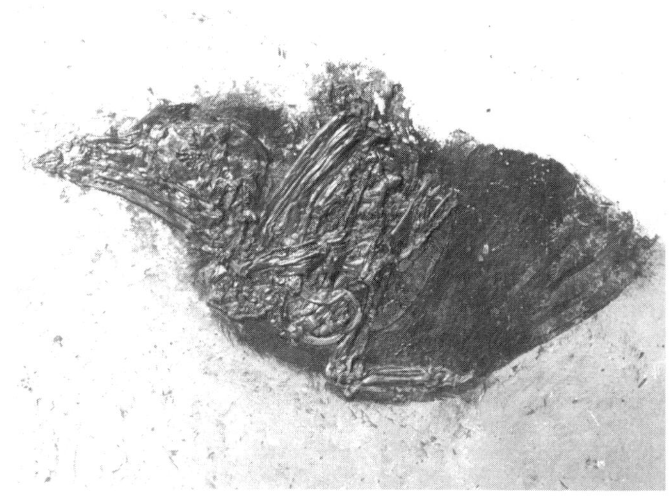

An undescribed medial Eocene fossil bird from the Messel oil shales of Germany that is quite reminiscent of the ground-rollers. The specimen is exceptionally well preserved and includes the pinions, but the hind parts are missing. × 0.46. (Courtesy S. Peters and Forschungsinstitut Senckenberg)

The relict cuckoo roller family, Leptosomidae, of Madagascar contains but a single species, the cuckoo roller (*Leptosomus discolor*). Like other coraciiforms, the cuckoo roller nests in cavities, primarily in tree hollows. (Drawing by George Miksch Sutton)

The Indian roller (*Coracias benghalensis*) is a member of the ancient Old World roller family the Coraciidae, which contains 12 species in 2 genera. (Drawing by George Miksch Sutton)

Eocene and Oligocene, enjoying dynamic evolution and widespread distribution. Few specific anatomical characteristics bind these birds together, but they do share rather small feet that have an unusual toe arrangement in which the three anterior toes are joined for part of their length, a condition known as syndactyly. Most coraciiform birds are carnivorous, eating small fish, amphibians and reptiles, small mammals, and insects. In addition, they are cavity nesters, digging holes in banks or rotten trees.

Percy Lowe summarized our knowledge of coraciiform evolutionary relations by saying that "the Coraciiformes have for many years been loaded with a heterogeneous collection of forms which custom has blindly accepted" (1948, 572). However, studies on the stapes (Feduccia 1977d, 1979), hindlimb musculature (Maurer and Raikow 1981), skeletal structure (Cracraft 1971c), and feeding apparatus (Burton 1985), DNA comparisons (Sibley and Ahlquist 1990), and study of recently discovered fossils permit us to distinguish between the ancient relicts and the derived forms and better understand the evolution and zoogeography of this fascinating group of birds.

Two groups of coraciiforms possess the primitive, reptilian form of the middle-ear ossicle, the various rollers and the hornbills. Centered in the Old World tropics are the twelve species of the Coraciidae, or rollers, so named for their aerial acrobatics. They are known from as early as the Eocene of the Northern Hemisphere, but the fossil record says little other than that they were once widespread. Even more primitive are two relict families now confined to Madagascar, the ground-rollers (Brachypteraciidae) and the cuckoo roller (Leptosomidae). Ancient forms resembling these relict groups may well have been ancestral to the passerine birds and, perhaps, the piciform and other land bird groups as well.

The oldest fossil referred to the Coraciiformes is *Halcyornis toliapicus*, known from a cranium from the Lower Eocene London Clay in England. It was described by Colin Harrison and Cyril Walker (1972) in its own family, Halcyornithidae, but without the postcranial skeleton its systematic allocation must remain tentative. Restudy of the varied North American Eocene fossils that were originally placed in the family Primobucconidae (Feduccia and Martin 1976) has shown that this group contains a variety of forms belonging to at least three separate groups. One of these, *Primobucco mcgrewi*, along with other undescribed Eocene Green River fossils, is close to the ground-rollers, Brachypteraciidae, and was apparently facultatively zygodactyl (Houde and Olson 1989, 2032). According to Stefan Peters (1991, 575), there are at least three unnamed species of rollers (Coraciidae) from the Middle Eocene Messel oil shales of Germany. In 1988 the tiny *Sylphornis bretouensis* was described in a new family, Sylphornithidae, by Cécile Mourer-Chauviré (1988a) from the Upper Eocene of Quercy, France, as a family of truly tiny coraciiform birds, showing close affinity with the Antillean todies, Todidae, and the Malagasy ground-rollers among the classical "coraciiforms," as well as with the family Bucconidae, the puffbirds. "At least one species . . . lived also in the forests around the Eocene lake of Messel. Its size was about that of the hummingbird *Chrysolampis mosquitus*" (Peters 1991, 575).

Rollers of the modern type appear to have been abundant in the Eocene. *Geranopterus alatus*, described by Alphonse Milne-Edwards in 1892, from the late Eocene to Oligocene Phosphorites du Quercy, "is a roller, . . . very close to the living genus *Coracias*" (Olson 1985a, 125). A beautifully preserved fossil from the Lower Eocene Green River Formation was considered to be similar in morphology to the living genus *Eurystomus* (Olson 1985a, 125). Olson (1987b) has since speculated that the Green River *Eurystomus*-like fossil might be a form of steatornithid and that the coraciiforms might be the wellsprings of the caprimulgiforms. This proposal certainly deserves further study.

The hornbills (Bucerotidae) of the Old World tropics, confined today to sub-Saharan Africa and parts of Asia east to the Philippines and New Guinea, are some of the most bizarre looking of all birds, with their enormous, down-curved bills, often surmounted by a large horny casque. The biology and adaptive radiation of hornbills has been reviewed by Alan Kemp (1978, 1995), and Kemp and Timothy Crowe (1985) have presented a detailed phylogenetic (cladistic) analysis of the Bucerotidae. Their diet is largely fruits and berries, but some may eat insects, small mammals, and snakes. Hornbills are somewhat convergent on their New World tropical equivalents, the toucans of the order Piciformes. The hornbill's bill, often equipped with a reinforcing dorsal casque, is used for feeding, preening, fighting, and sealing nests and comes in an amazing variety of sizes and colors, including a block of near solid ivory in *Buceros (Rhinoplax) vigil*, the helmeted hornbill, whose skull weighs in at 11 percent of the bird's body weight. Amazingly, the only fossil record of this intriguing family is that of a ground-hornbill, *Bucorvus brailloni*, from the Middle Miocene of Morocco (Brunet 1971; Olson 1985a, 137), far to the north of the modern range of ground-hornbills, which occur today only in sub-Saharan Africa.

Hoopoes (Upupidae, *Upupa*) and woodhoopoes (Phoeniculidae, *Phoeniculus* and *Rhinopomastus*) are yet another group of coraciiform birds. These birds are often placed in separate families, as they are here, and are considered by many to be related only distantly, if at all. Their unique, derived, anvil-shaped middle-ear bones tell us dif-

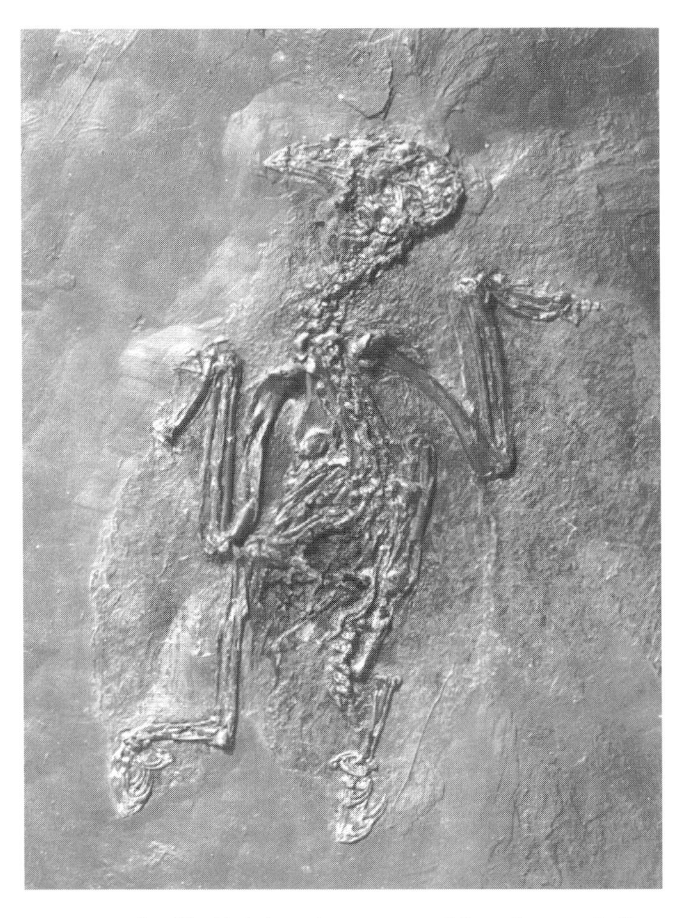

Numerous roller-like birds have been recovered from the Messel oil shales; this undescribed form has toes that are strongly reminiscent of those of falcons or owls. × 0.52. (Courtesy S. Peters and Forschungsinstitut Senckenberg)

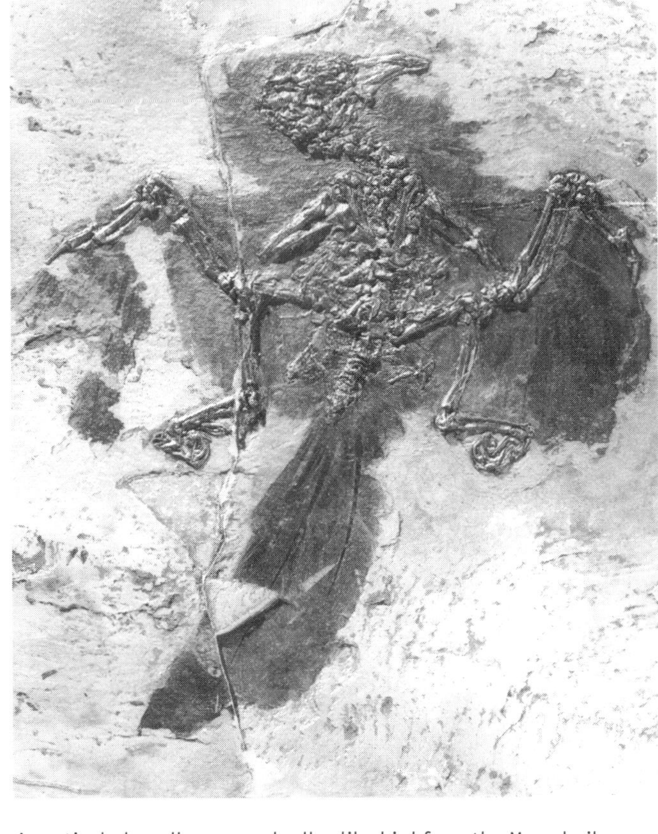

A particularly well-preserved roller-like bird from the Messel oil shales, with feet like those of a bird of prey. This bird had obviously molted, as revealed by one of the tail feathers, whose tip has just emerged from its quill. × 0.47. (Courtesy S. Peters and Forschungsinstitut Senckenberg)

The great hornbill (*Buceros bicornis*). The coraciiform family of hornbills, Bucerotidae (tropical Africa, Asia, and East Indies), contains 56 species in 9 genera. Hornbills, as the name implies, are noted for their oversized bills, as well as their habit of nesting in tree cavities and, during incubation, walling up the entrance with the female and her eggs inside until only a small feeding hole remains. (Drawing by George Miksch Sutton)

The southern ground-hornbill (*Bucorvus leadbeateri*) stalking the Loita Plains in Kenya. (Photo by M. P. Kahl/VIREO)

ferently, however; that both groups have such a bizarre structure would seem to argue strongly for a close evolutionary link, and Maurer and Raikow (1981) argued for such a link on the basis of myology (the study of muscles). In addition, Sibley and Ahlquist found that "the DNA evidence agrees with previous ideas about the relationships among the three genera of hoopoes; *Upupa* on one branch, *Phoeniculus* and *Rhinopomastus* on the other" (1990, 348).

The relation of hoopoes to other orders of birds is still another matter. As Percy Lowe noted of *Upupa*, "in some respects, especially as regards the palatal regions . . . [hoopoes are] characteristically Coraciiform; in others typically Passerine and in others Picine" (1946, 119). Many others who have worked on this strange group of African relicts share this view, but Sibley and Ahlquist concluded that "the hornbills and hoopoes are allied" (1990, 347), forming a clade distinctive from the other coraciiform birds.

Both families are from the Old World, and the eight species of woodhoopoes are presently confined to Africa south of the Sahara. Hoopoes, now represented by two species, extend into southern Europe, and the family Upupidae is present in the Eo-Oligocene deposits of Quercy, as are the other coraciiform families Alcedinidae, Todidae, Meropidae, and Coraciidae (Mourer-Chauviré 1989). A large flightless hoopoe, *Upupa antaios*, now extinct, lived on the South Atlantic island of Saint Helena during Pleistocene time or later (Olson 1975b). And a woodhoopoe is now known from the early Miocene of Bavaria and from the early Miocene of France (Ballmann 1969b). These finds, combined with reports of turacos and mousebirds from deposits of similar age in Europe (Ballmann 1969a, 1969b), indicate that Europe formerly had a much warmer climate, and that the present Ethiopian avifauna—hoopoes, mousebirds, and turacos—was once extensive in its range.

The Alcediniform Assemblage. The remaining coraciiform families include the bee-eaters (Meropidae), kingfishers (Alcedinidae), motmots (Momotidae), todies (Todidae), and trogons, which have usually been placed in their own order, the Trogoniformes. This assemblage is clearly united by peculiar middle-ear ossicles, characterized by a large, bulbous footplate region. These ossicles are somewhat similar to those of the suboscine passerines and led me at one time to conclude that the two groups might have a common ancestor (Feduccia 1975a, 1977d). Studies using scanning electron microscopy have since shown, however, that the two types are clearly distinct from each other. I have also suggested (1977d) that the bee-eater/kingfisher/motmot/tody/trogon assemblage be split off from the other Coraciiformes as the order Alcediniformes and follow this ordering here. This system, although it differs in arbitrary taxonomic treatment from others, does not depart significantly from the cladistic analysis based on "limb muscle synapomorphies" of Maurer and Raikow (1981) or from the DNA hybridization analysis of Sibley and Ahlquist that "the trogons are recognized as relatives of the coraciiforms; . . . the todies, motmots, bee-eaters, and kingfishers are allied"; and there exists a "trogon, roller, bee-eater, motmot, tody, kingfisher, clade (Coraciimorphae)" (1990, 347).

The twenty-six species of bee-eaters are very colorful, social birds that range throughout the Old World, particularly in tropical regions (Fry and Fry 1992). They are particularly fond of eating wasps and bees, and they forage by flying out to catch them on the wing. Bee-eaters nest colonially and, like many coraciiforms, dig their burrows in dirt banks. Mourer-Chauviré (1982, 1989) listed the Meropidae in the Eo-Oligocene of France, an occurrence that would not be at all surprising, given the paleodistribution of other coraciiform families. Bee-eaters are very closely allied with the kingfishers, and the two form a definitive clade within the advanced coraciiform assemblage.

The nearly cosmopolitan kingfishers appear to be the only birds in the "coraciiform" complex to have become well adapted physiologically to the variable climate of the temperate zone (Fry and Fry 1992). The modern forms would appear to be an Old World group that apparently recently invaded the New World via the Bering Strait land bridge that connected Eurasia and North America in the late Tertiary. Peters (1989, 2063) mentions Middle Eocene fossils from Messel, Germany, as showing similarities to the kingfishers, and Mourer-Chauviré (1982, 1989) lists the Alcedinidae as occurring in the Eo-Oligocene of France, both indicating an Old World origin. An undescribed kingfisherlike bird has been found in the Lower Eocene Green River Formation of Wyoming (Grande 1980, 212), and Houde and Olson have examined what appears to be the skeleton of a kingfisher from the Eocene London Clay (1989, 2034). These findings indicate a widespread Paleogene distribution of this family.

Like bee-eaters, kingfishers nest in burrows in dirt banks; some make their nests in tree cavities. The worldwide distribution of the kingfisher family is deceiving, for out of the ninety-five or so living species, only six species in two genera (all closely related) occur in the New World, all of them fishers, and there is only one European species, a form widespread over the Old World. Fifteen or so species occur in Africa, but the greatest density, some 60 percent of species, are found from Asia southward to Australia, occurring on Pacific islands—

Right, a common scimitar-bill (*Rhinopomastus cyanomelas;* Phoeniculiae, Africa, 2 genera, 8 species), and *left,* a Eurasian hoopoe (*Upupa epops;* Upupidae, Africa, Madagascar, and warm Eurasia, 1 genera, 2 species), with the unique anvil stapes possessed by both (scanning electron micrograph, *below,* × 40). (Drawings of birds by George Miksch Sutton; stapes from Feduccia 1977c)

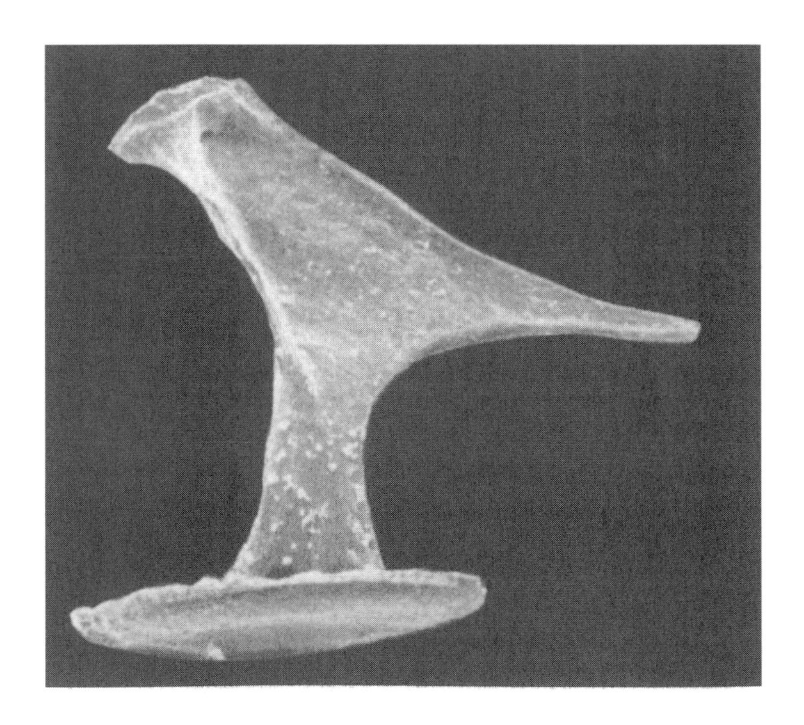

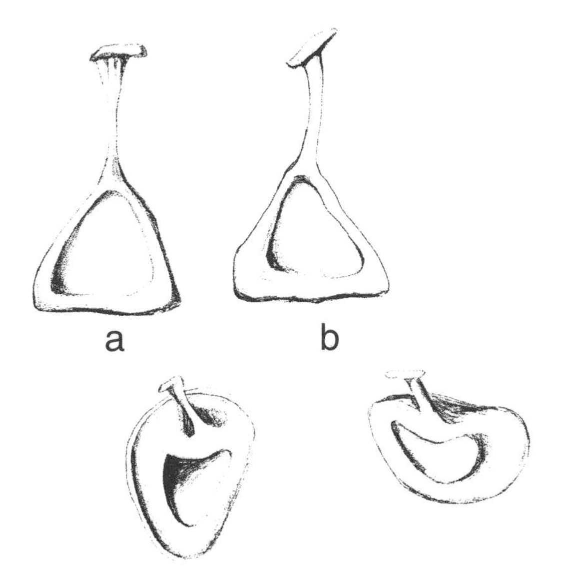

Posterior views (*above*) and views looking down on the footplate regions (*below*) of the stapes of *a,* a typical suboscine passerine bird, and *b,* a typical member of the bee eater/kingfisher/motmot/tody/trogon assemblage (Alcediniformes). All are drawn to approximately the same size. (From Feduccia 1977c)

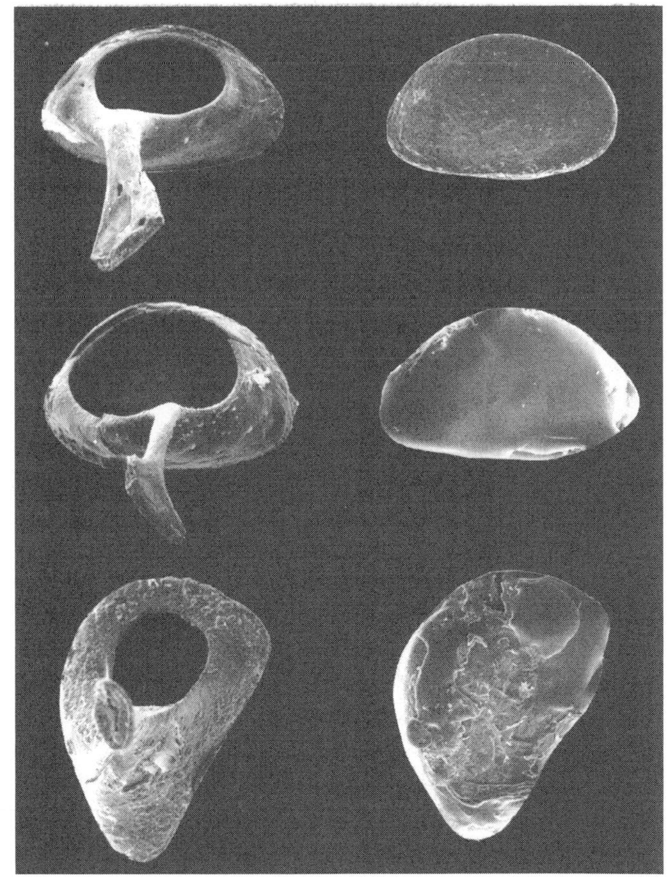

Left, scanning electron micrographs (SEM) of three views along the lengths of the bony stapes of: *right,* the Cuban trogon (*Priotelus temnurus*); *center,* the pygmy kingfisher (*Ceryle rudis*); and *left,* a suboscine passerine, the Andean cock-of-the-rock (*Rupicola peruviana*). The SEMS were taken so as to have all to approximately the same scale, × 25 to × 35, here reduced 50%. *Right,* views of the tops, left column, and bottoms of the footplates (region of insertion into the oval window); *from top to bottom,* trogon, kingfisher, and suboscine. The SEMS are scaled from × 45 to × 55, here reduced 50%. (From Feduccia 1979c)

Representatives of the living families of alcediniform birds, united by their possession of a unique type of middle-ear bone. *Left to right and top to bottom,* Bee-eaters (Meropidae, warm Old World, 3 genera, 26 species): European bee-eater (*Merops apiaster*). Motmots (Momotidae, Neotropics, 6 genera, 9 species): blue-crowned motmot (*Momotus momota*). Kingfishers (Alcedinidae, worldwide, 18 genera, 94 species): laughing kookaburra (*Dacelo novaeguineae*). Todies (Todidae, Greater Antilles, 1 genus, 5 species): Cuban tody (*Todus multicolor*). Trogons (Trogonidae [often an order Trogoniformes], pantropical except Australasia, 6 genera, 39 species): mountain trogon (*Trogon mexicanus*). (Drawings by George Miksch Sutton)

Eocene kingfisherlike bird from the Green River Formation of Wyoming under study by the author. Length about 15 cm (6 in.). (Photo courtesy Black Hills Institute)

among them Sumatra, Java, Borneo, the Philippines, Sulawesi, the Moluccas, the Lesser Sundas, and Papua New Guinea and its offshore islands—as far east as Samoa (Fry 1980, 152).

The dazzling, gemlike kingfishers provide a splendid example of adaptive radiation. Kingfishers range in size from very small forms about 13 centimeters (5 in.) in length to the crow-sized 43-centimeter-long (17 in.) laughing kookaburra (*Dacelo novaeguineae*) of Australia. Although they are best known for their invasion of the fishing niche, which has prevented them from competing with more advanced land birds, kingfishers in fact have diverse feeding habits. Many of the Malaysian kingfishers are primitive within the group, occurring in ancient rain forest habitat and foraging by surveillance from a branch followed by diving or swooping onto their prey of insects and other invertebrates and small vertebrates. These kingfishers take prey from the ground or near the surface of shaded water, often hovering briefly before they plunge on their prey. Interestingly, there are a number of woodland kingfishers, such as the aptly named woodland kingfisher (*Halcyon senegalensis*) of sub-Saharan Africa (Fry and Fry 1992, 158), that bathe by plunging recklessly, or diving headlong into wooded ponds, then shaking, and preening at the perch. This behavior may well presage the habits observed in the better-known, derived "fishermen."

Kingfishers range from exclusively dry-land predators (primarily insectivores), such as the Old World forest kingfishers that never go near water but hunt in arid zones for insects and small vertebrates, to land and water feeders to those that feed, like the familiar North American fishers, exclusively from the water. The highly specialized pied kingfisher (*Ceryle rudis*), the only "pelagic" fisher, which can catch two fish in a dive and eat a crab at sea, usually roosts gregariously, with as many as 220 individuals gathering in a few square meters of papyrus (Fry and Fry 1992, 239). At another end of the evolutionary spectrum are the Australian laughing kookaburra and related forms. This hook-billed predator of the Australian woodlands is fond of snakes and will even eat the young of other birds. Other bizarre forms, such as the "digging" hook-billed kingfisher (*Melidora macrorrhina*) and shovel-billed kookaburra (*Clytoceyx rex*), live in the humid lowland rain forests of New Guinea, where they feed on the ground, the shovel-billed kookaburra foraging by plowing or shoveling an area of leaf-littered earth with its huge bill, capturing a variety of food, primarily earthworms.

The nine species of motmots are moderate-sized forest birds restricted today to the Neotropics, from Mexico south to northern Argentina. The todies are tiny relatives of the motmots that survive today as five species restricted to the West Indies, the only family of birds confined to that area. Both todies and motmots nest in burrows and eat insects, though motmots are also known to eat small vertebrates and some fruit. Motmots have long tails with racket-shaped tips, and both motmots and todies have broad, flattened bills with serrated edges.

Todies and motmots were both probably widespread during the medial Tertiary, but in the late Tertiary a combination of climatic deterioration and competition with more advanced land birds entirely eliminated the tody-motmot assemblage from the Old World. The same changes in the New World forced the restriction of the range of motmots to Central America. They then spread into South America during the Great American Interchange, with the opening of the Central American land bridge, some 2.5 to 3 million years ago. Similar change no doubt affected the North American todies. A fossil tody, *Palaeotodus emryi*, described from the Oligocene of Wyoming (Olson 1976b), was characterized by larger size and proportionately longer wings than the living genus *Todus*. Interestingly, in 1966, James Bond had expressed reservations about todies' ability to cross even narrow water gaps because of their feeble flight, but *Palaeotodus* was a larger bird and showed proportions suggestive of greater powers of flight.

More recently, with the description of two new fossil species of *Palaeotodus*, Cécile Mourer-Chauviré (1985) has shown that the todies were a component of the late Eocene–Oligocene avifauna of the famous Phosphorites du Quercy. Three fossil species of todies are thus known: *Palaeotodus escampsiensis* and *P. itardiensis* from the Eo-Oligocene of France and *P. emryi* from the Oligocene of

Wyoming. Yet their ultimate origin is unknown, and Storrs Olson, in discussing *Palaeotodus*, has suggested of motmots and todies that "perhaps with material from earlier in the Oligocene it would not be possible to distinguish the two families, the Todidae having assumed its characteristics since that time" (1976b, 118).

Motmots appear to have originated in the Old World; a fossil named *Protornis glarniensis*, from the early Oligocene of Switzerland, seems to be a close ally. As was mentioned, the motmots could have crossed over the Bering land bridge, and, with the late Tertiary climatic deterioration and competition from advanced passerines, become restricted to the Central American region, only recently venturing into South America. It is equally likely, however, owing to their antiquity, that they dispersed into the New World across the North Atlantic land connection (De Geer route) that was present during the Eocene. At this time of "warm temperate" climate, according to Malcolm McKenna (1975), more than half of the known genera of fossil mammals were shared between North America and Europe. The recovery of an unnamed fossil motmot, quite close in morphology to the living motmots, from the late Miocene of Florida (Becker 1986c), is of particular interest. The presence of a motmot in Florida before a land connection between North and South America adds to the evidence that the Momotidae was not of South or Central American origin, as previously thought, but was widely distributed in the New World and no doubt entered Central and South America as part of the Great American Interchange, with the closure of the Panamanian Seaway, approximately 2.5 million years ago. This is particularly interesting when one examines the distribution of modern motmots: there are nine species in six genera, but only four species in three genera occur in South America, and all but one of the South American species are also widely distributed in Central America, indicating a recent arrival into South America (Feduccia 1977d), proposed as far back as 1923 by Frank M. Chapman.

Trogons. The thirty-nine species of trogons (Trogonidae), best known by the spectacular Central American quetzal, occur in the tropics of both the Old and New Worlds. Famous for their resplendent iridescent plumage, the trogons are among the most colorful of birds and are clearly the most divergent structurally of the alcediniform birds. In addition to a feathered tarsus, they have a unique type of yoke-toed foot called heterodactyl, in which the inner, or second, toe is shifted to the rear; in the zygodactyl foot of woodpeckers, cuckoos, and parrots, it is the outer, or fourth, toe that has migrated backward. Trogons feed in flycatcher fashion, eating insects extensively, along with some fruits. Like other alcediniforms, they are cavity nesters, but the nest is usually excavated in rotten wood or a termite nest rather than a dirt bank, and they will use preexisting tree excavations.

The trogons are of great zoogeographic interest because their remains, like those of so many other coraciiforms, indicate a much greater distribution in the past. With the removal of *Archaeotrogon*, mentioned earlier, the fossil record is represented by *Paratrogon gallicus*, from the early Miocene of France, and an unnamed putative trogon from the early Oligocene of Switzerland, this fossil with a heterodactyl foot (Olson 1976b). There is also a "trogon" fossil reported from the Oligocene of the Caucasus (Karkhu 1992). Trogons probably dispersed into the New World either early across the North Atlantic land connection or later via the Bering land bridge and were forced southward by late Tertiary climatic deterioration. Like motmots and todies, they spread into South America after the opening of a Central American land bridge. The present-day New World species of trogons, like the motmots, are found primarily in Central America and the West Indies (Feduccia 1977d). Following the zoogeographic pattern illustrated by the todies, two trogons are endemic to Cuba (*Priotelus temnurus*) and Hispaniola (*Priotelus* [=*Temnotrogon*] *roseigaster*). In the Old World tropics, the Ethiopian and Oriental zoogeographic regions, there are fourteen species in two genera.

The exact systematic position of the Trogonidae has been argued, although most authors have placed them somewhere near the coraciiform birds, and Max Fürbringer (1888) thought that the trogons were intermediate between the Coraciiformes and the "Pico-Passeriformes." Over the years trogons have been pushed back and forth from positions near the mousebirds to slots between the caprimulgiforms and coraciiforms, where Ernst Mayr and Dean Amadon (1951) placed them. In their cladistic analysis of the hindlimb musculature of the Coraciiformes, Maurer and Raikow concluded that their study supported "the inclusion of the Trogonidae in the Coraciiformes, specifically within the subinfraorder [!!] Alcedinides, thus allying the trogons most closely with the assemblage of todies, motmots, bee-eaters, and kingfishers" (1981, 418). This view conforms closely to data from the morphology of the stapes, which suggest that trogons were an early offshoot of the tody, motmot, bee-eater, and kingfisher assemblage. But the evidence from DNA hybridization is more equivocal; Sibley and Ahlquist concluded that "the trogons have no close living relatives and, unless they are more closely related to owls, nightjars, or turacos, they are members of the cluster that includes the rollers, todies, hornbills, kingfishers, motmots, and bee-eaters" (1990, 356).

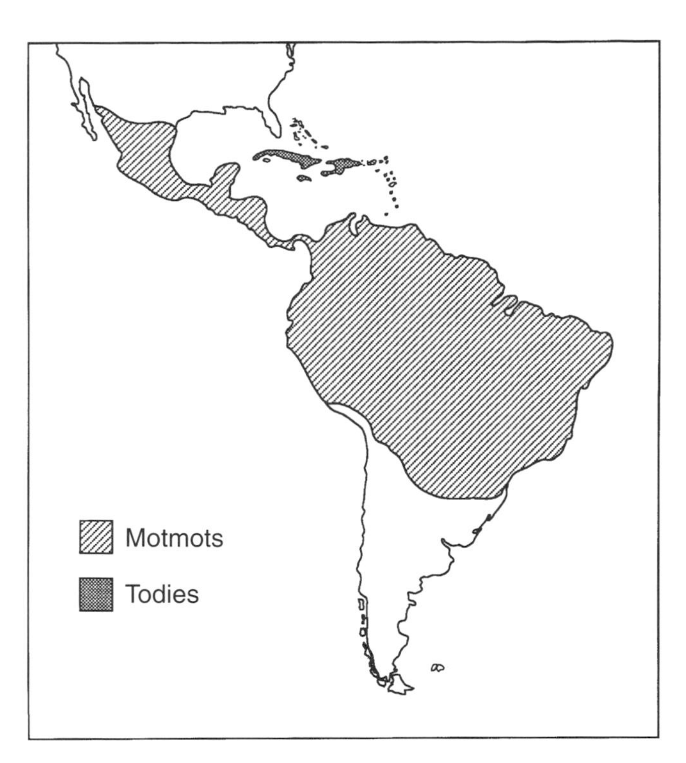

The present-day range of the coraciiform todies (Todidae) and motmots (Momotidae), both thought to have evolved in situ—todies in the West Indies and motmots in South America. However, fossil todies and motmots are now known from the Tertiary of North America and Europe, illustrating, along with many other groups, that bird biogeography cannot be deduced from current ranges. (After Austin 1961)

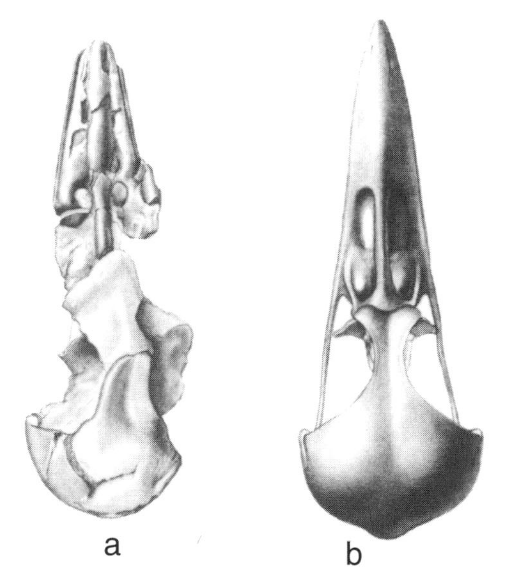

Skull of *a, Paleotodus emryi* (Middle Oligocene of Wyoming), in dorsal view compared to that of *b,* the Jamaican tody (*Todus subulatus*). (From Olson 1976)

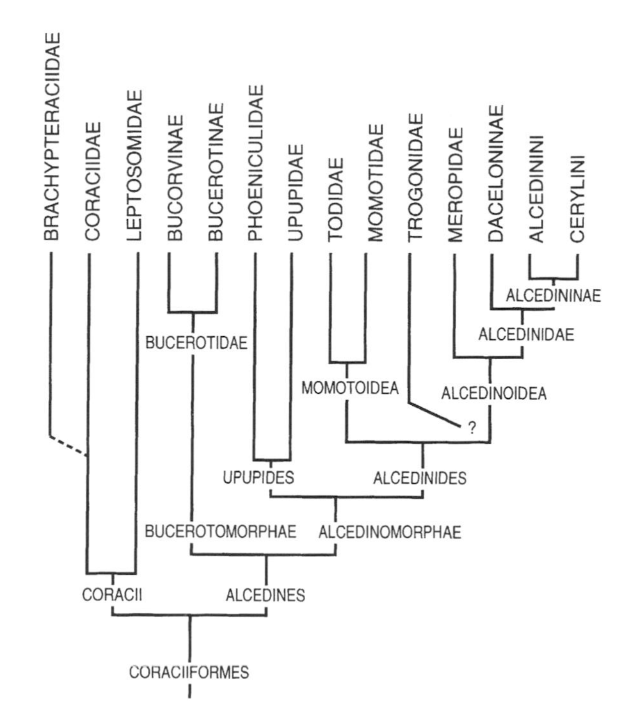

A phylogeny of the avian order Coraciiformes based on a study of appendicular myology (Maurer and Raikow 1981). The family Brachypteraciidae was not studied. The cladogram is constructed entirely on the basis of a cladistic analysis of 57 limb muscle characters and broadly agrees with many previous studies, providing a framework for further study. (Modified after Maurer and Raikow 1981)

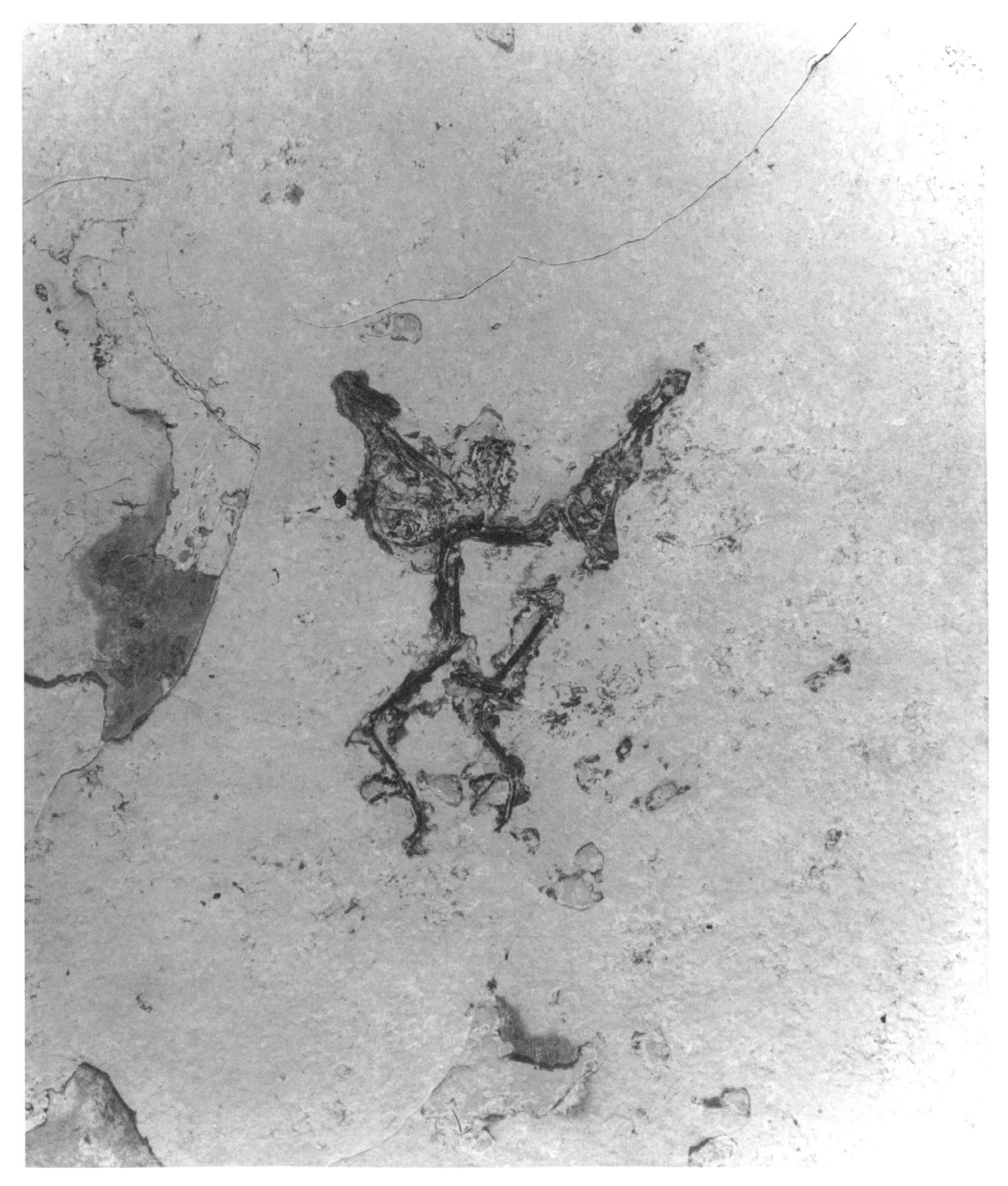

Fossil of the early Eocene primobucconid *Neanis kistneri* (Feduccia), a primitive perching piciform bird from the Green River Formation of Wyoming. In life it was 10–13 cm (4–5 in.) long. Note the zygodactyl foot. (From Feduccia 1977c)

Perching piciform birds are also known from Messel and, along with the sandcoleiforms and coraciiforms, were the predominant perching birds of the early Tertiary. Here a small, primitive Messel piciform has its zygodactyl feet preserved in position, with the first toe shorter than the others. × 0.81. (Courtesy S. Peters and Forschungsinstitut Senckenberg)

Piciform Birds. The order Piciformes is best known by the tree-creeping, or scansorial, woodpeckers (family Picidae). Yet the order is composed primarily of perching birds, and its characteristic feature is the zygodactyl foot, which evolved as a perching foot in the early members of the order and was later modified for foraging along tree trunks. In hitching up tree trunks, piciform birds spread the toes to the side to help them climb and use their stiffened tail feathers as a brace on the trunk.

Dating back well into Eocene time in the Northern Hemisphere, the Piciformes is, like Coraciiformes, an ancient order. One extinct family, the Zygodactylidae, here placed in the Piciformes for convenience, is known from the Miocene of France and Germany (Ballmann 1969b), and putative zygodactylids are present in the early Eocene London Clay material. These enigmatic birds show some similarities to the parrots and for the time being remain of uncertain affinity. Still other zygodactylous fossils are known from the Miocene of Morocco (Brunet 1961), but these birds are also of unknown affinity (Olson 1985a, 122). Widespread during Cenozoic time, the typical piciforms are now largely confined to the tropical regions of the Old and New Worlds, but are not represented in Australasia, Madagascar, or Oceania. Climate deterioration during the late Cenozoic, combined with competition from the more advanced perching birds, has left the woodpeckers as the piciforms' only representative in the temperate zone.

The primitive members of the classical order Piciformes are the puffbirds (Bucconidae) and jacamars (Galbulidae), which are typically separated into a suborder, the Galbulae, in recognition of their distinctive morphology, whereas the Pici, woodpeckers and allies, are usually allied with the Passeriformes. For example, Henry Seebohm (1890) erected a large order called the Pico-passeriformes, and the Passeriformes of Hans Friedrich Gadow (1892) contained two suborders, the Pici and Passeres. Unfortunately, the presence of a zygodactyl foot is insufficient to define the order, and as Walter Bock and Waldron De Witt Miller have noted, "There is little doubt that the Pici, the Psittaci and the Cuculidae have all acquired their zygodactyl foot independently of one another" (1959, 30). Corroborating this view are such birds as the Malagasy cuckoo roller *Leptosomus*, which is facultatively zygodactyl, capable of reversing the outer toe at will. The zygodactyl foot is therefore of doubtful value in uniting the Piciform families. In particular, as Sibley and Ahlquist have written, "It seems especially appropriate to question the alliance of the jacamars and puffbirds (our Galbuliformes) to the Piciformes because the two groups have little in common other than their zygodactyl feet" (1990, 327).

Because the puffbirds and jacamars show numerous distinctions from the Pici, the suborder containing the remaining forms—honeyguides, barbets, toucans, and woodpeckers—their position within the order Piciformes has been challenged in recent years. Edward Swierczewski and Robert Raikow (1981) performed a cladistic analysis of the hindlimb musculature with the classical Piciformes and proposed that the Piciformes is monophyletic, composed of two branches, or clades, the suborder Galbulae (Bucconidae and Galbulidae) and the suborder Pici, which itself was divided into two branches, the Capitonidae-Ramphastidae (barbets and toucans) and the Indicatoridae-Picidae (honeyguides and woodpeckers). In the same year Sharon Simpson and Joel Cracraft (1981) published another cladistic analysis, using skeletal characters, that agreed with Swierczewski and Raikow's phylogeny, and Cracraft (1981) cited the studies as supporting a "highly corroborated hypothesis of relationships" within the Piciformes. However, in 1983, Storrs Olson presented evidence that supported a polyphyletic Piciformes, placing the Galbulae in the coraciiform assemblage and the Pici near the Passeriformes, thus rebutting the arguments of

The primitive zygodactyl perching birds of the traditional piciform suborder Galbulae. Jacamars (Galbulidae, Neotropics, 5 genera, 18 species): black-chinned jacamar (*Galbula ruficola melanogenia*), *left.* Puffbirds (Bucconidae, Neotropics, 10 genera, 33 species): white-necked puffbird (*Notharcus macrorhynchos*), *right.* (Drawings by George Miksch Sutton)

Zygodactyl, advanced perching piciforms. *Left to right,* New World barbets (Capitonidae, Neotropics, 3 genera, 14 species): black-spotted barbet (*Capito niger*). Old World barbets (Megalaimidae, Asia and Africa, 3 genera, 26 species), not illustrated; African barbets (Lybiidae, Africa, 7 genera, 42 species), not illustrated. Honeyguides (Indicatoridae, Africa and tropical Asia, 4 genera, 17 species): greater honeyguide (*Indicator indicator*). Toucans, New World barbet derivatives (Ramphastidae, Neotropics, 6 genera, 41 species): Cuvier's toucan (*Ramphastos cuvieri*). (Drawings by George Miksch Sutton)

both Simpson and Cracraft and Swierczewski and Raikow. In Olson's view, "The structure of the zygodactyl foot in the Galbulae is very distinct from that in the Pici, and no unique shared derived characters of the tarsometatarsus have been demonstrated for these two taxa," and "There are fewer character conflicts with this hypothesis [of a polyphyletic Piciformes] than with the hypothesis that the Piciformes are monophyletic" (130). He concluded that the Galbulae are "closely related to the rollers, or Caracii of Maurer and Raikow (1981), which includes the Coraciidae, Brachypteraciidae [ground-rollers], and Leptosomatidae [cuckoo-rollers]. The Pici, on the other hand, are more closely related to the Passeriformes" (130). Of course, both Raikow and Cracraft rejected Olson's hypothesis in favor of their view that "a monophyletic origin of the Piciformes remains the hypothesis of choice" (1983, 134). Sibley and Ahlquist, weighing in on the basis of DNA evidence, erected an order Galbuliformes for the puffbirds and jacamars and concluded that the remaining Piciformes are closely related to one another but that "they are not closely related to the Passeriformes or to the Galbuliformes" (1990, 332). Although this conclusion might at first appear to support Olson's position, Sibley and Ahlquist's statement that the woodpeckers and allies are not closely allied with the passerines contravenes virtually all morphological evidence and casts doubt on the validity of DNA comparisons in this area. Nevertheless, given the DNA evidence, the generally poor track record of cladistic methodology in avian systematics, the zeal to establish "monophyletic" groups, and Olson's new evidence, the phylogeny remains unresolved, and only further research will settle the dust on this important issue in avian systematics. What emerges from the continuing controversy is a close nexus of the orders Coraciiformes and Piciformes through their basal members.

Primobucconidae, Puffbirds, and Jacamars. The problem of piciform phylogeny increases in significance when one considers the Eocene fossils that have been allocated to the extinct family Primobucconidae (Feduccia and Martin 1976; Feduccia 1976b), thought to be closely allied with the living puffbirds of Central and South America. As pointed out earlier, many of these fossils have been, with the study of additional fossil material and further preparation of the fossils, reallocated to several distinctive groups, including the Coraciiformes and the new order Sandcoleiformes. However, two truly zygodactyl species from the Eocene of Wyoming, *"Primobucco" olsoni* and *"Neanis" kistneri*, survived the reorganization. As Peter Houde and Storrs Olson remark, "Both are true members of the Galbulae, as evidenced by the structure of the sehnenhalter of the tarsometatarsus. . . . Characters shared with the Galbulae include limb proportions similar to those of the Bucconidae, as noted by Feduccia and Martin (1976); cranium with large postorbital process; posterior border of the sternum deeply notched . . . etc." (1989, 2031). They note similarities of *"N." kistneri* in the head and bill to the Galbulidae, and state that "The bill of *'P.' olsoni* is shorter and more robust and the caudal portion is very deep, as in the Bucconidae" (2032). Fossils belonging to the Galbulae are known from the Paleogene of Europe, but with the rapid revision of these groups, the fossils must await further study.

Both puffbirds and jacamars are today confined to the Neotropics. The thirty-three species of puffbirds are insectivorous forest birds living primarily in the Amazon Basin. They typically sit quietly on a perch and sally out from time to time to catch insects, or they may capture an occasional small frog or lizard from the ground. They come in a variety of forms, and one species, the swallow-wing (*Chelidoptera tenebrosa*), has become a bird of more open country, occurring along river banks, where it makes regular aerial sorties to capture insects in swallow fash-

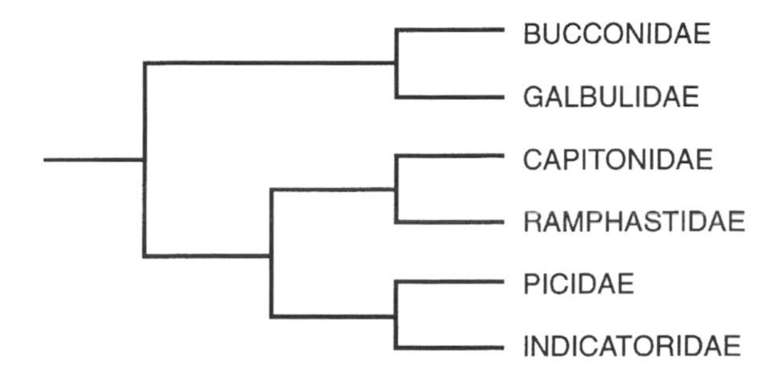

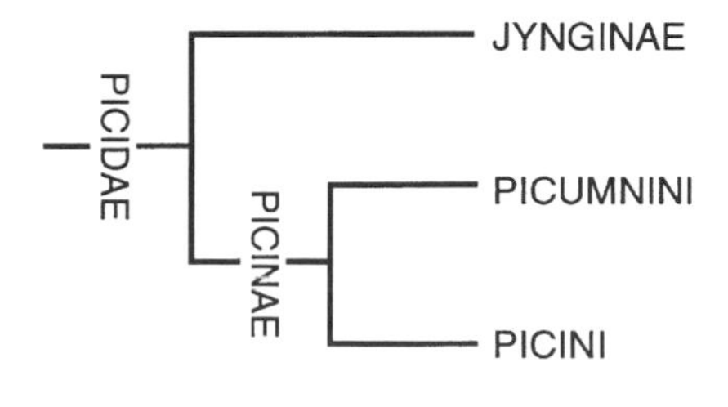

Left, cladogram showing higher-level relations within the traditional order Piciformes supported by Simpson and Cracraft (1981) and Swierczewski and Raikow (1981). Olson (1983) argued that the Galbulae were not of piciform but of coraciiform affinity. *Right,* cladogram illustrating generally accepted relations among the higher piciforms. (Modified after Swierczewski and Raikow 1981)

ion. Other species, such as the nunlets, *Nonnula*, are small puffbirds that move about the foliage in somewhat warbler fashion. The eighteen species of jacamars are more commonly found at forest edges. Like puffbirds, they are zygodactyl, but one species, the three-toed jacamar (*Jacamaralcyon tridactyla*), has lost one of its hind toes. Jacamars, like puffbirds, sally from their perch to catch insects on the wing, and both families excavate nest burrows in banks. However, jacamars have much more colorful plumage, often brightly iridescent green, and, unlike puffbirds, have long, pointed bills.

Barbets and Toucans. Of the "true" Piciformes, those contained in the suborder Pici, there are but three groups, the barbet-toucan clan, the intriguing Old World honeyguides, and the scansorial woodpeckers. The barbets (Capitonidae) are perching piciforms, some with large, broad bills, found in the tropics of both hemispheres. Most barbets are brightly colored, and some New World forest species may rank as among the world's most beautiful birds. They prefer to eat fruits and berries, but many species are primarily insectivorous, and Old World forms inhabit forests and woodlands but also arid brush country (Short and Horne 1980), unlike the forest-dwelling New World barbets. Like other members of the Pici, barbets nest in tree cavities. They are noted for their large bills, and some, such as the toucan barbet (*Semnornis ramphastinus*), have bills that so closely approach the morphology of a toucan bill that the separation of the New World barbets from the toucans into two families seems unjustified. In a phylogenetic analysis of the barbets and toucans in which morphology was analyzed and compared with DNA studies, Richard Prum (1988) of the University of Kansas concluded that the Capitonidae was not monophyletic but that the New World barbets and ramphastids formed a clade and that the sister-group to the ramphastids is the genus *Semnornis*. In other words, the New World barbets are more closely related to toucans than either group is to Old World barbets, a conclusion concordant with Sibley and Ahlquist's DNA hybridization studies (1990) and confirmed independently by Scott Lanyon and John Hall of the Field Museum of Natural History, using mitochondrial DNA sequence data. Prum noted that "the degree of congruence between morphology and the . . . DNA-DNA phylogeny of the ramphastoids is certainly among the greatest yet identified . . . for any avian group" (336). As he put it, "It now appears that the answer to the traditional question 'Why are there so few barbets in the New World?' is 'There aren't; there are toucans'" (339).

Barbets were once widespread, as discoveries of fossils in the Tertiary of Europe indicate. Stefan Peters has commented of several Middle Eocene Messel birds that, "although similar to the Capitonidae they seem to belong to a separate family" (1991, 574). Nevertheless, true barbets were first described as two species of a new genus *Capitonides* from the early Miocene of Bavaria by Peter Ballmann (1969b), who also referred a fossil from the late Miocene of France to the Capitonidae. In 1983, Ballmann described another species of *Capitonides* from the Middle Miocene of Germany and determined that his genus *Capitonides* is quite similar to the living genus *Trachyphonus*. This finding conforms to the view expressed by Swierczewski and Raikow (1981) that *Trachyphonus* is the most primitive genus of living barbets, a conclusion also reached by Prum's study. In addition to these early barbets, fossils of the Capitonidae are known from the early Miocene Thomas Farm local fauna of central Florida (Olson 1985a, 138), following a paleozoogeographic pattern that is repeated over and over again in birds.

The forty-one or so species of toucans (Ramphastidae) are a Neotropical group derived from barbet ancestors. Toucans are essentially enlarged barbets with oversized bills used for eating fruit, and there is a smooth morphological gradation in bill size and shape from such barbets as *Semnornis* through the seven species of toucanets (*Aulacorhynchus*) to the larger toucans and aracaris. Unlike the Capitonidae, there is no fossil record of any significance for the Ramphastidae. Like their small siblings the barbets, toucans nest in tree cavities, but unlike the hornbills, their convergent counterparts in the Old World, they do not seal the female in the nest cavity during incubation.

Honeyguides. A most unusual family of the suborder Pici is the Indicatoridae, the honeyguides (Friedmann 1954, 1955), which are drably colored, zygodactyl piciform birds with quite unusual habits. Confined to the Old World tropics, fifteen of the seventeen species occur in sub-Saharan Africa, whereas the Malaysian region and the western Himalayas are each home to one species. Honeyguides are unique among birds in a number of ways. First, of the six bird families that contain brood parasites, only the honeyguide family is wholly parasitic. No honeyguide is known to build its nest; all deposit their eggs in the nests of various host species. Newly hatched honeyguides have a calcareous hook on the tip of the bill that they use to kill their nestmates. Second, they are cerophagous, or wax-eaters, and must have special enzymes to help them break down the wax esters; they have even been known to enter churches to eat the candles. Cerophagy is, in fact, not restricted to honeyguides; many seabirds (particularly storm-petrels, Obst 1986) may depend on a substantial amount of wax in the diet, and wax-eating is known in other birds, particularly African bulbuls

Coevolution of bird and mammal is illustrated by the association of the greater honeyguide (*Indicator indicator*) and the honeybadger, or ratel (*Mellivora capensis*), as well as African people. The honeyguide leads the badger to a beehive, where the mammal reciprocates by tearing apart the beehive, giving the honeyguide access to the wax that is part of its diet. Africans attract the attention of honeyguides by grunting like the honeybadger and chopping on trees in imitation of the sound of opening a bees' nest. (From Friedmann 1954; drawing by Walter A. Weber; courtesy National Geographic Society)

The advanced scansorial piciforms. *Top,* woodpeckers and piculets (Picidae, including piculets, Picumninae, 27 genera, 214 species): imperial woodpecker (*Campephilus imperialis*). *Bottom,* wrynecks (Jynginae, Eurasia and Africa, 1 genus, 2 species): Eurasian wryneck (*Jynx torquilla*). (Drawings by George Miksch Sutton)

(Horne and Short 1990). Still other birds, including North American yellow-rumped warblers and tree swallows, eat waxy fruits (Place and Stiles 1992). Third, and perhaps most unusual, several species of African honeyguides have the extraordinary habit, as the name implies, of guiding mammals, particularly the honeybadger (*Mellivora capensis*) and African tribespeople, to bees' nests by flying back and forth with a loud, beckoning, chattering call. When the nest is broken open by the mammal, the honeyguide eats both bees and wax. This is a remarkable example of mutualism, with rewards for both parties. In association with this strange habit, honeyguides have unusually thick skin that is thought to protect them against stings; indeed, the honeyguides of the genus *Indicator* show a predilection for bees in their diet.

Picidae: Woodpeckers and Allies. Unlike the toucans, the woodpeckers and allies (Picidae) are not traceable back to a specific group within the Piciformes. As a scansorial or tree-trunk foraging group they are probably of late Tertiary origin, the oldest fossil woodpecker coming from the Middle Miocene of North America, and there are but a few other fossils, ranging from the late Miocene of Italy to a scattering of fossils from the North American Pliocene, but including woodpeckers of the flicker (*Colaptes*) and ivory-billed woodpecker (*Campephilus*) groups (Olson 1985a; Feduccia 1987). Woodpecker fossils are particularly difficult to identify, however, because of the extreme homogeneity in their postcranial morphology, and only distinctive woodpecker "types," such as *Colaptes*, are truly assignable to genus. There is, however, a piculet feather from early Miocene amber from Hispaniola that was identified by the Smithsonian feather expert Roxie Laybourne (Laybourne, Deedrick, and Hueber 1994). The success of the woodpeckers and their allies, the wrynecks and piculets (family Picidae), with some 215 species in twenty-eight genera, now spread throughout most of the world except Australia, is probably due to their acquisition of scansorial habits and their physiological adaptation to temperate climates.

The adaptive radiation of woodpeckers (Short 1982) has been restricted by the morphological uniformity demanded by the evolution of trunk-climbing adaptations, but within the family there is morphological variety ranging from the large "ivory-billed" types to the tiny piculets, and one form, *Picoides tridactylus*, as the Latin for three-toed woodpecker implies, has eliminated the first toe. The family contains two smaller, less specialized groups that lack a stiffened tail, the wrynecks (subfamily Jynginae, *Jynx*), confined to the Old World, and the piculets (subfamily Picumninae, *Picumnus*), which occur in the American as well as the Old World tropics. There are also the so-called ground woodpeckers, some twelve or so species that have become secondarily adapted for ground-foraging; they range from the ground- and trunk-foraging but tree-nesting flickers (*Colaptes*) to the fully terrestrial ground woodpecker (*Geocolaptes olivaceous*) of southern Africa (Short 1971). In addition, two high Andean flickers (*Colaptes rupicola* and *C. campestris*) occur above the tree line and so are primarily terrestrial but may nest in dirt banks or other suitable dirt walls and have therefore remained somewhat adapted for climbing.

If woodpeckers are, in fact, of mid-Tertiary origin, as the fossil record may or may not indicate, we should not necessarily assume that there were no trunk-foragers throughout the early Tertiary. There must have been numerous groups, now extinct, perhaps among the coraciiform birds and others, that attempted to live off the

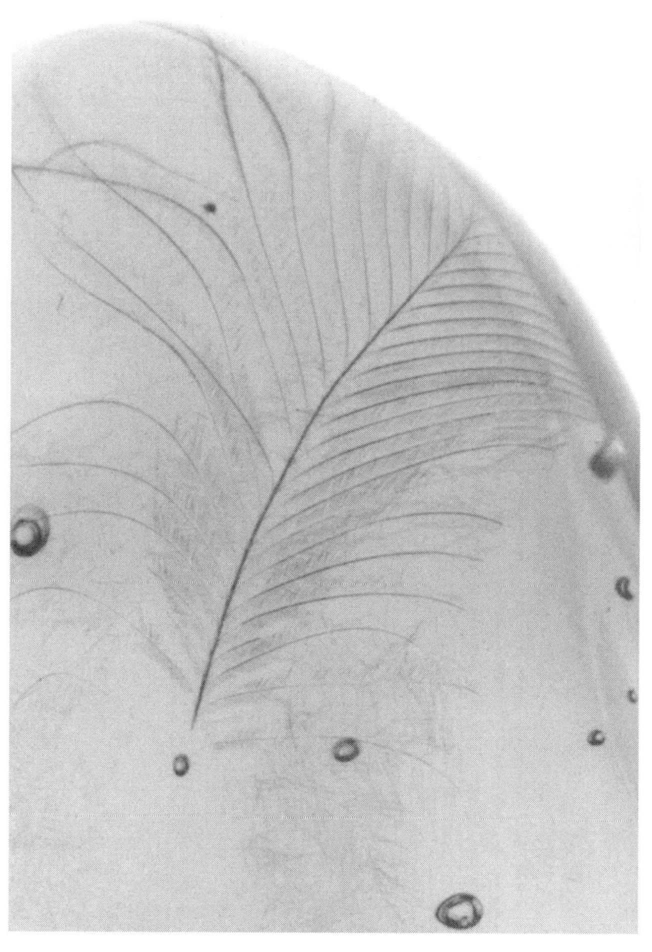

Amber from the Lower Miocene Palo Alto mine in the Dominican Republic reveals a partial feather, 17.5 mm long, that has been identified from plumaceous barbs as belonging to the woodpecker family Picidae by Roxie Laybourne of the Department of Vertebrate Zoology, Smithsonian Institution. Further study has revealed that the fossil was closely related to the Antillean piculet (*Nesoctites micromegas,* Picumninae) of Hispaniola. This feather provides evidence for a long presence of piculets in Hispaniola and is the earliest New World record of the Picidae. (Photo courtesy Roxie C. Laybourne)

trunks of trees and were ultimately supplanted by the more successful woodpeckers, which adapted their "perching" zygodactyl feet to a scansorial posture. In the early Cretaceous, as we saw in chapter 4, the small enantiornithine birds had claws equivalent in arc curvature to those of the modern nuthatches and were no doubt adapted for trunk-foraging, using the anisodactyl foot. Trunk-foraging has evolved independently in the modern passerines in numerous unrelated anisodactyl families, ranging from the scansorial suboscine woodcreepers (Dendrocolaptidae) to their oscine counterparts, the typical creepers, or brown creepers (Certhiidae) of the Northern Hemisphere. In all cases their morphology exhibits a set of convergent adaptations that are as predictable as those that evolve in birds that become foot-propelled or wing-propelled divers from disparate phylogenetic lineages.

The Passeriform or Modern Land Bird Radiation

By far the most numerous group of living birds are the passerines, order Passeriformes, often called songbirds. Passerines constitute approximately three-fifths of all living birds, or some 5,739 (59 percent) of the 9,702 species in 1,168 genera of birds recognized by Monroe and Sibley (1993). Most are small to medium-sized, but some are fairly large. According to Sibley and Ahlquist, "The largest passerine is the Greenland Raven (*Corvus corax principalis*), weighing up to 1700 grams (3.75 lbs.). The lyrebirds (*Menura*) may be the next largest. The smallest is the Pygmy Tit (*Psaltria exisis*) of Java" (1990, 577). The passerines were so successful during the late Tertiary that lines of demarcation among families and higher groups are very poorly defined, so that differences among many passerine families are not as great as those among nonpasserine genera. Instead of distinct evolutionary lines that can be traced by conventional methods, passerine phylogenies look like an upended head of an artist's camel hair paintbrush with the myriad single strands inextricably mixed. So poorly understood are passerine family boundaries that students of the group may use as few as 50 (Mayr and Amadon 1951) or 70 (Wetmore 1960) to as many as 104 families (Wolters 1975–1982) to accommodate the various species. The disparity is particularly and vividly illustrated by Fürbringer (1888), who recognized only 2 families of passerine birds.

Sibley and Ahlquist (1990, 552–577) have presented an excellent and exhaustive historical review of the systematics and classification of the passerines, one that reflects the group's morphological uniformity and hence confusion concerning their taxonomic affinities. Almost all recent workers have considered the order Passeriformes to be monophyletic, being defined by the following features: a special type of palate, termed aegithognathous; syringeal anatomy; an incumbent hallux, which is large and directed toward the back, creating an anisodactyl foot, with three anterior toes (digits 2, 3, and 4); the passerine type of plantar tendons; the passerine type of insertion of a forearm muscle, the tensor propatagialis brevis; bundled spermatazoa with a coiled head; and the distinctive construction of the passerine foot, which allows independent action of the hallux (also see Raikow 1982, 1986).

Why So Many Passerines? In a paper published in 1986, Robert Raikow asked, "Why are there so many kinds of passerine birds?" and searched, as many cladists would, for an explanation that could attribute the success of the group to the existence of "key adaptations," which would be identifiable as synapomorphies that define the monophyly of the group. However, the naïveté of the methodology, assuming that the sampling of synapomorphies has any systematic reality and is anything but an artifact of the sampling of the investigator, is revealed by the failure to consider such well-known passerine traits as vocal and song plasticity as possibly reflected in the complexity of the syrinx, learning ability, relative brain size, and other special neural and physiological parameters not definable as simple tabulational synapomorphies. So, for Raikow, all of the five so-called passerine synapomorphies "fail to provide any convincing key innovation that might explain the evolutionary success of the group." Storrs Olson notes

> that practically every one of the major postcranial bones in the Passeriformes has a distinctive and characteristic morphology that allows it to be recognized and identified as passerine regardless of size, systematic position within the order, or the superimposition of specialized locomotory adaptations. Regardless of how many of these features can be broken down into discrete, derived characters, there is much more to "passerineness" than the few characters outlined by Raikow (1986) would indicate. That this morphology has contributed to the success of the Passeriformes can hardly be doubted, as the basic passerine bauplan has apparently never been selected against, and is retained and recognizable, for example, in the highly aerially adapted swallows (Hirundinidae) or in the terrestrial passerines such as pittas (Pittidae). In no known instance have the Passeriformes given rise to a subsidiary group in which the passerine bauplan has been lost or significantly obscured. (Olson, unpubl. MS)

In the numerous rebuttals that followed Raikow's paper (Fitzpatrick 1988; Kochmer and Wagner 1988; and Vermeij 1988), however, a number of previously known explanations were nicely summarized. First, the relative size of organisms, not tabulated in cladistic methodology, is of considerable biological importance. In 1973, Leigh van Valen discovered that taxa that are composed of small organisms tend to be richer in species than those composed of large organisms; this holds true for angiosperms, birds, and mammals. Of the 4,004 species of the world's mammals, for example, 1,591 (40 percent) are in the order Rodentia and 950 (21 percent) are in the order Chiroptera, the bats; these two orders contain the majority of small living mammals. Passerine birds fit this general canon admirably.

In addition, as John Fitzpatrick appropriately notes, "Compared with other avian clades, passerines do appear to show superior morphological and neural capacity to experiment with novel physical environments, and to learn new behavioral techniques that permit them to enter and take advantage of these novel surroundings. . . . This behavioral plasticity, perhaps in consort with rapid population turnover, may endow passerine birds with a greater potential for rapid morphological evolution compared to non-passerines. Vocal complexity and dialect formation could accentuate such a potential among certain clades. Large brain size of song birds may be associated with this behaviorally mediated increase in potential for speciation" (1988, 74).

Raikow's arguments also carry the implicit assumption that the passerines represent an avian order that is as old as other avian clades recognized at the ordinal level, a contention that is contravened by the fossil evidence. As we have seen, most of the orders of living birds date from the Paleogene, but "some of the most significant paleontological information about passerines concerns their absence" (Olson 1985a, 139). Passerines are absent from the abundant fossil avifaunas now known from the Lower Eocene London Clay and Green River Formation of Wyoming, the Middle Eocene Messel oil shales of Germany, and the late Eocene to late Oligocene Phosporites du Quercy, France, and given the large avian assemblages from these deposits, including truly tiny coraciiform and apodiform birds, we must conclude that passerines simply were not present. Passerines were absent, at least as a formidable group, until after the Paleogene, when because of a variety of morphological, neurological, behavioral, and ecological factors, they adaptively radiated like no other group that had preceded them. Complex nest-building behaviors (see Collias and Collias 1964, 1984), attributes associated with more or less undefinable neurological and behavioral traits, in essence released the passerines from the limited resource of cavity nesting (a feature of almost all of the more primitive related land birds, the coraciiform and piciform birds), into an unexploited and unrestricted ecological zone to undergo their explosive adaptive radiation. And finally, as another indication of the uniqueness of the group, passerines have elevated basal metabolic rates compared to other birds. The basal (at rest) metabolic rates of passerines averages some 50 to 60 percent higher than nonpasserines of equivalent body size (Lasiewski and Dawson 1967; Calder and King 1974), indicating yet another parameter in which passerines have broken away from the normal avian mold and perhaps another indication, in some as yet unknown way, of their explosive success in the late Tertiary. Indeed, the debate over why there are so many kinds of passerines has provided insight into one major perspective of biological methodology—namely, the constrained and limited nature of the cladistic approach.

A Mid-Tertiary Explosive Radiation. As for the fossil record of passerines, what little there is provides scant information, and the lack of evidence on the passerine radiation is certainly one of the great voids in avian paleontology. Walter Boles of the Australian Museum in Sydney has described two bones of particular interest from an early Eocene site in southeastern Queensland dated at about 54 million years ago. The two small fragments are the proximal end of a carpometacarpus and the distal end of a tibiotarsus, the size of a finch and a thrush, respectively, and closely resemble those of the Passeriformes (Boles 1995a). If these fossils are in fact passerine, they would add fuel to the fire of the southern origin hypothesis and would push back the known age of the Passeriformes by a geological epoch. However, as we have learned from the numerous lessons of other early Tertiary avian mosaics, extreme caution must be exercised with single bones of this age, and these fossils, like other small bird fossils from Eocene localities in the Northern Hemisphere, show similarities with mousebirds, rollers, and piciforms. They could be passerine, but more evidence is needed before acceptance.

As for the Northern Hemisphere, the most famous North American putative passerine, *Palaeospiza bella*, from the Middle Oligocene Florissant shales of Colorado (Allen 1878), was reexamined by Storrs Olson (1985a, 139), who concluded that it was not a passerine but possibly a coraciiform. The oldest passerine fossils that can be identified with certainty are two unnamed fossils from the Uppermost Oligocene of Allier, France; they belong to typical oscines, "and probably to an advanced form within the group" (Mourer-Chauviré, Hugueney, and Jonet 1989, 844), and a nearly complete carpometacarpus belonging to a suboscine family from the beginning of the Upper Oligocene of the Phosporites du Quercy has been recovered but is unassignable to any Recent family (Mourer-Chauviré, pers. comm.). Before this discovery, the earliest known passerines were fossils described by Alphonse Milne-Edwards (1867–1871) from the Lower Miocene of France and a few undescribed passerines from the early Miocene Thomas Farm local fauna in Florida (Olson 1985a, 140). Peter Ballmann (1969b) made the extremely interesting finding that passerine fossils from the early Miocene deposits at Wintershof West, Bavaria, outnumber all the other birds found there combined!

From the Lower-Middle Miocene Pinturas Formation of southern Argentina, Jorge Noriega and Luis Chiappe (1991) have recovered passerine remains that belong probably to a suboscine passerine, whereas oscine passerines begin to appear in North America in the medial Miocene and become more numerous into the Pliocene, numbering in the hundreds in the late Pliocene Rexroad and Hagerman local faunas of Kansas and Idaho. Unfortunately, most are unidentifiable to genus or even family because of their extreme morphological uniformity. A swallow from the Upper Pliocene of Kansas has been described as a new species, *Hirundo aprica*, a form close to the living barn swallow, *H. rustica* (Feduccia 1967), swallows being one of the few distinctive families of oscine passerines.

Walter Boles (1993, 1995a) has identified a modern logrunner *Orthonyx*, a distinctive endemic Australo-Papuan group represented by two living species, from the Miocene of northwestern Queensland, along with a lyrebird described as *Menura tyewanoides*. Among the more interesting Miocene passerines are corvids from the medial Miocene of France, a medial to late Miocene nuthatch from France (Ballmann 1973), a fossil belonging to the Sylviidae from the Miocene of Italy (Ballmann 1973), a sparrow (*Ammodramus*) from the late Miocene of Kansas (Steadman 1982), and a bunting (*Passerina*) from the Pliocene of Mexico (Steadman and McKitrick 1982), some 4 million years old. As for suboscines, aside from the isolated find from the Miocene of Argentina, Ballmann (1969b) has identified an indeterminate species of broadbill, Eurylaimidae, a family now restricted to Africa and Asia, from the early Miocene of Wintershof West, Bavaria.

A picture thus emerges of passerines being absent in the Northern Hemisphere until after the Paleogene and beginning their remarkable and explosive adaptive radiation by the Miocene. As noted, oscine passerines outnumber all other birds in the Lower Miocene Wintershof West avifauna from Bavaria (Ballmann 1969b); before this time, in the Oligocene of the Northern Hemisphere, passerines are extremely scarce or absent, while the record of suboscines is so sparse as to prevent all but the most meager speculation. In 1982, Storrs Olson and I hypothesized that the entire passerine assemblage originated in the continents of the Southern Hemisphere and spread into the Northern Hemisphere during the mid-Tertiary. Although the many fossil finds of the past decade do not contravene this hypothesis, the extremely bizarre zoogeographic distribution illustrated by mousebirds, rollers, todies, motmots, and trogons means that the Southern Hemisphere origin hypothesis must remain a speculation and that solv-

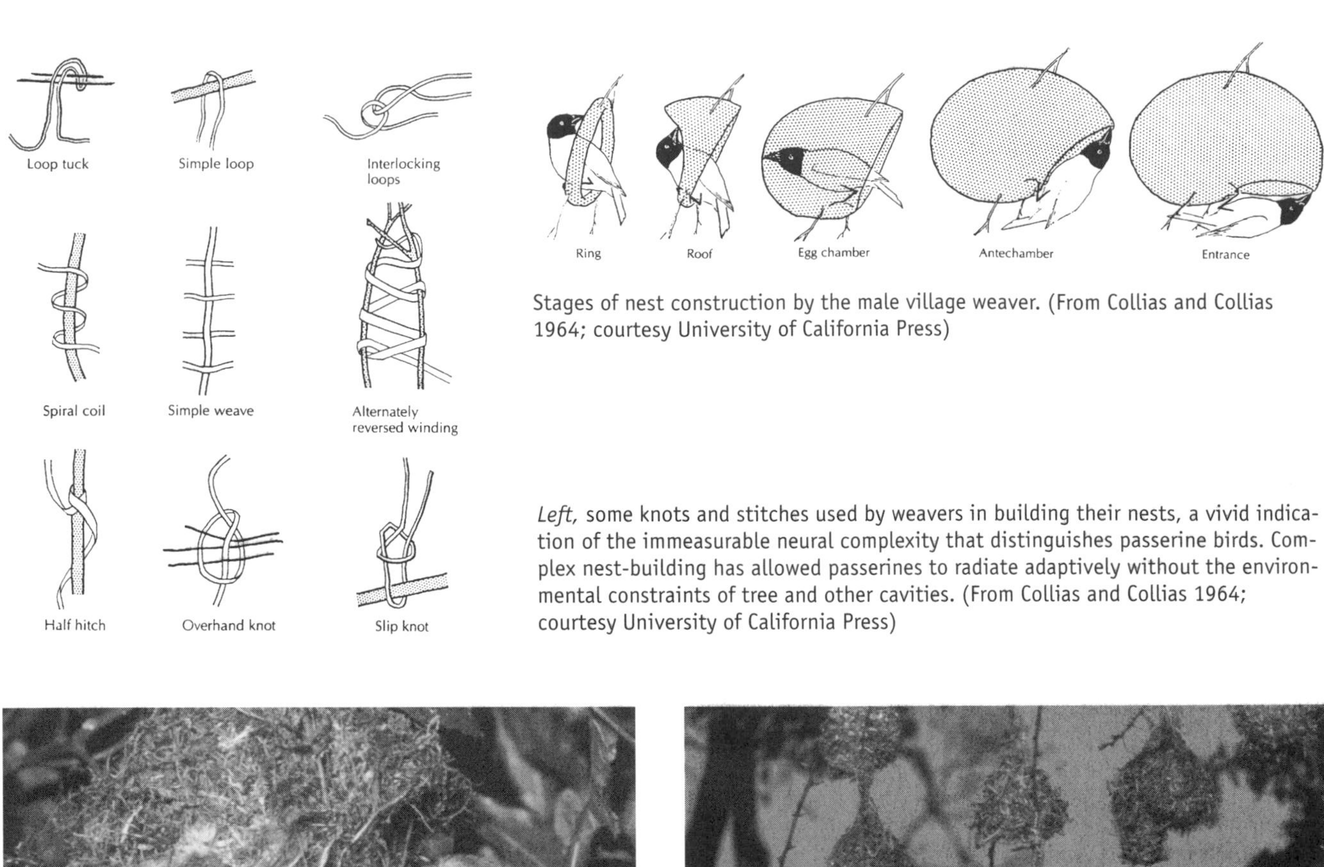

Stages of nest construction by the male village weaver. (From Collias and Collias 1964; courtesy University of California Press)

Left, some knots and stitches used by weavers in building their nests, a vivid indication of the immeasurable neural complexity that distinguishes passerine birds. Complex nest-building has allowed passerines to radiate adaptively without the environmental constraints of tree and other cavities. (From Collias and Collias 1964; courtesy University of California Press)

Complex nests of passerines from four families: *Clockwise from upper right,* village weaver (*Ploceus cucullatus,* Ploceinae); chestnut-headed oropendola (*Psarocolius wagleri,* Icteridae); cliff swallow (*Hirundo pyrrhonota,* Hirundinidae); brown thornbill (*Acanthiza pusilla,* Meliphagidae). (Photos: P. Davey—*Ploceus;* N. G. Smith—*Psarocolius;* D. and M. Zimmerman—*Hirundo;* R. Brown—*Acanthiza*/all VIREO)

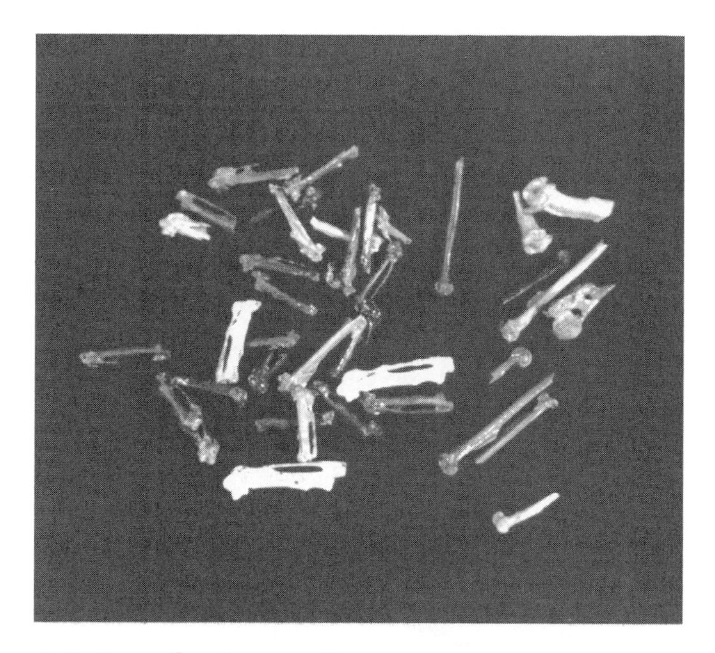

Oscine bones from the Upper Pliocene deposits of the Rexroad Formation in western Kansas. Hundreds of passerine fossils have been recovered from North American Pliocene deposits and all appear to represent oscines. (Photo by author)

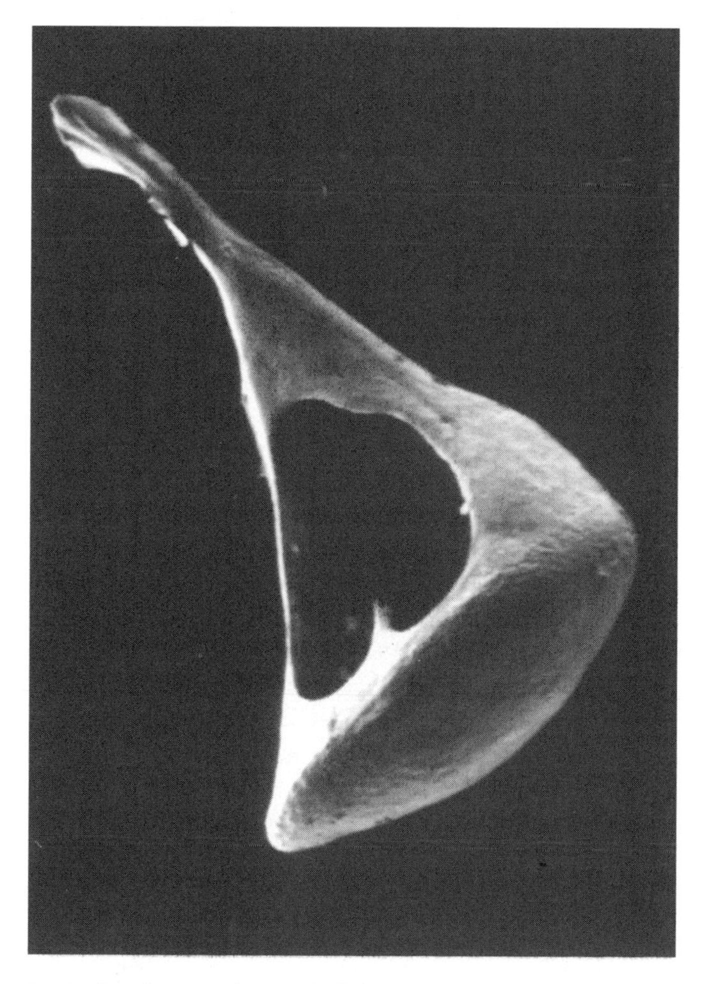

A scanning electron micrograph of the stapes of a South American ovenbird, the rufous hornero (*Furnarius rufus*), a New World suboscine, showing the middle-ear ossicle characteristic of all suboscine birds. Approximately × 50. (From Feduccia 1974b)

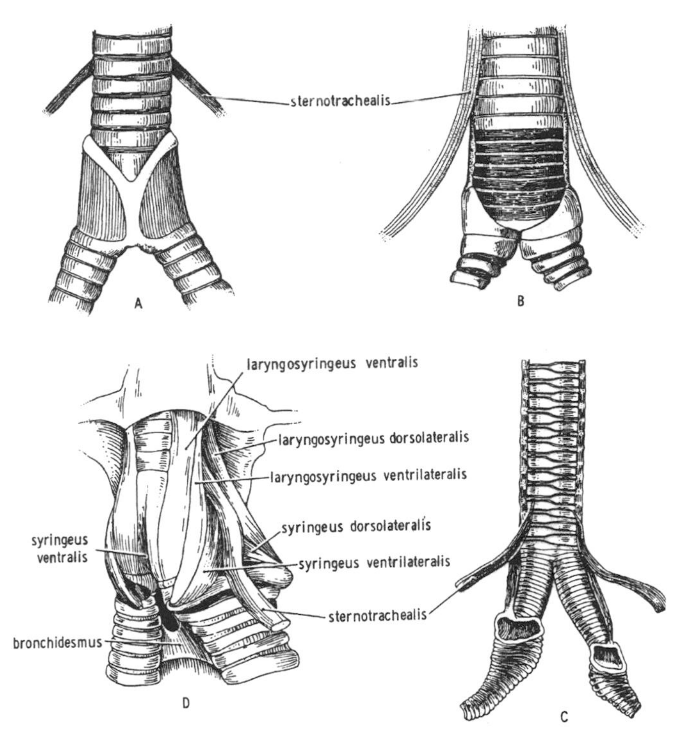

Types of syringeal anatomy: *A,* tracheobronchial syrinx of the suboscine "false sunbird" the sunbird asity (*Neodrepanis coruscans*); *B,* tracheal syrinx of the suboscine chestnut-billed gnateater (*Conopophaga aurita*); *C,* bronchial syrinx of the caprimulgiform oil-bird (*Steatornis caripensis*); *D,* tracheobronchial syrinx of the oscine true sunbird the little spiderhunter (*Arachnothera longirostra*). The avian voice box is not the larynx but the syrinx and is broadly divisible as (1) bronchial, a modification of the first few bronchial rings; (2) tracheal, a modification of the lower end of the windpipe, or trachea; or (3) tracheobronchial, involving both the trachea and the bronchi. The different arrangement of these muscles has been invaluable in dividing the order Passeriformes into natural groups. (From Van Tyne and Berger, *Fundamentals of ornithology,* copyright 1959, reprinted by permission of John Wiley and Sons)

ing the riddle of passerine origins will continue to be one of the great challenges for avian paleontology.

Dividing the Passerines: Suboscines and Oscines. The term "songbirds" was invented by people living in the temperate zones and referred initially only to the suborder Passeres of the order Passeriformes—the oscines—the noted songsters of the temperate zones. The oscines constitute about four-fifths of the passerine birds and are extremely uniform in their morphology. The rest, the suboscines, exhibit slightly less morphological uniformity, especially in the anatomy of the syrinx, and the vast majority are classified in the suborder Tyranni, a group of New World birds prominent in the Neotropics, especially in the deepest regions of Amazonia. The most familiar of these forms outside the tropics are the New World flycatchers (family Tyrannidae); many species of these flycatchers migrate north from the tropics, their ancestral homeland, to the temperate zone to nest in the summer, returning to the tropics in the winter. There are also several relict groups of Old World tropical suboscines, including the wide-ranging broadbills (Eurylaimidae) and pittas (Pittidae) and the Malagasy endemic philepittas (Philepittidae), whose classification is uncertain.

The two suborders of passerines have been distinguished from each other primarily by detailed anatomical analysis of the musculature of the syrinx. Suboscines all have simple syringes, whereas the oscines have more than three pairs of intrinsic syringeal muscles (Ames 1971). On the whole, oscine songs appear to be much more complex than the songs of the suboscines. A German anatomist, Johannes Müller, studied the muscles of the syrinx more than a century ago (1847) and subdivided the entire passerine assemblage on the basis of this structure, placing the suboscines into several suborders and the oscines into the single suborder Passeres. Peter Ames (1971) restudied the passerine syrinx in great detail and found that the oscine syrinx is complex but uniform throughout the suborder, suggesting a narrowly monophyletic group. He found suboscines to have simpler syringes, but he also found the structure to be highly variable throughout the group. Ames concluded that taxonomic decisions based on the syringes would have to be made with caution and that basing the unity of the suboscines on the simple syringes was not satisfactory. If simple syringeal morphology was, as it seemed, a retained primitive characteristic, it would not be suitable evidence for establishing evolutionary relations. However, more recent studies of the middle-ear ossicle have shown persuasively that the suboscines are, in fact, a tight, monophyletic group. All of the suboscines in both the Old and New Worlds are characterized by a stapes with a large, bulbous footplate region, resembling that of the alcediniform birds but differing in specific details (Feduccia 1974, 1979). Oscines, in contrast, have a simple stapes with a flat, oval footplate.

A number of characters of suboscines in addition to the simple syrinx are clearly primitive within the passerine assemblage: the lack of a sternal spine in some forms, a simple scapula, and the primitive single (as opposed to double) fossa of the humerus in all species. In addition, the sperm of the suboscines are not as complex as those of oscines (Henley et al. 1978; Feduccia 1979).

New Zealand Wrens and Australian Lyrebirds and Scrub-Birds: Primitive Living Passerines. Three groups of birds, thought by most authors to be primitive among the passerines, do not fit the classical definition of either suboscines or oscines; these are the New Zealand wrens (Acanthisittidae) and the Australian lyrebirds (Menuridae) and scrub-birds (Atrichornithidae). The three living New Zealand wrens are best known by the nuthatchlike rifleman (*Acanthisitta chloris*), which occurs commonly on both North and South Islands and is closely allied with the other forms, the South Island wren, bush wren, and extinct Stephens Island wren, all of the genus *Xenicus*. These birds were considered to be suboscines by most workers from 1882 to 1975, when stapes morphology illustrated that this could not be correct; in 1984, Raikow assigned them to the oscine branch on the basis of the loss of a part of one flexor muscle. As for the DNA evidence, Sibley and Ahlquist (1990, 582) concluded that "the acanthisittids are the survivors of an ancient passerine lineage with no close living relatives. We include them in the suborder Tyranni because they are not oscines (Passeri), but it is possible that they should be assigned to a third suborder as the sister group of the Tyranni and Passeri" (1990, 582). Sibley and Ahlquist's statement seems to me to be an accurate appraisal of the morphological evidence as well, and the group must remain incertae sedis, as Ernst Mayr proposed in 1979.

Another group of uncertain taxonomic status includes the primitive Australian passerines—the two species of lyrebirds (*Menura*) and two species of scrub-birds (*Atrichornis*)—which are commonly placed in their own suborder, the Menurae. Lyrebirds are noted for their beautiful, large, lyre-shaped tails and amazing vocal abilities; they are perhaps the best mimics within the class Aves, imitating everything from parrot flocks flying overhead to barking dogs and the blows of an ax. Scrub-birds, like lyrebirds, are primarily ground-dwellers but are nondescript small birds of predominantly brown color. Whereas most oscines have more than three pairs of intrinsic sy-

Primitive passerines of Australia and New Zealand. *Left to right,* Australian lyrebirds (Menuridae, 1 genus, 2 species): superb lyrebird (*Menura novaehollandiae*). Australian scrub-birds (Atrichornithidae, 1 genus, 2 species): noisy scrub-bird (*Atrichornis clamosus*). New Zealand wrens (Acanthisittidae [Xenicidae], 2 genera, 4 species): bush wren (*Xenicus longipes*). All three families have primitive syringeal anatomies and the primitive condition of the stapes, as opposed to the derived suboscine type of ear bone. (Drawings by George Miksch Sutton)

The Old World suboscines, the Pittidae, Philepittidae, and the Eurylaimidae, are a monophyletic group, as evidenced by numerous studies ranging from hindlimb myology to DNA-DNA hybridization. *Left to right,* asities and false sunbirds (Philepittidae, Madagascar, 2 genera, 4 species): velvet asity (*Philepitta castanea*). Broadbills (Eurylaimidae, Africa and southeastern Asia, 8 genera, 14 species): black-and-yellow broadbill (*Eurylaimus ochromalus*). Pittas (Pittidae, Old World tropics, 1 genus, 31 species): hooded pitta (*Pitta sordida*). (Drawings by George Miksch Sutton)

ringeal muscles, lyrebirds have a simpler syrinx with just three pairs, and scrub-birds also have three, plus a few fibers that may constitute a fourth (Sibley 1974, 73).

Most authors have concluded that *Menura* and *Atrichornis* are closely related primitive passerines (Raikow 1985b; Bock and Clench 1985). Charles Sibley (1974, 78), however, concluded that *Menura* is most closely related to the birds-of-paradise and bowerbirds, with scrub-birds as close allies, basing this rather bizarre conclusion on DNA hybridization studies (Sibley and Ahlquist 1990, 610). In sharp contrast to these findings, in 1982 Storrs Olson and I produced considerable morphological evidence to suggest that the Menurae, along with the Rhinocryptidae (the tapaculos, a New World suboscine family) are among the most primitive of all passerines and the likely remnants of the ancestral stock from which the remainder of the passerines arose. We found absolutely no morphological possibility of a close affinity of birds-of-paradise and bowerbirds with the Menurae, as did Raikow (1985b); the two groups are as dramatically different as any two of the most divergent passerines. Our conclusions are similar to those of Peter Ames, who wrote that "no single group of oscines can be considered syringeally primitive, in the sense that the Menurae can be considered so" (1971, 164). In fact, the primitive New World Rhinocryptidae and the Australian Menurae are so strikingly similar in morphology that a close affinity of the two must remain a possibility. Whatever the ultimate answer, it is beyond doubt that the Menurae are an extremely primitive group of passerines, that they do not fit the classical definitions of either suboscines or oscines.

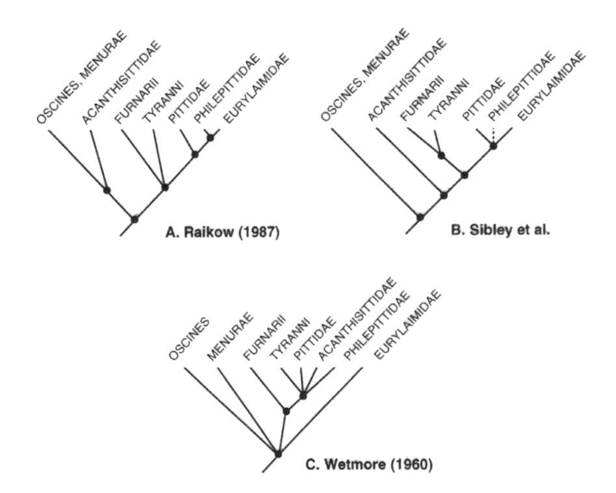

Comparison of results of three studies of passeriform relations: *A,* simplified cladogram of Robert Raikow's (1987) study of hindlimb myology; *B,* simplified cladogram of DNA hybridization studies of Charles Sibley et al. (1982) and Sibley and Jon Ahlquist (1985, 1990); *C,* hierarchical structure of the classification of Alexander Wetmore (1960). A single nomenclature is used for ease of comparison. The taxon Furnarii corresponds to the Furnarioidea of Sibley et al. (1982) and of Wetmore (1960) and to a corresponding but unnamed taxon of Sibley and Ahlquist (1985). The taxon Tyranni corresponds to the Tyrannoidea of Sibley et al. (1982), the Tyranni of Sibley and Ahlquist (1985), and to the New World families of Wetmore's (1960) taxon Tyrannoidea. Sibley and Ahlquist's classification (1990), based on the latest DNA hybridization studies, employs a parvorder Tyrannida, a parvorder Thamnophilida for typical antbirds, and a parvorder Furnariida for ovenbirds, woodcreepers, gnateaters, tapaculos, and the main groups of antbirds. (Modified after Raikow 1988)

Old World Suboscines. The true suboscine taxonomic categories are more clearly defined. In the Old World these birds are represented by three distinctive families of brightly colored perching birds that managed to hang on to their peculiar niches in the face of oscine competition. These relicts are now distributed erratically throughout the tropics. Among the most primitive are the broadbills (Eurylaimidae), fourteen brightly colored species of insectivorous, carnivorous (eating small frogs and lizards), and frugivorous birds. Broadbills construct a purselike hanging nest of grasses and other fibers with the entrance hole approximately in the center. They range erratically across tropical Africa and occur in Asia from the Himalayan foothills to the Philippines and on to Sumatra and Borneo. A fossil broadbill, mentioned previously, is known from the Miocene of Bavaria (Ballmann 1969b), indicating that the group was once much more widespread.

The other major group of Old World suboscine birds is the pittas (Pittidae). Chubby, primarily ground-dwelling birds with unusually short tails that are often not apparent at a distance, pittas are brightly colored, with different forms displaying a variety of vivid greens, reds, purples, and yellows. The thirty-one species are also placed in the same genus, *Pitta,* testifying to the homogeneity of the group. All inhabit the lower regions and floors of moist tropical forests, and in their habitat and body form they are highly convergent with the New World tropical antpittas and some of the antbirds (Formicariidae). Pittas feed almost entirely on earthworms and build a large globular nest with an entrance hole on one side. The center of their distribution is southwestern Asia and Malaysia, but two species are confined to central and east Africa, and four species have invaded Australia. Their relictual distribution across the Old World tropics is reminiscent of that of the broadbills, but unlike the broadbills, a number of pittas are migratory, nesting as far north as China and Japan and returning to the tropics for the winter.

The only other remnants of the Old World suboscine radiation are the asities of the family Philepittidae. Like many other primitive animals, they survive today only on the island refugium of Madagascar, where of the 201 extant avian species, 105, or 52 percent, are endemic (Lan-

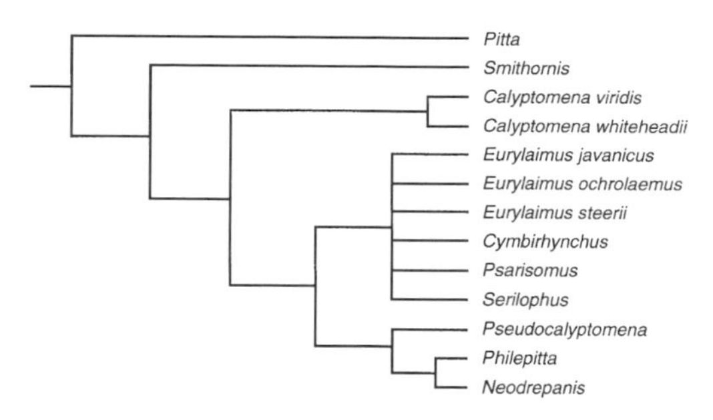

Most-parsimonious phylogenetic hypothesis for broadbills and asities, based on morphological study by Richard Prum (1993). Phylogenetic analysis of 21 characters; phylogenetic analysis of these data combined with 15 additional characters from Raikow (1987) yielded the same phylogenetic hypothesis. *Pitta* was the outgroup. (Modified after Prum 1993)

grand 1990; Goodman 1994b). The velvet asity and Schlegel's asity (*Philepitta*) are small, plump arboreal birds that superficially resemble pittas, whereas the two species of *Neodrepanis*, formerly called false sunbirds, are convergent with the more recently evolved oscine sunbirds of the family Nectariniidae, which contains 117 species and is widely distributed across Africa, Asia, Australia, and adjacent islands. Like true sunbirds, the sunbird asity and yellow-bellied asity dip their recurved bills into flowers to obtain nectar and small insects. Only the velvet asity is known to make a nest, a woven hanging structure with a side entrance.

The vicissitudes of Old World suboscine classification are detailed by Sibley and Ahlquist (1990). Among the more recent studies is that of Robert J. Raikow (1987), who concluded that the Philepittidae is the sister-group of the Eurylaimidae and that the Pittidae is the sister-group of the philepittid-eurylaimid clade, a finding that agrees broadly with evidence from DNA hybridization studies (Sibley and Ahlquist 1990, 584–585).

New World Suboscines. The New World suboscines have classically been divided on the basis of syringeal architecture into two superfamilies: the Furnarioidea, containing the ovenbirds and woodcreepers, antbirds, and tapaculos; and the Tyrannoidea, containing the flycatchers, sharpbill, plantcutters, cotingas, and manakins. Most numerous and best known of the New World suboscines are the 417 species of tyrant flycatchers (family Tyrannidae). These are primarily arboreal insect-eaters, with rather broad bills surrounded by special feathers modified as bristles that aid in capturing prey. In South America some forms have become largely terrestrial and have come to resemble larks or pipits. Many tyrant flycatchers are remarkably similar, at least superficially, to the Old World flycatchers of the large oscine family Muscicapidae. Old and New World flycatchers have similar habits, capturing insects on the wing by making short sallies from a prominent perch. A number of New World flycatchers migrate into North America to nest during the summer; most return to the tropics for the winter, though some winter in the southern United States.

Several groups of South American suboscines appear to be closely related to the flycatchers but are generally treated as separate families. The lone species of sharpbill (Oxyruncidae) is represented mostly by scattered specimens in museums. These specimens have been collected in six countries ranging from Costa Rica to Brazil, but no one has observed this species in its natural surroundings long enough to learn its habits! The three species of plantcutters (Phytotomidae) are so called because they use their saw-toothed bills for cutting plant material for food; they are among the few avian folivores (but not exclusively). Plantcutters range in temperate South America from western Peru to Patagonia.

The sixty-five species of cotingas (Cotingidae) are structurally primitive and bear a superficial resemblance to the Old World broadbills because both groups have evolved broadened bills adapted for eating fruit. This heterogeneous family includes such unusual forms as bellbirds, fruitcrows, umbrellabirds, cocks-of-the-rock, becards, and many more, most of them with striking and often gaudily colored plumage. Many are exceptionally large for passerines. The sixteen or so species of becards have been thought by some to be flycatchers. One form, the Jamaican becard (*Pachyramphus niger*), is the only cotinga to have entered the West Indies; and the northernmost species, the rose-throated becard (*P. aglaiae*), nests as far north as the southwestern United States. The two species of cocks-of-the-rock (*Rupicola*) are birds of the deep forest that have evolved an unusual method of courtship in which the gaudy males display for the drab females on special territories called leks, or arenas.

The fifty-three species of manakins (family Pipridae) are a homogeneous group of small suboscines, usually no bigger than 13 centimeters (5 in.) long (Prum 1992). Many authors have assigned manakins as close relations of the cotingas, probably because, like cotingas, manakin males are exceptionally brightly colored and display on leks to attract the drabber females, but the evidence for affinity is inconclusive. Many manakins have special modifications of the wing feathers that produce strange rattles and buzzing noises in flight.

The 279 species of ovenbirds (231) and woodcreepers or woodhewers (48) (Furnariidae and Dendrocolaptiae, often included in the single family Furnariidae) constitute an extremely diverse family of small to medium-sized, mainly brownish birds that range in habit from being com-

New World suboscines are thought to be monophyletic and are usually divided into two groups, the Furnarioidea and the Tyrannoidea. The Furnarioidea consists of ovenbirds, woodcreepers, antbirds, antthrushes, gnateaters, and tapaculos. *Left to right and top to bottom,* Ovenbirds (Furnariidae, Neotropics, 34 genera, 218 species): rufous hornero (*Furnarius rufus*). Woodcreepers (Dendrocolaptidae, Neotropics, 13 genera, 49 species): ivory-billed woodcreeper (*Xiphorhynchus flavigaster*). Antbirds (Thamnophilidae, Neotropics, 45 genera, 190 species), not illustrated. Antthrushes (Formicariidae, Neotropics, 7 genera, 60 species): black-faced antthrush (*Formicarius analis*). Gnateaters (Conopophagidae, Neotropics, 1 genus, 8 species): slaty gnateater (*Conopophaga ardesiaca*). Tapaculos (Rhinocryptidae, Neotropics, 12 genera, 29 species): chestnut-breasted huet-huet or turco (*Pteroptochos tarnii castaneus*). (Drawings by George Miksch Sutton)

pletely terrestrial to being completely scansorial (Feduccia 1973b; Raikow 1994). The woodcreepers spend their lives hitching up tree trunks in search of food and in appearance and habits closely resemble the unrelated temperate zone oscine family Certhiidae, the typical creepers. Ovenbirds are not related to the oscine wood warbler (*Seiurus aurocapillus*) that is often called an ovenbird; they are so-named for several species that build a domed, clay nest shaped like an old-fashioned earth oven. Most, however, dig tunnels in banks or build an open-cup nest. The ovenbirds are insectivorous, and a large number of species glean their food from deep-forest foliage. So morphologically different are the various members of the ovenbird family—earthcreepers, cinclodes, spinetails, canasteros, foliage-gleaners, treehunters, and many others—that it may be the most diverse family of all the passerine birds.

Another extremely large suboscine grouping consisting of two closely allied families of primarily deep-forest, insectivorous birds comprises the antbirds and antthrushes (Thamnophilidae, 190 species, and Formicariidae, 60 species), which, like ovenbirds and woodcreepers, are most abundant in the Amazon Basin. Antbirds come in a great variety of shapes and sizes, ranging from the wrenlike antwrens to the shrikelike antshrikes. The pittalike antpittas and ant-thrushes look very much like the fully terrestrial Old World suboscines. Most antpittas build a simple cup-shaped nest in branches near the ground. The 8 species of gnateaters (*Conopophaga*) are often included

The New World suboscines generally known as the Tyrannoidea include the tyrant flycatchers, cotingas, sharpbill, plantcutters, and manakins. Tyrant flycatchers (Tyrannidae, New World, 104 genera, 417 species): eastern kingbird (*Tyrannus tyrannus*). Cotingas (Cotingidae, Neotropics, 26 genera, 65 species): Amazonian umbrellabird (*Cephalopterus ornatus*). Sharpbills (Oxyruncidae [may be a cotinga], Neotropics, 1 genus, 1 species): sharpbill (*Oxyruncus cristatus*). Plantcutters (Phytotomidae [may be cotingas], South America, 1 genus, 3 species): white-tipped plantcutter (*Phytotoma rutila*). Manakins (Pipridae, Neotropics, 12 genera, 53 species): red-capped manakin (*Pipra mentalis*). (Drawings by George Miksch Sutton)

in the above grouping (Ames, Heimerdinger, and Warter 1968), but Monroe and Sibley (1993) place them in the Conophagidae; they are small, completely terrestrial forest birds that closely resemble the antpittas of the genus *Grallaria*. According to the DNA hybridization data of Sibley and Ahlquist (1990), the "Furnarioidea" of Alexander Wetmore (1960) is far more complex than previously thought. If their data are correct, the typical antbirds (Thamnophilidae) are the descendants of one lineage, and the sister-branch produced the birds now included in the Furnarioidea and Formicarioidea, the latter including the ground antbirds (Formicariidae), gnateaters (Conopophagidae), and tapaculos (Rhinocryptidae).

The twenty-nine species of tapaculos of the family Rhinocryptidae are a relict family of primitive suboscine birds that are now found mostly in the Chilean Andes but also occur in other parts of South and Central America. Alfred Newton noted that the name *tapaculo*, "Of Spanish origin, . . . is intended as a reproof to the bird for the shameless way in which, by erecting its tail, it exposes its hinder parts" (1896, 946). Tapaculos are insectivorous terrestrial or semiterrestrial nonmigratory birds that are largely confined to scrublands and the thick undergrowth of high mountain forests. Their nests range from tunnels in banks to bulky, domed constructions of grasses. Tapaculos are morphologically primitive, having a four-notched sternum and a humerus resembling that of primitive coraciiforms, and certain members have the primitive type of stapes. Most also have a poorly developed flight apparatus, often almost entirely lacking the sternal keel and having greatly reduced, unfused clavicles, some exhibiting ratite-like clavicular splints, and some species may be essentially flightless. As mentioned, tapaculos exhibit numerous similarities to the Old World Menurae; in fact, in 1859, Jean Cabanis and Ferdinand Heine included the lyrebird (*Menura*) in the family Rhinocryptidae (then Pteroptochidae).

Amazingly, a single humerus from the cave deposits from the Isle of Pines (off Cuba) and a tibiotarsus from a cave in Camaguey Province, Cuba, clearly belong to a tapaculo (Olson and Kurochkin 1987). When the humerus was first recovered, it presented such an insurmountable zoogeographic dilemma that Storrs Olson "set the specimen aside in disbelief" (353). Some five years later Kurochkin visited Olson at the Smithsonian, bringing with him the tibiotarsus from Camaguey, some 500 kilometers (310 mi.) from the other specimen, and it became apparent that the identification of the humerus had been correct. No member of the superfamily Furnarioidea had ever been recorded in the West Indies, and there is no evidence of any land connection between Cuba and either North or South America, so rafting must have been the mode of dispersal. Therefore, contrary to what one would surmise from the modern distribution of the family, tapaculos must once have been widespread and more capable of dispersal than was previously thought. Perhaps, in fact, the ability of all animals to disperse has been grossly underestimated; certainly ground sloths of South American affinity must have swum or were rafted directly from the continent to the Greater Antilles.

Tapaculo fossils from Cuba and the adjacent Isle of Pines present a zoogeographic dilemma and vivid reminders of how we tend to view the distribution of today's birds as static. This map shows the present-day distribution of the Rhinocryptidae, and arrows indicate the location of the fossil discoveries. The white-throated tapaculo (*Scelorchilus albicollis*) of Chile is also depicted. (From Newton 1896)

The DNA hybridization studies (Sibley and Ahlquist 1990, 590) indicate, as does a mountain of morphological evidence, that the New World suboscines are more closely related to one another than any of them are to any Old World suboscine group. Yet Sibley and Ahlquist (1990, 600–601) depart from classical systematics in placing the typical antbirds (separated as a family Thamnophilidae) in close alliance with the tyrant flycatchers and allies, the former Tyrannoidea, a conclusion that is not supported by most morphological evidence.

The Oscine Radiation. The oscines represent an absolute extreme among birds, and perhaps all living vertebrates, in their morphological uniformity. All have a nearly identical syringeal structure, as we have seen, a very similar pattern of feather tracts, and the simple, primitive type of middle-ear ossicle. Most are small land birds primarily adapted for feeding on insects, small fruits, and seeds, and therefore many of the differences we do see

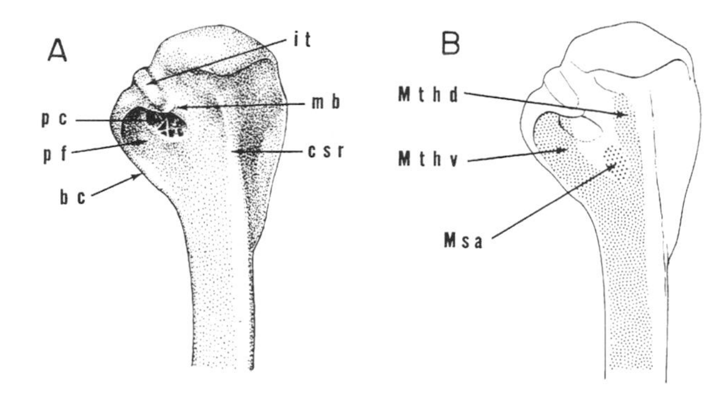

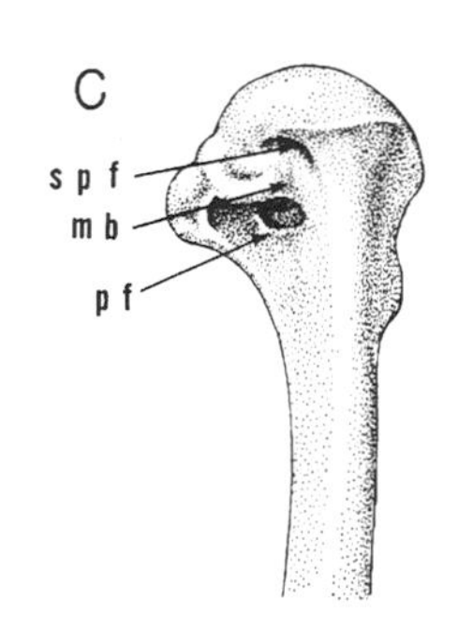

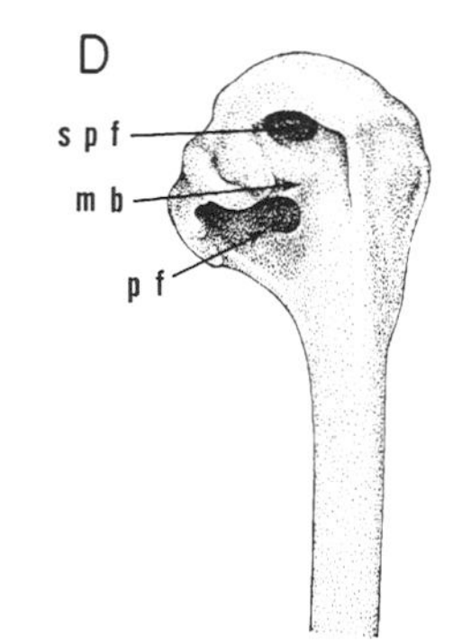

Humerus, right proximal end, of oscines: *A, Corvus* (crows), *C, Eulabes* (starlings), and *D, Sturnus* (starlings), to show the pneumatic fossa. *Corvus* has the single fossa condition, thought to be primitive. *Eulabes* starlings show a beginning stage in the development of the second, more anterio-proximal fossa, and the double-fossa condition is seen in starlings of the genus *Sturnus*. Abbreviations: *b c,* bicipital crest; *c s r,* capital-shaft ridge; *i t,* internal tuberosity; *m b,* medial bar; *p c,* pneumatic canal; *p f,* pneumatic fossa; and *s p f,* second pneumatic fossa. Attachments for muscles (*B*) are dorsal (*M t h d*) and ventral (*M t h v*) heads of the M. triceps humeralis (fine stippling); and insertion of the M. scapulohumeralis anterior (*M s a*), located at the distal end of the pneumatic fossa (heavy stippling). The pneumatic fossa is single in all coraciiforms and allies, except for the Todidae and Indicatoridae (it is divided in the todies, and a second fossa has reached a well-formed state in the honeyguides). The pneumatic fossa is single in all suboscines. The New World nine-primaried oscines appear to be the only larger subgroup within the oscines that uniformly has a fully developed double condition. (From Bock 1962; courtesy *Auk*)

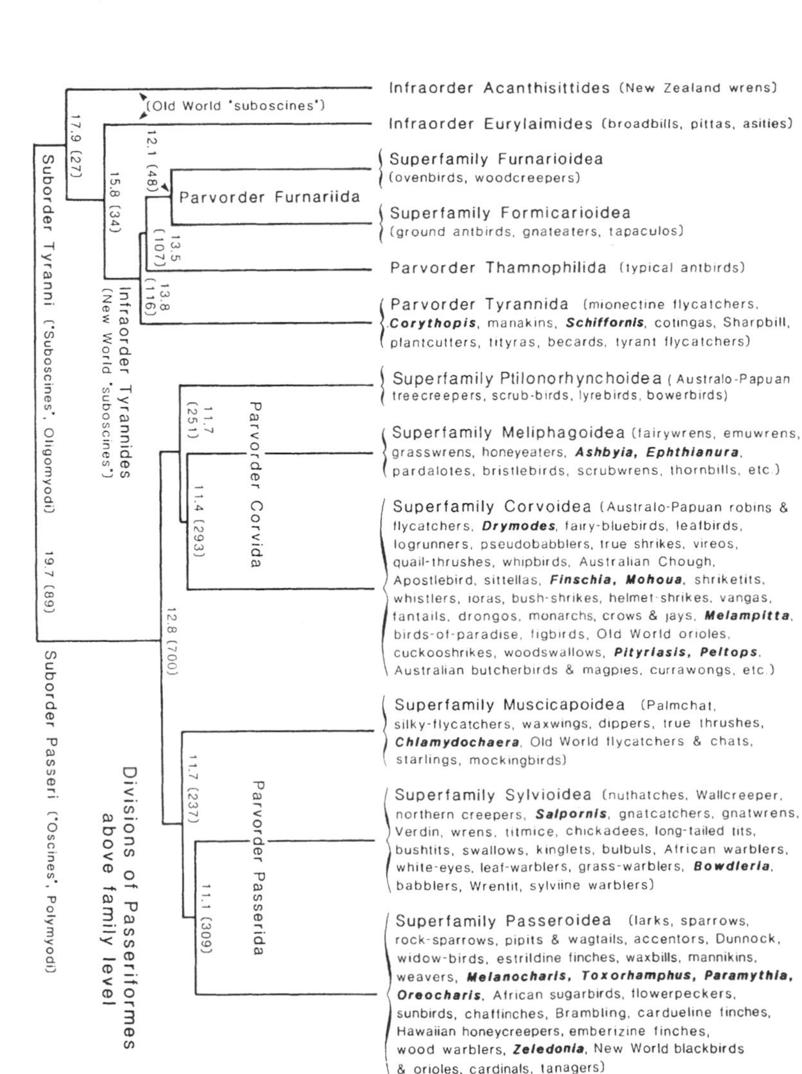

Division of the Passeriformes above the family level determined by average linkage clustering based on DNA-DNA hybridization. (From Sibley and Ahlquist 1990; courtesy Yale University Press)

among them are in the feeding mechanism and bill, structures that are often convergent rather than indicative of any evolutionary link. The lack of major gaps among the various oscine families suggests that the adaptive radiation of modern, advanced oscines is probably quite recent, perhaps dating to the medial to late Tertiary.

Two of the few important characteristics that have been used classically to separate the oscines are the form of the proximal end of the humerus and the number of primary feathers. The primitive type of humerus has only a single fossa, or excavation, at the proximal end; this style of humerus is found in many oscines (and in all of the suboscines, piciforms, and coraciiforms), but other oscine groups have the advanced double fossa. In flying birds the number of primary feathers varies from nine to twelve, but passerine birds typically have ten. Yet in some oscines, the tenth (or outermost) primary is reduced to a vestigial feather, and so these birds are termed "nine-primaried," a derived condition.

Among early studies of oscines, one of the most important was William Beecher's (1953) attempt to develop a phylogeny based on jaw musculature. Harrison Tordoff's study (1954) of the relations among nine-primaried oscines followed, as did a number of single character studies, including an examination of the pneumatic fossa of the humerus by Walter Bock (1962) and a study on the appendicular myology of New World nine-primaried oscines by Robert J. Raikow (1978). Charles Sibley and Jon Ahlquist (1990) have presented an excellent general summary of the history of the systematics of oscines.

Burt Monroe and Charles Sibley (1993) include within their suborder Passeri some 4,578 species in 875 genera, while Walter Bock and John Farrand (1980) count in their suborder Passeres some 4,177 species in 823 genera—in both cases this is about half of the world's birds. Yet because of oscines' presumed recent origin, as indicated by the amount of recent species formation, the lack of generic and familial definition among oscines, the lack of oscine fossils before the Oligocene, and the oscines' extreme morphological uniformity, it has been much harder to categorize the oscines than the suboscines. In 1888 the famed anatomist Max Fürbringer actually recognized just two oscine families; in 1934 the distinguished German ornithologist Erwin Stresemann recognized forty-nine; and in 1960 Alexander Wetmore, then dean of American ornithology, recognized fifty-four, a number that approximates today's usage. Charles Sibley accurately summarized our understanding of the New World nine-primaried oscines in 1970: "This cumbersome and not wholly accurate phrase is sanctioned by custom to designate a large assemblage of passerine birds. They are not confined to the New World, although more numerous there in terms of species, and some passerines with nine primaries are not included. The confusion is compounded because there is no agreement about the boundaries of the group" (98–99).

Some major groupings of oscines have classically included the corvine assemblage, which includes the crows and jays, bowerbirds, and birds-of-paradise. As an illustration of the confusion surrounding their classification, these birds have been placed both at the beginning and at the end of the oscines. They have been placed last on the basis of their purported superior mental capabilities and the highly complex behavior and plumage of bowerbirds and birds-of-paradise. Contrarily, they have been placed at the beginning of the oscine classification because of their primitive structural features: all corvids have ten primaries, a primitive feature, and the primitive single fossa of the humerus.

Another major classical grouping included the predominantly Old World group of ten-primaried oscines, including the Old World warblers, Old World flycatchers, babblers, and thrushes, and such New World groups as the wrens and mimids. Still a third traditional category, generally considered to be the most advanced grouping, consisted of the so-called New World nine-primaried oscines. This predominantly New World group includes the vireos, wood warblers, tanagers, icterids, and the emberizine and cardueline finches.

Sibley and Ahlquist's DNA hybridization studies revealed two major groupings within the Passeri, the parvorders Corvida and Passerida; each is subdivided into three superfamilies (1990, 603). The Corvida is composed primarily of oscines in Australasia and their allies; as Sibley and Ahlquist note, "It seems apparent that the group originated and radiated in Australia because the oldest elements occur today only, or mainly, in Australia and New Guinea" (603). Morphological evidence provides mixed support for this conclusion, and it awaits corroboration from other lines of evidence.

Because of the confusion surrounding oscine systematics, it is convenient here to follow, as a provisional standardization, the groupings of Sibley and Ahlquist (1990), which Frank Gill (1995) has adopted in his excellent text *Ornithology;* traditional family names are used here, however. Morphological studies corroborate many of their conclusions, and at present their groupings stand as a new framework for further investigations into the evolution of this complex assemblage. Sibley and Ahlquist group the oscines as follows: (1) Crow relatives, parvorder Corvida, including most of the families of Australasian birds and the crows, jays, drongos, shrikes, vangas, birds-of-paradise, and vireos (they also include the lyrebirds and scrub-birds in this division, but the morphological evidence is so contradictory to this conclusion [Feduccia and

Crows and allies (preceding two pages).
This group of oscines includes most of the Australasian passerine birds, as well as crows, jays, drongos, shrikes, vangas, birds-of-paradise, and vireos. (Sibley and Monroe 1990 and Monroe and Sibley 1993 designate these birds as a parvorder Corvida and all other oscines as a parvorder Passerida, divided into three superfamilies, illustrated in the next three figures.) Before Sibley and Ahlquist's work (1990), many of these Australasian families were placed into various Asian bird families that they most closely resemble in ecology and morphology, but they are now thought to be convergent on these Asian groups and closely allied instead to one another or to several non-Australian families. Representatives are, *from left to right, and top to bottom,* (*p. 359, row 1*) orange-winged sittella (*Daphoenositta chrysoptera,* Pachycephalidae); satin bowerbird (*Ptilonorhynchus violaceus,* Ptilonorhynchidae); regent honey-eater (*Xanthomyza phrygia,* Meliphagidae); (*row 2*) orange-bellied leafbird (*Chloropsis hardwickii,* Irenidae; formerly in Aegithinidae); loggerhead shrike (*Lanius ludovicianus,* Laniidae); yellow-throated vireo (*Vireo flavifrons,* Vireonidae); (*row 3*) rufous-browed peppershrike (*Cyclarhis gujanensis,* Vireonidae; Cyclarhidae of some authors); chestnut-sided shrike-vireo (*Vireolanius melitophrys,* Vireonidae, Vireolaniidae of some authors); white-throated magpie-jay (*Calocitta formosa,* Corvidae); (*row 4*) red bird-of-paradise (*Paradisaea rubra,* Paradisaeidae); hooded butcherbird (*Cracticus cassicus,* Artamidae); (*p. 360, row 1*) white-browed wood-swallow (*Artamus superciliosus,* Artamidae); Eurasian golden oriole (*Oriolus oriolus,* Oriolidae); large cuckooshrike (*Coracina macei,* Campephagidae); (*row 2*) greater racket-tailed drongo (*Dicrurus paradiseus,* Dicruridae); magpie-lark (*Grallina cyanoleuca,* Monarchidae, formerly in Grallinidae); curly-crested helmetshrike (*Prionops plumatus cristatus,* Prionopidae); (*row 3*) hook-billed vanga (*Vanga curvirostris,* Vangidae); nuthatch vanga (*Hypositta corallirostris*); huia (extinct) (*Heteralocha acutirostris,* Callaeatidae). (Drawings by George Miksch Sutton)

Family	Name	Distribution	Genera	Species
Climacteridae	Australo-Papuan treecreepers	Australia, New Guinea	2	7
Ptilonorhynchidae	Bowerbirds	Australasia	7	20
Maluridae	Fairywrens, emuwrens	Australia	5	26
Meliphagidae	Honeyeaters	Australasia	42	182
Pardalotidae	Pardalotes, bristlebirds	Australasia	2	7
Acanthizidae	Australian warblers	Australasia	14	61
Eopsaltriidae	Australian robins	Australasia	13	44
Irenidae	Fairy-bluebirds, leafbirds	Oriental	2	10
Aegithinidae	Ioras	Oriental	1	4
Orthonychidae	Logrunners, chowchillas	Australasia	1	2
Pomatostomatidae	Australo-Papuan babblers	Australasia	1	5
Laniidae	True shrikes	North America, Africa, Eurasia	3	30
Vireonidae	Vireos, peppershrikes	New World	4	52
Cinclosomatidae	Quail-thrushes, whipbirds	Australia	6	15
Corcoracidae	Australian chough, apostlebird	Australia	2	2
Pachycephalidae	Whistlers, sitellas	Australasia	14	59
Corvidae	Crows, jays, magpies	Worldwide	25	118
Paradisaeidae	Birds-of-paradise	Australasia	17	46
Artamidae	Wood-swallows, currawongs	Orient, Australasia	6	24
Oriolidae	Old World orioles, figbirds	Warm Old World	1	27
Campephagidae	Cuckooshrikes	Old World	7	84
Rhipiduridae	Fantails	Australasia	1	42
Dicruridae	Drongos	Old World tropics	2	24
Monarchidae	Monarch flycatchers, magpie-larks	Southeastern Asia, Australasia	18	98
Malacanotidae	Bushshrikes	Africa	8	49
Prionopidae	Helmetshrikes, batises	Africa	7	44
Vangidae	Vangas	Madagascar	12	14
Callaeatidae	New Zealand wattlebirds	New Zealand	3	3
Picathartidae	Chaetops, rockfowl	New Guinea, Africa	2	4

Note: The Acanthizidae and Maluridae were formerly included in the Sylviidae; the Orthonychidae, Cinclosomatidae, and Picathartidae were included in the heterogeneous family Timaliidae; and the Pachycephalidae was included in the Muscicapidae.

Thrushes and allies (superfamily Muscicapoidea of Sibley and Monroe 1990).
This group of oscines includes, in addition to thrushes, waxwings and allies, dippers, starlings, mockingbirds and allies, and Old World flycatchers and chats. Representatives are, *from left to right and top to bottom,* Bohemian waxwing (*Bombycilla garrulus,* Bombycillidae); phainopepla (*Phainopepla nitens,* Bombycillidae; Ptilogonatidae of some authors); palmchat (*Dulus dominicus,* Dulidae); white-throated dipper (*Cinclus cinclus,* Cinclidae); song thrush (*Turdus philomelos,* Turdidae); Asian paradise-flycatcher (*Terpsiphone paradisi,* Muscicapidae); brown thrasher (*Toxostoma rufum,* Mimidae); rosy starling (*Sturnus roseus,* Sturnidae). (Drawings by George Miksch Sutton)

Family	Name	Distribution	Genera	Species
Bombycillidae	Waxwings, silky-flycatchers, *Hypocolius*	Holarctic	4	7
Dulidae	Palmchat	Hispaniola	1	1
Cinclidae	Dippers	Holarctic, South America	1	5
Turdidae	Thrushes	Worldwide	21	179
Muscicapidae	Old World flycatchers and chats	Eurasia	48	273
Sturnidae	Starlings, mynas	Old World	27	114
Mimidae	Mimic thrushes, thrashers	New World	11	34

Sylvioid oscines and allies (superfamily Sylvioidea of Sibley and Monroe 1990).
This oscine group includes many Eurasian families, and some familiar Northern Hemisphere birds, the nuthatches, titmice, wrens, and swallows. It also encompasses the babblers, Old World warblers, and kinglets. Representatives are, *from left to right, and top to bottom,* red-breasted nuthatch (*Sitta canadensis,* Sittidae); Eurasian tree-creeper (*Certhia familiaris,* Certhiidae); grey-breasted wood-wren (*Henicorhina leucophrys,* Troglodytidae); green-backed tit (*Parus monticolus,* Paridae); barn swallow (*Hirundo rustica,* Hirundinidae); light-vented bulbul (*Pycnonotus sinensis,* Pycnonotidae); silvereye (*Zosterops lateralis,* Zosteropidae); blackcap (*Sylvia atricapilla,* Sylviidae); variegated laughingthrush (*Garrulax variegatus,* Sylviidae; sometimes Timaliidae). (Drawings by George Miksch Sutton)

Family	Name	Distribution	Genera	Species
Sittidae	Nuthatches, wallcreepers	Widespread, not South America	2	25
Certhiidae	Creepers	Holarctic, Africa, India	2	7
Troglodytidae	Wrens	New World; 1 species Holarctic	17	75
Polioptilidae	Gnatcatchers, verdin	New World	4	15
Paridae	Titmice	Eurasia, Africa, North America	3	53
Remizidae	Penduline-tits	Eurasia, Africa, western North America	4	12
Aegithalidae	Long-tailed tits	Eurasia, western North America	3	8
Hirundinidae	Swallows, martins	Worldwide	14	89
Regulidae	Kinglets	Palearctic	1	6
Hypocoliidae	*Hypocolius*	Southwest Asia	1	1
Cisticolidae	African warblers	Africa	17	120
Pycnonotidae	Bulbuls	Africa, southern Asia	21	138
Zosteropidae	White-eyes	Old World tropics and subtropics	13	95
Sylviidae	Old World warblers, laughingthrush, babblers, wrentit	Mostly Old World	101	552

Note: The Polioptilidae was previously included in the Sylviidae.

Oscine weavers and allies (superfamily Passeroidea of Sibley and Monroe 1990).
This predominantly Old World group comprises not only the familiar African weavers but also the ground-dwelling larks, sparrows, finches, wagtails, and accentors. Also included here are the nectar-feeding sunbirds and flowerpeckers, the cardueline finches, including crossbills and siskins, and the Hawaiian honeycreepers, which are derived from cardueline finches. According to Sibley and Ahlquist (1990), this group includes the nine-primaried oscines; here they are separated as a distinct grouping. Representatives are, *from left to right and top to bottom,* crested lark (*Galerida cristata,* Alaudidae); fire-breasted flowerpecker (*Dicaeum ignipectus,* Dicaeidae); orange-breasted sunbird (*Nectarinia violacea,* Nectariniidae); white wagtail (*Motacilla alba,* Motacillidae); rufous-striated accentor (*Prunella himalayana,* Prunellidae); eastern paradise-whydah (*Vidua paradisea,* Estrildidae; Ploceidae of some authors); Harris's sparrow (*Zonotrichia querula,* Fringillidae); Iiwi (*Vestiaria coccinea,* Drepanididae: derived from cardueline finches). (Drawings by George Miksch Sutton)

Family	Name	Distribution	Genera	Species
Alaudidae	Larks	Worldwide	17	91
Promeropidae	Sugarbirds	South Africa	1	2
Dicaeidae	Flowerpeckers	Oriental	2	44
Nectariniidae	Sunbirds	Old World tropics	5	123
Melanocharitidae	Berrypeckers, longbills	New Guinea	5	12
Passeridae	Old World sparrows, rock-sparrows	Old World	4	36
Motacillidae	Wagtails, pipits	Nearly worldwide	5	65
Prunellidae	Accentors or hedge-sparrows	Eurasia	1	13
Ploceidae	Weavers	Eurasia, Africa	17	118
Estrildidae	Estrildine finches, whydahs	Old World tropics	30	156
Fringillidae	Cardueline finches, chaffinch, olive warbler	Worldwide, except Australasia	22	141
Drepanididae	Hawaiian honeycreepers	Hawaiian Islands	18	30

Nine-primaried oscines (included in the superfamily Passeroidea in Sibley and Monroe 1990).
This diverse group of primarily New World birds, characterized by a strongly reduced tenth primary, includes the colorful New World warblers, tanagers, blackbirds and New World orioles, and New World "sparrows" and buntings. Numerous species nest in the temperate regions of North America and migrate south to the tropics in winter. Representatives are, *top row, left to right,* plushcap (*Catamblyrhynchus diadema,* Emberizidae; sometimes Catamblyrhynchidae); black-throated grey warbler (*Dendroica nigrescens,* Parulidae); wrenthrush (*Zeledonia coronata,* Parulidae; sometimes Zeledoniidae); *bottom row, left to right,* white-winged tanager (*Piranga leucoptera,* Thraupidae); swallow tanager (*Tersina viridis,* Thraupidae; sometimes Tersinidae); chestnut-headed oropendola (*Psarocolius wagleri,* Icteridae). DNA analyses have not supported Raikow's (1978) cladistic analysis of New World nine-primaried oscines, which bolstered the traditional concept of the group. Anthony Bledsoe (1988) measured dissimilarity of single-copy nuclear-DNA sequences for 13 species and showed that feeding specializations arose convergently and therefore that convergence characterizes the adaptive radiation of several levels of the nine-primaried assemblage. (Drawings by George Miksch Sutton)

Family	Name	Distribution	Genera	Species
Emberizidae	Buntings, longspurs, sparrows, towhees, brush-finches	Worldwide, except Australasia	32	156
Parulidae	New World warblers, wrenthrush	New World	25	115
Thraupidae	Tanagers, seedeaters, honeycreepers and allies	New World	104	413
Cardinalidae	Cardinal-grosbeaks and allies	New World	13	42
Icteridae	New World orioles, blackbirds, and allies	New World	26	97

Olson 1982] that at present it must remain in doubt); (2) thrush relatives, parvorder Passerida, superfamily Muscicapoidea, including waxwings, starlings, and mimic thrushes, plus the Old World flycatchers and chats; (3) Old World insect-eaters, parvorder Passerida, superfamily Sylvoidea, including swallows, titmice, nuthatches, white-eyes, babblers, and Old World warblers; and (4) weaver relatives, parvorder Passerida, superfamily Passeroidea, including crossbills, weavers, sunbirds, and larks, plus the New World nine-primaried oscines, the warblers, tanagers, blackbirds, and buntings (here the nine-primaried oscines are separated for historical reasons and to divide this cumbersome assemblage).

The Oscine-Suboscine Dichotomy. There have been two schools of thought concerning the competitive relations between the oscines and suboscines. One school, whose principal advocates were Paul Slud (1960) and Edwin Willis (1966), maintained that the suboscines are not actually primitive within the passerine assemblage and that they have held their own against competition from the more advanced oscines. These conclusions are based on the fact that suboscines are a very successful group of birds in South America in terms of adaptive radiation of morphological types and number of species; in addition, according to Willis, suboscines seem to be behaviorally adept in competing with winter oscine migrants in the tropics. However, this concept is based on the frequent observation of those, including myself, who have studied rain forest birds that the oscines are predominantly birds of the forest canopy and peripheral habitats; suboscines appear to rule the deep, inner forest niches. Yet when one considers how recent the Great American Interchange was and how limited the taxonomic diversity of South American oscines is (they are mostly "nine-primaried" oscines—more specifically, tanagers and allies and emberizid finches), this is not at all surprising. The other, more accepted school, which I follow, represents the classical view and has been championed by Ernst Mayr and Dean Amadon (1951; also see Amadon 1973 and Feduccia 1977d, 1980). It argues that suboscines are indeed more primitive and have been replaced by the more advanced oscines wherever the two groups have been in extended competition.

The oscines and suboscines surely had an ancient common ancestor. According to the classical view of Mayr and Amadon (1951), the suboscines represent a "first attempt" at becoming a highly advanced passerine bird, becoming widespread some time before the mid-Tertiary. The oscines, by contrast, did not undergo any extensive evolution until later in the Tertiary.

One of the main problems with current zoogeographic dogma is that most of the living orders of birds are thought by most authors (see Sibley and Ahlquist 1990) to be much older than the fossil record would indicate. The suboscines, or at least the ancestors of both Old and New World suboscines, are thought to have been ancient and to have been present in the southern continents of Gondwanaland before the separation of South America and Africa, say some 75 to 80 million years ago. As we have seen, this is certainly not the case, however, and in fact, what is more important than the simple separation of the continents is the width between them at a given time. The general idea (see Feduccia 1977d) goes as follows: after the continents began to drift apart in the late Cretaceous, the ancestral suboscines, if present in the early Tertiary, easily could have continued their dispersal across the Atlantic Ocean, which even as late as the Paleocene was very narrow, separating the two landmasses by perhaps no more than 600 kilometers (375 mi.)—the equivalent distance of the trans–Gulf of Mexico migration route followed today by the suboscine tyrant flycatchers. Also, at that time the mid-Atlantic ridge was probably dotted with volcanic islands that could have provided way stations for this dispersal (Raven and Axelrod 1975). There must have been some limited exchange, as evidenced by the presence of New World monkeys in South America. We could assume, then, an extensive Paleogene Southern Hemisphere adaptive radiation of suboscines. The discovery of innumerable fossil finds in the past decade has cast doubt on the existence of late Cretaceous suboscines as it has become obvious that the land birds of the Cretaceous were the enantiornithine birds and that in all probability the passerines, as well as perhaps almost all of the modern orders of living birds, evolved after the Cretaceous. Yet the zoogeographic proposal I have outlined may still warrant consideration, because current evidence points to passerines not being present, at least in any numbers, in the Northern Hemisphere until after the Paleogene, and if the southern continents were not too far apart, dispersal among a limited number of birds may still have been possible.

In the New World, the Central American land bridge had not yet come into being, and South America existed in isolation from North America. Even in mid-Tertiary times, when climates were quite mild, there were limited opportunities for New World suboscines to venture north. In the Old World, however, no geographic barriers existed, and the radiation of suboscines extended well into the north.

In the medial to late Tertiary, the oscines began an extensive adaptive radiation that edged out the suboscines wherever the two groups came into competition, as they frequently did in the Old World. In the late Tertiary the oscines apparently used the Bering land bridge to invade the

New World Northern Hemisphere. But until recently, the oscines remained isolated in North America, whereas the suboscines thrived in South America. It is this flourishing of suboscines in South America that led Slud and Willis to question the classical view of suboscine-oscine competition. A closer examination of the evidence, however, tends to lessen the impression of suboscine success. Clearly, suboscines did not break out of South America in any major way until the Great American Interchange during the late Pliocene. With the exception of tyrant flycatchers and a few isolated examples, such as the tapaculo from Cuba, suboscines are absent from the West Indies. The suboscines of Trinidad, for example, are not endemic but instead represent a recent invasion from South America when Trinidad was connected with that continent during the Pleistocene. The Central American suboscines are undoubtedly extralimital members of South American superspecies groups that have recently moved north. In addition, the fossil record indicates that there were no substantial numbers of suboscines in North America even as late as 3.8 million years ago—none are known as fossils. Given the moderate climates of the late Pliocene and Pleistocene in North America, it seems reasonable to assume that the suboscines, if present, would have undergone an extensive radiation.

During the late Pliocene advanced mammals crossed the land bridge into South America and precipitated a massive wave of extinction in the archaic fauna of the continent. The suboscines survived in spite of the late Pliocene invasion of oscines from North America, but the significance of this persistence diminishes when other factors are considered. In the very late Tertiary the major Andean uplift occurred, causing major topographic changes and creating new life zones, including deserts along the western coast, into which suboscines could radiate. Also, as Jürgen Haffer has shown (1969, 1974), Pleistocene fluctuations of humid and dry periods in the vast Amazon Basin resulted in refugia and therefore rapid speciation in the birds of the region, including the suboscines.

It is this recent extensive speciation, combined with the very recent and diverse speciation that occurred in the new habitats created by the Andean uplift, that has given the suboscines such an image of success when compared with the oscines. As Dean Amadon carefully summarized the situation, "All in all, I see no reason to abandon the classical view that the sub-Oscines are an early and in general less well-adapted group of Passeriformes that has persisted in South America longer than elsewhere because it was sheltered from the main wave of Oscine evolution" (1973, 274).

Though far from perfect, our knowledge of land birds does allow us to construct a very generalized history of their rise and dispersion. A number of waves of evolution seem to have characterized the appearance of at least the more advanced arboreal land birds, each one fleetingly successful and leaving a handful of relicts in the modern avifauna. Much of the zoogeography of modern birds has been attributed to continental drift and therefore to vicariance biogeography rather than to dispersal. Vicariance refers to the seemingly mysterious disjunct distributions of related organisms that are caused by geographic barriers that split the range of ancestral species or geologic events that separated once continuous ranges, as in the splitting of the southern continent Gondwanaland.

Cladists in particular have been enamored with continental vicariance biogeography, and Joel Cracraft (e.g., 1973a, 1974b, 1982a, 1982b, 1983; Rich 1975; Vickers-Rich 1991) was the first to attempt to explain most avian distributions by drifting continents. In 1974, as we saw in chapter 5, he attempted to explain the distribution of all flightless ratites as wanderers on drifting continents. Indeed, Sibley and Ahlquist (1990 and prior publications) used Cracraft's biogeographic hypothesis for the origin of ratites in Gondwanaland in their DNA studies to calibrate their molecular clock; unfortunately, as we saw in chapter 5, there is no evidence to indicate that any ratites are older than the Paleogene. If the fossil evidence is valid, and there is every reason to believe that it is, then ratites could not have been distributed by drifting continents. Rather, ratites evolved from flying paleognathous birds, the lithornithids, that dispersed to the various continents. As we saw, a Paleogene ratite, thought by Cracraft to have been derived from a southern continent ancestor that also gave rise to the South American rheas, has been located in the Eocene of Europe. Yet many authors have accepted wholesale Cracraft's vicariance biogeography to explain avian distributions; Sibley and Ahlquist (1990, 699–706), for example, depend largely on continental movements to explain the zoogeography of birds.

Surely the lesson of the ratites must be applied to the rise of land birds, for this, too, is a group for which continental vicariance has been invoked to explain present disjunct distribution patterns. Yet the only land birds known from the Cretaceous have turned out to be the enantiornithines and a few other primitive, also extinct ornithurines. There is simply no evidence that any of the living orders or families of land birds arose before the Paleogene. Drifting land masses may well explain a limited number of avian distribution patterns, but vicariance biogeography need not be invoked to explain any of them. One need only look to the migratory capabilities of certain birds.

Rails, reclusive marsh dwellers and "weak" fliers, devote only a relatively small percentage of their body

Breeding distribution (shaded areas) of Baillon's crake (*Porzana pusilla*) and its two apparent derivatives. (From Olson 1973, after Voous 1960)

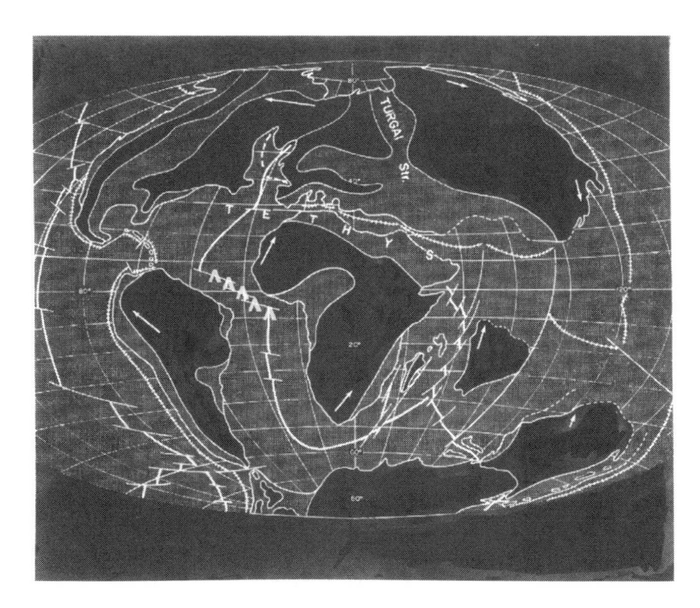

Reconstruction of the continents at the time of the late Cretaceous extinctions, approximately 65 million years ago. Note the volcanic island arcs indicated in the mid-Atlantic, halfway between mainland continents that are separated by a distance no more than that separating the Galápagos Islands from mainland Ecuador. (Adapted from Rich 1975)

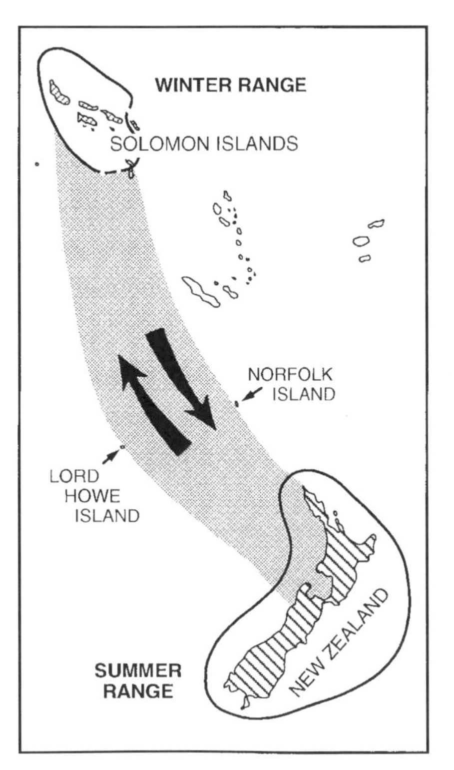

Migration path of one geographical race of the shining bronze cuckoo (*Chrysococcyx lucidus lucidus*), which traverses some 3,200 km (2,000 mi.) of ocean from its breeding grounds in New Zealand to its winter range in the Solomon Islands.

weight to the flight apparatus. Yet the rails are one of the most widespread of all avian families, perform remarkably long migrations over land and extensive stretches of ocean, and have developed new species, often flightless, in the South Pacific on almost all of the major and many of the minor oceanic islands. On the other side of the world, in the South Atlantic, the small Inaccessible Island rail (*Atlantisia rogersi*) occurs on an island in the Tristan de Cunha group, some 3,200 kilometers (2,000 mi.) east of South America and 2,900 kilometers (1,800 mi.) west of the Cape of Good Hope. The large, moorhenlike purple swamphen (*Porphyrio porphyrio*) is a clumsy-looking flier but ranges broadly from the Mediterranean basin and Africa across to Southeast Asia and on to Australia, New Zealand, and Micronesia. Although it is generally nonmigratory where it is found, it is capable of major journeys, apparently flying high at night. In New Zealand purple swamphens are known to cross the main mountain ranges because birds have been found dead on glaciers and snowfields (Ripley 1977, 297).

Then there is the lesson of the cattle egret (*Bubulcus ibis*), a native of southern Eurasia and northern Africa that has during the past half-century or so dramatically and explosively expanded its range. The species first appeared in the Americas in the coastal mangroves of British Guiana sometime between the First and Second World Wars, apparently having wandered or been blown across the South Atlantic. The cattle egret's range now includes South America, the West Indies, and North America up to Maine and Minnesota, and it has recently invaded Australia.

Among the land birds, the cuckoos undergo extensive migrations. The common cuckoo (*Cuculus canorus*), migrates from Europe to central Africa and from parts of Asia to the East Indies. The shining bronze-cuckoo (*Chrysococcyx lucidus*) of the Southern Pacific makes an almost incomprehensible migratory trip, traveling from New Zealand northward over some 3,200 kilometers (2,000 mi.) of ocean to the Solomon Islands. This trip, over the trackless South Pacific, is even more uncanny when one considers that the young birds make the same trip in the absence of their parents.

Even among the suboscines, considered to be remarkably sedentary, there are some remarkable revelations. For example, most of the plump-bodied and brightly colored pittas are year-round denizens of deep, moist tropical forest, yet the same species yearly migrate between nests in China and Japan and winter homes in the tropics, while the elusive African pitta (*Pitta angolensis*) may migrate some 1,280–1,920 kilometers (800–1,200 mi.) on an intra-Ethiopian migration (Moreau 1966, 236).

Among the New World suboscines, the tyrant flycatchers are remarkable migrants; several species nest as far north as Alaska, and many nest in North America and winter in Central and South America. One well-known species, the eastern kingbird (*Tyrannus tyrannus*), nests in North America as far north as central Canada and migrates thousands of miles to winter from Peru to Bolivia. The vermilion flycatcher (*Pyrocephalus rubinus*) likewise nests along the Gulf Coast of North America and winters in Argentina but has also managed to find its way to the Galápagos Islands, some 960 kilometers (600 mi.) off the coast of Ecuador. And we must keep in mind that positively the most sedentary group, the tapaculos, found their way to Cuba during the Tertiary, as we have seen. Likewise, the "sedentary" trogons and todies somehow arrived in the Greater Antilles, probably via northern dispersal routes.

Jared Diamond (1981) has emphasized that "fear of flying" is a barrier to bird distribution, but when one encompasses millions of years of dispersal time, even those birds that might, without our prior knowledge, be accused of a fear of flying, such as the rails, pittas, and tapaculos, may produce species capable of unexpected bursts of range expansion. Thus, fear of flying can be ascertained only ex post facto, and no group of birds, living or fossil, can be said to be incapable of wide-ranging dispersal based solely on anatomy. On the flip side, as Diamond points out, some species restricted to limited distributions are powerful fliers overland, and their failure to cross even narrow water gaps is through "choice," not inability. In addition, "for diverse animals and plants, the anatomical and behavioural bases of dispersal are subject to natural selection and may differ enormously between species within the same genus" (508). In the vast geologic past it takes just one avian species or family within an order to develop the migratory habit and disperse the group to parts unknown. After subsequent extinctions, the group may, like the West Indian endemic todies, give the impression of having evolved in situ within the confines of their current range. Surely the separation of the southern continent, say by some 1,600 kilometers (1,000 mi.), with partially submerged island arcs, could not have presented much of a barrier over millions of years for ancient passerine or other ancestors. The real biogeographic issue concerns not the continents' separation during the Cretaceous but how far apart various landmasses—South America and Africa, New Zealand and Australia, Madagascar and Africa—were during the early to medial Paleogene, when the explosive radiations of nonpasserines and passerines took place. I strongly suspect that the distribution of suboscines is related more to their early acquisition of restrictive tropical physiological adaptations than to any inability to expand their ranges. It was the oscines, with their advanced, more flexible neural parameters, physiological tolerance, and high reproductive potential features not measurable in cladistic analyses, that were capable of occupying both temperate and tropical zones.

Zoogeographic Realms. In 1858, Philip Lutley Sclater developed the concept that the world could be divided into major faunal "realms," and he used the current distribution of the major groups of birds to delimit the areas so defined. Thus, the Neotropical Realm has come to be characterized by such birds as the seriemas, oilbirds, barbets, toucans, trogons, puffbirds, jacamars, and suboscines, to mention a few. Yet when we look at the fossil record of birds we quickly discover that not only for South America but for all of the southern continents, almost all of the so-called characteristic taxa are in fact relicts of groups that were once widespread in the Paleogene or even later of the Northern Hemisphere (Olson 1989). Thus we find the pantropical parrots, trogons, and barbets in the Tertiary of Europe and North America and the Neotropical potoos in the Eo-Oligocene of France. We find secretarybirds, turacos, mousebirds, woodhoopoes, and broadbills in the Miocene of France and Germany, and African crowned-cranes (*Balearica*) in the Oligocene and Miocene of North America. We find the South American "endemic" oilbirds in the Eocene of North American and the Eo-Oligocene of France. We find motmots from the Oligocene of Switzerland and the late Miocene of Florida, and todies, now restricted to the Greater Antilles, in the Oligocene of Wyoming and the Eo-Oligocene of France. Even the most characteristic birds of the Australasian avifauna, the megapodes, frogmouths, and owlet-nightjars, are now known as fossils from the Eo-Oligocene of Quercy, France, and birds closely resembling the now endemic Malagasy ground-rollers (Brachypteraciidae) occur in the early Eocene of North America and the medial Eocene of Germany. And as if all of these examples were not enough, we now have a fully developed ratite, as a completely preserved and articulated fossil, from the Eocene of Germany! Sclater (1858) believed that the self-defining attributes of the zoogeographic provinces provided incontrovertible evidence that the organisms contained by each had their origin in the different parts of the world where they are now found. On the contrary, at least for birds, a quite different picture has emerged, with the southern continents, for example, containing an accumulation of relicts of the past, remnants of avian groups that are "restricted" to the southern continents, having failed to survive elsewhere in the face of either competition from advanced forms or a deteriorating climate during the late Tertiary.

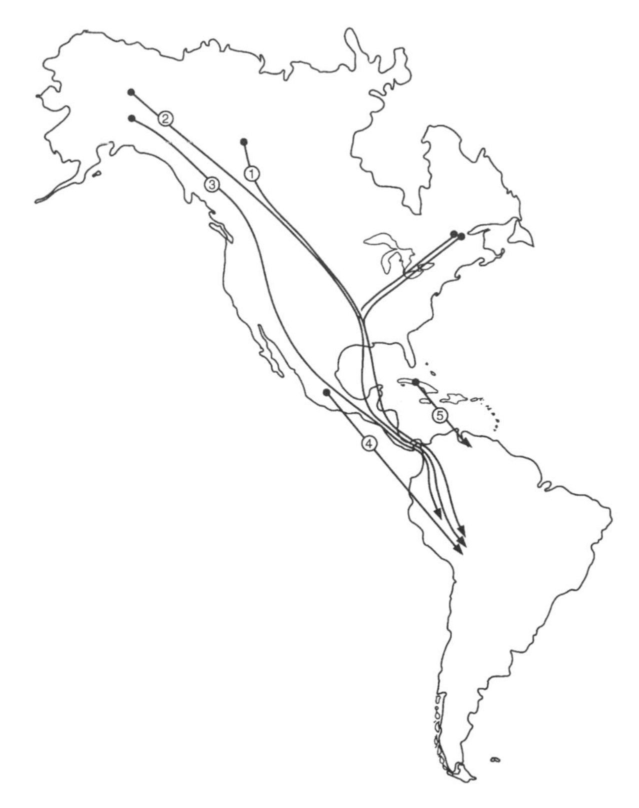

Map of the Western Hemisphere illustrating the post-mating migrations of some American suboscines of the flycatcher family Tyrannidae: (1) eastern kingbird (*Tyrannus tyrannus*), (2) willow flycatcher (*Empidonax traillii*); (3) western wood-pewee (*Contopus sordidulus*); (4) sulphur-bellied flycatcher (*Myiodynastes luteiventris*); (5) grey kingbird (*Tyrannus dominicensis*).

Current zoogeographic regions of the world. The double lines between the Oriental and Australian Regions enclose the transitional area of Wallacea. Given the mobility of birds, the validity of these regions through time, except for Australia, has been much less dramatic than with other groups. (From Serventy 1960; courtesy Academic Press)

Table 8.1. Avian Specialties of the Major Biogeographic Regions

Regions	Endemic Nonpasserine Families	Passerine Radiations
Nearctic and Palearctic (Holarctic): all of North America, Mexico, and the West Indies, plus all of Europe, northern Asia south to the Himalayas, and Africa north of the Sahara	Loons (Gaviidae) Auks (Alcidae)	Accentors (Prunellidae) [Palearctic] Buntings (Emberizidae) Cardueline finches (Fringillidae) Wood warblers (Parulidae) [Nearctic] Old World warblers (Sylviidae) [Palearctic]
Neotropical: all of South America plus Central America south of the Isthmus of Tehauntepec, Mexico	Rheas (Rheidae) Tinamous (Tinamidae) Curassows (Cracidae) Trumpeters (Psophiidae) Sunbittern (Eurypygidae) Seriemas (Cariamidae) Limpkin (Aramidae) Oilbird (Steatornithidae) Hoatzin (Opisthocomidae) Hummingbirds (Trochilidae) Motmots (Momotidae) Jacamars (Galbulidae) Puffbirds (Bucconidae) Toucans (Ramphastidae)	Tyrant flycatchers (Tyrannidae) Tanagers (Thraupidae) Antbirds (Thamnophilidae) Antthrushes (Formicariidae) Ovenbirds (Furnariidae) Woodcreepers (Dendrocolaptidae) Manakins (Pipridae) Cotingas (Cotingidae)
Ethiopian: Africa south of the Sahara	Ostrich (Struthionidae) Secretarybird (Sagitariidae) Guineafowl (Numididae) Mesites (Mesoenatidae) (M)* Turacos (Musophagidae) Mousebirds (Coliidae) Ground-rollers (Brachypteraciidae) (M) Cuckoo-roller (Leptosomidae) (M) Woodhoopoes (Phoeniculidae)	Larks (Alaudidae) Sunbirds (Nectariniidae) Weavers (Ploceidae)
Oriental: Southeast Asia from the Himalayas to northern Indonesia	None	Leafbirds (Irenidae) Pheasants (Phasianidae) Broadbills (Eurylaimidae) Pittas (Pittidae) Babblers (Timaliidae) Flowerpeckers (Dicaeidae)
Australasian: Australia and New Guinea from Lombok south plus the islands of the southwest Pacific	Emus (Dromiceidae) Cassowaries (Casuariidae) Kiwis (Apterygidae) Kagu (Rhynochetidae) Cockatoos (Cacatuinae) Lories (Loriinae) Owlet-nightjars (Aegothelidae)	Birds-of-paradise (Paradisaeidae) Whistlers (Pachycephalidae) Honeyeaters (Meliphagidae) Monarch flycatchers (Monarchidae) Australian warblers (Acanthizidae)

*(M) = Madagascar only
Source: After Gill 1995

Distributing the Passerines. As for passerines, like many groups of land birds, the end of the Oligocene and the beginning of the Miocene appears to have been a critical time in the history of their distribution and adaptive radiation. For until this time the southern continents were too far removed from the northern continents to make dispersal practical for all but the most exceptionally wide-ranging forms. This is the period when the first substantial interchange took place between northern and southern avifaunas. In the late Miocene, for example, the distance between North and South America was some 150 kilometers (94 mi.) or more (Webb 1978), yet a limited overwater exchange occurred, involving, among other fauna, the South American megalonychid and myoldontid sloths and a North American procyonid (Marshall et al. 1979). But it was not until the closing of the Panamanian seaway in the late Pliocene, some 2.5 to 3 million years ago, that the final isthmian link was solidified between North and South America, prompting the Great American Interchange. In the Old World no such dramatic isolation existed, and by the medial Tertiary the Old World oscines had outcompeted their suboscine counterparts wherever they came into contact, leaving today only a few suboscine relicts exhibiting a broad range of morphological types.

After the early Miocene, repeated invasions of oscines into the New World across the Bering land bridge produced the myriad families of oscines present in the Americas today, although the New World apparently did not reciprocate to any significant degree with its newly evolved oscine stock. Throughout the world, climatic deterioration in the late Tertiary, combined with competition from advanced oscine counterparts, forced the more primitive tropical arboreal land birds, such as the perching piciforms and coraciiforms, southward into the tropics. In the New World, the Great American Interchange led to the extensive radiation of many birds. The Central American alcediniform birds, such as the trogons and motmots, entered South America; the South American suboscines moved northward; and, perhaps most important, the oscines invaded the Neotropics, where they are now firmly established, as exemplified by the radiation of the resplendent tanagers.

A Final Thought

The entire history of birds is one of dramatic invasion of new ecological zones by diverse groups, with subsequent adaptive radiations, extinctions, and a temporal layering of new adaptive types. And each time disparate avian groups have entered similar ecological zones, parallel adaptations have resulted in massive convergence, often obscuring evolutionary affinities. The late Permian on through the Triassic was a period of arboreal experimentation in diverse reptile groups, protobirds moving through feather-assisted jumping and then parachuting, leading ultimately to gliding and actual flight. Birds descended from some form of small arboreal archosaur, and poorly known urvogels, with their three sharply clawed, unreduced hand digits, characterized the late Jurassic. By the early Cretaceous, the critical avian dichotomy had been breached, producing the sauriurine birds, the urvogels and opposite birds, on one hand, and, on the other, the more advanced fliers, the ornithurines. Sauriurine birds had fleshy tails, primitive pectoral regions, and were poor temperature regulators (probably quasi-ectothermic) and incapable of extended flight; but in the tropical setting of the Mesozoic, they, like the dinosaurs, prospered and adaptively radiated to become the dominant land birds of the Age of Reptiles. The early Cretaceous ornithurines, like their endothermic mammalian counterparts, played a relatively minor role during this period, but their advanced morphology tells of an endothermic bird capable of extended flight and migratory ability, vividly illustrated by finds of *Ichthyornis* hundreds of miles out to sea, and it is possible that during the Age of Reptiles the ornithurines were more or less confined to the shorelines and seas, for by late Cretaceous the group produced varied lineages of ancient shorebirds.

The cataclysm that terminated the Cretaceous period completely decimated the opposite birds, possibly because of their poor ability to regulate body temperature. Cretaceous extinctions also resulted in a narrow Cretaceous-Tertiary bottleneck that only a small number of ornithurine birds transcended—primarily "transitional shorebirds," along with perhaps a few other types, possibly including primitive paleognaths and some others, though we know little of these birds.

With the slate wiped clean and consequent ecological voids left by the Cretaceous extinctions, birds immediately began what can only be characterized as a dramatic and extraordinary explosive evolution emanating primarily, or exclusively, from a few types of ornithurine shorebirds, perhaps with initial neognathous shorebird and land bird lineages. This initial dichotomy quickly diverged into the modern orders within a period of some 10 million or so years to fill all available niches. A second wave of explosive adaptive radiation began in the late Oligocene but intensified during the Miocene, at least in the Northern Hemisphere, when passerines filled the trees with the seemingly endless variety of brilliant gems we know today as the songbirds.

Not only were birds devastated during the late Cretaceous cataclysm, but during the past two millennia Polynesians ate their way through the South Pacific, extirpating entire avian assemblages and their forest habitat, and modern Europeans followed suit on the continental mainlands. Birds, "the most splendid expression of creation," as Roger Tory Peterson has put it, have, from the Cretaceous upheaval to the present, been the miner's canary—the most sensitive indicators of environmental disturbance. With the rapid progression in the devastation of the world's rain forests and other critical environments, the welfare of birds stands singularly as a foreboding presage to the environmental future of our world.

REFERENCES

Abler, W. L. 1992. The serrated teeth of tyrannosaurid dinosaurs, and biting structures in other animals. *Paleobiology* 18:161–183.

Adams, D. F., S. J. Hessel, P. F. Judy, J. A. Stein, and H. L. Abrams. 1976. Potassium-argon ages from the Galápagos Islands. *Science* 192:465–468.

Alexander, R. McN. 1989. *Dynamics of dinosaurs and other extinct giants.* New York: Columbia University Press.

Allen, J. A. 1878. Description of a fossil passerine bird from the insect-bearing shales of Colorado. *Bulletin of the U.S. Geological and Geographical Survey* 4:443–445.

Allen, R. P. 1956. The flamingos: their life history and survival. *National Audubon Society Research Report* 5:1–285.

Allen, T. T. 1962. Myology of the limpkin. Ph.D. diss., University of Florida, Gainesville.

Alvarenga, H. M. F. 1982. Uma gigantesca ave fóssil do cenozóico brasileiro: *Physornis brasiliensis* sp. n. *Anais da Academia Brasileira Ciências* 54:697–712.

———. 1983. Uma ave ratita do paleoceno brasileiro: bacia calcária de Itaboraí, estado do Rio de Janeiro, Brasil. *Boletim do Museu Nacional (Rio de Janeiro), Geologia,* n.s., 41:1–11.

———. 1985a. Um novo Psilopteridae (Aves: Gruiformes) dos sedimentos terciários de Itaboraí, Rio de Janeiro, Brasil. *NME-DNPM Série Geologia 27 Paleontologia Estratigraphiá* 2:17–20.

———. 1985b. Notas sobre os Cathartidae (Aves) e descrição de um novo gênero do cenozóico brasileiro. *Anais Academia Brasileira de Ciências* 57:349–357.

———. 1995. A large and probably flightless anhinga from the Miocene of Chile. *Courier Forschungsinstitut Senckenberg,* 181:149–161.

Alvarenga, H. M. F., and J. F. Bonaparte. 1992. A new flightless landbird from the Cretaceous of Patagonia. *Los Angeles County Museum of Natural History, Science Series* 36:51–64.

Alvarez, R., and S. L. Olson. 1978. A new merganser from the Miocene of Virginia (Aves: Anatidae). *Proceedings of the Biological Society of Washington* 91:522–532.

Amadon, D. 1947. An estimated weight of the largest known bird. *Condor* 49:159–164.

———. 1973. Birds of the Congo and Amazon forests: a comparison. In *Tropical forest ecosystems in Africa and South America: a comparative review,* ed. E. S. Ayensu and W. D. Duckworth, 267–277. Washington, D.C.: Smithsonian Institution Press.

Ameghino, F. 1895. Sur les oiseaux fossiles de Patagonie. *Boletin Instituto Geográfico Argentina* 15:501–602.

American Ornithologists' Union. 1983. *Check-list of North American Birds.* 6th ed. [Washington, D.C.]: American Ornithologists' Union.

Ames, P. L. 1971. The morphology of the syrinx in passerine birds. *Bulletin of the Peabody Museum of Natural History, Yale University* 37:1–194.

Ames, P. L., M. A. Heimerdinger, and S. L. Warter. 1968. The anatomy and systematic position of the antpipits *Conopophaga* and *Corythopis*. *Peabody Museum of Natural History, Yale University, Postilla* 114:1–32.

Anderson, A. 1989. *Prodigious birds: moas and moa-hunting in New Zealand.* Cambridge: Cambridge University Press.

———. 1991. Early bird threatens *Archaeopteryx*'s perch. *Science* 253:35.

Andors, V. A. 1988. Giant groundbirds of North America (Aves, Diatrymidae). Ph.D. diss., Columbia University.

———. 1991. Paleobiology and relationships of the giant groundbird *Diatryma* (Aves: Gastornithiformes). *Proceedings of the Twentieth International Ornithological Congress,* 563–571.

———. 1992. Reappraisal of the Eocene groundbird *Diatryma* (Aves: Anserimorphae). *Los Angeles County Museum of Natural History, Science Series* 36:109–126.

Andrews, C. W. 1896. Note on a nearly complete skeleton of *Aptornis defossor* Owen. *Geological Magazine* 96:241–242.

———. 1899. On the extinct birds of Patagonia.—I. The skull and skeleton of *Phororhacos inflatus* Ameghino. *Transactions of the Zoological Society of London* 15(3):55-86.

Archey, G. 1941. The moa: a study of the Dinornithiformes. *Bulletin of the Auckland Institute and Museum* 1:1–145.

Arrendondo, O. 1976. The great predatory birds of the Pleistocene of Cuba. *Smithsonian Contributions to Paleobiology* 27:169–187.

———. 1982. Los Strigiformes fósiles del pleistoceno cubano. *Sociedad Venezolana de Ciencias Naturales Boletin* 37:33–55.

Arrendondo, O., and S. L. Olson. 1994. A new species of owl of the genus *Bubo* from the Pleistocene of Cuba (Aves: Strigiformes). *Proceedings of the Biological Society of Washington* 107:436–444.

Ashmole, N. P. 1971. Sea bird ecology and the marine environment. In *Avian biology*, vol. 1, ed. D. S. Farner and J. R. King, 233–286. New York: Academic Press.

Ashmole, N. P., and M. J. Ashmole. 1967. Comparative feeding ecology of sea birds of a tropical oceanic island. *Bulletin of the Peabody Museum of Natural History, Yale University* 24:1–131.

Atkinson, I. A. E., and P. R. Millener. 1991. An ornithological glimpse into New Zealand's pre-human past. *Proceedings of the Twentieth International Ornithological Congress*, 129–192.

Auffenberg, W. 1981. *The behavioral ecology of the Komodo monitor*. Gainesville: University of Florida Press.

Baer, J. G. 1957. Répartition et endémicité des cestodes chez les reptiles, oiseaux et mammifères. *International Union of Biological Sciences*, ser. B, 32:270–292.

Bailey, K. 1976. Potassium-argon ages from the Galápagos Islands. *Science* 192:465–466.

Bakker, R. T. 1975. Dinosaur renaissance. *Scientific American* 232(4):58–78.

———. 1986. *The dinosaur heresies*. New York: William Morrow.

Bakker, R. T., and P. M. Galton. 1974. Dinosaur monophyly and a new class of vertebrates. *Nature* 248:168–172.

Bakker, R. T., M. Williams, and P. J. Currie. 1988. *Nanotyrannus*, a new genus of pygmy tyrannosaur, from the latest Cretaceous of Montana. *Hunteria* 1(5):1–30.

Balda, R. P., G. Caple, and W. R. Willis. 1985. Comparison of the gliding to flapping sequence with the flapping to gliding sequence. In *The beginnings of birds*, ed. M. K. Hecht et al., 267–277. Eichstätt: Freunde des Jura-Museum.

Ballmann, P. 1969a. Les oiseaux miocènes de La Grive-Saint-Alban (Isère). *Géobios* 2:157–204.

———. 1969b. Die Vögel aus der altburdigalen Spaltenfüllung von Wintershof (West) bei Eichstätt in Bayern. *Zitteliana* 1:5–60.

———. 1970. Ein neuer Vertreter der Musophagidae (Aves) aus dem Chattium von Gaimersheim bei Ingolstadt (Bayern). *Mitteilungen Bayerischen Staatssammlung für Paläontologie und historische Geologie* 10:271–276.

———. 1972. Les oiseaux miocènes de Vieux-Collonges. *Documents des Laboratoires de Géologie, Lyon* 50:93–101.

———. 1973. Fossile Vögel aus dem Neogen der Halbinsel Gargano (Italien). *Scripta Geologica* 17:1–75.

———. 1976. Fossile Vögel aus dem Neogen der Halbinsel Gargano (Italien) zweiter Teil. *Scripta Geologica* 38:1–59.

———. 1983. A new species of fossil barbet (Aves: Piciformes) from the late Miocene of the Nördlinger Ries (southern Germany). *Journal of Vertebrate Paleontology* 3:43–48.

Balouet, J.-C., and S. L. Olson. 1987. A new extinct species of giant pigeon (Columbide: *Ducula*) from archaeological deposits on Wallis (Uvea) Island, South Pacific. *Proceedings of the Biological Society of Washington* 100:769–775.

Bang, B. G. 1960. Anatomical evidence for olfactory function in some species of birds. *Nature* 188:547–549.

———. 1967. The nasal organs of the black and turkey vultures: a comparative study of the cathartid species. *Journal of Morphology* 115:153–184.

———. 1971. Functional anatomy of the olfactory system in twenty-three orders of birds. *Acta Anatomica* 79 (suppl. 58):1–76.

Barreto, C., R. M. Albrecht, D. E. Bjorling, J. R. Horner, and N. J. Wilsman. 1993. Evidence of the growth plate and the growth of long bones in juvenile dinosaurs. *Science* 262:2020–2023.

Barrick, R. E., and W. J. Showers. 1994. Thermophysiology of *Tyrannosaurus rex:* evidence from oxygen isotopes. *Science* 265:222–224.

Barrowclough, G. 1992. Biochemical studies of the higher-level systematics of birds. *Bulletin of the British Ornithological Club* 112A:39–52.

Barrowclough, G. F., and K. W. Corbin. 1978. Genetic variation and differentiation in the Parulidae. *Auk* 95:691–702.

Barsbold, R. 1983. Carnivorous dinosaurs from the Cretaceous of Mongolia. *Transactions of the Joint Soviet-Mongolian Palaeontological Expedition*, 19. Moscow, 120 pp. [In Russian.]

Barthel, K. W., N. H. M. Swinburne, and S. C. Morris. 1990. *Solnhofen: a study in Mesozoic paleontology*. Cambridge: Cambridge University Press.

Baur, J. G. 1883. Der Tarsus der Vögel und Dinosaurier. *Morphologisches Jahrbuch* 8:417–456.

———. 1884. Dinosaurier und Vögel. *Morphologisches Jahrbuch* 10:446–454.

Beardsley, T. 1986. Fossil bird shakes evolutionary hypothesis. *Nature* 322:677.

Beauchamp, G. 1992. Diving behavior in surf scoters and Barrow's goldeneye. *Auk* 109:819–827.

Becker, J. J. 1986a. Reidentification of *"Phalacrocorax" subvolans* Brodkorb as the earliest record of Anhingidae. *Auk* 103:804–808.

———. 1986b. A new vulture (Vulturidae: *Pliogyps*) from the late Miocene of Florida. *Proceedings of the Biological Society of Washington* 99:502–508.

———. 1986c. A fossil motmot (Aves: Momotidae) from the late Miocene of Florida. *Condor* 88:478–482.

———. 1987a. Additional material of *Anhinga grandis*

Martin and Mengel (Aves: Anhingidae) from the Miocene of Florida. *Proceedings of the Biological Society of Washington* 100:358–363.
———. 1987b. Revision of *"Falco" ramenta* Wetmore and the Neogene evolution of the Falconidae. *Auk* 104:270–276.
Bédard, J. 1969. Adaptive radiation in the Alcidae. *Ibis* 111:189–198.
Beddard, F. E. 1898. *The structure and classification of birds*. London: Longmans, Green.
Beebe, C. W. 1915. A tetrapteryx stage in the ancestry of birds. *Zoologica* 2:39–52.
Beecher, W. J. 1953. A phylogeny of oscines. *Auk* 70:270–333.
Bellairs, A. D'A., and C. R. Jenkin. 1960. The skeleton of birds. In *Biology and comparative physiology of birds*, vol. 1, ed. A. J. Marshall, 241–300. New York: Academic Press.
Bennett, A. F. 1991. The evolution of activity capacity. *Journal of Experimental Biology* 160:1–23.
Bennett, A. F., and B. Dalzell. 1973. Dinosaur physiology: a critique. *Evolution* 127:170–174.
Bennett, A. F., and K. A. Nagy. 1977. Energy expenditure in free-ranging lizards. *Ecology* 58:697–700.
Bennett, A. F., and J. A. Ruben. 1986. The metabolic and thermoregulatory status of therapsids. In *The ecology and biology of mammal-like reptiles*, ed. N. Hotton, P. MacLean, J. J. Roth, and E. C. Roth, 207–218. Washington, D.C.: Smithsonian Institution Press.
Bennett, S. C. 1992. Sexual dimorphism of *Pteranodon* and other pterosaurs with comments on cranial crests. *Journal of Vertebrate Paleontology* 12:422–434.
———. 1993a. Year-classes of pterosaurs from the Solnhofen Limestone of southern Germany. *Journal of Vertebrate Paleontology* 13:26A (abstract).
———. 1993b. The ontogeny of *Pteranodon* and other pterosaurs. *Paleobiology* 19:92–106.
Benton, M. J. 1979. Ectothermy and the success of dinosaurs. *Evolution* 33:983–997.
———. 1985. Classification and phylogeny of the diapsid reptiles. *Zoological Journal of the Linnean Society* 84:97–164.
Benton, M. J., and J. M. Clark. 1988. Archosaur phylogeny and the relationships of the Crocodylia. In *The phylogeny and classification of the tetrapods*, vol. 1, ed. M. J. Benton, 295–338. Oxford: Clarendon Press.
Berger, A. E., and M. Simpson. 1986. Diving depths of Atlantic puffins and common murres. *Auk* 103:828–830.
Berger, A. J. 1960. The musculature. In *Biology and comparative physiology of birds*, vol. 1, ed. A. J. Marshall, 301–344. New York: Academic Press.
Berlioz, J. 1950. Systématique. In *Traite de Zoologie*, vol. 15: *Oiseaux*, ed. P. P. Grasse, 845–1055. Paris: Masson et Cie.
Berman, S. L., and R. J. Raikow. 1982. The hindlimb musculature of the mousebirds (Coliiformes). *Auk* 99:41–57.
Bickart, K. J. 1982. A new thick-knee, *Burhinus*, from the Miocene of Nebraska, with comments on the habitat requirements of the Burhinidae (Aves: Charadriiformes). *Journal of Vertebrate Paleontology* 1:273–277.
Blackburn, D. G., and H. E. Evans. 1986. Why are there no viviparous birds? *American Naturalist* 128:165–190.
Bledsoe, A. H. 1988a. Nuclear DNA evolution and phylogeny of the New World nine-primaried oscines. *Auk* 105:504–515.
———. 1988b. A phylogenetic analysis of postcranial skeletal characters of the ratite birds. *Annals of the Carnegie Museum* 57:73–90.
Bledsoe, A. H., and R. J. Raikow. 1990. A quantitative assessment of congruence between molecular and nonmolecular estimates of phylogeny. *Journal of Molecular Evolution* 30:247–259.
Block, B. A., J. R. Finnerty, A. F. R. Stewart, and J. Kidd. 1993. Evolution of endothermy in fish: mapping physiological traits on a molecular phylogeny. *Science* 260:210–214.
Boas, J. E. V. 1930. Über das Verhältnis der Dinosaurier zu den Vögeln. *Morphologisches Jahrbuch* 64:223–247.
Bock, W. J. 1956. A generic review of the Ardeidae (Aves). *American Museum Novitates* 1779:1–49.
———. 1962. The pneumatic fossa of the humerus in the Passeres. *Auk* 79:425–443.
———. 1963a. Evolution and phylogeny in morphologically uniform groups. *American Naturalist* 97:265–285.
———. 1963b. The cranial evidence for ratite affinities. *Proceedings of the Eighth International Ornithological Congress*, 39–54.
———. 1964. Kinetics of the avian skull. *Journal of Morphology* 114:1–42.
———. 1965. The role of adaptive mechanisms in the origin of the higher levels of organization. *Systematic Zoology* 14:272–287.
———. 1967. The use of adaptive characters in avian classification. *Proceedings of the Fourteenth International Ornithological Congress*, 61–74.
———. 1969. The origin and radiation of birds. *Annals of the New York Academy of Sciences* 167:147–155.
———. 1973. Philosophical foundations of classical evolutionary classification. *Systematic Zoology* 22:375–392.
———. 1985. The arboreal theory for the origin of birds. In *The beginnings of birds*, ed. M. K. Hecht et al., 199–207. Eichstätt: Freunde des Jura-Museum.
———. 1986. The arboreal origin of avian flight. *Memoires of the California Academy of Sciences* 8:57–72.
Bock, W. J., and P. Bühler. 1988. The evolution and biogeographical history of the paleognathous birds. *Current Topics in Avian Biology*. Proceedings of the One-hundredth Meeting, Deutsche Ornithologen-Gesellschaft, Bonn.
Bock, W, J., and M. H. Clench. 1985. Morphology of the noisy scrub-bird, *Atrichornis clamosus* (Passeriformes: Atrichornithidae): systematic relationships and summary. *Records of the Australian Museum* 37:243–254.
Bock, W. J., and J. Farrand, Jr. 1980. The number of species and genera of Recent birds: a contribution to

comparative systematics. *American Museum Novitates* 2703:1–29.
Bock, W. J., and W. De W. Miller. 1959. The scansorial foot of the woodpeckers, with comments on the evolution of perching and climbing feet in birds. *American Museum Novitates* 1931:10–45.
Boles, W. E. 1991. The origin and radiation of Australasian birds: perspectives from the fossil record. *Proceedings of the Twentieth International Ornithological Congress,* 383–391.
———. 1993. A logrunner *Orthonyx* (Passeriformes: Orthonychidae) from the Miocene of Riversleigh, northwestern Queensland. *Emu* 93:44–49.
———. 1995a. A preliminary analysis of the Passeriformes from Riversleigh, northwestern Queensland, Australia, with the description of a new species of lyrebird. *Courier Forschunginstitut Senckenberg* 181:163–170.
———. 1995b. The world's oldest songbird. *Nature* 374:21–22.
Bonaparte, J. F. 1970. *Pterodaustro guinazui* gen et sp. nov. Pterosaurio de la formación Lagarcito, provincia de San Luis, Argentina. *Acta Geologica Lilloana* 10:207–226.
———. 1975. Nuevos materiales de *Lagosuchus talampayensis* Romer (Thecodontia-Pseudosuchia) y su significado en el origen de los Saurischia. *Acta Geológica Lilloana* 13:5–90.
———. 1978. El Mesozoico de America del Sur y sus tetrapodos. *Opera Lilloana* 26:1–596.
———. 1984. Locomotion in rauisuchid thecodonts. *Journal of Vertebrate Paleontology* 3:210–218.
Bonaparte, J. F., and J. E. Powell. 1980. A continental assemblage of El Brete, northeastern Argentina. *Mémoires de la Société Géologique de France,* n.s., 139:19–28.
Bonaparte, J. F., J. A. Salfitty, G. Bossi, and J. E. Powell. 1977. Hallazgos de dinosaurios y aves cretácicas en la Formación Lecho de El Brete (Salta), proximo al limite con Tucumán. *Acta Geológica Lilloana* 14:5–17.
Bond, J. 1966. Affinities of the Antillean avifauna. *Caribbean Journal of Science* 6:173–176.
Bouvier, M. 1977. Dinosaur Haversian bone and endothermy. *Evolution* 31:449–450.
Brasil, L. 1914. *Grues.* Pt. 26 of *Genera Avium,* ed. W. Wytsman. Brussels: V. Verteneuil.
Brett-Surman, M. K., and G. S. Paul. 1985. A new family of bird-like dinosaurs linking Laurasia and Gondwanaland. *Journal of Vertebrate Paleontology* 5:133–138.
Briggs, D. E. G. 1991. Extraordinary fossils. *American Scientist* 79:130–141.
Brodkorb, P. 1963a. Catalogue of fossil birds, Part 1 (Archaeopterygiformes through Ardeiformes). *Bulletin of the Florida Academy of Sciences* 7(4):179–293.
———. 1963b. Birds from the Upper Cretaceous of Wyoming. *Proceedings of the Thirteenth International Ornithological Congress,* 55–70.
———. 1963c. A giant flightless bird from the Pleistocene of Florida. *Auk* 80:111–115.
———. 1964. Catalogue of fossil birds. Part 2 (Anseriformes through Galliformes). *Bulletin of the Florida State Museum, Biological Sciences* 8:195–335.
———. 1967. Catalogue of fossil birds. Part 3 (Ralliformes, Ichthyornithiformes, Charadriiformes). *Bulletin of the Florida State Museum, Biological Sciences* 11:99–220.
———. 1971. Origin and evolution of birds. In *Avian biology,* ed. D. S. Farner and J. R. King, 19–55. New York: Academic Press.
———. 1976. Discovery of a Cretaceous bird apparently ancestral to the orders Coraciiformes and Piciformes (Aves: Carinatae). *Smithsonian Contributions to Paleobiology* 27:67–73.
Brooks, A. 1945. The underwater actions of diving ducks. *Auk* 62:517–523.
Broom, R. 1906. On the early development of the appendicular skeleton of the ostrich, with remarks on the origin of birds. *Transactions of the South African Philosophical Society* 16:355–368.
———. 1913. On the South African pseudosuchian *Euparkeria* and allied genera. *Proceedings of the Zoological Society of London* 1913:619–633.
Brouwers, E. M., W. A. Clemens, R. A. Spicer, T. A. Ager, L. D. Carter, and W. V. Sliter. 1987. Dinosaurs on the North Slope, Alaska: high latitude, latest Cretaceous environments. *Science* 237:1608–1610.
Browne, M. W. 1993. Two clues back idea that birds arose from dinosaurs. *New York Times,* 28 December.
———. 1994. Study may shake birds from the dinosaur tree. *New York Times,* 17 March.
Brunet, J. 1961. Oiseaux. In *Le gisement de vértebrés miocènes de Beni Mellal (Maroc).* Etude systématique de la faune de mammifères et conclusions générales [by R. Lavocat]. *Notes et Mémoires, Service Géologique* (Morocco) 155:105–108.
———. 1971. Oiseaux miocènes de Beni-Mallal (Maroc): un complément à leur étude. *Notes et Mémoires, Service Géologique* (Morocco) 31:109–111.
Brush, A. H. 1993. The origin of feathers: a novel approach. In *Avian biology,* vol. 9, ed. D. S. Farner, J. R. King, and K. C. Parkes, 121–162. London: Academic Press.
Bryant, H. N., and A. P. Russell. 1993. The occurrence of clavicles within Dinosauria: implications for the homology of the avian furcula and the utility of negative evidence. *Journal of Vertebrate Paleontology* 12:171–184.
Bryant, L. J. 1983. *Hesperornis* in Alaska. *PaleoBios* 40:1–7.
Buffetaut, E., J. Le Loeuff, P. Mechin, and A. Mechin-Salessy. 1995. A large French Cretaceous bird. *Nature* 377:110.
Bühler, P. 1980. Zur Oberschnabelbeweglichkeit der Schnepfenvogel. *Ökologie der Vogel* 2(1):128–129.
———. 1985. On the morphology of the skull of *Archaeopteryx.* In *The beginnings of birds,* ed. M. K. Hecht et al., 135–140. Eichstätt: Freunde des Jura-Museum.
———. 1987. On the mobility of the upper jaw and the

segments of the braincase in the Mesozoic birds. *Documents des Laboratoires de Géologie, Lyon* 99:41–48.
———. 1992. Light bones in birds. *Los Angeles County Museum of Natural History, Science Series* 36:385–394.
Bühler, P., L. D. Martin, and L. M. Witmer. 1988. Cranial kinesis in the late Cretaceous birds *Hesperornis* and *Parahesperornis*. *Auk* 105:111–122.
Buisonjé, P. H. de. 1985. Climatological conditions during deposition of the Solnhofen limestones. In *The beginnings of birds*, ed. M. K. Hecht et al., 45–65. Eichstätt: Freunde des Jura-Museum.
Burchak-Abramovich, N. I. 1951. [*Urmiornis* (*Urmiornis maraghanus* Mecq.) ostrich-like bird of the *Hipparion* fauna of Transcaucasia and southern Ukraine.] *Izvestiya Akademii Nauk Azerbaidzhanskoi S.S.R.* 6:83–94. [In Russian.]
Burger, R., K. Ducate, K. Robinson, and H. Walter. 1975. Radiocarbon date for the largest extinct bird. *Nature* 258:709.
Burghardt, G. M., H. W. Greene, and A. S. Rand. 1977. Social behavior in hatchling green iguanas: life at a reptile rookery. *Science* 195:689–690.
Burt, W. H. 1929. Pterylography of certain North American woodpeckers. *University of California Publications in Zoology* 30(15):427–442.
Burton, J. A., ed. 1973. Owls of the world, their evolution, structure and ecology. Milan: A and W Visual Library.
Burton, P. J. K. 1974a. *Feeding and feeding apparatus in waders: a study of anatomy and adaptations in the Charadrii*. London: British Museum.
———. 1974b. Jaw and tongue features in Psittaciformes and other orders with special reference to the anatomy of the tooth-billed pigeon (*Didunculus strigirostris*). *Journal of Zoology* 174:255–276.
———. 1978. The intertarsal joint of the harrier-hawks *Polyboroides* ssp. and the crane hawk *Geranospiza caerulescens*. *Ibis* 120:171–177.
———. 1985. Anatomy and evolution of the feeding apparatus in the avian orders Coraciiformes and Piciformes. *Bulletin of the British Museum (Natural History), Zoological Series* 47:331–443.
Burton, R. 1990. *Bird flight*. New York: Facts on File.
Cabanis, J., and F. Heine. 1859. *Museum Heineanum*. Vol. 2. Halberstadt: R. Frantz.
Cade, T. J., and L. Greenwald. 1966. Nasal salt secretions in falconiform birds. *Condor* 68:338–350.
Calder, W. A., III. 1978. The kiwi. *Scientific American* 239(1):132–142.
———. 1979. The kiwi and egg design: evolution as a package deal. *Bioscience* 29:461–467.
———. 1983. Body size, mortality, and longevity. *Journal of Theoretical Biology* 102:135–144.
———. 1984. *Size, function, and life history*. Cambridge, Mass.: Harvard University Press.
Calder, W. A., and T. J. Dawson. 1978. Resting metabolic rates of ratite birds: the kiwis and the emu. *Comparative Biochemistry and Physiology*, ser. A, 60:479–481.
Calder, W. A., and J. R. King. 1974. Thermal and caloric relations of birds. In *Avian biology*, vol. 4, ed. D. S. Farner, J. R. King, and K. C. Parkes, 259–413. New York: Academic Press.
Calzavara, M., G. Muscio, and R. Wild. 1981. *Megalancosaurus preonensis* n.g., n. sp., a new reptile from the Norian of Fruili, Italy. *Gortania* 2:49–64.
Campbell, K. E. 1979. The non-passerine Pleistocene avifauna of the Talara Tar seeps, northwestern Peru. *Life Sciences Contributions of the Royal Ontario Museum* 118:1–203.
———. 1980. A review of the Rancholabrean avifauna of the Itchtucknee River, Florida. *Museum of Natural History of Los Angeles County, Contributions to Science* 330:119–129.
Campbell, K. E., and L. Marcus. 1992. The relationship of hindlimb bone dimensions to body weight in birds. *Los Angeles County Museum of Natural History, Science Series* 36:395–412.
Campbell, K. E., Jr., and E. P. Tonni. 1980. A new genus of teratorn from the Huayquerian of Argentina (Aves: Teratornithidae). *Los Angeles County Museum of Natural History, Contributions in Science* 330:59–68.
———. 1982. Preliminary observations on the paleobiology and evolution of teratorns (Aves: Teratornithidae). *Journal of Vertebrate Paleontology* 1:265–272.
———. 1983. Size and locomotion in teratorns (Aves: Teratornithidae). *Auk* 100:390–403.
Caple, G., R. P. Balda, and W. R. Willis. 1983. The physics of leaping animals and the evolution of preflight. *American Naturalist* 121:455–476.
———. 1984. Flap about flight. *Animal Kingdom* 87:33–38.
Carlquist, S. 1965. *Island Life*. New York: Garden City Press for the American Museum of Natural History.
Carroll, R. L. 1988. *Vertebrate paleontology*. New York, W. H. Freeman.
Carroll, R. L., and Dong Z.-M. 1991. *Hupehsuchus*, an enigmatic aquatic reptile from the Triassic of China, and the problem of establishing relationships. *Philosophical Transactions of the Royal Society of London*, ser. B, 331:131–153.
Case, J. A., M. O. Woodburne, and D. S. Chaney. 1987. A gigantic phororhacoid (?) bird from Antarctica. *Journal of Paleontology* 61:1280–1284.
Chandler, A. C. 1916. A study of the structure of feathers with reference to their taxonomic significance. *University of California Publications in Zoology* 13:243–446.
Chandler, R. M. 1994. The wing of *Titanis walleri* (Aves: Phorusrhacidae) from the late Blancan of Florida. *Bulletin Florida State Museum of Natural History, Biology Series* 36(6):175 180.
Chapman, F. M. 1923. The distribution of motmots of the genus *Momota*. *Bulletin American Museum of Natural History* 48:27–59.
Charig, A. J., F. Greenaway, A. C. Milner, C. A. Walker, and P. J. Whybrow. *Archaeopteryx* is not a forgery. *Science* 232:622–626.
Chatterjee, A. 1989. The oldest Antarctic bird. *Journal of Vertebrate Paleontology* 9(3):16A.

Chatterjee, S. 1982. Phylogeny and classification of the thecodontian reptiles. *Nature* 295:317–320.
———. 1985. *Postosuchus*, a new thecodontian reptile from the Triassic of Texas and the origin of tyrannosaurs. *Philosophical Transactions of the Royal Society of London*, ser. B, 309:395–460.
———. 1991. Cranial anatomy and relationships of a new Triassic bird from Texas. *Philosophical Transactions of the Royal Society of London*, ser. B, 332:277–342.
———. 1994. *Protoavis* from the Triassic of Texas: the oldest bird. *Journal für Ornithologie* 135:330.
———. 1995. The Triassic bird *Protoavis*. *Archaeopteryx* 13:15–31.
Cheneval, J. 1983a. Révision du genre *Palaelodus* Milne-Edwards, 1863 (Aves, Phoenicopteriformes) du gisement aquitanien de Saint-Gérand-le-Puy (Allier, France). *Géobios* 16:179–191.
———. 1983b. Les Anatidae (Aves, Anseriformes) du gisement aquitanien de Saint-Gérand-le-Puy (Allier, France). *Symposium International G. Cuvier* (Montbéliard, 1982), 85–98.
———. 1984. Les oiseaux aquatiques (Gaviiformes à Ansériformes) du gisement aquitanien de Saint-Gérand-le-Puy (Allier, France): révision systématique. *Palaeovertebrata* 14:33–115.
———. 1987. Les Anatidae (Aves, Anseriformes) du miocène de France: révision systématique et évolution. *Documents des Laboratoires de Géologie, Lyon* 99:137–157.
———. 1989. Fossil bird study, and paleoecological and paleoenvironmental consequences: example from the Saint-Gérand-le-Puy deposits (Lower Miocene, Allier, France). *Palaeogeography, Palaeoclimatology, Palaeoecology* 73:295–309.
Cheneval, J., and F. Escuillié. 1992. New data concerning *Palaelodus ambiguus* (Aves: Phoenicopteriformes: Palaelodidae): ecological and evolutionary interpretations. *Los Angeles County Museum of Natural History, Science Series* 36:209–224.
Chiappe, L. M. 1990. A flightless bird from the late Cretaceous of Patagonia (Argentina). *Archosaurian Articulations* 1(10):73–77.
———. 1991a. Cretaceous avian remains from Patagonia shed new light on the early radiation of birds. *Alcheringa* 15:333–338.
———. 1991b. Cretaceous birds of Latin America. *Cretaceous Research* 12:55–63.
———. 1992. Enantiornithine (Aves) tarsometatarsi and the avian affinities of the late Cretaceous Avisauridae. *Journal of Vertebrate Paleontology* 12:344–350.
———. 1993. Enantiornithine (Aves) tarsometatarsi from the Cretaceous Lecho Formation of northwestern Argentina. *American Museum Novitates* 3083:1–27.
———. 1995. The phylogenetic relationships of the Cretaceous birds of Argentina. *Courier Forschungsinstitut Senckenberg*, 181:55–63.
Chiappe, L. M., and J. O. Calvo. 1994. *Neuquenornis volans*, a new late Cretaceous bird (Enantiornithes: Avisauridae) from Patagonia, Argentina. *Journal of Vertebrate Paleontology* 14:230-246.
Chinsamy, A. 1995. Ontogenetic changes in the bone histology of the late Jurassic ornithopod *Dryosaurus lettowvorbeki*. *Journal of Vertebrate Paleontology* 15:96–104.
Chinsamy, A., L. M. Chiappe, and P. Dodson. 1994. Growth rings in Mesozoic birds. *Nature* 368:196–197.
Chinsamy, A., and P. Dodson. 1995. Inside a dinosaur bone. *American Scientist* 83:174–180.
Chu, P. C. 1995. Phylogenetic reanalysis of Strauch's osteological data set for the Charadriiformes. *Condor* 97:174–196.
Clark, B. D. 1977. Energetics of hovering flight and the origin of bats. In *Major patterns in vertebrate evolution*, ed. M. K. Hecht et al., 423–425. New York: Plenum Press.
Clark, J. M. 1992. Review of *Origins of the higher groups of vertebrates: controversy and consensus*, ed. H.-P. Schultze and L. Trueb. *Journal of Vertebrate Paleontology* 12:532–536
Clemens, W. A., and L. G. Nelms. 1993. Paleoecological implications of Alaskan terrestrial vertebrate fauna in latest Cretaceous time at high paleolatitudes. *Geology* 21:503–506.
Clench, M. H. 1970. Variability in body pterylosis, with special reference to the genus *Passer*. *Auk* 87:650–691.
Colbert, E. H. 1989. The Triassic dinosaur *Coelophysis*. *Bulletin of the Museum of Northern Arizona* 57:1–160.
Colbert, E. H., R. B. Cowles, and C. M. Bogert. 1946. Temperature tolerances in the American alligator and their bearing on the habits, evolution and extinction of the dinosaurs. *Bulletin of the American Museum of Natural History* 86:327–374.
Colbert, E. H., and M. Morales. 1991. *Evolution of the vertebrates*. 4th ed. New York: Wiley-Liss.
Collias, N. E., and E. C. Collias. 1964. Evolution of nest-building in the weaverbirds (Ploceidae). *University of California Publications in Zoology*, no. 73.
———. 1984. *Nest building and bird behavior*. Princeton, N.J.: Princeton University Press.
Collins, C. T. 1976a. Two new species of *Aegialornis* from France, with comments on the ordinal affinities of the Aegialornithidae. *Smithsonian Contributions to Paleobiology* 27:121–127.
———. 1976b. A review of the Lower Miocene swifts (Aves: Apodidae). *Smithsonian Contributions to Paleobiology* 27:129–132.
———. 1983. A reinterpretation of pamprodactyly in swifts: a convergent grasping mechanism in vertebrates. *Auk* 100:735–737.
Coombs, W. P., Jr. 1978. Theoretical aspects of cursorial adaptation in dinosaurs. *Quarterly Review of Biology* 53:393–418.
Cooper, A., G. K. Chambers, A. C. Wilson, and S. Paabo. 1991. Molecular studies of New Zealand's extinct ratites. *Proceedings of the Twentieth International Ornithological Congress*, 554.

Cooper, A., C. Mourer-Chauviré, G. K. Chambers, A. von Haeseler, A. C. Wilson, and S. Paabo. 1992. Independent origins of New Zealand moas and kiwis. *Proceedings of the National Academy of Sciences* 89:8741–8744.

Cope, E. D. 1876. On a gigantic bird from the Eocene of New Mexico. *Proceedings of the Academy of Natural Sciences of Philadelphia* 28(2):10–11.

Cott, H. 1961. Scientific results of an inquiry into the ecology and economic status of the Nile crocodile (*Crocodilius niloticus*) in Uganda and northern Rhodesia. *Transactions of the Zoological Society of London* 29:215–337.

Cottam, P. A. 1957. The pelecaniform characters of the skeleton of the shoe-bill stork *Balaeniceps rex*. *Bulletin of the British Museum (Natural History)* 5:51–72.

Courtice, G. 1987. Museum officials confident *Archaeopteryx* is genuine . . . but opponents renew demands for proof. *Nature* 328:657.

Cowen, R. 1981. Homonyms of *Podopteryx*. *Journal of Paleontology* 55:483.

Cowen, R., and J. H. Lipps. 1982. An adaptive scenario for the origin of birds and their flight. *Proceedings of the Third North American Paleontological Convention*, 1:109–112.

Cracraft, J. 1968. A review of the Bathornithidae (Aves, Gruiformes), with remarks on the relationships of the suborder Cariamae. *American Museum Novitates* 2326:1–46.

———. 1969. Systematics and evolution of the Gruiformes (Class Aves). 1. The Eocene family Geranoididae and the early history of the Gruiformes. *American Museum Novitates* 2388:1–41.

———. 1971a. Systematics and evolution of the Gruiformes (Class Aves). 2. Additional comments on the Bathornithidae, with descriptions of new species. *American Museum Novitates* 2449:1–14.

———. 1971b. Caenagnathiformes: Cretaceous birds convergent in jaw mechanism to dicynodont reptiles. *Journal of Paleontology* 45:805–809.

———. 1971c. The relationships and evolution of the rollers: families Coraciidae, Brachypteraciidae, and Leptosomatidae. *Auk* 88:723–752.

———. 1972. A new Cretaceous charadriiform family. *Auk* 89:36–46.

———. 1973a. Continental drift, paleoclimatology, and the evolution and biogeography of birds. *Journal of Zoology* 169:455–545.

———. 1973b. Systematics and evolution of the Gruiformes (Class Aves). 3. Phylogeny of the suborder Grues. *Bulletin of the American Museum of Natural History* 151:1–127.

———. 1974a. Phylogeny and evolution of ratite birds. *Ibis* 115:494–521.

———. 1974b. Continental drift and vertebrate distribution. *Annual Review of Ecology and Systematics* 5:215–261.

———. 1976. The species of moas (Aves: Dinornithidae). *Smithsonian Contributions to Paleobiology* 27:189–205.

———. 1981. Toward a phylogenetic classification of the recent birds of the world (class Aves). *Auk* 98:681–714.

———. 1982a. Phylogenetic relationships and monophyly of loons, grebes, and hesperornithiform birds, with comments on the early history of birds. *Systematic Zoology* 31:35–56.

———. 1982b. Geographic differentiation, cladistics, and vicariance biogeography: reconstructing the tempo and mode of evolution. *American Zoologist* 22:411–424.

———. 1983. Cladistic analysis and vicariance biogeography. *American Scientist* 71:273–281.

———. 1985. Monophyly and phylogenetic relationships of the Pelecaniformes: a numerical cladistic analysis. *Auk* 102:834–853.

———. 1986. The origin and early diversification of birds. *Paleobiology* 12(4):383–399.

———. 1987. DNA hybridization and avian phylogenetics. *Evolutionary Biology* 21:47–96.

———. 1988a. Early evolution of birds. *Nature* 331:389–390.

———. 1988b. The major clades of birds. In *The phylogeny and classification of the tetrapods*, vol. 1: *Amphibians, reptiles, birds*, ed. M. J. Benton, 339–361. Oxford: Clarendon Press.

Cracraft, J., and P. V. Rich. 1972. Systematics and evolution of Cathartidae in the Old World Tertiary. *Condor* 74:272–283.

Crile, G., and D. P. Quiring. 1940. A record of the body weight and certain organ and gland weights of 3,690 animals. *Ohio Journal of Science* 40:219–259.

Crowe, T. M., and L. L. Short. 1992. A new gallinaceous bird from the Oligocene of Nebraska, with comments on the phylogenetic position of the Gallinuloididae. *Los Angeles County Museum of Natural History, Science Series* 36:179–188.

Cruickshank, A. R. I. 1972. The proterosuchian thecodonts. In *Studies in vertebrate evolution*, ed. K. A. Joysey and T. S. Kemp, 89–119. Edinburgh: Oliver and Boyd.

Crusafont, P. M., and J. F. de Villalta Comella. 1955. Parte paleontológica. Apendice 1. Aves. In *Burdigaliense continental de la Cuenca del Vallés-Penedés*, by M. Crusafont, J. F. de Villalta, and Y. J. Truyols, 236–237. *Memorias y Comunicaciones, Instituto Geológico Diputación Barcelona* 12.

Currie, P. J. 1982. Bird footprints from the Gething Formation (Aptian, Lower Cretaceous) of northeastern British Columbia, Canada. *Journal of Vertebrate Paleontology* 1:187–194.

———. 1987. Bird-like characteristics of the jaws and teeth of troödontid theropods (Dinosauria, Saurischia). *Journal of Vertebrate Paleontology* 7:72–81.

———. 1989. Long distance dinosaurs. *Natural History* 6:60–65.

Dames, W. 1884. Über *Archaeopteryx*. *Paläontologische Abhandlungen* 2:119–198.

Daniels, M. 1994. [Report on birds from the Naze London Clay.] *Society of Avian Paleontology and Evolution Newsletter* 8:10–12.
Darwin, C. 1855. Letter to J. D. Hooker, March 7, 1855. In *The life and letters of Charles Darwin*, ed. F. Darwin, 1888. Reprint ed., New York: Johnson Reprint, 1969.
———. 1859. *On the origin of species by means of natural selection, or the preservation of favoured races in the struggle for life.* 2d ed. London: John Murray.
———. 1934. *Charles Darwin's diary on the voyage of the H.M.S. "Beagle."* Ed. Nora Barlow. London, Cambridge University Press.
Dawson, A., A. R. Goldsmith, and T. J. Nicholls. 1985. Thyroidectomy results in termination of photorefractoriness in starlings (*Sturnus vulgaris*) kept in long daylengths. *Journal of Reproductive Fertility* 74:527–533.
Dawson, A., F. J. McNaughton, A. R. Goldsmith, and A. A. Degen. 1994. Ratite-like neoteny induced by neonatal thyroidectomy of European starlings, *Sturnus vulgaris*. *Journal of Zoology* (London) 232:633–639.
De Beer, G. 1954. *Archaeopteryx lithographica: a study based on the British Museum specimen.* London: British Museum (Natural History).
———. 1956. The evolution of ratites. *Bulletin of the British Museum (Natural History)* 4:59–70.
Delacour, J., and D. Amadon. 1973. *Curassows and related birds.* New York: American Museum of Natural History Press.
Desmond, A. J. 1976. *The hot-blooded dinosaurs.* New York: Dial Press.
———. 1982. *Archetypes and ancestors.* Chicago: University of Chicago Press.
Diamond, J. 1991. A new species of rail from the Solomon Islands and convergent evolution of flightlessness. *Auk* 108:461–470.
Diamond, J. M. 1981. Flightlessness and fear of flying in island species. *Nature* 293:507–508.
———. 1991. Twilight of Hawaiian birds. *Nature* 353:505–506.
Dickson, D. 1987. Feathers still fly in row over fossil bird. *Science* 238:475–476.
Dietz, R. S., and J. C. Holden. 1970. The breakup of Pangaea. *Scientific American* 222(10):34–39.
Dodd, M. H. I., and J. M. Dodd. 1976. The biology of metamorphosis. In *Physiology of the amphibia*, ed. B. Lofts, 467–599. New York: Academic Press.
Dodson, P. 1985. International *Archaeopteryx* conference (conference report). *Journal of Vertebrate Paleontology* 5:177–179.
Dollo, L. 1882. Première note sur les dinosauriens de Bernissart. *Bulletin Museum d'Histoire Naturale Belgique* 1:161–180.
———. 1883a. Note sur la présence chez les oiseaux du "troisième trochanter" des dinosauriens et sur la fonction de celui-ci. *Bulletin Museum d'Histoire Naturale Belgique* 2:13–19.
Dong, Zhi-ming. 1993. A Lower Cretaceous enantiornithine bird from the Ordos Basin of Inner Mongolia, People's Republic of China. *Canadian Journal of Earth Sciences* 30:2177–2179.
Duff, R. 1949. *Pyramid Valley, Waikari, North Canterbury.* Christchurch, New Zealand.
Dumbacher, J. P., B. M. Beehler, T. F. Spande, H. M. Garraffo, and J. W. Daly. 1992. Homobatrachotoxin in the genus *Pitohui:* chemical defense in birds? *Science* 258:799–801.
Dyck, J. 1985. The evolution of feathers. *Zoologica Scripta* 14:137–154.
Edington, G. H., and A. E. Miller. 1941. The avian ulna: its quill-knobs. *Proceedings of the Royal Society of Edinburgh*, ser. B, 61:138–148, pls. 1–7.
Ellenberger, P., and J. F. De Villalta. 1974. Sur la presence d'un ancêtre probable des oiseaux dans le Muschelkalk supérieur de Catalogne (Espagne): note preliminaire. *Acta Geologica Hispanica* 9(5):162–168.
Elzanowski, A. 1974. Preliminary note on the palaeognathous bird from the Upper Cretaceous of Mongolia. *Palaeontologica Polonica* 30:103–109.
———. 1976. Palaeognathous bird from the Cretaceous of Central Asia. *Nature* 265:51–53.
———. 1977. Skulls of *Gobipteryx* (Aves) from the Upper Cretaceous of Mongolia. *Paleontologica Polonica* 37:153–165.
———. 1981. Embryonic bird skeletons from the late Cretaceous of Mongolia. *Paleontologica Polonica* 42:147–179.
———. 1983. Birds in Cretaceous ecosystems. *Acta Paleontologica Polonica* 28:75–92.
———. 1995. Cretaceous birds and avian phylogeny. *Courier Forschungsinstitut Senckenberg*, 181:41–59.
Elzanowski, A., and P. Galton. 1991. Braincase of *Enaliornis*, an early Cretaceous bird from England. *Journal of Vertebrate Paleontology* 11:90–107.
Elzanowski, A., and P. Wellnhofer. 1992. A new link between theropods and birds from the Cretaceous of Mongolia. *Nature* 359:821–823.
———. 1993. Skull of *Archaeornithoides* from the Upper Cretaceous of Mongolia. *American Journal of Science* 293-A: 235–252.
———. 1994. Cranial morphology of *Archaeopteryx:* evidence from the seventh specimen. *Journal für Ornithologie* 135:331.
Emslie, S. D. 1987. Age and diet of fossil California condors in Grand Canyon, Arizona. *Science* 237:768–770.
———. 1988a. An early condor-like vulture from North America. *Auk* 105:529–535.
———. 1988b. The fossil history and phylogenetic relationships of condors (Ciconiiformes: Vulturidae) in the New World. *Journal of Vertebrate Paleontology* 8:212–228.
———. 1995. A catastrophic death assemblage of a new species of cormorant and other seabirds from the late Pliocene of Florida. *Journal of Vertebrate Paleontology* 15:313–330.
Emslie, S. D., and G. S. Morgan. 1994. A catastrophic death assemblage and paleoclimatic implications of Pliocene seabirds in Florida. *Science* 264:684–686.

Evans, J. 1865. On portions of a cranium and of a jaw, in the slab containing the fossil remains of the *Archaeopteryx*. *Natural History Review*, n.s., 5:415–421.

Ewer, R. F. 1965. The anatomy of the thecodont reptile *Euparkeria capensis* Broom. *Philosophical Transactions of the Royal Society of London* 176:197–221.

Farlow, J. O. 1976. A consideration of the trophic dynamics of a late Cretaceous large-dinosaur community (Oldman Formation). *Ecology* 57:841–857.

———. 1990. Thermal energetics and thermal biology. In *The Dinosauria*, ed. D. B. Weishampel, P. Dodson, and H. Osmolska, 43–55. Berkeley: University of California Press.

Farlow, J. O., P. Dodson, and A. Chinsamy. 1995. Dinosaur biology. *Annual Review of Ecology and Systematics* 26:445–471.

Farlow, J. O., C. V. Thompson, and D. E. Rosner. 1976. Plates of the dinosaur *Stegosaurus:* forced convection heat loss fins? *Science* 192:1123–1125.

Feduccia, A. 1967. A new swallow from the Fox Canyon local fauna (Upper Pliocene) of Kansas. *Condor* 69:526–527.

———. 1972. Variation in the posterior border of the sternum in some tree-trunk foraging birds. *Wilson Bulletin* 84:315–328.

———. 1973a. Dinosaurs as reptiles. *Evolution* 27:166–169.

———. 1973b. Evolutionary trends in the Neotropical ovenbirds and woodhewers. *Ornithological Monographs* 13:1–69.

———. 1974. Morphology of the bony stapes in New and Old World suboscines: new evidence for common ancestry. *Auk* 91:427–429.

———. 1975a. Morphology of the bony stapes (columella) in the Passeriformes and related groups: evolutionary implications. *University of Kansas Museum of Natural History Miscellaneous Publications* 63:1–34.

———. 1975b. Morphology of the bony stapes in the Menuridae and Acanthisittidae: evidence for oscine affinities. *Wilson Bulletin* 87:418–420.

———. 1976a. Osteological evidence for shorebird affinities of the flamingos. *Auk* 93:587–601.

———. 1976b. *Neanis schucherti* restudied: another Eocene piciform bird. *Smithsonian Contributions to Paleobiology* 27:95–99.

———. 1977a. Neuer Stammbaum von Ente und Flamingo. *Bild der Wissenschaft, Akzent* 5(7):1.

———. 1977b. Hypothetical stages in the evolution of modern ducks and flamingos. *Journal of Theoretical Biology* 67:715–721.

———. 1977c. The whalebill is a stork. *Nature* 266:719–720.

———. 1977d. A model for the evolution of perching birds. *Systematic Zoology* 26:19–31.

———. 1978a. *Presbyornis* and the evolution of ducks and flamingos. *American Scientist* 66:298–304.

———. 1978b. Evolution von Enten und Flamingo. *Proceedings of the Seventeenth International Ornithological Congress*, 1243–1248.

———. 1979. Comments on the phylogeny of perching birds. *Proceedings of the Biological Society of Washington* 92:689–696.

———. 1985a. On why the dinosaur lacked feathers. In *The beginnings of birds*, ed. M. K. Hecht et al., 75–79. Eichstätt: Freunde des Jura-Museum.

———. 1985b. The morphological evidence for ratite monophyly: fact or fiction. *Proceedings of the Eighteenth International Ornithological Congress*, 184–190.

———. 1987. Two woodpeckers from the late Pliocene of North America. *Proceedings of the Biological Society of Washington* 100:462–464.

———. 1991. A preliminary study of skeletal pathology of birds in zoos and its implications. *Proceedings of the Twentieth International Congress*, 1930–1936.

———. 1993a. Evidence from claw geometry indicating arboreal habits of *Archaeopteryx*. *Science* 259:790–793.

———. 1993b. Aerodynamic model for the early evolution of feathers provided by *Propithecus* (Primates, Lemuridae). *Journal of Theoretical Biology* 160:159–164.

———. 1994a. The great dinosaur debate. *Living Bird* 13(4):28–33.

———. 1994b. Tertiary bird history: notes and comments. In *Major features of vertebrate evolution*, ed. D. P. Prothero and R. M. Schoch, 178–189. Knoxville: University of Tennessee Press.

———. 1995a. The aerodynamic model for the evolution of feathers and feather misinterpretation. *Courier Forschungsinstitut Senckenberg*, 181:73–86.

———. 1995b. Explosive evolution in Tertiary birds and mammals. *Science* 267:637–638.

Feduccia, A., and C. E. Ferree. 1978. Morphology of the bony stapes (columella) in owls: evolutionary implications. *Proceedings of the Biological Society of Washington* 91:431–438.

Feduccia, A., and L. D. Martin. 1976. The Eocene zygodactyl birds of North America (Aves: Piciformes). *Smithsonian Contributions to Paleobiology* 27:101–110.

Feduccia, A., and P. O. McGrew. 1974. A flamingolike wader from the Eocene of Wyoming. *Contributions to Geology, University of Wyoming* 113(2):49–61.

Feduccia, A., and B. McPherson. 1993. A petrel-like bird from the late Eocene of Louisiana: earliest record of the order Procellariiformes. *Proceedings of the Biological Society of Washington* 106:749–751.

Feduccia, A., and S. L. Olson. 1982. Morphological similarities between the Menurae and Rhinocryptidae, relict passerine birds of the Southern Hemisphere. *Smithsonian Contributions to Zoology* 366: 1–22.

Feduccia, A., and H. B. Tordoff. 1979. Feathers of *Archaeopteryx:* asymmetric vanes indicate aerodynamic function. *Science* 203:1021–1022.

Feduccia, A., and M. R. Voorhies. 1989. Miocene hawk converges on secretarybird. *Ibis* 131:349–354.

———. 1992. Crowned cranes (Gruidae: *Balearica*) in the Miocene of Nebraska. *Los Angeles County Museum of Natural History, Science Series* 36:239–248.

Feduccia, A., and R. Wild. 1993. Birdlike characters in

the Triassic archosaur *Megalancosaurus*. *Naturwissenschaften* 80:654–566.
Ferrer-Condal, L. 1954. Notice préliminaire concernant la présence d'une plume d'oiseau dans le jurassique supérieur du Montsec (Province de Lérida, Espagne). *Proceedings of the Eleventh International Ornithological Congress*, 268–269.
Fischman, J. 1993. A closer look at the dinosaur-bird link. *Science* 262:1975.
Fisher, D. C. 1981. Crocodilian scatology, microvertebrata concentrations and enamel-less teeth. *Paleobiology* 7:262–275.
Fisher, H. I. 1940. The occurrence of vestigial claws on the wings of birds. *American Midland Naturalist* 23:234–243.
Fisher, J., and R. T. Peterson. 1964. *The world of birds*. Garden City, N.J.: Doubleday.
Fitzpatrick, J. W. 1988. Why so many passerine birds? A response to Raikow. *Systematic Zoology* 37:72–77.
Fjeldså, J. 1976. The systematic affinities of sandgrouses, Pteroclididae. *Videnskabelige Meddelelser Fra Dansk Naturhistorisk Forening* 139:179–243.
Folger, T. 1993. Dinosaur update: the blood of the dinos. *Discover* (special issue):19.
Fox, R. C. 1974. A middle Campanian, nonmarine occurrence of the Cretaceous toothed bird *Hesperornis* Marsh. *Canadian Journal of Earth Sciences* 11:1335–1338.
———. 1984. *Ichthyornis* (Aves) from the early Turonian (late Cretaceous) of Alberta. *Canadian Journal of Earth Sciences* 21:258–260.
Friedmann, H. 1954. Honey-guide: the bird that eats wax. *National Geographic* 104:551–560.
———. 1955. The honey-guides. *Bulletin of the United States National Museum* 208:1–292.
Frith, H. J. 1967. *Waterfowl in Australia*. Honolulu: East-West Center Press.
Fry, C. H. 1980. The evolutionary biology of kingfishers (Alcedinidae). *Living Bird* 18:113–160.
Fry, C. H., and K. Fry. 1992. *Kingfishers, bee-eaters, and rollers: a handbook*. Princeton, N.J.: Princeton University Press.
Funston, S. 1992. *The dinosaur question and answer book*. Boston: Little, Brown.
Fürbringer, M. 1888. *Untersuchungen zur Morphologie und Systematik der Vögel*. 2 vols. Amsterdam: Holkema.
———. 1902. Zur vergleichenden Anatomie des Brustschulterapparates und der Schultermuskeln. Part 5. *Jenaische Zeitschrift für Naturwissenschaft* 36:289–736.
Gadow, H. 1892. On the classification of birds. *Proceedings of the Zoological Society of London* 1892:229–256.
———. 1893. Vögel. II.—Systematischer Theil. In *Klassen und Ordnungen des Thier-Reichs*, ed. H. G. Bronn, vol. 6(4). Leipzig: C. F. Winter.
———. 1896. Pterylosis. In *A dictionary of birds*, ed. A. Newton, 744–748. London: Adam and Charles Black.
Gaillard, C. 1939. Contribution a l'étude des oiseaux fossiles. *Archives Muséum d'Histoire Naturelle, Lyon* 15:1–100.
Galton, P. M. 1970. Ornithischian dinosaurs and the origin of birds. *Evolution* 24:448–462.
———. 1974. The ornithischian dinosaur *Hypsilophodon* from the Wealdon of the Isle of Wight. *Bulletin of the British Museum (Natural History), Geology Series* 251:1–152.
———. 1980. Avian-like tibiotarsi of pterodactyloids (Reptilia: Pterosauria) from the Upper Jurassic of East Africa. *Paläontologische Zeitschrift* 54:331–342.
Gans, C., I. Darevski, and L. P. Tatarinov. 1987. *Sharovipteryx*, a reptilian glider? *Paleobiology* 13:415–426.
Gardiner, B. G. 1982. Tetrapod classification. *Journal of the Linnean Society* 74:207–232.
Gardner, L. L. 1925. The adaptive modifications and the taxonomic value of the tongue in birds. *Proceedings of the United States National Museum* 67(19):1–49.
Gauthier, J. A. 1984. A cladistic analysis of the higher systematic categories of the Diapsida. Ph.D. diss., University of California, Berkeley.
———. 1986. Saurischian monophyly and the origin of birds. *Memoires of the California Academy of Sciences* 8:1–55.
Gauthier, J. A., A. G. Kluge, and T. Rowe. 1988. Amniote phylogeny and the importance of fossils. *Cladistics* 4:105–209.
Gauthier, J. A., and K. Padian. 1985. Phylogenetic, functional, and aerodynamic analyses of the origin of birds and their flight. In *The beginnings of birds*, ed. M. K. Hecht et al., 185–197. Eichstätt: Freunde des Jura-Museum.
George, J. C., and A. J. Berger. 1966. *Avian myology*. New York: Academic Press.
Ghiselin, M. T. 1984. Narrow approaches to phylogeny: a review of nine books of cladism. *Oxford Survey of Evolutionary Biology* 1:209–222.
Gill, F. B. 1995. *Ornithology*. 2d ed. New York: W. H. Freeman.
Gill, F. B., and F. H. Sheldon. 1991. Review of *Phylogeny and classification of birds*, by Charles Sibley and Jon E. Ahlquist. *Science* 252:1003–1005.
Gingerich, P. D. 1972. A new partial mandible of *Ichthyornis*. *Condor* 74:471–473.
———. 1973. Skull of *Hesperornis* and early evolution of birds. *Nature* 243:448–462.
———. 1976. Evolutionary significance of the Mesozoic toothed birds. *Smithsonian Contributions to Paleobiology* 27:23–33.
Gingerich, P. D., S. M. Raza, M. Arif, M. Anwar, and Z. Xiaoyuan. 1994. New whale from the Eocene of Pakistan and the origin of cetacean swimming. *Nature* 368:844–847.
Glenny, F. H., and H. Friedmann. 1954. Reduction of the clavicles in the Mesoenatidae, with some remarks concerning the relationship of the clavicle to flight-function in birds. *Ohio Journal of Science* 54(2):111–113.
Goedert, J. L. 1988. A new late Eocene species of Plotopteridae (Aves: Pelecaniformes) from northwestern Oregon. *Proceedings of the California Academy of Sciences* 45(6):97–102.

Goodman, S. M. 1994a. The enigma of antipredator behavior in lemurs: evidence of a large extinct eagle on Madagascar. *International Journal of Primatology* 15:97–102.

———. 1994b. Description of a new species of subfossil eagle from Madagascar: *Stephanoaetus* (Aves: Falconiformes) from the deposits of Amphasambazimba. *Proceedings of the Biological Society of Washington* 107:421–428.

Gould, J. 1852. On a new and most remarkable form in ornithology. *Proceedings of the Zoological Society of London* 1851:1–2.

Gould, S. J. 1977. *Ontogeny and phylogeny*. Cambridge, Mass.: Harvard University Press.

———. 1986a. The *Archaeopteryx* flap. *Natural History* (September):16–25.

———. 1986b. Play it again, life. *Natural History* (February):23–26.

———. 1989. Wonderful life. *The Burgess Shale and the nature of history*. New York: W. W. Norton.

Gould, S. J., and E. Vrba. 1982. Exaptation—a missing term in the science of form. *Paleobiology* 8:4–15.

Grajal, A., S. D. Strahl, R. Parra, M. G. Dominguez, and A. Neher. 1989. Foregut fermentation in the hoatzin, a Neotropical leaf-eating bird. *Science* 245:1236–1238.

Grande, L. 1980. Paleontology of the Green River Formation, with a review of the fish fauna. *Geological Survey of Wyoming Bulletin* 63:1–333.

Grant, B. R., and P. R. Grant. 1993. Evolution of Darwin's finches caused by a rare climatic event. *Proceedings of the Royal Society of London*, ser. B, 251:111–117.

Grayson, D. K. 1977. Pleistocene avifaunas and the overkill hypothesis. *Science* 195:691–693.

Greenaway, F., A. C. Milner, C. A. Walker, and P. J. Whybrow. 1986. *Archaeopteryx* is not a forgery. *Science* 232:622–626.

Greenwood, J. J. D. 1993. Theory fits the bill in the Galápagos Islands. *Nature* 362:699.

Gregory, J. T. 1951. Convergent evolution: the jaws of *Hesperornis* and the mosasaurs. *Evolution* 5:345–354.

———. 1952. The jaws of the Cretaceous toothed birds *Ichthyornis* and *Hesperornis*. *Condor* 54:73–89.

Hachisuka, M. 1953. *The dodo and kindred birds*. London: Witherby.

Haffer, J. 1969. Speciation in Amazonian forest birds. *Science* 65:131–137.

———. 1974. *Avian speciation in tropical South America, with a systematic survey of the toucans (Ramphastidae) and jacamars (Galbulidae)*. Publications of the Nuttall Ornithological Club, 14. Cambridge, Mass: Nuttall Ornithological Club.

Hall, M. L. 1983. Origin of Española Island and the age of terrestrial life on the Galápagos Islands. *Science* 221:545–547.

Halliday, T. 1978. *Vanishing birds*. New York: Holt, Rinehart and Winston.

Halstead, L. B. 1982. Evolutionary trends and the phylogeny of the Agnatha. In *Problems of phylogenetic reconstruction*, ed. K. A. Joysey and A. E. Friday, 159–196. New York: Academic Press.

Hammer, W. R., and W. J. Hickerson. 1994. A crested theropod dinosaur from Antarctica. *Science* 264:828–830.

Harland, W. B., R. L. Armstrong, A. V. Cox, L. E. Craig, A. G. Smith, and D. G. Smith. 1990. *A geologic time scale*. Cambridge: Cambridge University Press.

Harrison, C. J. O. 1973. The humerus of *Ichthyornis* as a taxonomically isolating character. *Bulletin of the British Ornithologists' Club* 93:123–126.

———. 1978. *Bird families of the world*. New York: Harry N. Abrams.

———. 1981. Re-assignment of *Amphipelagus* [sic] *majori* from Ciconiidae (Ciconiiformes) to Ergilornithidae (Gruiformes). *Tertiary Research* 3:111–112.

Harrison, C. J. O., and C. A. Walker. 1972. The affinities of *Halcyornis* from the Lower Eocene. *Bulletin of the British Museum (Natural History), Geology Series* 21:151–169.

———. 1973. *Wyleyia:* a new bird humerus from the Lower Cretaceous of England. *Paleontology* 16:721–728.

———. 1975a. The Bradycnemidae, a new family of owls from the Upper Cretaceous of Romania. *Paleontology* 18:563–570.

———. 1975b. A new swift from the Lower Eocene of Britain. *Ibis* 117:162–164.

———. 1976a. A reappraisal of *Prophaethon shrubsolei* Andrews (Aves). *Bulletin of the British Museum (Natural History), Geology Series* 27:1–30.

———. 1976b. Birds of the British Upper Eocene. *Zoological Journal of the Linnean Society* 59:323–351.

———. 1977. Birds of the British Lower Eocene. *Tertiary Research Special Papers* 3:1–32.

Harshman, J. 1994. Reweaving the tapestry: what can we learn from Sibley and Ahlquist (1990)? *Auk* 111:377–388.

Hartman, F. A. 1961. Locomotor mechanisms in birds. *Smithsonian Institution Miscellaneous Collections* 143:1–91.

Hasegawa, Y., S. Isotani, K. Nagai, K. Seki, T. Suzuki, H. Otsuka, M. Ota, and K. Ono. 1979. [Preliminary notes on penguinlike bird fossils of the Oligocene-Miocene era.] *Bulletin of the Kitakyushu Museum of Natural History* 1:41–60. [In Japanese.]

Hatch, D. E. 1970. Energy conserving and heat dissipating mechanisms of the turkey vulture. *Auk* 87:111–124.

Haubitz, B. M., W. Prokop, W. Dohring, J. H. Ostrom, and P. Wellnhofer. 1988. Computed tomography of *Archaeopteryx*. *Paleobiology* 14:206–213.

Haubold, H., and E. Buffetaut. 1987. Une nouvelle interprétation de *Longisquama insignis*, reptile énigmatique du trias supérior d'Asie centrale. *Académie des Sciences, Comptes Rendus*, ser. 2A, 305:65–70.

Hecht, M. K. 1976. Phylogenetic inference and methodology as applied to the vertebrate record. *Evolutionary Biology* 9:335–363.

Hecht, M. K., and B. M. Hecht. 1994. Conflicting developmental and paleontological data: the case of the bird manus. *Acta Palaeontologica Polonica* 38(3–4): 329–338.

Hecht, M. K., J. H. Ostrom, G. Viohl, and P. Wellnhofer, eds. 1985. *The beginnings of birds*. Eichstätt: Freunde des Jura-Museum.

Hecht, M. K., and S. Tarsitano. 1982. The paleobiology and phylogenetic position of *Archaeopteryx*. *Géobios Mémoires Spéciale* 6:141–149.

Hedges, S. B., and C. G. Sibley. 1994. Molecules vs. morphology in avian evolution: the case of the "pelecaniform" birds. *Proceedings of the National Academy of Sciences* 91:9861–9865.

Heilmann, G. 1926. *The origin of birds*. London: Witherby.

Heimerdinger, M. A., and P. L. Ames. 1967. Variation in the sternal notches of suboscine passeriform birds. *Peabody Museum of Natural History, Yale University, Postilla* 105:1–44.

Heinroth, O. 1923. Die Flügel von *Archaeopteryx*. *Journal für Ornithologie* 71:277–283.

Heller, F. 1959. Ein dritter *Archaeopteryx*-Fund aus den Solnhofener Plattenkalken von Langenaltheim/Mfr. *Erlanger Geologische Abhandlungen* 31:3–25.

Helms, J. 1982. Zur Fossilization der Federn des Urvögels (Berliner Exemplar). *Wissenschaftliche Zeitschrift, Humboldt Universität Mathematisch Naturwissenschaftliche Reihe* 31:185–199.

Henley, C., A. Feduccia, and D. P. Costello. 1978. Oscine spermatozoa: a light and electron microscopy study. *Condor* 80:41–48.

Hennig, W. 1966. *Phylogenetic systematics*. Urbana: University of Illinois Press.

Hertel, F. 1992. Morphological diversity of past and present New World vultures. *Los Angeles County Museum of Natural History, Science Series* 36:413–418.

———. 1994. Diversity in body size and feeding morphology within past and present vulture assemblages. *Ecology* 75:1074–1084.

Herzog, K. 1968. Anatomie und Flugbiologie der Vögel. Stuttgart: Gustav Fisher Verlag.

Hesse, A. 1988. Die Messelornithidae—ein neue Familie der Kranichartigen (Aves: Gruiformes: Rhynocheti) aus dem Tertiär Europas und Nordamerikas. *Journal für Ornithologie* 129:83–95.

———. 1990. Die Beschreibung der Messelornithidae (Aves: Gruiformes: Rhynocheti) aus dem Alttertiär Europas und Nordamerikas. *Courier Forschungsinstitut Senckenberg* 124:1–165.

———. 1992. A new species of *Messelornis* (Aves: Gruiformes: Messelornithidae) from the Middle Eocene Green River Formation. *Los Angeles County Museum of Natural History, Science Series* 36:171–178.

Higham, C. F. W. 1981. *The Maoris: Cambridge introduction to the history of mankind, topic book*. Cambridge: Cambridge University Press.

Hillenius, W. J. 1992. The evolution of nasal turbinates and mammalian endothermy. *Paleobiology* 12:450-458.

———. 1994. Turbinates in therapsids: evidence for late Permian origins of mammalian endothermy. *Evolution* 48:207–229.

Hinchliffe, J. R. 1985. "One, two, three" or "two, three, four": an embryologist's view of the homologies of the digits and carpus of modern birds. In *The beginnings of birds*, ed. M. K. Hecht et al., 141–147. Eichstätt: Freunde des Jura-Museum.

———. 1989a. An evolutionary perspective of the developmental mechanisms underlying the patterning of the limb skeleton on birds and tetrapods. *Géobios Mémoires Spéciale* 12:217–225.

———. 1989b. Reconstructing the archetype: innovation and conservatism in the evolution and development of the pentadactyl limb. In *Complex organismal functions: integration and evolution in vertebrates*, ed. D. Wake and G. Roth, 171–189. London: John Wiley and Sons.

———. 1991. Developmental approaches to the problem of transformation of limb structure in evolution. In *Developmental Patterning of the Vertebrate Limb*, NATO Advanced Study Institute Series A 205, ed. J. R. Hinchliffe, J. Jurle, and D. Summerbell, 313–323. New York: Plenum Press.

Hinchliffe, J. R., and M. K. Hecht. 1984. Homology of the bird wing skeleton: embryological versus paleontological evidence. *Evolutionary Biology* 18:21–39.

Ho, C. Y.-K., E. M. Prager, A. C. Wilson, D. T. Osuga, and R. E. Feeney. 1976. Penguin evolution: protein comparisons demonstrate phylogenetic relationship to flying aquatic birds. *Journal of Molecular Evolution* 8:271–282.

Hoch, E. 1980. A new middle Eocene shorebird (Aves: Charadriiformes, Charadrii) with columboid features. *Los Angeles County Museum of Natural History, Contributions to Science* 330:33–49.

Holdaway, R. N. 1989. New Zealand's pre-human avifauna and its vulnerability. In *Moas, mammals and climate in the ecological history of New Zealand*, ed. M. R. Rudge. *New Zealand Journal of Ecology* 12:11–25 (supplement).

Holmgren, N. 1955. Studies on the phylogeny of birds. *Acta Zoologica* 36:243–328.

Hopson, J. A. 1977. Relative brain size and behavior in archosaurian reptiles. *Annual Review of Ecology and Systematics* 8:429–448.

———. 1980. Relative brain size in dinosaurs: implications for dinosaurian endothermy. In *A cold look at the warm-blooded dinosaurs*, ed. R. D. K. Thomas and E. C. Olson, 287–310. American Association for the Advancement of Science Selected Symposium, 28. Boulder, Colo.: Westview Press.

Horne, J. F. M., and L. L. Short. 1990. Wax-eating by African common bulbuls. *Wilson Bulletin* 102:339–341.

Horner, J. R. 1984. The nesting behavior of dinosaurs. *Scientific American* 250(4):130–137.

Horner, J. R., and J. Gorman. 1990. *Digging dinosaurs*. New York: Harper and Row.

Horner, J. R., and P. Makela. 1979. Nest of juveniles pro-

vides evidence of family structure among dinosaurs. *Nature* 282:296–298.
Horner, J. R., and D. B. Weishampel. 1989. Dinosaur eggs: the inside story. *Natural History* (December):60–67.
Hotton, N., III. 1980. An alternative to dinosaur endothermy: the happy wanderers. In *A cold look at the warm-blooded dinosaurs*, ed. R. D. K. Thomas and E. C. Olson, 311–350. American Association for the Advancement of Science Selected Symposium, 28. Boulder, Colo.: Westview Press.
Hou, L.-H. 1980. [New form of the Gastornithidae from the Lower Eocene of the Xichuan, Honan.] *Vertebrata PalAsiatica* 18:111–115. [In Chinese with English abstract.]
———. 1994. A late Mesozoic bird from Inner Mongolia. *Chinese Science Bulletin* 32:264–266, 1 fig.
———. 1995. Morphological comparisons between *Confuciusornis* and *Archaeopteryx*. In *Sixth Symposium on Mesozoic Terrestrial Ecosystems and Biota, Short Papers*, ed. Aling Sun and Yuanging Wang, 193–201. Beijing: China Ocean Press.
Hou, L.-H., and Z. Liu. 1984. A new fossil bird from Lower Cretaceous of Gansu and early evolution of birds. *Scientia Sinica* 27:1296–1302.
Hou, L.-H., and J. Zhang. 1993. A new fossil bird from Lower Cretaceous of Liaoning. *Vertebrata PalAsiatica* 7:217–224.
Hou, L.-H., Z.-H. Zhou, Y.-C. Gu, and H. Zang. 1995. [*Confusiusornis sanctus*, a new late Jurassic sauriurine bird from China.] *Chinese Science Bulletin* 40(8):726–729. [In Chinese.]
Hou, L.-H., Z.-H. Zhou, L. D. Martin, and A. Feduccia. 1995. The oldest beaked bird is from the "Jurassic" of China. *Nature* 377:616–618.
Houck, M. A., J. A. Gauthier, and R. E. Strauss. 1990. Allometric scaling in the earliest fossil bird, *Archaeopteryx lithographica*. *Science* 247:195–198.
Houde, P. 1986. Ostrich ancestors found in the Northern Hemisphere suggest new hypothesis of ratite origins. *Nature* 324:563–565.
———. 1987a. Histological evidence for the systematic position of *Hesperornis* (Odontornithes: Hesperornithiformes). *Auk* 104:125–129.
———. 1987b. Critical evaluation of DNA hybridization studies in avian systematics. *Auk* 104:17–32.
———. 1988. *Paleognathous birds from the early Tertiary of the Northern Hemisphere*. Cambridge, Mass.: Nuttall Ornithological Club.
———. 1994. Evolution of the Heliornithidae: reciprocal illumination by morphology, biogeography and DNA hybridization (Aves: Gruiformes). *Cladistics* 10:1–9.
Houde, P., and H. Haubold. 1987. *Palaeotis weigelti* restudied: a small Eocene ostrich (Aves: Struthioniformes). *Paleovertebrata* 17:27–42.
Houde, P., and S. L. Olson. 1981. Paleognathous carinate birds from the early Tertiary of North America. *Science* 214:1236–1237.
———. 1989. Small arboreal nonpasserine birds from the early Tertiary of western North America. *Proceedings of the Nineteenth International Ornithological Congress*, 2030–2036.
———. 1992. A radiation of coly-like birds from the Eocene of North America (Aves: Sandcoleiformes New Order). *Los Angeles County Museum of Natural History, Science Series* 36:137–160.
Houston, D. C. 1976. Food searching behavior in griffon vultures. *East African Wildlife Journal* 12:63–77.
———. 1984. Does the king vulture *Sarcorhamphus papa* use a sense of smell to locate food? *Ibis* 126:67–69.
———. 1985. Evolutionary ecology of Afrotropical and Neotropical vultures in forests. In *Neotropical ornithology*, ed. P. Buckley, M. Foster, E. Morton, R. Ridgely, and F. Buckley, 856–864. *Ornithological Monographs*, no. 36.
———. 1988. Competition for food between Neotropical vultures in forest. *Ibis* 130:402–417.
Howard, H. 1932. Eagles and eagle-like vultures of the Pleistocene of Rancho La Brea. *Contributions to Paleontology, Carnegie Institution of Washington* 429:1–82.
———. 1933. A new species of owl from the Pleistocene of Rancho La Brea, California. *Condor* 35:66–69.
———. 1947. California's flightless birds. *Los Angeles County Museum Quarterly* 6:7–11.
———. 1957. A gigantic "toothed" marine bird from the Miocene of California. *Santa Barbara Museum of Natural History Bulletin* (Department of Geology) 1:1–23.
———. 1966a. A possible ancestor of the lucas auk (family Mancallidae) from the Tertiary of Orange County, California. *Los Angeles County Museum of Natural History, Contributions to Science* 101:1–8.
———. 1966b. Two fossil birds from the Lower Miocene of South Dakota. *Los Angeles County Museum of Natural History, Contributions to Science* 107:1–8.
———. 1968. Tertiary birds from Laguna Hills, Orange County, California. *Los Angeles County Museum of Natural History, Contributions to Science* 142:1–21.
———. 1969. A new avian fossil from Kern County, California. *Condor* 71:68–69.
———. 1970. A review of the extinct avian genus *Mancalla*. *Los Angeles County Museum of Natural History, Contributions to Science* 203:1–12.
———. 1976. A new species of flightless auk from the Miocene of California (Alcidae: Mancallinae). *Smithsonian Contributions to Paleobiology* 27:141–146.
———. 1978. Late Miocene marine birds from Orange Co. California. *Los Angeles County Museum of Natural History, Contributions to Science* 290:1–28.
Howgate, M. E. 1984. The teeth of *Archaeopteryx* and a reinterpretation of the Eichstätt specimen. *Zoological Journal of the Linnean Society* 82:159–175.
———. 1985. Back to the trees for *Archaeopteryx* in Bavaria. *Nature* 31:435–436.
Hoyle, F., and C. Wickramasinghe. 1986. *Archaeopteryx, the primordial bird: a case of fossil forgery*. London: Christopher Davies.
Huene, R. von. 1914. Beiträge zur Geschichte der Archosaurier. *Geologische und Paläontologische Abhandlungen*, n.s., 13:1–53.

Hull, D. L. 1970. Contemporary systematic philosophies. *Annual Review of Ecology and Systematics* 1:19–54.

Hurlbert, S. H., and C. C. Y. Chang. 1983. Ornitholimnology: effects of grazing by the Andean flamingo (*Phoenicoparrus andinus*). *Proceedings of the National Academy of Sciences* 80:4766–4769.

Hurst, C. H. 1893. The digits in a bird's wing. *Natural Science News* 3:274–281.

Hutchinson, G. E. 1950. Survey of contemporary knowledge of biogeochemistry. 3. The biogeochemistry of vertebrate excretion. *Bulletin of the American Museum of Natural History* 96:1–554.

———. 1954. Marginalia: the dodo and the solitaire. *American Scientist* 42:300–305.

Hutchinson, J. H. 1993. *Avisaurus:* a "dinosaur" grows wings. *Journal of Vertebrate Paleontology* 13(3):43A.

Huxley, T. H. 1867. On the classification of birds and on the taxonomic value of the modifications of certain of the cranial bones observable in that class. *Proceedings of the Zoological Society of London* 1867:415–472.

———. 1868a. Remarks upon *Archaeopteryx lithographica. Proceedings of the Royal Society of London* 16:243–248.

———. 1868b. On the animals which are most nearly intermediate between the birds and reptiles. *Annals and Magazine of Natural History* 2:66–75.

———. 1870a. Further evidence of the affinity between the dinosaurian reptiles and birds. *Quarterly Journal of the Geological Society of London* 26:12–31.

———. 1870b. On the classification of the Dinosauria, with observations on the Dinosauria of the Trias. *Proceedings of the Geological Society of London* 26:32–38.

———. 1882. On the respiratory organs of *Apteryx. Proceedings of the Zoological Society of London* 1882:560–569.

Ingold, J. L., J. C. Vaughn, S. S. I. Guttman, and L. R. Maxson. 1989. Phylogeny of the cranes (Aves: Gruidae) as deduced from DNA-DNA hybridization and albumin micro-complement fixation analyses. *Auk* 106:595–602.

Jacob, J., and H. Hoerschelmann. 1985. Klassifizierung der Flamingos (Phoenicopteriformes) durch vergleichende Analyse von Bürzeldrüsensekreten. *Zeitschrift für Zoologische Systematik und Evolutionsforschung* 23:49–58.

Jacob, J., and V. Ziswiler. 1982. The uropygial gland. In *Avian biology*, vol. 6, ed. D. S. Farner, J. R. King, and K. C. Parkes, 199–324. New York: Academic Press.

James, H. F., and S. L. Olson. 1983. Flightless birds. *Natural History* 92:30–40.

———. 1991. Descriptions of thirty-two new species of birds from the Hawaiian Islands. Part II: Passeriformes. *Ornithological Monographs* 46:1–88.

Janzen, D. H. 1995. Who survived the Cretaceous? *Science* 268:785.

Jehl, J. R., Jr. 1968a. Relationships in the Charadrii (shorebirds): a taxonomic study based on color patterns of the downy young. *Transactions of the San Diego Society of Natural History* 3:1–54.

———. 1968b. Foraging behavior of *Geranospiza nigra*, the blackish crane hawk. *Auk* 85:493–494.

Jenkin, P. M. 1957. The filter-feeding and food of flamingoes (Phoenicopteri). *Philosophical Transactions of the Royal Society of London, Biological Sciences*, ser. B, 240:401–493.

Jenkins, F. A., Jr. 1993. The evolution of the avian shoulder joint. *American Journal of Science* 293A:253–267.

Jenkins, F. A., Jr., K. P. Dial, and G. E. Goslow, Jr. 1988. A cineradiographic analysis of bird flight: the wishbone in starlings is a spring. *Science* 241:1495–1498.

Jenkins, F. A., and D. M. Walsh. 1993. An early Jurassic caecilian with limbs. *Nature* 365:246–250.

Jenkins, R. J. F. 1974. A new giant penguin from the Eocene of Australia. *Palaeontology* 17:291–310.

Jensen, J. A. 1981. Another look at *Archaeopteryx* as the "oldest" bird. *Encyclia* 58:104–128.

Jensen, J. A., and K. Padian. 1989. Small pterosaurs and dinosaurs from the Uncompahgre fauna (Brushy Basin Member, Morrison Formation: ?Tithonian), late Jurassic, western Colorado. *Journal of Paleontology* 63:364–373.

Jepsen, G. L. 1964. Riddles of the terrible lizards. *American Scientist* 52:227–246. Originally published in *Princeton Alumni Weekly* 64:6 (1963).

Jerison, H. J. 1973. *Evolution of the brain and intelligence.* New York: Academic Press.

J. L. M. 1978. The oldest fossil bird: a rival for *Archaeopteryx? Science* 199:284.

Johnston, P. A. 1979. Growth rings in dinosaur teeth. *Nature* 278:635–636.

———. 1980. Johnston replies. *Nature* 188:195.

Jollie, M. 1958. Comments on the phylogeny and skull of the Passeriformes. *Auk* 75:26–35.

———. 1976. A contribution to the morphology and phylogeny of the Falconiformes. *Evolutionary Theory* 1:285–298.

———. 1977. A contribution to the morphology and phylogeny of the Falconiformes. Part 4. *Evolutionary Theory* 2:1–141.

Jones, J. 1945. The banded stilt. *Emu* 45:1–36.

Jung, W. 1974. Die Konifere *Brachyphyllum nepos* Saporta aus den Solnhofener Plattenkalken (unteres Untertithon), ein Halophyt. *Mitteilungen Bayerischen Staatssammlung für Paläontologie und historische Geologie* 14:49–58.

Jurcsák, T., and E. Kessler. 1986. Evolutia avifaunei pe teritoriul Românei (I). *Crisia* 16:577–615.

Kahl, M. P. 1963. Thermoregulation in the wood stork with special reference to the role of the legs. *Physiological Zoology* 36:141–151.

———. 1970. East Africa's majestic flamingos. *National Geographic* 137:276–294.

———. 1975. Ritualized displays. In *Flamingos*, ed. J. Kear and N. Duplaix-Hall, 142–149. Berkhamsted: T. and A. D. Poyser.

Karkhu, A. A. 1988. [A new family of Apodiformes from the Palaeogene of Europe.] *Paleontological Journal* 3:78–88. [In Russian.]

———. 1992. Morphological divergence within the order Apodiformes as revealed by the structure of the humerus. *Los Angeles County Museum of Natural History, Science Series* 36:379–384.

———. 1992. An Oligocene trogon from the N-Caucasus. S.A.P.E. Symposium, 1992, not published.

Kemp, A. C. 1978. A review of the hornbills: biology and radiation. *Living Bird* 17:105–136.

———. 1995. The hornbills. Oxford: Oxford University Press.

Kemp, A. C., and T. M. Crowe. 1985. The systematics and zoogeography of Afrotropical hornbills (Aves: Bucerotidae). In *Proceedings of the International Symposium on African Vertebrates*, ed. K.-L. Schuchmann, 279–324. Bonn: Museum Koenig.

Kemp, T. 1986. Feathered flights of fancy. *Nature* 324:185.

Kessler, E. 1984. Lower Cretaceous birds from Cornet (Romania). In *Third symposium on Mesozoic terrestrial ecosystems*, ed. W.-E. Reif and R. Westphal, 119–121. Tübingen: Attempto Verlag.

———. 1987. New contributions to the knowledge about the Lower and Upper Cretaceous birds from Romania. *Occasional Papers of the Tyrrell Museum of Paleontology* 3:133–135.

Kessler, E., and T. Jurcsák. 1984. Fossil bird remains in the bauxite from Cornet (Romania, Bihor county). *Travaux du Muséum d'Histoire Naturelle "Grigore Antipa"* 24:393–401.

———. 1986. New contributions to the knowledge of the Lower Cretaceous bird remains from Cornet (Romania). *Travaux du Muséum d'Histoire* Naturelle *"Grigore Antipa"* 28:289–295.

Kingsolver, J. G., and M. A. R. Koehl. 1985. Aerodynamics, thermoregulation, and the evolution of insect wings: differential scaling and evolutionary change. *Evolution* 39:488–504.

Kinsky, F. C. 1960. The yearly cycle of the northern blue penguin (*Eudyptula minor novaehollandiae*) in the Wellington harbour area. *Dominion Museum Record* (New Zealand) 3:145–218.

Kitto, G. B., and A. C. Wilson. 1966. Evolution of malate dehydrogenase in birds. *Science* 153:1408–1410.

Knudsen, E. I., and M. Konishi. 1979. Mechanisms of sound localization in the barn owl (*Tyto alba*). *Journal of Comparative Physiology* 133:13–21.

Kochmer, J. P., and R. H. Wagner. 1988. Why are there so many kinds of passerine birds? Because they are small. A reply to Raikow. *Systematic Zoology* 37:69–70.

Konishi, M. 1993. Listening with two ears. *Scientific American* 268(4):66–73.

Koolos, J. G. M., A. R. Kraaijeveld, G. E. J. Langenbach, and G. A. Zweers. 1989. Comparative mechanics of filter feeding in *Anas platyrhynchos*, *Anas clypeata* and *Aythya fuligula* (Aves, Anseriformes). *Zoomorphology* 108:269–290.

Koolos, J. G. M., and G. A. Zweers. 1991. Integration of pecking, filter feeding and drinking in waterfowl. *Acta Biotheoretica* 39:107–140.

Kooyman, G. L., R. W. Davis, J. P. Croxall, and D. P. Costa. 1982. Penguins. *Science* 217:726–727.

Kooyman, G. L., and T. G. Kooyman. 1995. Diving behavior of emperor penguins nurturing chicks at Coulman Island, Antartica. *Condor* 97:536–549.

Krajewski, C. 1989. Phylogenetic relationships of the cranes (Gruiformes: Gruidae) based on DNA hybridization. *Auk* 106:603–618.

Krajewski, C., and J. W. Fetzner, Jr. 1994. Phylogeny of cranes (Gruiformes: Gruidae) based on cytochrome-*B* DNA sequences. *Auk* 111:351–365.

Kruuk, H. 1967. Competition for food between vultures of East Africa. *Ardea* 55:171–193.

Kurochkin, E. N. 1976. A survey of the Paleogene birds of Asia. *Smithsonian Contributions to Paleobiology* 27:75–86.

———. 1981. [New representatives and evolution of two archaic gruiform families in Eurasia.] *Transactions of the Joint Soviet-Mongolian Paleontological Expedition* 15:59–85. [In Russian.]

———. 1982. [New order of birds from the Lower Cretaceous of Mongolia.] *Doklady Akademii Nauk S.S.R.* 262:452–455. [In Russian.]

———. 1985a. A true carinate bird from Lower Cretaceous deposits in Mongolia and other evidence of early Cretaceous birds in Asia. *Cretaceous Research* 6:271–278.

———. 1985b. [Birds of central Asia in the Pliocene.] *Transactions of the Joint Soviet-Mongolian Paleontological Expedition* 26:1–120. [In Russian.]

———. 1988. Cretaceous Mongolian birds and their significance for the study of bird phylogeny. English summary of fossil reptiles and birds of Mongolia. *Transactions of the Joint Soviet-Mongolian Paleontological Expedition* 34:108.

———. 1994. Synopsis and evolution of Mesozoic birds. *Journal für Ornithologie* 135:332.

———. 1995. Synopsis of Mesozoic birds and early evolution of class Aves. *Archaeopteryx* 13:47–66.

Kurochkin, E. N., and A. N. Lungu. 1970. [A new ostrich from the Middle Sarmatian of Moldavia.] *Palaeontological Journal* 1970:103–111. [English translation of *Palaeontological Journal* 1:118–126.]

Kurochkin, E. N., and N. Mayo. 1973. Las lechuzas gigantes del Pleistoceno Superior de Cuba. *Academica Ciencias, Cuba Instituto Geologico* 3:56–60.

Kuroda, N. 1954. On the classification and phylogeny of the order Tubinares, particularly the shearwaters (*Puffinus*), with special considerations on their osteology and habit differentiation. Tokyo: Herold.

Kurtén, B. 1971. *The age of mammals*. New York: Columbia University Press.

Lacasa Ruiz, A. 1985. Nota sobre las plumas fósiles del yacimiento eocretácico de "La Pedrera-Cabrúa" en la Sierra del Montsec (Prov. Lérida, España). *Llerda* 47:227–238.

———. 1986. Nota preliminar sobre el hallazgo de restos óseos de un ave fósil en el yacimiento Neocomiense del Montsec Prov. Lérida, España. *Llerda* 47:203–206.

Lamb, J. P., Jr., L. M. Chiappe, and P. G. P. Ericson. 1993. A marine enantiornithine from the Cretaceous of Alabama. *Journal of Vertebrate Paleontology* 13(3):45A.

Lambrecht, K. 1929. Mesozoische und tertiäre Vögelreste aus Siebenbürgen. *Comptes Rendus, Congrés International de la Zoologie (Budapest, 1927)* 10, pt. 2:1262–1275.

———. 1931a. *Protoplotus beauforti* n.g. n.sp., ein Schlangenhalsvogel aus Tertiär von W.-Sumatra. *Dienst van den Mijnbouw in Nederlandsch-Indië, Wetenschap Meded* 17:15–24.

———. 1931b. *Gallornis straeleni* n.g. n.sp., ein Kreidevogel aus Frankreich. *Bulletin du Musée Royale d'Histoire Naturelle de Belgique* 7(30):1–6.

———. 1933. *Handbuch der Paläornithologie*. Berlin: Gebrüder Bornträger.

Langrand, O. 1990. *Guide to the birds of Madagascar.* New Haven and London: Yale University Press.

Langston, W. 1981. Pterosaurs. *Scientific American* 244(2):122–136.

Lanham, U. N. 1947. Notes on the phylogeny of the Pelecaniformes. *Auk* 64:65–70.

Lanyon, S. M. 1992. Review of *Phylogeny and Classification of Birds*, by Charles G. Sibley and Jon E. Ahlquist. *Condor* 94:304–307.

Lanyon, S. M., and J. G. Hall. 1994. Re-examination of barbet monophyly using mitochondrial-DNA sequence data. *Auk* 111:389–397.

Lanyon, W. E. 1963. *Biology of birds*. New York: Garden City Press.

Lasiewski, R. D., and W. R. Dawson. 1967. A re-examination of the relation between standard metabolic rate and body weight in birds. *Condor* 69:13–23.

Lawick-Goodall, J. Van. 1968. Tool-using bird: the Egyptian vulture. *National Geographic* 1968:631–641.

Laybourne, R. C., D. W. Deedrick, and F. M. Hueber. 1994. Feather in amber is the earliest New World fossil of Picidae. *Wilson Bulletin* 106:18–25.

Lemoine, V. 1881. Sur le *Gastornis edwardsii* et le *Remiornis heberti* de l'éocène inférieur des environs de Reims. *Académie des Sciences, Comptes Rendus* 93:1157–1159.

Lewin, R. 1982. Adaptation can be a problem for evolutionist. *Science* 216:1212–1213.

———. 1983. How did vertebrates take to the air? *Science* 221:38–39.

———. 1985. On the origin of insect wings. *Science* 230:428–429.

———. 1987. Domino effect invoked in Ice Age extinctions. *Science* 238:1509–1510.

———. 1988a. Conflict over DNA clock results. *Science* 241:1598–1600.

———. 1988b. DNA clock conflict continues. *Science* 241:1756–1759.

Ligon, J. D. 1967. Relationships of the cathartid vultures. *Occasional Papers of the Museum of Zoology, University of Michigan*, 651.

———. 1970. Still more responses of the poor-will to low temperatures. *Condor* 72:496–498.

———. 1993. The role of phylogenetic history in the evolution of contemporary avian mating and parental care systems. *Current Ornithology* 10:1–46.

Linnaeus, C. 1758. *Systema naturae per regna tria naturae.* 10th ed., rev. 2 vols. Holmiae: L. Salmii.

Livezey, B. C. 1986. A phylogenetic analysis of recent anseriform genera using morphological characters. *Auk* 103:737–754.

———. 1988. Morphometrics of flightlessness in the Alcidae. *Auk* 105:681–698.

———. 1989a. Flightlessness in grebes (Aves, Podicipedidae): its independent evolution in three independent genera. *Evolution* 43:29–54.

———. 1989b. Morphometric patterns in Recent and fossil penguins (Aves, Sphenisciformes). *Journal of Zoology, London* 219:269–307.

———. 1991. A phylogenetic analysis and classification of recent dabbling ducks (tribe Anatini) based on comparative morphology. *Auk* 108:471–507.

———. 1992. Flightlessness in the Galápagos cormorant (*Compsohalieus* [*Nannopterum*] *harrisi*): heterochrony, gigantism, and specialization. *Zoological Journal of the Linnean Society* 105:155–224.

———. 1993a. Morphology of flightlessness in *Chendytes*, fossil seaducks (Anatidae: Mergini) of coastal California. *Journal of Vertebrate Paleontology* 13:185–199.

———. 1993b. An ecomorphological review of the dodo (*Raphus cucullatus*) and solitaire (*Pezophaps solitaria*), flightless Columbiformes of the Mascarene Islands. *Journal of Zoology, London* 230:247–292.

Livezey, B. C., and L. D. Martin. 1988. The systematic position of the Miocene anatid *Anas*[?] *blanchardi* Milne-Edwards. *Journal of Vertebrate Paleontology* 8:196–211.

Lockley, M. 1991. *Tracking dinosaurs*. Cambridge: Cambridge University Press.

Løvtrup, S. 1985. On the classification of the taxon Tetrapoda. *Systematic Zoology* 34:463–470.

Lowe, P. R. 1928. A description of *Atlantisia rogersi*, the diminutive flightless rail of Inaccessible Island (South Atlantic), with some notes on flightless rails. *Ibis*, 12th ser., 4:99–131.

———. 1935. On the relationships of the Struthiones to the dinosaurs and to the rest of the avian class, with special reference to the position of *Archaeopteryx*. *Ibis*, 13th ser., 5:298–429.

———. 1944. Some additional remarks on the phylogeny of the Struthiones. *Ibis* 86:37–42.

———. 1946. On the systematic position of the woodpeckers (Pici), honey-guides (*Indicator*), hoopoes and others. *Ibis* 88:103–127.

———. 1948. What are the Coraciiformes? *Ibis* 90:572–582.

Lucas, A. M., and P. R. Stettenheim. 1972. *Avian anatomy: Integument*. 2 vols. Agricultural Handbook no. 362. Washington, D.C.: U. S. Government Printing Office.

Lucas, F. A. 1901. *Animals of the past*. New York: McClure Phillips.

Lucas, S. G., and R. E. Sullivan. 1982. *Ichthyornis* in the late Cretaceous Mancos Shales (Juana Lopez Member), northwestern New Mexico. *Journal of Paleontology* 56:545–547.

Maddox, J. 1993. Bringing the extinct dodo back to life. *Nature* 365:291.

Maderson, P. F. A. 1972. On how an archosaurian scale might have given rise to an avian feather. *American Naturalist* 146:424–428.

Madsen, C. S., K. P. McHugh, and S. R. de Kloet. 1988. A partial classification of waterfowl (Anatidae) based on single-copy DNA. *Auk* 105:452–459.

Madsen, J. H., Jr. 1976. *Allosaurus fragilis:* a revised osteology. *Utah Geological and Mineralogical Survey Bulletin* 109:1–163.

Marden, J. H. 1987. Maximum lift production during take off in flying vertebrates. *Journal of Experimental Biology* 130:235–258.

Marsh, O. C. 1873. On a new subclass of fossil birds (Odontornithes). *American Journal of Science*, 3d ser., 5:161–162.

———. 1877. Introduction and succession of vertebrate life in America. *American Journal of Science*, 3d ser., 14:337–378.

———. 1880. *Odontornithes: a monograph on the extinct toothed birds of North America.* Report of the U.S. Geological Exploration of the Fortieth Parallel, no. 7, Washington, D.C.

———. 1881. Discovery of a fossil bird in the Jurassic of Wyoming. *American Journal of Science*, 3d ser., 21:341–342.

———. 1896. *The dinosaurs of North America.* Sixteenth Annual Report of the U.S. Geological Survey, Washington, D.C.

Marshall, L. G. 1978. The terror bird. *Bulletin of the Field Museum of Natural History* 49:6–15.

———. 1988. Land mammals and the Great American Interchange. *American Scientist* 76:380–388.

———. 1994. The terror birds of South America. *Scientific American* 270(2):90–95.

Marshall, L. G., R. F. Butler, R. E. Drake, G. H. Curtis, and R. H. Tedford. 1979. Calibration of the Great American Interchange. *Science* 204:272–279.

Marshall, L. G., D. D. Webb, J. J. Sepkoski, and D. M. Raup. 1982. Mammalian evolution and the Great American Interchange. *Science* 215:1351–1357.

Martin, L. D. 1983a. The origin of birds and of avian flight. *Current Ornithology* 1:105–129.

———. 1983b. The origin and early radiation of birds. In *Perspectives in Ornithology*, ed. A. H. Brush and G. A. Clark, Jr., 291–338. Cambridge: Cambridge University Press.

———. 1984. A new hesperornithid and the relationships of the Mesozoic birds. *Transactions of the Kansas Academy of Science* 87(3/4):141–150.

———. 1985. The relationship of *Archaeopteryx* to other birds. In *The beginnings of birds*, ed. M. K. Hecht et al., 177–183. Eichstätt: Freunde des Jura-Museum.

———. 1987. The beginning of the modern avian radiation. *Documents des Laboratoires de Géologie, Lyon* 99:9–20.

———. 1988. Review of *The origin of birds and the evolution of flight*, edited by Kevin Padian. *Auk* 105:596–597.

———. 1991. Mesozoic birds and the origin of birds. In *Origins of the higher groups of tetrapods*, ed. H.-P. Schultze and L. Treube, 485–540. Ithaca, N.Y.: Cornell University Press.

———. 1992. The status of the late Paleocene birds *Gastornis* and *Remiornis*. *Los Angeles County Museum of Natural History, Science Series* 36:97–108.

———. 1995a. A new skeletal model of *Archaeopteryx*. *Archaeopteryx* 13:33–40.

———. 1995b. The Enantiornithes: Terrestrial birds of the Cretaceous in avian evolution. *Courier Forschungsinstitut Senckenberg*, 181:23–36.

Martin, L. D., and R. M. Mengel. 1975. A new species of Anhinga (Anhingidae) from the Upper Pliocene of Nebraska. *Auk* 92:137–140.

Martin, L. D., and A. M. Neuner. 1978. The end of the Pleistocene in North America. *Transactions of the Nebraska Academy of Sciences* 6:117–126.

Martin, L. D., and J. D. Stewart. 1977. Teeth in *Ichthyornis* (class: Aves). *Science* 195:1331–1332.

———. 1982. An ichthyornithiform bird from the Campanian of Canada. *Canadian Journal of Earth Sciences* 19:324–327.

———. 1985. Homologies in the avian tarsus. *Nature* 315:159.

Martin, L. D., J. D. Stewart, and K. N. Whetstone. 1980. The origin of birds: structure of the tarsus and teeth. *Auk* 97:86–93.

Martin, L. D., and J. Tate. 1976. The skeleton of *Baptornis advenus* (Aves: Hesperornithiformes). *Smithsonian Contributions to Paleobiology* 27:35–66.

Martin, P. S. 1973. The discovery of America. *Science* 179:969–974.

Martin, P. S., and R. G. Klein, eds. 1984. *Quaternary extinctions: a prehistoric revolution.* Tucson: University of Arizona Press.

Martins-Neto, R. C., and A. W. A. Kellner. 1988. Primeiro registro de pena na Formação Santana (cretáceo inferior), Bacia do Araripe, nordeste do Brasil. *Anais da Academia Brasileira de Ciências* 60:61–68.

Marx, J. L. 1978. Warm-blooded dinosaurs: evidence pro and con. *Science* 199:1424–1426.

Matthew, W. D., and W. Granger. 1917. The skeleton of *Diatryma*, a gigantic bird from the Lower Eocene of Wyoming. *Bulletin of the American Museum of Natural History* 37:307–326.

Maurer, D., and R. J. Raikow. 1981. Appendicular myology, phylogeny, and classification of the avian order Coraciiformes (including Trogoniformes). *Annals of the Carnegie Museum of Natural History* 50:417–434.

Mayr, E. 1960. The emergence of evolutionary novelties. In *The evolution of life*, ed. S. Tax, 349–380. Chicago: University of Chicago Press.

———. 1979. *Check-list of birds of the world.* Vol. 8. Ed.

M. A. Traylor, Jr. Cambridge, Mass.: Museum of Comparative Zoology.

———. 1981. Biological classification: toward a synthesis of opposing methodologies. *Science* 214:510–516.

Mayr, E., and D. Amadon. 1951. A classification of recent birds. *American Museum Novitates* 1496:1–42.

Mayr, E., and W. J. Bock. 1994. Provisional classifications v. standard avian sequences: heuristics and communication in ornithology. *Ibis* 136:12–18.

Mayr, F. X. 1973. Ein Neuer *Archaeopteryx*-fund. *Paläontologica Zeitschrift* 47:17–24.

McDowell, S. 1948. The bony palate of birds. Part I. The Paleognathae. *Auk* 65:520–549.

McGilp, J. N., and A. M. Morgan. 1931. The nesting of the banded stilt (*Cladorhynchus leucocephalus*). *South Australian Ornithologist* 11(2):37–53.

McGowen, C. 1984. Evolutionary relationships of ratites and carinates from the ontogeny of the tarsus. *Nature* 307:733–735.

———. 1985. Homologies in the avian tarsus, McGowan replies. *Nature* 315:159–160.

———. 1989. Feather structure in flightless birds and its bearing on the question of the origin of feathers. *Journal of Zoology, London* 218:537–547.

———. 1991. *Dinosaurs, spitfires, and sea dragons*. Cambridge, Mass.: Harvard University Press.

McGrew, P. O., and A. Feduccia. 1973. A preliminary report on a nesting colony of Eocene birds. *Wyoming Geological Association Twenty-fifth Conference Guidebook*, 163–164.

McKenna, M. 1975. Fossil mammals and the early Eocene North Atlantic land continuity. *Annals of the Missouri Botanical Garden* 62:335–353.

McKinney, M. L., ed. 1988. *Heterochrony in evolution*. New York: Plenum Press.

McKinney, M. L., and K. J. McNamara. 1991. *Heterochrony: the evolution of ontogeny*. New York: Plenum Press.

McNab, B. K. 1994. Energy conservation and the evolution of flightlessness in birds. *American Naturalist* 144:628–642.

McNamara, J. A. 1976. Aspects of the feeding ecology of *Cladorhynchus leucocephalus* (Recurvirostridae). B.S. thesis, University of Adelaide.

Meinke, D. K., K. Padian, and J. Kappelman. 1980. Growth rings in dinosaur teeth. *Nature* 288:193–194.

Meyer, H. von. 1857. Beiträge zur näheren Kenntniss fossiler Reptilien. *Neues Jahrbuch für Mineralogie, Geologie, und Paläontologie* 1857:532–543.

———. 1861. Vögel-Federn und *Palpipes pricus* von Solnhofen. *Neues Jahrbuch für Mineralogie, Geologie, und Paläontologie* 1861:561.

Mikhailov, K. E., and E. N. Kurochkin. 1988. The eggshells of Struthioniformes from the Palearctic and its position in the system of views on ratite evolution. English summary in *Fossil reptiles and birds of Mongolia. Transactions of the Joint Soviet-Mongolian Paleontological Expedition* 34:108–109.

Millard, R. A. 1995. The body temperature of *Tyrannosaurus rex*. *Science* 267:1666.

Millener, P. R. 1982. And then there were twelve: the taxonomic status of *Anomalopteryx oweni* (Aves: Dinornithidae). *Notornis* 29:165–170.

Miller, A. H. 1931. An auklet from the Eocene of Oregon. *University of California Publications, Department of Geological Sciences* 20:23–26.

———. 1953. A fossil hoatzin from the Miocene of Colombia. *Auk* 70:484–489.

———. 1963. Fossil ratite birds of the late Tertiary of South Australia. *Records of the South Australian Museum* 14:413–420.

Miller, A. H., and L. V. Compton. 1939. Two fossil birds from the Lower Miocene of South Dakota. *Condor* 41:153–156.

Miller, H. W. 1968. Invertebrate fauna and environment of deposition of the Niobrara Formation (Cretaceous) of Kansas. *Fort Hays Studies, Sciences Series* 8:1–90.

Miller, L. H. 1916. Two vulturid raptors from the Pleistocene of Rancho La Brea. *University of California Publications, Department of Geology* 9:105–109.

———. 1925. The birds of Rancho La Brea. *Carnegie Institution of Washington Publication* 349:63–106.

———. 1935. New bird horizons in California. *University of California, Los Angeles, Publications in the Biological Sciences* 1(5):73–80.

Miller, L. H., and I. S. DeMay. 1942. The fossil birds of California: an avifauna and bibliography with annotations. *University of California Publications in Zoology* 47:47–142.

Miller, R. S., and J. P. Hatfield. 1974. Age ratios of sandhill cranes. *Journal of Wildlife Management* 38:234–242.

Milne-Edwards, A. 1867–1871. *Recherches anatomiques et paléontologiques pour servir à l'histoire des oiseaux fossiles de la France*. 4 vols. Paris: Victor Masson et Fils.

———. 1892. Sur les oiseaux fossiles des dépots éocènes de phosphate de chaux du sud de la France. *Comptes Rendus, Congrés d'Ornithologie International (Budapest, 1891)* 2:60–80.

Milner, A. R. 1985. *Cosesaurus*—the last proavian? *Nature* 315:544.

———. 1993. Ground rules for early birds. *Nature* 362:589.

Milner, A. R., and S. E. Evans. 1991. The Upper Jurassic diapsid *Lisboasaurus estesi*—a maniraptoran theropod. *Palaeontology* 34:503–513.

Mindell, D. P. 1992. DNA-DNA hybridization and avian phylogeny. *Systematic Biology* 41:126–134.

Mlikovsky, J. 1989a. A new guineafowl (Aves: Phasianidae) from the late Eocene of France. *Annalen des Naturhistorischen Museums in Wien* 90A:63–66.

———. 1989b. A new swift (Aves: Apodidae) from the late Eocene of France. *Annalen des Naturhistorischen Museums in Wien* 90A:59–62.

Molnar, R. E. 1980. Australian late Mesozoic tetrapods: some implications. *Mémoires de la Société Géologique de France* 139:131–143.

———. 1986. An enantiornithine bird from the Lower

Cretaceous of Queensland, Australia. *Nature* 322:736–738.

Monastersky, R. 1990. Chinese bird fossil: mix of old and new. *Science News* 138:246–247.

———. 1991. The lonely bird. *Science News* 140:104–105.

———. 1993. The accidental reign. *Science News* 143:60–62.

Monroe, B. L., Jr., and C. G. Sibley. 1993. *A world checklist of birds*. New Haven and London: Yale University Press.

Montevecchi, B. 1994. The great auk cemetery. *Natural History* (August):6–8.

Mooers, A., and P. Cotgreave. 1994. Sibley and Ahlquist's tapestry dusted off. *Trends in Ecology and Evolution* 9:458–459.

Moreau, R. E. 1966. *The bird faunas of Africa and its islands*. New York: Academic Press.

Morell, V. 1994. Warm-blooded dino debate blows hot and cold. *Science* 265:188.

Morse, E. S. 1872. On the tarsus and carpus of birds. *Annals of the Lyceum of Natural History* (New York) 10:1–22.

Mourer-Chauviré, C. 1978. La pouche à phosphate de Sainte-Neboule (Lot) et sa faune de vertébrés du Ludien supérieur. 6. Oiseaux. *Paleovertebrata* 8:217–229.

———. 1980. The Archaeotrogonidae from the Eocene and Oligocene Phosphorites du Quercy (France). *Los Angeles County Museum of Natural History, Contributions to Science* 330:17–31.

———. 1981. Première indication de la présence de Phorusracidés, famille d'oiseaux géants d'Amerique du Sud, dans le tertiaire européen: *Ameghinornis* nov. gen. (Aves, Ralliformes des Phosphorites du Quercy, France). *Géobios* 14:637–647.

———. 1982. Les oiseaux fossiles des Phosphorites du Quercy (éocène supérieur à oligocène supérieur): implications paléobiogéographiques. *Géobios Mémoires Spéciale* 6:413–426.

———. 1983. Les Gruiformes (Aves) des Phosphorites du Quercy (France). I. Sous-ordre Cariamae (Cariamidae et Phorusrhacidae) systématique et biostratigraphie. *Palaeovertebrata* 13:83–143.

———. 1985. Les Todidae (Aves, Coraciiformes) des Phosphorites du Quercy (France). *Proceedings of the Koninklijke Nederlandse Akademie van Wetenschappen*, ser. B, 88:407–414.

———. 1987. Les Strigiformes (Aves) des Phosphorites du Quercy (France): systématique, biostratigraphie et paléobiogéographie. *Documents des Laboratoires de Géologie, Lyon* 99:89–135.

———. 1988a. Le gisement du Bretou (Phosphorites du Quercy, Tarn-et-Garonne, France) et sa faune de vertébrés de l'éocène supérieur. II. Oiseaux. *Palaeontographica*, ser. A, 205:29–50.

———. 1988b. Les Aegialornithidae (Aves: Apodiformes) des Phosphorites du Quercy: comparaison avec la forme de Messel. *Courier Forschungsinstitut Senckenberg* 107:369–381.

———. 1989. Les Caprimulgiformes et les Coraciiformes de l'éocène et de l'oligocène des Phosphorites du Quercy et description de deux genres nouveaux de Podargidae et Nyctibiidae. *Proceedings of the Nineteenth International Ornithological Congress*, 2047–2055.

———. 1991. Les Horusornithidae nov. fam., Accipitriformes (Accipitriformes, Aves) a articulation intertarsienne hyperflexible de l'éocène du Quercy. *Géobios* 13:183–192.

———. 1992a. Un ganga primitif (Aves, Columbiformes, Pteroclidae) de très grande taille dans le paléogène des Phosphorites du Quercy (France). *Académie des Sciences, Comptes Rendus*, ser. 2A, 314:229–235.

———. 1992b. Une nouvelle famillie de perroquets (Aves, Psittaciformes) dans l'éocène supérieur des Phosphorites du Quercy (France). *Géobios* 14:169–177.

———. 1992c. The Galliformes (Aves) from the Phosphorites du Quercy (France): systematics and biostratigraphy. *Los Angeles County Museum of Natural History, Science Series* 36:67–96.

———. 1993. Les gangas (Aves, Columbiformes, Pteroclidae) du paléogène et du miocène inférieur de France. *Palaeovertebrata* 22:73–98.

———. 1994. A large owl from the Palaeocene of France. *Paleontology* 37(2):339–348.

———. 1995. The Messelornithidae (Aves: Gruiformes) from the Paleogene of France. *Courier Forschungsinstitut Senckenberg* 181:95–105.

Mourer-Chauviré, C., R. Bour, F. Moutou, and S. Ribes. 1994. *Mascarenotus* nov. gen. (Aves, Strigiformes), genre endemique eteint des Mascareignes et *M. grucheti* n. sp., espece eteinte de la Réunion. *Académie des Sciences, Comptes Rendus*, ser. 2A, 318:1699–1706.

Mourer-Chauviré, C. R. Bour, and S. Tibes. 1995a. Position systématique du solitaire de la Réunion: nouvelle interprétation basée sur les restes fossiles et les récits des anciens voyageurs. *Académie de Sciences, Comptes Rendus*, ser. 2a, 320:1125–1131.

———. 1995b. Was the solitaire of Réunion an ibis? *Nature* 373:568.

Mourer-Chauviré, C., and J. Cheneval. 1983. Les Sagittariidae fossiles (Aves, Accipitriformes) de l'oligocène des Phosphorites du Quercy et du miocène inférieur de Saint-Gérand-le-Puy. *Géobios* 16:443–459.

Mourer-Chauviré, C., M. Hugueney, and P. Jonet. 1989. Découverte de Passeriformes dans l'oligocène supérieur de France. *Académie des Sciences, Comptes Rendus*, ser. 2A, 309:843–849.

Mourer-Chauviré, C., and A. S. Marco. 1988. Présence de *Tyto balearica* (Aves, Strigiformes) dans des gisements continentaux du pliocène de France et d'Espagne. *Géobios* 21:639–644.

Mourer-Chauviré, C., and F. Poplin. 1985. Le mystère des tumulus de Nouvelle-Calédonie. *La Recherche* 16:1094.

Moynihan, M. 1976. *The New World primates*. Princeton, N.J.: Princeton University Press.

Mudge, B. F. 1879. Are birds derived from dinosaurs? *Kansas City Review of Science* 3:224–226.

Müller, G. B., and G. P. Wagner. 1991. Novelty in evolution: restructuring the concept. *Annual Reviews of Evolution and Systematics* 22:229–256.

Müller, J. 1847. Über die bisher unbekannten typischen Verschiedenheiten der Stimmorgane der Passerinen. *Abhandlungen Königlich Akademische Wissenschaft, Berlin,* 1–71.

Murphy, R. C. 1936. *Oceanic birds of South America.* 2 vols. New York: American Museum of Natural History.

Mutschler, Von O. 1927. Die Gymnospermen des Weissen Jura von Nusplingen. *Jahresberichte und Mitteilungen des Oberrheinischer Geologische "Vereines"* (January):25–50.

Nachtigall, W. 1977. Zur Bedeutung der Reynoldszahl in der Schwimmphysiologie und Flugbiophysik. *Fortschritte der Zoologie* 24(2/3):13–56.

Nelson, C. H., and K. R. Johnson. 1987. Whales and walruses as tillers of the sea floor. *Scientific American* 256(2):112–117.

Nessov, L. A. 1984. [Upper Cretaceous pterosaurs and birds from Central Asia.] *Paleontological Journal* 1984(1):47–57. [In Russian.]

———. 1986. [The first find of the Late Cretaceous bird *Ichthyornis* in the Old World and some other bird bones from the Cretaceous and Paleogene of Soviet Middle Asia.] *USSR Academy of Sciences, Proceedings of the Zoological Institute* 147:31–38. [In Russian.]

———. 1990. [Small *Ichthyornis* and other findings of bird bones from the Bissekty Formation (Upper Cretaceous) of central Kizylkum Desert.] *USSR Academy of Sciences, Proceedings of the Zoological Institute* 210:59–62. [In Russian.]

———. 1992a. Mesozoic and Paleogene birds of the USSR and their paleoenvironments. *Los Angeles County Museum of Natural History, Science Series* 36:465–478.

———. 1992b. [Record of the localities of Mesozoic and Paleogene with avian remains in the USSR, and the description of new findings.] *Russian Journal of Ornithology* 1:7–50. [In Russian.]

———. 1992c. Nonflying birds of meridional late Cretaceous sea straits of North America, Scandinavia, Russia and Kazakhstan as indicators of features of oceanic circulation. *Bulletin Moscow Society of National Geology* 67:78–83.

Nessov, L. A., and L. J. Borkin. 1983. [New records of bird bones from the Cretaceous of Mongolia and Soviet Middle Asia.] *USSR Academy of Sciences, Proceedings of the Zoological Institute,* 116:108–110. [In Russian.]

Nessov, L. A., and A. A. Jarkov. 1989. [New Cretaceous-Paleogene birds of the USSR and some remarks on the origin and evolution of the class Aves.] *Proceedings of the Zoological Institute of Leningrad* 197:78–97. [In Russian with English summary.]

———. 1993. [Hesperornithiforms in Russia.] *Russian Journal of Ornithology* 2:37–54. [In Russian.]

Nessov, L. A., and A. V. Pantheleev. 1993. On the similarity of the late Cretaceous ornithofauna of South America and western Asia. *Russian Academy of Sciences, Proceedings of the Zoological Institute* 252:84–94.

Nessov, L. A., and B. V. Prizemlin. 1991. [Big advanced seabirds of the order Hesperornithiformes of the late Senonian of the Turgaj Strait: the first findings of the group in the USSR.] *Proceedings of the Zoological Institute of Leningrad* 23:85–107. [In Russian with English summary.]

Newton, A. 1896. *A dictionary of birds.* London: Adam and Charles Black.

Noble, J. W. 1993. Fossils suggest a turkey with a dinosaur's bite. *New York Times,* 15 April.

Nopcsa, F. von. 1907. Ideas on the origin of flight. *Proceedings of the Zoological Society of London* 1907:223–236.

———. 1923. On the origin of flight in birds. *Proceedings of the Zoological Society of London* 1923:463–477.

Norberg, R. Å. 1981. Why foraging birds in trees should climb and hop upwards rather than downwards. *Ibis* 123:281–288.

———. 1983. Optimum locomotion modes of foraging birds in trees. *Ibis* 125:172–180.

———. 1985. Function of vane asymmetry and shaft curvature in bird flight feathers: inferences on flight ability of *Archaeopteryx.* In *The beginnings of birds,* ed. M. K. Hecht et al., 303–318. Eichstätt: Freunde des Jura-Museum.

———. 1995. Feather asymmetry in *Archaeopteryx. Nature* 374:221.

Norberg, U. M. 1985. Evolution of flight in birds: aerodynamic, mechanical, and ecological aspects. In *The beginnings of birds,* ed. M. K. Hecht et al., 293–302. Eichstätt: Freunde des Jura-Museum.

———. 1986. On the evolution of flight and wing forms in bats. In *Bat flight—Fledermausflug,* ed. W. Nachtigall, 13–26. Biona Report 5. Stuttgart: Gustav Fischer Verlag.

———. 1990. *Vertebrate flight.* Berlin: Springer-Verlag.

Norell, M. A., E. S. Gaffney, and L. Dingus. 1995. *Discovering dinosaurs in the American Museum of Natural History.* New York: Alfred A. Knopf.

Noriega, J. I., and L. M. Chiappe. 1991. El mas antiguo Passeriformes, Aves, de America del Sur. *Ameghiniana* 28:410.

Norman, D. B. 1992. Dinosaurs past and present. *Journal of Zoology* (London) 228:173–181.

Novacek, M. J. 1994. Whales leave the beach. *Nature* 368:804.

Novas, F. C. 1993. New information on the systematics and postcranial skeleton of *Herrerasaurus ischigualastensis* (Theropoda: Herrerasauridae) from the Ischigualasto Formation (Upper Triassic) of Argentina. *Journal of Vertebrate Paleontology* 13: 400–423.

Obst, B. 1986. Wax digestion in Wilson's storm-petrel. *Wilson Bulletin* 98:189–195.

Oliver, W. R. B. 1949. The moas of New Zealand and Australia. *Dominion Museum Bulletin* (New Zealand) 15:1–206.

Olson, S. L. 1973. Evolution of the rails of the South Atlantic islands (Aves: Rallidae). *Smithsonian Contributions to Zoology* 152:1–53.

———. 1975a. *Ichthyornis* in the Cretaceous of Alabama. *Wilson Bulletin* 87:103–105.

———. 1975b. Paleornithology of St. Helena Island, South Atlantic Ocean. *Smithsonian Contributions to Paleobiology* 23:1–49.

———. 1976a. A new species of *Milvago* from Hispaniola, with notes on other fossil caracaras from the West Indies (Aves: Falconidae). *Proceedings of the Biological Society of Washington* 88:355–366.

———. 1976b. Oligocene fossils bearing on the origins of the Totidae and Momotidae (Aves: Coraciiformes). *Smithsonian Contributions to Paleobiology* 27:111–119.

———. 1977a. A great auk, *Pinguinis,* from the Pliocene of North Carolina (Aves: Alcidae). *Proceedings of the Biological Society of Washington* 90:690–697.

———. 1977b. A Lower Eocene frigatebird from the Green River Formation of Wyoming (Pelecaniformes: Frigatidae). *Smithsonian Contributions to Paleobiology* 35:1–33.

———. 1977c. A synopsis of the fossil Rallidae. In *Rails of the world: a monograph of the family Rallidae,* ed. D. S. Ripley, 509–525. Boston: Godine.

———. 1979. Multiple origins of the Ciconiiformes. *Proceedings of the Colonial Waterbird Group* 1978:165–170.

———. 1980. A new genus of penguin-like pelecaniform bird from the Oligocene of Washington (Pelecaniformes: Plotopteridae). *Los Angeles County Museum of Natural History, Contributions to Science* 330:51–57.

———. 1982a. A critique of Cracraft's classification of birds. *Auk* 99:733–739.

———. 1982b. The generic allocation of *Ibis pagana* Milne-Edwards, with a review of fossil ibises (Aves: Threskiornithidae). *Journal of Vertebrate Paleontology* 1:165–170.

———. 1983. Evidence for a polyphyletic origin of the Piciformes. *Auk* 100:126–133.

———. 1984a. A hammerkop from the early Pliocene of South Africa (Aves: Scopidae). *Proceedings of the Biological Society of Washington* 97:736–740.

———. 1984b. A very large enigmatic owl (Aves: Strigidae) from the late Pleistocene at Ladds, Georgia. *Carnegie Museum of Natural History, Special Publication* 8:44–46.

———. 1985a. The fossil record of birds. In *Avian biology,* vol. 8, ed. D. S. Farner, J. R. King, and K. C. Parkes, 79–252. New York: Academic Press.

———. 1985b. Early Pliocene Procellariiformes (Aves) from Langebaanweg, South-western Cape Province, South Africa. *Annals of the South African Museum* 95:123–145.

———. 1985c. An early Pliocene marine avifauna from Duinefontein, Cape Province, South Africa. *Annals of the South African Museum* 95:147–164.

———. 1987a. Review of *The beginnings of birds,* ed. M. K. Hecht et al. *American Scientist* 75:74–75.

———. 1987b. An early Eocene oilbird from the Green River Formation of Wyoming (Caprimulgiformes: Steatornithidae). *Documents des Laboratoires de Géologie, Lyon* 99:56–70.

———. 1989. Aspects of the global avifaunal dynamics during the Cenozoic. *Proceedings of the Nineteenth International Ornithological Congress,* 2023–2029.

———. 1992a. *Neogaeornis wetzeli* Lambrecht, a Cretaceous loon from Chile. *Journal of Vertebrate Paleontology* 12:122–124.

———. 1992b. A new family of primitive landbirds from the Lower Eocene Green River Formation of Wyoming. *Los Angeles County Museum of Natural History, Science Series* 36:127–136.

———. 1994. A giant *Presbyornis* (Aves: Anseriformes) and other birds from the Paleocene Aquia Formation of Maryland and Virginia. *Proceedings of the Biological Society of Washington,* 107:429–435.

———. 1995. Redescription of *Thiornis sociata* Navas, a nearly complete Miocene grebe from Spain (Aves: Podicipedidae). *Courier Forschungsinstitut Senckenberg,* 181:131–140.

Olson, S. L., J. C. Balouet, and C. T. Fisher. 1987. The owlet-nightjar of New Caledonia, *Aegotheles savesi,* with comments on the systematics of the Aegothelidae. *Le Gerfaut* 77:341–352.

Olson, S. L., and A. Feduccia. 1979. Flight capability and the pectoral girdle of *Archaeopteryx. Nature* 278:247–248.

———. 1980a. Relationships and evolution of flamingos (Aves: Phoenicopteridae). *Smithsonian Contributions to Zoology* 316:1–73.

———. 1980b. *Presbyornis* and the origin of the Anseriformes (Aves: Charadriomorphae). *Smithsonian Contributions to Zoology* 323:1–24.

Olson, S. L., and Y. Hasegawa. 1979. Fossil counterparts of giant penguins from the North Pacific. *Science* 206:688–689.

Olson, S. L., and H. F. James. 1982. Fossil birds from the Hawaiian Islands: evidence for wholesale extinction by man before Western contact. *Science* 21:633–635.

———. 1991. Descriptions of thirty-two new species of birds from the Hawaiian Islands. Part I. non-Passeriformes. *Ornithological Monographs* 45:1–88.

Olson, S. L., and E. N. Kurochkin. 1987. Fossil evidence of a tapaculo in the Quaternary of Cuba (Aves: Passeriformes: Scytalopodidae). *Proceedings of the Biological Society of Washington* 100:353–357.

Olson, S. L., and D. C. Parris. 1987. The Cretaceous birds of New Jersey. *Smithsonian Contributions to Paleobiology* 63:1–22.

Olson, S. L., and D. T. Rasmussen. 1986. The paleoenvironment of the earliest hominoids: new evidence from the Oligocene avifauna of Egypt. *Science* 233:1202–1204.

Olson, S. L., and D. W. Steadman. 1977. A new genus of flightless ibis (Threskiornithidae) and other fossil birds from cave deposits in Jamaica. *Proceedings of the Biological Society of Washington* 92:23–27.

———. 1981. The relationships of the Pedionomidae (Aves: Charadriiformes). *Smithsonian Contributions to Zoology* 337:1–25.

Olson, S. L., and A. Wetmore. 1976. Preliminary diagnoses of extraordinary new genera of birds and Pleistocene deposits in the Hawaiian Islands. *Proceedings of the Biological Society of Washington* 89:247–258.

Osborn, H. F. 1900. Reconsideration of the evidence for a common dinosaur-avian stem in the Permian. *American Naturalist* 34:777–799.

———. 1916. Skeletal adaptations of *Ornitholestes, Struthiomimus, Tyrannosaurus*. *Bulletin of the American Museum of Natural History* 35:733–771.

Ostrom, J. H. 1969a. Osteology of *Deinonychus antirrhopus*, an unusual theropod from the Lower Cretaceous of Montana. *Bulletin of the Peabody Museum of Natural History, Yale University* 30:1–165.

———. 1969b. Terrestrial vertebrates as indicators of Mesozoic climates. *Proceedings of the North American Paleontological Convention*, pt. D:347–376.

———. 1970. *Archaeopteryx:* notice of a "new" specimen. *Science* 170:537–538.

———. 1973. The ancestry of birds. *Nature* 242:136.

———. 1974. *Archaeopteryx* and the origin of flight. *Quarterly Review of Biology* 49:27–47.

———. 1975. The origin of birds. *Annual Review of Earth and Planetary Science* 3:55–77.

———. 1976a. *Archaeopteryx* and the origin of birds. *Biological Journal of the Linnean Society* 8:91–182.

———. 1976b. Some hypothetical anatomical stages in the evolution of avian flight. *Smithsonian Contributions to Paleobiology* 27:1–21.

———. 1976c. On a new specimen of *Deinonychus antirrhopus*. *Brevoria* 439:1–21.

———. 1978a. The osteology of *Compsognathus longipes* Wagner. *Zitteliana* 4:73–118.

———. 1978b. New ideas about dinosaurs. *National Geographic* 154:152–185.

———. 1979. Bird flight: how did it begin? *American Scientist* 67:46–56.

———. 1985a. The Yale *Archaeopteryx:* the one that flew the coop. In *The beginnings of birds*, ed. M. K. Hecht et al., 359–367. Eichstätt: Freunde des Jura-Museum.

———. 1985b. The meaning of *Archaeopteryx*. In *The beginnings of birds*, ed. M. K. Hecht et al., 161–176. Eichstätt: Freunde des Jura-Museum.

———. 1986a. The cursorial origin of avian flight. *Memoires of the California Academy of Sciences* 8:73–81.

———. 1986b. The Jurassic "bird" *Laopteryx priscus* reexamined. *Contributions to Geology, University of Wyoming, Special Paper* 3:11–19.

———. 1987a. *Protoavis*, a Triassic bird? *Archaeopteryx* 5:113–114.

———. 1987b. Romancing the dinosaurs. *The Sciences* (May–June):56–63.

———. 1991a. The bird in the bush. *Nature* 353:212.

———. 1991b. The question of the origin of birds. In *Origins of the higher groups of tetrapods: controversy and consensus*, ed. H.-P. Schultze and L. Trueb, 467–484. Ithaca, N.Y.: Cornell University Press.

———. 1992. Comments on the new (Solnhofen) specimen of *Archaeopteryx*. *Los Angeles County Museum of Natural History, Science Series* 36:25–27.

———. 1994. On the origin of birds and of avian flight. In *Major features of vertebrate evolution*, ed. D. P. Prothero and R. M. Schoch, 160–177. Knoxville: University of Tennessee Press.

———. 1995. Wing biomechanics and the origin of bird flight. *Neues Jahrbuch für Geologie und Paläontologie Abhandlungen* 195:253–266.

Owen, R. 1836. Aves. In *Todd's cyclopaedia in anatomy and physiology*, 1:265–358.

———. 1839. On the bone of an unknown Struthious bird from New Zealand. *Proceedings of the Zoological Society of London* 1839:169–170.

———. 1862. On the fossil remains of a longtailed bird, *Archaeopteryx macrura* from the lithographic slate of Solnhofen. *Proceedings of the Zoological Society of London* 12:272–273.

———. 1863. On the *Archaeopteryx* of von Meyer, with a description of the fossil remains of a long-tailed species from the lithographic stone of Solnhofen. *Philosophical Transactions of the Royal Society of London* 153:33–47.

———. 1866. *On the anatomy of vertebrates*. Vol. 2: *Birds and mammals*. London.

———. 1875. Monographs of the British fossil Reptilia of the Mesozoic formations. Part II. (Genera *Bothriospondylus, Cetiosaurus, Omosaurus*). *Palaeontographical Society, Monographs* 29:15–93.

———. 1879. *Memoirs on the extinct wingless birds of New Zealand; with an appendix* . . . . 2 vols. London: John van Voorst.

Owen-Smith, N. 1987. Pleistocene extinctions: the pivotal role of megaherbivores. *Paleobiology* 13:351–362.

Padian, K. 1982. Macroevolution and the origin of major adaptations: vertebrate flight as a paradigm for the analysis of patterns. *Proceedings of the Third North American Paleontological Convention*, 2:387–392.

———. 1983. A functional analysis of flying and walking in pterosaurs. *Paleobiology* 9:218–239.

———. 1984. The origin of pterosaurs. *Third symposium on Mesozoic terrestrial ecosystems*, ed. W.-E. Reif and R. Westphal, 163–168. Tübingen: Attempto Verlag.

———. 1985. The origins and aerodynamics of flight in extinct vertebrates. *Paleontology* 28:413–433.

Palmer, R. S. 1962. *Handbook of North American birds*. Vol. 1. New Haven: Yale University Press.

Parker, W. K. 1887. On the morphology of birds. *Proceedings of the Royal Society of London* 42:52–58.

———. 1888a. On the structure and development of the wing in the common fowl. *Transactions of the Zoological Society of London*, ser. B, 179:385–395.

———. 1888b. On the presence of claws in the wings of birds. *Ibis* 5:124–128.

———. 1891. On the morphology of the reptilian bird *Opisthocomus cristatus*. *Philosophical Transactions of the Royal Society of London*, ser. B, 13:45–83.

Parkes, K. C. 1966. Speculations of the origin of feathers. *Living Bird* 5:77–86.

Parkes, K. C., and G. A. Clark, Jr. 1966. An additional character linking ratites and tinamous, and an interpretation of their monophyly. *Condor* 68:459–471.

Parris, D. C., and J. Echols. 1992. The fossil bird *Ichthyornis* in the Cretaceous of Texas. *Texas Journal of Science* 44:201–212.

Patterson, B., and J. L. Kraglievich. 1960. Sistemáticas y nomenclatura de las aves fororracoideas del plioceno argentino. *Publicaciones del Museo Municipal de Ciencias Naturales y Tradicional de Mar del Plata* 1(1):1–51.

Patterson, C. 1964. A review of Mesozoic acanthopterygian fishes with special reference to those of the English Chalk. *Philosophical Transactions of the Royal Society of London*, ser. B, 247:213–482.

———. 1982. Morphological characters and homology. In *Problems of phylogenetic reconstruction*, ed. K. A. Joysey and A. E. Friday, 21–74. New York: Academic Press.

Patterson, C., and P. V. Rich. 1987. The fossil history of the emus, *Dromaius* (Aves: Dromaiinae). *Records of the South Australian Museum* 36:63–126.

Paul, G. S. 1988. *Predatory dinosaurs of the world*. New York: Simon and Schuster.

Payne, R. B., and C. J. Risley. 1976. Systematics and evolutionary relationships among the herons (Ardeidae). *Miscellaneous Publications of the Museum of Zoology, University of Michigan* 150:1–115.

Pennycuick, C. J. 1972. Animal flight. *Studies in biology*, 33. Southampton: Camelot Press.

———. 1986. Mechanical constraints on the evolution of flight. *Memoires of the California Academy of Sciences* 8:83–98.

Perle, A., L. M. Chiappe, R. Barsbold, J. M. Clark, and M. A. Norell. 1994. Skeletal morphology of *Mononykus olecranus* (Theropoda: Avialae) from the late Cretaceous of Mongolia. *American Museum Novitates* 3105:1–29.

Perle, A., M. A. Norell, L. M. Chiappe, and J. M. Clark. 1993a. Flightless bird from the Cretaceous of Mongolia. *Nature* 362:623–626.

———. 1993b. Flightless bird from the Cretaceous of Mongolia [correction]. *Nature* 362:188.

Peters, D. S. 1983. Die "Schnepfenralle" *Rhynchaeites messelensis* Wittich 1898 ist ein Ibis. *Journal für Ornithologie* 124:1–27.

———. 1987a. *Juncitarsus merkeli*, n. sp. stützt die Ableitung der Flamingos von Regenpfeifervögeln (Aves: Charadriiformes: Phoenicopteridae). *Courier Forschungsinstitut Senckenberg* 97:141–155.

———. 1987b. Ein "Phorusrhacide" aus dem Mittel-Eozän von Messel (Aves: Gruiformes: Cariamae). *Documents des Laboratoires de Géologie, Lyon* 99:71–88.

———. 1989. Fossil birds from the oil shale of Messel (Lower Middle Eocene, Lutetian). *Proceedings of the Nineteenth International Ornithological Congress*, 2056–2064.

———. 1991. Zoogeographical relationships of the Eocene avifauna from Messel (Germany). *Proceedings of the Twentieth International Ornithological Congress*, 572–577.

———. 1992a. A new species of owl (Aves: Strigiformes) from the Middle Eocene Messel oil shale. *Los Angeles County Museum of Natural History, Science Series* 36:161–170.

———. 1992b. Messel birds: a land-based assemblage. In *Messel: an insight into the history of life and of the earth*, ed. S. Schaal and W. Ziegler, 135–151. Oxford: Clarendon Press.

———. 1994. *Messelastur gratulator* n. gen. n. spec., ein Griefvogel aus der Grube Messel (Aves: Accipitridae). *Courier Forschungsinstitut Senckenberg* 170:3–9.

———. 1995. *Idiornis tuberculata* n. spec., ein weiterer ungewohnlicher Vogel aus der Grube Messel (Aves: Gruiformes: Cariamidae: Idiornithinae). *Courier Forschungsinstitut Senckenberg* 181:107–119.

Peters, D. S., and E. Gorgner. 1992. A comparative study on the claws of *Archaeopteryx*. *Los Angeles County Museum of Natural History, Contributions to Science* 36:29–37.

Peterson, R. T. 1963. *The birds*. New York: Time-Life Nature Library.

Petronievics, B. 1925. Über die Berliner *Archaeornis*. *Geologica Balkanskoga poluostrova* 8:37–84.

———. 1927. Nouvelles recherches sur l'ostéologie des Archaeornithes. *Annales Paleontologica* 16:39–55.

Piatt, J. F., and D. N. Nettleship. 1985. Diving depths of four alcids. *Auk* 102:293–297.

Place, A. R., and E. W. Stiles. 1992. Living off the wax of the land: bayberries and yellow-rumped warblers. *Auk* 109:334–345.

Place, A. R., N. Stoyan, R. G. Butler, and R. R. Ricklefs. 1989. The physiological basis of stomach oil formation in Leach's storm-petrel, *Oceanodroma leucorhoa*. *Auk* 106:687–699.

Polack, J. S. 1838. *New Zealand*. London: R. Bentley.

Pomeroy, D. E. 1980. Growth and plumage changes of the grey crowned crane *Balearica regulorum gibbericeps*. *Bulletin of the British Ornithologists' Club* 100(4):219–222.

Poplin, F., and C. Mourer-Chauviré. 1985. *Sylviornis neocaledoniae* (Aves, Galliformes, Megapodiidae), oiseau géant éteint de l'île des Pins (Nouvelle-Calédonie). *Géobios* 18:73–97.

Portmann, A., and W. Stingelin. 1961. The central nervous system. In *Biology and comparative physiology of birds*, vol. 2, ed. A. J. Marshall, 1–36. New York: Academic Press.

Prager, E. M., and A. C. Wilson. 1980. Phylogenetic relationships and rates of evolution in birds. *Proceedings of the Seventeenth International Ornithological Congress*, 1209–1214.

Prager, E. M., A. C. Wilson, D. T. Osuga, and R. E. Feeney. 1976. Evolution of flightless land birds on

southern continents: transferrin comparison shows monophyletic origin of ratites. *Journal of Molecular Biology* 8:283–294.
Prange, H. D., J. F. Anderson, and H. Rahn. 1979. Scaling of skeletal mass to body mass in birds and mammals. *American Naturalist* 113:103–122.
Proctor, N. S., and P. J. Lynch. 1993. *Manual of ornithology: avian structure and function.* New Haven and London: Yale University Press.
Prohaska, F. 1976. The climate of Argentina, Paraguay and Uruguay. In *Climates of Central and South America: world survey of climatology*, vol. 12, ed. W. Schwerdtfeger, 13–112. New York: Elsevier.
Prum, R. O. 1988. Phylogenetic interrelationships of the barbets (Aves: Capitonidae) and toucans (Aves: Ramphastidae) based on morphology with comparisons to DNA-DNA hybridization. *Zoological Journal of the Linnean Society* 92:313–343.
———. 1992. Syringeal morphology, phylogeny, and evolution of the Neotropical manakins (Aves: Pipridae). *American Museum Novitates* 3043:1–65.
Pycraft, W. P. 1894. Wing of *Archaeopteryx. Journal of the Oxford University Junior Science Club* 1:172–176.
———. 1900. On the morphology and phylogeny of the Palaeognathae (Ratitae and Crypturi) and Neognathae (Carinatae). *Transactions of the Zoological Society of London* 15:149–290.
———. 1903. The claws on the wings of birds. *Knowledge and Scientific News* 26:221–224.
Raath, M. A. 1985. The theropod *Syntarsus* and its bearing on the origin of birds. In *The beginnings of birds*, ed. M. K. Hecht et al., 219–227. Eichstätt: Freunde des Jura-Museum.
Rabinowitz, P. D., M. F. Coffin, and D. Falvey. 1983. The separation of Madagascar and Africa. *Science* 220:67–69.
Raikow, R. J. 1970. Evolution of diving adaptations in the stifftail ducks. *University of California Publications in Zoology* 94:1–52.
———. 1978. Appendicular myology and relationships of the New World nine-primaried oscines (Aves: Passeriformes). *Bulletin of the Carnegie Museum of Natural History* 7:1–52.
———. 1982. Monophyly of the Passeriformes: test of a phylogenetic hypothesis. *Auk* 99:431–445.
———. 1984. Hindlimb myology and phylogenetic position of the New Zealand wrens. *American Zoologist*, Abstract 446, 24:3.
———. 1985a. Locomotor system. In *Form and function in birds*, vol. 3, ed. A. S. King and J. McLelland, 57–193. New York: Academic Press.
———. 1985b. Systematic and functional aspects of the locomotor system of the scrub-birds, *Atrichornis*, and lyrebirds, *Menura* (Passeriformes: Atrichornithidae and Menuridae). *Records of the Australian Museum* 37:211–228.
———. 1986. Why are there so many kinds of passerine birds? *Systematic Zoology* 35:255–259.
———. 1987. *Hindlimb myology and evolution of the Old World suboscine passerine birds (Acanthisittidae, Pittidae, Philepittidae, Eurylaimidae).* Ornithological Monographs, 41. Washington, D.C.: American Ornithologists' Union.
———. 1994. A phylogeny of the woodcreepers (Dendrocolaptinae). *Auk*, 111:104–114.
Raikow, R. J., and J. Cracraft. 1983. Monophyly of the Piciformes: a reply to Olson. *Auk* 100:134–138.
Randolph, S. E. 1994. The relative timing of the origin of flight and endothermy: evidence from the comparative biology of birds and mammals. *Zoological Journal of the Linnaean Society* 112:389–397.
Räsänen, M. E., A. M. Linna, J. C. R. Santos, and F. R. Negri. 1995. Late Miocene tidal deposits in the Amazonian foreland basin. *Science* 269:386–389.
Rasmussen, D. T., S. L. Olson, and E. L. Simons. 1987. Fossil birds from the Oligocene Jebel Qatrani Formation, Fayum Province, Egypt. *Smithsonian Contributions to Paleobiology* 62:1–20.
Rasmussen, D. T., and R. F. Kay. 1992. A Miocene anhinga from Colombia, and comments on the zoogeographic relationships of South America's Tertiary avifauna. *Los Angeles County Museum of Natural History, Science Series* 36:225–230.
Rautian, A. S. 1978. A unique bird feather from Jurassic lake deposits in the Karatau. *Paleontological Journal* 1979:520–528. [English translation of *Paleontological Journal* 12:106–114.]
Raven, P. H., and D. I. Axelrod. 1975. History of the flora and fauna of Latin America. *American Scientist* 63:420–429.
Rawles, M. E. 1960. The integumentary system. In *Biology and comparative physiology of birds*, vol. 1, ed. A. J. Marshall, 189–240. New York: Academic Press.
Rayner, J. M. V. 1981. Flight adaptations in vertebrates. *Symposium of the Zoological Society of London* 48:137–172.
———. 1985a. Mechanical and ecological constraints on flight evolution. In *The beginnings of birds*, ed. M. K. Hecht et al., 279–288. Eichstätt: Freunde des Jura-Museum.
———. 1985b. Cursorial gliding in proto-birds: an expanded version of a discussion contribution. In *The beginnings of birds*, ed. M. K. Hecht et al., 289–292. Eichstätt: Freunde des Jura-Museum.
———. 1986. Vertebrate flapping flight mechanics and aerodynamics, and the evolution of flight in bats. In *Bat Flight—Fledermausflug*, ed. W. Nachtigall, 27–74. Biona Report 5. Stuttgart: Gustav Fischer Verlag.
———. 1988. The evolution of vertebrate flight. *Biological Journal of the Linnean Society* 34:269–287.
———. 1991. Avian flight evolution and the problem of *Archaeopteryx*. In *Biomechanics in evolution*, ed. J. M. V. Rayner and R. J. Wooton, 183–212. Cambridge: Cambridge University Press.
Rea, A. 1983. Cathartid affinities: a brief overview. In *Vulture biology and management*, ed. S. A. Wilbur and J. A. Jackson, 26–54. Berkeley: University of California Press.

Regal, P. J. 1975. The evolutionary origin of feathers. *Quarterly Review of Biology* 50:35–66.
Reichel, M. 1941. *L'Archaeopteryx:* un ancètre des oiseaux. *Nos Oiseaux* 16:93–107.
Reid, R. E. H. 1981. Lamellar-zonal bone with zones and annuli in the pelvis of a sauropod dinosaur. *Nature* 292:49–51.
———. 1984. The histology of dinosaurian bone, and its possible bearing on dinosaur physiology. *Symposium of the Zoological Society of London* 52:629–662.
Renesto, S. 1994. *Megalancosaurus,* a possibly arboreal archosauromorph (Reptilia) from the Upper Triassic of northern Italy. *Journal of Vertebrate Paleontology* 14:38–52.
Rice, W. R. 1982. Acoustical location of prey by the marsh hawk: adaptation to concealed prey. *Auk* 99:403–413.
Rich, P. V. 1972. A fossil avifauna from the Upper Miocene Beglia Formation of Tunisia. *Notes du Service Géologique* [Tunisia] 35:29–66.
———. 1975. Changing continental arrangements and the origin of Australia's non-passeriform continental avifauna. *Emu* 75:97–112.
———. 1976. The history of birds on the island continent Australia. *Proceedings of the Eighteenth International Ornithological Congress,* 53–65.
———. 1979. *The Dromornithidae.* Bureau of Mineral Resources (Geology and Geophysics) Bulletin 184. Canberra: Australian Government Publishing.
———. 1980a. The Australian Dromornithidae: a group of large extinct ratites. *Los Angeles County Museum of Natural History, Contributions to Science* 330:93–103.
———. 1980b. "New World vultures" with Old World affinities? *Contributions to Vertebrate Ecology* 5:1–115.
———. 1983. The fossil history of vultures: a world perspective. In *Vulture biology and management,* ed. S. R. Wilbur and J. A. Jackson, 3–25. Berkeley: University of California Press.
Rich, P. V., and D. J. Bohaska. 1976. The world's oldest owl: a new strigiform from the Paleocene of southwestern Colorado. *Smithsonian Contributions to Paleobiology* 27:87–93.
———. 1981. The Ogygoptyngidae, a new family of owls from the Paleocene of North America. *Alcheringa* 5:95–102.
Rich, P. V., and P. J. Haarhoff. 1985. Early Pliocene Coliidae (Aves, Coliiformes) from Langebaanweg, South Africa. *Ostrich* 56:20–41.
Rich, P. V., and A. McEvey. 1977. A new owlet-nightjar from the early to mid-Miocene of eastern New South Wales. *Memoires of the National Museum of Victoria* 38:247–253.
Rich, P. V., T. H. Rich, B. E. Wagstaff, J. M. Mason, E. B. Douthitt, R. T. Gregory, and E. A. Felton. 1988. Evidence for low temperatures and biologic diversity in Cretaceous high latitudes of Australia. *Science* 242:1403–1406.
Rich, P. V., and R. J. Scarlett. 1977. Another look at *Megaegotheles,* a large owlet-nightjar from New Zealand. *Emu* 77:1–8.
Rich, P. V., and Van Tets, G. F. 1982. Fossil birds of Australia and New Guinea: their biogeographic, phylogenetic and biostratigraphic input. In *The fossil vertebrate record of Australasia,* ed. P. V. Rich and E. M. Thompson, 235–384. Clayton: Monash University.
Rich, P. V., and C. A. Walker. 1983. A new genus of Miocene flamingo from East Africa. *Ostrich* 54:95–104.
Richmond, N. D. 1965. Perhaps juvenile dinosaurs were always scarce. *Journal of Paleontology* 39:503–505.
Ricklefs, R. E. 1983. Avian postnatal development. In *Avian biology,* vol. 7, ed. D. S. Farner, J. R. King, and K. C. Parkes, 1–83. New York: Academic Press.
Ricqles, A. de. 1969. L'histologie osseuse envisagée comme indicateur de la physiologie thermique chez les tétrapodes fossiles. *Académie des Sciences, Comptes Rendus,* ser. 2A, 268:782–785.
———. 1974. Evolution of endothermy: histological evidence. *Evolutionary Theory* 1:51–80.
———. 1980. Tissue structures of dinosaur bone: functional significance and possible relation to dinosaur physiology. In *A cold look at the warm-blooded dinosaurs,* ed. R. D. K. Thomas and E. C. Olson, 103–139. American Association for the Advancement of Science Selected Symposium, 28. Boulder, Colo.: Westview Press.
Rieppel, O. 1989. *Helveticosaurus zollingeri* Peyer (Reptilia, Diapsida) skeletal paedomorphosis, functional anatomy and systematic affinities. *Palaeontographica,* ser. A, 208:123–152.
———. 1993. Studies on skeletal formation in reptiles. IV. The homology of the reptilian (amniote) astragalus revisited. *Journal of Vertebrate Paleontology* 13:31–47.
Rietschel, S. 1985. Feathers and wings of *Archaeopteryx,* and the question of her flight ability. In *The beginnings of birds,* ed. M. K. Hecht et al., 251–260. Eichstätt: Freunde des Jura-Museum.
Ripley, S. D. 1977. *Rails of the world.* Boston: Godine.
Rogers, D. 1987. Description and interpretation of a Lower Cretaceous fossil feather from Koonwarra, South Gippsland, Victoria. 3d year thesis. Clayton: Monash University.
Romer, A. S. 1966. *Vertebrate paleontology.* 3d ed. Chicago: University of Chicago Press.
———. 1972. The Chanares (Argentina) Triassic reptile fauna. X. Further remains of the thecodonts *Lagerpeton* and *Lagosuchus. Brevoria* 394:1–7.
Rootes, W. L., and R. H. Chabreck. 1993. Cannibalism in the American alligator. *Herpetologica* 49:99–107.
Rooth, J. 1965. The flamingos on Bonaire (Netherlands Antilles): habitat, diet and reproduction of *Phoenicopterus ruber ruber. Natuurwetenschappelijke Studiekring voor Suriname en de Nederlandse Antillen* (Utrecht) 41:1–151.
Ross, C. A., ed. 1989. *Crocodiles and alligators.* New York: Facts on File.
Ruben, J. 1991. Reptilian physiology and the flight capacity of *Archaeopteryx. Evolution* 45:1–17.

———. 1993. Powered flight in *Archaeopteryx*: response to Speakman. *Evolution* 47:935–938.
———. 1995. The evolution of endothermy in mammals and birds: from physiology to fossils. *Annual Review of Physiology* 57:69–95.
———. 1996. The evolution of endothermy in mammals, birds and their ancestors. *Journal of Experimental Biology*, in press.
Rudge, M. R., ed. 1989. Moas, mammals and climate in the ecological history of New Zealand. *New Zealand Journal of Ecology* 12 (suppl.):1–169.
Russell, D. A. 1967. Cretaceous vertebrates from the Anderson River, Northwest Territories. *Canadian Journal of Earth Sciences* 4:21–38.
———. 1969. A new specimen of *Stenonychosaurus* from the Oldman Formation (Cretaceous) of Alberta. *Canadian Journal of Earth Sciences* 6:595–612.
———. 1972. Ostrich dinosaurs from the late Cretaceous of western Canada. *Canadian Journal of Earth Sciences* 9:375–402.
———. 1989. *An odyssey in time: the dinosaurs of North America*. Toronto: University of Toronto Press.
Russell, D. E. 1982. Tetrapods of the northwestern European Tertiary Basin. *Geologisches Jahrbuch*, ser. A, 60:5–74.
Ryan, P. G., B. P. Watkins, and W. R. Siegfried. 1989. Morphometrics, metabolic rate and body temperature of the smallest flightless bird: the Inaccessible Island rail. *Condor* 91:465–467.
Sanz, J. L., and J. F. Bonaparte. 1992. A new order of birds (class Aves) from the Lower Cretaceous of Spain. *Los Angeles County Museum of Natural History, Science Series* 36:39–50.
Sanz, J. L., J. F. Bonaparte, and A. Lacasa. 1988. Unusual early Cretaceous birds from Spain. *Nature* 331:433–435.
Sanz, J. L., and A. D. Buscalioni. 1992. A new bird from the early Cretaceous of Las Hoyas, Spain, and the early radiation of birds. *Paleontology* 35:829–845.
Sanz, J. L., J. M. Chiappe, and A. D. Buscalioni. 1995. The osteology of *Concornis lacustris* (Aves: Enantiornithes) from the Lower Cretaceous of Spain and a reexamination of its phylogenetic significance. *American Museum Novitates* 3133:1–23.
Sanz, J. L., and N. López-Martínez. 1984. The prolacertid lepidosaurian *Cosesaurus aviceps* Ellenberger and Villalta, a claimed "protoavian" from the Middle Triassic of Spain. *Géobios* 17:747–753.
Sarich, V. M., C. W. Schmid, and J. Marks. 1989. DNA hybridization as a guide to phylogenies: a critical analysis. *Cladistics* 5:3–32.
Sauer, E. G. F. 1976. Aepyornithoid eggshell fragments from the Miocene and Pliocene of Anatolia, Turkey. *Palaeontographica* 153:62–115.
Sauer, E. G. F., and P. Rothe. 1972. Ratite eggshells from Lanzarote, Canary Islands. *Science* 176:43–45.
Savile, D. B. O. 1957. Adaptive radiation of the avian wing. *Evolution* 11:212–224.
———. 1962. Gliding and flight in the vertebrates. *American Zoologist* 2:161–166.
Scarlett, R. J. 1968. An owlet-nightjar from New Zealand. *Notornis* 15:254–266.
Schinz, H. R., and R. Zangrel. 1937. Beiträge zur Osteogenese des Knochensystems beim Haushuhn, bei der Haustaube und beim Haubenseissfuss. *Denkshriften der Schweizerischen Naturforschenden Gesellschaft* 77:117–164.
Schlee, D. 1973. Harzhonservierte fossile Vogelfedern aus der untersten Kreide. *Journal für Ornithologie* 114:207–219.
Schlesinger, W. H., J. M. H. Knops, and T. H. Nash, III. 1993. Arboreal sprint failure: lizardfall in a California oak woodland. *Ecology* 74:2465–2467.
Scholey, K. D. 1986. An energetic explanation for the evolution of flight in bats. In *Bat flight—Fledermausflug*, ed. W. Nachtigall, 1–12. Biona Report 5. Stuttgart: Gustav Fischer Verlag.
Schorger, A. W. 1947. The deep diving of the loon and old-squaw and its mechanism. *Wilson Bulletin* 59:151–159.
Schultze, H.-P., and L. Trueb, eds. 1991. *Origins of the higher groups of tetrapods: controversy and consensus*. Ithaca, N.Y.: Cornell University Press.
Sclater, P. L. 1858. On the general geographical distribution of members of the class Aves. *Journal of the Proceedings of the Linnean Society, Zoological Series* 2:130–145.
Seebohm, H. 1890. *The classification of birds: an attempt to classify the subclasses, orders, suborders, and some of the families of existing birds*. London: R. H. Porter.
Seeley, H. G. 1866. An epitome of the evidence that Pterodactyles are not reptiles, but a new subclass of vertebrate animals allied to birds (Saurornia). *Annals and Magazine of Natural History*, 3d ser., 17:321–331.
———. 1870. *The Ornithosauria: an elementary study of the bones of Pterodactyles*. Cambridge: Cambridge University Press.
———. 1881. Prof. Carl Vogt on the *Archaeopteryx*. *Geological Magazine*, 2d ser., 8:300–309.
Sereno, P. C. 1990. Clades and grades in dinosaur systematics. In *Dinosaur systematics: perspectives and approaches*, ed. K. Carpenter and P. J. Currie, 9–20. Cambridge: Cambridge University Press.
———. 1991. Basal archosaurs: phylogenetic relationships and functional implications. *Society of Vertebrate Paleontology Memoires* 2:1–53.
———. 1993. The pectoral girdle and forelimb of the basal theropod *Herrerasaurus ischigualastensis*. *Journal of Vertebrate Paleontology* 13:425–450.
Sereno, P. C., and A. B. Arcucci. 1990. The monophyly of crurotarsal archosaurs and the origin of bird and crocodile joints. *Neue Jahrbücher für Geologie und Paläontologie Abhandlungen* 180:21–52.
———. 1993. Dinosaurian precursors from the Middle Triassic of Argentina: *Lagerpeton chanarensis*. *Journal of Vertebrate Paleontology* 13:385–399.

Sereno, P. C., C. A. Forster, R. R. Rogers, and A. M. Monetta. 1993. Primitive dinosaur skeleton from Argentina and the early evolution of Dinosauria. *Nature* 361:64–66.
Sereno, P. C., and F. E. Novas. 1992. The complete skull and skeleton of an early dinosaur. *Science* 258:1137–1140.
———. 1993. The skull and neck of the basal theropod *Herrerasaurus ischigualastensis*. *Journal of Vertebrate Paleontology* 13:451–476.
Sereno, P. C., and C. Rao. 1992. Early evolution of avian flight and perching: new evidence from the Lower Cretaceous of China. *Science* 255:845–848.
Sereno, P. C., and R. Wild. 1992. *Procompsognathus*: theropod, "thecodont" or both? *Journal of Vertebrate Paleontology* 12:435–458.
Sharov, A. G. 1970. An unusual reptile from the Lower Triassic of Fergana. *Paleontological Journal* 1970(1):127–130. [Russian translation.]
———. 1971. [New flying reptiles from the Mesozoic deposits of Kasakhstan and Kirgizia.] *Trudy Paleontologichesky Institut Akademiya Nauk S.S.R.* (Moscow) 130:104–113. [In Russian.]
Sheldon, F. H. 1987. Phylogeny of herons estimated from DNA-DNA hybridization data. *Auk* 104:97–108.
Sheldon, F. H., and A. H. Bledsoe. 1993. Avian molecular systematics, 1970s to 1990s. *Annual Review of Ecology and Systematics* 24:243–278.
Sheldon, F. H., B. Slikas, M. Kinnarney, F. B. Gill, E. Zhao, and B. Silverin. 1992. DNA-DNA hybridization evidence of phylogenetic relationships among major lineages of *Parus*. *Auk* 109:173–185.
Shilov, I. A., and B. Stephan. 1976. [Were the flying reptiles homeotherms?] *Zoological Journal* 55(7):1038–1045. [In Russian.]
Short, L. L. 1971. The evolution of terrestrial woodpeckers. *American Museum Novitates* 2467:1–23.
———. 1982. The woodpeckers of the world. *Delaware Museum of Natural History Monographs*, no. 4.
Short, L. L., and J. F. M. Horne. 1980. Ground barbets of East Africa. *Living Bird* 18:179–186.
Shubin, N. 1991. The implications of "the bauplan" for development of the tetrapod limb. In *Developmental patterning of the vertebrate limb*, ed. J. R. Hinchliffe, J. Jurle, and D. Summerbell, 411–421. New York: Plenum Press.
Shubin, N. H. 1994. History, ontogeny, and evolution of the archetype. In *Homology, the hierarchical basis for comparative biology*, ed. B. K. Hall, 249–269. San Diego, Calif.: Academic Press.
Sibley, C. G. 1970. A comparative study of the egg-white proteins of passerine birds. *Bulletin of the Peabody Museum of Natural History, Yale University* 32:1–131.
———. 1974. The relationships of the lyrebirds. *Emu* 74(2):65–79.
Sibley, C. G., and J. E. Ahlquist. 1972. A comparative study of the egg-white proteins of non-passerine birds. *Bulletin of the Peabody Museum of Natural History, Yale University* 39:1–276.
———. 1985. The phylogeny and classification of the Australo-Papuan passerine birds. *Emu* 85:1–14.
———. 1990. *Phylogeny and classification of birds: a study in molecular evolution*. New Haven and London: Yale University Press.
Sibley, C. G., and B. L. Monroe, Jr. 1990. *Distribution and taxonomy of birds of the world*. New Haven and London: Yale University Press.
Sill, W. D. 1974. The anatomy of *Saurosuchus galilei* and the relationships of the rauisuchid thecodonts. *Bulletin of the Museum of Comparative Zoology* 146:317–362.
Simonetta, A. M. 1963. Cinesi e morfologia del cranio negli uccelli non-passeriformi: studio su varie tendenze evolutive. Part II. Striges, Caprimulgiformes ed Apodiformes. *Archivo Zoologico Italiano* (Turin) 52:1–36.
Simpson, G. G. 1946. Fossil penguins. *Bulletin of the American Museum of Natural History* 87:1–100.
———. 1961. *Principles of animal taxonomy*. New York: Columbia University Press.
———. 1980. *Splendid isolation: the curious history of South American mammals*. New Haven and London: Yale University Press.
Simpson, S. F., and J. Cracraft. 1981. The phylogenetic relationships of the Piciformes (class Aves). *Auk* 98:481–494.
Slud, P. 1960. The birds of Finca "La Selva," Costa Rica: a tropical wet forest locality. *Bulletin of the American Museum of Natural History* 128:49–148.
Snell, R. R. 1985. Underwater flight of long-tail duck (oldsquaw) *Clangula hyemalis*. *Ibis* 127:267.
Snow, E., ed. 1978. *An atlas of speciation in African non-passerine birds*. London: British Museum.
Sokal, R. R., and P. H. A. Sneath. 1963. *Principles of animal taxonomy*. San Francisco: W. H. Freeman.
Speakman, J. R. 1993. Flight capabilities in *Archaeopteryx*. *Evolution* 47:336–340.
Speakman, J. R., and S. C. Thompson. 1994. Flight capabilities of *Archaeopteryx*. *Nature* 370:514.
———. 1995. Speakman and Thompson reply. *Nature* 374:221–222.
Spearman, R. I. C., and J. A. Hardy. 1985. Integument. In *Form and function in birds*, vol. 3, ed. A. S. King and J. McLelland, 1–56. London: Academic Press.
Spotila, J. R. 1980. Constraints of body size and environment on the temperature regulation of dinosaurs. In *A cold look at the warm-blooded dinosaurs*, ed. R. D. K. Thomas and E. C. Olson, 233–252. American Association for the Advancement of Science Selected Symposium, 28. Boulder, Colo.: Westview Press.
Spotila, J. R., P. W. Lommen, G. S. Bakken, and D. M. Gates. 1973. A mathematical model for body temperatures of large reptiles: implications for dinosaur ecology. *American Naturalist* 107:391–404.
Stager, K. E. 1964. The role of olfaction in food location by the turkey vulture (*Cathartes aura*). *Los Angeles*

County Museum of Natural History, Contributions to Science 81:1–63.
Starck, M. J. 1995. Comparative anatomy of the external and middle ear of palaeognathous birds. *Advances in Anatomy, Embryology and Cell Biology* 131:1–137.
Steadman, D. W. 1980. A review of the osteology and paleontology of turkeys (Aves: Meleagridinae). *Los Angeles County Museum of Natural History, Contributions to Science* 330:131–207.
———. 1981. Birds of the British Lower Eocene (Review). *Auk* 98:205-207.
———. 1982. A re-examination of *Palaeostruthus hatcheri* (Shufeldt), a late Miocene sparrow from Kansas. *Journal of Vertebrate Paleontology* 1:171–173.
———. 1986. Two new species of rails (Aves: Rallidae) from Mangaia, southern Cook Islands. *Pacific Studies* 40:27–43.
———. 1995. Prehistoric extinctions of Pacific Island birds: biodiversity meets zooarchaeology. *Science* 267:1123–1131.
Steadman, D. W., and P. S. Martin. 1984. Extinction of birds in the late Pleistocene of North America. In *Quatarnary extinctions*, ed. P. Martin and R. Klein, 466–477. Tucson: University of Arizona Press.
Steadman, D. W., and M. C. McKitrick. 1982. A Pliocene bunting from Chihuahua, Mexico. *Condor* 84:240–241.
Steadman, D. W., and N. G. Miller. 1987. California condor associated with spruce–jack pine woodland in the late Pleistocene of New York. *Quaternary Research* 28:415–426.
Stegmann, B. C. 1969. Über die systematische Stellung der Tauben und Flughuhner. *Zoologische Jahrbücher Systematik* 96:1–51.
———. 1978. *Relationships of the superorders Alectoromorphae and Charadriomorphae (Aves): a comparative study of the avian hand.* Publications of the Nuttall Ornithological Club, 17. Cambridge, Mass: Nuttall Ornithological Club.
Steiner, H. 1917. Das problem der diastataxie des vögelflugels. *Jenaische Zeitschrift für Naturwissenschaften* 55:221–496.
———. 1938. Der *Archaeopteryx*—Schwanz der Vögelembryonen. *Vierteljahrsschrift der Naturforschenden Gesellschaft in Zürich* 83:279–300.
Stejneger, L. 1885. *The standard natural history*, vol. 4: *Aves*. Ed. J. S. Kingsley. Boston: S. E. Cassino.
Stephan, B. 1985. Remarks on the reconstruction of the *Archaeopteryx* wing. In *The beginnings of birds*, ed. M. K. Hecht et al., 261–265. Eichstätt: Freunde des Jura-Museum.
———. 1987. *Urvögel: Archaeopterygiformes.* Wittenberg Lutherstadt: A. Ziemsen Verlag.
———. 1992. Vorkommen und Ausbildung der Fingerkrallen bei rezenten Vögeln. *Journal für Ornithologie* 133:251–277.
Stettenheim, P. 1976. Structural adaptations in feathers. *Proceedings of the Sixteenth International Ornithological Congress*, 385–401.
Stolpe, M. 1932. Physiologisch-anatomische Untersuchungen über die hintere Extremität der Vögel. *Journal für Ornithologie* 80:161–247.
———. 1935. *Columbus, Hesperornis, Podiceps:* ein Vergleich ihrer hinteren Extremitat. *Journal für Ornithologie* 83:115–128.
Storch, G. 1993. "Grube Messel" and African–South American faunal connections. In *The African–South America connections*, ed. W. George and R. Lavocat, 76–86. Oxford Monographs on Biogeography, no. 7. Oxford: Clarendon Press.
Storer, R. W. 1956. The fossil loon *Colymboides minutus. Condor* 58:413–426.
———. 1960. Evolution in the diving birds. *Proceedings of the Twelfth International Ornithological Congress*, 694–707.
———. 1971. Adaptive radiation of birds. In *Avian biology*, vol. 1, ed. D. S. Farner and J. R. King, 149–188. New York: Academic Press.
Strauch, J. G., Jr. 1978. The phylogeny of the Charadriiformes (Aves): a new estimate using the method of character compatibility analysis. *Transactions of the Zoological Society of London* 34:263–345.
———. 1985. The phylogeny of the Alcidae. *Auk* 102:520–539.
Stresemann, E. 1932. La structure des remiges chez quelques rales physiologiquement aptères. *Alauda* 4:1–5.
———. 1934. *Handbuch de Zoologie*. Vol. 7, pt. 2: *Aves*. Ed. W. Kukenthal and T. Krumbach. Berlin: Walter de Gruyter.
Strickland, H. E. 1849. Supposed existence of a giant bird in Madagascar. *Annals and Magazine of Natural History*, 2d ser., 4:338–339.
Strickland, H. G., and A. G. Melville. 1848. *The dodo and its kindred.* London: Reeve, Benham and Reeve.
Struhsaker, T. T., and M. Leakey. 1990. Prey selectivity by crowned hawk-eagles on monkeys in the Kibale Forest, Uganda. *Behavioral Ecology and Sociobiology* 26:435–443.
Svec, P. 1980. Lower Miocene birds from Dolnice (Cheb basin), western Bohemia. *Casopis pro Mineralogii a Geologii* 25:377–387.
———. 1982. Two new species of diving birds from the Lower Miocene of Czechoslovakia. *Casopis pro Mineralogii a Geologii* 27:243–260.
Swafford, D. L. 1991. PAUP: phylogenetic analysis using parsimony, version 3.0s. Computer program distribution by Illinois Natural History Survey, Champaign.
Swierczewski, E. V., and R. J. Raikow. 1981. Hindlimb morphology, phylogeny, and classification of the Piciformes. *Auk* 98:466–480.
Swinton, W. E. 1970. *The dinosaurs*. London: Allen and Unwin.
Sy, M. 1936. Funktionall-anatomische Untersuchungen am Vögelflugel. *Journal für Ornithologie* 84:199–296.
Taborsky, M., and B. Taborsky. 1993. The kiwi's parental burden. *Natural History* 102(12):50–57.
Talent, J. A., P. M. Duncan, and P. L. Handley. 1966. Early Cretaceous feathers from Victoria. *Emu* 66:81–86.

Tambussi, C. 1995. Fossil Rheiformes of Argentina. *Courier Forschungsinstitut Senckenberg,* 181:121–129.
Tarsitano, S. 1985. The morphological and aerodynamic constraints on the origin of avian flight. In *The beginnings of birds,* ed. M. K. Hecht et al., 319–332. Eichstätt: Freunde des Jura-Museum.
———. 1991. *Archaeopteryx: Quo Vadis?* In *Origins of the higher groups of tetrapods: controversy and consensus,* H.-P. Schultze and L. Trueb, 541–576. Ithaca, N.Y.: Cornell University Press.
Tarsitano, S., and M. K. Hecht. 1980. A reconsideration of the reptilian relationships of *Archaeopteryx. Zoological Journal of the Linnean Society of London* 69:257–263.
Taylor, E. L., T. N. Taylor, and N. R. Cuneo. 1992. The present is not the key to the past: a polar forest from the Permian of Antarctica. *Science* 257:1675–1677.
Taylor, J. R. E. 1986. Thermal insulation of the down and feathers of pygoscelid penguin chicks and the unique properties of penguin feathers. *Auk* 103:160–168.
Temple, R. C., and L. M. Anstey, eds. 1936. *The travels of Peter Mundy in Europe and Asia, 1608–1667.* Vol. 5. London: Hakluyt Society.
Temple, S. A. 1977. Plant-animal mutualism: coevolution with dodo leads to near extinction of plant. *Science* 197:885–886.
Thomas, A. L. R. 1993. On the aerodynamics of birds' tails. *Philosophical Transactions of the Royal Society of London,* ser. B, 340:361–380.
Thomas, R. D. K., and E. C. Olson, eds. 1980. *A cold look at the warm-blooded dinosaurs.* American Association for the Advancement of Science Selected Symposium, 28. Boulder, Colo.: Westview Press.
Thorington, R. W., and T. L. Heaney. 1981. Body proportions and gliding adaptations of flying squirrels (Petauristinae). *Journal of Mammalogy* 62:101–114.
Thouless, C. R., J. H. Fanshawe, and B. C. R. Bertram. 1989. Egyptian vultures *Neophron percnopterus* and ostrich *Struthio camelus* eggs: the origins of stone-throwing behavior. *Ibis* 131:9–15.
Thulborn, R. A. 1982. Significance of ankle structures in archosaur phylogeny. *Nature* 299:657.
———. 1984. The avian relationships of *Archaeopteryx,* and the origin of birds. *Zoological Journal of the Zoological Society of London* 82:119–158.
Thulborn, R. A., and T. L. Hamley. 1985. A new palaeoecological role for *Archaeopteryx.* In *The beginnings of birds,* ed. M. K. Hecht et al., 81–89. Eichstätt: Freunde des Jura-Museum.
Tome, M. W., and D. A. Wrubleski. 1988. Underwater foraging behavior of canvasbacks, lesser scaups, and ruddy ducks. *Condor* 90:168–172.
Tonni, E. P. 1980a. Un pseudodontornítido (Pelecaniformes, Odontopterygia) de gran tamaño, del terciario de Antártida. *Ameghiniana* 17:273–276.
———. 1980b. The present state of knowledge of the Cenozoic birds of Argentina. *Los Angeles County Museum of Natural History, Contributions to Science* 330:105–114.
Tordoff, H. B. 1954. Relationships in the New World nine-primaried oscines. *Auk* 71:273–284.
Tordoff, H. B., and J. R. Macdonald. 1957. A new bird (family Cracidae) from the early Oligocene of South Dakota. *Auk* 74:174–184.
Tracy, C. R. 1976. Tyrannosaurs: evidence for endothermy? *American Naturalist* 110:1105–1106.
Unwin, D. M., and N. N. Bakhurina. 1994. *Sordes pilosus* and the nature of the pterosaur flight apparatus. *Nature* 371:62–64.
Vakhrameev, V. A. 1991. *Jurassic and Cretaceous floras and climates of the Earth.* Cambridge: Cambridge University Press.
Van der Leeuw, A., and G. A. Zweers. 1994. Evolution and development of Anseriform feeding mechanisms. *Journal für Ornithologie* 135:41.
Van Tyne, J., and A. J. Berger. 1976. *Fundamentals of ornithology.* 2d ed. New York: John Wiley and Sons.
Van Valen, L. 1973. Body size and numbers of plants and animals. *Evolution* 27:27–35.
Van Valkenburgh, B. 1995. Tracking ecology over geological time: evolution within guilds of vertebrates. *TREE* 10:71–76.
Varricchio, D. J. 1993. Bone microstructure of the Upper Cretaceous theropod dinosaur *Troödon formosus. Journal of Vertebrate Paleontology* 13:99–104.
Varricchio, D. J., and L. M. Chiappe. 1995. A new enantiornithine bird from the Upper Cretaceous Two Medicine Formation of Montana. *Journal of Vertebrate Paleontology* 15:201–204.
Vasquez, R. J. 1992. Functional osteology of the avian wrist and the evolution of flapping flight. *Journal of Morphology* 211:259–268.
Vaughn, T. A. 1978. *Mammalogy.* 2d ed. Philadelphia: W. B. Saunders.
Vermeij, G. J. 1988. The evolutionary success of passerines: a question of semantics? *Systematic Zoology* 37:70–72.
Vickers-Rich, P. 1991. The Mesozoic and Tertiary history of birds on the Australian Plate. In *Vertebrate palaeontology of Australasia,* ed. P. Vickers-Rich, T. M. Monaghan, R. R. Baird, and T. H. Rich, 721–808. Melbourne: Pioneer Design Studio and Monash University Publications Committee.
Vickers-Rich, P., and T. H. Rich. 1993. Australia's polar dinosaurs. *Scientific American* 269(1):50–55.
Villee, C. A., W. F. Walker, Jr., and R. D. Barnes. 1984. *General zoology.* 6th ed. Philadelphia: W. B. Saunders.
Viohl, G. 1985a. Geology of the Solnhofen lithographic limestone and the habitat of *Archaeopteryx.* In *The beginnings of birds,* ed. M. K. Hecht et al., 31–44. Eichstätt: Freunde des Jura-Museum.
———. 1985b. Carl F. and Ernst O. Häberlein, the sellers of the London and Berlin specimens of *Archaeopteryx.* In *The beginnings of birds,* ed. M. K. Hecht et al., 349–357. Eichstätt: Freunde des Jura-Museum.
Vogt, C. 1880. *Archaeopteryx macrura,* an intermediate form between birds and reptiles. *Ibis* 4:434–456.
Voorhies, M. R. 1981. Ancient ashfall creates a Pom-

peii of prehistoric animals. *National Geographic* 159:66–75.
Voorhies, M. R., and J. R. Thomasson. 1979. Fossil grass anthoecia within Miocene rhinoceros skeletons. *Science* 206:331–333.
Voous, K. H. 1960. *Atlas of European birds*. London: Thomas Nelson and Sons.
Wagner, J. A. 1861a. Über ein neues, augeblich mit Vögelfedern versehenes Reptil aus dem Solenhofener lithographischen Schiefer. *Sitzungsberichte der Bayerischen Akademie der Wissenschaften* 2:146–154.
———. 1861b. Neue Beiträge zur Kenntnis der urweltlichen Fauna des lithographischen Schiefers; *Compsognathus longipes* Wagner. *Abhandlungen, Bayerische Akademie der Wissenschaften* 9:30–38.
———. 1862. On a new fossil reptile supposed to be furnished with feathers. *Annals and Magazine of Natural History*, 3d ser., 9:261–267.
Waldman, M. 1970. A third specimen of a Lower Cretaceous feather from Victoria, Australia. *Condor* 72:377.
Walker, A. D. 1964. Triassic reptiles from the Elgin area: *Ornithosuchus* and the origin of carnosaurs. *Philosophical Transactions of the Royal Society of London*, ser. B, 248:53–134.
———. 1969. The reptile fauna of the "Lower Keuper" Sandstone. *Geological Magazine* 106:470–476.
———. 1972. New light on the origin of birds and crocodiles. *Nature* 237:257–263.
———. 1977. Evolution of the pelvis in birds and dinosaurs. In *Problems in vertebrate evolution*, vol. 4, ed. S. M. Andrews et al., 319–358. London: Linnean Society.
———. 1985. The braincase of *Archaeopteryx*. In *The beginnings of birds*, ed. M. K. Hecht et al., 123–134. Eichstätt: Freunde des Jura-Museum.
Walker, C. A. 1981. New subclass of birds from the Cretaceous of South America. *Nature* 292:51–53.
Wallace, A. R. 1876. The geographical distribution of animals. London: Macmillan.
Warheit, K. I. 1992. A review of the fossil seabirds from the Tertiary of the North Pacific: plate tectonics, paleoceanography, and faunal change. *Paleobiology* 18:401–424.
Watson, G. E. 1975. *Birds of the Antarctic and Sub-Antarctic*. Washington, D.C.: American Geophysical Union.
———. 1976. . . . And birds took wing. In *Our continent, a natural history of North America*, ed. E. H. Colbert and S. L. Fishbein, 98–107. Washington, D.C.: National Geographic Society.
Webb, M. 1957. The ontogeny of the cranial bones, cranial peripheral and cranial parasympathetic nerves, together with a study of the visceral arches of *Struthio*. *Acta Anatomica* 38:81–203.
Webb, S. D. 1974. *Pleistocene mammals of Florida*. Gainesville: University of Florida Press.
———. 1976. Mammalian faunal dynamics of the Great American Interchange. *Paleobiology* 2:216–234.
———. 1978. A history of savannah vertebrates in the New World. Part II: South America and the Great American Interchange. *Annual Review of Ecology and Systematics* 9:393–426.
———. 1991. Ecogeography and the Great American Interchange. *Paleobiology* 17:266–280.
———. 1995. Biological implications of the middle Miocene Amazon seaway. *Science* 269:361–362.
Weber, E., and A. Hesse. 1995. The systematic position of *Aptornis*, a flightless bird from New Zealand. *Courier Forschungsinstitut Senckenberg* 181:293–301.
Weishampel, D. B., P. Dodson, and H. Osmólska, eds. 1990. *The Dinosauria*. Berkeley: University of California Press.
Welles, S. P., and R. A. Long. 1974. The tarsus of theropod dinosaurs. *Annals of the South African Museum* 64:191–218.
Wellnhofer, P. 1974. Das fünfte Skelettexemplar von *Archaeopteryx*. *Paläontographica*, ser. A, 147:169–216.
———. 1985. Remarks on the digit and pubis problems of *Archaeopteryx*. In *The beginnings of birds*, ed. M. K. Hecht et al., 113–122. Eichstätt: Freunde des Jura-Museum.
———. 1988a. A new specimen of *Archaeopteryx*. *Science* 240:1790–1792.
———. 1988b. Ein neues Exemplar von *Archaeopteryx*. *Archaeopteryx* 6:1–30.
———. 1991. *The illustrated encyclopedia of pterosaurs*. New York: Crescent Books.
———. 1992. A new specimen of *Archaeopteryx* from the Solnhofen limestone. *Los Angeles County Museum of Natural History, Science Series* 36:3–23.
———. 1993. Das siebte Exemplar von *Archaeopteryx* aus den Solnhofener Schichten. *Archaeopteryx* 11:1–48.
Wetmore, A. 1926. Fossil birds from the Green River deposits of eastern Utah. *Annals of the Carnegie Museum* 16:391–402.
———. 1944. A new terrestrial vulture from the Upper Eocene deposits of Wyoming. *Annals of the Carnegie Museum* 30:57–69.
———. 1951. A revised classification for the birds of the world. *Smithsonian Miscellaneous Collections* 117(4):1–22.
———. 1960. A classification for the birds of the world. *Smithsonian Miscellaneous Collections* 39:1–37.
———. 1967. Re-creating Madagascar's giant extinct bird. *National Geographic* 132:488–493.
Whetstone, K. N. 1983. Braincase of Mesozoic birds: I. New preparation of the "London" *Archaeopteryx*. *Journal of Vertebrate Paleontology* 2:439–452.
Whetstone, K. N., and L. D. Martin. 1979. New look at the origin of birds and crocodiles. *Nature* 279:234–236.
———. 1981. Reply to McGowan and Baker. *Nature* 289:98.
Whetstone, K. N., and P. J. Whybrow. 1983. A "cursorial" crocodilian from the Triassic of Lesotho (Basutoland), South Africa. *Occasional Papers of the Museum of Natural History, University of Kansas* 106:1–37.

Whittow, G. C., and H. Tazawa. 1991. The early development of thermoregulation in birds. *Physiological Zoology* 64:1371–1390.

Wiedersheim, R. 1882. *Lehrbuch der vergleichenden Anatomie der Wirbelthiere.* Jena: Fischer Verlag.

Wieland, G. R. 1942. Too hot for the dinosaur. *Science* 96:359.

Wild, R. 1978. Die Flugsaurier (Reptilia, Pterosauria) aus der Oberen Trias von Cene bei Bergamo, Italien. *Bolletino Società Paleontologica Italiana* 17(2):176–256.

———. 1983. Über den Ursprung der Flugsaurier. In Weltenburger Akademie, *Erwin Rutte-Festschrift*, 231–238. Weltenburg: Kelheim.

———. 1984. Flugsaurier aus der Obertrias von Italien. *Naturwissenschaften* 71:1–11.

Wiley, E. O. 1981. *Phylogenetics*. New York: John Wiley and Sons.

Willis, E. O. 1966. The role of migrant birds at swarms of army ants. *Living Bird* 5:187–231.

Williston, S. W. 1879. Are birds derived from dinosaurs? *Kansas City Review of Science* 3:457–460.

———. 1898. *Birds*. University Geological Survey, University of Kansas 4(2):43–64.

———. 1925. *The osteology of the reptiles.* Cambridge, Mass.: Harvard University Press.

Witmer, L. M. 1987. The nature of the antorbital fossa of archosaurs: shifting the null hypothesis. In *Fourth symposium on Mesozoic terrestrial ecosystems, short papers*, ed. P. J. Currie and E. H. Koster, 230–235. Tyrrell Museum of Paleontology, Alberta, Occasional Papers, 3.

———. 1990. The craniofacial air sac system of Mesozoic birds (Aves). *Zoological Journal of the Linnean Society* 100:327–378.

———. 1991. Perspectives on avian origins. In *Origins of the higher groups of tetrapods: controversy and consensus*, ed. H.-P. Schultz and L. Trueb, 427–466. Ithaca, N.Y.: Cornell University Press.

Witmer, L. M., and K. D. Rose. 1991. Biomechanics of the jaw apparatus of the gigantic Eocene bird *Diatryma:* implications for diet and mode of life. *Paleobiology* 17:95–120.

Woitkevitsch, A. A. 1966. *The feathers and plumage of birds.* London: Sidgwick and Jackson.

Wolters, H. E. 1975–82. *Die Vogelarten der Erde.* Hamburg: Paul Parey.

Woodward, A. S. 1907. On a new dinosaurian reptile (*Scleromochlus taylori*, gen. et sp. nov.) from the Trias of Lossiemouth, Elgin. *Proceedings of the Geological Society of London* 63:140–144.

Woolfenden, G. E. 1961. Postcranial osteology of the waterfowl. *Bulletin of the Florida State Museum, Biological Sciences* 6(1):1–129.

Worthy, T. H. 1991. An overview of the taxonomy, fossil history, biology and extinction of moas. *Proceedings of the Twentieth International Ornithological Congress*, 555–562.

Wyles, J. S., J. G. Kunkel, and A. C. Wilson. 1983. Birds, behavior, and anatomical evolution. *Proceedings of the National Academy of Sciences* 80:4394–4397.

Yalden, D. W. 1985. Forelimb function in *Archaeopteryx*. In *The beginnings of birds*, ed. M. K. Hecht et al., 91–97. Eichstätt: Freunde des Jura-Museum.

Zhou, Z.-H. 1995a. Discovery of new Cretaceous birds in China. *Courier Forschungsinstitut Senckenberg,* 181:9–23.

———. 1995b. Discovery of a new enantiornithine bird from the early Cretaceous of China. *Chinese Science Bulletin* 33:108–113.

———. 1995c. Discovery of a new enantiornithine bird from the early Cretaceous of Liaoning, China. *Vertebrata PalAsiatica* 33(2):99–113.

Zhou, Z.-H., F. Jin, and J. Zhang. 1992. Preliminary report on a Mesozoic bird from Liaoning, China. *Chinese Science Bulletin* 37:1365–1368.

Zimmer, C. 1992. Ruffled feathers. *Discover* 13(5):44–54.

———. 1994. Masters of an ancient sky. *Discover* 15(2):42–54.

Zink, R. M., and P. R. Baverstock. 1991. Modern biochemical approaches to avian systematics. Symposium 7. *Proceedings of the Twentieth International Ornithological Congress*, 589–636.

Zinsmeister, W. J. 1982. Review of the Upper Cretaceous–Lower Tertiary sequence on Seymour Island, Antarctica. *Journal of the Geological Society of London*, ser. B, 139:779–785.

———. 1985. Seymour Island expedition. *Antarctic Journal* 20:41–42.

Zusi, R. L. 1974. An interpretation of skull structure in penguins. In *The biology of penguins*, ed. B. Stonehouse, 59–84. London: Macmillan.

———. 1984. A functional and evolutionary analysis of rhynchokinesis in birds. *Smithsonian Contributions to Zoology* 395:1–40.

Zusi, R. L., and K. I. Warheit. 1992. On the evolution of intraramal mandibular joints in pseudodontorns (Aves: Odontopterygia). *Los Angeles County Museum of Natural History, Science Series* 36:351–360.

Zweers, G. A. 1974. Structure, movement, and myography of the feeding apparatus of the mallard (*Anas platyrhynchos* L.): a study in functional anatomy. *Netherlands Journal of Zoology* 24(4):323–467.

Zweers, G. A., H. Berkhoudt, and J. C. Vanden Berge. 1994. Behavioral mechanisms of avian feeding. *Advances in Comparative and Environmental Physiology* 18:241–279.

Zweers, G. A., F. de Jong, H. Berkhoudt, and J. C. Vanden Berge. 1995. Filter feeding in flamingos (*Phoenicopterus ruber*). *Condor* 97:297–324.

Zweers, G. A., A. F. C. Gerritsen, and P. J. van Kranenburg-Voogd. 1977. Mechanics of feeding of the mallard (*Anas platyrhynchos* L.; Aves, Anseriformes). *Contributions to Vertebrate Evolution* 3:1–109.

The following references are among the most recent available at the time of publication. The italic notations at the end of each entry indicate the section of the text to which the entries apply.

Alvarenga, H. M. F. 1995. Um primitivo membro da ordem Galliformes (Aves) do terciário médio da Bacia de Taubaté, Estado de São Paulo, Brasil. *Anais da Academia Brasiliera de Ciências* 67:33–44. The first gallinaceous bird described from the Tertiary of South America—*Ameripodius silvasantosi*, from the Upper Oligocene or Lower Miocene of southeastern Brazil—is a primitive galliform belonging to the extinct family Quercymegapodiidae, described initially from the Paleogene of France. It is closest to the Megapodiidae among the extant families of Galliformes. *New Tertiary galliform, chapter 6*

Averianov, A. O., et al. 1991. [Bony-toothed birds (Aves: Pelecaniformes: Odontopterygia) of the late Paleocene and Eocene of the western margin of ancient Asia.] *USSR Academy of Sciences, Proceedings in Zoology of the Institute of Ecology and Fauna of Eurasiatic Birds* 239:3–12. Fossils of bony-tooths recovered from the Paleocene–Eocene of the western margin of Asia. [In Russian.] *Range extension for bony-toothed birds, chapter 5*

Avise, J. C., W. S. Nelson, and C. G. Sibley. 1994. DNA sequence support for a close phylogenetic relationship between some storks and New World vultures. *Proceedings of the National Academy of Sciences* 91:5173–5177. Mitochondrial cytochrome-*B* gene nucleotide sequences confirm the distant relatedness of New and Old World vultures, which are similar through convergent evolution, and suggest a close phylogenetic alliance between at least some New World vultures and storks, a conclusion reached previously through DNA-DNA hybridization. *Storks and New World vultures, chapter 7*

Baskin, J. A. 1995. The giant flightless bird *Titanis walleri* (Aves: Phorusrhacidae) from the Pleistocene of Texas. *Journal of Vertebrate Paleontology* 15:842–844. Pedal phalanx of *Titanis* recovered from the latest Pleistocene (Rancholabrean) of San Patricio County, Texas. *Giant phorusrhacid from Texas, chapter 6*

Bennett, S. C. 1995. A statistical study of *Rhamphorhynchus* from the Solnhofen Limestone of Germany: Year-classes of a single large species. *Journal of Vertebrate Paleontology* 69:569–580. All species of the toothed, tailed Solnhofen pterosaur *Rhamphorhynchus* represent a single species that did not have determinate growth but was equivalent in growth to extant crocodilians. Specimens fall into discrete size classes (year-classes) resulting from seasonal mortality or preservation of specimens. *Jurassic pterosaurs, chapters 1–2*

Björklund, M. 1994. Phylogenetic relationships among Charadriiformes: Reanalysis of previous data. *Auk* 111:825–832. Reanalysis of data from Strauch (1978) using cladistic techniques shows scolopacine and charadriine waders to be a monophyletic group, with the alcids forming a basal group and the thick-knees a sister-group to all waders and larids. *Shorebird phylogeny, chapter 5*

Cheneval, J. 1995. A fossil shearwater (Aves: Procellariiformes) from the Upper Oligocene of France and the Lower Miocene of Gemany. *Courier Forschungsinstitut Senckenberg* 181:187–198. *Frigidafons brodkorbi* is a new genus and species of Oligocene–Miocene shearwater. *New genus of Oligocene–Miocene shearwater, chapter 4*

Chiappe, L. M. 1995. The first eighty-five million years of avian evolution. *Nature* 378:349–355. Reviews the Jurassic and Cretaceous evolution of birds and supports the view that birds are theropod dinosaurs, that flight evolved from the ground up, possibly twice, that *Archaeopteryx* can be treated as the ancestor of all subsequent birds, and that birds originated late in the Mesozoic. Maintains *Mononykus* as a bird (see Zhou 1995, below). Views the modern radiation of bird orders in the classical, gradualistic fashion. However, the New Jersey greensands birds, thought to be Cretaceous, have been shown to be Paleocene in age, and there are no clear examples of modern bird orders, save general shorebird morphs, in the Mesozoic. Also, hesperornithiforms reached the latest Cretaceous (Maastrichtian) of North America and Mongolia (Kurochkin 1995:53). *Avian evolution, chapters 2–4*

Chiappe, L. M., and A. Chinsamy. 1996. *Pterodaustro*'s true teeth. *Nature* 379:211–212. Sectioned teeth from the mandible show that the filter-feeding apparatus is composed of true dental structures not comparable to the epidermally derived sieving structures of flamingos, ducks, and baleen whales. Modified filtering teeth are also found in the Permian mesosaurs, the pterosaurs *Ctenochasma* and *Gnathosaurus*, and the crab-eating seal *Lobodon carcinophagus*. *Pterosaur filter-feeding, chapter 5*

Chinsamy, A., L. M. Chiappe, and P. Dodson. 1995. Mesozoic avian bone microstructure: Physiological implications. *Paleobiology* 21:561–574. Enantiornithines and *Patagopteryx* show growth rings that indicate alternating periods of slow and fast growth, suggesting slower growth rates and a different physiology from modern ornithurines. The appearance of growth rings in a large number of theropods casts doubt on their proposed endothermic status, along with the view that endothermic birds evolved from theropods that were already hot-blooded. *Growth rings in opposite birds, chapters 3–4, 6*

Davis, P. G., and D. E. K. Briggs. 1995. Fossilization of feathers. *Geology* 23:783–786. Lacustrine settings provide the most important settings for feather preservation, and therefore a bias may exist in the avian fossil record in favor of inland fresh water habitats. *Feather fossilization, chapter 1*

Demes, B., E. Forchap, and H. Herwig. 1991. They seem to glide: Are there aerodynamic effects in leaping prosimian primates? *Zeitschrift für Morphologie und Anthropologie* 78(3):373–385. Leaping primates often assume a horizontal position while airborne, with limbs spread out and skin folds between upper limbs and trunk exposed. Coefficients of lift and drag for flying primates and calculated values give the animals a 5 percent gain or loss in leaping distance, and therefore a significant influence of aerodynamic forces on the flight path can be assumed. The smaller-bodied species (e.g., galagos) are more strongly influenced by their great surface areas, and air speed gains importance in the larger-bodied species (e.g., sifakas). The absolute amounts of lift and drag can be determined only by wind-tunnel testing. *"Flying" primates, chapter 3*

Elzanowski, A., and M. K. Brett-Surman. 1995. Avian premaxilla and tarsometatarsus from the uppermost Cretaceous of Montana. *Auk* 112:762–767. Fragmentary premaxilla and tarsometatarsus from the uppermost Cretaceous Hell Creek Formation of Montana conform to the charadriiform-gruiform assemblage and the anseriform lineage, respectively, the latter agreeing with the "transitional shorebird"–anseriform mosaic *Presbyornis.* Both fossils fit the mold for transitional shorebirds. *Late Cretaceous ornithurine bird fossils, chapter 4*

Elzanowski, A., and P. Wellnhofer. 1996. Cranial morphology of *Archaeopteryx:* Evidence from the seventh specimen. *Journal of Vertebrate Paleontology* 16:81–94. Avian features of the skull show that *Archaeopteryx* is a bird rather than a "feathered nonavian archosaur." The palatine is distinctively avian; the ectopterygoid has a hook-shaped jugal process remotely similar to that of certain theropods but also approached in the thecodont *Ornithosuchus. Skull of new* Archaeopteryx, *chapters 1–2*

Gatesy, S. M., and K. P. Dial. 1996. Locomotor modules and the evolution of avian flight. *Evolution* 50:331–340. An analysis of anatomical subregions ("locomotor modules"), which are highly integrated functional units during locomotion, and the sequence of modifications of the primitive tetrapod locomotor system through time. Decoupling of the pelvic and caudal locomotor modules freed the tail to attain a new affiliation with the forelimb during flight. Flight resulted from the origin and novel association of locomotor modules in ground-dwelling theropod dinosaurs, and flight evolved from the ground up. *Ground-up origin of bird flight, chapter 3*

Geist, N. R., and T. D. Jones. 1996. Juvenile skeletal structure and the reproductive habits of dinosaurs. *Science,* in press. Skeletal anatomy of perinatal extant archosaurs (birds and crocodilians) and perinatal dinosaurs suggests that the known dinosaur hatchlings were all precocial, a finding consistent with overall similarity in nesting behavior of dinosaurs and crocodilians. *No altricial dinosaurs, chapter 3*

Griffiths, C. S. 1994. Monophyly of the Falconiformes based on syringeal morphology. *Auk* 111:787–805. Cladistic analysis of data derived from syringeal morphology supports the monophyly of the Falconiformes, as well as the monophyly of three clades within the order: the Cathartidae, the Falconidae, and an Accipitrinae-*Sagittarius-Pandion* cluster, with the Cathartidae positioned as basal to the other two clades. Data support inclusion of the cathartids within the Falconiformes (contra Avise, Nelson, and Sibley 1994). *Raptor phylogeny, chapter 7*

Grimaldi, D. A. 1996. Captured in amber. *Scientific American,* April 1996:84–91. Incredible specimens preserved in amber include a feather, some 90 million to 94 million years old, from the Cretaceous of central New Jersey, discovered some five years ago. It is the oldest record of a terrestrial bird from North America. *Cretaceous feather from North America, chapter 4*

Hedges, S. B., M. D. Simmons, M. A. M. Van Dijk, G.-J. Caspers, W. W. DeJong, and C. G. Sibley. 1995. Phylogenetic relationships of the hoatzin, an enigmatic South American bird. *Proceedings of the National Academy of Sciences* 92:11662–11665. DNA sequences of hoatzins and thirteen other birds show that hoatzins are most closely related to cuckoos among living avian orders; they are the sister-group to the other cuculiforms. Turacos, colies, and coraciiforms, which may have some affinity, were excluded from the comparisons. *Hoatzin relations, chapter 8*

Herman, A. B., and R. A. Spicer. 1996. Palaeobotanical evidence for a warm Cretaceous Arctic Ocean. *Nature* 380:330–333. Fossil leaves indicate that the Arctic Ocean was relatively warm, remaining above 0°C even during winter months, implying that there was significant poleward heat transport during all seasons and helping to explain the presence of Arctic dinosaurs (p. 118). *Warm Cretaceous Arctic Ocean, chapter 3*

Hou, Lian-hai. Personal communication. More specimens of *Confuciusornis* have been recovered from the late Jurassic of China, and there is a newly discovered small bird, possibly ornithurine, from the *Confuciusornis* locality, with a keeled sternum. Sternum is well developed, like a high-footed cup, and bears a relatively high keel. Forelimb is robust. Tarsometatarsus (16 mm) is short and not quite one-half the length of the tibiotarsus (34 mm). Digit 3 together with the claw is the longest, 20 mm; digit 1 together with the claw is 10 mm long. If a true ornithurine, it will confirm the early dichotomy of sauriurine and ornithurine birds indicated in the diagram on p. 172. *Late Jurassic Chinese birds, chapter 4*

Hunt, G. R. 1996. Manufacture and use of hook-tools by Caledonian crows. *Nature* 379:249–251. New Caledonian crows *Corvus moneduloides* manufacture and use two types of hook tool, characterized by a high degree of standardization, to capture prey. This high degree of technical capability is remarkable, especially given that use of hooks and imposition of form in tool shaping first appeared in the stone and bone tool-using cultures of early humans after the Lower Paleolithic. Study confirms the extraordinarily high intelligence within the corvine birds, adding to the enigma surrounding their structurally primitive position within the passerines. *Tool use by crows, chapter 8*

Kellner, A. W. A. 1996. Fossilized theropod soft tissue. *Nature* 379:32. Well-preserved soft tissue from an early Cretaceous theropod from Brazil was devoid of "any structure covering the skin, . . . such as feathers, which should be present if they were originally present." *Theropod skin preserved, chapter 3*

Krause, David. Personal communication. Several new taxa of primitive birds discovered in the late Cretaceous (Campanian) of Madagascar are described by L. Chiappe, C. Forster, and D. Krause. *Sauriurine birds from the Cretaceous of Madagascar, chapter 4*

Ligon, David. Personal communication. The critical limiting factor for green woodhoopoes is the availability of roost cavities, including loose bark. Although predation is high in such settings, woodhoopoes are intolerant of low nighttime temperatures. The African honeybee is a major competitor for woodhoopoe roosting holes, and Ligon observed a woodhoopoe trying to enter a roosting hole occupied by bees. The bird became very nervous and finally sought refuge in a bark crevice. Illustrates that many primitive living birds exist in an unperfected world of temperature regulation, and therefore the portrayal of urvogels and opposite birds as ectotherms or quasi-ectotherms becomes much more believable. *Thermoregulation, chapters 3, 4, 8*

Norell, M. A., J. M. Clark, L. M. Chiappe, and D. Dashzeveg. 1995. A nesting dinosaur. *Nature* 378:774–776. A fossil of the late Cretaceous theropod *Oviraptor* preserved over a nest with dinosaur eggs is assumed to belong to the dinosaur and is interpreted as suggesting an "avian type of nesting behavior in oviraptorids." However, more than a hundred species of lepidosaurs (snakes and lizards), as well as crocodilians, brood eggs or maintain a nest, and bird eggs and dinosaur eggs are fundamentally different (see K. F. Hirsch, K. L. Stadtman, W. E. Miller, and J. H. Madsen, Jr., Upper Jurassic dinosaur egg from Utah, *Science* 243 [1989]:1711–1713). *Nesting dinosaur, chapter 3*

Noriega, J. I. 1995. The avifauna from the "Mesopotamian" (Ituzaingo Formation; Upper Miocene) of Entre Rios Province, Argentina. *Courier Forschungsinstitut Senckenberg* 181:141–148. Avifauna includes the large swimming flamingo *Megapaloelodus*, as well as the probably flightless snakebird *Macranhinga paranensis*, similar in size to the modern *Anhinga anhinga*. *Miocene aquatic avifauna, chapters 4–5*

Noriega, J. I., and C. P. Tambussi. 1995. A late Cretaceous Presbyornithidae (Aves: Anseriformes) from Vega Island, Antarctic Peninsula: Paleobiogeographic implications. *Ameghiniana* 32:57–61. Fossil humerus, tibiotarsus, femur, and coracoid of a *Presbyornis*-like bird ("transitional shorebird") recovered from the Maastrichtian of Antarctica. *Presbyornis in Antarctica, chapters 4–5*

Ruben, John. Personal communication. Without the long posterior abdominal extension of the sternum as seen in ornithurine, but not enantiornithine birds, opposite birds would have lacked the capacity for higher rates of oxygen consumption during flight and therefore a modern avian-style flow-through lung. Uncinate processes hinging the rib cage are found in all ornithurine birds but not in any enantiornithine bird or urvogel, facts that argue for modern air-sac breathing in ornithurine birds but against it in all sauriurine birds, including *Archaeopteryx*, as they would be unable to ventilate the abdominal air-sac system. Indicates that the avian dichotomy was characterized not only by divergent locomotory architecture but also by thermoregulatory and activity physiology. *Activity physiology and thermoregulation, chapters 3–4*

Storch, G., B. Engesser, and M. Wuttke. 1996. Oldest fossil record of gliding in rodents. *Nature* 379:439–441. A well-preserved late Oligocene gliding rodent, *Eomys quercyi*, belongs to the extinct family Eomyidae, which now constitutes the fourth family of rodents with representatives capable of flying. Others are the squirrels (Sciuridae), scaly-tailed flying squirrels (Anomaluridae), and dormice (Gliridae). *Gliding rodent, chapter 3*

Tambussi, C. P., and J. I. Noriega. 1995. Falconid bird from the Middle Eocene La Meseta Formation, Seymour Island, west Antarctica. *Journal of Vertebrate Paleontology* 15:55A. Tarsometatarsus belongs to the Polyborinae (Falconidae). *Earliest New World falconiform fossil, chapter 7*

Tambussi, C. P., J. I. Noriega, A. Gaždzicki, A. Tatur, M. A. Reguero, and S. F. Vizcaino. 1994. Ratite bird from the Paleogene La Meseta Formation, Seymour Island, Antarctica. In *XXI Polar Symposium*, ed. S. M. Zaleqki, pp. 45–48. Warsaw: Polish Academy of Sciences. Fragment of distal tarsometatarsus with two trochleae reported from the upper part of the La Meseta Formation, Antarctica, most likely ranging from medial to late Eocene or early Oligocene. Phorusrhacids not excluded. *Large bird (ratite?) from Antarctica, chapter 6*

Tambussi, C. P., and E. P. Tonni. 1988. Un Diomedeidae (Aves: Procellariiformes) del Eoceano tardio de la Antartida. *5th Jornadas Argentinas de Paleontologia de Vertebrados*, 4. Late Eocene albatross reported from the La Meseta Formation of Antarctica. *Albatross from the Eocene of Antarctica, chapter 4*

Welman, J. 1995. *Euparkeria* and the origin of birds. *South African Journal of Science* 91:533–537. Braincase of various archosaurs and *Archaeopteryx* illustrates synapomorphies in common between the medial Triassic primitive archosauromorph *Euparkeria*, *Archaeopteryx*, and modern birds. Theropods are too specialized in the braincase to be bird ancestors. *Thecodont origin of birds, chapter 2*

Zhou, Z. 1995. Is *Mononykus* a bird? *Auk* 112(4): 958–963. Morphology of *Mononykus* refutes its avian status and shows that it is a small theropod dinosaur that converged on birds and moles in certain general characters. *Mononykus is a theropod, chapters 2–3*

Index

Page numbers followed by the letter F indicate an illustration